Oil Well Stimulation

ROBERT S. SCHECHTER
W. A. (Monty) Moncrief Centennial Endowed Chair
in Petroleum Engineering
Professor of Chemical Engineering
Professor of Petroleum Engineering
The University of Texas at Austin

PRENTICE HALL, Englewood Cliffs, New Jersey 07632

Library of Congress Cataloging-in-Publication Data

Schechter, Robert Samuel.
Oil well stimulation / Robert S. Schechter.
p. cm.
Includes bibliographical references and index.
ISBN 0-13-949934-2
1. Oil fields—Production methods. I. Title.
TN870.S37 1992
622'.3382—dc20 91-8063
CIP

Acquisitions editor: *Michael Hays*
Editorial/production supervision: *Raeia Maes*
Manufacturing buyers: *Kelly Behr/Susan Brunke*
Cover designer: *Karen Salzbach*

Printed in the United States of America

10 9 8 7 6 5 4 3 2

ISBN 0-13-949934-2

Prentice-Hall International (UK) Limited, *London*
Prentice-Hall of Australia Pty. Limited, *Sydney*
Prentice-Hall Canada Inc., *Toronto*
Prentice-Hall Hispanoamericana, S.A., *Mexico*
Prentice-Hall of India Private Limited, *New Delhi*
Prentice-Hall of Japan, Inc., *Tokyo*
Simon & Schuster Asia Pte. Ltd., *Singapore*
Editora Prentice-Hall do Brasil, Ltda., *Rio de Janeiro*

Contents

PART 6
SAND CONTROL METHODS: GRAVEL PACKING AND CONSOLIDATION TECHNIQUES 549

Preface

This book is intended as a text for both undergraduate and graduate courses in petroleum engineering, with particular emphasis on the subject of well stimulation. There are ample materials and sufficient end-of-chapter homework problems of varying difficulty to satisfy the needs of both courses. Parts of the book have also served as a basis for industrial short courses, and it is hoped that some of the material presented will also be of value to researchers engaged in developing new stimulation processes or products. However, the primary audience is thought to be university students interested in petroleum engineering. There is no other comprehensive book currently available for this purpose. There are monographs containing the subject matter covered, but these generally do not attempt to delve into the more fundamental aspects of the problem nor are the processes quantified. The presentation is most often qualitative and offered without examples. These books are only intended for use in industrial short courses, and apparently the authors are hesitant to model stimulation processes because they recognize their enormous complexity and fear that the degree of samplification needed to reduce them so that they are amenable to simple models will lead to useless, or at best misleading, results. On the other hand, when a process is necessarily described by a number of interacting variables, some dependent on both time and position, it is not easy for students to grasp the importance of these variables when the discussion is qualitative nor can they sense the relative importance of the relevant variables. Thus, the viewpoint adopted in writing this book is that models are essential. For pedagogical purposes, it seems better to oversimplify certain aspects of a process, hopefully unimportant ones, in order to obtain a mathematical representation. This representation is the main feature of this book and the one which sharply distinguishes it from any other currently available text. The models developed can all be evaluated using simple hand calculators, and yet they retain the essential features of the process. When possible, predictions using these relativelty simple models have been compared to experimental results or to results obtained by sophisticated numerical techniques to demonstrate their limitations and to emphasize the real complexity of the processes.

Since the approach taken here is in many cases new and different, it is to be expected that new concepts will emerge. This is true. It is hopes that these new concepts will lead the way to new directions for research and that ultimately new processes or new products will emerge. If such results do occur, then this will be an added feature not anticipated at the onset.

The book is complete with examples and end-of-chapter homework problems that have been assigned a degree of difficulty. Level one problems, designated by a single asterisk (*) are intended to reinforce concepts presented in a chapter and they are often similar to examples that are developed within the text. Level two problems, designated by two asterisks (**), are more comprehensive and may require the unification of several concepts, numerical evaluation, or even the derivation of new but related equations. Level two problems are intended for use primarily in graduate classes, but undergraduates may want to attempt some of them.

Early in the writing, the author decided to discard the use of field units and to adhere strictly to SI units. Occasionally, variables are expressed in field units together with their corresponding SI values to help the reader more familiar with field units to visualize the magnitudes of the terms. Because field units are not used, the equations presented here are not cluttered with the distracting constants often found in the petroleum engineering literature but are in fact valid in any consistent set of units. All examples are developed in SI units. It is hoped that the American petroleum industry will make the transition to SI units in the near future. This industry is, after all, truly international and the use of SI units is actually commonplace throughout most of the world. In solving problems, the recommended procedure is to take all of the data given in field units and convert them to SI units. A table of conversion factors is presented for this purpose in the Appendix at the end of the text.

There are topics considered in this book which some practitioners will characterize as being outside the scope of well stimulation; however, these, in particular Perforating Methods (Part 3, Chapter 7) and Sand Control (Part 6, Chapter 19) are very much related to maintaining good well performance and cannot really be completely divorced from concepts fundamental to the subject. Properties of Formation Materials and Treatment Fluids (Part 1, Chapters 1, 2, and 3) is a study of the chemical, thermal, and mechanical properties of reservoir materials as well as stimulation fluids. These properties are pertinent when evaluating any of the stimulation processes, so that considerable economy is gained by developing an understanding of them at the beginning of one's studies.

Formation damage is often the origin of the need for well stimulation. The particular well treatment recommended and its ultimate success will critically depend on correctly recognizing and understanding the mechanisms giving rise to the damage to be removed. Any book that deals with the subject matter presented here but omits consideration of formation damage is seriously flawed. Thus, in recognition of its importance, the coverage on Formation Damage (Part 2, Chapters 4, 5, and 6) is extensive. Many aspects of formation damage not found in any other monograph are included.

The primary methods for well stimulation are Hydraulic Fracturing (Part 4, Chapters 8, 9, 10, 11, and 12) and Matrix Acidizing (Part 5, Chapters 13, 14, 15, 16, 17, and 18). The coverage of these two subjects ranges from fundamental principles to design techniques and new concepts, especially in acid fracturing and matrix acidizing. These topics will hopefully interest even experienced practitioners. These two parts are also suitable and have been used for industrial short courses. The examples worked out in the text are relevant to industrial practice as well as for promoting a fundamental understanding of well stimulation.

I have greatly benefitted from the critical comments provided by Professor A. D. Hill (The University of Texas at Austin) and his students. I also wish to acknowledge the contributions of Ms. Joye Johnson who has helped edit and has typed parts of the book. Most thanks, however, are due my wife, Mary Ethel Schechter, who has typed, retyped, and re-retyped this manuscript. This laborious procedure might have seriously strained any marriage save for the modern miracle of word processors.

R. S. Schechter

PART 1
Properties of Formation Materials and Treatment Fluids

1
Chemical Properties of Oil-Bearing Formations and Resident Fluids

Oil and gas well completion and stimulation practices have become increasingly sophisticated as our understanding of the complex nature of the physical and chemical systems we are dealing with has improved. Knowledge about these systems has grown enormously since World War II, when many major oil and service company research facilities devoted strictly to production research were founded. It is not possible to grasp the significance of a particular well treatment formulation and the procedure used to administer it without having a full appreciation of the mechanical, chemical, and thermal properties of the system. In this chapter we shall focus on the chemical properties of the naturally occurring materials, with subsequent chapters being devoted to their physical properties and the properties of the materials used in treatments.

To facilitate our discussion it is convenient to divide it into the two categories of chemical and physical properties. The material presented here is necessarily quite condensed but hopefully precise. References will be provided to assist the interested reader in finding more detailed information as needed.

The chemical composition of both solid and fluid materials will be considered. Two general types of formations will be of interest—namely, sandstone and limestone—but it may be difficult to draw this distinction with any degree of accuracy since formations put into one of these two categories may, in fact, have many constituents in common with those formations placed in the other category. In general, it is better to have chemical, petrographic, mechanical, and other types of analyses to characterize the particular formation of interest rather than to use simple classification schemes. The differences in composition between different oil-bearing formations arise because of differences in the weathering, transportation, deposition, diagenesis, and lithification processes. Postlithification changes can also be significant. We shall, therefore, begin with a brief review of the composition and mineralogy of sedimentary rocks.

1.1 ROCK COMPOSITION AND MINERALOGY

The minerals of the common sedimentary rocks can be divided in terms of origin into four major groups: (1) minerals that survive weathering and transportation (detrital minerals), (2) new minerals formed during weathering and transportation (secondary minerals), (3) minerals that form directly from solutions owing to chemical or biochemical reactions (precipitated minerals), and (4) minerals formed in sediments during and after deposition (authigenic minerals).

The major detrital minerals are quartz, orthoclase, microcline, and plagioclase. The clay minerals make up the bulk of the secondary minerals, as do calcite and aragonite for the precipitated minerals. All of these can be authigenic minerals. Dolomite mainly forms as an authigenic mineral.

Sandstones can be considered to represent a segregation of quartz and, similarly, limestones represent a segregation of CaO. Most limestones have a rather low content of detrital and secondary silicate minerals. Both sandstone and limestone formations may undergo significant changes in mineralogy due to the for-

TABLE 1.1 Chemical Formulae of the Major Minerals [1]

Minerals	Composition
Quartz	SiO_2
Orthoclase (feldspar)	$KAlSi_3O_8$
Microcline	$KAlSi_3O_8$
Plagioclase	$NaAlSi_3O_8$–$CaAl_2Si_2O_8$
Muscovite	$KAl_3Si_3O_{10}(OH)_2$
Biotite	$K(Mg, Fe)_3AlSi_3O_{10}(OH)_2$
Kaolinite	$Al_4Si_4O_{10}(OH)_8$
Montmorillonite (smectite)	$(Al, Mg, Fe)_4(Al, Si)_8O_{20}(OH)_4 \cdot nH_2O$
Illite	$K_{0-2}Al_4(Al, Si)_8O_{20}(OH)_4$
Chlorite	$(Mg, Fe, Al)_3(Al, Si)_4O_{10}(OH)_2 \cdot (Mg, Al)_3(OH)_6$
Calcite	$CaCO_3$
Aragonite	$CaCO_3$
Dolomite	$CaMg(CO_3)_2$

mation of authigenic materials. Table 1.1 shows the composition of major minerals found in sedimentary rock.

Sandstones

Orthoquartzites are composed almost entirely of quartz grains and chert (mostly chalcedonic quartz) fragments. The sand grains are usually cemented by silica or calcium carbonate. These sediments are often more than 98% SiO_2. Other sandstones contain both detrital quartz and detrital feldspars. These arkose sandstones contain significant amounts of Al_2O_3, K_2O, Na_2O, and CaO. In addition to quartz and one or more feldspars, many of these sandstones contain clay minerals, a variety of other minerals, and rock fragments. Terms that have been applied to these rocks include lithic sandstone, graywacke, and subgraywacke. Many of the trace metals such as TiO_2, P_2O_3, and MnO may come from clay minerals, from precipitation of minerals such as celestite during diagenesis, or from accessory detrital minerals such as rutile, zircon, monazite, etc.

Limestones

All limestones are composed primarily of $CaCO_3$ in the form of calcite and aragonite. The $CaCO_3$ occurs as (1) discrete fragments and particles of either inorganic or organic origin, (2) microcrystalline material formed from a carbonate mud, and (3) coarsely or finely crystalline cementing material formed by inorganic or biochemical precipitation. A wide array of textures results from variations in the nature of the discrete fragments and particles. Lithographic limestones have, in general, been formed from carbonate mud. Fossiliferous limestone is composed of fossil fragments cemented by coarsely crystalline calcite. Cemented, inorganic, or biochemical rounded masses (ooliths) make up oolitic limestone. For all three limestones, although quite different in texture, $CaCO_3$ is the only major component. Because calcite and aragonite are never compositionally pure, small amounts of magnesium, iron, manganese, barium, and strontium may be found in a particular analysis. Magnesium is the most abundant of these and if MgO exceeds 1 wt%, then the mineral dolomite is likely to be present. Also present in trace or significant quantities are silicate impurities.

Many carbonate rocks, particularly Paleozoic and Precambrian rocks, have a significant amount of dolomite. Any rock containing more than 50% of the mineral is known as the rock dolomite.

Shales

The major minerals of shales are the clay minerals (kaolinite, montmorillonite, illite, chlorite) and quartz. Thus, the most abundant oxides are SiO_2 and Al_2O_3. A particularly high silica content results when silica-bearing shells or volcanic ash occurs in a rock along with detrital quartz. A high K_2O content may be due to detrital feldspar, authigenic feldspar, detrital muscovite, illite, or potassium adsorbed by the clay minerals. Shales may also contain some organic matter.

Composition of Sedimentary Rocks

As discussed previously, sandstones and carbonates represent segregations of the oxides SiO_2 and CaO, respectively. Variations in the extent of segregation are reflected in the large overall compositional variations primarily due to differences in the amount of quartz, chert, and clay minerals in sandstones and carbonates. Since shales represent less segregation of SiO_2 and CaO during weathering processes and since the majority of sedimentary rocks are shales, the average composition for sedimentary shales is close to the average composition for igneous rocks.

It will be important to keep these chemical compositions in mind when attempting to design a workover treatment or an initial stimulation program. Chemical composition will very often be of primary concern.

1.2 CLAYS

To design an adequate well treatment, it is essential to consider the compatibility of the treatment fluids and the formation materials as well as the many other parameters that may directly influence the ultimate performance of the treated well. All too often one hears of cases where the injectivity of the treating fluids decreases to such an extent that the project is abandoned, or that fractured wells have not recovered to the extent expected, or that the productivity of a well declined rapidly following a treatment. In all of these cases, it is possible that fluid interacting with the clays present in the formation may be responsible. In this section we examine the different types of clay minerals we expect will be present in an oil-bearing formation, try to understand their interactions with fluids and with each other, and finally, explore the effect of these interactions on the permeability of the formation. The mechanisms giving rise to permeability reduction are discussed in Part 2.

There are many types of clays, and they each exhibit a different response upon interaction with fluids of different compositions. Because of the complexities of the systems involved, the coverage in this section is necessarily abbreviated. Comprehensive treatments are available [2, 3].

In this section the chemical structure of clay minerals is briefly discussed, because an understanding of their structure is needed to appreciate their impact on well productivity and well stimulation. Once having considered the structure of clays, their interaction with water and the electrolytes contained in both naturally occurring waters and those used in stimulations fluids will be of primary interest.

Structure of Clay Minerals

There are basically two structural units involved in the atomic lattices of most of the clay minerals. One unit consists of a sheet of closely packed oxygens or hydroxyls in which aluminum, iron, or magnesium atoms are embedded in octahedral coordination so that they are equidistant from six oxygens or hydroxyls. A typical structure is shown in Fig. 1.1. When aluminum is present, only two thirds of the possible positions are filled to balance the structure gibbsite, which has the formula $Al_2(OH)_6$. When magnesium is present, all the positions are filled to balance the structure brucite, which has the formula $Mg_3(OH)_6$ [2].

The second type of sheet is composed of silica tetrahedra, which can be described as follows: In each tetrahedron a silicon atom is equidistant from four oxygens (or hydroxyls if needed to balance the structure). The silica tetrahedral groups are arranged to form a hexagonal network, which is repeated indefinitely to form a sheet of composition $Si_4O_6(OH)_4$ as shown in Fig. 1.2. The tetrahedra are arranged so that all their tips point in the same direction and all their bases are in the same plane. A sheet containing the silicon atoms with each silicon in the tetrahedral cavity, when viewed from above, would reveal the silicon atoms to be arrayed in a hexagonal network. Above each silicon is an oxygen or a hydroxyl, positioned at the tip of the tetrahedron.

There are four clays abundant in sandstones: kaolinite, smectite (or montmorillonite), illite, and chlorite. The structure of each will be described briefly.

Kaolinite. Clays can be classified according to the number and the type of sheets that combine to form a layer. Thus, a 1:1 clay is formed when one tetrahedral sheet combines with an octahedral one. Kaolinite is a 1:1 clay [2]. A layer consisting of the two sheets combined is shown in Fig. 1.3. The silicon tetrahedra and the octahedral aluminum sheets are linked by shared oxygens, whereas the lower surface of the octahedral sheet is composed of hydroxyl ions as shown in Fig. 1.3. Thus, each aluminum atom is linked to two oxygen atoms and a hydroxyl group. The silicon atoms are linked to oxygen atoms, one above in the octahedral sheet along with three others at the top of the lattice.

Since the alumina and silica sheets are covalently bonded (they share the same oxygen atoms), hydrodynamic, capillary, electrical, or solvation forces will

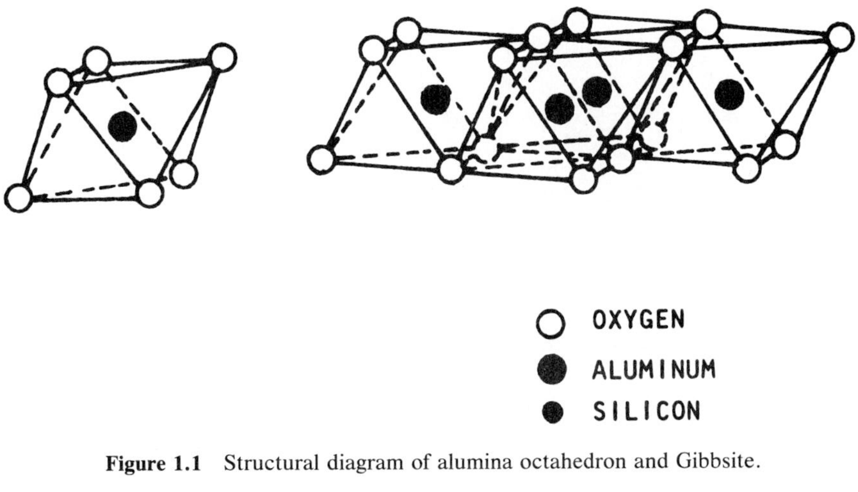

Figure 1.1 Structural diagram of alumina octahedron and Gibbsite.

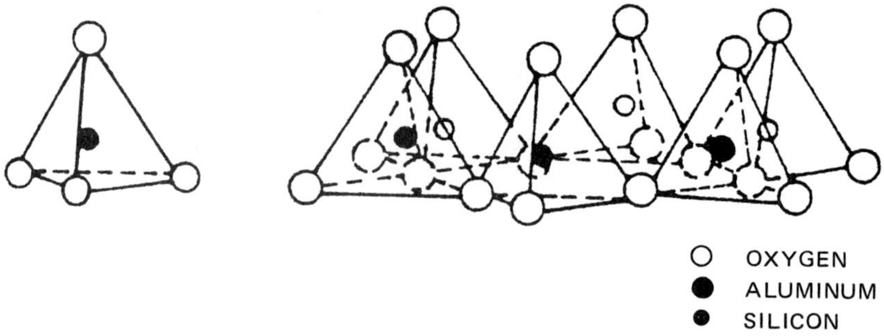

Figure 1.2 Structural diagram of silica tetrahedron and hexagonal network.

not be sufficient to separate the two sheets forming a single layer. However, kaolinite particles that occur naturally are composed of many layers stacked one atop the other as depicted by Fig. 1.4. The forces binding the layers together are hydrogen bonds supplemented by van der Waals forces. The strength of this binding is sufficient to prevent water penetration between the layers when a kaolinite particle is immersed in water [4]. For this reason kaolinite is classified as a nonswelling clay. It will become evident this does not at all mean that kaolinite in oil or gas reservoirs will never be a source of formation damage. Swelling clays are often thought to be the only source of particles that clog the pores and give

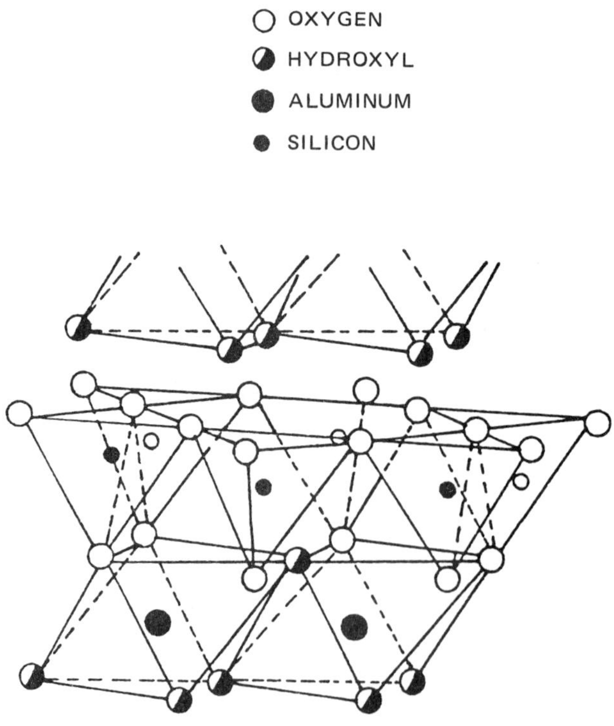

Figure 1.3 Lattice structure of kaolinite. It is composed of one tetrahedral sheet linked to an octahedral one.

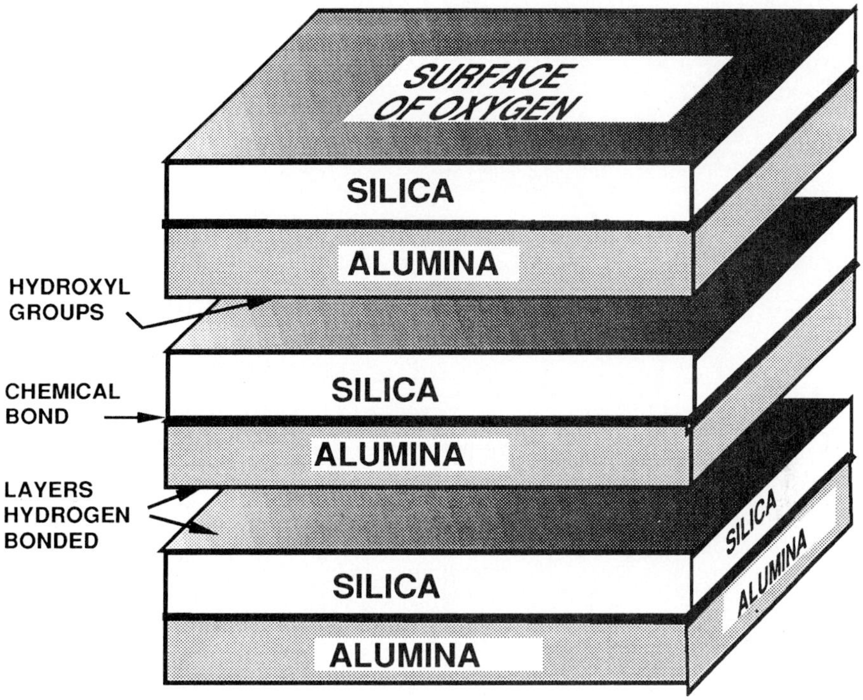

Figure 1.4 Stacking of two-sheet layers to form kaolinite crystals.

rise to formation damage, but this is not the case. Indeed sandstones, which do not contain measurable quantities of swelling clays, may be prone to formation damage (see Part 2).

Montmorillonites. Montmorillonites (smectites) are composed of layers consisting of two silica tetrahedral sheets surrounding a central alumina octahedral sheet. All tips of the tetrahedra point in the same direction toward the center of the unit. The tetrahedral and octahedral sheets are combined so that the tip of the tetrahedra and one of the hydroxyl units of the octahedral sheet form a common surface.

In the stacking of the silica–alumina–silica sheets, the oxygen layers of each unit are adjacent to corresponding oxygen layers of the neighboring units resulting in a very weak bond and an excellent cleavage between them. The outstanding feature of the montmorillonite structure is that water and other polar molecules can enter between the unit layers as shown by Fig. 1.5, causing the lattice to expand [2, 4]. The spacing between layers is therefore not fixed but varies from about 0.96 nm* when no polar molecules are between the unit layers to substantially complete separation of the individual layers in some cases.

* Units and unit conversions are SI units and these are discussed in the Appendix. It will be noted that nm refers to a length equal to 10^{-9} meters.

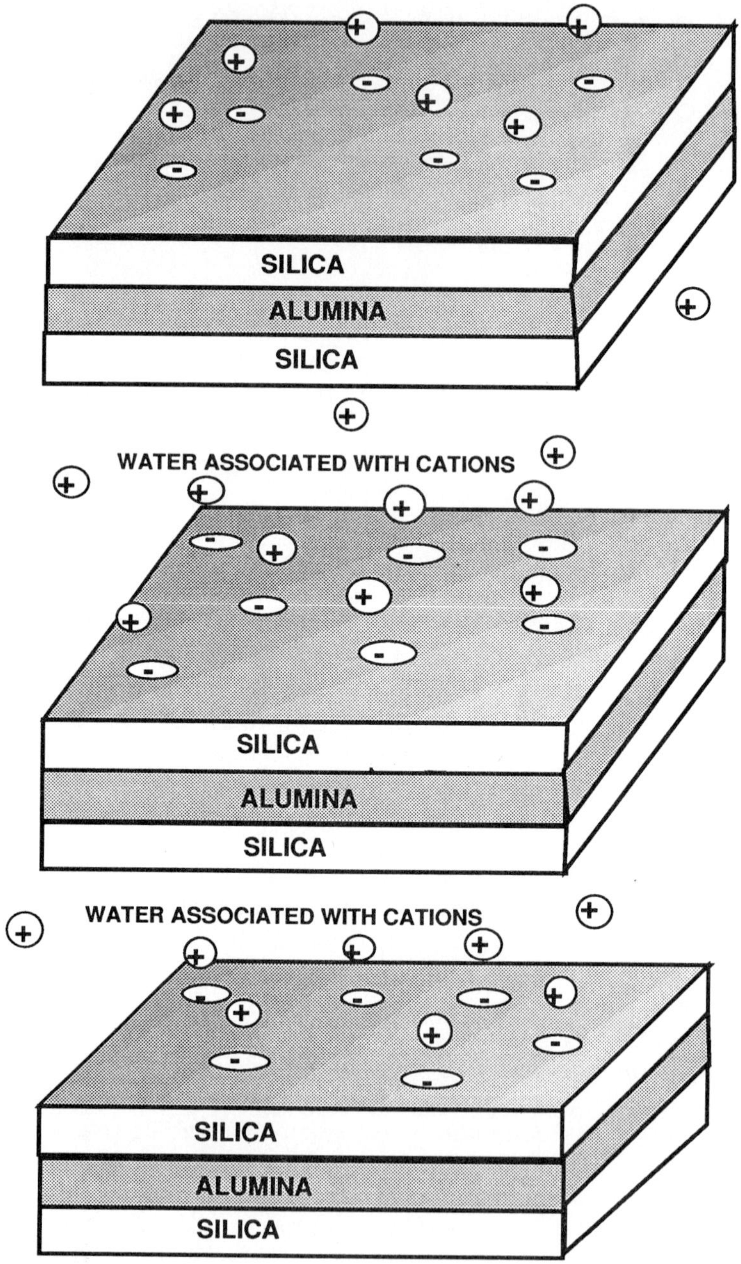

Figure 1.5 Section showing three layers of montmorillonite. Each layer is composed of three sheets separated by water associated with cations.

Montmorillonite is classified here as a 2:1 clay and the three sheets, which together form a layer, are covalently bonded as depicted in Fig. 1.5. Each layer is separated by a thin film which may contain cations, organic matter, and water.

The thickness of the liquid film increases as the dipole moment of the reacting fluid increases. It appears that the thickness of water layers between successive silicate layers is an integral number of molecules [2]; that is, the water layer is one, two, three, etc., molecular water layers thick. A natural montmorillonite may have a regular ordering of a single thickness of water layers, or it may be a random mixture of different *hydrates*. The tendency of montmorillonites to expand in the presence of water and other polar liquids has been identified as one reason for the water sensitivity of reservoir rocks or shales. However, the swelling of the clays is probably not the prime cause of permeability damage. A more likely difficulty is the clogging of pore throats by deflocculated (dispersed) clays (see Part 2).

Illite. The basic structure of illite minerals is similar to montmorillonite, each layer being composed of two silica tetrahedral sheets with a central octahedral sheet. The tips of the tetrahedra in each silica sheet point toward the center of the unit and are combined with the octahedral sheet in a single layer with a suitable replacement of OH by O. The unit is the same as that for montmorillonite, except some of the silicons are always replaced by aluminums [2] and the resultant charge deficiency is balanced by potassium ions positioned between the layers.

In some natural illites, there may be less substitution of aluminum for silicon and the potassium ions between layers may be partially replaced by other cations—possibly calcium, magnesium, or even hydrogen. Depending on the degree to which aluminum has been substituted for silicon and other cations—perhaps sodium has replaced potassium—illites may, somewhat like montmorillonites, exhibit some swelling upon contact with water. This penetration between layers by water might vary from layer to layer. Thus, in some cases illites may tend to reduce formation permeability by swelling.

Chlorites. Chlorites generally consist of alternate kaolinite-like and brucite-like layers, although there is usually considerable substitution within the structure. The bonding between layers is partly electrostatic in character as a consequence of the substitution and partly due to the interaction between adjacent oxygen and hydroxyls. This latter hydrogen bonding mechanism is similar to that found in kaolinites. The chlorite structure is depicted in Fig. 1.6.

Occurrence of Clays in Reservoirs

The clays of primary importance to our discussion are those formed within the rock pore system and generally attached to rock mineral surfaces. Dispersed clays in rocks are authigenic in origin, having developed subsequent to sediment deposition by precipitation of clay crystals from pore fluids. Particular clay mineral species develop in response to changes in the water chemistry brought about by changing temperature, pressure, and groundwater conditions during burial and

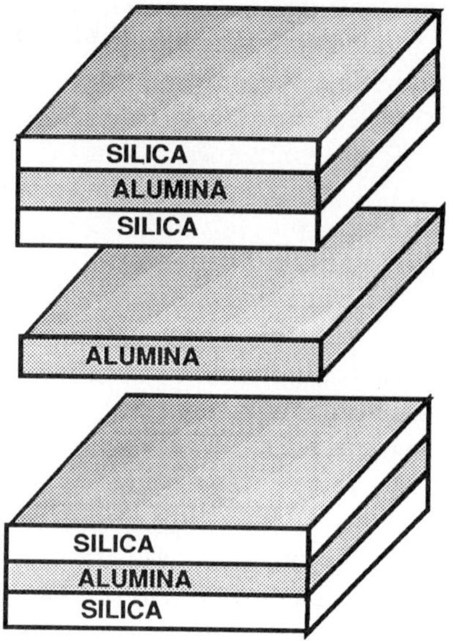

Figure 1.6 Diagrammatic sketch showing the structure of a mixed layer clay chlorite. The alumina or Gibbsite layer is often replaced by brucite ($MgOH_2$).

compaction. Since dispersed clays generally occur as a rock pore-filling component and have a variety of crystal sizes and shapes, they exhibit a broad spectrum of adverse effects on rock fluid flow and fluid saturation properties.

Neasham [5] has proposed three different categories of dispersed clays.

1. *Discrete particle clays* reflect the typical mode of occurrence of kaolinite in sandstones. Kaolinite usually develops as pseudohexagonal, platy crystals that are either attached to pore walls or occupy intergranular pores. A sketch of discrete kaolinite booklets is shown in Fig. 1.7(a). Kaolinite crystals that extensively fill pores have a random arrangement with respect to one another and affect rock petrophysical properties primarily by reducing the intergranular pore volume and behaving as migrating fines in the pore system.

2. *Pore lining clays* are attached to the pore walls, often forming a rather continuous and thin (~12 μm) clay mineral coating, as shown in Fig. 1.7(b). The clay crystals are oriented either parallel or perpendicular to the wall surface. If the crystals are oriented perpendicularly, then the crystals may be intergrown to form a continuous clay layer containing abundant micropore spaces with pore diameters in the range of 2 μm. Illite, chlorite, and montmorillonites have been observed with pore lining morphologies.

3. *Pore bridging clays* also include illite, chlorite, and montmorillonites. As shown in Fig. 1.7(c), the extensive development of intergrown and/or intertwined clay crystals within the pore system creates both microporosity and tortuous fluid flow pathways.

It has been found by Neasham that systems having sands of comparable grain size exhibit lower permeabilities to both air and brine if the clays are the

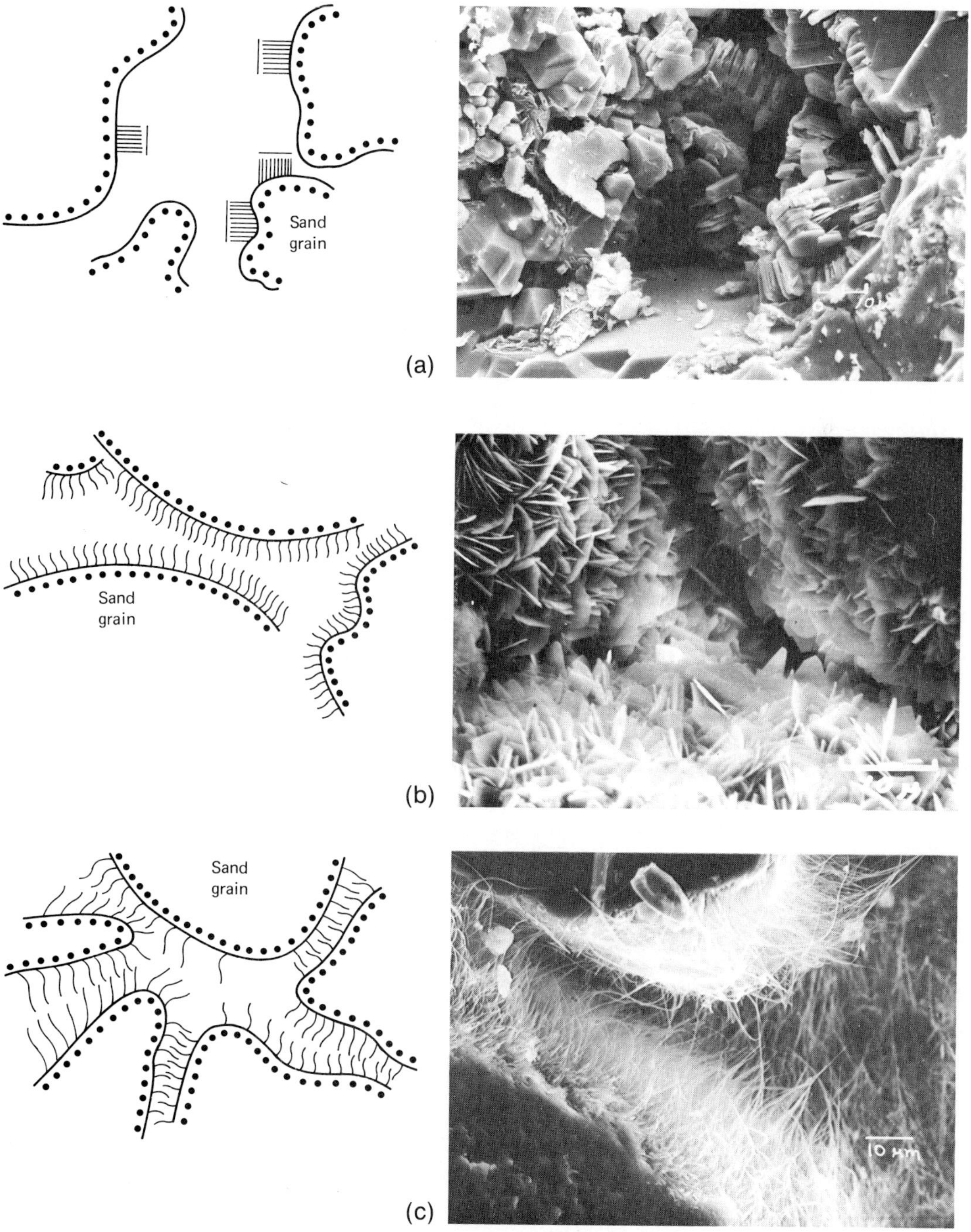

Figure 1.7 (**a**) Discrete particle kaolinite. (**b**) Pore lining chlorite. (**c**) Pore bridging illite. (Photographs courtesy Haliburton Co.)

pore bridging type. Sands with significant quantities of both pore lining and discrete clays can still have large permeabilities.

The pore lining and pore bridging systems generally have higher connate water saturations, but all have relative permeabilities to oil of about unity indicating that the larger quantities of water are held in the very fine pores which do not contribute materially to the flow. On the other hand, the relative permeability to water at residual oil is small for all systems and decreases as the system evolves from discrete to pore lining to pore bridging types.

1.3 SURFACE CHARGE OF CLAYS AND OTHER MINERALS

Most minerals when placed in contact with aqueous electrolyte solutions exhibit a surface charge, which can be detected by simply imposing an electric field across an aqueous suspension. It will be observed that the suspended particles will migrate toward one of the electrodes and when the polarity of the electrodes is reversed, the direction of migration will also be reversed. This movement under the influence of an electric field of particles suspended in an aqueous solution is known as *electrophoresis*, and the motion of the particle signals the presence of a fixed surface charge on the particle surface. Since the chemicals used in well treatment often bear a net charge (they are ions), then it is clear that adsorption will be greatly enhanced when the ion in aqueous solution has a charge opposite that on the surface. Thus, surface charge is an important factor.

Origin of Surface Charge

There are at least three different mechanisms responsible for the existence of a fixed surface charge. The first one relates primarily to clay minerals, but the other mechanisms will apply quite generally.

When clay minerals are formed during diagenesis, the "wrong" ion may, from time to time, be incorporated into the lattice sites. If this does not occur often, then the crystal structure of the particular clay mineral may not be changed. Hence, the substitution of an aluminum ion into the tetrahedral silicon lattice (see Fig. 1.2) can take place without altering the fundamental layer structure. This type of substitution is termed *isomorphic* since the morphology (form) of the mineral is unaltered [6]. However, the substitution of an Al^{+3} ion for Si^{+4} leaves an excess negative charge within the lattice structure. Similar substitutions, such as Fe^{+2} and Mg^{+2} for Al^{+3}, will also leave the lattice structure with a net negative charge. This charge is fixed or *frozen* in the lattice structure and will not be altered by changes in the solution composition [6].

Another important contribution to surface charge is a result of broken bonds at the edges of the clay crystals or of dislocations along the crystal faces [4, 6].

Finally, mineral oxides and silicates tend to be hydroxylated in water [7, 8]. This reaction with water takes place on the exposed surface of the mineral and can be visualized as follows:

$$H_2O + M^+ + MO^- \rightleftharpoons 2MOH \tag{1.1}$$

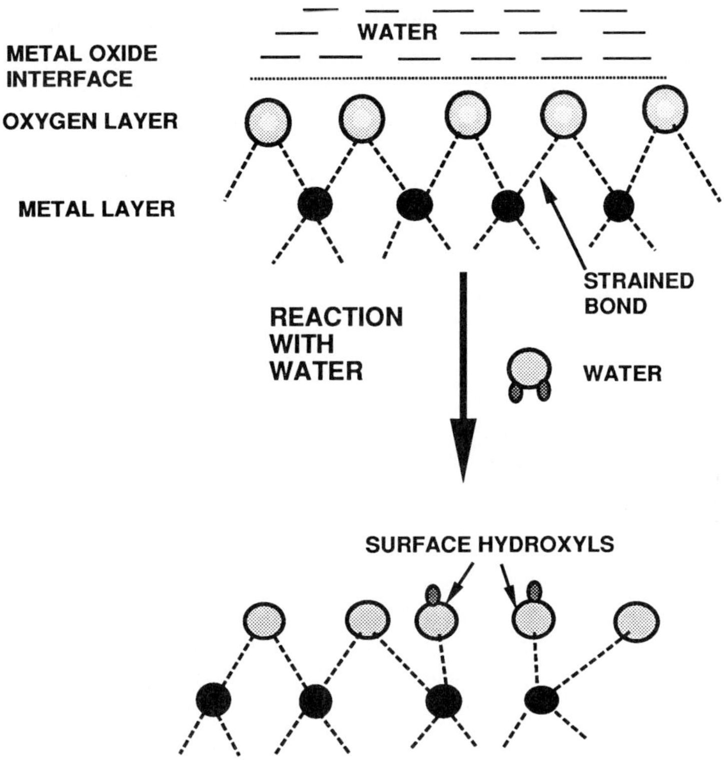

Figure 1.8 Schematic representation of a portion of a mineral oxide surface showing a mechanism by which surface hydroxyls form. An oxygen–metal bond is disrupted in the process.

where M^+ represents a metal ion incorporated into the crystal lattice, but near the surface and exposed to water, and MO^- denotes an oxide ion. This process is depicted in Fig. 1.8. Note that the breaking of one metal–oxygen bond produces upon reaction two surface hydroxyls. These hydroxylated sites, which are denoted here as MOH, are amphoteric and they can in turn receive or donate a proton (H^+) depending on the pH of the water in contact with the mineral. Thus, in an aqueous environment a surface hydroxyl (MOH) can associate with a proton to form MOH_2^+ or it can donate a proton leaving the oxide ion (MO^-) [9]. In low pH solutions where there are many protons it is more likely to accept a proton rather than donate one. Thus, the presence of acids in treatment fluids is likely to increase the degree of proton association thereby rendering the surface positive.

Similarly, at high pH in the presence of a caustic, which might for example be present in certain drilling fluids, the inverse reaction is more probable; that is, the surface MOH group may yield a proton leaving an MO^- group. Thus by accepting or donating protons, surface hydroxyl groups create a surface charge which unlike the fixed charge associated with clays will depend on the solution composition.

Davis et al. [10] represent these two reactions as follows:

$$MOH_2^+ \rightleftharpoons MOH + H^+ \tag{1.2}$$

and

$$MOH \rightleftharpoons MO^- + H^+ \tag{1.3}$$

where MOH_2^+ simply represents the surface hydroxyl site upon which a proton has adsorbed and MO^- is the group remaining when a proton is drawn into solution. These reactions make quantitative the concept that surface hydroxyls can accept or donate a proton. It would be productive to pursue this concept further by considering the process to be a chemical one which attains equilibrium with the ions in solution [10]; however, to proceed would take us beyond the scope of this text.

The adsorption or desorption of protons can be measured by simply adding an amount of acid or base to a solution containing a suspension of the mineral in question. The pH resulting from the addition of a known quantity of acid can be measured and compared with the pH that would be expected if none of the acid (protons) adsorb. This difference is, of course, a measure of the proton adsorption or desorption. Figure 1.9 shows the surface charge of kaolinite determined by this titration procedure as a function of the pH. Note that surface charge depends on both pH and electrolyte concentration. As the pH increases, the amount of negative charge per unit of surface area increases. The number MO^- groups as given by Eq. (1.3) increases. This is the expected trend.

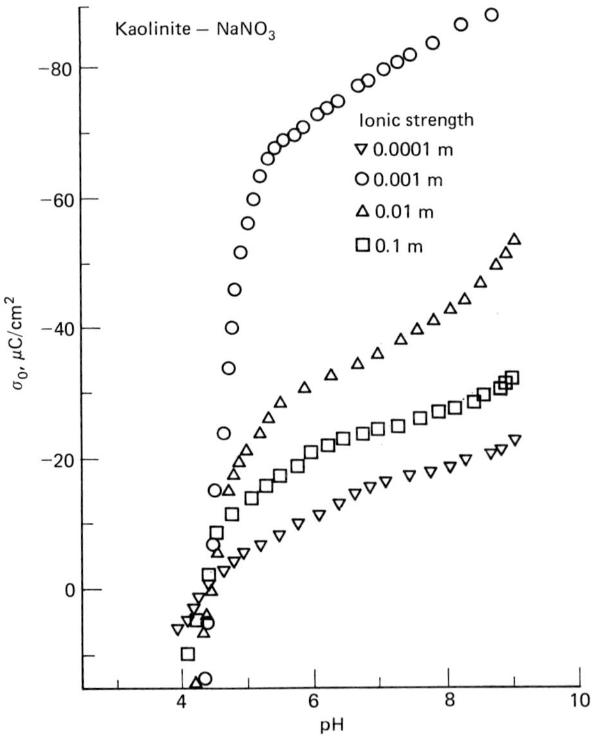

Figure 1.9 Surface charge development of kaolinite as a function of pH and ionic strength in $NaNO_3$ electrolyte solutions [11].

The surface charge at a given pH is clearly a function of the number of surface hydroxyls actually present on the mineral surface. The determination of the site density is complex. The difficulties have been discussed by Yates et al. [12]. Riese reports 6×10^{14} sites/cm^2 for the density of hydroxyl sites on kaolinite [11]. This means that if fully charged, the surface charge would be 96 $\mu C/cm^2$. Thus for a surface charge of 48 $\mu C/cm^2$ in Fig. 1.9, one-half of the hydroxyl sites have been converted to the MO^- form.

An interesting feature shown by Fig. 1.9 is that surface charge also increases as electrolyte concentration is increased. Thus, the pH and the electrolyte concentration of the water in contact with the reservoir solids are both important factors in determining surface charge.

Another interesting feature is that there appears to be a certain value of the pH for which the surface charge vanishes. For kaolinite this pH is about 4.5. The fact that the surface charge vanishes can be confirmed by suspending small kaolinite particles in an aqueous solution and imposing an electric field. If the particles do not respond, then the net charge on the particle is zero. Plotted in Fig. 1.10 is the electrophoretic mobility of kaolinite particles, where mobility is defined as the particle velocity divided by the applied electric field. The negative sign implies that the particle moves in a direction opposite that expected of a positive charge. Thus as shown by Fig. 1.10, the surface charge is negative for all pH values above 4.5. The electrophoretic mobility does apparently vanish as the pH approaches 4.5. This pH is known as the *point of zero charge* (pzc). It is an important quantity. If the solution pH is greater than the pzc, the surface charge is negative.

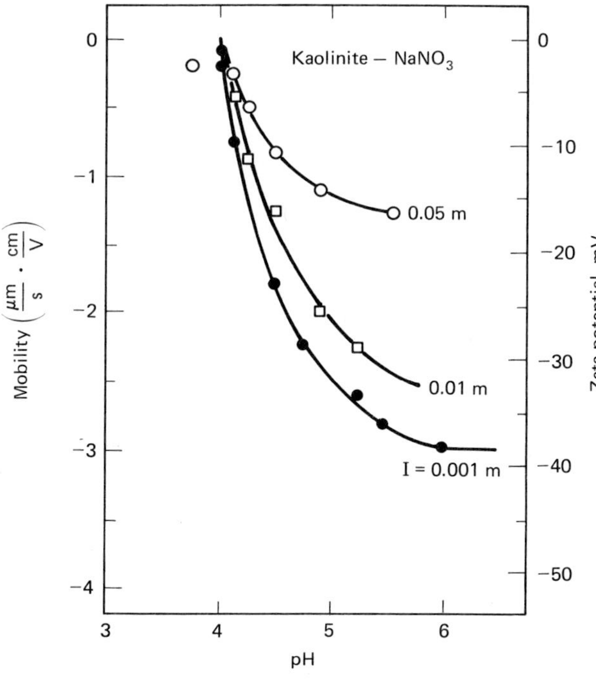

Figure 1.10 Electrophoretic mobility and zeta potential of kaolinite as a function of pH and ionic strength in $NaNO_3$ electrolyte solutions [11].

TABLE 1.2 The Point of Zero Charge (pzc) of Some Representative Reservoir Materials [14]

Material	pzc
Quartz (SiO_2)	2.7
Corundun (Al_2O_3)	9
Hematite (Fe_2O_3)	5
Calcite ($CaCO_3$)	9.5
Kaolinite ($Al_4Si_4O_{10}(OH)_8$)	4.6
Montmorillonite (Al, Mg, Fe)$_4$(Al, Si)$_8$$O_{20}(OH)_4 \cdot nH_2O$	2.0
Albite ($NaAlSi_3O_8$)	2.0

(*Note:* The values listed should be considered to be representative, but not precise. Actually, the pzc is quite sensitive to small differences in overall composition and for the same mineral may vary over a wide range. For example, the range of values for quartz is between 2.0 and 2.7).

Mobility which depends on the viscosity of the solution is often reported in terms of an electrical potential characteristic of surface charge and electrolyte concentration. This potential, which is called the *zeta potential*, appears again in Chapter 4. It is defined here for convenience. The zeta potential, $\mathcal{L}$, is defined by the Smoluchowski equation as [13]

$$\frac{U}{\mathcal{E}} = \frac{\epsilon \mathcal{L}}{\mu} \tag{1.4}$$

where U is the velocity of a particle in an electric field $\mathcal{E}$. The ϵ is the dielectric permittivity of the solvent (see Ex. 1.1) and μ is the fluid viscosity.

The pzc depends on the mineral. Table 1.2 gives selected values. These are important when selecting molecules for use in well treatments or in oil recovery processes. For example, polymer adsorption in sandstones may be minimized by using negatively charged polymers. This is true because at near neutral values of the pH, the quartz particles and many other minerals are negatively charged. It is not surprising therefore, that all polymers considered to be candidates for mobility control agents are anionic polyelectrolytes (negatively charged). Thus in selecting chemicals for use in well treatments, it is prudent to keep the values given in Table 1.2 in mind.

Electrical Double Layer

The origin of surface charge has been attributed to any one or a combination of mechanisms described in the previous section. Furthermore, it is known that electrical charges of one particular sign strongly attract charges of an opposite sign. In general, the systems of interest here are overall electrically neutral because of this strong attractive force. To balance the surface charge on mineral surfaces, ions of an opposite sign will gather in the aqueous solution near the charged surface. These ions will not be drawn immediately adjacent to the surface because of thermal fluctuations. Instead, they will remain in a thin layer surrounding the mineral surface with increasing concentration as the distance from the surface decreases. This situation is very much like the atmosphere surrounding the earth. When viewed from a spaceship, our atmosphere appears as a thin layer much smaller than the diameter of the earth. It exists because each molecule in the air is attracted to the earth's surface by gravity. However, due to random molecular motions caused by temperature, the molecules do not rest on the surface of the earth but instead remain near it. The density of the atmosphere is greater near the earth's surface and decreases with increasing distance (altitude) away from it.

Similarly, the concentration of ions having a charge opposite to that of the surface (called *counterions* to denote their opposite charge) is determined by a balance of two *forces*. One is the electrostatic force, which attracts the ions of opposite charge toward the surface. The electrical forces will be counteracted by thermal fluctuations, which will tend to mix these counterions with others in solution. The result of the competition between these two forces is a cloud of ions surrounding the charged surface and the ions in this cloud will be primarily, but not solely, counterions. Figure 1.11 depicts this situation. The electrical double layer consists of both a highly localized charge, the surface charge, and a diffuse

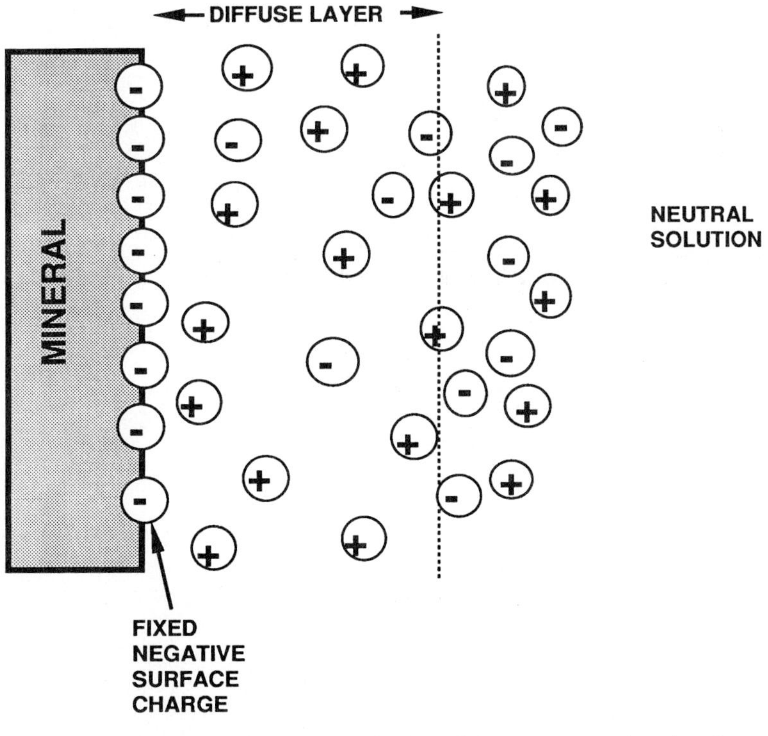

Figure 1.11 Sketch showing the distribution of charge near a mineral surface.

charge. It should be noted that the diffuse region contains more positive ions, the counterions, than negative ions. This is due to the attraction of the negatively charged surface. However, the diffuse region also contains, as depicted in Fig. 1.11, some negative ions. This is a result of diffusion from the neutral solutions toward the diffuse region, which contains fewer negative ions than the neutral solution. The neutral solution contains an equal number of negative and positive charges per unit volume.

An analysis of the diffuse region will reveal that its thickness is about five times the Debye length, λ, which is given as [13]

$$\lambda = \sqrt{\frac{\epsilon RT}{4\pi F^2 \sum Z_i^2 c_{i,0}}} \tag{1.5}$$

where ϵ is the dielectric constant of the fluid medium (for water at 25°C use 8.73 $\times$ 10^{-9} C/V-m), F is Faraday's number which is 9.65 $\times$ 10^7 C/kg equ, Z_i is the valence of the ith ionic species and has units of kg equ/kg mole, RT is the thermal energy per kg mole which is $(8.3147 \times 10^3)T$ in V-C/kg mole, and T is the temperature in °K. The concentration of the ionic species $c_{i,0}$ is the concentration in the neutral solution, not in the diffuse region, and should be expressed in kg moles/m^3.

The Debye length, λ, is expressed in meters.

Example 1.1 Thickness of the Diffuse Region

If a mineral surface attains a surface charge when immersed in an aqueous solution containing 10^{-3} kg moles sodium chloride (NaCl)/m^3, estimate the thickness of the diffuse region. The temperature of the solution is 298°K.

Solution Using Eq. (1.5), the Debye length is calculated as follows: The first step is to calculate the ionic strength, which is defined as

$$I = \sum Z_i^2 c_{i,0}$$

For sodium chloride, $Z_{Na^+} = 1$ kg equ/kg mole and $Z_{Cl^-} = -1$ kg equ/kg mole. Thus for the concentrations given

$$I = [(1)^2 10^{-3} + (-1)^2 10^{-3}] = 2 \times 10^{-3} (\text{kg equ})^2/\text{kg mole-m}^3$$

Thus

$$\lambda^2 = \frac{8.73 \times 10^{-9} \text{ C/V-m}(8.3147 \times 10^3 \text{ V-C/kg mole-°K})298 \text{ °K}}{4\pi(9.65 \times 10^7 \text{ C/kg equ})^2 \, (2 \times 10^{-3} \text{ (kg equ)}^2/\text{kg mole-m}^3)}$$

or

$$\lambda = 9.6 \text{ nm}$$

Thus at a distance of approximately 50 nm, approximately 5λ, measured from the surface the diffuse part of the electrical double layer will have merged into the bulk solution. At this point there will be as many negative as there are positive charges in a unit volume of solution. It is important to note that this thickness is perhaps not too much smaller than many of the pores in sedimentary rock. Thus, it should be anticipated that the electrical forces will play a prominent role in determining the movement of fine particles within the pore of a sedimentary rock. This is discussed in Chapter 4.

The thickness of the diffuse layer is estimated to be about five times the Debye length. The electrical potential is an exponentially decreasing function measured from the surface [13], but for all practical purposes, the solution can be considered to be neutral for distances greater than five times the Debye length.

Practical Importance of Surface Charge

When two clay particles or when a clay particle approaches a fixed surface, then there are two forces of prime importance. The particles may repel one another because of their electrostatic charge. The range of this interaction is related to λ. If λ is small, then the forces of repulsion (if two particles have the same charge) do not come into play until the particles approach very closely. At this point other forces may also become important; namely, the attractive van der Waals forces (see Chapter 4).

As a consequence, if λ is small, flocculation, agglomeration or adhesion is probable. On the other hand, for large λ, particles are likely to remain dispersed.

The importance of the composition of water in the flocculation or peptization of clays can now be at least partially understood. Equation (1.5) shows that if the electrolyte concentration is increased, λ decreases. But decreasing λ implies a decreased diffuse layer thickness, thus making it easier for particles of like charge to approach one another. This aids flocculation.

Also it becomes evident that adding ions with a large valence (Z_i) will reduce λ.

Example 1.2 Thickness of the Diffuse Region in Calcium Chloride Solutions

Repeat the calculation shown in Example 1.1, replacing the sodium chloride with 1×10^{-3} kg moles/m^3 solution of calcium chloride ($CaCl_2$).

Solution Note that $Z_{Ca^{++}} = +2$ kg equ/kg mole since calcium is divalent and furthermore, since there are two chloride ions (Cl^-) per mole of calcium chloride, $c_{Cl^-} = 2(1 \times 10^{-3}$ kg moles/m^3).

Thus
$$I = \sum Z_i^2 c_{i,0} = Z_{Ca^{++}}^2 c_{Ca^{++}} + Z_{Cl^-}^2 c_{Cl^-}$$

or
$$I = (2)^2(1 \times 10^{-3}) + (1)^2(2 \times 10^{-3}) = 6 \times 10^{-3}(\text{kg equ})^2/\text{kg mole-m}^3$$

From Eq. (1.5)
$$\lambda^2 = \frac{8.73 \times 10^{-9}(8.3147 \times 10^3)298}{4\pi(9.65 \times 10^7)^2(6 \times 10^{-3})}$$

or
$$\lambda = 5.6 \text{ nm}$$

Thus, the thickness of the diffuse layer is reduced when calcium chloride replaces sodium chloride even though the concentrations are the same. This can be observed by comparison with the Debye length obtained in Example 1.1.

1.4 CATION OR BASE EXCHANGE

It seems evident that those cations in the diffuse layer (see Fig. 1.11) are the same ones that exist in the bulk water immediately adjacent to the surface. If the electrolyte composition of the bulk water phase is changed by, for example, replacing a sodium chloride solution with one composed entirely of calcium chloride, then the sodium ions initially present in the diffuse layer will be replaced by calcium ions. This exchange process is called *ion exchange* and since mineral particles taken together with the thin diffuse layer surrounding them are electrically neutral, then the number of equivalents of Na^+ originally present must be exchanged for precisely the same number of equivalents of Ca^{++}. Since $Z_{Ca^{++}} = 2$, this means that one mole of sodium would be replaced by one-half mole of calcium to maintain electrical neutrality.

There are really two different issues of great interest. The first relates to the quantity of ions present within the diffuse part of the electrical double layer. This can be measured by first converting the mineral to sodium form by repeatedly contacting the mineral with a fresh NaCl solution. After a number of such contacts, all of the exchangeable cations other than Na^+ will have been stripped from the surface. The surface is then said to be in sodium form. Taking a mineral in sodium form and then contacting repeatedly with calcium chloride will ultimately remove all of the sodium from the surface. By measuring the amount of Na^+ removed from the surface by this procedure, one can assign a value to the cation-exchange capacity which is expressed in kg equ per 100 kg of the mineral A detailed discussion of ion-exchange capacity has been presented by Helfferich [15].

The exchangeable ions are held around the outside of the clay mineral structural unit and it is thought that the exchange reaction does not affect these structural units. The cations may penetrate between the structural layers resulting in the expansion or collapse of the distances separating the layers, or the exchange sites may be located at the edges of the layers where there are broken bonds. In montmorillonites it has been estimated that about 80% of the exchange sites are located along the basal planes and the remainder at the edges. A good review of the data now available relating to the collapse of clay layers and the subsequent fixation of the ion has been presented by Sawhney [16].

A range of cation-exchange capacities for clay minerals is shown in Table 1.3. These data indicate that montmorillonite, when present in the reservoir rock, will have a substantial influence on the total cation-exchange capacity of the rock.

TABLE 1.3 Cation-Exchange Capacity of Clay Minerals Expressed in kg equivalents per 100 kg,* $\overline{Q}_v$

Kaolinite	3–15
Montmorillonite	80–150
Illite	10–40
Chlorite	10–40

* These are fundamental values of the cation-exchange capacity for various minerals.

Source: Adapted from Grim [2] and van Olphen [6].

TABLE 1.4 Comparison of the Properties of Two Sandstones [17], Q_v

	Sandstone	
	Berea	Second Wall Creek
Permeability (mD)	195	32
Porosity (%)	22.4	21.4
Cation-Exchange Capacity* (kg equ/m^3 of pore volume)	.041	0.147

*Cation-exchange capacities expressed in these units are called *system values*.

For example, the data presented in Table 1.4 show the cation-exchange capacity of a Berea sandstone core and a core (Second Wall Creek) taken from the Salt Creek field, Natrona County, Wyoming [17]. Although both are sandstones having very similar porosities, the Second Wall Creek core has a cation-exchange capacity three or four times that of the Berea core. X-ray analysis revealed the presence of montmorillonite in the Second Wall Creek core not found in Berea [17].

Comparing the cation-exchange capacities given in Table 1.3 with those in Table 1.4, we see that the units are different. Those in Table 1.3 may be thought of as being fundamental values intrinsic to the clay and independent of the method of determination. This is called the *fundamental cation-exchange capacity* and is represented by the symbol, $\overline{Q}_v$. The system cation-exchange capacity is denoted Q_v and represents the cation-exchange capacity per unit of pore volume. The two are related by the equation

$$Q_v = \overline{Q}_v \rho_{\text{rock}} \left(\frac{1 - \phi}{\phi} \right) \tag{1.6}$$

This relationship is derived by noting that a unit volume of formation rock contains a mass of rock equal to the multiple of the rock density, ρ_{rock}, and the volume of the rock per unit volume of formation, $1 - \phi$, where ϕ is the formation porosity. The pore volume per unit of volume is ϕ. The cation-exchange capacity per unit volume of reservoir is found by multiplying $\overline{Q}_v$ by $\rho_{\text{rock}}(1 - \phi)$. The system cation-exchange capacity is computed by dividing by ϕ, the volume of the pores. It is this value which will ultimately enter all of the calculations related to well treatments. It is important to note that the system Q_v depends on the porosity, which is a property of the formation and therefore a function of the system as well as the fundamental cation-exchange capacity.

Example 1.3 The Significance of the Cation-Exchange Capacity

Suppose the formation water in the pores of a Second Wall Creek core (see Table 1.4) contains 0.5 kg moles sodium chloride/m^3 (approximate composition of seawater). How many kg equivalents of sodium cation are present per unit volume of rock and what proportion of these are associated with the mineral surface charge?

Solution Consider a 1 m^3 sample of Second Wall Creek rock. The pore volume is 0.214 m^3 and the cation-exchange capacity is 0.147 kg equ/m^3 of pore volume. The negative surface charges must be counterbalanced by sodium cations. Thus the total kg equivalents of sodium is given by

$$(0.214)(0.5 + 0.147) = 0.138 \text{ kg equ}$$

The fraction on the surface equals 0.147/0.647 = 0.23. This calculation shows that a significant fraction of the cations contained in the rock are associated with the charged surfaces. These cations are bound to the surface by electrostatic forces. If a liquid sample is drawn from the formation, only the ions present in the aqueous phase will be produced. To obtain full knowledge of the total cation inventory, one must measure the cation-exchange capacity of a formation rock sample.

Although knowledge of the cation-exchange capacity will prove to be essential in designing some well treatments, it is also often necessary to know which cations present in the formation waters are associated with the clays and more particularly the quantities of each. The issue is to determine which cations are present within the ionic atmosphere shown in Fig. 1.11. For example, if the bulk water near the surface contains both calcium and sodium cations then so will the diffuse layer. If the composition of the water is changed, some ions leave and other ions will enter the diffuse layer. The only requirement imposed on this exchange is that the net surface charge plus the charge in the diffuse region remain zero.

If an aqueous solution containing two cations A and B is contacted with a clay, the process of ion exchange can be formally represented as a chemical reaction [15] as follows:

$$Z_B\text{A-clay} + Z_A\text{B(solution)} \rightleftharpoons Z_A\text{B-clay} + Z_B\text{A(solution)} \tag{1.7}$$

Thus for example if A represents calcium and B sodium, then $Z_A = Z_{Ca^{++}} = 2$ and $Z_B = Z_{Na^+} = 1$ and the reaction shows that calcium associated with the clay can *react* with sodium in the aqueous phase to produce a sodium-clay and a calcium in solution. The reaction is, furthermore, shown to be reversible; that is, cation B attached to the clay can also exchange with cation A in solution. The appearance of the valences Z_A and Z_B in Eq. (1.7) assures that equal quantities of electrical charge are exchanged.

Since the reaction is reversible, an equilibrium will be obtained for a fixed solution composition such that [18, 19]

$$\overline{K}_{AB} = \left[\frac{\theta_B^{Z_A} c_A^{Z_B}}{\theta_A^{Z_B} c_B^{Z_A}}\right]\left[\frac{f_B^{Z_A} \gamma_A^{Z_B}}{f_A^{Z_B} \gamma_B^{Z_A}}\right] \tag{1.8}$$

where θ_A and θ_B represent the fraction of the cation sites occupied by A and B, respectively, and c_A and c_B represent their aqueous phase concentrations, normally expressed as kg moles/m^3; however, here it is more convenient to express electrolyte concentrations in terms of kg equ/m^3. The quantities in the brackets are activity coefficients, which are required to convert concentrations to thermodynamically relevant terms. The f_i refers to the activity coefficient of ions associated with the clays, and γ_i refers to the solution-phase activity coefficients and θ_A equals the total number of equivalents of ion A associated with a given mass of clay divided by the cation-exchange capacity. Thus, the concentration of adsorbed ion A expressed as kg equ/m^3 of pore volume is

$$\bar{c}_A = \theta_A Q_v \tag{1.9}$$

The $\overline{K}_{AB}$ shown in Eq. (1.8) is a thermodynamic property which depends on temperature and pressure but does not depend on the composition of either the aqueous phase or the ion-exchange sites. Unfortunately, $\overline{K}_{AB}$ is difficult to obtain primarily because surface activity coefficients are poorly understood. Gaines and Thomas [20] assume that a clay mineral has a fixed number of ion-exchange sites permeable to water but not to anions. Based on this approximation, they were able to obtain an expression for f_A and f_B but extensive data are required for their evaluation and few systems have been systematically studied.

The aqueous-phase activity coefficients can be determined using a variety of techniques [21], but the procedures are complex in multicomponent electrolyte solutions such as those represented by natural waters. Because of the difficulty in determining the activity coefficients, they are often ignored. It is more efficient to use a selectivity coefficient, defined as follows [18]:

$$K_{AB} = \frac{\theta_B^{Z_A} c_A^{Z_B}}{\theta_A^{Z_B} c_B^{Z_A}} \tag{1.10}$$

There is no thermodynamic basis for treating K_{AB} as though it is independent of composition, and careful studies show that it is not. However, here K_{AB} will be treated as a constant. Experiments show that for some cases over a narrow range of concentrations this is an adequate approximation except perhaps at very small aqueous-phase electrolyte concentrations, which are seldom encountered in any application considered here.

Figure 1.12 shows a plot of $K_{Na,Ca}$, the selectivity coefficient for the ion-exchange reaction of sodium and calcium, as a function of the fraction of the surface covered by sodium cations [22]. Although considerable scattering of the data is evident, the selectivity coefficient appears roughly independent of the fractional coverage for ionic strengths $[I = (1/2)\sum Z_i^2 c_i]$ of 0.1 and 1.0 (kg equ)^2/kg mole-m^3. However, for a small value of $I = 0.01$, $K_{Na,Ca}$ differs considerably from the other two values. $K_{Na,Ca}$ also appears to depend on the fractional coverage. Note that $I = 0.01$ corresponds to about 0.17 wt% sodium chloride. For comparison, seawater contains about 3 wt%. Thus, such small ionic strengths will rarely be encountered in field situations. An acceptable approximation is to assume $K_{Na,Ca}$ to be a constant.

Figure 1.12 shows that for $I = 0.1$ and 1.0 [(kg equ)^2/kg mole-m^3], the average value of $K_{Na,Ca} \approx 6.4$. This selectivity coefficient applies when the ion-exchange process starts with the clay in sodium form (reactant) and ends with the clay in calcium form (a product). If the reaction is reversed, it is not difficult to argue that $K_{Ca,Na} = 1/K_{Na,Ca}$ and therefore, for the system considered in Fig. 1.12, $K_{Ca,Na} = 1/6.4 = 0.16$.

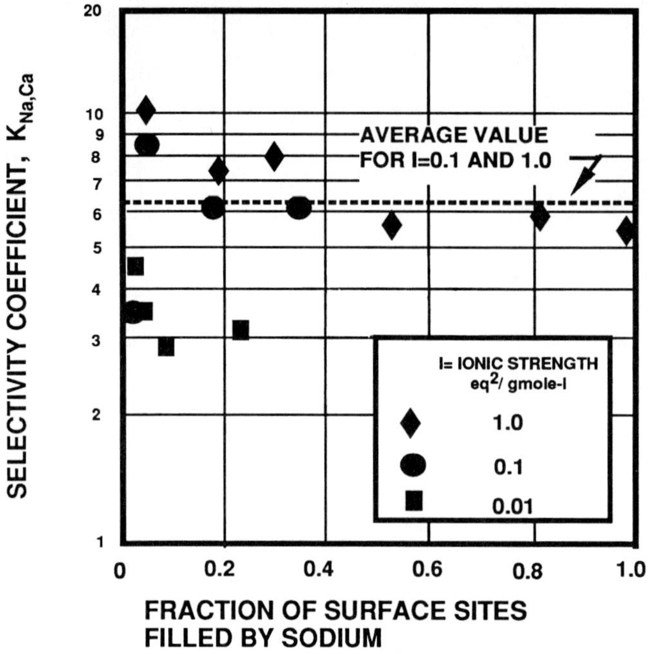

Figure 1.12 The selectivity coefficient for calcium–sodium ion exchange as a function of the fractional coverage of sodium for various ionic strengths. (Data recalculated from Rodgers et al. [22]).

Example 1.4 Ion-Exchange Equilibria

The cation-exchange capacity of the Wyoming clay used to obtain the selectivity coefficients plotted in Fig. 1.12 is reported to be 0.95 kg equ/kg clay [22]. Suppose that 2×10^{-3} kg of the clay in sodium form is mixed with 1 kg of solution containing 10^{-3} kg moles of $CaCl_2$. The density of the solution is 1010 kg/m^3. What will the final solution composition be? What fraction of the surface cation-exchange sites will be taken up by sodium cations. Take $K_{Ca,Na} = 0.16$.

Solution In the clay suspension there must be a total of (2×10^{-3} kg clay)(0.95 kg equ/kg clay) $= 1.9 \times 10^{-3}$ kg equ Na^+. These are the sodium cations that were associated with the clay surface when the clay was mixed with water to form the suspension. The total amount of calcium in the suspension is the amount originally present in the water. This amount is (1×10^{-3} kg moles $CaCl_2$)(2 kg equ Ca^{++}/kg mole $CaCl_2$) $= 2 \times 10^{-3}$ kg equ Ca^{++}.

If c_{Na} and c_{Ca} are the final aqueous concentrations of Na^+ and Ca^{++}, respectively, and θ_{Na} and θ_{Ca} are the fraction surface coverages, then a sodium balance is

$$1.9 \times 10^{-3} = c_{Na}\left(\frac{1}{1010}\right) + \theta_{Na}(0.95)(2 \times 10^{-3})$$

since c_{Na} has the dimensions kg equ/m^3 and the calcium concentrations must satisfy the following equation:

$$2 \times 10^{-3} = c_{Ca}\left(\frac{1}{1010}\right) + \theta_{Ca}(0.95)(2 \times 10^{-3})$$

Also it is evident that

$$\theta_{Na} + \theta_{Ca} = 1$$

The equilibrium equation, Eq. (1.10), is

$$0.16 = \frac{\theta_{Na}^2 c_{Ca}}{\theta_{Ca} c_{Na}^2}$$

There are four equations and four unknowns. These can be solved and it is easy to verify that

$$\theta_{Na} = 0.394$$

$$\theta_{Ca} = 0.606$$

$$c_{Na} = 1.163 \text{ kg equ/m}^3$$

$$c_{Ca} = 0.857 \text{ kg equ/m}^3$$

satisfy the four equations. An additional check is needed to ensure that the final aqueous composition is electrically neutral. The concentration of chloride must equal the sum of the two cation concentrations. Thus, $c_{Cl} = c_{Na} + c_{Ca}$ and the solution is electrically neutral.

The calculations shown in the previous example can be repeated for a number of different solution compositions. Typical results are presented in graphical form in Fig. 1.13. Plotted is a fraction of calcium ions in solution y_{Ca} defined as

$$y_{Ca} = c_{Ca}/c_{AN}$$

where c_{AN} is the total anion concentration (in this case $c_{AN} = c_{Cl}$), versus the fraction of the surface covered by calcium ions. This graph shows two striking features. For small anion concentrations, c_{AN}, the surface is covered almost en-

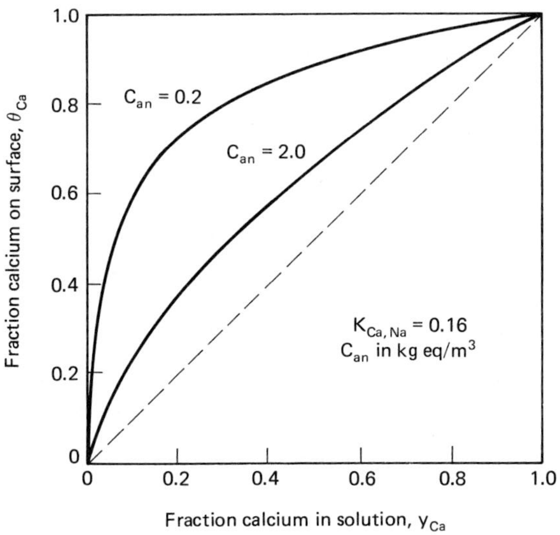

Figure 1.13 Ion exchange isotherms showing the fraction surface coverage of calcium plotted against the fraction in solution for various values of electrolyte concentration.

tirely by calcium ions even when the solution in contact with the clays contains very few calcium ions relative to the number of sodium ions. Secondly, regardless of the anion concentration, the surface exhibits a strong preference for calcium although this preference is less pronounced as the total anion concentration is increased. This behavior has important ramifications with respect to formation damage (see Chap. 4).

The montmorillonites have a strong preference for divalent ions as opposed to monovalent ions. This preference remains true for most clays and becomes even more pronounced as the surface charge density of the clay increases. For example, van Olphen reports that the charge density of illites is larger than that of montmorillonites [6]. We would therefore expect illites to show a stronger preference for calcium than does montmorillonite. This stronger preference would be reflected as an increased selectivity coefficient when the calcium form of the clay is the product. Consistent with this hypothesis Rogers et al. [22] found that $K_{Na,Ca}$ for illites is, as anticipated, greater than the corresponding value for either montmorillonites or attapulgites. Thus, the selectivity coefficient is a function of both the degree and type of isomorphic substitution since illites and montmorillonites are both three-layer clays, as well as the clay structure, temperature, solution pH, and many other factors. For critical cases such as the design of expensive and complex oil recovery processes, the selectivity coefficients of representative formation samples should be measured.

The same concepts used in Example 1.4 to determine the equilibrium state of a solution containing two cations can be applied to a multicomponent cation mixture. Suppose c_i^T represents the total concentration of cation i expressed in kg equ/m^3. Then a material balance expressing the fact that the cations are present either in the aqueous solution or associated with the clays can be written as follows:

$$c_i^T = c_i + \theta_i Q_v \qquad i = 1, 2, \ldots, (N) \tag{1.11}$$

Note that the system cation-exchange capacity is used here to maintain all terms in Eq. (1.11) on the same basis, namely, per volume of aqueous phase. The c_i and θ_i are unknowns and the c_i^T are generally known quantities. Thus there are $2N$ unknowns but only N material balance equations. An additional N equations are required. The equilibrium conditions provide $N - 1$ of these additional equations. Thus Eq. (1.10)

$$K_{ij} = \frac{\theta_j^{Z_i} c_i^{Z_j}}{\theta_i^{Z_j} c_j^{Z_i}} \tag{1.12}$$

provides $N - 1$ additional independent equations. One can obviously write $N(N - 1)/2$ equations for all i and j excluding $i = j$; however, only $N - 1$ of these will be independent. The remaining equations can be derived by multiplying and dividing combinations of any $N - 1$ independent equilibrium relationships. Thus for a 3-component system, K_{12} and K_{13} are independent. This is clear because components 2 and 3 appear just once. However, K_{23} can be expressed as $K_{13}/$

TABLE 1.5 Selectivity Coefficients for Various Cation-Exchange Reactions*

Ion-exchange reaction	Substrate	Selectivity coefficient	Notes
Na - mont + K	Wyoming benonite < 1 μ	$K_{Na,K} = 1.8$	
Na - kaolinite + K	Georgia kaolinite < 2 μ	$K_{Na,K} = 5.2$	Average over a wide range.
Na - illite + K	Winsum illite, Neth. <2 μ	$K_{Na,K} = 400$	
Na - mont + Ca	Clay spur <2 μ	$K_{Na,Ca} = 3.6$	Average over a wide range.
Na - verm + Ca	Transvaal "world"	$K_{Na,Ca} = 1.4$	
Na - illite + Ca	Illite API 35	$K_{Na,Ca} = 8.5$	Average over a wide range.
Na - mont + Ba	Chambers API 23 0.1–0.5 μ	$K_{Na,Ba} = 3.4$	
K - illite + Mg	Winsum illite, Neth. <2 μ	$K_{K,Mg} = 0.1$	
K - mont + Ca	Clay spur <2 μ	$K_{K,Ca} = 0.61$	
K - illite + Ca	Winsum illite, Neth. <2 μ	$K_{K,Ca} = 0.11$	
K - kaolinite + Ca	Chodau kaolinite <2 μ	$K_{K,Ca} = 0.031$	
K - chlorite + Ca	Rimpfischwang chlorite <2 μ	$K_{K,Ca} = 0.045$	
Ca - mont + Ba	Utah bentonite <1 μ	$K_{Ca,Ba} = 1.5$	
Mg - mont + Ca	Utah bentonite <1 μ	$K_{Mg,Ca} = 1.17$	

*Average values adapted from data compiled by Bolt [23].

K_{12} and is therefore not independent. The last equation needed to solve for the $2N$ unknowns is

$$\sum \theta_i = 1 \tag{1.13}$$

and as a check any solution should be electrically neutral so that

$$\sum c_i = c_{AN} \tag{1.14}$$

where c_{AN} is the anion concentration.

Table 1.5 lists a few selectivity constants taken from an extensive listing offered by Bolt [23]. Since the values given are often averages of a range of values, the reader should not confer an unjustified degree of precision upon them. They merely serve to indicate trends. The scatter of values seen in Fig. 1.12 is, unfortunately, the rule rather than the exception. Nevertheless, certain trends can be observed by examining the results presented in Table 1.5. Generally when potassium cations (K^+) are exchanged for sodium cations (Na^+), $K_{Na,K}$ is greater than unity indicating a preference of the clays for potassium cations. The Winsum illite is seen to preferentially adsorb K^+ almost to the complete exclusion of Na^+.

Calculations will show that divalent cations are generally, but not always, preferentially selected over monovalent ions, whereas the exchange between Ca^{++} and Ba^{++} or Ca^{++} and Mg^{++} indicates that the clay systems studied have little preference for one divalent ion as compared to another one.

1.5 CHEMISTRY OF FORMATION WATERS

Cations

The composition of the dissolved solids found in oil field water is dependent on a number of factors: the composition of the water in the depositional environment of the sedimentary rock, subsequent changes by rock/water interaction during sediment compaction, changes by water/rock interaction during migration, and changes by mixing with infiltrating waters. Since most oil-bearing sediments are marine in origin, it is not surprising that the major cation in oil field brines is sodium (Na^+).

Calcium (Ca^{++}) is also generally present in oil field waters, but its concentration is quite variable, ranging from less than 100 ppm (parts per million) to values in excess of 30,000 ppm although its concentration is usually much less than 30,000 ppm. In general, waters associated with limestones, dolomites, or gypsum will contain more calcium than waters associated with sandstones [24]. The presence of divalent ions such as calcium will be shown to be an important consideration in designing well treatments. As the total amount of dissolved solids increases, the divalent ion concentration may also be expected to increase [26].

Magnesium (Mg^{++}) concentrations in oil field brines, like calcium concentrations, are dependent upon the origin of the waters and the rock/water interaction. Perhaps the most important reaction affecting the magnesium content is the formation of dolomite.

The other cations often present in oil field brines in concentrations greater than 10 ppm are potassium (K^+), strontium (Sr^{++}), lithium (Li^+), and barium (Ba^{++}). These alkali and alkali earth metals react similarly (e.g., with respect to ion exchange) to sodium and calcium.

Potassium is generally depleted relative to sodium in oil field waters as compared to the ratio of the two in seawater. This is probably caused by mineral formation, which uses potassium ions in preference to sodium [25].

Strontium is often present in concentrations ranging from a trace to 3000 ppm. Barium may also occur in concentrations exceeding 100 ppm; often its concentration is much less.

Other cations that may occur in concentrations greater than 10 ppm are aluminum, ammonium, iron, lead, manganese, and zinc. Aluminum is present in most clays but its solubility is a complex function of the solution pH and composition. Low pH waters (less than 4) will generally exhibit large aluminum concentrations.

Some investigators believe that the presence of ammonium (NH_4^+) in formation waters is an indicator of petroleum that has been acted upon by bacteria.

The solubility of iron is a strong function of pH, the iron compound involved, and the amounts and types of other ions present in solution. Iron in the +3 state of oxidation (Fe^{+++}) tends to precipitate more readily (e.g., at lower pH) than iron in the +2 state (Fe^{++}). Formation rocks contain iron carbonates, hydroxides, oxides, and sulfides as well as iron that has been substituted into tetrahedral and octahedral sheets of the clays. Therefore, when the waters are in equilibrium with the formation minerals, they generally contain iron.

Substantial amounts of lead or manganese may be found in a few oil field waters, but normally only trace quantities are present.

Anions

Water containing quantities of cations (Na^+, Ca^{++}, etc.) must necessarily contain similar quantities of anions (Cl^-, SO_4^{--}, etc.) so that the solution is electrically neutral. The major anion in most formation waters is chloride. It is also the predominate anion in seawater. Chloride ions do not readily form precipitates. Chlorides are sometimes used as tracers since they are relatively unreactive and do not tend to adsorb on clays in substantial quantities.

Bromide ranks as the next most predominant anion in many formation waters. As a general rule, formation water samples are more likely to be analyzed for bicarbonate or sulfate than for bromide. Thus, there is a substantial lack of data documenting bromide concentrations; however, their concentrations range from less than 50 ppm to more than 6000 ppm.

Bicarbonates or carbonates (or both) are almost always present in formation waters. Waters in equilibrium with formations containing carbonate minerals will contain carbonate in some form, which depends on the pH and other factors including concentration of ions, temperature, etc.

Seawater contains about 900 ppm of sulfur as sulfate, whereas formation waters may contain quantities ranging from almost none to several thousand parts per million. The amount of sulfate ion present depends on many factors, including bacterial activity.

Formation waters may also contain iodide, fluoride, phosphates, arsenates, and numerous other anions present in trace quantities [24].

The primary constituents of oil field waters are listed in Table 1.6.

Silica

The solubility of silica in water is a function of temperature, pressure, pH, and other ions in solution. Most silica in natural water probably is in the form of silicic acid, H_4SiO_4. The solubility increases with increasing temperature up to about

TABLE 1.6 Primary Constituents and Properties of Oil Field Waters [27]

Cations	Anions	Other properties
Calcium (Ca^{++})	Chloride (Cl^-)	pH
Magnesium (Mg^{++})	Carbonate (CO_3^{--})	Suspended solids
Sodium (Na^+)	Bicarbonate (HCO_3^-)	Turbidity
Iron (Fe^{++} or Fe^{+++})	Sulfate (SO_4^{--})	Temperature
Barium (Ba^{++})	Bromide (Br^-)	Specific gravity
Strontium (Sr^{++})		Dissolved oxygen
		Dissolved carbon dioxide
		Sulfide as H_2S
		Bacterial population

125°C. Between 125°C and 200°C, the solubility decreases with increasing temperature. Formation waters usually contain less than 30 ppm dissolved silica [24].

Representation of Water Composition

The analysis of formation waters is often presented graphically. The diagram or pattern obtained by graphically plotting the results of water analysis will often highlight important points about the analysis that may be missed by simply examining a table of data. There are many different diagrams in use. However, the

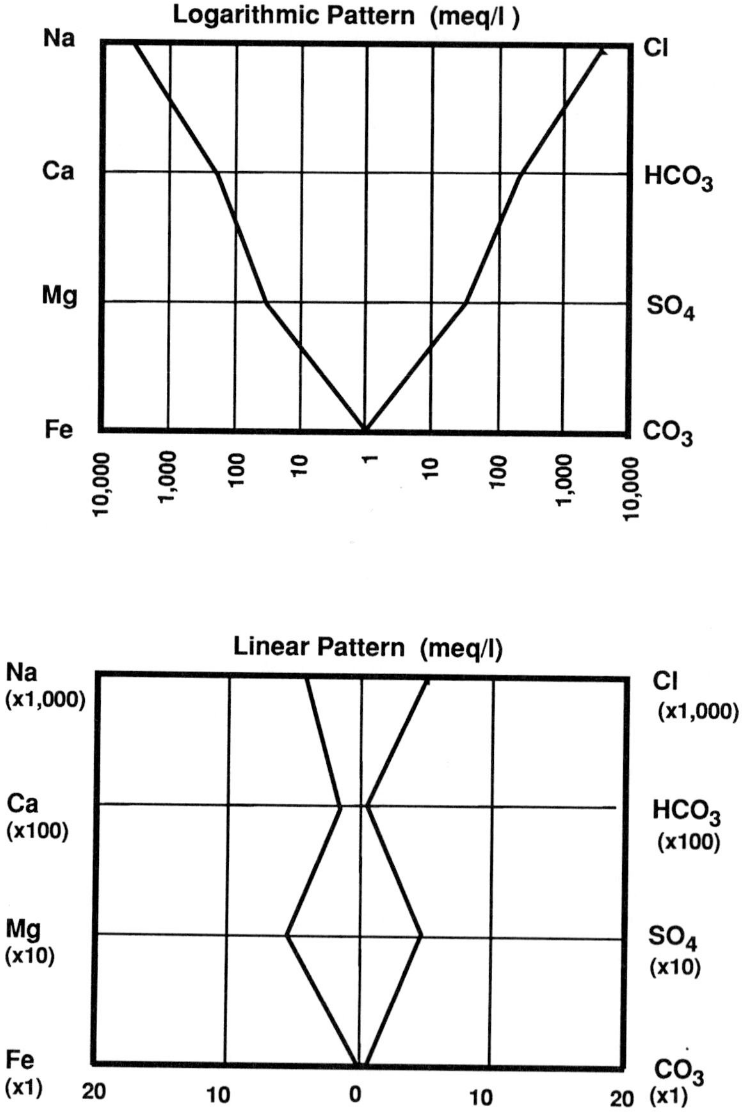

Figure 1.14 Graphical presentation of water analysis Stiff method.

Stiff method [25] has been adopted by the API and is the one most likely to be encountered by the practicing petroleum engineer. A typical pattern is shown in Fig. 1.14. The concentrations of the cations are plotted to the left of the centerline and the concentrations of the anions to the right of the centerline. Since the solutions must be electrically neutral, the number of equivalents of cations must equal that of the anions. The analysis shown in Fig. 1.14 is tabulated in Table 1.7.

Table 1.8 gives values of the equivalent weights of some selected compounds. Note that one equivalent weight of cation in an aqueous solution must be balanced by an equivalent weight of anions. Thus if a given volume of solution contains 20 kg of calcium, it would contain 35.5 kg of chloride.

Example 1.5 Verification of Water Analysis

Each formation water sample must contain the same total equivalents of cations (positive) as it does of anions (negative). Given the seawater analysis shown in Table 1.9, determine if it satisfies the requirement of electrical neutrality.

Solution The weight of one equivalent is given in Table 1.8 and thus the total cation kg equ/m^3 is given by

$$\text{Total cation} = (10.8/23) + (1.3/12.2) + (0.4/20) = 0.596 \text{ kg equ/m}^3$$

Similarly,

$$\text{Total anion} = (19.4/35.5) + (2.7/48) + (1.4/61) = 0.626 \text{ kg equ/m}^3$$

Thus there is a small discrepancy that may be attributable to experimental error or the failure to analyze for a cation, such as potassium, which may actually be present.

Total Alkalinity

The total alkalinity is formally defined as the equivalent sum of the bases that are titratable with strong acid. All anions converted to unchanged species as the solution pH is progressively lowered are included in the total alkalinity [28]. Thus bicarbonates (HCO_3^-), carbonates (CO_3^{--}), borate ($B(OH)_4^-$), silicic acid

TABLE 1.7 Water Analysis Shown in Fig. 1.14

Ion	meq/liter	mg/liter
Na^+	4,053	93,230
Ca^{++}	258	5,173
Mg^{++}	51	620
$Fe^{(total)}$	0.6	12
Cl^-	4,320	153,200
HCO_3^-	3	195
SO_4^{--}	40	1,910
CO_3^{--}	0	0

TABLE 1.8 Selected Ions, Compounds, and Equivalent Weights

Cations	Symbol	Equivalent weight	Anions	Symbol	Equivalent weight
Barium	Ba^{++}	68.7	Bicarbonate	HCO_3^-	61
Calcium	Ca^{++}	20	Carbonate	CO_3^{--}	30
Hydrogen	H^+	1	Chloride	Cl^-	35.5
Iron			Hydroxyl	OH^-	17
Ferrous	Fe^{++}	27.9	Oxide	O^{--}	8
Ferric	Fe^{+++}	18.6	Phosphate	PO_4^{--}	31.6
Magnesium	Mg^{++}	12.2	Sulfate	SO_4^{--}	48
Sodium	Na^+	23	Sulfide	S^{--}	16
Strontium	Sr^{++}	43.8	Sulfite	SO_3^{--}	40

($H_3SiO_4^-$), bisulfide (HS^-), organic anions, and hydroxyl ions (OH^-) which are converted to H_2CO_3, $B(OH)_3$, H_4SiO_4, H_2S, organic acid, and H_2O, respectively, all contribute to the alkalinity. They are included in the titration of natural waters with a strong acid. The sum [28] shown in Eq. (1.14) represents moles per volume of acid required to neutralize a natural water:

$$\text{Alkalinity} = c_{\text{HCO}} + 2c_{\text{CO}_3} + c_{\text{B(OH)}_4} + c_{\text{H}_3\text{SiO}_4} + c_{\text{HS}} + c_{\text{organic anion}} + c_{\text{OH}} - c_{\text{H}} \tag{1.14}$$

In most formation waters the bicarbonate and carbonate ions are by far the largest contributors to the alkalinity. It is important to note that the alkalinity is independent of temperature and pressure. It does not change even if carbon dioxide (CO_2) is dissolved or removed from the solution since CO_2 is a neutral species carrying no net electrical charge. For example, removal of one mole of CO_2 at the expense of HCO_3^- will increase the concentration of OH^- by one mole. Thus alkalinity titrations obtained from surface measurements, assuming no precipitation of solid (which does change the alkalinity because, e.g., both positive charges Ca^{++} and negative charges CO_3^{--} are removed from solution), are useful. The total alkalinity is very useful in determining the dangers of calcium carbonate precipitation if, for example, the solution pH changes.

TABLE 1.9 Approximate Seawater Analysis

Cations	kg/m^3	Anions	kg/m^3
Na^+	10.8	Cl^-	19.4
Mg^{++}	1.3	SO_4^{--}	2.7
Ca^{++}	0.4	HCO_3^-	1.4

1.6 CHEMICAL FACTORS IN CRUDE OILS

Crude oil is not a uniform substance and its appearance and characteristics will vary from reservoir to reservoir. It can contain dissolved gases, solids, and colloidal suspensions. The main constituents of crude oil are, however, hydrogen and carbon present in the form of hydrocarbons; sulfur, nitrogen, and oxygen also appear in the form of organic compounds. Metals may also appear either as complexes with the organic matter or as salts of organic acids.

The API gravity has been one method of characterizing oils, but early observations that the residues from atmospheric distillation of Pennsylvanian crudes separated solid waxes on cooling whereas residues from California or Texas did not separate waxes but contained asphaltic materials led to a differentiation between *wax-bearing* and *wax-free* crude oils. Intermediate between these two types of crude are many oils, which upon distillation, yield residues containing both paraffin waxes and asphaltic materials. These are called *intermediate crudes.*

Taking into account the three types of crudes and their specific gravities a classification scheme evolved, which is in common use today. The modern scheme includes a fourth classification. *Naphthenic crudes* are those which also have wax and asphatenic residues but are more paraffinic than intermediate crudes.

The important aspects of concern are those chemical factors which can lead to production problems. Paraffin crudes, for example, may tend to precipitate waxes as they are cooled from reservoir temperatures. These waxes can clog the matrix near the wellbore, the perforations, or the tubing. Well treatments to remove these waxes may be required to restore adequate production.

Asphaltic crudes may contain colloidal particles which will agglomerate and precipitate upon addition of a nonpolar compound such as heptane, pertane, or carbon dioxide. In some cases this tendency can result in formation damage (see Chap. 4).

REFERENCES

1.1. Brownlow, A. H., *Geochemistry*, Chap. 7, Englewood Cliffs: Prentice Hall, 1982.

1.2. Grim, R. E., *Clay Mineralogy*, 2nd ed., Manchester: McGraw-Hill, 1968.

1.3. Velde, B., *Clay Minerals: A Physico-Chemical Explanation of Their Occurrence*, Amsterdam: Elsevier, 1985.

1.4. Bennett, R. H., and Hulbert, M. H., *Clay Microstructure*, International Human Resources Development Corporation, Boston (1986), 25.

1.5. Neasham, J. J., "The Morphology of Dispersed Clays in Sandstone Reservoirs and its Effect on Shaliness, Pore Space and Fluid Properties," SPE 6858, presented at the 52nd Annual Fall Technical Conference and Exhibition of the Society of Petroleum Engineers, Denver, Colorado, 1977.

1.6. Van Olphen, H., *Introduction to Clay Chemistry*, 2nd ed., London: Interscience, 1977.

1.7. Parfitt, G. D., and Rochester, C. H., in *Characterization of Powder Surfaces*, G. D. Parfitt and K. S. Sing (eds.), San Francisco: Academic Press, 1976.

1.8. Klier, K., and Zettlemoyer, A. C., *J. Colloid and Interface Sci., 58* (1977) 216.

1.9. James, R. O., and Parks, G. A., in *Surface and Colloid Science*, Vol. 12, E. Matijevic (ed.), New York: Plenum Press, 1981.
1.10. Davis, J. A., James, R. O., and Leckie, J. D., *J. Colloid and Interface Sci., 63* (1978) 480.
1.11. Riese, A. C., "Adsorption of Radium and Thorium onto Quartz and Kaolinite: A Comparison of Solution/Surface Equilibria Models," PhD dissertation, Colorado School of Mines, Golden, 1982.
1.12. Yates, D. E., Levine, S., and Healy, T. W., *J. Chem. Soc., Faraday Trans. I, 70* (1974) 1807.
1.13. Hiemenz, P. C., *Principles of Colloid and Surface Chemistry*, 2nd ed., New York: Marcel Dekker (1986) 690.
1.14. Strumm, W., and Morgan, J. J., *Aquatic Chemistry*, New York: John Wiley, 1970, 583, and Drever, J. I., *The Geochemistry of Natural Waters*, Englewood Cliffs: Prentice Hall, 1982.
1.15. Helfferich, F., *Ion Exchange*, New York: McGraw-Hill, 1962.
1.16. Sawhney, B. L., *Clay and Clay Minerals, 20* (1972) 93.
1.17. Smith, F. W., *J. Pet. Tech., 30* (1978) 959.
1.18. Reddy, M. M., in *Ion Exchange and Solvent Extraction*, Vol. 7, J. A. Marinsky and Y. Marcus (eds.), New York: Marcel Dekker (1977) 165.
1.19. Barrier, R. M., and Klinowski, J., *J. Chem. Soc., Faraday Transactions I, 70* (1974) 2080.
1.20. Gaines, G. L., and Thomas, H. C., *The J. Chem. Phys., 21* (1953) 714.
1.21. Harned, H. S., and Owens, B. B., *The Physical Chemistry of Electrolyte Solutions*, 3rd ed., New York: Reinhold Book Company, 1958.
1.22. Rogers, W. J., Shiao, S. Y., Meyer, R. E., Westmoreland, C. G., and Lietzke, M. H., in *Surface Phenomena in Enhanced Oil Recovery*, D. O. Shah (ed.), New York: Plenum Press (1982) 695.
1.23. Bolt, G. H., *Soil Chemistry*, Chap. 3, Amsterdam: Elsevier, 1979.
1.24. Collins, A. G., and Wright, C. C., "Enhanced Oil Recovery Injection Waters," U.S. Department of Energy Final Report DOE/BETC RI-82/5, 1982.
1.25. Collins, A. G., *Geochemistry of Oilfield Waters*, Amsterdam: Elsevier (1975) 138.
1.26. Gash, B. H., Griffith, T. D., and Chan, A. F., "Phase Behavior Effects on the Oil Displacement Mechanism of Broad Equivalent Weight Surfactants Systems," SPE 9812, presented at the Second Symposium on Enhanced Oil Recovery of Society of Petroleum Engineers, Tulsa, Oklahoma, 1981.
1.27. Patton, C. C., *Oilfield Water Systems*, Campbell Petroleum Services, Norman, Oklahoma, 1977.
1.28. Drever, J. I., *The Geochemistry of Natural Waters*, Englewood Cliffs: Prentice Hall, 1982.

PROBLEMS

***1.1.** If the fixed surface charge on a kaolinite particle can be entirely attributed to the substitution of Al^{+3} for Si^{+4} in the silica sheet, what fraction of silicon atoms have been replaced to give a surface charge of $-20\ \mu C/m^2$? The area per silicon atom in the silica sheet can be taken to be approximately 30 nm^2.

* Level one problems designated by a single asterisk are intended to reinforce concepts presented in a chapter and they are often similar to examples that are developed within the text.

***1.2.** What is the Debye length when a charged solid surface is contacted with an aqueous solution containing 10^{-3} kg moles/m^3 of KCl? The temperature is 18°C.

Is this length changed if NaCl is substituted for KCl keeping the concentration constant? Is it changed if Na_2SO_4 is substituted for KCl? If the answer in either or both cases is yes, calculate the new Debye length.

****1.3.** The model for creation of surface charge on mineral oxide surfaces as depicted by Fig. 1.8 envisions protonation and deprotonation reactions. These are represented by Eqs. (1.2) and (1.3). If the equilibrium constant for reaction (1.2) is K_1 and that for reaction (1.3) is K_2, then show that the pH of zero charge (pzc) is given by

$$\text{pzc} = -\log \sqrt{K_1 K_2}$$

****1.4.** A formation water contains 30,000 ppm $CaCl_2$ and 100,000 ppm NaCl. If the cation-exchange capacity of the reservoir rock is 1.5×10^{-3} kg equ/kg of rock, calculate the total amount of Ca^{++} (in kg) present in 1 m^3 of formation. This amount includes the Ca^{++} in the solution contained within the pores and that associated with the clays. The density of the rock is 2500 kg/m^3, the density of the formation water is 1100 kg/m^3, and the formation porosity is 0.15.

For this formation $K_{Na,Ca}$ is 7.5.

***1.5.** The cation-exchange capacity of a reservoir rock is 1.5 equ/kg of rock. If the formation water contains only calcium chloride ($CaCl_2$) at a concentration of 30,000 ppm, calculate the total amount of Ca^{++} (in kg) that would be associated with the ion-exchange sites of the clays and in the pores of 1 m^3 of reservoir rock. The density of the rock is 2500 kg/m^3 and that of the formation water is 1010 kg/m^3. The porosity of formation is 0.20.

***1.6.** A formation water contains 1.2 wt% sodium chloride and 0.3 wt% calcium chloride. How many kg of calcium are retained on the clays per m^3 of formation material? There are no other ions in solution and the formation water density is 1018 kg/m^3.

DATA

Well pattern: 40-acre spacing

Formation thickness: 12 m

Cation-exchange capacity: 10 meq/kg of rock

Rock density: 2700 kg/m^3

Formation porosity: 0.18

Selectivity coefficient: $K_{Ca,Na} = 0.2$

Permeability: 25 mD

****1.7.** One liter of an aqueous electrolyte solution is blended with 2.0 g of kaolinite ($\overline{Q}_v$ = 15 g equ/100 g clay) which is in the potassium form prior to mixing. The *final* (after mixing) electrolyte concentration in the water is 12 g/liter of calcium, 23 g/liter of potassium, and 42.1 g/liter of chloride. Answer the following:

(a) Prove that the final solution is electrically neutral.

(b) Calculate the original solution composition; that is, the composition that existed before mixing clay with the aqueous solution, given that $K_{Ca,K} = 0.25$.

** Level two problems designated by two asterisks are more comprehensive and may require the unification of several concepts, numerical evaluation, or even the derivation of new but related equations. Level two problems are intended for use primarily in graduate classes, but undergraduates may want to attempt some of them.

1.8. One hundred grams of a clay in sodium form are blended with 1 liter of an aqueous solution containing 13 meq of calcium chloride ($CaCl_2$) and 12 meq of sodium chloride (NaCl). After sufficient time has elapsed so that the clay and solution are in equilibrium, analysis of the solution shows that it contains 15 meq/liter NaCl. It is known that $K_{Ca,Na} = 2$. What is the cation-exchange capacity of the clay expressed in meq per 100 g?

*1.9. Plot the water analysis shown in the accompanying table using the Stiff linear pattern representation. Does the analysis as presented represent an electrically neutral solution?

Ion	Concentration (ppm)
Sodium, Na^+	69,000
Calcium, Ca^{++}	20,000
Magnesium, Mg^{++}	120
Sulfate, SO_4^{--}	164,000
Chloride, Cl^-	64,000

*1.10. Given in Table P1.10 are formation water analyses from Ghawar field in Saudi Arabia. Waterflooding with seawater is considered after having waterflooded for some years with Wasia water. Construct a Stiff plot showing a composition of high-salinity Arab-D. Use the logarithmic scale. Comment on possible difficulties which might arise using seawater [see J. C. Lindlof and K. G. Stoffer, "A Case Study of Seawater Injection Incompatibility," *J. Pet. Tech.* (July 1983) 1256–1262].

TABLE P1.10 Analyses of Water Associated with Seawater Flood at Uthmaniyah

	Low-Salinity Arab-D	High-Salinity Arab-D	Wasia	Seawater
Na^+	29,680	51,187	1,504	18,043
Ca^{++}	13,574	29,760	392	652
Mg^{++}	1,575	4,264	66	2,159
Ba^{++}	8	10	0	0
Sr^{++}	557	1,035	5	11
HCO_3^-	369	351	192	119
Cl^-	73,861	143,285	2,577	31,808
SO_4^{--}	404	108	700	4,450
CO_3^{--}	0	0	0	27
TDS, mg/dm^3	120,000	230,000	5,436	57,269

Tabulate values are given in mg/dm^3.

2
Physical Properties of Formation Materials

The mechanical properties of rock are still subjects of active research, indicating both their complexity and importance. It is not possible to completely consider these matters in the brief description given here in Sections 2.1 and 2.2. For additional information the reader should consult Jaeger [1], which remains an excellent primer, and perhaps Boubrie et al. [2], for a modern description of wave propagation in porous media. The information presented here is, however, sufficient for our purposes.

Thermal properties are also important whenever temperature represents a significant factor in the design of a well treatment. These properties are summarized in Sections 2.3–2.5.

2.1 MECHANICAL PROPERTIES OF FORMATION ROCKS

Linear Elasticity

The classical laws of elasticity have generally been developed for homogeneous and isotropic materials. Their application to porous media cannot be fully justified. The materials of interest here are by their very essence heterogeneous on all scales. At the pore level, one finds a complex topology formed by grains composed of many different minerals and on a larger scale, cracks, layers, and other chemical and structural features defying description except in some average sense are widely prevalent. Thus, the chemical and physical states of rocks are quite varied and these states have not been related in any fundamental way to mechanical behavior [3].

Most of the applications of concern to us requiring the response of rocks to an applied force consider them to be ideal elastic bodies. Although the dangers of such a gross simplification should be recognized, more complex phenomena such as *creep*, known to occur under certain conditions, have not yet been included in most production engineering calculations. Even more disconcerting is the fact that predictions defining the limitations of the linear laws of elasticity are not yet available, so we cannot adequately define the valid range of the equations developed in this text.

For an ideal elastic body characterized by its Young's modulus, E, and its Poisson's ratio, ν, the compressive stresses σ_x, σ_y, and σ_z required to yield strains ϵ_x, ϵ_y, and ϵ_z are given as follows [2]:

$$\epsilon_i = \frac{1 + \nu}{E}\sigma_i - \frac{\nu}{E}(\sigma_x + \sigma_y + \sigma_z) \qquad (i = x, y, \text{ or } z) \tag{2.1}$$

where the σ_i are stresses defined as the ith component of the force per unit area that material at lessor values of the coordinate axis i exerts on the material at greater values of i across a plane drawn perpendicular to i. Thus, a positive value of σ_i ($i = x, y,$ or z) denotes that the material on one side of an imaginary plane drawn perpendicular to the i-axis tends to compress the material on the side at greater i or equivalently, the force on the material at lessor values of i is a tension. The ϵ_i are strains in the i-direction. They represent the decrease (compression) in distance between two points along the i-axis after the forces have been applied divided by the original distance between them. We note that the definitions given here embody a degree of arbitrariness with regard to the signs and others may define σ_i and ϵ_i so that they are negative when our values are positive [1].

Young's modulus and Poisson's ratio are the mechanical properties of rock most often reported (see Table 2.1); however, Eq. (2.1) can be written in other entirely equivalent forms. For example, the following equation is one such form:

$$\sigma_i = 2G\left\{\epsilon_i + \frac{\nu}{1 - 2\nu}e\right\} \tag{2.2}$$

where e is the decrease in volume per unit volume given by

$$e = \epsilon_x + \epsilon_y + \epsilon_z \tag{2.3}$$

The modulus of rigidity, G, is related to Young's modulus by the equation

$$G = \frac{E}{2(1 + \nu)} \tag{2.4}$$

Equations (2.1) and (2.2) apply in any coordinate system; however, generally shearing forces as well as compressive forces exist. There is one coordinate system, called the *principal coordinate system*, for which shearing forces do not exist [1]. Since all of the problems considered here can be solved in principal coordinates, shearing stresses are not defined.

The equations relating stress and strain given here represent the most general

TABLE 2.1 Properties of Various Formations [12]

Formation name	Depth (m)	Location	Young's modulus *($\times 10^{-7}$kPa)	Poisson's ratio	Permeability to air ($\times 10^{15}$/m^2)	Porosity (%)
Austin Chalk	731	Caldwell Co., Tex.	1.6	0.276	0.03	4.8
Bartelsville	2408	Cleveland Co., Okla.	4.7	0.124	13.0	16.6
Basal Quartz	1487	Calgary, Canada	1.4	0.147	0.85	13.4
Buda Lime	1829	Frio Co., Tex.	2.6	0.245	0.01	2.1
Cotton Valley	2743	Claiborne Co., La.	1.8	0.016	0.27	8.7
Devonian	3660	Glasscock Co., Tex.	4.3	0.088	0.01	2.1
Grayburg	875	Crockett Co., Tex.	3.3	0.342	2.4	16.9
McClosky Dolomite	914	Edwards Co., Ill.	4.9	0.370	34	9.7
Mississippian	2738	San Juan Co., N. Mex.	5.4	0.110	0.43	5.5
Muddy	2105	Campbell Co., Wyo.	1.4	0.143	10.6	18.3
Red Oak	1972	Latimer Co., Okla.	1.4	1.125	0.80	13.1
San Andres	1048	Ward Co., Tex.	4.4	0.260	0.03	0.3
Smackovev	2896	Navarro Co., Tex.	3.7	0.310	4.9	20.1
Wilcox	2682	Bee Co., Tex.	3.4	0.181	0.11	9.0

* This notation means that each entry in this column has been multiplied by 10^{-7}. Thus, for Austin Chalk, $E = 1.6 \times 10^{+7}$kPa.

linear constituitive relationship for an isotropic and homogeneous material. Whether or not such a relationship actually applies to real systems is a matter requiring experimental justification. Figure 2.1 shows the behavior of rock during a simple compression test [2]. For stresses less than σ_x^S, the main impact of the compressive stress on the rock is to close any existing small cracks and fissures. Since few new cracks are created during this initial stage of compression, it is reversible; that is, if the stress is decreased, the stress–strain curve will be retraced. During this initial phase the stress–strain curve is not linear and the material tends to stiffen due to the closure of microscopic cracks.

Once the applied stress exceeds σ_x^S the crack closure phase is complete and additional forces act to strain the solid. Linear behavior is often observed for stresses exceeding σ_x^S.

The linear range is, however, limited. Once the stresses are sufficient to initiate new cracks, as, for example, at σ_x^F, the process is both nonlinear and irreversible since relaxing the compressive force does not heal the newly created cracks. Finally, above σ_x^L the rapid growth of cracks is observed. Failure occurs at σ_x^M.

Based on the trends shown by Fig. 2.1, the region of stresses between σ_x^S

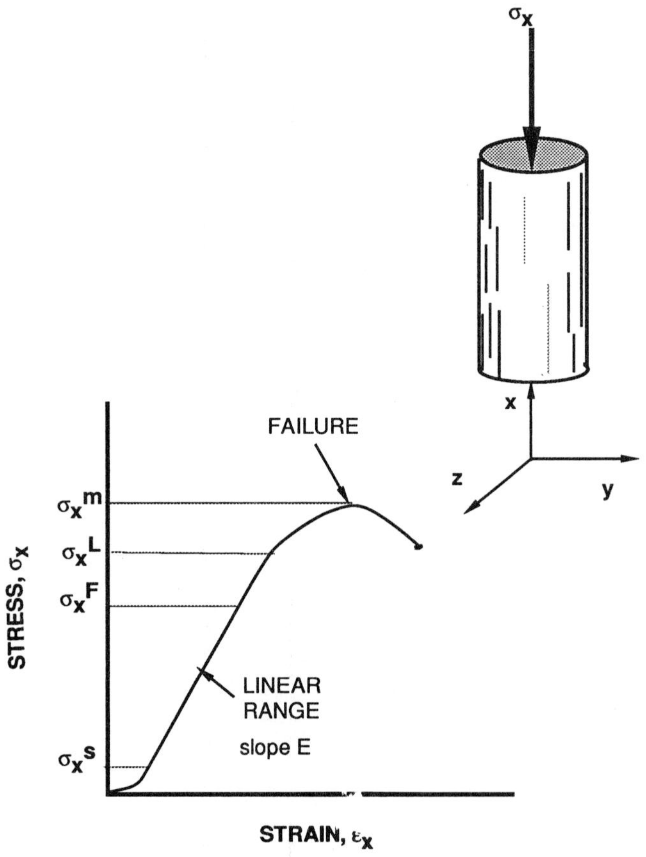

Figure 2.1 Rock behavior during a simple compression test.

and σ_x^F is linear and a meaningful Young's modulus and Poisson's ratio can be defined. The lithostatic pressure of in-situ rocks generally lies between σ_x^S and σ_x^F and thus over a limited range of conditions we would expect the equations of linear elasticity to be useful [2].

Porous Media

The materials of primary interest here are porous and at least some of the pores are interconnected, thereby permitting the flow of fluid. When these pores are filled with a fluid, then the fluid pressure helps to support compressive stresses exerted on an element of rock volume. The representation of the mechanical properties of rock must therefore recognize the existence of a pore pressure, p. One such representation is as follows [4]:

$$\sigma_i - \alpha p = 2G\left(\epsilon_i + \frac{\nu}{1 - 2\nu}e\right) \qquad (i = x, y, z) \tag{2.5}$$

where α is a material property and $\sigma_i - \alpha p$ can be interpreted as that part of the total stress supported by the rock skeleton. We will generally take $\alpha = 1$. There is evidence to indicate that in some cases this is a reasonable approximation [5, 6].

The quantities G and ν in Eq. (2.5) may be considered to be properties of the rock skeleton.

Wave Propagation in Elastic Media

The mechanical properties of formation rock can be estimated based on sonic logs if the porosity is known based on an independent measurement [7, 8]. The concept is that the velocities of the compression (P-wave) and the shear (S-wave) waves contain the relevant [2, 9, 10] information. The two different waves are illustrated by Fig. 2.2. Consider a long rod that has been struck by a hammer at one of its ends. If the blow is delivered along the rod's axis, the material at that end will be compressed and this compression will propagate with a certain velocity along its length. This is the P-wave.

If, on the other hand, a blow is delivered perpendicular to the rod's axis, it

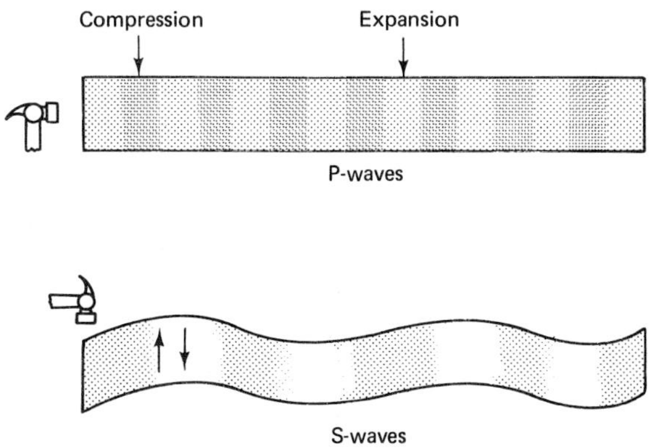

Figure 2.2 P-WAVES and S-WAVES in a solid rod. The S-WAVE velocity is generally slower than the P-WAVE velocity.

results in a shear; that is, a change of shape without a change of volume. This deformation will also propagate along the axis with a certain velocity depending on how strongly the material resists a shearing force. This is the S-wave (see Fig. 2.2). If the bar is struck at a certain angle, then both waves are generated. The velocities of the two waves are in general different.

Both velocities depend on the lithology of the rock, the porosity, and the fluid filling the pores. For reservoir rocks the range of P-wave velocities varies from 8 km/sec to 1.5 km/sec [2, 7]. The S-wave velocity is slower, roughly one half that of the P-wave. Thus, the S-wave velocity must be determined from a recording of the wave train which contains the signal mixed with other waves stemming from reflections from interfaces, anisotropic media, boundary waves, etc. Considerable experience may be required to interpret a sonic log and identify the S-wave. Tixier et al. [8] describe the measurement of the shear wave velocity in friable tertiary sediments as *elusive*.

If both the P- and S-wave velocities are known, then [8]

$$G = \rho_b v_P^2 \tag{2.6}$$

and

$$\nu = \frac{1}{2}\,\frac{\left(\dfrac{v_P}{v_S}\right)^2 - 2}{\left(\dfrac{v_P}{v_S}\right)^2 - 1} \tag{2.7}$$

where v_p and v_s are the P-wave and S-wave velocities, respectively, and ρ_b is the bulk density. These equations apply to a homogeneous elastic material. They are only approximately correct for porous formation rock. G and ν can be calculated given v_p, v_s, and the porosity. The porosity is needed to estimate the bulk density.

Figure 2.3 shows a correlation that can sometimes be used to estimate Young's modulus based on the sonic travel time (v_p^{-1}). This graph does not, however, reflect differences to clay content, fluid properties, etc., which are all known to influence the wave velocities [8].

Example 2.1 Deformation of a Rock Sample

A cubic-shaped sample of Austin Chalk taken from a depth of 731 m is subjected to an axial compressive load of 1000 kPa. No confining pressure is applied to the lateral sides. What is the strain in the axial direction and the volume change of the sample?

Solution Table 2.1 gives the following information for this sample:

$$E = 1.6 \times 10^7 \text{ kPa}$$

$$\nu = 0.276$$

We also note that the porosity is small (4.8%) and can be ignored. From Eq. (2.4) the modulus of rigidity is calculated to be

$$G = \frac{E}{2(1+\nu)} = \frac{1.6 \times 10^7}{2(1.276)} = 6.27 \times 10^6 \text{ kPa}$$

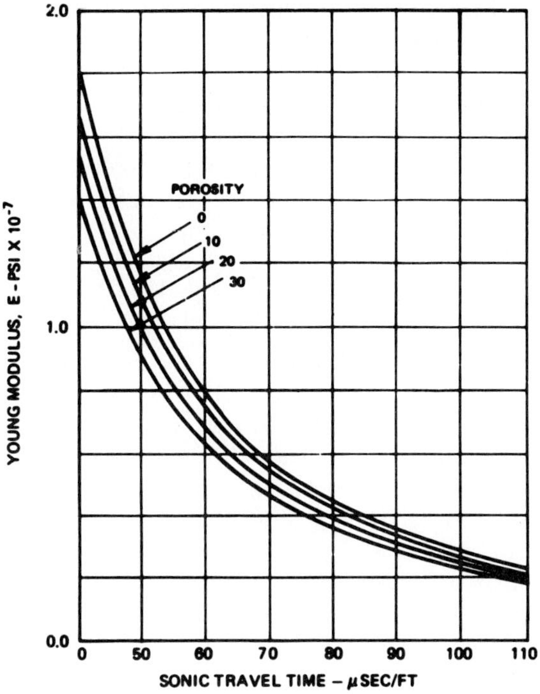

Figure 2.3 Graph showing correlation of Young's modulus (E) versus acoustic travel time. (After B. B. Williams et al., *Acidizing Fundamentals*, Monograph Series, 6, SPE, Richardson, Texas, 1979.)

Based on Eq. (2.2) and considering the coordinate system shown in Fig. 2.1 we have

$$\sigma_x = 1000 = 2(6.27 \times 10^6)\left(\epsilon_x + \frac{0.276}{1 - 2(0.276)}e\right)$$

for the axial load. (Note ϵ_x is positive because the force exerted on the sample at the upper surface is in the positive x-direction.) Along lateral sides there is no stress. Thus,

$$0 = \epsilon_z + \frac{0.276}{1 - 2(0.276)}e \qquad \text{and} \qquad 0 = \epsilon_y + \frac{0.276}{1 - 2(0.276)}e$$

Since $e = \epsilon_x + \epsilon_y + \epsilon_z$ and $\epsilon_z = \epsilon_y$, we can solve these equations to obtain the axial strain

$$\epsilon_x = +6.25 \times 10^{-5}$$ (the positive sign denotes a decrease in length per unit length)

and the fraction volume decrease

$$e = +2.8 \times 10^{-5}$$ (the positive sign denotes a net decrease in volume)

Rock Embedment Strength

Another mechanical property of interest in both propped and acidized fractures is the rock embedment strength, S_{RES}. In this test a high-speed steel ballpoint of 0.05-inch diameter is used [13]. This is mounted on the upper platen of a hydraulic testing machine that loads the specimen hydraulically. The preferred size of rock specimen is a piece of core 3.5 inches in diameter and 6 or more inches long. The additional length is desirable to reduce the possibility of rock breakage. The ballpoint is brought into contact with the rock specimen and the steel ballpoint is embedded to a depth of 0.0125 inch. When the desired penetration is achieved, the load is recorded. The indentation is inspected under a microscope and its diameter in the plane of the rock surface is measured. This diameter is used to calculate a *projected* area, which is then divided into the load to obtain the rock embedment strength. This relationship is

$$S_{\mathrm{RES}} = \frac{4W_p}{\pi d_1^2} \tag{2.8}$$

where W_p is the load and d_1 is the diameter of the indentation. Table 2.2 shows a number of embedment strengths for various formations. A more complete compilation can be found in the *Hydraulic Fracturing Monograph* [13]. Note S_{RES} varies between 7×10^5 kPa and 4×10^6 kPa in laboratory tests.

TABLE 2.2 Embedment Strengths for Various Formations [13]

Formation name	Formation type	Field name	State	County	Approx. depth (m)	S_{RES} (kPa $\times 10^6$)
Atoka (Red Oak)	Sand	Red Oak	Okla.	Latimer	2298	1.4
Grayburg	Lime	Hobbs	N. Mex.	Lea	1250	2.1
Mesa Verde	Sand	Beaver Crk.	Wyo.	Freemont	1088	1.7
San Andres	Lime	Slaughter	Tex.	Gaines	1524	1.1
Sprayberry	Sand	South Wells	Tex.	Dawson	2804	1.1
Walters (Prickly)	Sand	Walters	Okla.	Cotton	664	0.12

2.2 MECHANICAL PROPERTIES OF FLUIDS

We shall be interested in both the isothermal compressibility and viscosity of fluids found in hydrocarbon-bearing formations. These fluids include formation waters, oil, and gas.

Isothermal Compressibility

Isothermal compressibility is defined as

$$\kappa = -\frac{1}{V}\left(\frac{\partial V}{\partial p}\right)_T = \frac{1}{\rho}\left(\frac{\partial \rho}{\partial p}\right)_T \tag{2.9}$$

where V is volume and T is temperature. The κ as defined is always positive. For

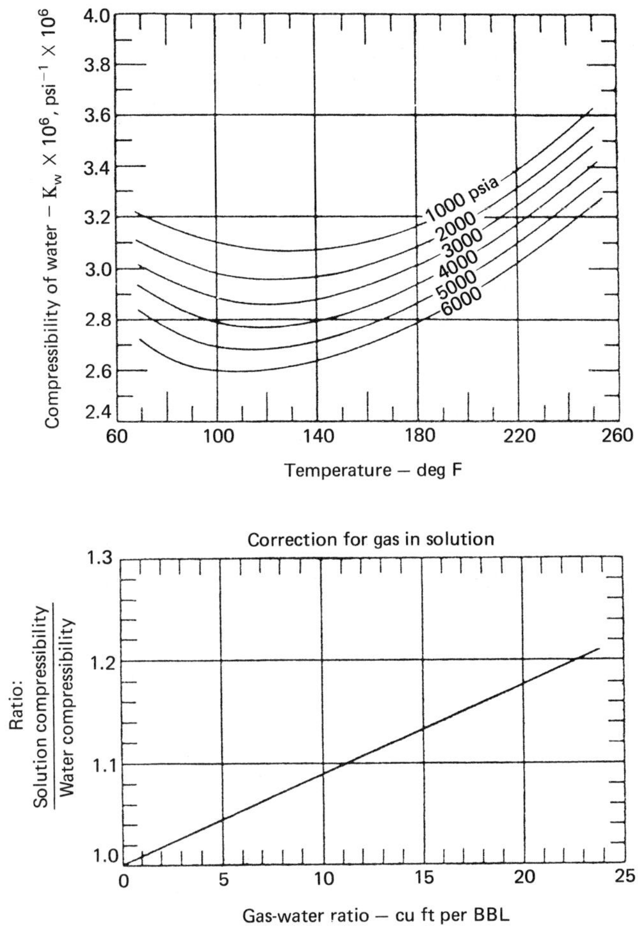

Figure 2.4 Isothermal compressibility of water. (From C. R. Dodson and M. B. Standing, *Drilling and Production Practice*, Washington, D.C.: American Petroleum Institute, 1944, p. 173.)

water, the values shown by Fig. 2.4 will be adequate for all of our purposes. However, we note that the presence of dissolved solids will reduce the isothermal compressibility of water [15] and dissolved gases will increase it [14]. If a more accurate estimate of the isothermal compressibility is required, then extensive data and correlations compiled by Earlougher are available [16].

The isothermal compressibility of an undersaturated oil should, where possible, be computed from laboratory PVT data for the oil existing in the reservoir. If data are not available, Trube's correlation for compressibility (Fig. 2.5) may be used [17]. To use this correlation an estimate of both the pseudocritical pressure and pseudocritical temperature is required. Figure 2.6 shows one correlation [17] for these parameters. The estimated oil compressibility based on Figs. 2.5 and 2.6 is calculated from the equation

$$\kappa_{\text{oil}} = \frac{C_{\text{pr}}}{p_{\text{pc}}} \tag{2.10}$$

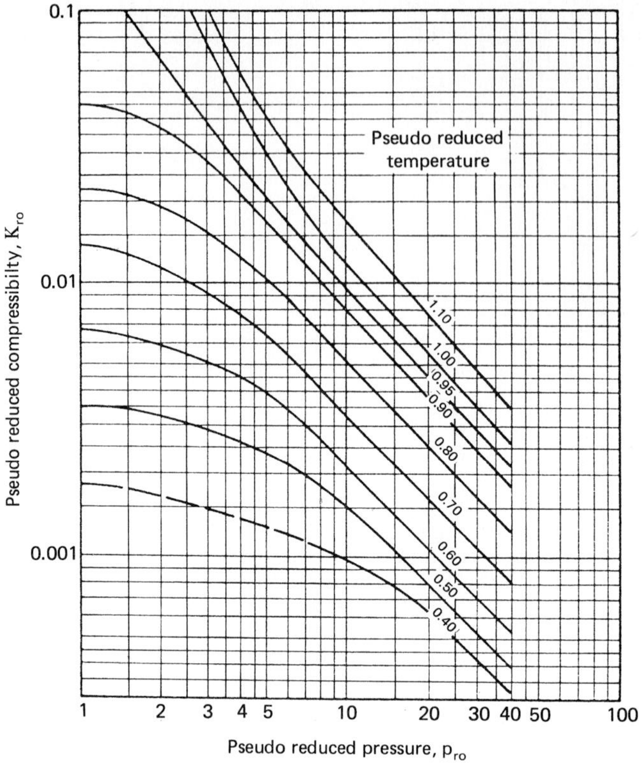

Figure 2.5 Graph showing correlation of the pseudoreduced compressibility for an undersaturated oil versus pseudoreduced pressure. [From A. C. Trube, *Transactions of the American Institute of Mining and Metallurgical Engineering, 210* (1957), p. 341.]

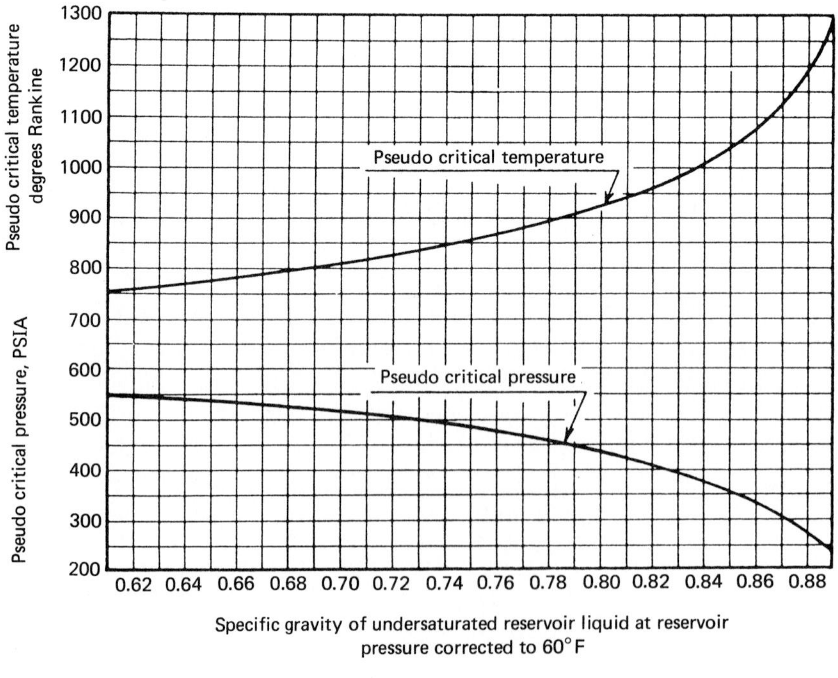

Figure 2.6 Approximate correlation of liquid pseudocritical properties with specific gravity. [From A. C. Trube, *Transactions of the American Institute of Mining and Metallurgical Engineering, 210* (1957), p. 341.]

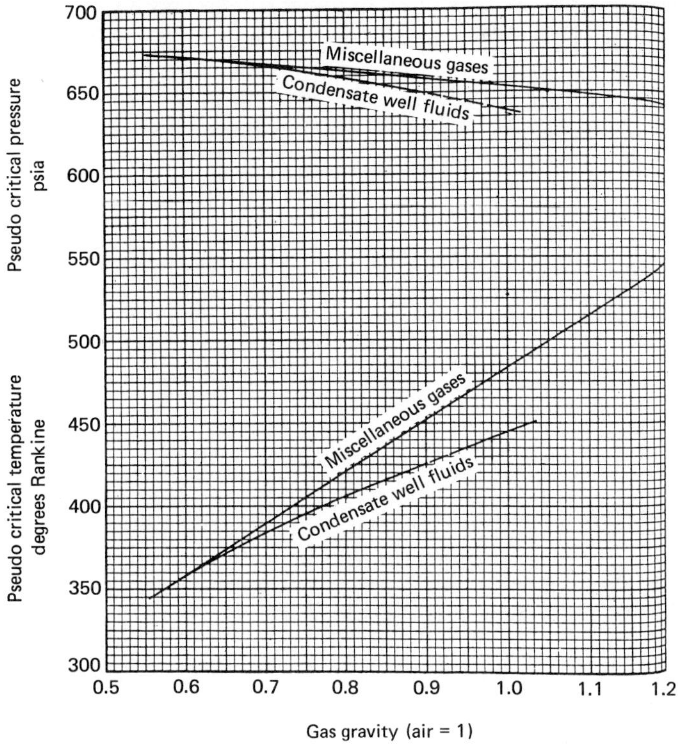

Figure 2.7 Correlation of pseudocritical properties of condensate well fluids and miscellaneous natural gas with fluid gravity. (From G. Brown et al., "Natural Gasoline and the Volatile Hydrocarbons," Natural Gasoline Association of America, Tulsa, 1948.)

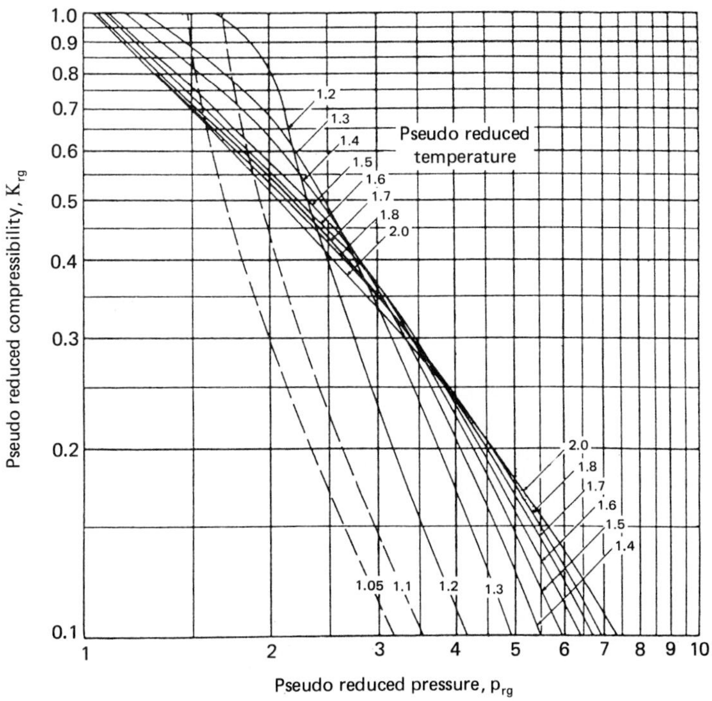

Figure 2.8 Graph showing correlation of pseudoreduced compressibility for natural gases versus pseudoreduced pressure. [From H. C. Trube, *Transactions of the American Institute of Mining Engineers, 210* (1957), p. 355.]

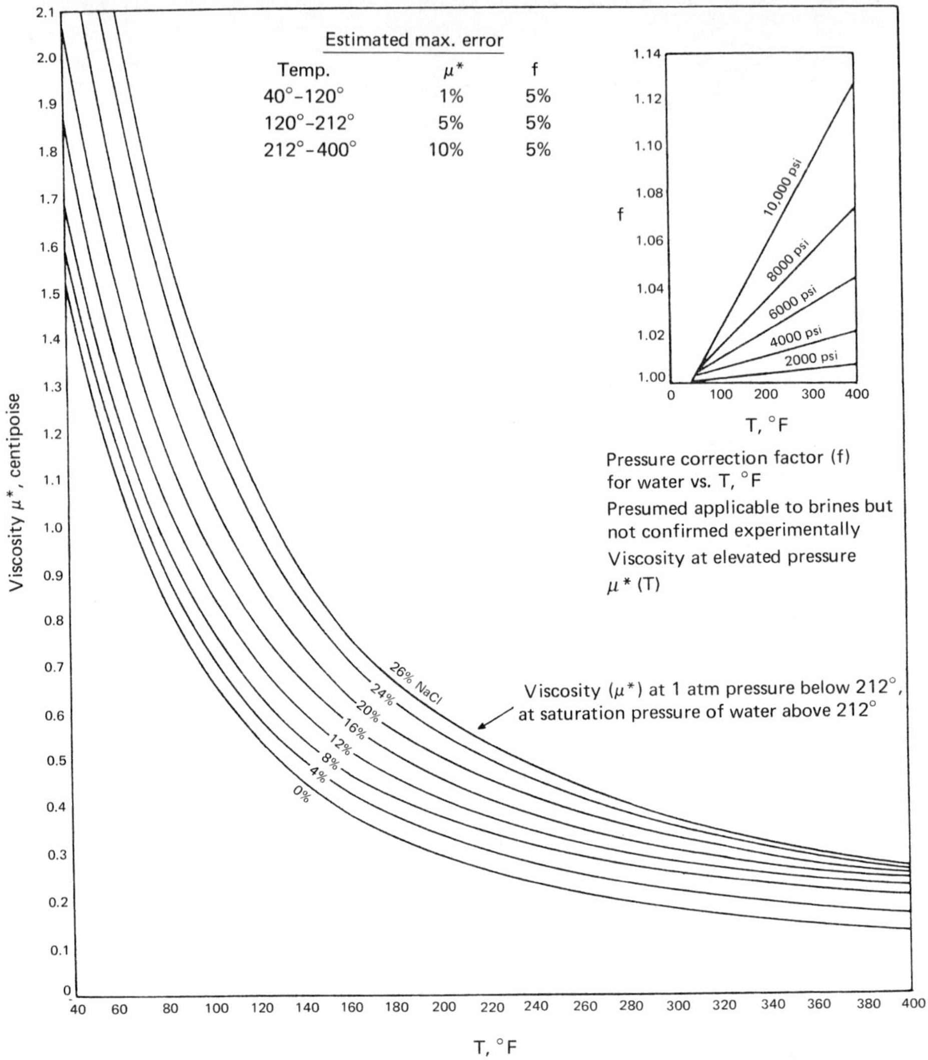

Figure 2.9 Water viscosity at various salinities and temperatures. (With permission of Shell Oil Company.)

where C_{pr} is the pseudoreduced compressibility and p_{pc} is the pseudocritical pressure. Again it should be emphasized that the oil compressibility obtained using the procedure outlined here is an estimate and can be substantially in error. In particular, the presence of dissolved gas will alter the compressibility [18]. For calculations requiring accurate values of the compressibility, these additional factors should be considered.

The isothermal compressibility of natural gas can be estimated using a procedure quite similar to that recommended for undersaturated oils. Figure 2.7 provides an estimate of the pseudocritical pressure and temperature, and the pseudoreduced compressibility is then given by Fig. 2.8 [18].

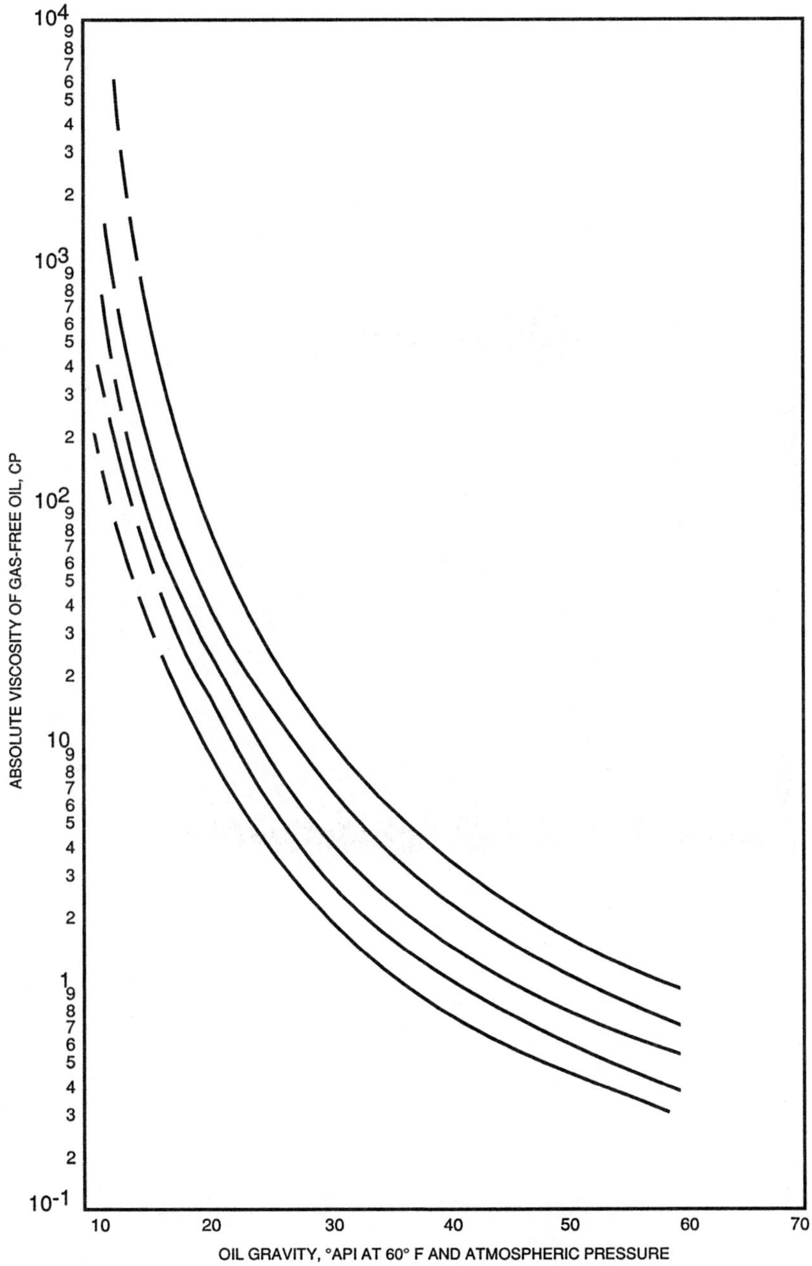

Figure 2.10 Dead-oil viscosity at reservoir temperature and atmospheric pressure. [From C. Beale, *Transactions of the American Institute of Mining and Metallurgical Engineers, 165* (1946), p. 94.]

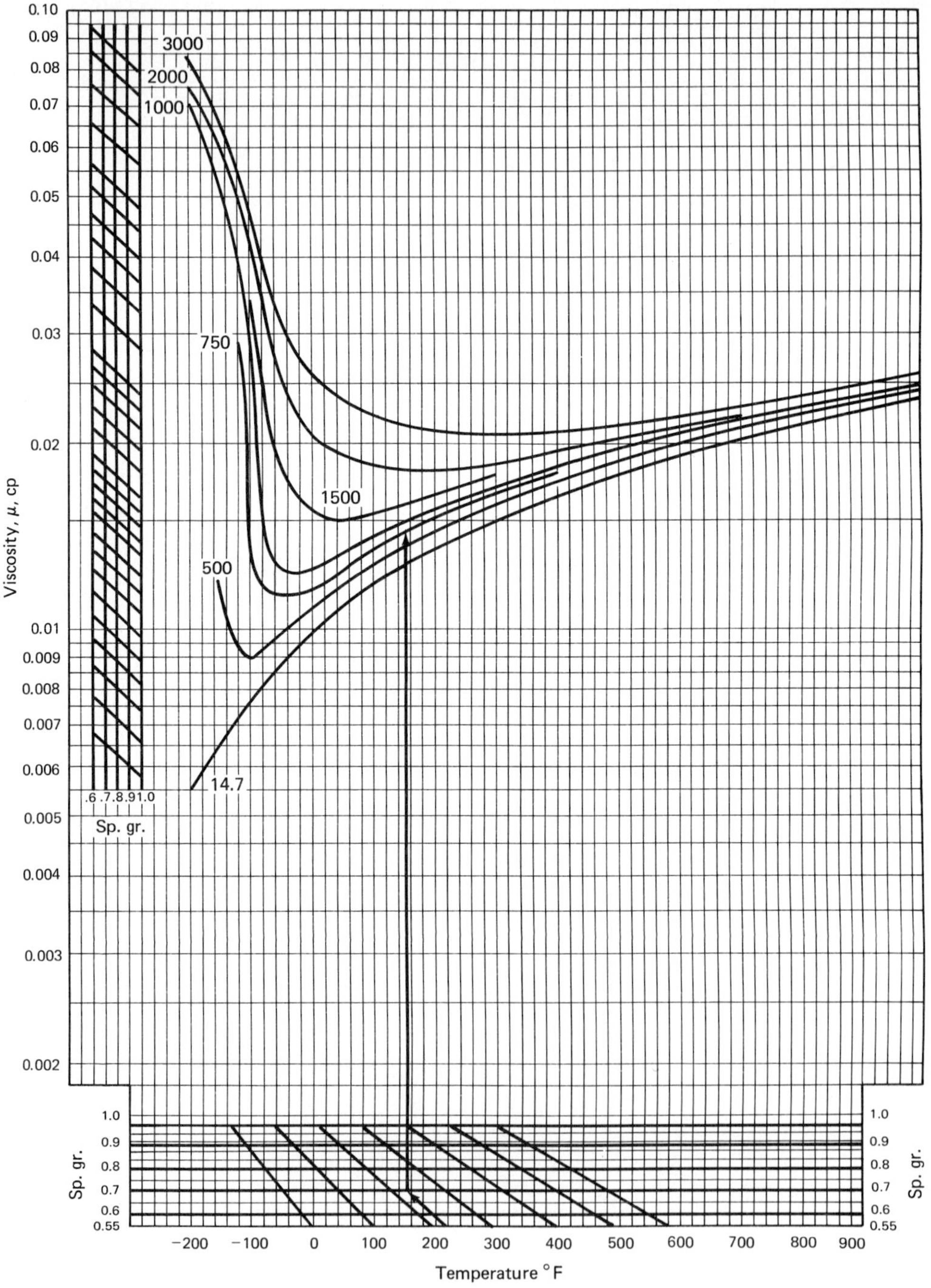

Figure 2.11 Viscosity of hydrocarbon gases expressed in centipoise. (From Engineering Data Book, 9th ed., Section 16, Gas Suppliers Association, Tulsa, Oklahoma, 1972.)

Fluid Viscosity

The viscosities of water, oil, and gas are given by Figs. 2.9, 2.10, and 2.11. Again we note that the values for crude oils are rough estimates only and laboratory data are preferable. The liquid viscosities are not a strong function of pressure but depend on the amount of gas dissolved. Correlations accounting for the amount of dissolved gas are summarized by Earlougher [16].

2.3 THERMAL PROPERTIES OF POROUS ROCK

The temperature within the earth's crust increases with depth approximately in accordance with the equation

$$T = T_s + \omega D \tag{2.11}$$

where T_s is the average yearly temperature and D is the depth. The geothermal gradient, ω, varies from place to place but is approximately 3°C per 100 m. A simple calculation will show that a formation at a depth of 3000 m may have a temperature in excess of 100°C.

There are a number of factors related to the design of well treatments, such as fluid viscosity, solubility, or rate of chemical reaction, which depend on the fluid temperature. It will often be necessary to calculate the fluid temperature in contact with reservoir rock. This calculation requires knowledge of two factors—heat capacity and thermal conductivity of both the fluid and the rock.

Heat Capacity

Heat capacity is a measure of the ability of a body to store heat. C_p is defined as the heat [expressed here in kilocalories (kcal)] required to increase the temperature of 1 kg of mass by 1°C under conditions of constant pressure.

Somerton [23] has shown that heat capacities of sandstones and limestones can be accurately predicted from heat capacities of constituent mineral oxides. He also indicated that heat capacities of α-quartz and calcite represent with good accuracy the heat capacities of the respective reservoir types.

Thus, the heat capacity of a sandstone can be approximated by the following equation [24]:

$$C_{p\ \text{rock}} = 0.1812 + 1.452 \times 10^{-4} T - \frac{1.495 \times 10^3}{T^2} \tag{2.12}$$

where T must be expressed in °K. Here C_p has the units kcal/kg-°C. Similarly, for limestones [24]

$$C_{p\ \text{rock}} = 0.1968 + 1.189 \times 10^{-4} T - \frac{3.076 \times 10^3}{T^2} \tag{2.13}$$

These heat capacities apply to the rock skeleton material. For fluid-filled rocks, the fluid heat capacity, $C_{p\ \text{fluid}}$, must contribute. The presence of the fluid in the pores can be taken into account as follows:

$$\rho_f C_{pf} = (1 - \phi)\rho_{rock} C_{p\ rock} + \phi \rho_{fluid} C_{p\ fluid} \tag{2.14}$$

Here, ρ_f is the bulk density of the fluid-filled rock, that is the density of the formation, and C_{pf} is its heat capacity. In almost all applications ρ_f and C_{pf} are the most useful quantities.

Thermal Conductivity

Thermal conductivity, K_f, is defined by the equation

$$Q_x = -K_f \frac{\partial T}{\partial x} \tag{2.15}$$

where Q_x is the flux of heat in the x-coordinate direction [kcal/m^2-sec]. In all calculations relating to well treatment the fluid and the rock are assumed to have the same temperature at each point. Because of the very large surface-to-volume ratios normally found in subterranean formations, this is an excellent approximation. The units for K_f are expressed in kcal/m-sec-°C.

Somerton [23] has measured the thermal conductivities of a few fluid-filled cores at 32°C. He found considerable variation between sandstones and stated that his results, presented here in Table 2.3, could not be correlated in any obvious way. Thus, for applications requiring precise values of K_f, Somerton recommended that they be measured at field temperatures and pressures using the formation rock and formation fluids.

The data given in Table 2.3 were obtained at 32°C. It is expected that these

TABLE 2.3 Measured Thermal Conductivities at 32°C [23]. (Thermal Conductivity, K_f, kcal/m^2-sec-°C)

Sample	Porosity	Air filled ($\times 10^4$)	Oil filled[b] ($\times 10^4$)	Water filled[c] ($\times 10^4$)	Oil and water filled[d] ($\times 10^4$)
1. Sandstone	0.196	2.1	3.25	6.58	5.89
2. Sandstone	0.4	1.18	2.39	4.34	—
3. Silty sand[a]	0.4	1.18	—	—	—
4. Silty sand[a]	0.43	1.08	2.58	4.59	—
5. Siltstone	0.36	1.40	2.29	4.29	—
6. Siltstone	0.196	1.64	—	—	—
7. Shale	0.071	2.49	—	4.03	—
8. Limestone	0.186	4.06	5.15	8.74	6.99

[a] Disaggregated sample.

[b] The thermal conductivity of the oil was 0.318×10^{-4}kcal/m-sec-°C.

[c] The thermal conductivity of water is 1.46×10^{-4}kcal/m-sec-°C.

[d] The water saturation was about 35%.

values will be too large at higher temperatures. Birch and Clark [25] reported that the thermal conductivities of most minerals decrease markedly with temperature.

The values of K_f reported in Table 2.3 vary as a function of the particular fluid contained within the pores. This fact should be borne in mind when selecting an appropriate K_f for a given application. Prats, in his monograph dealing with thermal recovery processes, has collected extensive thermal data and up-to-date correlations for these properties [26]. If data are not available, the reader should consult the Appendices in Prats's monograph.

2.4 THERMAL PROPERTIES OF FORMATION FLUIDS

The heat capacity of water is approximately 1 kcal/kg-°C. This value will apply with sufficient accuracy to the applications considered in this book.

The only oil and gas thermal property of interest here is heat capacity. For estimating the heat capacities of crude oils, Fallon and Watson [28] recommend the correlation

$$\begin{aligned} C_{p\,\text{oil}} &= [0.355 + 0.128 \times 10^{-2}(\text{API})] \\ &\quad + [0.503 + 0.117 \times 10^{-2}(\text{API})] \times 10^{-3}T(0.05\tilde{K} + 0.41) \end{aligned} \tag{2.16}$$

where API is related to the specific gravity as follows:

$$\text{API} = \frac{141.5}{\text{Sp. Gr.}\left(\dfrac{60°\text{F}}{60°\text{F}}\right)} - 131.5 \tag{2.17}$$

where Sp. Gr. (60°F/60°F) is the oil density at 60°F divided by the density of water at the same temperature and T is in °F. $\tilde{K}$ is the Universal Oil Products Characterization Factor.

$$\tilde{K} = \frac{(\text{Normal boiling point in °R})^{1/3}}{\text{Sp. Gr.}\left(\dfrac{60°\text{F}}{60°\text{F}}\right)} \tag{2.18}$$

Typical values of $\tilde{K}$ are given in Table 2.4, which indicates their range is not large.

The heat capacity of natural gas can be assumed to be similar to that of

TABLE 2.4 $\tilde{K}$ Values

Origin	Type of Oil	$\tilde{K}$
Pennsylvania	Crude	12.2–12.5
Midcontinent	Crude	11.8–12.0
Gulf Coast	Crude	11.0–11.8
East Texas	Crude	11.9
California	Crude	10.8–11.8

methane, which is given by [27]

$$C_{p\ gas} = 8.2 + 1.307 \times 10^{-2}T + 8.75 \times 10^{-7}T^2 - 2.63 \times 10^{-9}T^3 \tag{2.19}$$

where T is in °C and the heat capacity is expressed in kcal/kg mole-°C.

Example 2.2 Steady-State Temperature Profile

A rectangular sandstone core (0.1 m in length and 0.02 m × 0.02 m in cross section) is carefully insulated along its lateral sides so that heat can only flow along the length of the core. If one face of the core is maintained at a temperature of 100°C while the other is kept at 25°C, ultimately a steady-state temperature distribution will be established. Here *steady state* implies that the temperature at each point is time independent.

Find the temperature as a function of position within the sandstone core once steady state has been established and determine the amount of energy (heat) stored in the core under steady-state conditions. Consider only that amount of heat necessary to elevate the temperature above 25°C as being *stored*. The sandstone has a porosity of 15% and is filled with water.

Solution A sketch of the system is shown in Fig. 2.12. Take the x-direction to be along the length of core with the point $x = 0$ being at the inlet face where the temperature is 100°C. Since the system is at steady state, then as much heat is removed from each segment as enters. Thus, steady state is characterized by

$$Q_x = \text{Constant}$$

Note that since there is no heat flow in the y- or z-directions, $Q_y = Q_z = 0$. From Eq. (2.15) we see that

$$T = -\frac{Q_x}{K_f}x + B \tag{2.20}$$

where B is a constant. Since $T = 373$°K at $x = 0$, $B = 373$. Also, $T = 298$°K at $x = 0.1$ m. Thus,

$$T = 373 - 750x \tag{2.21}$$

This is the temperature variation as a function of x produced because of the steady-state heat flow. It varies linearly with x.

To calulate the heat stored, we multiply a small length, dx, having a temperature $T(x)$ by the cross-sectional area (0.02)(0.02) m². This gives a volume. The

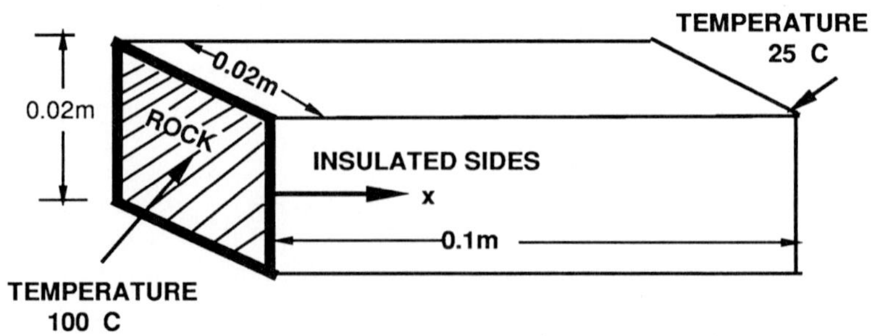

Figure 2.12 An insulated rock with imposed temperature difference.

volume multiplied by $\rho_f C_{pf}$ [see Eq. (2.14)] yields the heat capacity of this little element. Therefore, the amount of heat stored is

$$\text{Stored heat} = \int_0^{0.1} (0.02)(0.02)\rho_f C_{pf}(T(x) - 298°\text{K})\, dx \tag{2.22}$$

Based on the linear relationship between $T(x)$ and x, $dx = -1/750\ dT$,

$$\text{Stored heat} = \int_{373}^{298} -\frac{(0.02)(0.02)}{750}(T - 298)(\rho_f C_{pf})\, dT \tag{2.23}$$

Based on Eqs. (2.12) and (2.14) we can write for a water-filled sandstone

$$\rho_f C_{pf} = (0.85)\left(2640\ \frac{\text{kg}}{\text{m}^3}\right)\left(0.1812 + 1.452 \times 10^{-4}\,\text{T} - \frac{1.495 \times 10^3}{T^2}\right)$$
$$\frac{\text{kcal}}{\text{kg-°C}} + (0.15)\left(1000\ \frac{\text{kg}}{\text{m}^3}\right)\left(1\ \frac{\text{kcal}}{\text{kg-°C}}\right) \tag{2.24}$$

Substituting Eq. (2.24) into Eq. (2.23) and integrating between limits, we find

$$\text{Stored heat} = 0.963\ \text{kcal}$$

REFERENCES

2.1. Jaeger, J. C., *Elasticity, Fracture and Flow*, Chap. 1, London: Chapman and Hall, 1978.
2.2. Boubrie, T., Coussy, O., and Zinszner, B., *Acoustics of Porous Media*, Houston: Gulf Publishing Company, 1987.
2.3. Jaeger, J. C., and Cook, N. G. W., *Fundamentals of Rock Mechanics*, 2nd ed., London: Chapman and Hall (1977) 7.
2.4. Jaeger, J. C., and Cook, N. G. W., *Fundamentals of Rock Mechanics*, 2nd ed., London: Chapman and Hall (1977) 199.
2.5. Handin, J., Hager, R. V., Friedman, M., and Feather, J. N., *Bull. Amer. Assoc. Pet. Geol., 47* (1953) 717.
2.6. Murrel, S. A. F., *Geophys. J., 10* (1965) 231.
2.7. Kowalski, J. J., "Formation Strength Parameters from Well Logs," *SPWLA Symp. Trans.* (1975) and *SPWLA Reprint Vol.—Acoustic Logging* (1978).
2.8. Tixier, M. P., Loveless, G. W., and Anderson, R. A., *J. Pet. Tech., 27* (1975) 283.
2.9. Biot, M. A., *J. Acoust. Soc. Amer., 28* (1956) 168.
2.10. Biot, M. A., *J. Acoust. Soc. Amer., 34* (1962) 1254.
2.11. Williams, B. B., Gidley, J. L., and Schechter, R. S., *Acidizing Fundamentals,* Mono. Ser., *6*, Society of Petroleum Engineers, Richardson, Texas, 1979.
2.12. Haliburton, *The Fracbook*, Duncan: Haliburton Services, 1971.
2.13. Howard, G. C., and Fast, C. R., *Hydraulic Fracturing*, Mono. Ser., *2*, Society of Petroleum Engineers, Richardson, Texas, 1970.
2.14. Dodson, C. R., and Standing, M. B., *Drilling and Production Practices*, Washington, D.C.: American Petroleum Institute (1944) 173.
2.15. Long, G., and Chierici, G., *Pet. Eng. B25-B51* (July 1961).
2.16. Earlougher, R. C., Jr., *Advances in Well Test Analysis*, Mono. Ser., *5*, Society of Petroleum Engineers, Richardson, Texas, 1977.
2.17. Trube, A. C., *Trans. AIMME, 210* (1957) 341.

2.18. Trube, A. C., *Trans. AIMME, 210* (1957) 355.
2.19. Brown, G., Katz, D. L., Oberfell, G. G., and Alden, R. C., "Natural Gasoline and the Volatile Hydrocarbons," Natural Gasoline Association of America, Tulsa, 1948.
2.20. Matthews, C. S., and Russell, D. G., *Pressure Buildup and Flow Tests in Wells*, Mono. Ser., *1*, Society of Petroleum Engineers, Richardson, Texas, 1967.
2.21. Beal, C., *Trans. AIMME, 165* (1946) 94.
2.22. *Engineering Data Book*, 9th ed., Sec. 16, Gas Suppliers Association, Tulsa, Oklahoma, 1972.
2.23. Somerton, W. H., *Trans. AIMME, 213* (1958) 375.
2.24. Keley, K. L., "Contributions to the Data on Theoretical Meterology," *Bull. USBM, 476* (1949).
2.25. Birch, F., and Clark, H., *Ameri. J. Sci., 238*(8), Pt. I (1940) 529 and *238*(9), Pt. II (1940) 613.
2.26. Prats, M., *Thermal Recovery*, Mono. Ser., *7*, Society of Petroleum Engineers, Richardson, Texas, 1982.
2.27. Himmelblau, D. M., *Basic Principles and Calculations in Chemical Engineering*, Englewood Cliffs: Prentice Hall, 1974, Table E.1.
2.28. Fallon, J. F., and Watson, D. M., National Petroleum News Technology Section, June 7, 1944.

PROBLEMS

***2.1.** A cubic sample of Grayburg sandstone ($E = 3.3 \times 10^7$ kPa, $\nu = 0.342$, $\phi = 0.169$) is subjected to a stress of 3×10^4 kPa axial load to simulate overburden forces. A confining pressure of 1×10^4 kPa is maintained along the lateral sides of the sample and since the sample is permeable, the fluid pressure within the sample is also 1×10^4 kPa.

Calculate the axial strain and the fraction volume change of the sample which occurs in response to these loads.

***2.2.** A cubic sample of Grayburg sandstone ($E = 3.3 \times 10^7$ kPa, $\nu = 0.342$, $\phi = 0.169$) is subjected to a stress of 3×10^4 kPa axial load. A confining pressure of 1×10^4 kPa is maintained along the lateral sides of the sample, which is covered with a flexible rubber sheath to prevent the confining fluid from entering the sample (see Fig. P2.2). The fluid pressure (pore pressure p_L) shown in the drawing is 1×10^3 kPa.

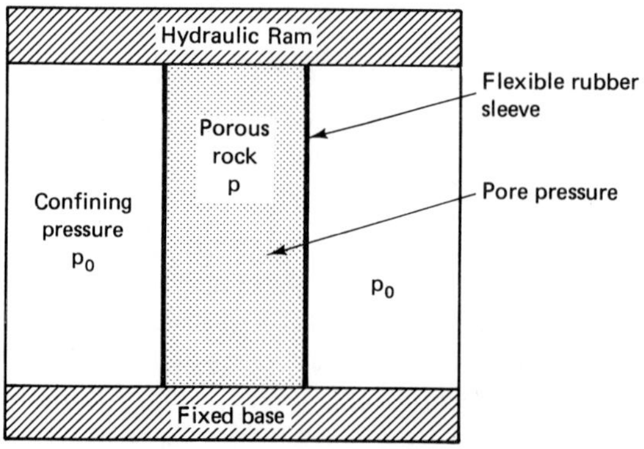

Figure P2.2

Calculate the axial strain and the fraction volume change of the sample in response to these loads.

****2.3.** It is well known that wave motion is often described by a partial differential equation having the form

$$\frac{\partial^2 u}{\partial x^2} = -\frac{1}{c^2}\frac{\partial^2 u}{\partial t^2}$$

where c is the wave velocity. Let $u(x, t)$ be the displacement of a material point along the axis of the solid rod depicted in Fig. 2.2 representing the propagation of P-waves and neglecting the stresses in the directions perpendicular to the rod's axis show, by making a force balance about a differential element of the rod, that

$$-\frac{\partial \sigma_x}{\partial x} = \rho_b \frac{\partial^2 u}{\partial t^2}$$

where x is distance measured along the axis of the rod. Prove, therefore, that the velocity of the compression wave is approximately

$$v_p^2 = \frac{E}{\rho_b}$$

Why is this result an approximation and hence somewhat different from that expressed by Eq. (2.6)?

***2.4.** Based on the correlations presented in this chapter, estimate the isothermal compressibility of an undersaturated API 35 crude oil at a pressure of 21 MPa and temperature of 40°C.

***2.5.** If 1 m^3 of water is compressed from an initial pressure of 1 atmosphere (atm) to a final pressure of 400 atm while maintining the temperature constant at 100°F, what will be its final volume?

****2.6.** It has been suggested that considerable energy savings can be obtained if the co-current flow of a low API° gravity crude oil and water can be arranged so that the water forms an annular ring coating the surface of the pipe, as shown by Fig. P2.6. If the oil is a dead crude (API = °18) flowing at a volumetric rate of 10^4 cm^3/sec in a tube of radius 1 cm, what fraction of the pumping energy would be saved if a water layer 0.1 cm in thickness existed along the tube wall (R_o = 0.9 cm in Fig. P2.6) as compared to the case where no water layer is present?

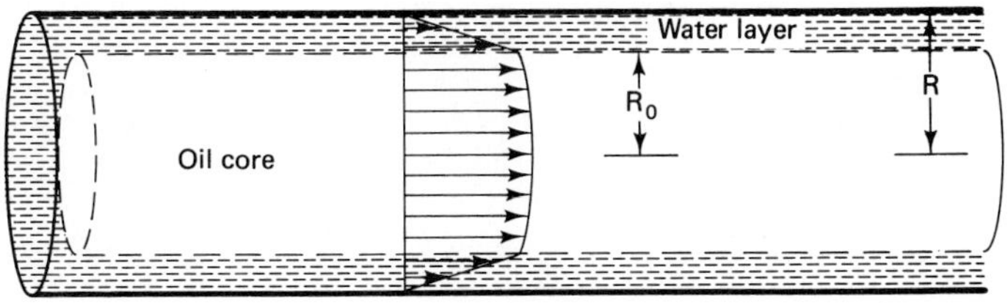

Figure P2.6

What water-to-oil ratio is needed to sustain the thin film? The system temperature is 38°C.

***2.7.** A long sandstone air-filled core (thermal conductivity = 0.023 kcal/hr-°C-m) has a cylindrical hole of radius 1 cm bored along the axis. If the core, which has a radius of 10 cm, is immersed in an air bath that is maintained at 25°C and if heaters inserted into the hole maintain the inner surface at 80°C, how many watts per centimeter of core length must the heater provide?

****2.8.** A long cylindrical sandstone core (permeability 200 mD, porosity = 0.199, and air-filled thermal conductivity = 0.023 kcal/hr-°C-m) has a cylindrical hole bored along its axis (radius = 1 cm). If the core, which has a radius of 10 cm, is immersed in an air bath at 100°C and if air is introduced into the centerhole at 25°C and forced to flow radially through the core at a rate of 20 g/hr-cm of core length, estimate the temperature within the core at the inner radius.

The heat capacity of air is 0.17 cal/g-°C.

****2.9.** A careful measurement shows that over the 50-m thick permafrost zone (frozen water cements the sand grains) the temperature of the crude oil changes from 60°C to 56°C as it is produced from a well on the North slope of Alaska. The production rate is 0.75 kg/sec and the heat capacity of the crude oil is 0.45 kcal/kg-°C. If the wellbore radius through this zone is 0.3 m and the thermal conductivity of melted permafrost is known to be 2.7×10^{-4} kcal/m-sec-°C, to what distance around the wellbore has the permafrost melted? Take the average temperature of the fluid in the wellbore to be 58°C. The permafrost melts at 0°C (melting point of ice).

***2.10.** A heterogeneous sandstone core sample can be considered to be made up of relatively clean sandstone interbedded with small vertical shale layers, as shown in Fig. P2.10. If the thermal conductivity of the shale is

$$K_{\text{shale}} = 0.1 \frac{\text{kcal}}{\text{hr-m-°C}}$$

and the thermal conductivity of the sand is

$$K_{\text{sand}} = 0.5 \frac{\text{kcal}}{\text{hr-m-°C}}$$

what *average* thermal conductivity would we measure for the sample shown in Fig. P2.10? (*Note: There are three vertical shale layers, each one occupying 10% of the length of the core.*) Heat flow is along the axis of the core. The walls of the sample are very well insulated, as shown in Fig. P2.10.

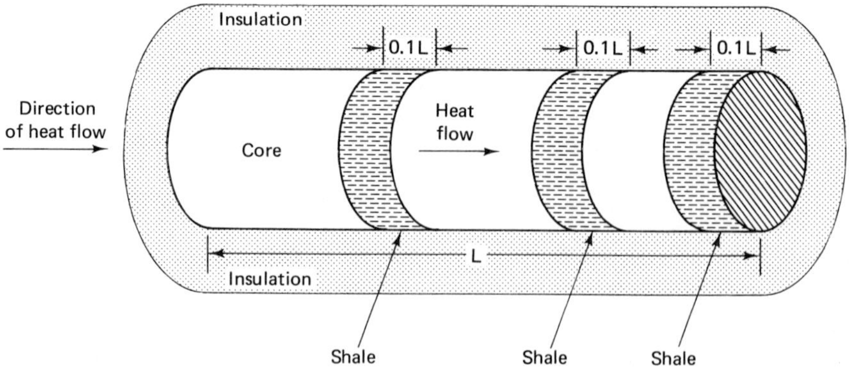

Figure P2.10

3
Chemical and Mechanical Properties of Injected Fluids

In this chapter the chemical and mechanical properties of injected fluids are considered. These fluids are formulated for specific purposes; for example, to create wide fractures during hydraulic fracturing, or to dissolve sandstone rock in a matrix acid stimulation, or to provide well control and be nondamaging when used as a workover fluid. Special formulations such as these are considered in the appropriate chapter dealing with that application; however, many of the treatment fluids have certain properties in common which can most conveniently be considered in a single chapter. For example, polymers, electrolytes, and surfactants are added to almost all treatment fluids. Also, foams and emulsions are used in a variety of applications and their properties will be described in this chapter. In the final section, acids are discussed. Since acids are used primarily to dissolve reservoir rock, their chemical reactions with rock are emphasized.

3.1 AQUEOUS POLYMER SOLUTIONS: CHEMICAL ASPECTS

Polymers are added to both water and oil to increase the apparent viscosity, to reduce the friction losses attending turbulent flow, or to increase the solids-carrying capacity. Water-soluble polymers play a major role and are emphasized in this chapter.

A polymer is a large molecule (macromolecule) composed of many simple units, called *structural units*, which are chemically linked together [1]. When the structural units are connected in a linear, chain-like fashion, the polymer is called

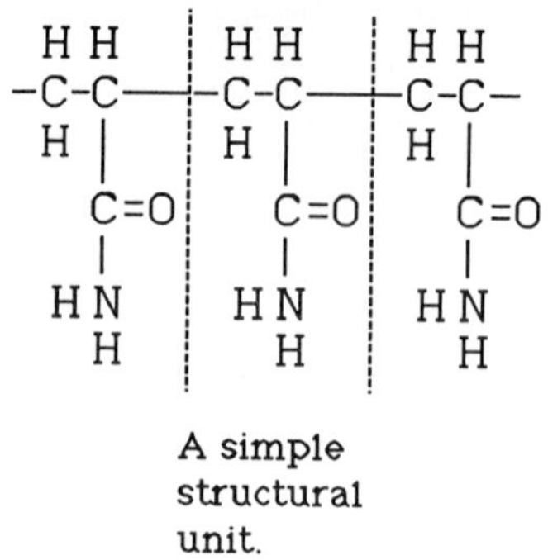

Figure 3.1 Polyacrylamide is composed of a series of repeated structural units.

a *homopolymer*. Polyacrylamide [1], shown in Fig. 3.1, and Xanthan gum [2], shown in Fig. 3.2, are examples of linear polymers. The Xanthan polymer does have a complicated structural unit, but is linear because the polymer is composed of repeated units of the one shown.

Sometimes two, rather than a single structural unit, are used to build a polymer. These are called *copolymers* and can be either random, block, or graft copolymers. As the name implies, random copolymers are those for which the sequence of the two structural units is random whereas block copolymers exhibit considerable structural order. In block copolymers the probability that adjacent structural units are identical is very high.

In graft copolymers one structural unit is connected in a linear fashion, whereas the second one is attached at random positions along the linear chain. Partially hydrolyzed polyacrylamide [1], as evident from its structure presented in Fig. 3.3, is a random copolymer. Because the carboxylic group dissociates from the sodium counterion in aqueous solution, this polymer is sometimes referred to as a *polyelectrolyte*.

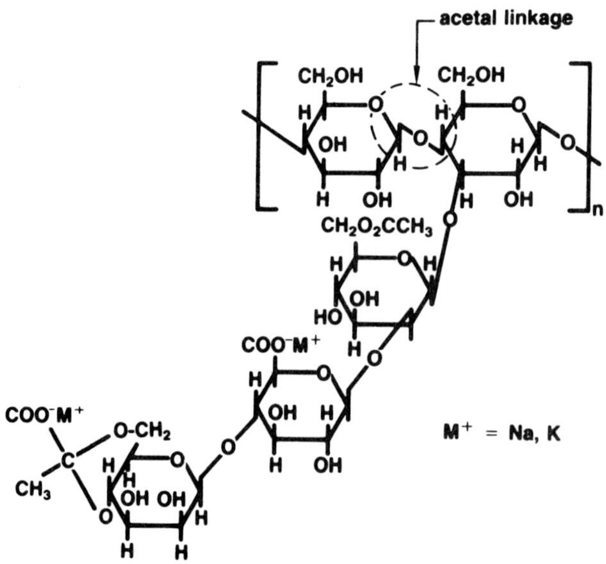

Figure 3.2 Structure of Xanthan gum. One of the repeated units is shown.

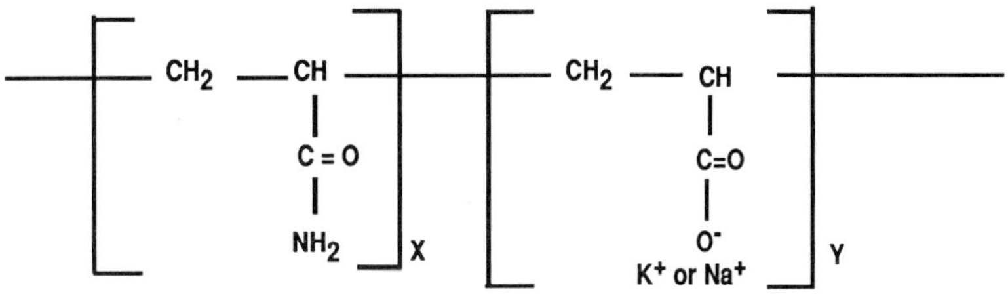

Figure 3.3 Structure of partially hydrolyzed polyacrylamide.

Example 3.1 The Size of Polymer Molecules

The bond length between adjacent singly bonded carbon atoms is 0.154 nm. What then is the total length of a polyacrylamide polymer molecule with a molecular weight of 950,000?

Solution First determine the molecular weight of a single structural unit. Since, as shown by Fig. 3.1, a unit is composed of three carbons, one oxygen, one nitrogen, and five hydrogens, the molecular weight is 71. Thus, the number of structural units is given by (950,000)/71 or 13,400 units. Since each unit includes two carbon–carbon bonds (one half of the two bonds linking one unit to adjacent units is counted), then

$$\text{Length} = (13{,}400)(2)(0.154 \text{ nm}) = 4.1 \times 10^3 \text{ nm}$$

Thus, the length of a single polymer molecule is 4.1 μm. This is roughly the size of many of the larger pores in a medium permeability sandstone. It may be expected, therefore, that some molecules will be filtered when an aqueous polymer solution is injected into a porous medium.

Natural and Synthetic Polymers

The classifications noted previously are convenient from the viewpoint of imparting some notion as to the arrangement of the structural units in the makeup of a polymer molecule, but for our purpose there is a second type of classification perhaps more useful. It will be discussed in this section.

Many polymers occur in nature. Guar gum [3] is an example (see Fig. 3.4). It is derived from the seed of the guar plant. Grinding the endosperm of the guar bean produces relatively pure guar. We will call guar and other polymers *natural polymers* when they are derived from natural products.

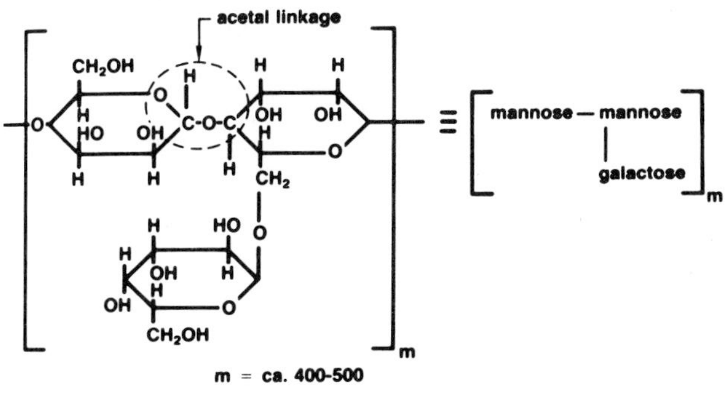

Figure 3.4 Chemical structure of guar gum.

Cellulose is the most abundant of all the natural polymers, but it is not soluble in water or in oils [4]. To be useful in treatment applications, cellulose must be modified chemically so that its water solubility is enhanced. Natural polymers such as cellulose, which are modified to enhance their solubility, are simply called *modified polymers*.

Polymers such as the polyacrylamides, which are manufactured by linking basic structural units together, are called *synthetic*.

Many of the natural or modified polymers are characterized by an acetal linkage in the polymer repeat unit [5]. In this linkage, shown in Fig. 3.5, a carbon atom is bonded with a single bond to two oxygen atoms while also bonded to at least one hydrogen atom. This linkage can be seen in the structure of Xanthan gum (Fig. 3.2) which is a natural polymer. The acetal carbon is linked by way of an oxygen to a nonacetal carbon of another monosaccharide (the basic structural unit) to form the chain. The existence of this linkage is an important feature of natural polymers since it is somewhat unstable and can be hydrolyzed (broken and reacted with water) in acidic solutions or broken by enzymes, cleaved by free radicals generated from catalysts such as ammonium persulfate, or broken oxidatively by redox agents [6, 7, 8].

Many of the applications of polymers in well treatment are best served if the polymer can be cleaved (broken) into smaller fragments after completion of the well treatment. This breaking of the polymer tends to reduce the viscosity of the solution and aids well production when flow is resumed following the treatment. Thus, the instability of the acetal linkage is sometimes a useful feature.

Synthetic polymers generally are formed by linking the structural units through carbon to carbon bonds. Considerable energy (80 kcal/mole) is required to break this bond, and these polymers tend to be quite stable. This will limit their application in well treatment. Shupe [9] has reported an extensive study of the chemical stability of polyacrylamide polymers and Wellington [10] has studied Xanthan. The stability of these polymers is greatly enhanced if oxygen is totally excluded from the system.

The acetal linkage is one important feature of the polymer structure. There are others. All water-soluble polymers, whether natural or synthetic, must have either polar groups or ionic groups incorporated into the basic structural unit. As was noted, partially hydrolyzed polyacrylamide dissociates in water and is, therefore, classified as an *ionic polymer*. The solubility of ionic polymers in water depends on this mechanism of dissociation and, if additional salt is added to the water, the solubility of ionic polymers is normally reduced. Thus, we would expect ionic polymers to be quite sensitive to the composition and concentration of the

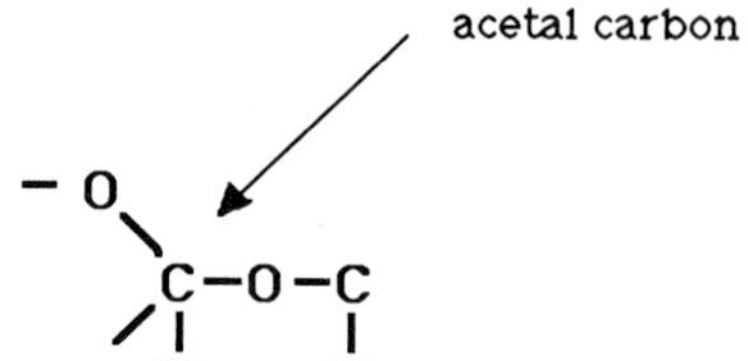

Figure 3.5 Acetal linkage prominent in many natural polymers.

electrolytes contained in the water in which the polymer molecules are dissolved. The solution viscosity of partially hydrolyzed polyacrylamides is, for example, quite salt sensitive because the attached carboxylates are ionic in character [11] (see Fig. 3.3).

Polar groups are those which interact strongly with water generally by hydrogen bonding, but do not dissociate. Polar groups tend to be less sensitive to added electrolytes. Guar gum has a number of polar groups (OH) attached to the basic structural unit which promotes its water solubility, but none of these dissociate. Thus, guar can be used in saline waters. The interaction by hydrogen bonding is, however, sensitive to temperature. Guar or modified guar may be readily soluble in cold water but may become progressively less soluble as the temperature is increased [3]. Thus, guar may have limited application in situations for which fluid stability over prolonged periods at elevated temperatures is required.

Modified Polymers

Cellulose is a linear polymer formed by linking glucose molecules. Even though the cellulose molecule is built of structural units that do have polar segments, they are not water soluble in their natural form. Since cellulose is abundant, it is of practical interest to modify these molecules to make them useful. This is done by substituting for the OH groups (there are three per structural unit). The product of the substitution reaction can have a maximum degree of substitution of 3; however, it should be noted that manufacturers often report the degree of substitution per ten structural units so that a degree of substitution equal to 7 in effect implies that on the average 0.7 of the three OH groups has been substituted.

Several substituted celluloses are available for oil field application. Sodium carboxymethylcellulose is prepared by the reaction of cellulose with chloroacetic acid in the presence of sodium hydroxide. The structure of the modified polymer (CMC) is shown in Fig. 3.6. Notice that while the original cellulose is polar, but not ionic in character, the CMC molecule does contain groups that dissociate in water. We would anticipate a degree of salt sensitivity with regard to the properties of aqueous solutions of CMC.

Cellulose can also be substituted by polar compounds, which will increase the water solubility but remain nonionic. Hydroxyethyl cellulose (HEC) and hydroxypropyl cellulose are examples of modified celluloses that do not dissociate in water.

Sequential treatment of sodium cellulose with ethylene oxide and then with chloroacetic acid forms sodium carboxymethylhydroxyethyl cellulose (CMHEC).

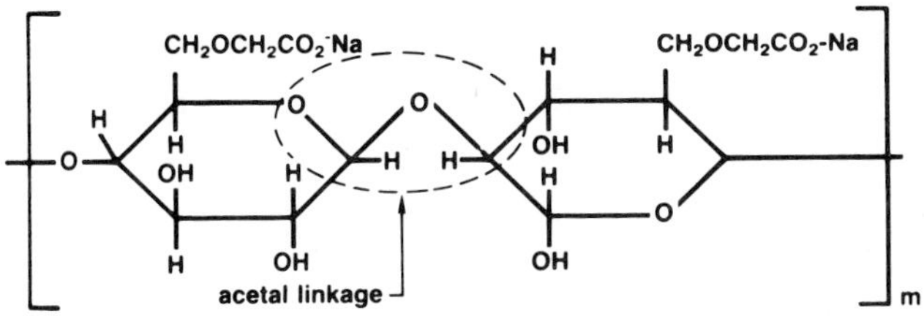

Figure 3.6 Structure of sodium carboxymethylcellulose (CMC) [5].

It has anionic groups attached but, because there are fewer of them than are generally substituted into CMC, its tolerance for more saline waters is better. The polar or nonionic molecules tend to be less soluble as the water temperature is increased. Thus, HEC may precipitate in hot water.

Which of the modified celluloses is selected will depend on the particular application, the quality of the water that is available, the other additives, and the formation temperature. All of these factors will prove to be quite important.

The widely used polymer in hydraulic fracturing is a modified guar gum called *hydroxypropyl guar*. This modified polymer has some advantages when compared to natural guar. One of the most often cited is the small residue (generally less than 2%) [5] which remains when hydroxypropyl guar is depolymerized (broken). This compares to the 10–14% residue found when natural guar breaks [5]. It has been suggested that this residue may result in damage following a well treatment, thereby reducing the effectiveness of the treatment. Thus, when it appears that residue in the broken polymer can be damaging to the success of the treatment, guar gum is not the polymer of choice even though it is relatively inexpensive.

Acid Stability

Synthetic polymers such as polyacrylamides are relatively stable in low pH solutions and are generally used in applications where breaking of the polymer is not required or desirable, as, for example, in drilling fluids. On the other hand, many of the natural or modified polymers degrade rapidly in acid solutions. Figure 3.7 shows the viscosity of several polymers as a function of time in 5 wt% HCl.

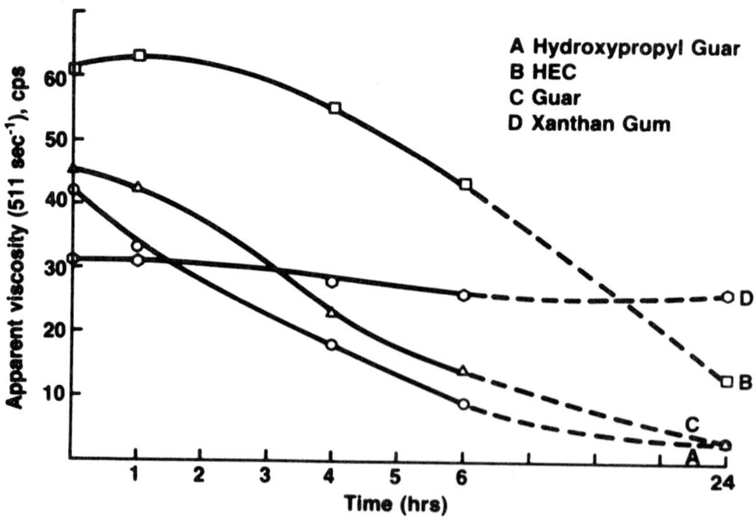

Figure 3.7 Viscosity versus time at 80°F (6 kg/m^3 or 50 lb$_m$/1000 gal) for various gelling agents (5 wt% HCl, 2 wt% KCl). [From J. Chatterji and J. K. Borchardt, *Journal of Petroleum Technology, 33* (1981), p. 2042.]

Further increase in the acid strength will only serve to accelerate the rate of degradation.

Xanthan gum appears to be more resistant to acid attack as compared to the other natural polymers tested even though acetal linkages are also prevalent in Xanthan. The reason for this added stability is not fully understood but appears to be related to the conformance of the Xanthan molecules in aqueous solution. They are believed to form helical structures in which Xanthan molecules are intertwined.

Crosslinking

Crosslinking is that process whereby polymer molecules dissolved in aqueous solutions are joined together or *linked* by a low molecular weight additive (cross-linking agent) which interacts either through hydrogen bonding or electrostatically with two polymer molecules at the same time [12, 13]. If a sufficient quantity of cross-linking agent is present, many of the molecules in solution may be linked, thereby forming a solution containing essentially a few very large molecular weight molecules. Crosslinking can significantly increase the fluid viscosity. Cross-linked fluids are highly non-Newtonian and their viscosity is difficult to characterize.

The most widely used types of crosslinking agent are the metal ions such as boron, chromium, titanium, aluminum [12–14], etc. A metal ion in solution usually associates with one or more groups that stabilize the ion. The associated groups are often called *ligands* and the metal cations plus the associated groups are sometimes called *complex ions*. There are no hard and fast rules determining the number of ligands that can associate with a metal ion (the maximum coordination number); however, this number is very often twice the ionic charge on the metal ion. Thus, for Al^{+++}, the maximum coordination number is 6. In the case of crosslinking, the metal ion coordinates with the hydroxyl groups of polymers to form a complex ion. These complex ions can form and dissociate quite rapidly. Those which do form and break rapidly are termed *labile*. If a complex ion is labile, there will be a rapid exchange rate, that is, association between a metal ion and a polymer may last for a short time; however, once an association is disrupted by, for example, shear or thermal fluctuations of the molecules, new associations form rapidly. The actual rates of these exchange reactions is quite sensitive to the type of metal ion, the structure of the polymer, fluid pH and temperature, and ionic composition and concentration [14].

Figure 3.8 shows a metal ion (M^{n+}) with a valence n^+ associated with two polymer molecules. Each of the metal ions has its own range of applicability. For example, antimony ($n^+ = 5$) forms a cross-linked gel at a pH of 2, but is not useful for pH greater than 8. Antimony ($n^+ = 3$), on the other hand, works well at high pH [14].

Synthetic polymers such as polyacrylamides can be crosslinked using metal cations. Trivalent aluminum or chromium has been used [15, 16]. Polyacrylamides can also be crosslinked using a dealkehyde such as glyoral [15] or aldehydes such as formaldehyde [15].

It will be evident that cross-linked fluids now play a primary role in fracturing

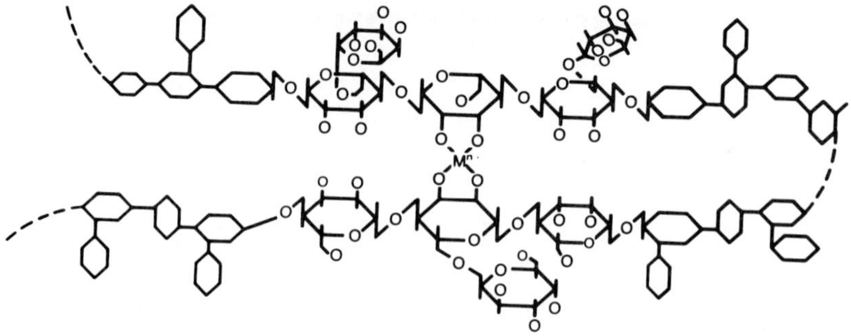

Figure 3.8 Intramolecular crosslinking of hydroxypropyl guar.

and their use is growing in gravel-packing operations and acidizing. It is important to have a general understanding of the mechanism causing the formation of complex ions and the exchange rates, all of which govern the behavior of these fluids under conditions of shear.

3.2 MECHANICAL PROPERTIES OF POLYMER SOLUTIONS

This section is concerned with two important mechanical properties of treatment fluids. The first, important for those flows which are laminar such as in fracture treatments, is the relationship between shear rate and stress. The mathematical model now used most often to describe the shear rate as a function of the applied stress is the power-law model [17]. This is a two-parameter model. It will be of interest and importance to examine the stability of various treatment fluids by studying the variation of the two parameters with temperature, pH, etc.

The second mechanical feature, which applies to turbulent flows, is that of drag or friction reduction. The pressure drop attending turbulent flow can sometimes be reduced by adding polymer to the fluid. This reduction can be important. For example, if the surface pressures tend to become excessive during a well treatment, drag-reducing additives can be added in quite small quantities to reduce this pressure.

The Power-Law Model

The experiment depicted by Fig. 3.9, showing a lower plate set in motion at a constant velocity, v, while the upper plate is held fixed, is most useful for understanding viscosity (apparent) and the methods for measuring this important mechanical property. As shown in Fig. 3.9, the lower plate will require a force, F, to maintain a steady velocity v. If the force required to attain a certain velocity is large, then the fluid contained between the plates is said to be *viscous*. Very viscous fluids are those which are difficult to shear. The viscosity can be defined mathematically by the equation

$$F/A = \mu v/w \tag{3.1}$$

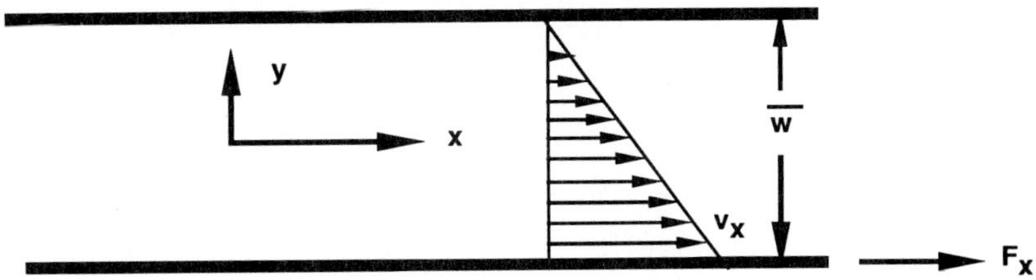

Figure 3.9 A flow system consisting of two parallel plates. The lower one is moving and the upper one is fixed.

The constant of proportionality, μ, is called the *viscosity* of the fluid, w is the spacing between the parallel plates, and A is the area of one of the plates.

It is useful for future work to write Eq. (3.1) in a more general form. Let us designate the shear stress in the x-direction across a plane oriented perpendicular to y as τ_{yx}. The sign convention chosen is such that τ_{yx} is positive if the x-component of force exerted by the fluid at the larger values of y on the fluid at smaller values of y is positive. For the case shown, the fluid at greater values of y (see Fig. 3.9) is moving at a slower rate than the fluid at lesser values of y and would tend to retard its motion. Thus, τ_{yx} is expected to be negative.

We can write Eq. (3.1) as

$$\tau_{yx} = \mu \frac{dv_x}{dy} \tag{3.2}$$

where v_x is the x-component of the fluid velocity.

Clearly, dv_x/dy is negative for the case depicted in Fig. 3.9 and since μ is always a positive number, τ_{yx} is negative. Equation (3.2) is more valuable than Eq. (3.1) since it is a differential equation and will apply even if the velocity is not a linear function of y. If dv_x/dy depends on y, then, as shown by Eq. (3.2), so will τ_{yx}. In the remaining chapters of the book, Eq. (3.2) will be used. If the fluid velocity is not restricted to a single direction, then more complex relationships between the stress and the strain rate will apply. These more complex situations are not encountered in this book; however, more general problems are considered by Bird et al. [17].

We have previously called μ the viscosity. If the same proportionality constant, μ, applies regardless of the velocity of the lower plate, the fluid is called *Newtonian*. In this case, a plot of τ_{yx} versus dv_x/dy is a straight line passing through the origin. Almost all reservoirs fluids are Newtonian. Their viscosities will depend on temperature, pressure, and fluid composition, but do not depend on the shear rate. The viscosity as defined by Eq. (3.2) is therefore a property of the fluid and does not depend on motion. In this case, the μ is a true viscosity.

Many of the fluids used for well treatment are not Newtonian. As shown in Fig. 3.10, the relationship between the stress and the strain rate for a polymer solution is not a linear one. For this fluid, the viscosity as calculated based on Eq. (3.2) is therefore a variable. It will depend on the shear rate. Such viscosities

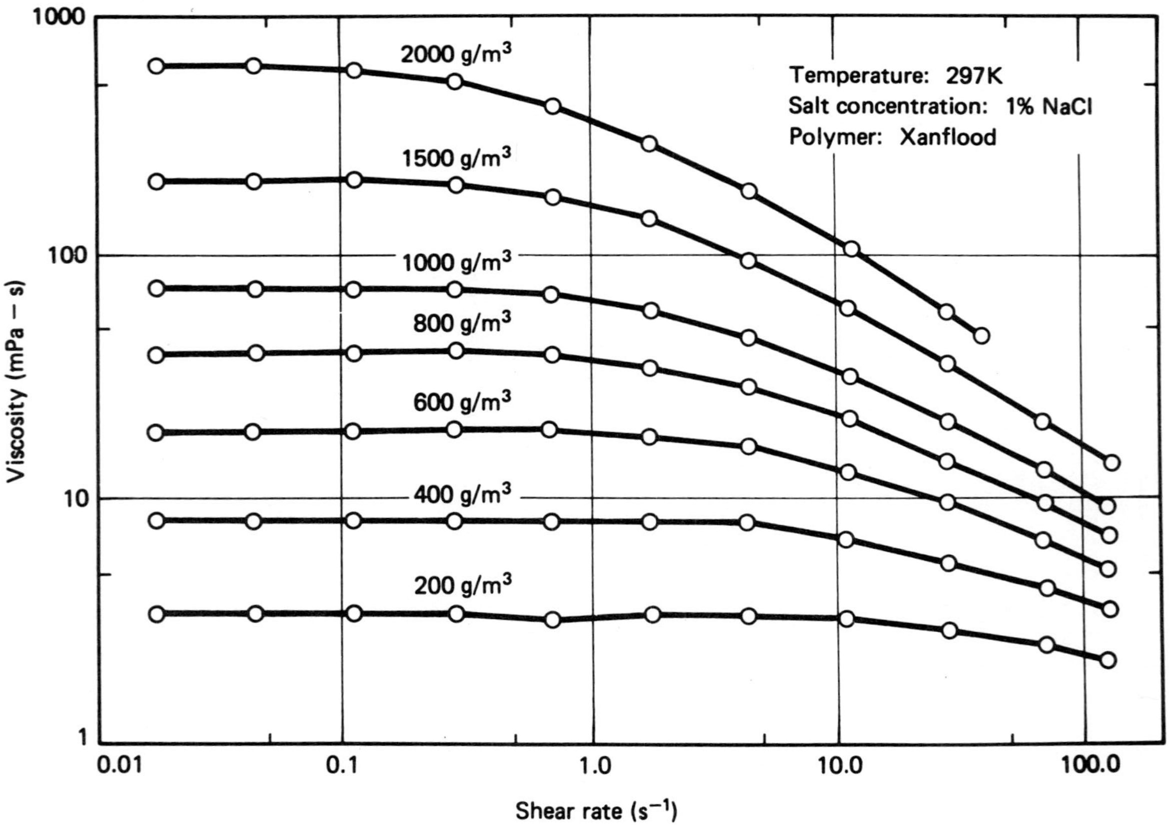

Figure 3.10 Viscometer comparison (3.6 kg/m^3 or 30 lb$_m$/1000 gal) for gel not crosslinked [18].

are apparent viscosities and to be complete, one must specify the shear rate at which any reported apparent viscosity has been measured.

Shear thinning fluids are those for which the viscosity decreases as the shear rate increases. One model used to define the apparent viscosity of shear thinning fluids is the power-law model. For this model, the apparent viscosity is given by

$$\mu_{app} = m \left| \frac{dv_x}{dy} \right|^{n-1} \tag{3.3}$$

where the vertical lines denote the absolute value of the shear rate and m and n are parameters. The flow index, n, measures deviations from Newtonian behavior ($n = 1$ gives a Newtonian fluid). For shear thinning fluids $n < 1$. The quantity m is sometimes called the *consistency index*.

The dimensions of the consistency index are kg/m-sec^{2-n} in SI units. Many published results give values of m as lb$_f$-sec^n/ft^2 without specifying that these are the units. Table 3.1 shows data for a fracture fluid taken from a service company brochure.

Example 3.2 Conversion of Units

Convert to SI units the values of the consistency index for hydroxypropyl guar given in Table 3.1 and calculate the apparent viscosity at a shear rate of 175 sec^{-1}.

$$m = \left(0.098\,\frac{\text{lb}_f\text{-sec}^{0.535}}{\text{ft}^2}\right)\left(\frac{32.2\ \text{lb}_m\text{-ft}}{\text{sec}^2\text{-lb}_f}\right)\left(\frac{\text{kg}}{2.2\ \text{lb}_m}\right)\left(\frac{3.28\ \text{ft}}{\text{m}}\right) = 4.70\ \text{kg/m-sec}^{1.465}$$

The other values of m expressed in SI units are tabulated in Table 3.1. The viscosity at a shear rate of 175 sec^{-1} can be calculated from Eq. (3.3).

$$\mu_{app} = 4.70(175)^{(0.535-1)} = 0.426\ \text{kg/m-sec}$$

This can be expressed in poise as follows:

$$\mu_{app} = \left(\frac{0.426\ \text{kg}}{\text{m-sec}}\right)\left(\frac{1000\ \text{g}}{\text{kg}}\right)\left(\frac{1\ \text{m}}{100\ \text{cm}}\right) = 4.26\,\frac{\text{g}}{\text{cm-sec}} = 4.26\ \text{poise}$$

The values of m and n for most non-Newtonian fluids are functions of temperature. The index n given tends to approach unity (Newtonian behavior) as the temperature increases. Also, and of considerable importance in fracturing, m and

TABLE 3.1 Rheological Parameters of Hydroxypropyl Guar Crosslinked (3 kg/m^3 of water or 25 lb/1000 gal)

Temp (°F)	Time (hr)	n (Flow index)	m (lb$_f$-sec^n/ft^2)	m (kg/m-sec^{2-n})
175	1	0.535	0.098	4.70
175	2	0.553	0.078	3.74
175	3	0.572	0.056	2.69
175	4	0.602	0.035	1.67
175	5	0.647	0.020	0.958
200	5	0.695	0.009	0.432

n both vary with time at elevated temperatures. This is due to the degradation of the polymer molecules at elevated temperatures.

Example 3.3 Power-Law Parameters

The force required to move the lower plate shown in Fig. 3.9 at a velocity of 1×10^{-3} m/sec is measured to be 1 Pa. To increase the velocity to 3×10^{-3} m/sec, a force of 2 Pa is required. The plates are spaced a distance 10^{-3} m apart. Based on these data, calculate the apparent viscosity expressed in poise that would be observed at a shear rate of 200 sec^{-1}. Assume that the fluid obeys the power-law model.

Solution From Eq. (3.1) the apparent viscosity is

$$\mu_{app} = \frac{Fw}{Av} = \frac{1 \text{ Pa } (10^{-3} \text{ m})}{10^{-3} \text{ m/sec}} = 1 \frac{\text{kg}}{\text{m-sec}}$$

when

$$\frac{v}{w} \equiv \frac{dv_x}{dy} = 1 \text{ sec}^{-1}$$

Similarly,

$$\mu_{app} = 0.667 \frac{\text{kg}}{\text{m-sec}} \quad \text{for} \quad \frac{dv_x}{dy} = 3 \text{ sec}^{-1}$$

Based on the definition of m and n, Eq. (3.3), we have

$$\ln \mu_{app} = \ln m + (n - 1) \ln \left| \frac{dv_x}{dy} \right|$$

At

$$\frac{dv_x}{dy} = 1; \qquad \ln m = 0$$

or

$$m = \frac{1 \text{ kg}}{\text{m-sec}^{2-n}}$$

At $dv_x/dy = 3$ sec^{-1}

$$\ln(0.667) = (n - 1) \ln 3$$

Solving for n we find

$$n = 0.631$$

For $dv_x/dy = 200$ sec^{-1} the apparent viscosity is given by

$$\mu_{app} = (1)(200)^{(0.631-1)} = 0.142 \frac{\text{kg}}{\text{m-sec}}$$

Drag Reduction in Turbulent Flow

If very small quantities of certain polymeric materials are added to a fluid flowing in turbulent flow, a substantial reduction in the friction loss attending flow may be observed. This reduction in pressure loss was apparently first observed by Toms and is sometimes called the *Toms effect*. Savins [19] has reviewed the early research of this seemingly paradoxical phenomenon. More recent reviews, which summarize the theories proposed to explain the mechanism responsible for drag reduction, are also available [20, 21]. Drag reduction results in pipes may be conveniently presented in terms of the Fanning friction factor, f,

$$f = \frac{1}{2}\left(\frac{D}{L}\right)\frac{\Delta p}{\rho \bar{v}_z^2} \tag{3.4}$$

where $\Delta p/L$ is the pressure drop per unit length of horizontal tube, D is the tube diamcter, and $\bar{v}_z$ is the average velocity in the tube. Figure 3.11 shows data presented by Bird et al. [22] for water with and without small quantities of polyethylene oxide polymer added. It is clear from Fig. 3.11 that drag reduction is strictly a turbulent flow phenomenon. It does not result from maintaining laminar flow past the usual transition region. Note that the addition of only 20 ppm to water can reduce the friction factor by one half. This means, according to Eq. (3.4), that for a given flow rate the pressure drop is halved. This is quite significant.

The mechanism producing drag reduction is not yet fully understood, although a number of polymeric characteristics for potentially good reducers have been empirically determined. A long-chain backbone and flexibility are important characteristics of good drag-reducing agents. If, for example, two polymers have the same molecular weight, the linear one will generally be more effective than

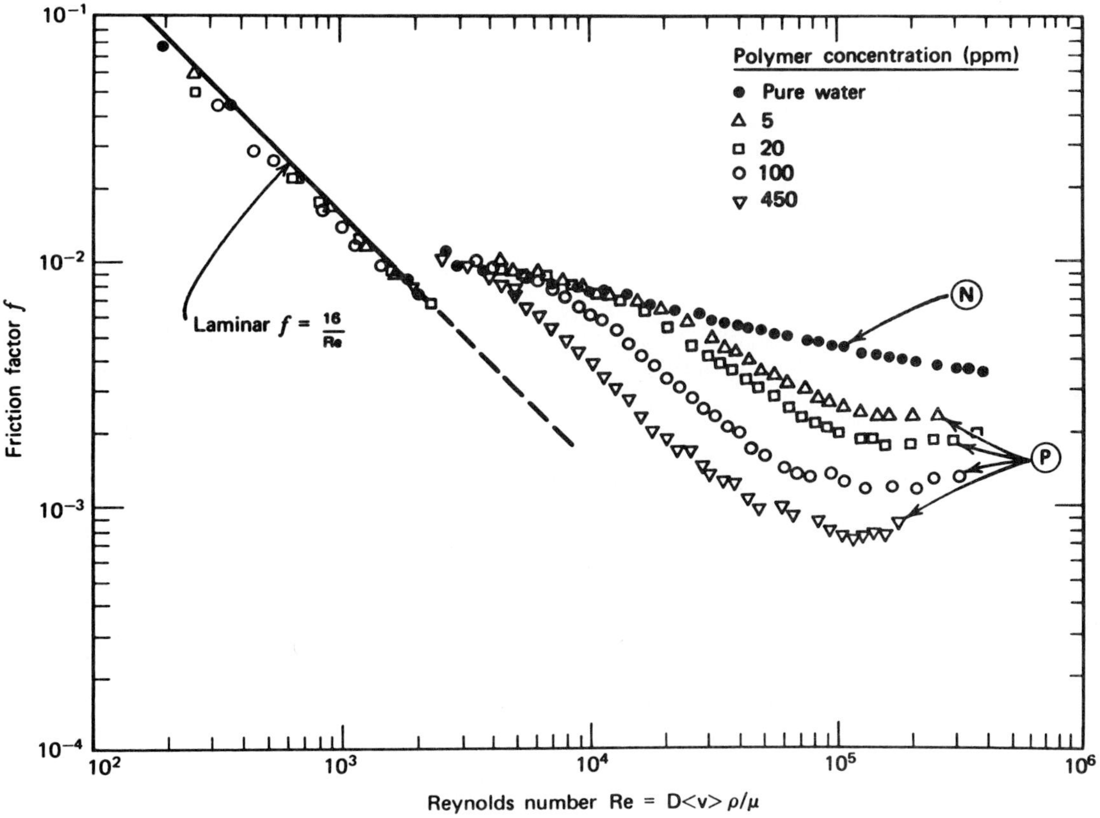

Figure 3.11 Friction factor for dilute aqueous solutions of polyethylene oxide. (From R. B. Bird et al., *Dynamics of Polymeric Liquids, Vol. 1*, New York: John Wiley, 1977.)

one which is highly branched. Similarly, the polymer composed of the lower molecular weight monomer will generally be the most effective if applied at the same solution weight fraction and if all polymer molecular weights are equal [22].

Although the mechanism by which drag reduction occurs has not been firmly established, it appears that viscoelasticity plays an important role. Let us consider the simplest possible model for a viscoelastic fluid, namely, a Maxwell fluid, which for small deformations may be defined by the following equation [22]:

$$\tau_{yx} + \lambda_t \frac{\partial \tau_{yx}}{\partial t} = \mu \frac{\partial v_x}{\partial y} \tag{3.5}$$

where λ_t represents a relaxation time. If λ_t vanishes, Eq. (3.5) reduces to the relationship between the shear stress and the strain rate for a Newtonian fluid. Thus, λ_t is a measure of the viscoelasticity or *memory* of the fluid. It only contributes to the shear stress when the motion of the fluid depends on time. As polymer is initially added to a solvent such as water, we would expect λ_t to increase much faster than μ; however, at higher polymer concentrations, the viscosity increases rapidly. The net result of increasing polymer concentration is complex, but the trends can partly be understood by considering a simplistic, but instructive, model for energy dissipation in turbulent flows.

The effect of the fluid elastic properties on the dissipation of energy can be illustrated by considering a fluid element undergoing an oscillatory small amplitude shear deformation in a large fluid body. Let the frequency of the motion be ω and the amplitude of the oscillatory motion be v_o. The oscillatory motion will produce a shear wave whose propagation is represented by

$$\rho \frac{\partial v_x}{\partial t} = \frac{\partial \tau_{yx}}{\partial y} \tag{3.6}$$

where y is measured in a direction perpendicular to the oscillating fluid element. The shear field is generated by the following motion:

$$v_x = v_o \exp(j\omega t) \qquad (\text{at } y = 0) \tag{3.7}$$

where $j = \sqrt{-1}$. Also the motion is damped so that

$$v_x \to 0 \qquad (\text{as } y \to \infty) \tag{3.8}$$

We will assume that the fluid is viscoelastic and satisfies the Maxwell relationship represented by Eq. (3.5). Using this model, our aim is to calculate the energy dissipated over one period $T = 2\pi/\omega$ as a function of the fluid properties λ_t and μ. We will find that the rate of energy dissipation generally decreases as λ_t increases at a fixed value of μ and it increases as μ increases at a fixed value of λ_t.

To see this trend, assume that the fluid velocity is represented by the equation

$$v_x(y, t) = A(y) \exp(j\omega t) \tag{3.9}$$

where $A(y)$ is a function of y but does not depend on time. Substituting Eq. (3.9) into Eq. (3.5), gives

$$\tau_{yx} = \mu^* \frac{dA}{dy} \exp(j\omega t) \tag{3.10}$$

where the complex viscosity μ^* is given by

$$\mu^* = \frac{\mu}{1 + \omega^2\lambda_t^2} - j\frac{\omega\lambda_t\mu}{1 + \omega^2\lambda_t^2}$$

Substituting Eq. (3.9) and Eq. (3.10) into the momentum balance, Eq. (3.6), it is seen that the function $A(y)$ must satisfy the differential equation

$$\frac{d^2A}{dy^2} - M^2A = 0 \tag{3.11}$$

where

$$M = \sqrt{\frac{j\omega\rho}{\mu^*}}$$

Thus we find

$$A(y) = v_o \exp(-My) \tag{3.12}$$

where we choose the root of M which has a positive real part; that is, Re(M) > 0. Equation (3.12) shows that the function $A(y)$ is an exponential one. This function can now be substituted into Eq. (3.10) to determine the shear stress as a function of both time and position. In this analysis we are, however, only interested in a measure of the energy required to maintain the oscillatory motion of the fluid element. At any time the rate energy dissipation is given by $-v_x\tau_{yx}|_{y=0}$. As a measure of energy dissipation we take the magnitude of this oscillatory function. Thus

$$E = |(v_x\tau_{yx})|_{y=0} \tag{3.13}$$

where the vertical lines denote the absolute value of the complex function. After some calculation, we find this absolute value to be

$$E = v_o^2\left[\frac{\mu\omega\rho}{\sqrt{1 + \omega^2\lambda_t^2}}\right]^{1/2} \tag{3.14}$$

This equation shows that based on our simplistic model increasing the viscosity μ increases the turbulent energy dissipation rate, whereas increasing λ_t decreases the rate. As polymer is added to a solvent, λ_t increases rapidly while μ does not change at first. Thus, the addition of small concentrations of polymer often decreases friction. This situation is depicted in Fig. 3.11. The friction factor decreases as the polymer concentration increases. In the range of concentrations for which μ does not change, a decreasing friction factor is a direct indication of decreased friction. As the polymer concentration is increased and both μ and λ_t increase substantially [actually one would not expect Eq. (3.5) to describe the behavior of such solutions], then it is not possible, based on the friction factor alone, to predict whether or not the pressure drop will increase or decrease when a fixed amount of fluid is pumped through a pipe. Measurements show that often a maximum in drag reduction is observed as the polymer concentration is increased [23, 24]. This can be interpreted based on Eq. (3.14) as the point at which the viscosity increases E faster than λ_r tends to decrease it. For a partially hydrolyzed polyacrylamide the maximum drag reduction corresponds to a concentration somewhat less than 100 ppm [23].

Drag reduction is a most important consideration in the design of fracturing fluids. It will become evident that high rates are desirable in achieving wide fractures and the addition of drag-reducing fluids will help keep the well head pressure at acceptable levels even at high injection rates.

3.3 SURFACTANTS

Surfactants (sometimes called *surface active agents*) are used in almost every phase of well treatment. Virtually every fluid injected into hydrocarbon-rich formations contains surfactants as an additive. Their function may be to help stabilize oil and water emulsions or, on the contrary, they may be designed to break undesirable emulsions. Surfactants may be used to alter wettability, to increase the solubility of oil in water, or to disperse additives either in oil or water. The list of applications is a seemingly endless one. Many of the uses to which surfactants are intended require careful selection of an appropriate molecule; not just any surfactant will suffice. Remarkably, in the design of well treatments most surfactants are selected for use with little or no laboratory data to support the choice and sometimes without full knowledge of their properties under the conditions at which they will be applied. This can lead to results contrary to those which are intended and the addition of the surfactant may hinder the success of the treatment rather than help it. Unless the surfactant has a verified specific purpose, it should not be added to treatment fluids. In this section we shall try to describe some of the properties of surfactants which the practitioner should be aware of and which will also be useful in our subsequent discussions of treatment design.

Surfactants have two essential properties of great practical value. They are extremely surface active at both water/air and water/oil interfaces where surface activity is defined as the tendency to concentrate at an interface and to decrease the interfacial or surface tension. This large degree of surface activity is a result of the molecular structure of a surfactant molecule as depicted in Fig. 3.12. There are two distinct, separate portions of a surfactant molecule—one which is called the *hydrophilic moiety* (the "water loving" part) and a second which is called the *hydrophobic moiety* (the "water hating" part). One part of the molecule, the ionic $SO_3^- Na^+$ group, prefers to be in the aqueous phase whereas the second part, the hydrocarbon, has a greater affinity for oil than for water.

By altering one part of the surfactant molecule or both parts, one can either increase or decrease its affinity for water. It is this balance between the hydrophilic (surfactant's "head") and the hydrophobic (surfactant's "tail") moieties which is crucial in governing the properties of a surfactant; however, as will be observed, this balance is but one factor and not the only consideration.

1.

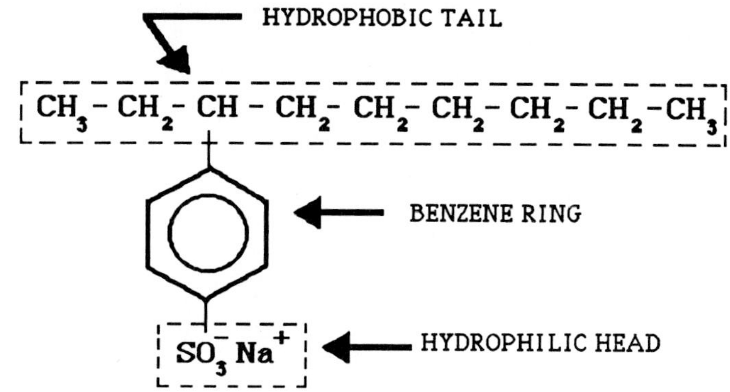

Sodium P (1-butyl heptyl) Benzene Sulfonate (3-ϕ-C_9ABS)

2.

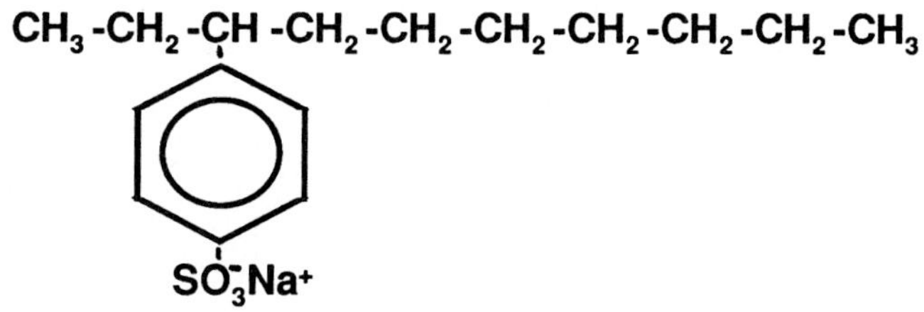

Sodium P (1-butyl octyl) Benzene Sulfonate (3-ϕ-C_{10}ABS)

Figure 3.12 Typical surfactant structures showing two distinct, separate portions of the molecule.

Surface activity is not the only property that may be attributed to surfactants. Many other types of molecules, such as alcohols, exhibit a large degree of surface activity. Alcohols can greatly reduce the surface tension at water/gas interfaces even though they are present in only small quantities. However, alcohols are not considered here to be typical surfactants. A surfactant is considered here to be a molecule that is surface active by virtue of its structure and which forms micelles in aqueous solutions, although this distinction is admittedly somewhat artificial [25]. A micelle is an aggregate of a large number of surfactant molecules in which the hydrophilic portions of the molecules are arranged so that they are in contact with water and the hydrophobic portions are collected together so that they form a separate phase which excludes water molecules. One possible arrangement is shown in Fig. 3.13. Surfactant molecules are shown arranged in a spherical form with the tails of the surfactant (hydrophobic part) collected together. In this ar-

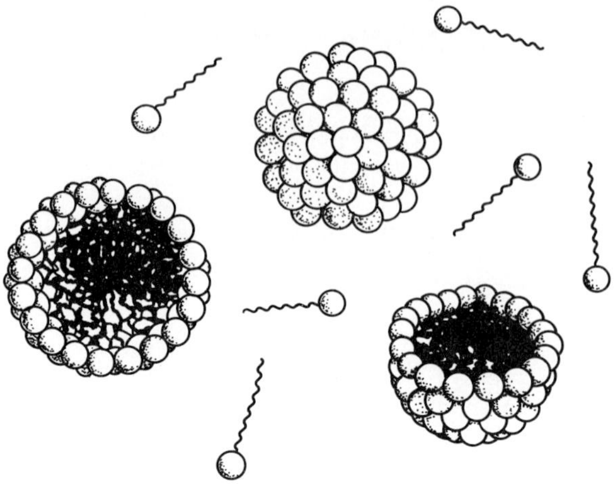

Figure 3.13 Sketch depicting a spherical micelle in aqueous solution.

rangement the interior of the sphere closely resembles a very small drop of oil and indeed in many applications it is useful to consider this to be the case. It has been found, for example, that the solubility of certain oily materials not normally very soluble in water can be enhanced to a remarkable extent by the addition of surfactants [25]. It seems likely that these oily materials are incorporated into the interior of the micelles. The micelles are quite small and are invisible to the eye. Indeed, the radius of the micelle shown in Fig. 3.13 is roughly the length of the surfactant's tail, which may range from 2 to 4 nm. The presence of these small structures can be discerned by light-scattering experiments or inferred from their solution properties, but micellar solutions are often quite transparent. They will easily pass through most pores in sedimentary rocks, so micellar solutions can be injected as treatment fluids.

It is first useful to examine briefly the molecular structure of the most often applied commercial surfactant molecules. These can be divided into four groups, depending on the nature of the hydrophilic moiety.

The surfactant head group can be classified as anionic, cationic, nonionic, or amphoteric. *Anionic* surfactants ionize in aqueous solutions with the long chain (surfactant) carrying a negative charge [25]. *Cationic* surfactants also ionize in solution with the long chain bearing a positive charge. *Nonionic* surfactants have a polar head group, which has a strong affinity for water but does not ionize [25]. Some authors recognize a fourth classification in which the long chain carries a positive or a negative charge, depending on the pH of the solution. These surfactants are called *amphoteric* [26]. Each of these will be discussed in the next section.

Anionic Surfactants

Anionic surfactants are frequently added to well treatment fluids. These surfactants bear a negative charge when they ionize in aqueous solution and since most reservoir minerals are also negatively charged at near neutral and higher values of the solution pH (see Chap. 1), anionic surfactants exhibit minimal adsorption.

$CH_3(CH_2)_{10}CH_2OSO_3^-Na^+$

Figure 3.14 The molecular structure of one isomer of sodium dodecyl sulfate.

Alkylbenzene sulfonates (Fig. 3.12), salts of carboxylic acids, petroleum sulfonates, olefin sulfonates, and alkyl sulfates (see Fig. 3.14) all enjoy application. The hydrocarbon tail of the surfactant molecule is an important feature and by varying its length and degree of branching, the surfactant can be made to serve a number of functions. Rosen [26] provides a good summary of the influence of chain structure on the surfactant properties.

The surfactant sodium 4-(4-dodecyl) benzene sulfonate has 12 carbons in the hydrophobic chain. As this surfactant is added to an aqueous solution, the surface tension initially decreases linearly with the logarithm of the surfactant concentration. This trend is shown by Fig. 3.15. The surface tension of water is 0.076 N/m and with very small concentrations of surfactant this is observed as rapidly decreasing to 0.035 N/m. It is interesting to note that when the surfactant concentration has been increased to about 100 μmole/l, further increases in the surfactant concentration do not significantly lower the surface tension. In fact, the curve shown indicates a sharp transition between the linearly decreasing surface tension portion of the curve and the plateau region where the surface tension does not change further. This sharp transition between the two regions is thought to be the solution concentration at which aggregates of surfactant, the micelles, begin to form. The concept is that once micelles begin to form, the addition of more surfactant to the solution only results in the creation of more micelles and that the surfactant concentration at the interface between the air and the water

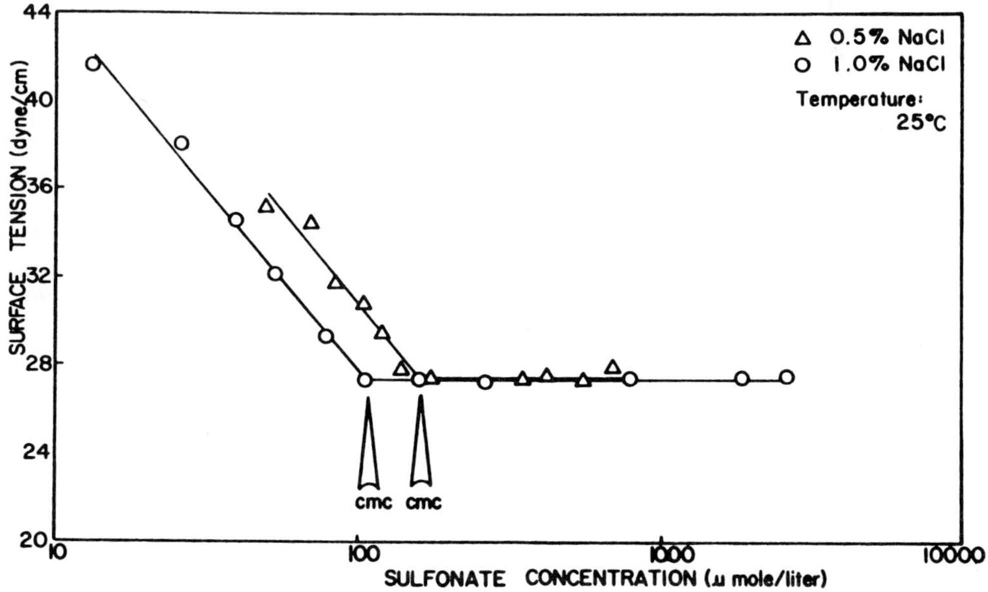

Figure 3.15 Critical micelle concentration of sodium 4-(4-dodecyl) benzene sulfonate from surface tension measurements [27].

remains constant. Thus rather than adsorbing at the interface, the surfactant molecules take collective action to form their own interface in the form of micelles, as depicted by Fig. 3.13, thereby removing the tails from the water. This explains why the surface tension does not further decrease when additional surfactant is added to the solution.

The concentration at which micelles first form is, for all practical purposes, a distinct one and can be used to characterize a surfactant. The concentration at the sharp break shown in Fig. 3.15 is called the *critical micelle concentration*. There are many properties that show a sharp break at or very near the critical micelle concentration (CMC) [25, 26]. The osmotic pressure, solution electrical conductivity, detergency, and many other properties can all be used to determine the surfactant CMC. The CMC values of a few representative surfactants are given in Table 3.2. The important feature to note is that the CMC is related to the surfactant tail length. As the length of the surfactant's tail is increased, the solution concentration at which micelles form steadily decreases. At some point, however, the surfactant molecules will precipitate as a crystal or a gel before reaching the concentration needed to form micelles. If this occurs the surfactant is said to be at a temperature that is less than its Krafft temperature. Increasing the temperature will reduce the tendency for the surfactant to crystallize and at the Krafft temperature micelles rather than crystals form. The phase diagram shown in Fig. 3.16 shows the Krafft temperature for a dodecyl benzene sulfonate. It also shows the CMC as a function of the temperature. For temperatures less than the Krafft temperature, the surfactant precipitates as solid crystals. The

TABLE 3.2 Critical Micelle Concentration of Anionic Surfactants [26]

Surfactant type	n	CMC (kg moles/m^3)	Temperature (°C)
Fatty acid soaps			
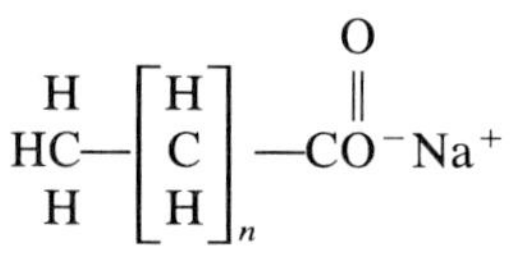	9	0.050	25
	11	0.012	25
	12	0.0072	25
Alkyl sulfonate			
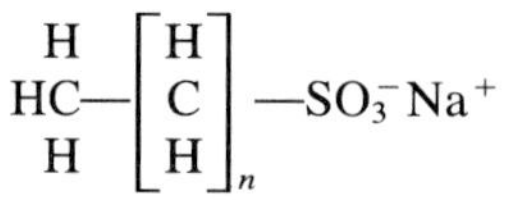	11	0.023	40
	11	0.025	50
	12	0.012	40
	13	0.0059	40
Alkyl sulfate			
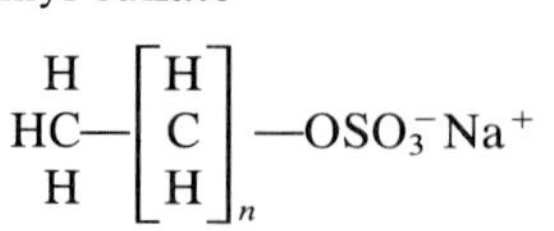	12	0.0076	45
	13	0.0033	45
	14	0.0020	45

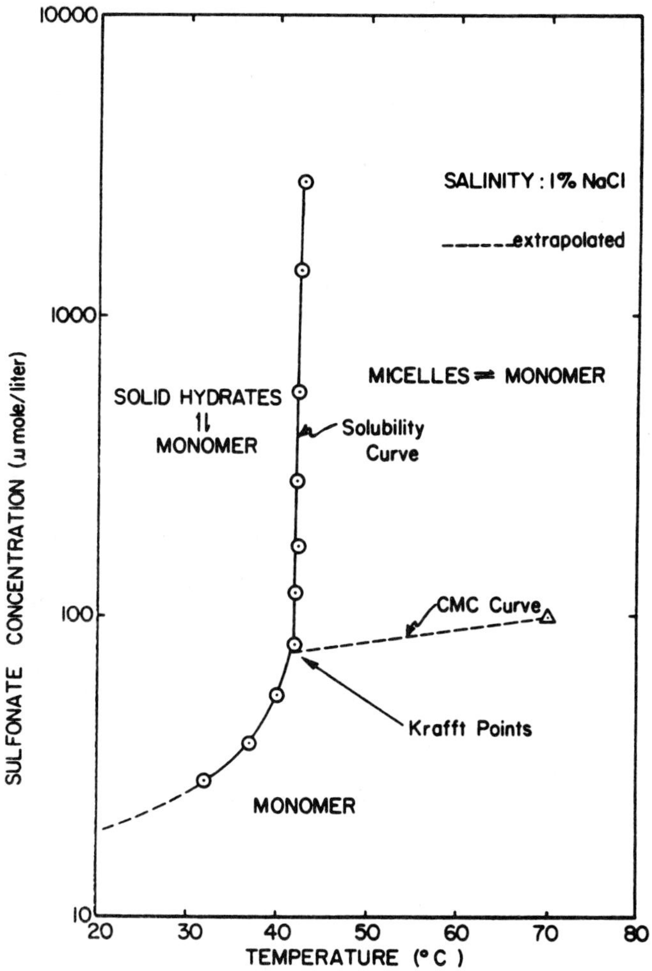

Figure 3.16 Phase diagram for sodium 4-(2-dodecyl) benzene sulfonate [27].

Krafft temperature of the dodecyl benzene sulfonate is observed to be 45°C. This particular surfactant cannot be used in many applications for which the aqueous phase temperature is less than 45°C since precipitated surfactant will settle out or be filtered out if the fluid is forced into the formation. Thus, the Krafft temperature of a surfactant is important since it marks a lower limit at which the surfactant can be applied. It should be noted, however, that if an oil phase is present as would be the case of an oil/water emulsion, then the surfactant molecules may partition into the oil rather than crystallize. In this case, it might be possible to use an ionic surfactant at temperatures less than the Krafft temperature.

The critical micelle concentration is also a very important property since below this concentration micelles do not exist. To disperse a corrosion inhibitor which has limited water solubility, it is, for example, necessary to have micelles since the inhibitor will be solubilized into their interior, oil-like environment. Thus, in this application concentrations in excess of the CMC are required.

Both the CMC and the Krafft temperature are sensitive functions of the composition of dissolved solids present in the water. This is important to remember since, as shown in Fig. 3.17, the Krafft temperature increases rapidly as the salt concentration in the water is increased. The values of the CMC shown in Fig. 3.15 demonstrate the sensitivity of the CMC to the salt concentration. Not shown, but of significance, is the particular sensitivity of anionic surfactants to multivalent ions such as Ca^{++} and Mg^{++}. These tend to precipitate anionic surfactants, although this tendency can be overcome to some extent by increasing the surfactant concentration.

Since both the CMC and the Krafft temperature vary with the salt content of the waters into which the surfactant are to be dissolved, the use of an anionic surfactant in waters of unknown or uncontrolled quality should be approached with caution. In general as the concentration of either dissolved solids or multivalent ions increases, the tail length of the surfactant used must be shortened. This often leads to a decreased efficiency. Thus, surfactant selection is an optimization problem as are most choices in engineering design. It is most important to recognize that the surfactant of choice for application in one well may not be the optimum choice for the same treatment in another field.

The alkyl benzene sulfonates are not the only anionic surfactants produced commercially. Alkyl sulfates derived from coconut oil or fatty alcohols are also widely applied. Petroleum sulfonates, made by sulfonating the aromatic fractions of certain refinery streams, are also manufactured in commerical quantities. These are also used in well treatment as wetting agents, dispersants, and cosurfactants.

Cationic Surfactants

There are two general categories of cationic surfactants. The first of these consists of long-chain primary, secondary, and tertiary amines. These are water soluble only in acidic solutions, where they ionize to form a long-chain cation and simple

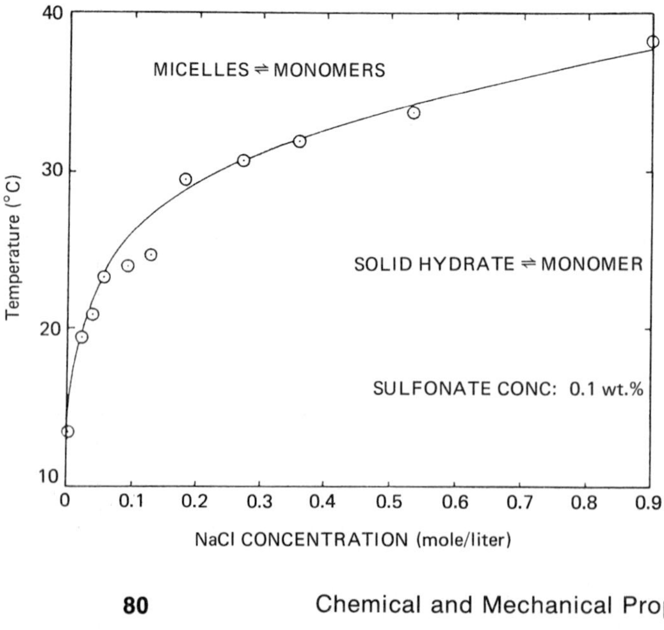

Figure 3.17 Effect of salt addition on the Krafft temperature of aqueous solutions of sodium 4-(4-dodecyl) benzene sulfonate [27].

anion salt. Amines are used as corrosion inhibitors to protect the tubular goods during acid treatments. Their application for this purpose will be described in Part 5. Figure 3.18 shows some molecular structures that are cationic.

The second important category of cationic surfactant is the quaternary ammonium compounds. These ionize to form long-chain cations over a wide range of solution pH. This class of cationic surfactant is depicted in Fig. 3.18.

Cationic surfactants experience the same sensitivity to multivalent ions or increased concentrations of dissolved solids as do anionic surfactants and therefore the same care must be exercised in their application as with anionic surfactants.

Cationic and anionic surfactants are generally incompatible. When mixed they tend to precipitate in aqueous solutions. These two classes of surfactant should not be blended in well treatment fluids without special reason and at sufficiently high concentration of either the anionic or the cationic surfactant (not both) so that the complexes of the two surfactants can be solubilized into the micelles of whichever surfactant is in excess. This is a particularly important point that should be carefully noted. An exception may be when mixing two surfactants of opposite charge but one having a fluorinated lipophile. Since hydrocarbon lipophiles apparently do not readily blend with fluorinated ones to form mixed micelles [28], it may be that such mixtures will not exhibit the same tendency to precipitate. Such mixtures may very well then have application in well treatments.

All of the long-chain cationic surfactants are germicidal. Quaternary ammonium compounds are among the most potent bactericides known and are widely used for this purpose.

As noted in Chapter 1, most mineral surfaces in contact with water carry a negative charge. Cationic surfactants are strongly adsorbed on negatively charged surfaces and cationic surfactants will be more readily exhausted from solution when in contact with reservoir materials than will be other types of surfactants. The exception to this rule occurs when the aqueous solution is strongly acidic

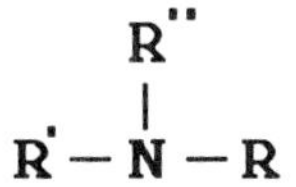

tertiary amine
(cationic in acid medium)

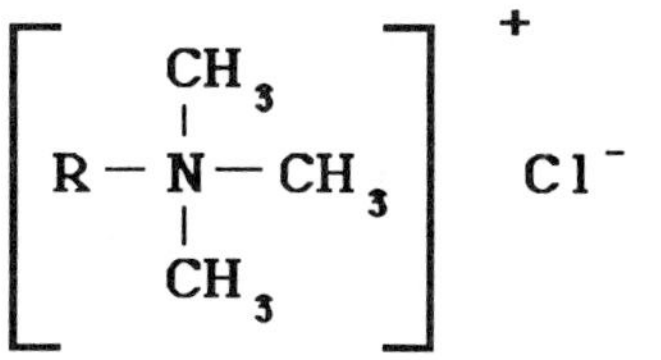

Quaternary Ammonium

Figure 3.18 Several types of cationic surfactants.

(low pH). Under these conditions mineral oxide surfaces tend to become positively charged (see Chap. 1) and cationic surfactant adsorption is significantly reduced.

If, on the other hand, surfactant adsorption on reservoir surfaces is desired to increase oil wettability, then cationic surfactants are the molecules of choice.

Nonionic Surfactants

We have seen that the Krafft temperature and the CMC of both anionic and cationic surfactants are sensitive to the ionic composition of water. In some cases where the quality of the water is unknown or uncontrolled, this sensitivity can be a decided disadvantage. The source of the sensitivity is, of course, the ionic nature of the surfactants. Both anionic and cationic surfactants ionize in aqueous solution. Indeed, it is this process of ionization which gives the head group of the surfactant its hydrophilic character. Water solubility can be achieved in another way, namely, by enhancing the polarity of the head group. The resulting surfactant will have both hydrophilic and hydrophobic segregated structures, but the hydrophilic head group will not become ionized in water. Such surfactants, which play on the hydrogen bonding properties of water, will be expected to be more tolerant of the composition of water. This will indeed turn out to be the case; however, nonionic surfactant performance is much more temperature sensitive than ionic surfactant performance. Thus, one may trade one type of difficulty for a second one.

The most prevalent class of commercially available nonionic surfactant is derived by reacting ethylene oxide with an appropriate material containing a reactive hydrogen. Since the product of this reaction is an alcohol which also contains an active hydrogen, the reaction will proceed extending the length of the ethylene oxide chain until in many cases a water soluble compound is formed. If the initial material is a hydrophobic compound of suitable molecular weight, it will become a surface active agent after a sufficient number of ethoxy groups have been added. Thus, the reaction of a long-chain alcohol will proceed as follows:

$$\underset{\text{(alcohol)}}{R\text{—}OH} + \underset{\text{(ethylene oxide)}}{\underset{\diagdown O \diagup}{CH_2\text{—}CH_2}} \longrightarrow \underset{\text{(ethoxylated alcohol)}}{ROCH_2CH_2OH} \tag{3.15}$$

$$\begin{array}{ccc} ROCH_2CH_2OH + \underset{\diagdown O \diagup}{CH_2\text{—}CH_2} & \longrightarrow & ROCH_2CH_2OCH_2CH_2OH \\ \vdots & & \vdots \end{array} \tag{3.16}$$

$$RO(CH_2CH_2O)_xH + \underset{\diagdown O \diagup}{CH_2\text{—}CH_2} \longrightarrow RO(CH_2CH_2O)_{x+1}H \tag{3.17}$$

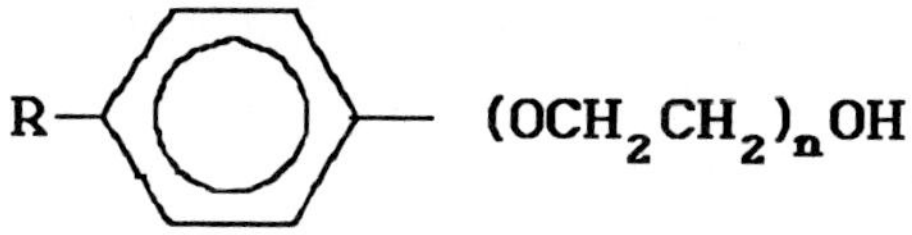

Alkylphenol ethoxylate

R $(OCH_2CH_2)_nOH$ Alcohol ethoxalate

Figure 3.19 Typical structures of nonionic surfactants.

The final product shown is a molecule containing $x + 1$ ethylene oxide units (we will say EO $= x + 1$).

Figure 3.19 shows the structure of a number of nonionic surfactant molecules. Both are ethoxylated compounds. Such compounds comprise the overwhelming commercial production of nonionic surfactants. Other types of nonionic surfactants are produced and may find limited application in well treatment; however, the ethoxylated compounds are most widely used because of their ready availability.

The critical micelle concentrations of nonionic surfactants relative to their anionic or cationic counterparts are quite small. The reason for this is that the repulsive effect of a surface of like charges, as shown by the spherical micelle depicted in Fig. 3.13, does not enter into the determination of the stability of a nonionic micelle. Thus, nonionic micelles are able to form at very low solution concentrations. Consequently, the Krafft temperatures are quite low and are not a factor in the selection of a nonionic surfactant.

CMCs of representative nonionic surfactant molecules are shown in Table 3.3. These can be compared with values of anionic surfactants having similar tail

TABLE 3.3 Critical Micelle Concentrations of Some Nonionic Surfactants at 25°C [26]

Surfactant	Average mole ratio of ethylene oxides	Average molecular weight	CMC (moles/l)
Nonyl phenol	9	616	70
Nonyl phenol	10.5	682	70
Nonyl phenol	15	880	130
Nonyl phenol	20	1100	150
Nonyl phenol	30	1540	175
Octyl phenol	1	250	49.5
Octyl phenol	2	294	76.5
Octyl phenol	3	338	250
Octyl phenol	4	382	268
Octyl phenol	5	426	283
Octyl phenol	10	602	304

structures (see Table 3.2) and it will be observed that nonionic CMCs are one to two orders of magnitude smaller than those of anionic surfactants.

The cloud point temperature is another important property of nonionic surfactants which differs from that of either anionic or cationic surfactants. As the temperature of an aqueous solution is increased, the degree of hydrogen bonding between the surfactant's hydrophilic head and surrounding water molecules decreases and at an elevated temperature, the surfactant forms a separate phase—not a solid crystalline phase, but some type of liquid crystalline phase which generally remains suspended in the aqueous solution giving it a cloudy appearance. This upper consolute temperature is a function of the number of ethylene oxide adducts (EO units), of the hydrophilic tail structure of the surfactant molecule and the salt content of the solution. The sensitivity of the cloud point to these variables is illustrated by Figs. 3.20 and 3.21. It is important to recognize that although nonionic surfactants may be effective for application at one temperature, their strong temperature-dependent behavior may render them much less effective at elevated temperatures. Thus, the performance of nonionic surfactants must be established at reservoir temperatures when selecting one for use.

Increasing the number of EO units does increase the cloud point temperature. Thus, for high-temperature applications those nonionic with larger values

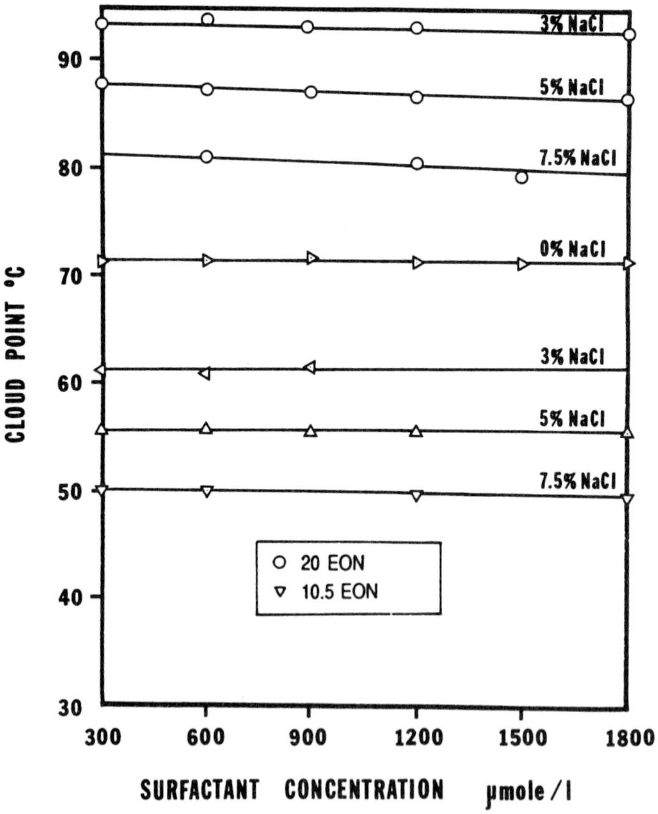

Figure 3.20 Cloud points of ethoxylated nonylphenols, one with 20 ethylene oxide units (20 EON) and one with 10.5 units in aqueous solutions of various salinities (NaCl) [29].

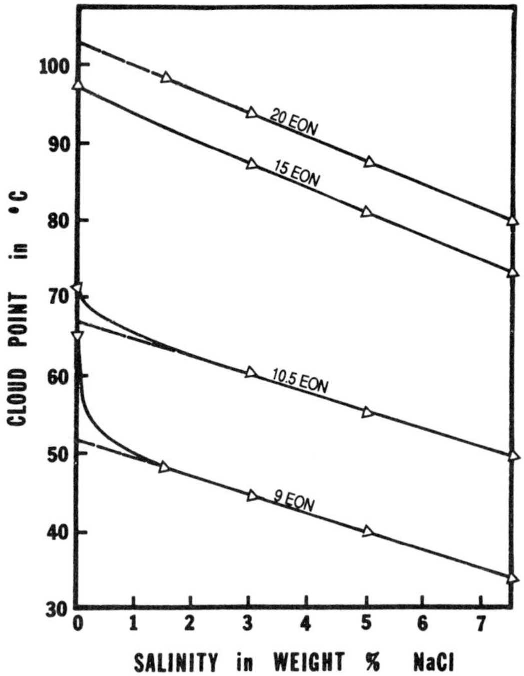

Figure 3.21 Dependence of cloud points of various ethoxylated nonylphenols as a function of salinity. The EON number refers to the number of ethylene oxide numbers [29].

of EO would be expected to be most useful. Increasing the number of EO units does, however, reach a point of diminishing return and it is difficult to obtain water solubility at temperatures in excess of 90°C without reducing the number of carbon atoms in the chain regardless of the number of EO units added. Thus application of nonionic surfactants in reservoirs of higher temperature should be done with care and full knowledge that their water solubility is limited.

The cloud point temperature is almost independent of surfactant concentration. However, in the presence of salt the cloud point temperature is depressed as shown in Fig. 3.21. Although the effect is not large, the ionic composition of the water in which the surfactant is to be dissolved should be taken into account. For many applications of surfactants, the balance between the tendencies to dissolve in oil and in water are critical and a small alteration of the affinity of the surfactant for either phase may change the result dramatically [25].

Fluorocarbon Surfactants

Fluorocarbons form surfaces of lower free energy than hydrocarbon surfaces. Consequently, fluorocarbon surfactants lower the surface tension of solutions to a greater extent than hydrocarbon surfactants. Surface tensions as low as 0.030 N/m can be obtained using surfactants with a hydrocarbon tail, whereas values as low as 0.017 N/m have been reported using fluorocarbon surfactants [30].

Fluorocarbons are commercially available in anionic, cationic, or nonionic form.

3.4 FOAMS

Foams are used as fracturing, gravel transporting, and drilling fluids and it is of interest, therefore, to consider briefly their physical properties. A foam consists of gas bubbles separated by walls of liquid film. Remarkably foams may exhibit elasticity, rigidity, and non-Newtonian behavior. Depending on the thickness of the liquid films and the size of the gas bubbles, the foam density can range from that nearly equivalent to the liquid down to the gas density. A parameter often cited to define the state of a foam is its *quality*, Γ, which is defined as the volume of gas per volume of foam. Quality is a volume fraction, and gas volumes are very sensitive to both pressure and temperature. Therefore, when stating the quality of a foam it is necessary to also provide the conditions under which this quality applies.

For qualities less than approximately 50%, the foam may be considered to be a dispersion of gas bubbles in a liquid. Under these conditions the bubbles need not be in contact and the rheological behavior will be governed primarily by the properties of the liquid, with small corrections to allow for the presence of the gas bubbles. Such low-quality foams are not particularly useful in well stimulation and attention will be focused primarily on systems having qualities greater than 55% and usually greater than 70%.

Foam Stability

As noted previously, low-quality foam consists of nearly spherical bubbles separated by thick liquid walls. As liquid drains from the walls under the influence of gravity, the gas bubbles deform and become more or less structured. Foams of very high quality exhibit a polyhedral shape.

Foam stability is dependent upon a number of factors, one of the most important being the rate at which the liquid drains from between the walls separating gas bubbles causing the films to become thinner and thereby increasing the probability it will rupture. Thus, one method of enhancing foam stability is to increase the viscosity of the liquid phase. This may be accomplished through the addition of polymer to the liquid to increase its viscosity prior to blending the liquid and the gas to create the foam [31–34]. The drainage may be by gravity or due to the pressure differences created by differences in curvature. This latter effect occurs at Plateau borders [35].

If two foam bubbles are brought together as shown in Fig. 3.22, the pressures in each will differ because the radii of the two bubbles differ. The pressure inside each bubble is given by Laplace's equation written to account for two gas interfaces

$$\Delta p = \frac{4\gamma}{r} \tag{3.18}$$

where r is the average radius (actually the reciprocal of the mean curvature), Δp is the pressure difference between the gas inside the bubble and that outside, and γ is the surface tension. It follows that the pressure in the smaller bubble is larger than that in the larger one. Because of this pressure difference there will be a

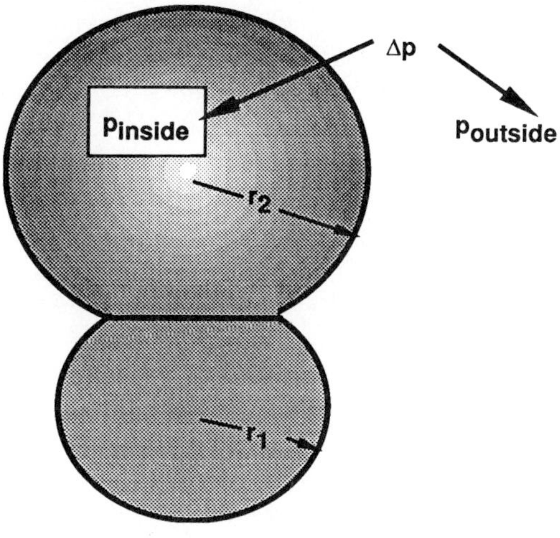

Figure 3.22 Two foam bubbles in contact. Mass transfer will result from the small one to the larger one because the pressure inside the smaller one is larger.

diffusion of gas through the liquid film from the smaller bubble to the larger one [36–38]. The pressure difference is the driving force for this diffusion and can, as shown by Eq. (3.18), be reduced by decreasing the surface tension. Decreasing the surface tension is accomplished by the addition of surfactants, which for several reasons are necessary to achieve reasonable foam lifetimes unless the viscosity of the liquid is large.

Surfactants play a second role in enhancing foam stability. Once the drainage process has resulted in the walls of the foam thinning to very small proportions (on the order of 15 nm), rupture of the walls occurs quite rapidly in the absence of surfactants. If surfactants are added, the films can become extremely thin and persist for some time (days) before rupture occurs. This is thought to be related to electrostatic forces when ionic surfactants are used and due to structural forces when nonionic surfactants are used [39]. Thus, it is of great importance to add surfactants when attempting to create foams with good stability.

Different surfactants provide different degrees of foam stability under different conditions. Those surfactants which may yield foams having long lifetimes under one set of conditions of temperature and salt concentration may be entirely unsatisfactory under others. The particular surfactant selected for a given application should be matched to the quality of the water and the temperature of the formation.

Example 3.4 Surface Area of Foam [36, 40]

An apparatus for measuring the surface area of foam is shown in Fig. 3.23. The pressure is measured in the vapor space above the foam column. Assuming that the pressure varies in accordance with the empirical equation

$$p - p_0 = \beta[1 - \exp(-t/T_R)]$$

where the parameters β and T_R that best fit the pressure measured as a function of time are 5 cm H_2O and 60 min., respectively. Show that $\beta = 4\gamma A_0/3$ where A_0 is the initial foam area per unit volume of foam. Calculate the area of the foam per unit of volume, A, after 100 minutes have elapsed. Assume that at any time the foam

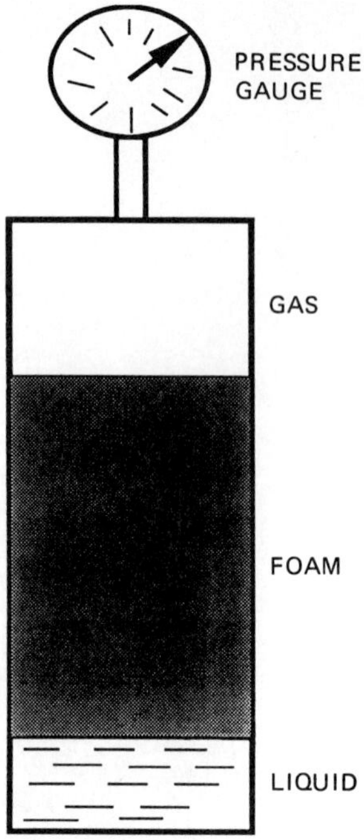

Figure 3.23 Apparatus for measuring the surface area of a foam as a function of time.

consists of a large number of spherical bubbles each having the same radius, r. The surface tension is 0.03 N/m.

Solution In a volume V there is a volume V_g of free gas and a volume $V - V_g$ of foam, neglecting the volume occupied by the liquid films. The pressure in the free gas is p and the pressure in the foam is $p + 4\gamma/r$. Thus the total moles of gas is given by

$$pV_g + \left(p + \frac{4\gamma}{r}\right)(V - V_g) = n_T RT$$

where n_T is total number of moles of gas present, a constant. This equation can be simplified to

$$pV + \frac{4\gamma}{r}(V - V_g) = n_T RT$$

But
$$V - V_g = N_b \frac{4\pi r^3}{3}$$

where N_b is the number of bubbles. Since the total area of foam is the number of bubbles times the area per bubble, we have

$$VA = N_b(4\pi r^2) \qquad \text{and} \qquad pV + \frac{4\gamma AV}{3} = n_T RT$$

Since the total number of moles is constant,

$$p + \frac{4\gamma A}{3} = \frac{n_T RT}{V} = \left(p_0 + \frac{4\gamma A_0}{3}\right) = \text{constant}$$

Thus $$p - p_0 = \frac{4\gamma}{3}(A_0 - A) = \frac{4\gamma A_0}{3}\left(1 - \frac{A}{A_0}\right)$$

and by comparing this to the empirical equation, it is seen that $\beta = 4\gamma A_0/3$.

After 100 minutes

$$p - p_0 = (5 \text{ cm } H_2O)\,(1 - \exp[-100/60])$$

or $$p - p_0 = 4.06 \text{ cm } H_2O = (4.06 \text{ cm})\,(980 \text{ cm/sec}^2)\,(1 \text{ g/cm}^3)$$

Thus $$p - p_0 = 3975 \text{ dynes/cm}^2$$

Using this value

$$A_0 - A = 3(3975)/4(30) = 99.4 \text{ cm}^{-1}$$

Since $$A_0 = \frac{3(5 \text{ cm})(980 \text{ cm/sec}^2)\,(1 \text{ g/cm})}{4(30 \text{ dynes/cm})} = 122.5 \text{ cm}^{-1}$$

we find $$A = 23.1 \text{ cm}^{-1}$$

Foam Rheology

Generally foams of high quality can flow only if the gas bubbles are deformed. Princen [41] and Prud'homme [42] have calculated the stress needed to induce a deformation of a two-dimensional hexagonal foam, as depicted by Fig. 3.24, which occurs at high gas qualities (> 90°). A shear deformation of the two-dimensional hexagons causes an increase in surface area which is resisted by the interfacial tension. An elastic deformation occurs below a certain yield stress. Above this value a more stable configuration can be reached in which each row of the two-dimensional foam cells has been shifted in the direction of the force by one cell width. This two-dimensional model does not really apply to three-dimensional foams; however, foams do exhibit a yield stress as shown by the data plotted in Fig. 3.25, where the yield is shown to increase as the quality of the foam is increased. The two-dimensional yield stress has been hypothesized to be given by [41]

$$\tau_{\text{yield}} \cong \frac{2\gamma}{r} \tag{3.19}$$

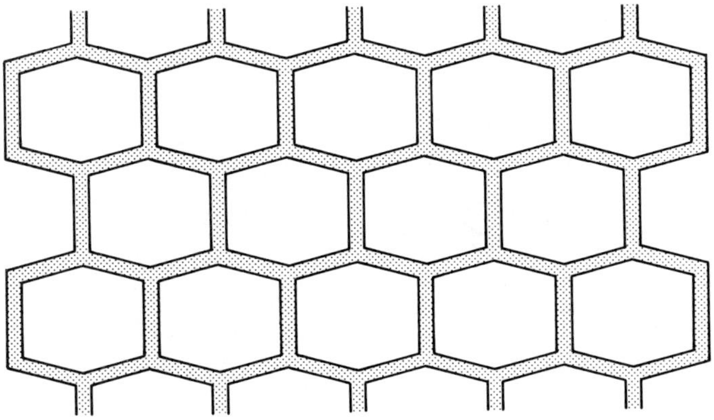

Figure 3.24 A two-dimensional honeycomb foam.

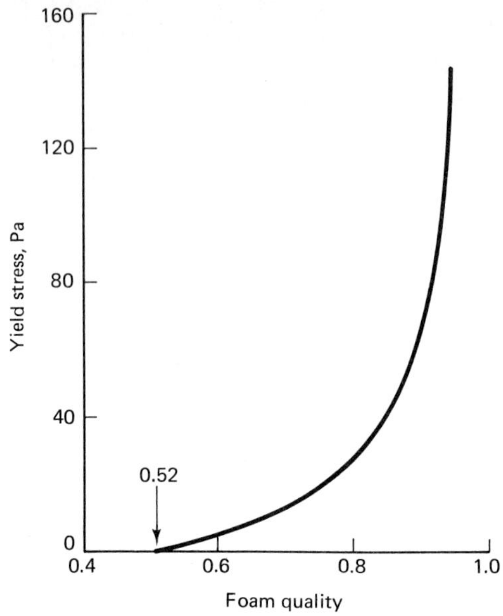

Figure 3.25 Yield stress of foam [43].

This expression can only apply at high qualities, for as shown by Fig. 3.25, the yield stress is a strong function of quality. Other more recent and systematic studies have confirmed the trends shown by Fig. 3.25. Reidenbach et al. [31] found the yield stress to be an exponential function of quality. Their empirical equation is

$$\tau_{\text{yield}} = C_i \exp 9\Gamma \tag{3.20}$$

where C_i is a property of the surfactants and the gas (normally CO_2 or N_2).

Example 3.5 Estimation of Bubble Size

Assuming Eq. (3.19) does actually represent three-dimensional foam behavior, determine the bubble radius that corresponds to a yield stress of 62.4 Pa. (For the specific foam represented by Fig. 3.25 this corresponds to a 93% quality.) The surface tension is 3.3×10^{-2} N/m.

Solution Based on Eq. (3.19)

$$r = \frac{2\gamma}{\tau_{\text{yield}}} = \frac{2(3.3 \times 10^{-2})}{62.4} = 1.06 \times 10^{-3} \text{ m} = 1.06 \text{ mm}$$

The average bubble radius was not reported by Blauer et al. [43]. This value, which appears quite reasonable, cannot therefore be confirmed; however, Kraynik [44] has reported a foam radius of 0.125 mm at the same quality.

The fact that foams do exhibit a yield stress is important. Many of the applications of foam in both well stimulation and drilling depend on this property for particle-carrying capacity.

In addition to the yield stress, the apparent viscosity is also of interest [45]. Mitchell [46] found that the pressure drop attending the flow of foam in pipes could be approximated by considering foam to behave in shear as a Bingham plastic [17]. Thus,

$$\tau_{xy} = \pm \tau_{\text{yield}} + \mu_{\text{p}} \frac{dv_x}{dy} \tag{3.21}$$

where μ_{p} is the plastic viscosity and the sign preceding the yield stress must be selected so that it is the same as the sign of τ_{xy}. The yield stress is a positive quantity.

The flow of a Bingham plastic between parallel plates is depicted by Fig. 3.26. There is a central core moving at a constant velocity surrounded by a region of varying shear rate. The extent of the central region is determined by both the imposed pressure gradient and the yield stress. A force balance on the element of fluid shown in Fig. 3.27 gives

$$\tau_{yx}\,\Delta x + p(x)y - p(x + \Delta x)y = 0$$

or dividing by Δx and taking the limit as Δx vanishes, one obtains

$$\tau_{yx} = y \frac{dp}{dx} \tag{3.22}$$

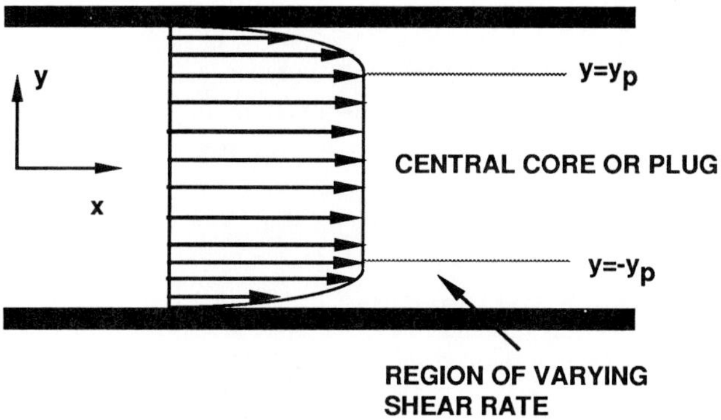

Figure 3.26 Velocity profile of a Bingham plastic flowing between parallel plates. The shear stress exerted throughout the central core is smaller than the yield stress.

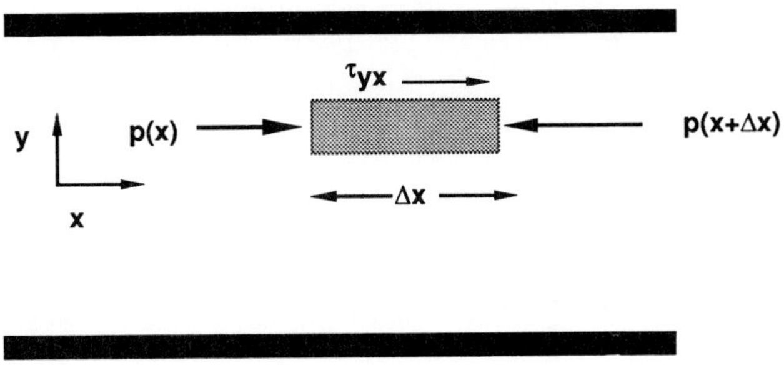

Figure 3.27 Forces acting on an element of fluid. These include the frictional force and the pressure forces.

Based on this equation, it is seen that

$$y_p = \frac{\tau_{yield}}{-\frac{dp}{dx}} \tag{3.23}$$

where the negative sign in Eq. (3.21) has been selected since dp/dx is negative. Equation (3.23) shows the relationship between the extent of the central core and the yield stress.

Apparently characteristic of most foam flow is the presence of a thin liquid film coating the wall. Indeed for some applied pressure gradients, it appears that the entire region of varying shear rate may be devoid of foam and consist of a liquid film [44, 47]. At higher pressure gradients, the foam may also be sheared, but the liquid film may still be the origin of most of the friction loss. To understand the empirical equations that have been used to express foam flow, it is helpful to recognize the existence of the liquid film.

Figure 3.28 shows the plastic viscosity plotted against the foam quality [43]. Again the more interesting region is that of high quality where μ_p becomes large. Hirasaki and Lawson [48] show that the viscosity is also a function of the foam texture (bubble size) and that specifying the quality alone is insufficient. Harris [49] also found texture to be important, but not as important as quality or the rheological properties of the liquid. Thus, two foams both having the same quality but different bubble size distributions will exhibit different viscosities. Therefore the results, shown by Fig. 3.28, are then an indication of the trends that can be expected, but should not be accepted as sufficient for design purposes.

Example 3.6 The Flow of Foam between Parallel Plates

Find the pressure gradient required to cause a foam (90% quality) to flow at an average velocity of 0.165 m/sec through a slit between closely spaced parallel plates

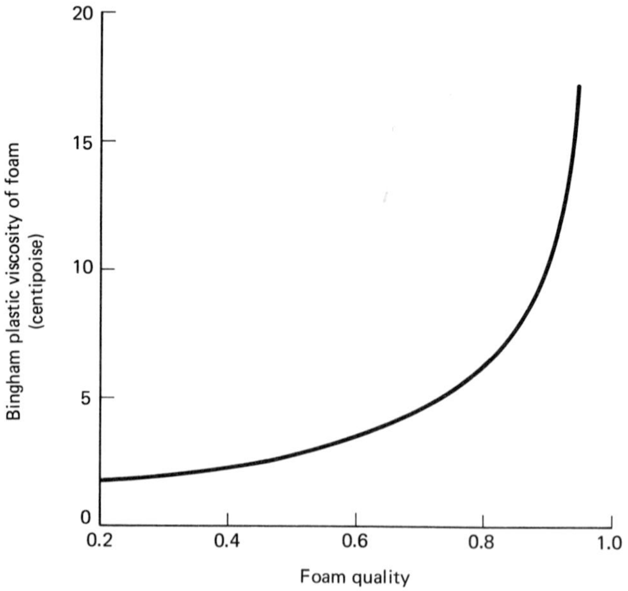

Figure 3.28 Bingham plastic viscosity of foam [43].

separated by a total distance of 1 cm. Assume that the rheological properties of the foam are defined by Figs. 3.25 and 3.28.

Solution To resolve this problem, an equation relating the average velocity to the pressure gradient is needed. Equation (3.22) is general. It applies irrespective of the rheological model so that substituting Eq. (3.21) into Eq. (3.22) one obtains

$$\mu_p \frac{dv_x}{dy} = \tau_{\text{yield}} + y\frac{dp}{dx} \qquad (\text{for } y \leqslant y_p) \tag{3.24}$$

Integrating this differential equation and noting that $v_x = 0$ at $y = w/2$ yields

$$v_x = -\frac{1}{2\mu_p}\left(\frac{w}{2}\right)^2 \frac{dp}{dx}\left[1 - \left(\frac{2y}{w}\right)^2\right] - \left(1 - \frac{2y}{w}\right)\frac{w\tau_{\text{yield}}}{2\mu_p} \tag{3.25}$$

This equation gives the distribution of velocity in the region of variable shear. Furthermore, the plug velocity is given by Eq. (3.25) by substituting y_p for y. Thus

$$V_{\text{plug}} = -\frac{1}{2\mu_p}\left(\frac{w}{2}\right)^2 \frac{dp}{dx}\left[1 - \left(\frac{2y_p}{w}\right)^2\right] - \left(1 - \frac{2y_p}{w}\right)\frac{w\tau_{\text{yield}}}{2\mu_p} \tag{3.26}$$

Finally, the average velocity is given by

$$\langle v_x \rangle = \frac{2}{w}\int_0^{w/2} v_x\, dy = \frac{2}{w}\int_0^{y_p} V_{\text{plug}}\, dy + \frac{2}{w}\int_{y_p}^{w/2} v_x\, dy \tag{3.27}$$

Integrating

$$\langle v_x \rangle = \frac{2y_p}{w} V_{\text{plug}} - \frac{w\tau_{\text{yield}}}{2\mu_p}\left[\frac{1}{2} - \left(\frac{2y_p}{w}\right) + \frac{1}{2}\left(\frac{2y_p}{w}\right)^2\right] - \frac{1}{2\mu_p}\left(\frac{w}{2}\right)^2\frac{dp}{dx}\left[\frac{2}{3} - \frac{2y_p}{w} + \frac{1}{3}\left(\frac{2y_p}{w}\right)^3\right] \tag{3.28}$$

Equation (3.28) relates the average velocity to the pressure gradient; however, since y_p also depends on the pressure gradient, it is easiest to use a trial-and-error procedure to calculate it. The yield stress at 90% quality appears to be about 57.5 Pa (from Fig. 3.25) and the Bingham plastic viscosity is about 11 cp (from Fig. 3.28). Also w is given as 10^{-2} m. Substituting these values into Eq. (3.26) together with an estimate for the value of dp/dx permits the velocity of the centerplug to be calculated. Take as first approximation $dp/dx = -1.28 \times 10^4$ Pa/m. With this value, from Eq. (3.23) $2y_p/w = 2(57.5)/10^{-2}(1.28 \times 10^4) = 0.9$. Thus,

$$V_{\text{plug}} = -\frac{(5 \times 10^{-3})(57.5)}{(1.1 \times 10^{-2})}(1 - 0.9) + \frac{1}{2}\frac{(5 \times 10^{-3})^2(1.28 \times 10^4)}{(1.1 \times 10^{-2})}(1 - 0.9^2)$$

or

$$V_{\text{plug}} = 0.15 \text{ m/sec}$$

The average velocity is determined by Eq. (3.28) as

$$\langle v_x \rangle = (0.9)(0.15) - \frac{(5 \times 10^{-3})(57.5)}{(1.1 \times 10^{-2})} \left[\frac{1}{2} - 0.9 + \frac{1}{2}(0.9)^2 \right] + \frac{(5 \times 10^{-3})^2}{2(1.1 \times 10^{-2})}(1.28 \times 10^4) \left[\frac{2}{3} - 0.9 + \frac{1}{3}(0.9)^3 \right]$$

or

$$\langle v_x \rangle = 0.145 \text{ m/sec}$$

Thus, our guess for the pressure gradient is too small since this is smaller than the desired velocity. Continuing this trial-and-error procedure, the pressure gradient is finally found to be $-1.29 \times 10^{+4}$ Pa/m. The final value of $2y_p/w$ is 0.89, indicating that the sheared region is quite small for this case.

Foams that are used in well stimulation often include polymer as well as surfactant to enhance stability and to modify foam rheology. Since much of the viscous behavior of foams is related to shearing thin liquid films, especially those coating or near the wall, then the addition of polymer might be expected to convert the Bingham fluid to a *yield pseudoplastic* [50], which is defined by replacing the constant Bingham plastic viscosity, μ_p, in Eq. (3.21) by a power-law relationship so that

$$\tau_{xy} = \pm\tau_{\text{yield}} + m_f \left| \frac{dv_x}{dy} \right|^{n-1} \frac{dv_x}{dy} \tag{3.29}$$

Studies of foam flow in pipes by Reidenbach et al. [31] and Cawiezel and Niles [51] have shown that m_f, the consistency index of the foam solution, and the value of n (flow index) are both essentially the same as that of the aqueous polymer solution. Apparently, the properties of the aqueous polymer solution that is blended with the gas to form the foam dominate the viscous behavior of a foam.

Harris [49] has studied foam rheology in a flow loop measuring the time-dependent pressure gradient and the bubble distribution at the same time. He confirmed again that the rheological behavior is governed primarily by the foam quality. The liquid phase rheological properties are also, as noted here, important. The texture (bubble size) was found to be of lesser importance, but finer texture foams result from higher pressures, higher shear rates, and higher surfactant concentrations. All of these factors will, to some extent, influence the rheological properties, but for design purposes Eq. (3.29) with the appropriate values of m_f, n, and τ_{yield} will suffice.

3.5 EMULSIONS

Stability

An emulsion can be defined as a dispersion of liquid drops in a second immiscible liquid. The discontinuous or internal phase is the liquid that is broken into small droplets, whereas the continuous or external phase is the surrounding liquid. As depicted by Fig. 3.29, when water is the external phase the emulsion is said to

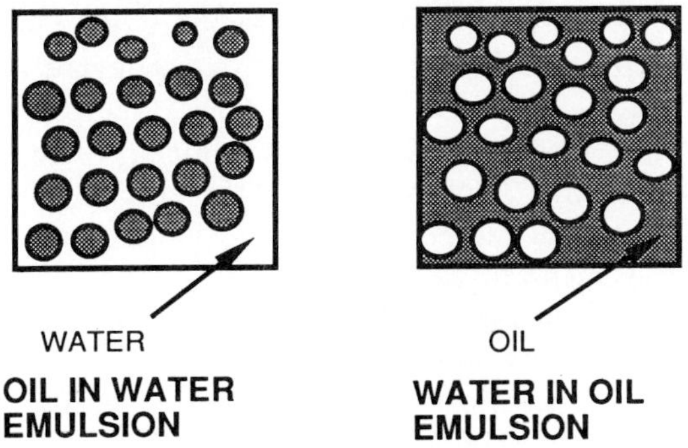

Figure 3.29 Diagram showing W/O and O/W emulsions. Emulsions consist of one liquid phase dispersed into a second immiscible liquid phase.

be an O/W type (oil-in-water), and when water is the discontinuous phase the emulsion is W/O (water-in-oil).

Mixtures of oil and water such as those depicted in Fig. 3.29 are unstable. The droplets tend to rise (or fall depending on the relative densities), a process which is called *creaming*. As the droplets compress into a smaller and smaller volume, coalescence to form larger droplets then tends to occur. If emulsifiers (surfactants) have not been added to the mixture, the creaming and coalescence processes will take place rapidly. After a short time, the system will consist of a continuous oil phase separated from the aqueous phase by a sharp interface.

The addition of appropriately balanced surfactant molecules to the blend of immiscible liquids can greatly stabilize emulsions [25, 52, 53]. The surfactant molecules strongly adsorb at the liquid/liquid interface reducing the interfacial tension. Depending on how the emulsion is produced, reducing the interfacial tension will normally reduce the drop sizes thereby slowing the rate of creaming. This will, of course, tend to stabilize the emulsion. The interfacial tension plays a key role in fixing the distribution of drop sizes because the pressure inside of a drop increases from its value given by Eq. (3.18) to a larger value whenever a drop is distorted from its spherical shape, as, for example, in a shear field. This increase in pressure will tend to restore the original spherical shape and oppose the break-up of the drop. The magnitude of the restoring pressure depends directly on the interfacial tension.

The most important role of the surfactant molecule is to reduce the coalescence rate between emulsion droplets. There are a number of mechanisms that contribute to the increased stability. Ionic surfactants adsorbed at the interface will normally give rise to a surface charge causing a repulsion when drops approach one another [54]. This repulsive force is similar to the one experienced by a particle approaching a wall, as described in Chapter 4. Nonionic surfactants stabilize emulsions by sterically hindering the coalescence process [54].

Shinoda and Friberg [53] discuss the stabilizing influence of liquid crystals. These viscous phases composed of surfactant, oil, and water can form a *shell* surrounding the dispersed phase drops mechanically hindering their coalescence.

Perhaps the most important role of surfactant in stabilizing emulsions is the creation of interfacial tension gradients. When two drops are forced to approach one another, the film of continuous phase separating the two drops is thinned. The continuous phase flows outward from between the drops pushing surfactant to the periphery of the film as shown by Fig. 3.30. This action transporting surfactant from the region separating the drops causes a concentration gradient of surfactant along the interface.

The interfacial tension between oil and water is reduced, often substantially [25], by the presence of surfactant. Thus because of the concentration gradient, there also exists a gradient of interfacial tension. This provides an additional stress [55, 56] as shown by Fig. 3.31. The interfacial tensions are depicted as tensions and the resultant is a pull on a surface element. The side of the fluid element where the surfactant concentration is least experiences the greatest tension. The resultant of the interfacial tensions is a force opposing the motion of the fluid. Thus, the interfacial tension gradient hinders the drainage of the continuous phase from the region separating the two drops and thereby reduces the rate of coalescence.

Whenever two immiscible liquids are vigorously blended to form an emulsion, it seems reasonable to suppose that the phase which will appear as the discontinuous one will be the one for which the coalescence rate is slowest. For example, if in a given system oil drops in water coalesce more rapidly than do water drops in oil, then the emulsion that forms will be the W/O. Bancroft's rule states that the phase containing the surfactant will in general be the continuous

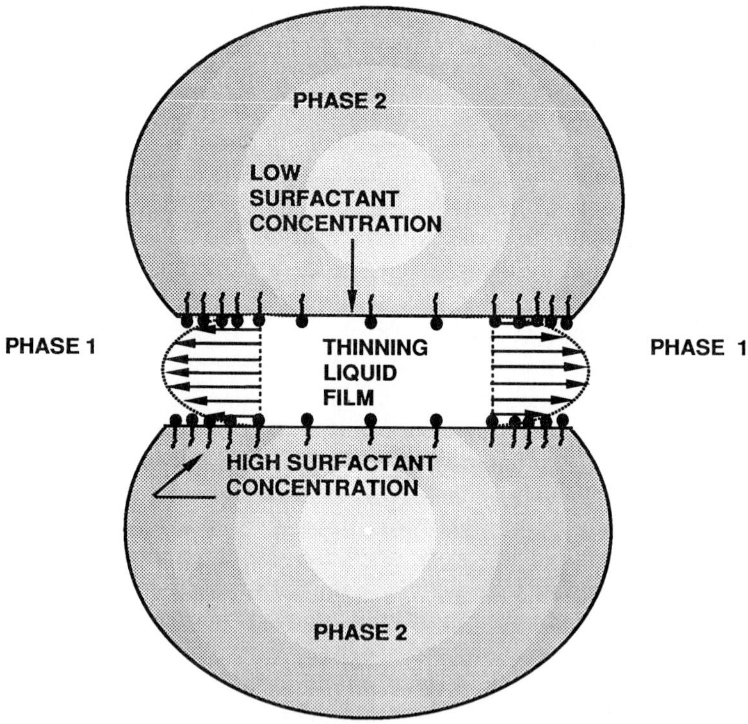

Figure 3.30 Diagram showing the mechanism by which a concentration gradient of surfactant is produced.

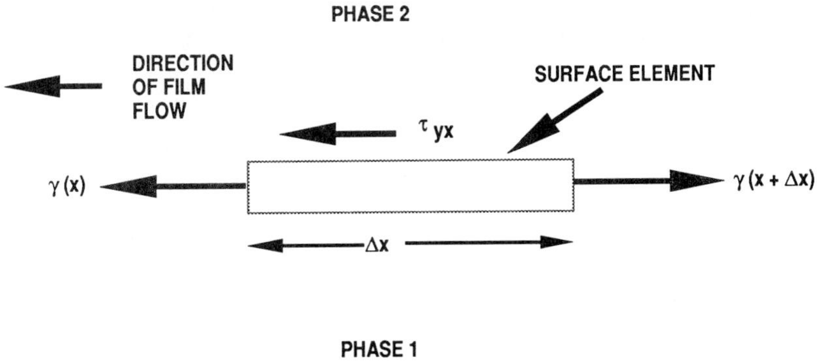

Figure 3.31 Diagram showing an element of interfacial area subjected to two different forces: shear forces and interfacial tension forces.

phase [57]. It can be seen that the interfacial tension gradient is also largest when the surfactant is in the continuous phase.

Figure 3.32 depicts the two different cases. In Case A the surfactant, which is in the continuous phase, is prevented or at least hindered from reentering the region separating the drops by the hydrodynamic motion of the fluid. Thus, this motion helps to maintain large interfacial tension gradients. Conversely, in Case B when the surfactant is in the discontinuous phase, then the hydrodynamic motion of the fluid circulating within the drop tends to transport surfactant to the interfacial region thereby decreasing the gradients of interfacial tension. Thus the surfactant, which is swept along the surface by the flow of the continuous phase, is more readily replenished when the surfactant is in the drop rather than in the continuous phase. Emulsion stability is in part determined by the phase in which the surfactant is found.

Figure 3.33 shows a series of test tubes containing surfactant, oil, and water. There are at least two phases in each of the tubes but in some (the three central tubes) there are three phases. That phase containing the bulk of the surfactant is called a *micellar solution* (this phase is sometimes called a *microemulsion* [25]). In the tubes on the left, the surfactant is in the lower phase and the micellar solution is in equilibrium with an oil phase. Such systems are sometimes labelled as Type I and if the two phases are vigorously blended, the oil will generally be the discontinuous phase.

Type II systems are those for which the micellar solution is in equilibrium with water. The emulsion formed will be W/O.

When the micellar solution is in equilibrium with both water and oil (Type III systems), the emulsions formed are unstable and break rapidly. The temperature at which the emulsions break rapidly is called the *phase inversion temperature* [53].

There are important implications of Bancroft's rule which can now be understood. In breaking an emulsion at the production site, one should add a water soluble surfactant to break a W/O emulsion because the presence of surfactant in the discontinuous phase will tend to result in the phase inversion point being

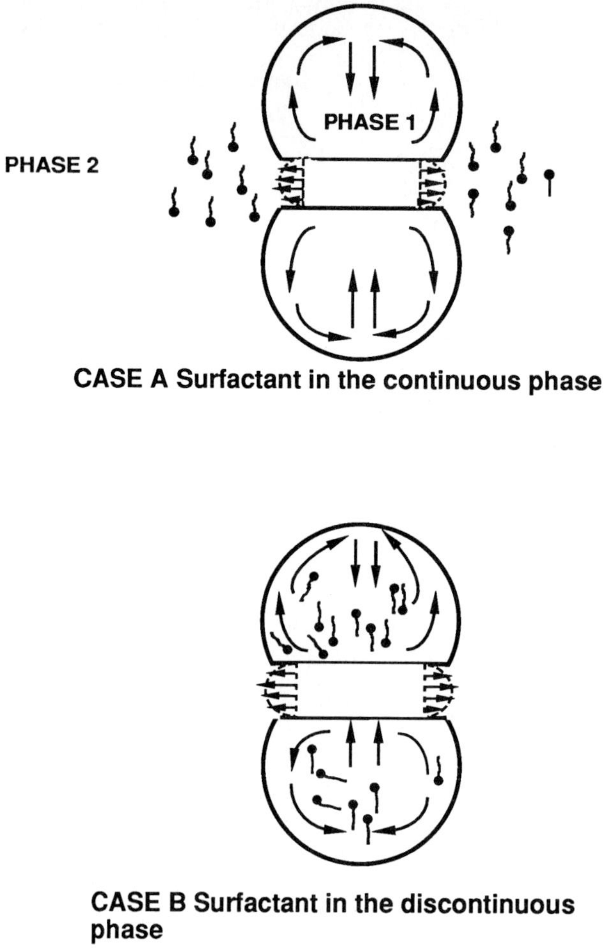

Figure 3.32 Sketch showing surfactant in either the continuous or the discontinuous phase. Flow tends to transport surfactant away from the interface in Case A and toward the interface in Case B.

approached and will tend to destabilize the emulsion. To break an O/W emulsion, adding a nonionic surfactant and heating may be effective. Figure 3.33 shows that heating increases the partitioning of nonionic surfactants into oil. This is because the affinity of the surfactant for water decreases as evidenced by the existence of a cloud point (see Fig. 3.20).

Figure 3.33 shows that many variables influence the partitioning of surfactants. Increasing the molar volume of the oil will, for example, generally favor partitioning of surfactant into the aqueous phase. Thus a surfactant balanced to break an emulsion composed of one crude oil at a given temperature will, in general, not be so balanced for another crude oil or the same crude oil at some other temperature. A surfactant which deemulsifies one system may possibly stabilize another. Thus, adding emulsifiers or deemulsifers to well treatment fluids should not be permitted unless the particular system has been laboratory tested under realistic field conditions. Otherwise, the result is almost certain to be different than the desired one.

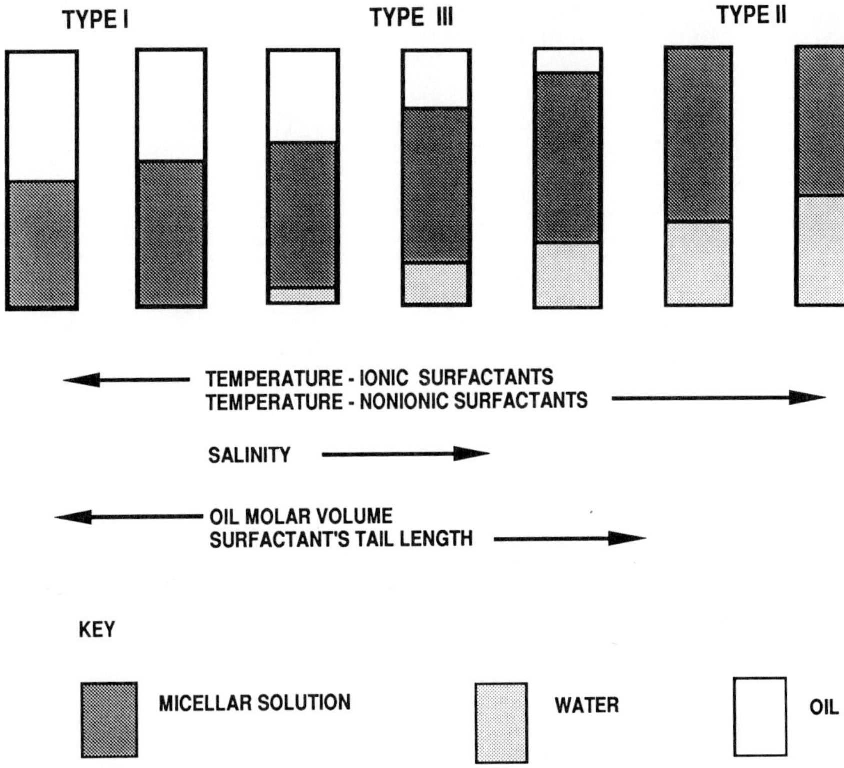

Figure 3-33 The factors determining surfactant partitioning are depicted here by showing a series of test tubes containing oil/water/surfactant. The surfactant is in the micellar phase.

Example 3.7 Selection of a Surfactant

Table 3.4 shows partition coefficients defined as the concentration of surfactant in an aqueous phase divided by the concentration in an isooctane phase. Which surfactant would you select to create a stable O/W emulsion? If the temperature is increased steadily, what would you expect to happen?

Solution To create an O/W emulsion, a surfactant soluble in water is required. If, however, it is too soluble, then surfactant adsorption at the interface will not be sufficient to stabilize the emulsion. Thus, choose a surfactant with 9 or 10 ethylene oxide units. Experiments will show that 9 is best.

Heating will cause the surfactant to partition into the oil. At some temperature normally less than the cloud point, which is roughly 70°C (see Fig. 3.20), the O/W emulsion will break and the phases rapidly separate. Thus, an emulsifier at one temperature becomes a deemulsifier at another temperature. This shows the importance of matching the surfactant to the particular system for a particular purpose.

Experiments have been performed to determine the emulsion stability of isooctane/water emulsions stabilized by ethoxylated octyl phenols [59]. Unfor-

TABLE 3.4 Physical Properties of the Ethoxylated Octyl Phenols [58]

Number of ethylene oxide units	Partition coefficient (Water/isooctane @25°C)
3	3.13×10^{-3}
4	9.83×10^{-3}
5	2.46×10^{-2}
6	5.92×10^{-2}
7	1.02×10^{-1}
8	5.0×10^{-1}
9	1.39
10	3.84
11	10.6

tunately, Graciaa et al. demonstrated that there are complicating factors, which reduces the criterion for emulsifier selection espoused here to the status of a "rule of thumb." It is not exact. The selection of a surfactant to stabilize an emulsion cannot be adequate based simply on consideration of a partition coefficient. Laboratory tests are required.

The importance of Bancroft's rule is that it shows the difficulty of selecting an appropriate surfactant and stresses the importance of testing them for each specific application.

A completely different mechanism for stabilizing emulsions is the adsorption of solid particles at the oil/water interface. This mechanism is particularly important when reservoir fines are being produced together with oil and water. This type of emulsion is known as a *Pickering emulsion* [54] and clearly requires a finite contact angle at confluence of the solid–oil–water. If the particles are hydrophilic, they tend to protrude into the water side of the oil/water interface tending to stabilize O/W emulsions. For a contact angle sharp on the oil side, W/O emulsions tend to be stabilized. The solid particles act as a mechanical barrier and may also alter the interfacial tension since they are often charged and repel one another.

Polymers may also be used to help stabilize emulsions, and many emulsion systems used for well treatment (e.g., emulsified acid fracturing fluids) contain added polymer. The polymer will reduce the rate of creaming by increasing the viscosity of the external phase and may in some cases adsorb at the oil/water interface reducing the coalescence rate. Both mechanisms will lead to greater emulsion stability [54].

Rheological Properties of Emulsions

The viscosity of an emulsion is an important property that may be crucial in its application. Very dilute emulsions (small volume fraction of the discontinuous phase) exhibit Newtonian behavior and the viscosity is given to a good approximation by the Einstein equation [60]

$$\mu = \mu_0(1 + 2.5\phi) \tag{3.30}$$

where μ is the viscosity of the dispersion, μ_0 is the viscosity of the continuous phase, and ϕ is the volume fraction of the discontinuous phase assumed to be small. The increase in viscosity above the value μ_0 results from energy dissipation when drops of an immiscible liquid are introduced into the continuous phase and the flow pattern is modified by their presence. Equation (3.30) is also limited to very small drops, which behave essentially like rigid spheres, and to small values of ϕ. Equation (3.30) must be modified if the drops do not behave like rigid spheres [63, 64].

For concentrated emulsions, which is the case of practical interest, there is no universally applicable theory. Concentrated emulsions are generally non-Newtonian and viscoelastic [61, 62]. At high shear rates the following relationship is sometimes satisfactory [62]:

$$\mu = \mu_0 \exp\left(\frac{2.5\phi}{1 - k\phi}\right) \tag{3.31}$$

where k depends on the hydrodynamic interaction between drops and increases as the drop size decreases [62]. Often $k \sim 1.5$; however, in any situation of practical importance, it should be measured.

Equation (3.31) is valid when most of the droplets have the same size. Near the phase inversion point the drop size distribution may become quite disperse and large emulsion viscosities may be obtained.

3.6 ACIDS

Acids are used as fracture fluids (Chap. 10), for scale removal, and for matrix treatments (Chaps. 16 and 17). Acids are also used to clean up gravel packs once they are positioned or as cleansing agents to preflush the formation prior to administering a near-wellbore chemical treatment. To be able to select the appropriate acid for a specific application, an understanding of the chemical reactions of different acids with minerals is necessary. In this section we discuss the types of acids available, the stoichiometry of their reactions with minerals, and some important thermodynamic limitations.

Acid Systems

Acid systems in current use can be classified as mineral acids, dilute organic acids, powdered organic acids, hybrid (or mixed) acids, or retarded acids. The most common members of each category are given in Table 3.5. All of the acids with the exception of hydrochloric–hydrofluoric and formic–hydrofluoric acid mixtures are used to treat carbonate formations. It is, with few exceptions, generally necessary to include hydrofluoric acid (HF) in the treatment of sandstones.

Mineral acids. Most acid treatments of carbonate formations employ hydrochloric acid (HCl). Usually it is used as a 15 wt% solution of hydrogen chloride

TABLE 3.5 Categories of Acids and Example Systems

Category	Examples of acids used in well treatment
Mineral acids	Hydrochloric acid Hydrochloric–hydrofluoric acid
Organic acids	Formic acid Acetic acid
Powdered acids	Sulfamic acid Chloroacetic acid
Mixed acids	Acetic–hydrochloric acid Formic–hydrochloric acid Formic–hydrofluoric acid
Retarded acids	Gelled acids Emulsified acids

gas in water. This concentration, sometimes called *regular acid*, was originally chosen because of inadequacies in early corrosion inhibitors and the difficulty of preventing corrosion of well tubulars by more concentrated solutions. With the development of improved inhibitors, high concentrations have become practical and in some cases concentrations to 30 wt% are used.

In addition to concentrations higher than 15 wt%, lower concentrations are commonly available. As an example, 5 to 7.5 wt% HCl are often used to displace connate water and remove carbonates ahead of HCl–HF mixtures to prevent the formation of sodium and potassium fluorosilicate materials capable of plugging the formation.

The continued use of HCl results from its moderate cost and soluble reaction products [calcium chloride ($CaCl_2$) and carbon dioxide (CO_2)]. The principal disadvantage of HCl is its high corrosivity on wellbore tubular goods. This high corrosivity is especially significant and expensive to control at temperatures above 120°C. Also, aluminum or chromium-plated metals, often found in pumps, are easily damaged. Sometimes, mixed acids or organic acids are used in place of HCl to reduce the corrosion.

HCl–HF mixtures are used almost exclusively for sandstone stimulation. HF is available commercially as a relatively pure material in anhydrous form or as a concentrated (40 to 70 wt%) aqueous solution. As it is used in the petroleum industry for well stimulation, HF is most often a dilute solution in HCl. It may be formed from dilution of concentrated solutions of HF or, more frequently, from the reaction of ammonium bifluoride NH_4F_2 with HCl. Often, 15 wt% HCl is used, and enough NH_4F_2 is added to create a solution containing 3 wt% HF. Consumption of hydrogen chloride by this reaction leaves 12 wt% HCl remaining in solution. Similarly, 6 wt% HF is often generated from 15 wt% HCl solutions and the final HCl concentration is approximately 9%. More concentrated solutions ranging from 12 wt% HCl and 10 wt% HF to 25 wt% HCl and 20 wt% HF are prepared by dilution of high-strength HF with an aqueous HCl solution. The corrosion characteristics of the HF–HCl mixture are comparable with those of HCl alone, and similar corrosion inhibitors are required.

Organic acids. The principal virtues of the organic acids are their lower corrosivity and easier inhibition at high temperatures. They have been used primarily in operations requiring a long acid–pipe contact time, such as a perforating fluid, or where aluminum or chrome-plated parts unavoidably will be contacted. Athough many organic acids are readily available, only two, acetic and formic, are used to any great extent in well stimulation.

Acetic acid was the first of the organic acids to be used in appreciable volumes in well stimulation. It is commonly available as a 10 wt% solution of acetic acid in water. At this concentration, the products of reaction (calcium and magnesium acetates) are generally soluble in spent acid. In addition to being used as a perforating fluid or as a fluid of low corrosivity in the presence of metals that corrode easily, acetic acid is often used in mixture with HCl in the hybrid acids.

On the basis of cost per unit of dissolving power, acetic acid is more expensive than either hydrochloric or formic acids. This greater expense generally dictates its use in small quantities only for the special application described.

Acetic acid reacts incompletely in the presence of its reaction products. The equilibrium set up between the products and reactants of the system often has been misinterpreted as evidence of a retarded reaction rate. Because of this, organic acids sometimes have been sold as retarded acids for regular acidizing operations. A discussion of acid equilibrium reactions is given later in this chapter.

Of the organic acids used in acidization, formic acid has the lowest molecular weight and, correspondingly, the lowest cost per volume of rock dissolved. It is a substantially stronger acid than acetic acid, though appreciably weaker than HCl. Like acetic acid, it reacts to an equilibrium concentration in the presence of its reaction products.

The principal advantage of formic over acetic acid is cost, although this is partially offset by the greater difficulty of inhibiting corrosion with this acid. Although more corrosive than acetic acid, formic acid corrodes uniformly and with less pitting than HCl, and effective inhibitors are available for its use at temperatures as high as 400°F. In high-temperature applications, the cost discrepancy with HCl narrows because of the high inhibitor concentrations required for HCl.

Powdered acids. Sulfamic and chloroacetic acids have only limited use in well stimulation, most of which is associated with their portability to remote locations in powdered form. They are white crystalline powders that are readily soluble in water. Generally, they are mixed with water at or near the wellsite. Sometimes these acids are cast in useful shapes such as "acid sticks," a form convenient for introduction into the wellbore.

Both sulfamic and chloroacetic acids are substantially more expensive than HCl on an equivalent-dissolving-power basis. Significant savings accompany their use when transportation and pumping charges can be eliminated.

Chloroacetic acid is stronger and more stable than sulfamic acid and is generally preferred when a powdered acid is appropriate. Sulfamic acid decomposes at about 80°C and is not recommended for applications where formation temperatures are above 70°C.

Acid mixtures. Acetic–hydrochloric and formic–hydrochloric acid mixtures, useful on carbonates, generally have been designed to exploit the dissolving-power economies of HCl while attaining the lower corrosivity (especially at high temperatures) of the organic acids. Therefore, their application is almost exclusively in high-temperature formations where corrosion inhibition costs greatly affect the overall treatment cost. The mixed acids sometimes have been sold as retarded acids because of the presence of the organic acid. It is important to recognize that, under formation conditions, carbon dioxide (CO_2) evolved by the HCl reaction greatly reduces the extent of reaction of organic acids and, in extreme cases, may prevent the organic acid from reacting.

A formic–hydrofluoric acid mixture, useful on sandstones, is sometimes employed in high-temperature applications because it is less corrosive than the comparable inorganic acid mixture, HF–HCl.

Retarded acids. The acid reaction rate can be slowed or retarded in a number of ways. The viscosity of the acid can be increased by gelling thereby slowing the diffusion of acid to the rock surface, the surface of the rock can be coated with an adsorbed layer of inert molecules thereby decreasing the rate at which acid reaches the rock surface, or the acid can be emulsified with an oil creating an acid-in-oil emulsion which reacts with the rock only when a drop of acid contacts the surface. All of these methods have disadvantages. Gelling the acid increases the injection pressures required for matrix acidization and as noted in the previous section, the gelling agents, often water-soluble polymers, tend to degrade rapidly in acid solutions.

Retarding the acid reaction by coating the rock is difficult under dynamic conditions. The use of surfactants which adsorb onto carbonate rock surfaces has been tried, but the results are not now encouraging. Emulsified acids have two disadvantages. They are viscous and they do not dissolve as much rock per unit of volume as does the acid alone.

Because of the inherent difficulties associated with each of the three methods of retarding the acid reaction rate that has been described, other methods that provide effectively decreased acid reaction rates have been developed. These are described in Chapter 16.

Stoichiometry of Acid–Carbonate Reactions

Stoichiometry refers to the proportions of the various reactants that participate in the reaction. While these proportions are easy to identify in the reaction of calcium carbonate or dolomite with HCl, reactions with naturally occurring carbonates are often complicated by the presence of other minerals that react with HCl. In addition to HCl those organic acids listed in Table 3.5 are also used to treat carbonate formations. These acids all react with carbonates to form carbon dioxide (CO_2), water (H_2O), and a calcium or magnesium salt ($CaCl_2$ or $MgCl_2$). Typical reactions are

$$\underset{}{2HCl} + \underset{\text{(calcite)}}{CaCO_3} \rightleftharpoons CaCl_2 + H_2O + CO_2 \tag{3.32}$$

and

$$4HCl + \underset{\text{(dolomite)}}{CaMg(CO_3)_2} \rightleftharpoons MgCl_2 + CaCl_2 + 2H_2O + 2CO_2 \tag{3.33}$$

These equations indicate the stoichiometry of the reaction. For example, Eq. (3.32) indicates that 2 moles of hydrochloric acid (HCl) react with 1 mole of limestone [calcium carbonate ($CaCO_3$)] to create 1 mole of calcium chloride ($CaCl_2$), 1 mole of water (H_2O), and 1 mole of carbon dioxide (CO_2).

We now introduce the concept of dissolving power. The gravimetric dissolving power, β, is defined as the mass of rock per mass of acid. The purpose of acid in oil field applications is to dissolve rock, and the dissolving power β is an important quantity. Given the reaction stoichiometry, β can easily be calculated. From Eq. (3.32) for HCl reacting with $CaCO_3$

$$\beta = \left[\frac{1 \text{ mole } CaCO_3}{2 \text{ moles HCl}}\right]\left[\frac{\text{molecular weight } CaCO_3}{\text{molecular weight HCl}}\right]\left[\frac{\text{mass HCl}}{\text{mass acid solution}}\right]$$

Example 3.8 The Gravimetric Dissolving Power

Calculate β for the dissolution of dolomite using a 30 wt% solution of HCl.

Solution The molecular weights of HCl and $CaMg(CO_3)_2$ (dolomite) are given in Tables 3.6 and 3.7. Thus

$$\beta = \frac{1(184.3)30}{4(36.47)100} = 0.379 \text{ mass of dolomite/mass of acid}$$

Perhaps an even more useful concept is that of the volumetric dissolving power, X, which is defined as the volume of rock dissolved per volume of acid reacted. Volumetric relationships are more interesting than gravimetric ones because acids are generally purchased by volume and most often we are interested

TABLE 3.6 Acids Used in Carbonate Acidization

Category	Molecular weight
Mineral Acids	
Hydrochloric (HCl)	36.47
Organic acids	
Formic (HCOOH)	46.03
Acetic (CH_3COOH)	60.05

TABLE 3.7 Molecular Weight of Components in HCl Reaction with Carbonates

Compound	Chemical formula	Molecular weight
Calcium carbonate (limestone)	$CaCO_3$	100.09
Calcium magnesium carbonate (dolomite)	$CaMg(CO_3)_2$	184.3
Calcium chloride	$CaCl_2$	110.99
Magnesium chloride	$MgCl_2$	95.3
Carbon dioxide	CO_2	44.01
Water	H_2O	18.02

in the volume of rock that can be dissolved. The volumetric dissolving power is simply related to the gravimetric dissolving power by the equation

$$X = \beta \frac{\rho_{\text{acid}}}{\rho_{\text{rock}}} \tag{3.34}$$

The specific gravities of HCl solutions are given in Table 3.8 and the dissolving powers are given in Table 3.9. The gravimetric dissolving power, β_{100}, given in Table 3.9 refers to a value for an acid having 100% strength. To find β for lesser strength solutions, one need only multiply β_{100} by the weight fraction of acid in the solution.

Also given in Table 3.9 are values of the dissolving power (X) for the organic acids acetic and formic. These values of X have been calculated with the assumption that acetic acid and formic acid react to completion upon contact with a large excess of carbonaceous materials. This, however, is not generally true. The acid reaction may cease, leaving live, unreacted acid in solution. This is a thermodynamic limitation and is discussed in a subsequent section of this chapter.

Example 3.9 Application of the Dissolving Power

Calculate the volume of 10 wt% formic acid required to increase the permeability of a limestone formation 10 m thick by a factor of 10 in a zone 1.5 m in radius around the wellbore. The wellbore radius is 0.2 m and the permeability response of the

TABLE 3.8 Specific Gravity of Aqueous Hydrochloric Acid Solutions [64] (at 20°C)

Percent HCl	Specific gravity
1	1.0032
2	1.0082
4	1.0181
6	1.0279
8	1.0376
10	1.0474
12	1.0574
14	1.0675
16	1.0776
18	1.0878
20	1.0980
22	1.1083
24	1.1187
26	1.1290
28	1.1392
30	1.1493
32	1.1593
34	1.1691
36	1.1789
38	1.1885
40	1.1980

TABLE 3.9 Dissolving Power of Various Acids

Formation	Acid	β_{100}	X 5%	X 10%	X 15%	X 30%
Limestone:	Hydrochloric (HCl)	1.37	0.026	0.053	0.082	0.175
$CaCO_3$	Formic (HCOOH)	1.09	0.020	0.041	0.062	0.129
$\rho_{CaCO_3} = 2.71$ g/cm^3	Acetic (CH_3COOH)	0.83	0.016	0.031	0.047	0.096
Dolomite:	Hydrochloric	1.27	0.023	0.046	0.071	0.152
$CaMg(CO_3)_2$	Formic	1.00	0.018	0.036	0.054	0.112
$\rho_{CaMg(CO_3)_2} = 2.87$ g/cm^3	Acetic	0.77	0.014	0.027	0.041	0.083

Data for organic acids have not been corrected for equilibrium.
β_{100} denotes gravimetric dissolving power for 100% strength acids.

limestone is given by

$$\frac{k}{k_0} = \left(\frac{\phi}{\phi_0}\right)^{10}$$

where k_0 and ϕ_0 are the original permeability and porosity, respectively. In this example $\phi_0 = 0.1$.

Solution Since

$$\frac{k}{k_0} = 10 = \left(\frac{\phi}{\phi_0}\right)^{10} = \left(\frac{\phi}{0.1}\right)^{10}$$

we find

$$\phi = 0.1266.$$

Thus to increase the porosity of the rock from $\phi_0 = 0.1$ to $\phi_0 = 0.126$, the volume of rock to be dissolved is

$$V = \pi[(1.5)^2 - (0.2)^2](0.126 - 0.1)(10) \quad \text{or} \quad V = 1.80 \text{ m}^3 \text{ of rock}$$

From Table 3.9 we see that the volumetric dissolving power is

$$X = 0.041 \text{ m}^3 \text{ of rock/m}^3 \text{ of acid}$$

Therefore,

$$\text{Acid volume} = \frac{1.80}{0.041} = 44 \text{ m}^3$$

Equilibrium in Acid–Carbonate Reactions

Along with the concepts of reaction stoichiometry and rate, it is important to understand equilibrium as it relates to acid reactions. When an acid reaction reaches equilibrium, the dissolution of formation material by the acid stops, even though acid molecules still may be present. Equilibrium is attained when the chemical activity (consider this to be a driving force for change) of the reaction products balances the activity of the reactants.

A general definition of equilibrium can be obtained from thermodynamic arguments [65]. The final result is that at equilibrium, the ratio of the chemical activity of products to reactants, with each activity raised to a power equal to its stoichiometric coefficient, is equal to a constant, called an *equilibrium constant*. For example, consider the generalized reaction

$$\mathrm{A + B \rightleftharpoons C + D}$$

The equilibrium constant for this reaction is defined as

$$K = \frac{a_\mathrm{C} a_\mathrm{D}}{a_\mathrm{A} a_\mathrm{B}} \tag{3.35}$$

The quantity a_i is the activity of the component i. These activities are thermodynamic potentials and are not easy to evaluate. Therefore, experimental data are generally required to obtain accurate values. The activity of a substance increases with its concentration in solution, but the relationship between activity and concentration is generally not linear. More detailed discussions of activities are given in texts on chemical thermodynamics [65].

Dissociation equilibrium. An important property of acids is that, in aqueous solution, they dissociate (ionize) by the reaction

$$\mathrm{HA \rightleftharpoons H^+ + A^-} \tag{3.36}$$

In this equation, the generalized acid is denoted as HA, with the ionized species being H^+ and A^- ions. The dissociation equilibrium is described by

$$K_D = \frac{a_{\mathrm{H}^+} a_{\mathrm{A}^-}}{a_\mathrm{HA}} \tag{3.37}$$

The equilibrium constant, K_D, in this case is called the *dissociation constant*. If at equilibrium an acid is highly dissociated, K_D will be a large number; if the acid is only slightly dissociated, K_D will be small.

The dissociation constants depend on temperature and can be calculated using Eq. (3.38) as follows:

$$-\log_{10} K_D = \frac{A_1}{T} - A_2 + A_3 T \tag{3.38}$$

In this equation, T is the temperature in °K and the constants A_1, A_2, and A_3 can be obtained from Table 3.10.

TABLE 3.10 Typical Values for Acid Dissociation Constants [66]

Acid	A_1	A_2	A_3
Acetic	1170.48	3.1649	0.013399
Formic	1342.85	5.2743	0.015168
Propionic	1213.28	3.3860	0.014055
Chloroacetic	1229.13	6.1714	0.016486

Acetic and formic acids have small dissociation constants compared with HCl. This means that under comparable conditions a much smaller quantity of acetic or formic acid will be dissociated into the reactive ionized state. Therefore, they are often called *weak acids*.

Reaction equilibrium. Under reservoir conditions, organic acids do not react to completion with either limestone or dolomite formations because of the limitations imposed by chemical equilibrium. Equilibrium occurs in the reservoir because CO_2 (one reaction product) is held in solution by reservoir pressure and not allowed to escape from the solution [67]. At low pressures, where the CO_2 can escape, the acid will, however, react to completion.

Results of tests relating the fraction of acid reacted, temperature, and acid composition at 6895 kPa are given in Figs. 3.34 and 3.35 [67]. These factors should be used to correct the dissolving power of organic acids. For example, Fig. 3.34 shows that at 150°F and 6895 kPa (1000 psia), only about 50% of a 10-wt% acetic

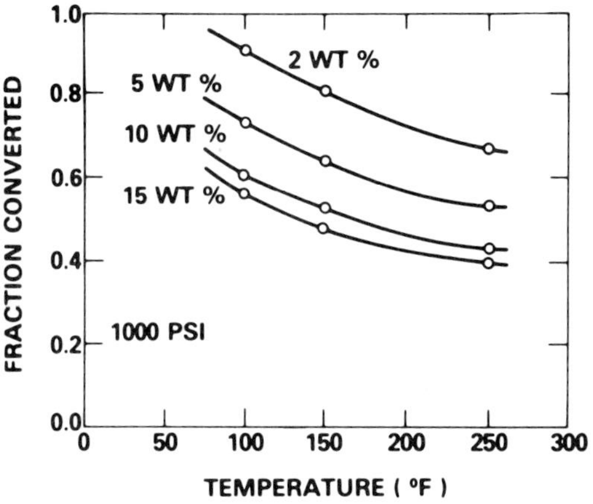

Figure 3.34 Fraction acid reacted at equilibrium versus temperature for the system acetic-acid/calcium-carbonate. [From J.C. Chatelain et al., *Society of Petroleum Engineers Journal* (Aug. 1976), p. 189.]

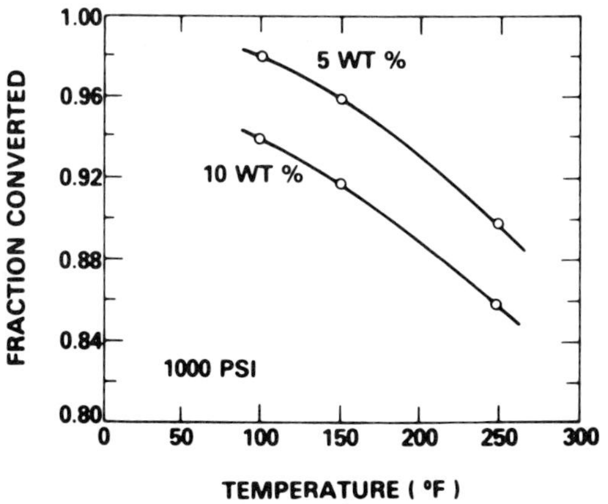

Figure 3.35 Fraction acid reacted at equilibrium versus temperature for the system formic-acid/calcium-carbonate. [From J. C. Chatelain et al., *Society of Petroleum Engineers Journal* (Aug. 1976), p. 189.]

acid solution will react. The dissolving power of acetic acid under these conditions, therefore, should be reduced by 50% from the value given in Table 3.9.

It has been found that the equilibrium state can be predicted approximately by the empirical equation [68]

$$1.6 \times 10^4 K_D = \frac{c_{CaA_2} c_{CO_2}}{c_{HA}} \tag{3.39}$$

where c_i is the concentration of component i in g moles/1000 g of H_2O. Experimental evidence indicates that equilibrium conditions with dolomite are very nearly the same as with limestone [67]; therefore, the extent of acid reaction can be roughly calculated by assuming that all carbonate formations behave at chemical equilibrium as though they are composed of $CaCO_3$. Equation (3.39) only applies to reactions of weak acids with $CaCO_3$, but knowledge of this is sufficient to estimate the extent of reaction with dolomite. Once the extent of reaction is determined, the correct dissolving power can be obtained by multiplying those values in Table 3.9 by the fraction of the acid which reacts.

Example 3.9 Equilibrium Limitations

Calculate that fraction of a 10 wt% of acetic solution which will react in the presence of an excess of dolomite at 150°F and at a pressure sufficient to retain all of the CO_2 produced by the reaction in the aqueous solution. This pressure will depend on the amount of acid that reacts.

Solution The fraction of the acid that will react is the same whether the formation rock is limestone or dolomite. Thus we will make the calculation as though the formation is $CaCO_3$. Equation (3.39) then applies.

Since there are 100 g of acid and 900 g of water in 1000 g of acid solution, the acid concentration is

$$c_{HAC} = \frac{(100)(1000)}{(60)(900)} = 1.85 \text{ moles/1000 g of } H_2O$$

where 60 is the molecular weight of acetic acid.

It we let ψ be the number of moles of acetic acid/1000 g of H_2O that react, then at equilibrium (neglecting the small quantity of H_2O produced by the reaction)

$$c_{CO_2} = \psi/2 \text{ (carbon dioxide produced)}$$

$$c_{Ca(CH_3COO)_2} = \psi/2 \text{ (calcium acetate produced)}$$

$$c_{H(CH_3COO)} = 1.85 - \psi \text{ (unreacted acid remaining)}$$

The value of K_D at 150°F is 1.48×10^{-5} found using Eq. (3.38). Substituting these values in Eq. (3.39) we find

$$(1.6 \times 10^4)(1.48 \times 10^{-5}) = \frac{(\psi/2)\,(\psi/2)}{1.85 - \psi}$$

Solving this quadratic equation yields

$$\psi = 0.93 \text{ mole of acid reacted/1000 g of } H_2O$$

The fraction of acid converted is then $0.93/1.85 = 0.50$.

This use of Eq. (3.39) assumes that both the CO_2 and the calcium–acid anion salt remain in the aqueous solution. Although the reservoir pressures are generally sufficient to force the CO_2 to remain in solution when modest acid concentrations are used, a separate CO_2-rich phase may form if the initial concentration of the acid is high enough. A second possibility is that CO_2 may escape from the aqueous phase into an intermingled oil phase if one is present. In either case, the concentration of CO_2 to use in Eq. (3.39) is that actually remaining in the aqueous phase. Clearly, reducing the concentration of CO_2 in the aqueous phase shifts the reaction so that more acid reacts.

It is useful to note again that the reaction of an organic acid in a hybrid acid mixture is also subject to thermodynamic limitations since the HCl will generally react first, producing a large quantity of CO_2. In accordance with Eq. (3.39), the added CO_2 will reduce the quantity of organic acid that can react before Eq. (3.39) is satisfied. Therefore, there is likely to be more unreacted organic acid than is indicated by Figs. 3.34 and 3.35 when hybrid acids are used.

The solubility of CO_2 in H_2O, given in Table 3.11, can be used to approximate solubility values needed for use in equilibrium calculations although the solubility in spent acid solutions is less [70].

If the CO_2 concentration is predicted to exceed that given in Table 3.11, the solubility from the table should be used in Eq. (3.39). This consideration is often important at low pressures, for high concentrations of organic acids, for hybrid acids, or when the formation contains appreciable CO_2.

The solubilities of calcium salts in water are given in Table 3.12. If the predicted salt concentration exceeds the value shown, the solubility from the table should be used in Eq. (3.39). Perhaps of even greater practical significance than the equilibrium conversion is the possible plugging of the formation with the precipitated salt from the organic acid if the salt concentration exceeds the solubility limit.

TABLE 3.11 Solubility of Carbon Dioxide in Water [69]

Pressure (kPa)	kg moles of CO_2/1000 kg of H_2O		
	95°F	167°F	212°F
2,533	0.58	0.31	0.24
5,068	1.00	0.57	0.46
7,602	1.25	0.77	0.64
10,136	1.31	0.92	0.79
12,670	1.37	1.10	1.02
15,203	1.43	1.20	1.15
20,271	—	1.32	1.33
30,407	1.60	1.43	1.45
40,543	1.70	—	—
50,678	—	1.69	1.73

TABLE 3.12 Solubilities of Organic Acid Calcium Salts in Water [26]

Temperature (°F)	kg/100 kg H_2O	
	Calcium formate	Calcium acetate
32	16.15	37.4
50	—	36.0
68	16.60	34.7
86	—	33.8
104	17.05	33.2
140	17.50	32.7
176	17.95	33.5
194	—	31.1
211	18.40	29.7

Stoichiometry of Acid–Sandstone Reactions

Acidizing treatments in sandstone formations normally employ a mixture of HCl and HF. This acid mixture is used because HF is reactive with clay minerals and feldspars that may be restricting the permeability near the wellbore. The origins of permeability damage in sandstones are discussed in Chapter 4. HCl alone is sometimes used to treat sandstones, but these treatments are normally successful only when the sandstone has a high carbonate content.

Chemical reactions between HF and silica (SiO_2) or calcite ($CaCO_3$), two important mineral constituents of sandstones, are comparatively simple. The reactions of HF with aluminosilicates, such as clays or feldspars, are complex. This complexity stems from two difficulties. Clays and feldspars may not be represented by a single stoichiometric equation. As discussed in Chapter 1, there is often substantial substitution into the crystalline structure of the mineral lattices. Thus the chemical formulae of clays and feldspars are often empirical, representing the average ratios of their constituent elements. A second difficulty is that the distribution of aqueous species such as AlF_3, AlF_2^-, SiF_6^{--}, SiF_4, etc. (see following section on equilibrium) depends on the solid-to-acid ratio. Thus, an equation representing a reaction of HF with a mineral should be considered to be an example rather than a precise description of the stoichiometry.

Equations describing the reaction of HF with SiO_2 are as follows:

$$SiO_2 + 4HF \rightleftharpoons SiF_4 + 2H_2O \tag{3.40}$$

and

$$SiF_4 + 2HF \rightleftharpoons H_2SiF_6 \tag{3.41}$$

The reaction products SiF_4 and H_2SiF_6 are highly soluble in water. In the presence of a large excess of HF, as compared to SiO_2, the reaction product is primarily H_2SiF_6; whereas, excess amounts of silica lead to SiF_4. Depending on which product predominates, the dissolving power may vary.

Example 3.10 Dissolving Power of HF Solutions

Calculate the volumetric dissolving power of a 3 wt% HF and 12 wt% HCl solution, assuming that the formation is composed of quartz. Consider two different cases. In the first case, take the predominate reaction product to be SiF_4; in the second case, consider it to be H_2SiF_6.

Solution When SiF_4 is the reaction product, the gravimetric dissolving power is given by

$$\beta_{100} = \left[\frac{1 \text{ mole } SiO_2}{4 \text{ moles HF}}\right]\left[\frac{1 \text{ mole HF}}{20 \text{ kg HF}}\right]\left[\frac{60 \text{ kg } SiO_2}{\text{mole } SiO_2}\right]$$

giving

$$\beta_{100} = 0.75 \text{ kg } SiO_2/\text{kg HF}$$

Taking $\rho_{SiO_2} = 2420 \text{ kg/m}^3$ and $\rho_{HF} = 1070 \text{ kg/m}^3$ (same as 15 wt% HCl, see Table 3.8), then

$$X = (0.75)\left(\frac{3}{100}\right)\left(\frac{1070}{2420}\right) = 0.010 \text{ m}^3 \text{ of rock/m}^3 \text{ of acid}$$

If H_2SiF_6 is the reaction product, then

$$\beta_{100} = \frac{1}{6}\left(\frac{60}{20}\right) = 0.5 \text{ kg } SiO_2/\text{kg HF}$$

and

$$X = (0.5)\left(\frac{3}{100}\right)\left(\frac{1070}{2420}\right) = 0.0066 \text{ m}^3 \text{ of rock/m}^3 \text{ of acid}$$

Comparison of these two numbers shows that the dissolving power does depend on the reaction products that are formed.

HF reacts with $CaCO_3$ as follows:

$$CaCO_3 + 2HF \rightleftharpoons CaF_2 + H_2O + CO_2 \tag{3.42}$$

The reaction product, calcium fluoride (CaF_2), is not very soluble in water and its formation can result in permeability loss. To avoid this difficulty and other possible precipitates, a preflush of HCl is recommended before introducing HF into the formation (see Chap. 16).

As an example of the reaction of HF and HCl mixtures with aluminosilicates such as feldspars and clays, we select albite ($NaAlSi_3O_8$). Thus,

$$NaAlSi_3O_8 + 14HF + 2H^+ \rightleftharpoons Na^+ + AlF_2^+ + 3SiF_4 + 8H_2O \tag{3.43}$$

The products AlF_2^+ and SiF_4 will be the predominate ones formed for a limited range of solid-to-acid ratios. For ratios outside of this limited range, other products are formed. This important aspect is discussed in the next section.

Assuming that Eqs. (3.40) and (3.43) represent the predominate ones, the dissolving power for HF solutions can be computed. Representative results, which can be used for crude estimates, are given in Table 3.13. Note that HCl is not appreciably reactive with silica and clay minerals and therefore is not included

TABLE 3.13 Dissolving Power for Hydrofluoric Acid

Acid concentration (wt %)	Quartz (SiO_2)		Albite ($NaAlSi_3O_8$)	
	β	X	β	X
2	0.015	0.006	0.019	0.008
3	0.023	0.010	0.028	0.011
4	0.030	0.018	0.037	0.015
6	0.045	0.019	0.056	0.023
8	0.060	0.025	0.075	0.030

Note: β = Mass of rock dissolved/mass of acid reacted.
X = Volume of rock dissolved/volume of acid reacted.

in the dissolving power calculation. Also it is assumed that the reactions go to completion. In the following section it will be shown that the reaction, Eq. (3.43), is unlikely to be representative of an equilibrium state in the presence of excess albite.

Equilibrium in Acid–Sandstone Reactions

Solutions of hydrofluoric and hydrochloric acids contain many different species including H^+, F^-, HF_2^-, HF, and of course Cl^-. Which of these will be present in dominate concentration depends on the concentrations of HF and HCl present. This is an important consideration because, for example, at high concentrations (in excess of 2 wt% HF), HF begins to behave as a strong acid forming H^+ and HF_2^- species in solution.

Association of HF to form various species in aqueous solution is illustrated by the following:

$$HF \rightleftharpoons H^+ + F^- \tag{3.44}$$

and

$$2HF \rightleftharpoons HF_2^- + H^+ \tag{3.45}$$

The equilibrium constant for reaction (3.44) defined as

$$K_D = \frac{[H^+]\,[F^-]}{[HF]} \tag{3.46}$$

is about 6.6×10^{-4} at 25°C [72]. In dilute solutions, Eq. (3.44) is the primary one, but as the HF concentration is increased, the second reaction, represented by Eq. (3.45), becomes important. The equilibrium constant for this reaction is

$$K_D = \frac{[H^+]\,[HF_2^-]}{[HF]^2} \tag{3.47}$$

where $K_D = 2.2 \times 10^{-3}$ at 25°C [71].

Aluminosilicates, such as clays and feldspars, may react with HF to form silicon and aluminum fluoride species, as indicated by Eq. (3.43). As noted previously, however, the reaction products shown by Eq. (3.43) are not the only ones that form. SiF_4, $HSiF_5$, and H_2SiF_6 may also exist [71], while aluminum may exist in aqueous solutions as Al^{+++}, AlF^{++}, AlF_2^+, . . . AlF_6^{---}. The aluminum ions compete with silicon for fluorine. The equilibrium constants for all of these reactions are presented in Chapter 16 and their implication with respect to the design of matrix acidizing treatment is discussed in some detail. Here we note that in the presence of an excess amount of clay, the aluminum fluoride compounds are more stable than are the silicon fluorides and the initial reaction product (H_2SiF_6), formed in the presence of a large excess of HF, begins to lose fluorines to the more stable aluminum compounds. At first the H_2SiF_6 loses two fluorines to form SiF_4, but ultimately as the reaction proceeds $Si(OH)_4$ is formed. This precipitates as a voluminous hydrated colloid ($Si(OH)_4 \cdot nH_2O$) [73, 74] that may be damaging to rock permeability. Although it is difficult to avoid the occurrence of these precipitates, techniques for calculating and minimizing the amount of colloidal silica which comes out of the solution are discussed in Chapter 16.

REFERENCES

3.1. Stevens, M. P., *Polymer Chemistry: An Introduction*, London: Addison-Wesley Publishing Co., 1975.

3.2. Gerad, T., in *Handbook of Water-Soluble Gums and Resins*, R. L. Davidson (ed.), New York: McGraw-Hill, 1980.

3.3. Whistler, R. L., *Industrial Gums*, New York: Academic Press, 1973.

3.4. Oh, E., Spurlin, H. M., and Graffin, M. W., *High Polymers Cellulose, 5*, Pt. 3, New York: Wiley-Interscience, 1955.

3.5. Chatterji, J., and Borchardt, J. K., *J. Pet. Tech., 33* (1981) 2042.

3.6. Newell, T. P., in *Cellulose Chemistry and Its Applications*, T. P. Newell and S. H. Zeronian (eds.), Chichester, England: Ellis Harwood Ltd., 1985.

3.7. Finch, P., and Roberts, J. C., in *Celluose Chemistry and Its Applications*, T. P. Newell and S. H. Zeronian (eds.), Chichester, England: Ellis Harwood Ltd., 1985.

3.8. Roberts, J. D., and Caserio, M. C., *Basic Principles of Organic Chemistry*, New York: W. A. Benjamin, Inc., 1964.

3.9. Shupe, R. D., "Chemical Stability of Polyacrylamide Polymers," SPE 9299, presented at the 55th Fall Technical Conference and Exhibition of the Society of Petroleum Engineers, Dallas, Texas, 1980.

3.10. Wellington, S. L., "Biopolymer Solution Viscosity Stabilization–Polymer Degradation and Antioxidant Use," SPE 9296, presented at the 55th Fall Technical Conference and Exhibition of the Society of Petroleum Engineers, Dallas, Texas, 1980.

3.11. Volk, H., and Friedrich, R. E., in *Handbook of Water-Soluble Gums and Resins*, R. L. Davidson (ed.), New York: McGraw-Hill, 1980.

3.12. Shultz, A. R., "Crosslinking," in *Encyclopedia of Polymer Technology, 4* (1966) 331.

3.13. Martell, A. E., "Complex Agents," in *Encyclopedia of Chemical Technology, 6* (1966) 1.

3.14. Conway, M. J., Almond, S. W., Briscoe, J. E., and Harris, L. E., *J. Pet. Tech., 35* (1983) 315.

3.15. Barnhart, J. E., and Sawyer, J. C., "Gelled Ammonium Nitrate Blasting Explosives," U.S. Patent No. 3,072,059 (1963).
3.16. Crisp, J. D., "Gelled Compositions Containing Galactomannan Gums," U.S. Patent No. 3,301,723 (1967).
3.17. Bird, R. B., Stewart, W. E., and Lightfoot, E. N., *Transport Phenomena*, New York: John Wiley, 1960.
3.18. Rogers, R. E., Veatch, R. W., and Nolte, K. G., "Pipe Viscometer Study of Fracturing Fluid Rheology," SPE 10258, presented at the 56th Fall Technical Conference and Exhibition of the Society of Petroleum Engineers, San Antonio, Texas, 1981.
3.19. Savins, J. G., *J. Inst. Pet., 47* (1961) 329.
3.20. Sellin, R. H. J., Hoyt, J. W., and Scrivener, O., *J. Hydraul. Res., 20* (1982) 29.
3.21. Berman, N. S., *Ann. Rev. Fluid Mech., 10* (1978) 47.
3.22. Bird, R. B., Armstrong, R. C., and Hassager, O., *Dynamics of Polymeric Liquids*, Vol. 1, New York: John Wiley, 1977.
3.23. Hoyt, J. W., *Polymer Letters, 9* (1971) 851.
3.24. Darby, R., *Encyclopedia of Fluid Mechanics*, N. P. Cheremisinoff (ed.), Houston: Gulf Publishing Co., 1986.
3.25. Bourrel, M., and Schechter, R. S., *Microemulsions and Related Systems*, New York: Marcel Dekker, 1988.
3.26. Rosen, M. J., *Surfactants and Interfacial Phenomena*, New York: John Wiley, 1978.
3.27. Fernandez, M., "Adsorption of Sulfonates from Aqueous Solutions onto Mineral Surfaces," MS thesis, The University of Texas at Austin, 1977.
3.28. Funasaki, N., and Hada, S., *J. Phy. Chem., 87* (1983) 342.
3.29. Thach, S., "The Mechanism of Nonyl Phenol Polyoxyethylene Adsorption on Some Mineral Surfaces," MS thesis, The University of Texas at Austin, 1979.
3.30. Clark, M. B., Pike, M. T., and Rengel, G. L., *J. Pet. Tech., 34* (1982) 1565.
3.31. Reidenbach, V. G., Harris, P. C., Lee, Y. N., and Lord, P. L., *Soc. Pet. Eng. J., 38* (1986) 31.
3.32. Harris, P. C., and Reidenbach, V. G., "High Temperature Rheological Study of Foam Fracturing Fluids," SPE 13177, presented at the 59th Annual Fall Technical Conference and Exhibition, Houston, Texas, 1984.
3.33. Holcomb, D. L., Galaway, E., and Curry, L., "The Chemistry, Physical Nature and Rheology of an Aqueous Stimulation Foam," SPE 9530, presented at the Eastern Regional Meeting of the Society of Petroleum Engineers, Morgantown, Virginia, 1980.
3.34. Rand, P. B., and Kraynik, A. M., *Soc. Pet. Eng. J., 35* (1983) 152.
3.35. Adamson, A. W., *Physical Chemistry of Surfaces*, 4th ed., New York: John Wiley, 1982.
3.36. Monsolve, A., and Schechter, R. S., *J. Colloid and Interface Sci., 97* (1984) 327.
3.37. Selecki, A., and Wasiak, R., *J. Colloid and Interface Sci., 102* (1984) 557.
3.38. Cheng, H. C., and Lemlich, R., *Ind. and Eng. Chem. Fund., 24* (1985) 44.
3.39. Israelachvili, J. N., *Intermolecular and Surface Forces*, New York: Academic Press, 1985.
3.40. Nishioka, G., and Ross, S., *J. Colloid and Interface Sci., 81* (1981) 1.
3.41. Princen, H. M., *J. Colloid and Interface Sci., 91* (1983) 160.
3.42. Prud'homme as reported by Heller and Kuntamukkla (see [47]).
3.43. Blauer, R. E., Mitchell, B. J., and Kohlhass, C. A., "Determination of Laminar, Turbulent and Transitional Foam Flow Losses in Tubes," SPE 4885, presented at the California Regional Meeting of the Society of Petroleum Engineers, San Francisco, California, 1974.

3.44. Kraynik, A. M., paper B10, presented at the Annual Meeting of the Rheological Society, Evanston, Illinois, 1982.
3.45. Raza, S. H., and Marsden, S. S., *Soc. Pet. Eng. J., 7* (1967) 4.
3.46. Mitchell, B. J., "Viscosity of Foam," PhD dissertation, The University of Oklahoma, Norman, 1969.
3.47. Heller, J. P., and Kuntamukkla, M. S., *Ind. and Eng. Chem., 26* (1987) 318.
3.48. Hirasaki, G. J., and Lawson, J. B., "Mechanism of Foam Flow in Porous Media—Apparent Viscosity in Smooth Capillaries," SPE 12129, presented at the 58th Annual Fall Technical Conference and Exhibition, San Francisco, California, 1983.
3.49. Harris, P. C., "Effects of Texture on Rheology of Foam Fracture Fluids," SPE 14257, presented at the 60th Annual Fall Technical Conference and Exhibition, Las Vegas, Nevada, 1985.
3.50. Govier, G. W., and Aziz, K., *The Flow of Complex Mixtures in Pipes*, Malabur, Florida: Robert E. Krieger Publishing Co., 1982.
3.51. Cawiezel, K. E., and Niles, T. D., "Rheological Properties of Foam Fracturing Fluids Under Downhole Conditions," SPE 16191, presented at the Hydrocarbon Economics and Evaluation Symposium of the Society of Petroleum Engineers, Dallas, Texas, 1987.
3.52. Walstra, P., in *Encyclopedia of Emulsion Technology Volume I Basic Theory*, P. Becher (ed.), New York: Marcel Dekker, 1983.
3.53. Shinoda, K., and Friberg, S., *Emulsions and Solubilization*, New York: John Wiley, 1986.
3.54. Tadros, T. F., and Vincent, B., in *Encyclopedia of Emulsion Technology Volume I Basic Theory*, P. Becher (ed.), New York: Marcel Dekker, 1983.
3.55. McKay, G. D. M., and Mason, S. G., *J. Colloid Sci., 18* (1963) 674.
3.56. Wason, D. T., McNamara, J. J., Shaw, S. M., and Sampath, K., *J. Rheology, 23* (1979) 181.
3.57. Peterson, R., Hamil, R. D., and McMahon, J. D., *J. Amer. Pharmacy Assoc., 55* (1964) 651.
3.58. Crook, E. H., Fordyce, D. B., and Trebbi, G. F., *J. Colloid Sci., 20* (1965) 191.
3.59. Graciaa, A., Lachaise, J., Marion, G., and Schechter, R. S., "A Study of the Required HLB for Emulsification," *Langmiur, 5* (1989) 1315.
3.60. Einstein, A., *Ann. Phys., 24* (1911) 591.
3.61. Darby, R., in *Emulsions and Emulsion Technology, Volume 3*, K. J. Lissant (ed.), New York: Marcel Dekker, 1984.
3.62. Sherman, P., in *Encyclopedia of Emulsion Technology, Volume 1*, P. Becher (ed.), New York: Marcel Dekker, 1983.
3.63. Lothian, G. F., and Chappel, F. D., *J. Appl. Chem, 1* (1951) 475.
3.64. *Handbook of Chemistry and Physics*, 48th ed., Cleveland: The Chemical Rubber Co., 1967.
3.65. Denbigh, K. G., *The Principles of Chemical Equilibrium*, Cambridge: Clarendon Press, 1981.
3.66. Robinson, R. A., and Stokes, R. H., *Electrolyte Solutions*, London: Butterworths, 1965.
3.67. Chatelain, J. C., Silberberg, I. H., and Schechter, R. S., *Soc. Pet. Eng. J., 28* (1976) 189.
3.68. Williams, B. B., Gidley, J. L., and Schechter, R. S., *Acidizing Fundamentals*, Mono. Ser., *6*, Society of Petroleum Engineers, Richardson, Texas, 1979.
3.69. Perry, J. H., *Chemical Engineers Handbook*, 3rd ed., New York: McGraw-Hill, 1950.

3.70. Malinis, S. D., and Savelyera, N. I., *Geochemical International* (1972) 410.
3.71. *International Critical Tables*, New York: McGraw-Hill (1933) 233.
3.72. Fogler, H. S., Lund, K., and McCune, C. C., *Chem. Eng. Sci., 30* (1975) 1325.
3.73. Labrid, J. C., *Soc. Pet. Eng. J., 27* (1975) 117.
3.74. Walsh, M. P., Lake, L. W., and Schechter, R. S., *J. Pet. Tech., 34* (1982) 2097.

PROBLEMS

***3.1.** A non-Newtonian fluid ($n = 0.7$, $m = 1$ kg/m-sec$^{1.3}$) is contained between parallel plates. The lower plate is fixed and the upper plate is free to move. If a force is exerted on the upper plate, it will move at an equilibrium velocity of v. If the force is doubled, will the new velocity be $2v$? If not, what will the relationship between the two velocities be?

***3.2.** A Fann viscometer consists of two closely spaced concentric cylinders with the fluid to be tested contained in the small gap between the cylinders. The outer cylinder is rotated while the inner cylinder is fixed. The torque exerted on the stationary inner cylinder is measured. Because the spacing between the cylinders is small, the strain rate ($\dot{\gamma}$) is approximately constant and equal to the velocity of the outer cylinder divided by the width of the gap between the cylinders.

Given the following data, calculate the torque that is exerted on the inner cylinder:

Radius of outer cyclinder: 3 cm
Radius of inner cyclinder: 2.8 cm
Height of inner cylinder: 4 cm
Rotational speed of outer cyclinder: 15 rev/min

The fluid is a power-law fluid for which $n = 0.45$ and $m = 0.3$ kg/m-sec$^{1.55}$.

***3.3.** The system of two parallel plates, each of which is 1 m in length, and spaced 0.1 cm apart, is shown in Fig. P3.3. This system can be used as a simple viscometer, since the velocity of the upper plate relative to a fixed lower plate can be measured as a function of the mass that is placed in the pan. Given the data shown in Table P3.3, calculate the parameters m and n which apply to the power-law fluid contained between the plates.

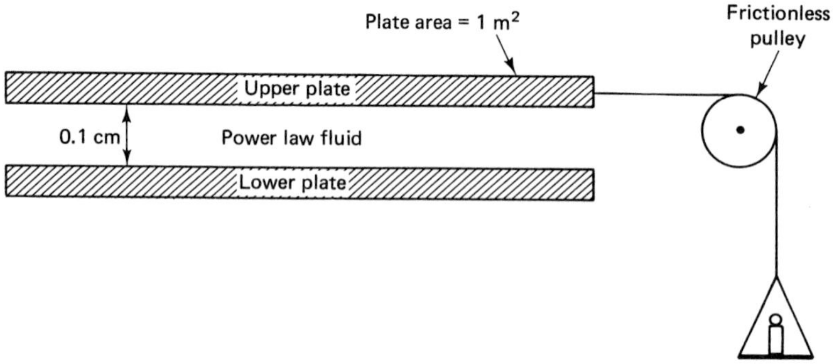

Figure P3.3

TABLE P3.3 Data

Mass in pan (kg)	Velocity of upper plate (m/sec)
0.1	1×10^{-3}
0.2	3×10^{-3}

***3.4.** The friction factor for turbulent flow of a Newtonian fluid is approximately given by the Blasius equation

$$f = \frac{0.0791}{\text{Re}^{+/4}}$$

where Re is the Reynolds number defined as $\rho \bar{v} D/\mu$. A fluid with a viscosity of 1 centipoise is pumped through a pipe (inside diameter = 0.05 m) that is 1000 m long. The pressure at the top of the pipe, which is inclined at an angle of 30° with respect to the horizontal, is 2000 kPa and is 7224 kPa at the bottom. Is the flow upward or downward in the pipe? Does the fluid contain drag-reducing agents? If so, prove your conclusion. The volumetric flow rate is 9.08×10^{-2} m^3/min and the fluid density is 1038 kg/m^3.

****3.5.** **(a)** Prove that the velocity of power-law fluid flowing between infinite parallel plates spaced a distance w apart is given by

$$v_x(y) = \left(-\frac{1}{m}\frac{dp}{dx}\right)^{1/n} \left(\frac{n}{n+1}\right) \left[\left(\frac{w}{2}\right)^{n+1/n} - y^{n+1/n}\right]$$

where y is a position coordinate measured from the centerline to the wall (for $y \geqslant 0$) and v_x is the velocity through the parallel plate system. The constants m and n are parameters in the power-law model.

(b) Plot $v_x(y)/v_{\max}$, where $v_{\max}$ is the maximum fluid velocity, as a function of $2y/w$ for $n = 1$ and $n = 0.5$.

(c) Show that the strain rate $\dot{\gamma} = |\, dv_x/dy \,|$ is given by

$$\dot{\gamma} = \left(-\frac{1}{m}\frac{dp}{dx}\right)^{1/n} y^{1/n}$$

(d) Prove that the average velocity is given by

$$\langle v_x \rangle = \left(-\frac{1}{m}\frac{dp}{dx}\right)^{1/n} \frac{n}{2n+1} \left(\frac{w}{2}\right)^{n+1/n}$$

****3.6.** Assume that a foam behaves as a Bingham plastic and prove that the following equation describes the volumetric flow rate q in a circular tube as a function of the pressure drop per unit length

$$q = -\left(\frac{dp}{dZ}\right)\frac{\Pi R^2}{8\mu}\left[1 - \frac{4}{3}\frac{r_p}{R} + \frac{1}{3}\left(\frac{r_p}{R}\right)^3\right]$$

Start your analysis by proving that

$$\tau_{rz} = -\frac{r}{2}\frac{dp}{dZ}$$

represents a force balance about a cylindrical volume such as the one shown in Fig. P3.7.

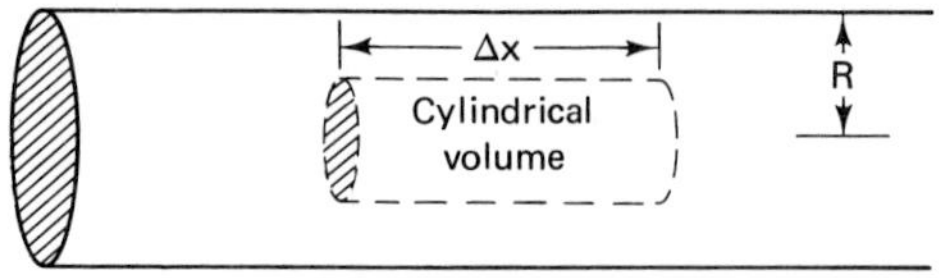

Figure P3.7

Argue that the fluid moves as a rigid plug within the region

$$\frac{r}{R} \leqslant \tau_{\text{yield}} \frac{4L}{D\,\Delta p}$$

***3.7.** A mixture containing 5 wt% hydrochloric acid (HCl) and 5 wt% acetic acid (CH_3CO_2H) is used to treat a limestone formation ($CaCO_3$), which is at 225°F. Assuming that all of the HCl reacts and that the carbon dioxide (CO_2) produced remains in the acid solution, calculate the percentage of CH_3CO_2H which is unreacted when the solution comes to equilibrium. What fraction of the CH_3CO_2H in a 5 wt% solution will be unreacted if the HCl is not added to the solution?

***3.8.** An inventor claims to have formulated an acid, HA, that has a dissociation constant $K_D = 10^{-5}$ and that this is the same as the corresponding value for acetic acid at the same temperature. Furthermore, the inventor states that because the calcium salt, CaA_2, has a limited solubility in water (about 10^{-2} g moles/1000 g of H_2O) this acid will react to a greater extent than will a 10 wt% solution of acetic acid when both acids have some initial concentration expressed in g moles/1000 g of H_2O.

(a) Compare the extent of conversion of the new acid with acetic acid.

(b) The inventor is seeking money to help develop this project. Would you be interested in investing? If so, what further information would you require? If not, why?

***3.9.** 2.5 m^3/m of a 6 wt% formic acid solution is used to matrix acidize a dolomite formation. The formation temperature is 175°F and the initial porosity is 11%. If the acid reacts to equilibrium within a radius of 0.3 m, what increase in porosity would you expect? The wellbore radius is 0.07 m.

***3.10.** In acidized carbonate formations an approximation relating the initial formation permeability, k_i, and the final one, k_f, to the change in porosity is

$$\frac{k_f}{k_i} = \left(\frac{\phi_f}{\phi_i}\right)^m$$

where m is a large number, say 10. If a carbonate formation ($\phi_i = 0.15$, $T = 250$°F) is treated with 2.5 m^3/m of a 10 wt% acetic acid solution and if the acid reacts to chemical equilibrium within a radius of 1 m measured from the centerline of the wellbore, what increase in permeability would you expect? The wellbore radius is 0.15 m.

PART 2
Formation Damage

4
The Origins of Formation Damage

4.1 INTRODUCTION

Flow surveys almost invariably show that a significant fraction of the zone open to the wellbore does not contribute to the total production. Since vertical permeabilities tend to be smaller than horizontal ones, these restricted production rates will ultimately result in considerable hydrocarbon being left in place when the well has reached its economic limit. Those portions of the open zone that are unproductive may well have been damaged as a result of careless drilling, completion, or production practices and are therefore avoidable. Thus, it is worthwhile to explore those mechanisms which can give rise to a reduction in the formation permeability. Even if it should turn out that the damage is not avoidable, an understanding of the nature of formation damage will be crucial in the selection of proper methods for stimulation. Perhaps the remedy might be as simple as selection of the appropriate perforating gun or so complex as to warrant a continuous, long-term series of well treatments.

By understanding the origins of formation damage, it will be possible to minimize their effect, and any consideration of oil well stimulation procedures will be incomplete without considering them. Most formation damage occurs in the region near the wellbore and results primarily, although not entirely, from the pores in this region becoming clogged with debris. This debris can have a physical, chemical, or biological origin. It can be produced by hydrodynamic forces which strip loosely attached fine particles from the pore walls (physical process), it can be produced by interaction of injected fluids with the reservoir rock (chemical), or it can be a product of bacterial activity. All of these mechanisms are discussed here. Irrespective of the origin of the debris, the mechanism by which the permeability is reduced is the same. The pore throats become clogged. However, not all debris or fine particles which enter the pore spaces or are generated there will clog the pore throats. Some, in fact most, will be captured on the surface of the pores in a nondamaging mode. Thus, the first issue to be resolved here relates to the ultimate fate of fine particles suspended in a fluid flowing in a porous medium. Following this discussion, the mechanisms responsible for producing the fine particles are then considered.

The clogging of pore throats is not, however, the only mechanism which can lead to reduced production. There are others including in-situ emulsification and wettability modifications which in turn lead to smaller relative permeabilities to oil. Both of these mechanisms are discussed here.

Having understood the origins of formation damage, it is then of interest to model these processes so as to be able to predict their effect on the formation permeability and the depth to which the damage extends into the formation around the wellbore. In fact, it is this latter consideration which is of greatest importance, for the depth of damage determines the type of well stimulation treatment that can be used to remove it. Thus, the acid stimulation of a limestone may be limited to depths of 50–100 cm. If the damaged zone extends into the formation at a much greater depth, then acidization will clearly not be the proper stimulation method. The modeling of formation damage is discussed in Chapter 5.

4.2 CAPTURE OF SUSPENDED PARTICLES

Particles suspended in a fluid flowing through a sandstone or limestone formation do not migrate far (at most a few pore diameters) without becoming attached to the pore walls, unless there exists a repulsive force between the particle and the wall. There are at least four mechanisms that tend to bring a suspended particle into contact with the pore wall [1, 2] and all of them are generally operative in systems of interest here. These mechanisms are depicted in Fig. 4.1, which shows a single sand grain positioned in a flowing stream of fluid containing suspended particles. It is immediately evident that those particles positioned on streamlines that approach the surface will be brought into contact. There are, in addition, three mechanisms that cause suspended particles to move in a direction different from the motion of the fluid and come into contact with the sand grain.

One of these mechanisms depicted in Fig. 4.1 is *sedimentation*. If the particle has a density different from that of the fluid, gravity will influence the particle motion. In Fig. 4.1 the particle has a greater density than the fluid and drifts across fluid streamlines to collide with the sand grain.

Inertia is another factor that will increase the collision rate. As the fluid swerves to bypass the sand grain, the denser particles will tend to move in a straight line toward it.

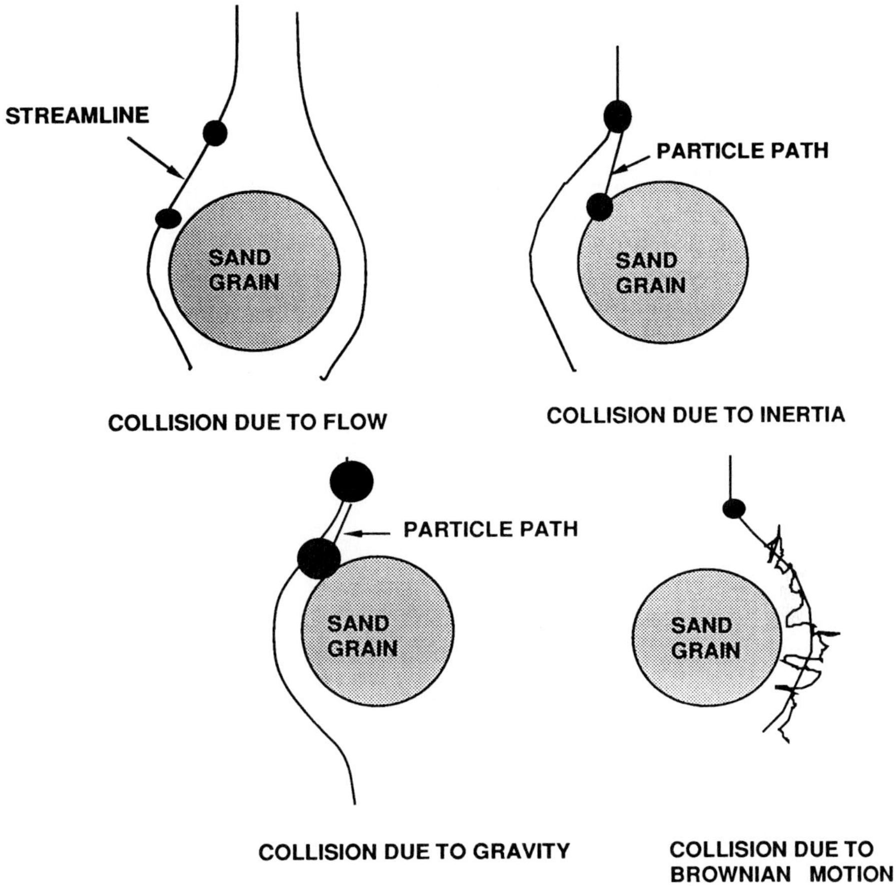

Figure 4.1 Various particle trajectories that result in a collision with the sand grain.

The final mechanism depicted in Fig. 4.1 is *Brownian motion*. The trajectory of a Brownian particle is suggested by the wavy line surrounding the streamlines near the surface. This essentially random movement, which is especially pronounced for small particles ($a < 1$ μm), has been attributed by Einstein [3] to the random impact of very energetic molecules with suspended fine particles. Because these impacts are random, the position of a given particle cannot be predicted; however, Einstein showed that the movement of a number of small spherical particles can be characterized by an effective diffusion coefficient expressed as follows:

$$D = \frac{kT}{6\pi\mu a} \tag{4.1}$$

where μ is the viscosity of the fluid and a is the particle radius. This Stokes–

Einstein equation shows that the movement is important for small particles and that the phenomenon becomes more important at elevated temperatures.

All of these collision mechanisms taken together with the complex topologies that typify the pore spaces of oil-bearing formations lead one to conclude that, excluding any repulsive forces, collisions between suspended particles and the pore walls are quite frequent and that a particle will on the average travel only a few pore diameters before colliding with the wall.

This assertion can be shown to be true by considering the collisions due entirely to Brownian motion. Although rather detailed analyses have been reported, including the interaction forces between the particles and the wall [4, 5], our purposes will be served by simply considering particles suspended in a fluid moving through a series of parallel capillaries. On the average a particle during a time (t) diffuses a distance proportional to $\sqrt{Dt}$. For a simple parallel capillary model, Dullien [6] cites an equation relating the capillary diameter (d_c) to the permeability

$$d_c = \sqrt{\frac{32k}{\phi}} \tag{4.2}$$

Thus, the average time required for a suspended particle to reach the capillary boundary starting from the centerline can be estimated from the equation

$$\sqrt{D\bar{t}} = \frac{d_c}{2} = \sqrt{\frac{8k}{\phi}} \tag{4.3}$$

An average particle moves a distance ($L_{cap} = u\bar{t}/\phi$) before colliding with the wall. Thus,

$$L_{cap} = \frac{8uk}{D\phi^2} \tag{4.4}$$

This average distance of movement can be shown to be on the order of the pore dimensions. Equation (4.4), therefore, confirms the assertion that, in the absence of repulsive forces, suspended particles do not move any great distance without experiencing multiple collisions with the wall.

Example 4.1 The Penetration of Small Particles into Porous Media

Determine the average distance a clay particle ($a = 0.1\ \mu$m) from a drilling fluid penetrates into a formation ($k = 5$ mD and $\phi = 0.15$) if the flux of drilling filtrate into the formation is 2×10^{-5} cm/sec in the near wellbore region. The fluid viscosity is 0.01 poise and the formation temperature is 80°C. Express your answer in pore diameters; that is, as L_{cap}/d_c.

Solution The Einstein equation for the diffusivity, Eq. (4.1), gives

$$D = \frac{(1.38 \times 10^{-20})(273 + 80)\ \text{J}}{6\pi\left(0.001\ \frac{\text{kg}}{\text{m-sec}}\right)(1 \times 10^{-7}\ \text{m})} = 2.6 \times 10^{-9}\ \frac{\text{m}^2}{\text{sec}}$$

The permeability in m^2 is

$$k = (5\ \text{mD})\left(\frac{D}{1000\ \text{mD}}\right)\left(9.87 \times 10^{-13}\ \frac{\text{m}^2}{D}\right) = 4.9 \times 10^{-15}\ \text{m}^2$$

Thus,
$$L_{\text{cap}} = \frac{8(2 \times 10^{-5})(4.9 \times 10^{-15})}{(0.15)^2(2.6 \times 10^{-9})} = 1.34 \times 10^{-8}\ \text{m}$$

Furthermore,
$$d_c = \sqrt{\frac{(32)(4.9 \times 10^{-15})}{(0.15)}} = 1.02 \times 10^{-6}\ \text{m}$$

so that
$$\frac{L_{\text{cap}}}{d_c} = 1.31 \times 10^{-2}$$

Thus, in moving one pore diameter, an average suspended clay particle will collide with the surface about a hundred times. Clearly, excluding any force preventing these collisions, suspended fines will be rapidly attached to the surface.

The conclusion that suspended fines are rapidly attached to pore surfaces does not seem to square entirely with practical experience. For example, produced fluids often contain fine particles which then settle in surface holding tanks forming a sludge. These particles are continuously produced and evidently have migrated with the produced fluids for some distance. These produced fine particles have avoided collisions with the wall because of a repulsive force or because, as is discussed in Chapter 5, particles can migrate large distances by a sequential process of attachment followed by detachment.

Produced particles are generally very fine, normally less than one third the size of the smaller pore throats. Particles larger than this minimum size will tend to be captured by a "log jam" mechanism at the pore throats as well as by attachment to the pore walls. These two modes of capture are depicted by Fig. 4.2. The "log jam" or plugging mode is due entirely to the hydrodynamic motion of the fluid and is clearly the more damaging of the two modes depicted. If a particle attaches itself to the pore wall, it will not alter the permeability by much; however, clogging the pore throats will have a profound effect on the permeability.

Thus, a crucial issue is the ultimate mode of capture. This is in turn determined primarily by the forces that act when a particle approaches a wall and by the size of the particles relative to the size of the pores (see Chap. 5). Based on these observations, a study of the forces that arise between a particle and a surface seems necessary.

Long-Range Forces: Van der Waals

Experience indicates that small particles often adhere tenaciously to surfaces. Dust particles cling to vertical surfaces despite the force of gravity. Similarly, vigorous blowing or shaking will often fail to dislodge dust particles. What force holds these particles to a surface? How does it vary with the particle size and shape? Is the force a function of the chemical composition of either the particle or the surface? How does the force vary with the spacing between the particle and the surface? All of these issues relate to the mode of particle capture.

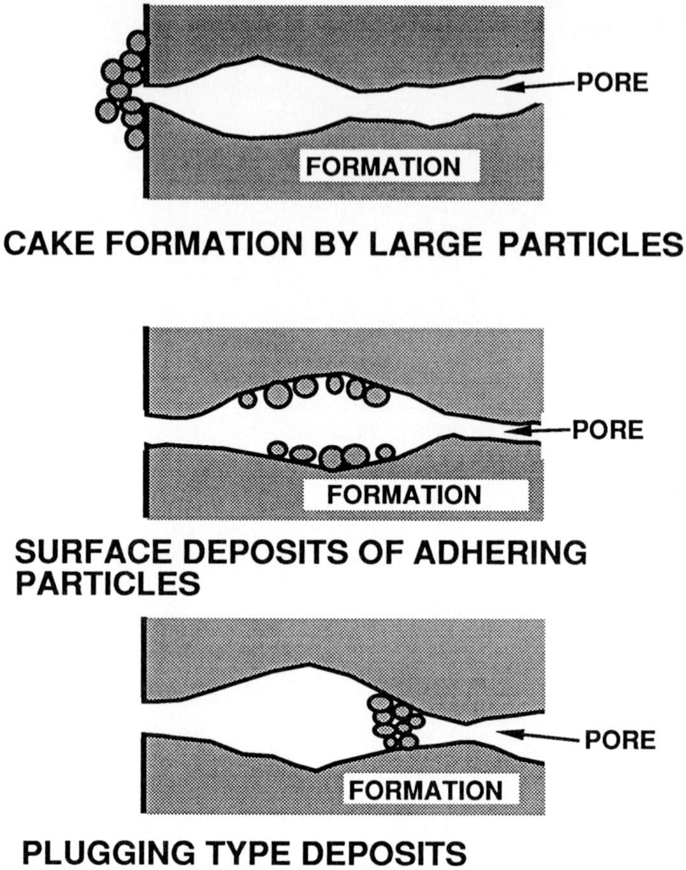

Figure 4.2 Illustration depicting three different types of deposits.

Surprisingly, the attractive force between microscopic particles and a surface is due to the forces between molecules. This is a surprising assertion since it is well known that the force between molecules decreases rapidly when two molecules are separated beyond a few molecular diameters. Indeed, the attractive potential between two molecules, $U_{\text{mol}}^{(A)}$, varies approximately as

$$U_{\text{mol}}^{(A)} = -\frac{\alpha}{r^6} \tag{4.5}$$

where r is the distance of separation and α is a constant [7]. This expression applies to polar molecules as well as to nonpolar ones (i.e., London or dispersion forces [7]). The negative sign appearing in Eq. (4.5) signifies that the force is attractive. [The reader will recall that a force is obtained from a potential by differentiating potential with respect to distance. Thus, the attractive force acting along a line between two molecules is $-d/dr\ U_{\text{mol}}^{(A)}$.] This force, called *van der*

Waals or *dispersion force*, is the cohesive force which binds liquids and many solids together.

The van der Waals force between two molecules decreases rapidly ($\sim r^{-7}$) as the distance between them increases. At first thought, one might believe that these forces will be entirely negligible as compared to hydrodynamic forces when considering the motion of a suspended particle.

Remarkably, despite the incredibly small values of the molecular attractive force between two molecules separated by a distance of 2 nm, the resultant force when all of the pairs of molecular interactions are summed is substantial. A microscopic body is composed of so many molecules that this large number multiplied by even an exceedingly small force per molecule yields a finite value. This can be better understood by considering Fig. 4.3. The number of molecules in volume dV_A is $\rho_A \, dV_A$, where ρ_A is the molecular density. Similarly, $\rho_B \, dV_B$ is the number of molecules in volume element dV_B. The attractive force between the molecules contained within these two volume elements is given by

$$\text{Component of force in normal direction} = -\frac{d}{dr}\left(\frac{\alpha}{r^6}\right) \cos\theta \, \rho_A \rho_B \, dV_A \, dV_B$$

Here, we see that the force between two molecules, one in each of the bodies, is multiplied by the two molecular densities which are large numbers ($\sim 10^{21}$ molecules/cm^3). Furthermore, to obtain the total force, one must integrate over both

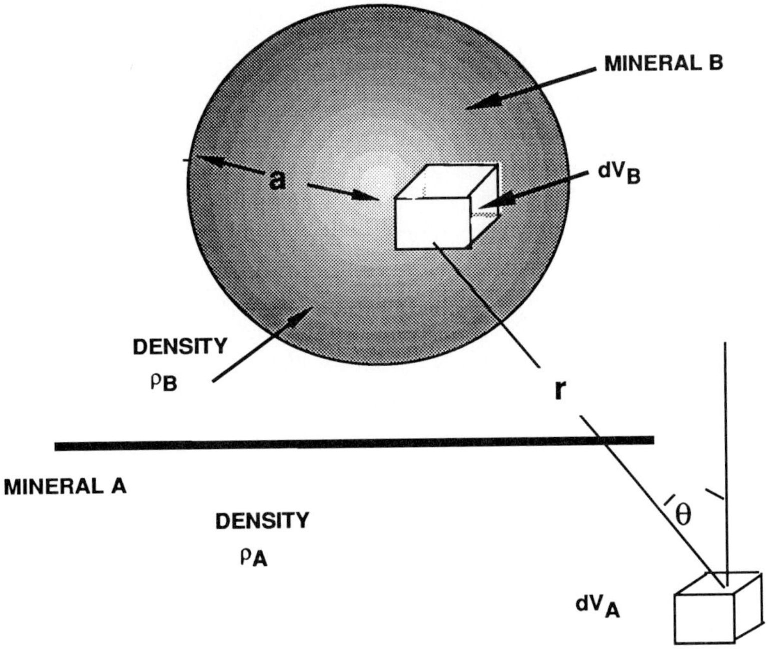

Figure 4.3 Molecules in volume element A interact with molecules in volume element B, which is part of a spherical particle of mineral B.

volumes V_A and V_B. Thus,

$$\begin{aligned} \overline{F}_N^{(A)} &= \text{Component of attractive force in normal direction} \\ &= -\iint_{V_A V_B} \frac{d}{dr}\left(\frac{\alpha}{r^6}\right) \cos\theta\, \rho_A \rho_B\, dV_A\, dV_B \end{aligned} \tag{4.6}$$

Integrating this equation over the volumes of both the spherical particle and the thick slab yields [7]

$$\overline{F}_N^{(A)} = \frac{A_H a}{6h^2} \quad \text{or} \quad \overline{U}_A = -\frac{A_H a}{6h} \tag{4.7}$$

where A_H is known as Hamaker's constant, h is the distance of separation shown in Fig. 4.3, and $\overline{U}_A$ *is the potential energy of interaction. Typical values of Hamaker's constant are about 10^{-19} J for interactions of two bodies separated by a vacuum. Taking* $A_H = 10^{-19}$, the force acting on a sphere having a radius of 1 μm at a distance (h) of 1 nm from a wall is 1.6×10^{-8} N. To obtain some understanding of the importance of this force, let us compare it to the gravitational force acting on the same small sphere.

The density of the particle (ρ_B) can be taken to be roughly 2500 kg/m^3 so that

$$F_g = \frac{4}{3}(\pi a^3)\rho_B g \tag{4.8}$$

is 1×10^{-13} N. The van der Waals force between the surface and the particle is 5 orders of magnitude larger! For suspended particles flowing through porous media, gravity can generally be neglected. The long-range forces are much more important.

Most mineral particles are not spheres and perhaps in some cases the force between two walls may be a better approximation. Israelachvili [7] gives F_N (expressed per unit area of surface) as

$$\overline{F}_N^{(A)} = \frac{A_H}{6\pi h^3} \quad \text{or} \quad \overline{U}_A = -\frac{A_H}{12\pi h^2} \tag{4.9}$$

where h is now the distance separating the two parallel surfaces.

The calculation of the force between a sphere and a wall illustrated by Fig. 4.3 tacitly assumes that the interaction between any pair of molecules is uninfluenced by the presence of surrounding molecules and the fluid separating the particle from the wall. This assumption is certainly not valid and both the presence of surrounding molecules and those in the intervening medium must be taken into account. These effects are both considered by the Lifshitz theory [8], which requires modification of the Hamaker constant but does not change the form of Eq. (4.7) or Eq. (4.9). Israelachvili [7] gives an approximate (but apparently accurate) expression for the Hamaker constant in terms of the dielectric constants and the refractive indices of the materials involved. Calculations using this equation will show that the Hamaker constant is reduced by about two orders of magnitude (to $\sim 10^{-21}$ J) by these corrections. In the considerations presented here, the value $A_H = 10^{-21}$ J will be of sufficient accuracy for our qualitative purposes and will be assumed in every case.

Long-Range Forces: Double-Layer Repulsion

In Chapter 1, the electrical double layer surrounding a mineral oxide surface is described. Except under certain very specific conditions (e.g., the pzc, see Section 1.3), mineral oxide surfaces are charged. It is not difficult to visualize, therefore, that when two minerals approach one another, an electrical force will arise when two double layers overlap. This force depicted by Fig. 4.4 will act in addition to the van der Waals forces and must be considered separately.

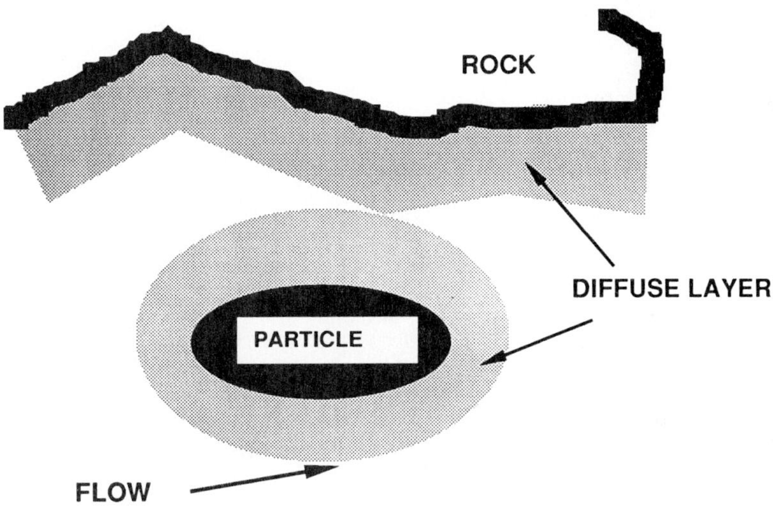

Figure 4.4 As a particle approaches a surface the diffuse part of their double layers will overlap. This leads to an additional force not included as part of the Hamaker constant.

Figure 4.5 is a schematic representation of the case for which the charge per unit area (σ) on the two surfaces is the same. Therefore, the electrical potential is symmetrical as shown. Since the two surfaces shown are similarly charged, the ions contained in the diffuse part of the double layers are also the same. Thus, the overlapping of the diffuse double layers will result in a repulsive force between the two surfaces. This is an additional force and must be added to the van der Waals attractive forces considered in the previous section. This repulsive force expressed per unit area, $\overline{F}_N^{(R)}$, is an effective pressure in excess of the normal hydrodynamic pressure, and it is given by the following equation for the particular

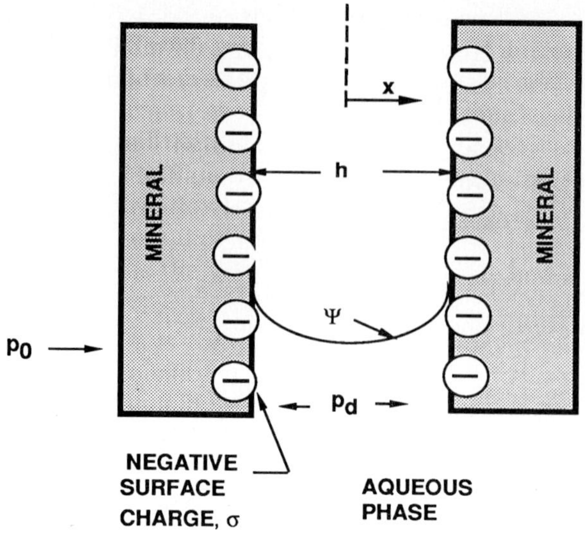

Figure 4.5 Parallel mineral surfaces are positioned a distance h apart. Since the plates have a fixed charge shown here as negative, a diffuse layer will surround each mineral. The electrical potential is denoted as ψ.

case where the cations and anions both have the same valence, Z [9]:

$$\overline{F}_{\mathrm{N}}^{(\mathrm{R})} = -\frac{2\pi\sigma^2}{\epsilon \sinh^2\left(\dfrac{h}{2\lambda}\right)} \tag{4.10}$$

In this equation, σ is the surface charge per unit area, λ is the Debye length, and ϵ is the dielectric permittivity. These terms all appear in the discussion of surface charge presented in Section 1.3 and it might be wise to review that information. $\overline{F}_{\mathrm{N}}^{(\mathrm{R})}$ is the force acting on one of the plates shown in Fig. 4.5 in excess of the fluid pressure.

Equation (4.10) is an approximation that will be satisfactory for the purposes of understanding formation damage; however, a comprehensive discussion and rigorous calculations (requiring numerical integration) are given by Verwey and Overbeek [10]. This model assumes both plates have a similar charge. If the surface and the particle are different minerals, they will not have the same charge. The repulsive force between surfaces of unlike charge has been considered [11, 12, 13], but these results are not cited here.

Example 4.2 The Excess Pressure Caused by Double-Layer Overlap

Kaolinite having one silicon atom per one thousand replaced isomorphically by an aluminum atom will have a surface charge of approximately -4×10^{-2} C/m^2. If two kaolinite surfaces with this charge are positioned 2 nm apart and separated by a water film, what excess force per unit area will be required to maintain this position? The water contains NaCl at a concentration of 10^{-2} kg moles/m^3 and its temperature is 80°C. At this temperature $\epsilon = 6.72 \times 10^{-9}$ C/V-m.

Solution Eq. (1.5) for the Debye length is

$$\lambda = \sqrt{\frac{\epsilon RT}{4\pi F^2 \sum Z_i^2 c_{i,0}}}$$

$$= \left[\frac{(6.72 \times 10^{-9})(8.3147 \times 10^3)(273 + 80)}{4\pi(9.65 \times 10^7)^2 2(0.01)}\right]^{1/2} = 2.9 \text{ nm}$$

The excess pressure is

$$\overline{F}_N^{(R)} = \frac{-2\pi(4 \times 10^{-2})^2}{(6.72 \times 10^{-9}) \sinh^2\left(\frac{2}{2(2.9)}\right)} = -1.21 \times 10^7 \text{ Pa}$$

This is a large repulsive force, but it must be added to the van der Waals force (which will be opposite in sign) to obtain a full picture of the particle behavior.

Equation (4.10) has been derived starting with a more general result. It has been simplified to the form presented here by expanding the equation so that it applies to small values of the electrical potential (ψ) divided by the thermal energy (RT) [9]. This is called the *Debye–Hückel approximation.*

The terms appearing in Eq. (4.10) have been presented in connection with the discussion of surface charge and the electrical double layer in Chapter 1; however, since Eq. (4.10) is presented without proof, it is worthwhile to discuss the terms to ascertain their plausibility. The reader will recall that the Debye length (λ) is a measure of the double-layer thickness. According to Eq. (4.10), increasing the Debye length increases the force tending to push the plates shown in Fig. 4.5 apart. This is reasonable because increasing the double-layer thickness causes the two layers to overlap at larger distances of separation.

A second trend, intuitively predictable, concerns the influence of the surface charge (σ). The repulsive force is proportional to σ^2. This assumes that the two surfaces shown in Fig. 4.5 have the same surface charge and, based on Coulomb's law, one would expect the force to be a multiple of the two charges. Thus, the force is repulsive independent of whether the charge is positive or negative.

Finally, the excess pressure is observed to increase as the distance of separation (h) is decreased. At small h the force tends to infinity.

Often it is more convenient to consider potential energies rather than forces. Recalling that a potential is defined as

$$\overline{F}_N^{(R)} = -\frac{d\overline{U}^{(R)}}{dh} \tag{4.11}$$

where $\overline{U}_R$ is the repulsive energy per unit of area, then assuming σ to be independent of h, one finds

$$\overline{U}_R = \frac{4\pi\sigma^2\lambda}{\epsilon}\left\{\coth\left(\frac{h}{2\lambda}\right) - 1\right\} \tag{4.12}$$

The surface charge will not remain constant as the two surfaces approach one another [14]; however, if the approach is rapid and the ions do not have time to diffuse, then this is perhaps the best approximation one can make without further knowledge of the process.

Long-Range Structural Forces

Experiments show that as two surfaces are brought together, additional forces other than the van der Waals and electrostatic forces become important [7]. These are apparently related to the arrangement of molecules in the liquid film or, rather, the difficulty of arranging them to fit precisely into the space between two walls. Because of the great importance of thin films in physical sciences, the underlying nature of these structural forces is under active investigation. For the purpose of discussion here, however, these forces are neglected.

Short-Range Repulsive Forces

At very small interatomic distances the electronic clouds of atoms overlap and there arises a strong repulsive force that determines just how closely two surfaces can approach one another. There is no rigorous theory that defines the form of these forces, but they are generally represented by a power-law dependence [7]. These forces will not be discussed further here, but it is important to remember that h, the distance of separation, cannot become zero. The repulsive forces will prevent this. Thus, the van der Waals forces given, for example, by Eq. (4.9) cannot become infinite which would imply, of course, that it would not be possible to separate the two bodies once they became attached.

The Derjaguin–Landau–Verwey–Overbeek (DLVO) Theory

The net force acting on a particle as it approaches a pore surface is the sum of the van der Waals and the electrostatic forces. Figure 4.6 shows the potentials plotted as a function of the separation distance between the two surfaces (remember the force is proportional to the slope of the potential curve). The attractive van der Waals potential is independent of the salt concentration in the aqueous phase, but the electrostatic potential is reduced as the salt concentration is increased. This is predicted by Eq. (4.12). The potentials are additive and Figs. 4.7(a)–(c) show their sum. For Case 4.7(a), there exists a large potential barrier that will tend to prevent the fine particle from approaching the pore wall (see Fig. 4.4). At the top of the potential barrier the repulsive force vanishes and for small values of the distance of separation, the net force becomes attractive.

For Case 4.7(b), the repulsive forces are much smaller and only a small energy barrier is indicated. If a particle approaches the wall with a sufficient velocity, it will "jump" this barrier and be attracted to the wall. Case 4.7(c) depicts a situation for which there is no barrier and each interaction will lead to particle attachment.

These different cases illustrate the crucial importance of the salt concentration in reducing the repulsive energy barrier. This barrier, when it exists, will

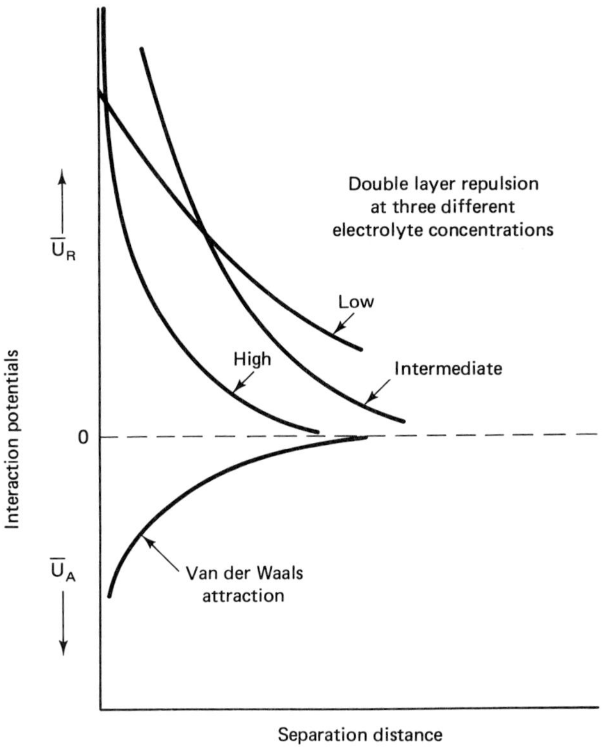

Figure 4.6 Repulsive and attractive energy as a function of particle separation at three electrolyte concentrations.

tend to increase the probability of pore plugging by migrating fines and therefore, in order to reduce formation damage, it is imperative that all injected fluids have a minimal electrolyte concentration. This concentration depends on a number of factors as will become evident and is generally too complex to calculate in advance. Well treatment fluids that are water based will normally contain at least 2 wt% potassium chloride (KCl) to minimize formation damage. In waterflooding projects, it may not be feasible to maintain this level of electrolyte. In some cases, the injection wells may have to be periodically stimulated or fractured so as to maintain a viable flooding rate.

Example 4.3 Minimum Electrolyte Concentration

What minimum concentration of KCl (expressed in wt%) is required to eliminate the potential barrier between a particle (assumed to be a flat slab) having a surface charge of 2×10^{-3} C/m^2 and a similarly charged surface? The formation temperature is 80°C. (See Ex. 4.2 for further data.)

Solution The potentials are additive so that

$$\overline{U}_{\text{TOTAL}} = \overline{U}_{\text{A}} + \overline{U}_{\text{R}}$$

Now to eliminate the potential barrier the electrolyte concentration must be adjusted so that at the peak value of $\overline{U}_{\text{TOTAL}}$ it is just zero. Translated into mathematical

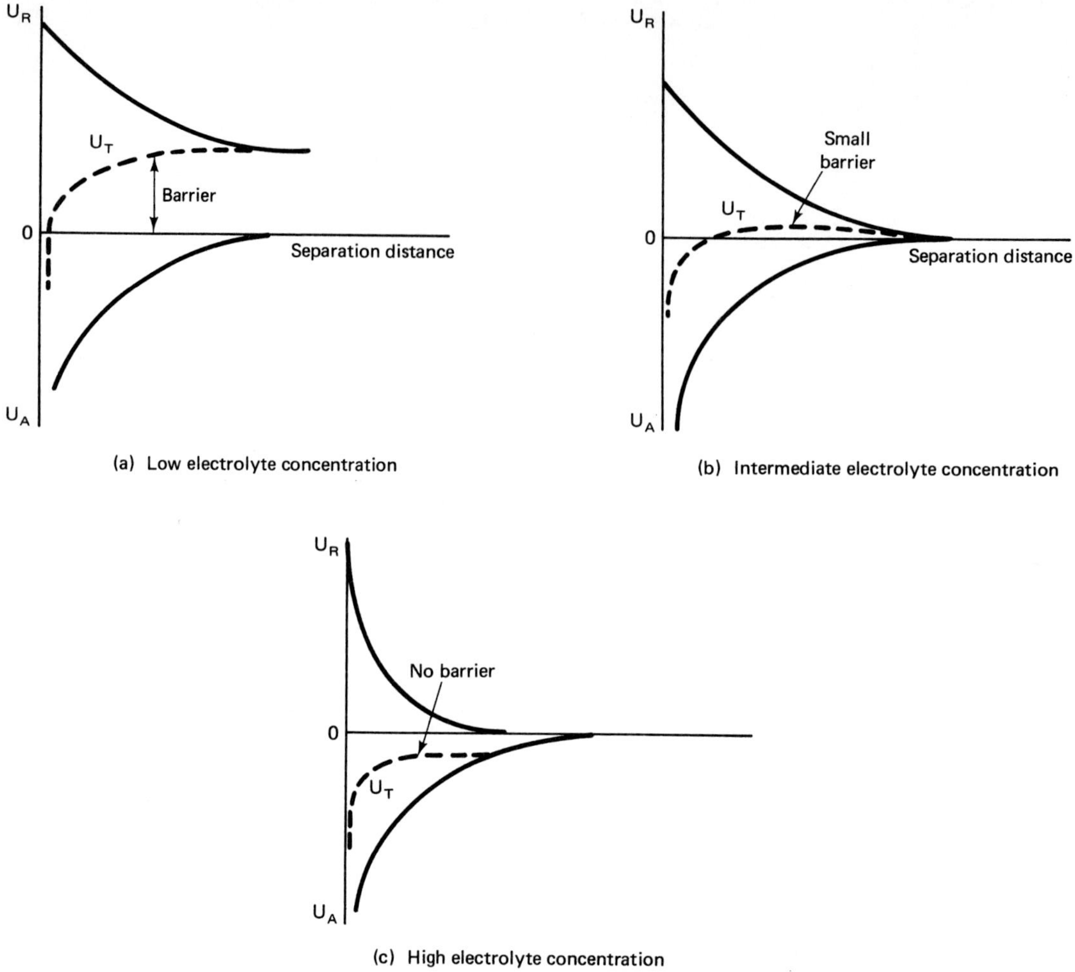

Figure 4.7 Net interaction energy as a function of particle separation. **(a)** Low electrolyte concentration. **(b)** Intermediate electrolyte concentration. **(c)** High electrolyte concentration.

terms, this means that $d\overline{U}_{\text{TOTAL}}/dh = \overline{U}_{\text{TOTAL}} = 0$ at the same point $h_{\max}$. Thus,

$$\frac{d\overline{U}_{\text{TOTAL}}}{dh} = \frac{A_{\text{H}}}{6\pi h^3} - \frac{2\pi\sigma^2}{\epsilon \sinh^2\left(\dfrac{h}{2\lambda}\right)} = 0$$

and

$$\overline{U}_{\text{TOTAL}} = -\frac{A_{\text{H}}}{12\pi h^2} + \frac{4\pi\sigma^2\lambda}{\epsilon}\left[\coth\left(\frac{h}{2\lambda}\right) - 1\right] = 0$$

Eliminating A_H, the resulting equation is

$$x = 2(\cosh x \sinh x - \sinh x \sinh x)$$

where $x = h_{max}/2\lambda$. This can be further simplified so that

$$\exp(-2x) = 1 - x$$

Solving for x gives

$$\frac{h_{max}}{2\lambda} = 0.795$$

This is the value of h for which the maximum of $\overline{U}_{TOTAL}$ vanishes. Substituting this value into the first equation

$$0 = \frac{A_H(8\lambda^3)}{6\pi h_{max}^3} - \frac{16\pi\sigma^2\lambda^3}{\epsilon \sinh^2\left(\frac{h_{max}}{2\lambda}\right)}$$

Rearranging

$$\lambda^3 = \frac{A_H}{6\pi x^3}\left(\frac{\epsilon \sinh^2 x}{16\pi\sigma^2}\right)$$

or

$$\lambda^3 = \frac{(10^{-21})(6.72 \times 10^{-9})}{96\pi^2(0.795)^3(2 \times 10^{-3})^2} \sinh^2(0.795) = 2.74 \times 10^{-27}$$

Thus,

$$\lambda = \sqrt{\frac{\epsilon RT}{4\pi F^2 \sum Z_i^2 c_{i,0}}} = 1.4 \times 10^{-9} \text{ m}$$

Solving for c_0, the minimum KCl concentration

$$c_0 = \left(\frac{1}{1.4 \times 10^{-9}}\right)^2 \left(\frac{(6.72 \times 10^{-9})(8.3147 \times 10^3)(273 + 80)}{4\pi(9.65 \times 10^7)^2(2)}\right)$$

or

$$c_0 = \frac{0.043 \text{ kg mole}}{\text{m}^3}$$

This is a small concentration and taking the density of the aqueous solution to be 1000 kg/m^3, then the weight percent is given by

$$\text{wt\%} = 100\left[\left(\frac{0.043 \text{ kg mole}}{\text{m}^3}\right)\left(\frac{74.5 \text{ kg KCl}}{\text{kg mole}}\right)\left(\frac{\text{m}^3}{1000 \text{ kg}}\right)\right] = 0.32 \text{ wt\%}$$

Thus, only a small salt concentration, in this case potassium chloride (KCl), is required to suppress the electrostatic repulsive barrier. The 2 wt%, which is normal industry practice, is well above this minimum value.

The capture of fine particles within the pore spaces of a porous media will generally be by the pore plugging mechanism unless the salt concentration exceeds a critical value. This has been verified experimentally [15], as will be discussed in a subsequent section. It should be understood that maintaining a sufficient electrolyte concentration will help reduce formation damage but will certainly not entirely prevent it. To understand the difficulty involved, the origins of fine particles suspended in the flowing fluid must be understood.

4.3 PHYSICAL PROCESSES THAT PRODUCE FORMATION DAMAGE

Drilling and Completion Fluids

Fine particles contained in either drilling or completion fluids will be damaging if there is fluid loss. Generally, it is advisable to use carefully filtered completion fluids and to minimize fluid loss during drilling. Substantial fluid loss during drilling should be recorded, because knowing the extent of the intrusion and the zones affected will help in the selection of a stimulation technique.

Detachment of Particles from Pore Walls

In this section, those forces which act to dislodge particles from the surface of the pore walls and hence increase the potential for formation damage are considered. An examination of sedimentary rock samples under an electron microscope will immediately reveal the rich diversity of pore filling minerals of diagenetic origin. These particles represent an indescribable variety of crystal shapes and sizes and have a wide range of chemical compositions. Many of the particles adhering to the pore walls are clay minerals, but other minerals such as silica are also present. It is a mistake to think only in terms of clays when considering the potential for formation damage.

The particles of greatest interest range in size from 1 to 100 μm. Particles smaller than 1 μm are difficult if not impossible to dislodge and pores are seldom much larger than 100 μm in diameter.

Particles in this size range adhere to surfaces with great tenacity. The adhesion force is much larger than the gravitational force so that their orientation with respect to the gravitational field is not a relevant issue. The force of adhesion is primarily the van de Waals forces and is roughly proportional to the particle diameter as shown by Eq. (4.7). On the other hand, the weight of a particle decreases as a function of the diameter cubed. For particles that are roughly 1 mm in diameter, the adhesion and gravitational forces are roughly balanced. For 1 μm particles, the force of adhesion greatly exceeds the gravitational force.

Particles once attached to the surface of the pore walls can be dislodged by hydrodynamic or capillary forces. The mechanism by which particles in the size range considered here are dislodged appears, as will be observed, to be one of peeling; however, to date, there are no physical models relating the removal process to established fields of physics. The difficulties arise from the fact that adhesive strength is a result of complex chemical and physical forces and mechanical strains and stresses which must all be considered simultaneously. The problem cannot be broken into a series of simpler subproblems. All of the factors must be considered at once. Consider a single spherical particle adhering to a surface.

Isolated sphere attachments to a planar surface. Figure 4.8 shows an isolated sphere attached to a plane surface with a fluid moving past. Remote from the particle, the fluid velocity increases linearly with distance measured from

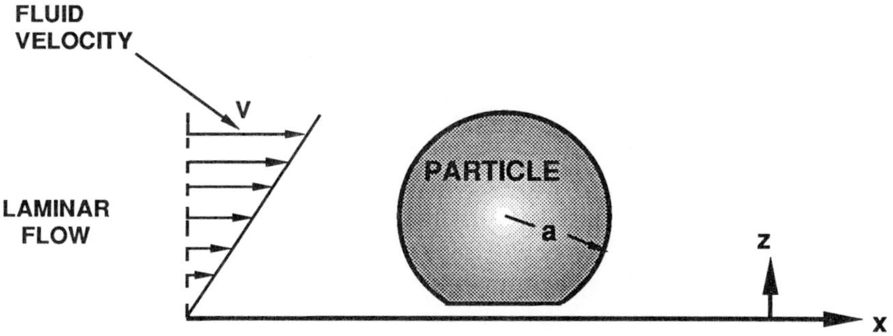

Figure 4.8 A spherical particle attached to a plane surface in a velocity field.

the surface. The hydrodynamic problem depicted by Fig. 4.8 has been considered by O'Niel [16] who concluded that there is no net lift force acting on the particle but that there is a tangential force (in the x-direction) given by

$$F_t = 10.2\ \pi \mu a \frac{dv}{dz} \tag{4.13}$$

where dv/dz is the change of fluid velocity with distance from the plane.

In addition to the tangential force, there is a torque exerted on the particle. When the moments are taken about the center of the sphere, this torque amounts to [16]

$$T_q = 7.55\ \pi \mu a^2 \frac{dv}{dz} \tag{4.14}$$

Equation (4.13) shows that the tangential force exists for all velocity gradients and a spherical particle resting on a smooth surface should be displaced by even small flows.

Experiments carried out by placing small (10 to 40 μm) polyethylene particles on a smooth glass surface reveal that, contrary to expectations, a substantial force (fluid velocity) is required to displace these particles from the surface. Figure 4.9 shows typical results [17]. There is apparently a critical force necessary to displace a particle from the surface. Experiments reveal that

1. The fluid velocity gradient, dv/dz, required to displace particles decreases with particle radius. Thus, larger particles will tend to be displaced from the pore surface first.

2. Although the velocity gradient required to displace small particles is greater than for larger particles, the actual force as defined by Eq. (4.13) is less.

3. Particles having a smaller Young's modulus are more difficult to displace than those having larger moduli.

4. There is an increase in the force required to displace a particle as the salt concentration is increased presumably because of the compression of the

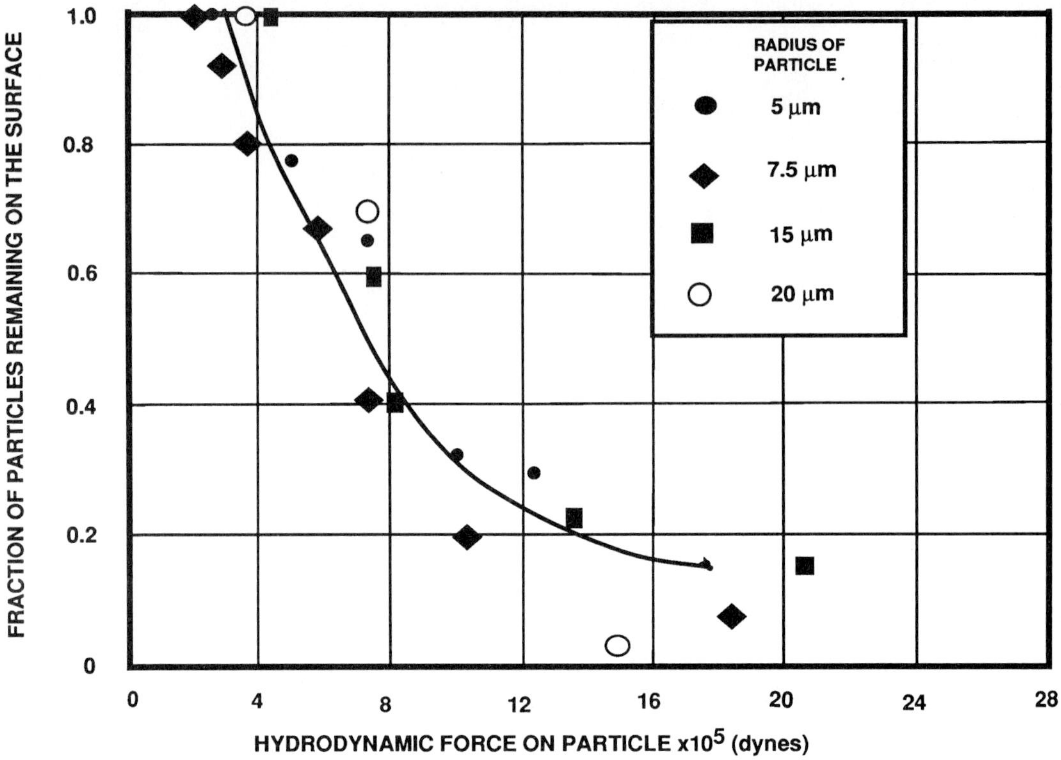

Figure 4.9 The fraction of particles displaced from a flat surface by a flowing fluid as a function of the total force acting on the particles [17].

double layer and consequent reduction of the electrostatic repulsive forces which tend to assist in displacing the particle from the surface.

5. A small effect of pH was observed [17] on particle displacement, but the addition of ethanol to the flowing aqueous stream in some cases substantially reduced the force needed to displace particles from a surface. Thus, there are chemical additives that can either exacerbate or even perhaps ameliorate [18] formation damage.

These observations are quite remarkable in view of the near-perfect sphericity of the particles used in the study. As noted, one would at first thought predict that the slightest fluid motion should disturb spherical particles resting on a plane surface but, in fact, a substantial force is required. The reason for this is that the van der Waals forces tend to crush the particle, deforming both the surface and the particle at the point of contact. Thus, the particle in Fig. 4.8 is depicted as being flattened at the point of contact. This is precisely the physical picture that will apply when an elastic sphere is positioned on a rigid surface. Derjaguin et al. [19] have derived an expression defining the deformation of a spherical particle by an applied force, which is equated to the van der Waals attractive forces giving

$$F_N^{(A)} = \frac{4a^2E}{3(1 - \nu^2)} \alpha^{3/2} \tag{4.15}$$

where α is the angle measured from the vertical to the point at which the deformed sphere first contacts the rigid boundary (see Fig. 4.10). E and ν are Young's modulus and the Poisson ratio, respectively (see Chap. 2). Since according to Eq. (4.7) since $F_N^{(A)}$ is proportional to the particle radius (a), α decreases with increasing particle radius and increasing Young's modulus. If it is presumed, as seems reasonable [17], that increasing α makes a particle more difficult to detach, then the observed trends can be understood.

Example 4.4 Relative Ease of Detaching Particles

Assuming that spherical particles with smaller values of α are the easiest to detach from a rigid plane surface where they are held by van der Waals forces, which will be more readily displaced: a 3 μm silica particle ($E = 4 \times 10^{10}$ Pa, $\nu = 0.15$) or a 3 μm carbonate particle ($E = 6 \times 10^{10}$ Pa, $\nu = 0.20$)?

Solution From Eq. (4.7)

$$\overline{F}_N^{(A)} = -\frac{A_H a}{6h^2}$$

Assuming A_H and h to be the same for both particles, then

$$\left(\frac{\alpha_{\text{quartz}}}{\alpha_{\text{carbonate}}}\right)^{2/3} = \frac{E_{\text{carbonate}}}{E_{\text{quartz}}} \cdot \frac{a_{\text{carbonate}}}{a_{\text{quartz}}} \cdot \frac{(1 - \nu^2_{\text{quartz}})}{(1 - \nu^2_{\text{carbonate}})}$$

or

$$\left(\frac{\alpha_{\text{quartz}}}{\alpha_{\text{carbonate}}}\right)^{2/3} = \left(\frac{6 \times 10^{10}}{4 \times 10^{10}}\right)\left(\frac{1 - (0.15)^2}{1 - (0.20)^2}\right) = 1.5$$

Thus, the carbonate particle would, according to this hypothesis, be dislodged from the pore surface by a slower motion than is required to displace the quartz particles. As evident from Eq. (4.7), size is the more relevant variable.

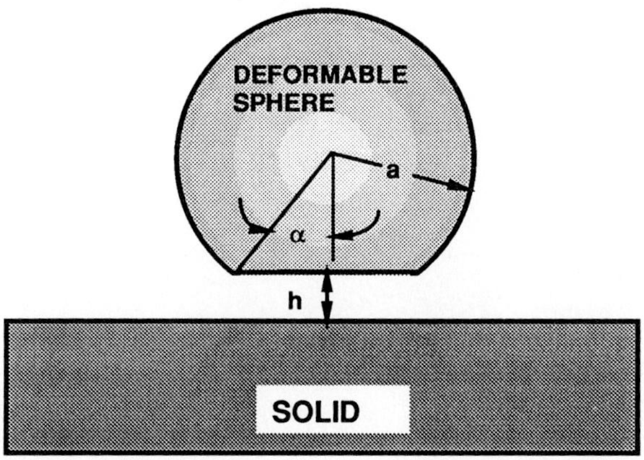

Figure 4.10 Sketch depicting deformation of a sphere by van der Waals forces attracting the sphere to a rigid plate.

Smaller particles are more difficult to displace than larger ones. This seems to be completely in correspondence with our intuition and yet when the particle sizes in produced fluids are measured, they are generally quite small (<1 μm). This is perhaps due to the fact that some of the attached minerals are actually a composite of smaller particles which break up upon detachment. It is certainly clear that removing a very small particle (<0.1 μm) by applying hydrodynamic forces of the magnitude associated with oil production or water injection operations is virtually impossible.

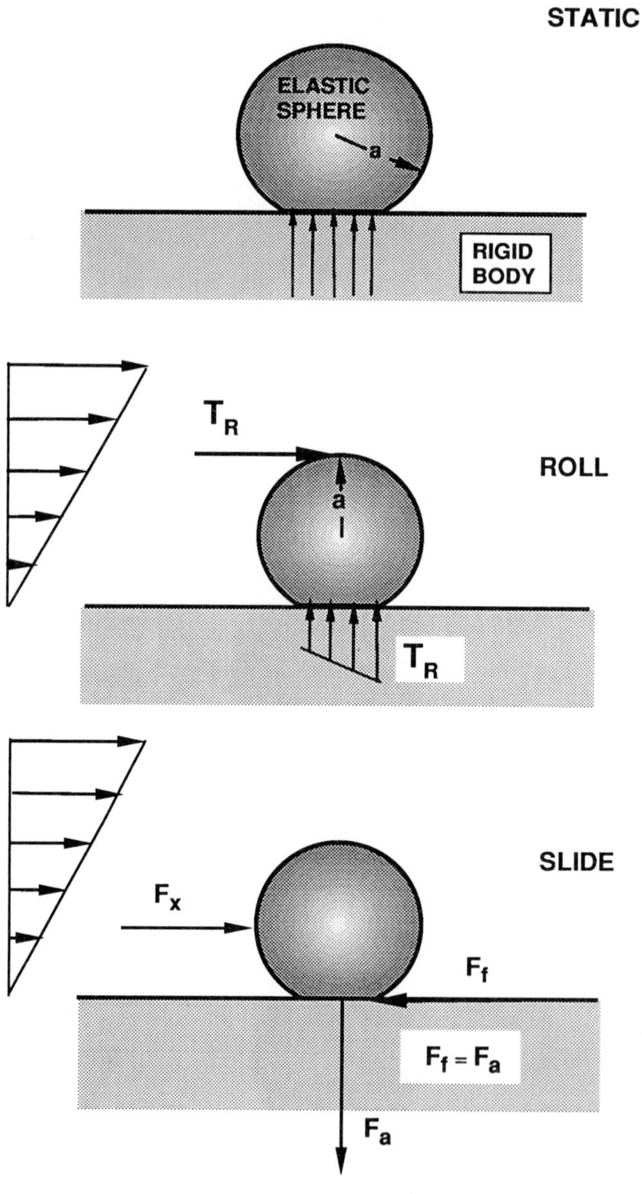

Figure 4.11 The three possible modes of particle detachment. The torque tends to roll or "peel" the particle from the surface. The tangential force tends to cause it to slide.

Mechanism of detachment. The hydrodynamic force has no component of lift. Thus, the attached particle must either slide or rotate and "peel" from the surface. Both mechanisms are indicated in Fig. 4.11. Chamoun [17] has demonstrated experimentally that 5 to 20 μm glass and polyethylene particles are peeled from the surface rather than removed by sliding. The torque is then the crucial feature [see Eq. (4.14)]. Particles having nonspherical shapes, which protrude into the flowstream a distance that is large compared to the square root of the contact area, will be dislodged more readily than those which hug the surface more closely. Pore filling clays (e.g., see Chap. 1) would be expected to be a rich source of fine particles. Dislodged particles are a primary source of formation damage.

Table 4.1 shows the overall composition of a Berea sandstone core and compares this composition with that of the fines defined as particles having a size less than 5 μm. It is evident that a substantial portion of the fines is composed of clays but by no means are clays the only fine minerals. Indeed, for Berea, quartz is listed as the major component. It is advantageous to remember these values, because in some cases stabilization of clays may not remove the possibility of formation damage. This is especially true for Berea [20], where clay stabilizers may not entirely control the damage caused by fines.

Thus, as shown by Table 4.1, there is a rich variety of different minerals which can be regarded as fines. These include secondary quartz precipitates, feldspars, carbonates, clays, or other minerals precipitated during diagenesis and lithification. Hydrodynamic forces can result in the dislodging and entrainment of these particles, which can then cause formation damage.

Studies reported by Muecke [21], Gruesbeck and Collins [22], and Gabriel and Inamder [20] all indicate the existence of a critical velocity for formation damage. Based on the discussion here, it is likely that particles are dislodged from the pore surface at all injection and production rates; however, at low velocities only large ones, perhaps too large to fit into pore throats and clog them, are dislodged. Of course, aggregates of smaller ones may also be dislodged, but their concentration in the flowing fluid may not be large enough to cause pore clogging jams.

Whatever the mechanism, experiments tend to show a critical velocity for

TABLE 4.1 Comparison of the Composition of Fines with the Bulk Core [20]

Mineral	Bulk Core (%)	Fines (<5 μm) (%)
Quartz	86	51
Feldspar	5	11
Dolomite	1	—
Siderite	1	1
Illite	4	15
Kaolinite	3	20
Chlorite	—	2

formation damage. Figure 4.12 shows representative data. It is observed that the fraction of the permeability lost for a fixed number of pore volumes of fluid depends of the velocity at which it is injected.

Figure 4.12 shows that oils will also dislodge particles that create formation damage. For example, isopar M is a clean paraffinic oil and its critical velocity is compared with 2 wt% KCl brine in Fig. 4.12. The critical velocity is larger than that of the KCl solution even though the oil viscosity is greater. One might expect that if all factors are equal, the fluid having the greatest viscosity would exhibit the smallest critical velocity. This rule is certainly not consistent with the data presented in Fig. 4.12. The reason for this discrepancy is perhaps related to the absence of electrical forces when the fluid is an oil.

It is interesting to note and perhaps of some practical interest that by slowing the velocity to a value less than the critical velocity, the decrease in the permeability may be arrested. Gabriel and Inamder [20] have shown that even if fluid is injected at a rate exceeding the critical velocity, if the flow is then slowed to a value less than the critical velocity, the permeability no longer declines; whereas, if the flow is continued at a rate exceeding the critical velocity, then the permeability decline continues. Finally, Fig. 4.12 shows the oil critical velocity is greatly increased in the presence of connate water.

To see just how a small quantity of water can help cement particles to pore walls, consider the particle shown in Fig. 4.13. Because water is presumed here to be the wetting phase, it is shown coating the particle. Furthermore, because there is an interfacial tension between oil and water, a force pointing in the direction of the arrows shown is exerted. This force is proportional to the perimeter

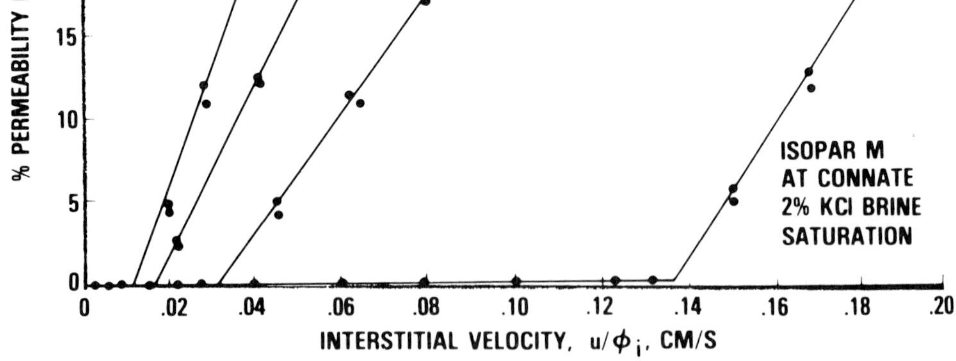

Figure 4.12 Permeability reduction versus interstitial velocity (computed from data compiled by Gruesbeck and Collins [22]). (Reprinted by permission of the Society of Petroleum Engineers.)

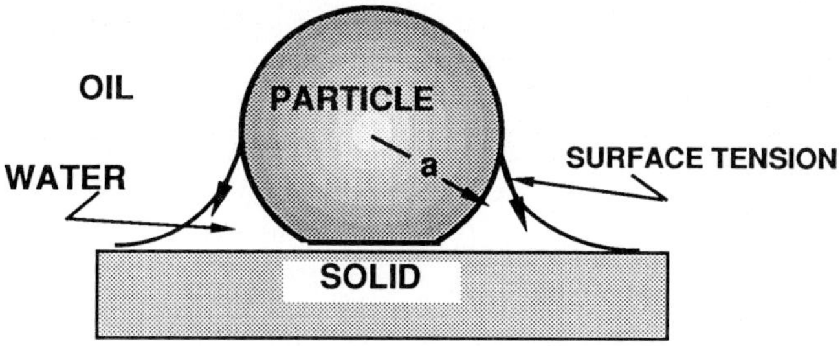

Figure 4.13 The interfacial tension on the particle shown acts along the oil/water interface thereby creating a force having a downward vertical component.

of the particle and to the interfacial tension. Thus,

$$F_{\text{capillary}} \propto 2\pi a\gamma \tag{4.16}$$

This force is large since $\gamma \sim 10$ mN/m and for a 10 μm particle, $F_{\text{capillary}} \cong -3 \times 10^{-7}$ N. This is compared to van der Waals forces $F_N^{(A)} \cong -2.5 \times 10^{-8}$ N when h in Eq. (4.7) is 0.18 nm, roughly the distance of approach permitted by the repulsive forces. Thus, the capillary forces can be important. If the water is removed from the near wellbore region by a stimulation fluid, for example, then the potential for formation damage is enhanced.

Fresh Water Shock of Reservoir Cores

A number of investigators have reported the rapid loss in formation permeability associated with the sudden injection of fresh water into a sandstone [24–30]. The results of a typical experiment consist of the formation permeability divided by the initial permeability to electrolyte solution, the permeability ratio, plotted as a function of the pore volumes of fresh water injected. Figure 4.14 is a typical graph. The permeability is shown to decrease to a small percentage of its original value after only 6 to 8 pore volumes have been injected.

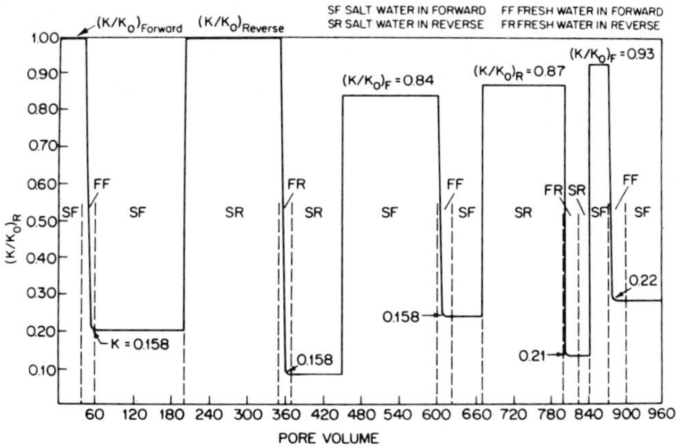

Figure 4.14 Permeability of Berea sandstone shown as a function of the number of pore volumes of fresh water. (From K. C. Khilar and H. S. Fogler, in *Surface Phenomena in Enhanced Oil Recovery,* D. O. Shah, ed., New York: Plenum Press, 1981. Used by permission.)

In these experiments, the flow velocities were maintained below the critical velocity so that the permeability decline is a function of the water composition and not the flow rate.

Figure 4.14 also presents data which show that if the flow is reversed by injecting 3 wt% NaCl solution into the outlet end, then most of the permeability lost during fresh water injection is regained. If fresh water is injected into the outlet end, the sudden loss in permeability is again observed.

Early theories suggested that this fresh water sensitivity is due to clay swelling [26, 29]. It is certainly true that in the presence of swelling clays, such as montmorillonite and to some extent illite (see Chap. 1), some permeability loss will be incurred by the expansion of clay layers [30]; however, experiments have definitely established that clay swelling is not the most important factor [31]. The dislodging of particles from the surface and the subsequent pore blocking are the essential mechanisms at work here.

Khilar and Fogler [15] have demonstrated the existence of a critical salt concentration. Concentrations greater than the critical one are nondamaging, whereas at concentrations less than the critical, substantial damage is observed. The critical salt concentrations have been measured for Berea sandstone and are reported in Table 4.2. These values will, no doubt, vary from sandstone to sandstone and with the solution pH. The value for KCl given in Table 4.2 is about the same as that found in Example 4.3. This can be considered to be an excellent agreement since this critical concentration will depend on both surface charge and particle shape. It is clear that the critical concentrations listed in Table 4.2 are essentially the electrolyte concentrations needed to suppress the electrostatic repulsive forces so as to permit dislodged particles to reattach to the pore surface rather than clogging the pore throats.

It should be recognized that the electrolyte effect on the Debye length, as might be suggested by Example 4.3, is not the full explanation. Example 4.3 would suggest that the same critical concentration applies to all one-one ($Z = 1$) electrolytes, but Table 4.2 shows that the critical concentration depends on the particular cation used in the experiment. This would seem to be associated with the specific influence of the cation on the surface charge. For example, critical concentrations for divalent cations such as calcium (Ca^{++}) are not given in Table

TABLE 4.2 Critical Salt Concentrations of Monovalent Cations (Berea Sandstones at 25°C [15])

Cation	Critical salt concentration (CSC)		Permeability reduction at the CSC
	CSC (kg moles/m^3)	*CSC (wt %)	
Sodium (Na^+)	0.07	0.41	46%
Lithium (Li^+)	0.07	0.27	43%
Potassium (K^+)	0.04	0.30	10%
Ammonium (NH_4^+)	0.01	0.05	8%
Cesium (Cs^+)	0.006	0.08	10%

* Expressed as chloride salt.

4.2 because for these cations, the salt concentration can be reduced to practically zero without reducing the sandstone permeability [15]. Jones [25] noted the importance of maintaining minimum levels of divalent ions to minimize permeability loss. He postulated that as long as the divalent ion concentration is high enough (at least one tenth of the salt concentration in the native waters) formation damage due to insufficient electrolyte concentrations will not occur. There are at least two problems with the practical implementation of this proposal. Waters containing divalent ions are more likely to yield precipitates when injected. Secondly, because of ion exchange with the formation minerals, divalent cations may be stripped from the injected water and replaced by other more damaging cations such as sodium. As discussed in Section 4.5, even though the injected water may contain an ample supply of divalent cations, the waters that interact with the formation may not.

The differences in critical concentration between monovalent and divalent ions is certainly due to differences in surface charge. Riese [32] has shown that even very small quantities of Ca^{++} in water reduce the surface charge on both kaolinite and quartz markedly as compared to concentrations of Na^+.

Since the surface charge is decreased in the presence of multivalent ions, one would also expect the zeta potential to be similarly reduced. This trend is shown by Fig. 4.15. As expected, trivalent ions more effectively reduce the zeta potential than equivalent amounts of divalent ions. Divalent or trivalent cations are not, however, good candidates for use in treatment fluids since they tend to be precipitated. Some well treatments to fix clays do, however, use multivalent ions.

It is perhaps worth noting that there is a remarkable correspondence between the zeta potential of clays in the presence of monovalent ions and the critical salt concentrations obtained by observing permeability reduction in Berea sandstone (Table 4.2). Based on the trends shown by Fig. 4.15, one would expect that of the monovalent ions, cesium should exhibit the smaller critical salt concentration. This is precisely what is observed.

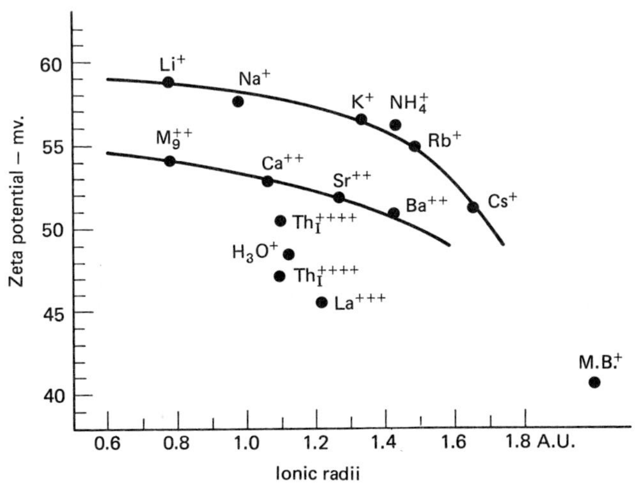

Figure 4.15 Dependence of the zeta potential of colloidal clay suspensions on charge and ionic radii of counterions [33]. The dot labelled M.B.$^+$ represents a cation of methylene blue. [From H. Jenny and R. F. Reitmeier, "Ionic Exchange in Relation to the Stability of Colloidal Systems," *J. Phys. Chem.*, *39* (1935), 593.]

It seems reasonable to ask why cesium chloride is not used in well treatment fluids rather than potassium chloride. The answer is purely an economic one. Potassium chloride costs less.

In conclusion, the following observations apply when fresh water is injected into formations:

1. After injecting large volumes of NaCl, the sudden injection of fresh water invariably results in substantial formation damage whatever the fluid velocity. This is understood in terms of the repulsion particles dislodged from the pore surface experience when they approach the wall (see Fig. 4.4). Particle trapping is then by the pore clogging mode.

2. After injecting large volumes of $CaCl_2$, sudden injection of fresh water at low velocities is nondamaging [25] because Ca^{++} reduces the surface charge.

3. Backflushing a freshwater damaged core with a concentrated salt solution removes the damage. Particles are displaced from the pore throats and are reattached to the pore walls where they are nondamaging. It should be noted that occasionally the permeability of a core may increase with pore volumes of concentrated salt solution injected because pore filling clays may be dislodged and reattached in positions that allow easier fluid passage.

4. After injection of large volumes of NaCl solution, injection of water with slowly decreasing concentrations of NaCl is often not damaging even when fresh water is ultimately injected [24]. One must understand ion-exchange processes to understand this result. The difference between slowly and suddenly decreasing the NaCl concentration is that suddenly decreasing the concentration depletes the small quantities of Ca^{++} normally present because of the small solubility of calcite ($CaCO_3$). Thus, it is the concentration of divalent ions which is the key variable in preventing fresh water damage.

The impact of fresh water damage is most relevant when considering waterflooding or waste disposal in deep wells. Here, it may not be economically feasible to maintain the desired level of electrolyte. Stimulation fluids, however, always contain an ample electrolyte concentration to minimize formation damage.

The Effect of pH on Fresh Water Shock

Experiments have shown that low pH solutions are nondamaging even if they do not contain added electrolytes [24, 34, 35]. Kia et al. [35], for example, report that acid solutions ($pH < 4.5$) do not materially damage Berea cores. Minerals containing multivalent cations (Fe^{++}, Ca^{++}, Mg^{++}, etc.) have a higher solubility in acidic solutions and these cations also adsorb, thereby reducing surface charge. It is for this reason that acidic solutions are less damaging than near neutral or basic solutions. It has recently been suggested that hydrochloric acid leaches aluminum from the clay lattices, thereby reducing the ion-exchange capacity of clays [38].

These observations imply that special clay stabilizing compounds may not be required when stimulating with acids. This matter is discussed in some detail in Chapter 16.

Sydansk [34] has proposed a high pH, potassium hydroxide treatment for stabilizing clays. A long soaking time is recommended with this treatment.

In-Situ Emulsification

An emulsion is generally regarded as a dispersion of liquid drops in a second, immiscible liquid. The drops most often range in size from 1 to 10 μm (see Chap. 3). Considerable mechanical energy and low interfacial tensions are generally required to create drops as small as 1 μm. This sort of emulsion is often encountered when oil and water are simultaneously produced.

When stable oil and water emulsions are produced and when production rates fall below expectation, then it is natural to suspect that these same emulsions exist within the interstices of the rock and, because of their high effective viscosity (Chap. 3), are flowing at a reduced rate. It is, however, unlikely that the same emulsions flowing into the production tank exist in situ. Emulsion drop sizes are as large as or larger than the pore sizes and thus cannot be expected to flow freely through the pores.

Although it seems unlikely that a dispersion of drops will exist within the pore interstices, there is a mechanism whereby distinct drops of nonwetting phase separated by lamella of wetting phase are created thereby leading to increased pressure drops for a constant production rate.

Figure 4.16 depicts an idealized pore. It is composed of two parts: a body that is a capillary tube of radius R_c and a pore throat characterized by two radii, R_T and R_1. Present within the pore are both wetting and nonwetting phases as shown. Because the interfaces are curved there is a capillary pressure. In the pore body

$$p_{cap} = p_N - p_W^B = \frac{\gamma}{R_c} \tag{4.17}$$

where p_N and p_W are the pressures in the nonwetting and wetting phases, re-

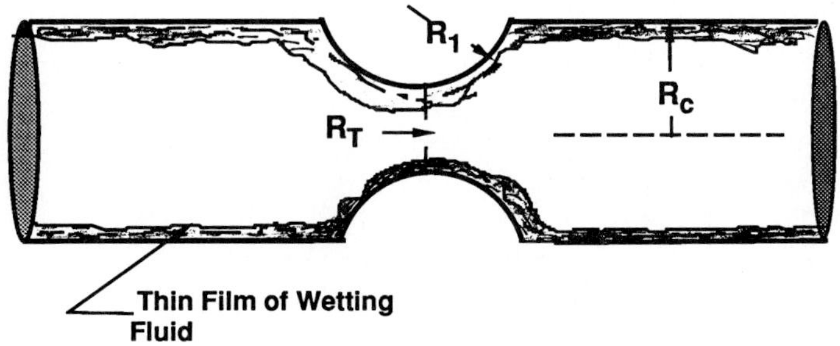

Figure 4.16 A highly idealized single pore filled with nonwetting fluid, except for a thin wetting-phase film coating the pore walls.

spectively. The interfacial tension is γ and $p_N - p_W^B$ is the capillary pressure in the pore body.

The capillary pressure in the pore throat is

$$p_{cap} = p_N - p_W^T = \gamma\left(\frac{1}{R_T} - \frac{1}{R_1}\right) \tag{4.18}$$

Now, if the nonwetting phase is stationary or flowing slowly, then p_N is approximately the same through the entire pore. For certain special geometries to be defined, $p_W^B > p_W^T$ and the wetting phase will tend to flow from the body of the pore into the pore throat. Such pores are called *generating sites* because, as shown by the time sequence of events depicted by Fig. 4.17, lamella are generated by virtue of this flow into the pore throats and, if these lamella are stable when displaced from the pore throat, the result is a train of discrete nonwetting phase droplets separated by thin films of wetting phase. These films each increase the pressure drop slightly, but if a sufficient number of them exist at a given time their total contribution can be quite significant.

The concept of a generating site as a mechanism for producing residual oil drops is apparently first due to Roof [39]. It has been extended by Mohanty et al. [40] and is now envisioned as a mechanism for forming foam in situ [41, 42].

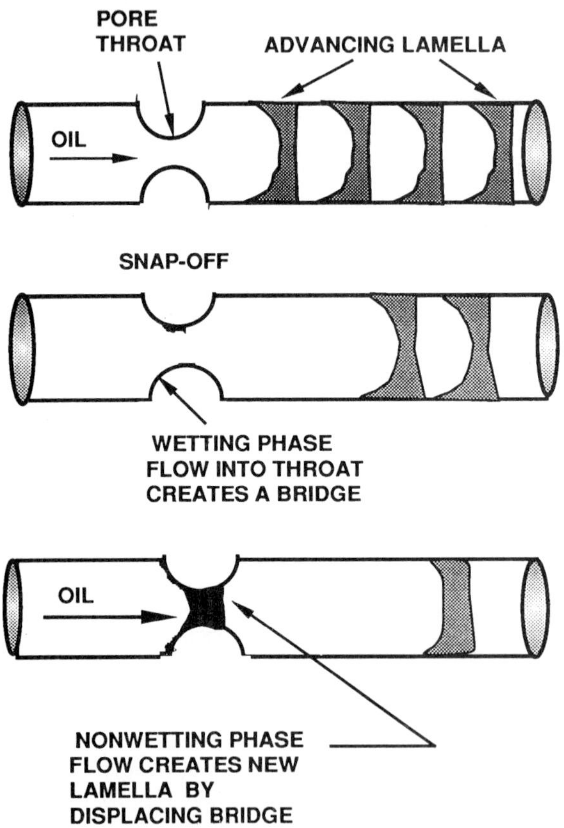

Figure 4.17 The same idealized capillary shown at different times depicting the sequence of events leading to formation of discrete nonwetting phase droplets.

For the idealized pore depicted by Fig. 4.16, a generating site is thus a pore for which

$$\frac{1}{R_T} > \frac{1}{R_1} + \frac{1}{R_c} \tag{4.19}$$

since curvatures $1/R_1$ and $1/R_c$ tend to draw wetting phase from the capillary and $1/R_T$ has the opposite effect.

To a very crude approximation, the flow of the wetting phase into the capillary can be imagined to be the same as a laminar flow between parallel walls and, as given in Chapter 8, the average velocity is

$$\bar{v}_x = \frac{\Delta p}{L} \frac{w^2}{3\mu} \tag{4.20}$$

where Δp is the pressure driving force and w is the wetting phase film thickness. L is a characteristic distance over which the fluid moves under the influence of Δp. Here, with an accuracy sufficient for qualitative purposes, take $L \cong R_1$ (more precise analyses have been presented [43, 44]).

Example 4.5 Rate of Lamella Generation

Consider the idealized pore shown by Fig. 4.16 with $R_T = 0.3\ \mu\text{m}$, $R_1 = 5\ \mu\text{m}$, and $R_c = 3\ \mu\text{m}$. If $10^{-15}\ \text{m}^3$ of wetting fluid are required to obtain snap-off (see Fig. 4.17), how many lamella are generated per unit of time? If the nonwetting phase velocity is 1×10^{-2} m/sec within the body of the capillary, how long are the discrete nonwetting phase drops? Take the wetting phase film thickness to be 0.05 μm, its viscosity to be 10^{-3} kg/m-sec, and the interfacial tension to be 2×10^{-2} N/m.

Solution The force driving wetting fluid into the throat from the body is

$$p_W^B - p_W^T = \gamma \left\{ \frac{1}{R_T} - \frac{1}{R_1} - \frac{1}{R_C} \right\}$$

or

$$\Delta p = 2 \times 10^{-2} \frac{\text{N}}{\text{m}} \left\{ \frac{1}{3 \times 10^{-7}} - \frac{1}{5 \times 10^{-6}} - \frac{1}{3 \times 10^{-6}} \right\} \cdot \frac{1}{\text{m}}$$

and

$$\Delta p = 5.6 \times 10^4\ \text{Pa}$$

The average wetting phase velocity is represented by Eq. (4.20) as

$$\bar{v}_x = \left(\frac{5.6 \times 10^4}{5 \times 10^{-6}} \right) \left(\frac{(5 \times 10^{-8})^2}{3(10^{-3})} \right) = 9.3 \times 10^{-3} \frac{\text{m}}{\text{sec}}$$

The volume flow rate into the throat is $2(2\pi R_C w)\bar{v}_x$, where $2\pi R_C w$ is the area available for flow into the pore throat from one side. The extra factor of 2 accounts for flow from both sides. Thus,

$$\text{time to snap-off} = \frac{10^{-15}}{2(9.3 \times 10^{-3})(2\pi)(5 \times 10^{-8})(3 \times 10^{-6})} = 5.7 \times 10^{-2}\ \text{sec}$$

and the length of a drop is

$$(1 \times 10^{-2})(5.7 \times 10^{-2}) = 5.7 \times 10^{-4}\ \text{m}$$

This example points out some interesting features that are pertinent to understanding the formation of blockages by this mechanism. The drop length found in Example 4.5 is not unreasonable. Residual oil blobs are often 5 to 10 times the size of a sand grain [45]. This length is, as will be observed, too long to create enough lamella to effectively decrease production. Equation (4.19) indicates that the most effective generating sites are those where very small pores empty into large ones. Thus, at an interface between two layers, one of low permeability with one of high permeability, the generation of drops will be rapid. Similarly, a vugular carbonate, which exhibits a significant contrast in size between the radius of the pore throat and the radius of the pore body, should also contain effective generating sites.

If a large number of lamella are generated, they would contribute to the pressure drop. Here, the pressure drop will be estimated based entirely on capillary forces although the fluid viscosity and surface tension gradients will both contribute [46]. The best approach would be to take into account the difference between advancing and receding contact angles [47, 48]; however, these depend on the fluid velocity. Instead, the moving interfaces will be assumed to take the form depicted by Fig. 4.17, flat at the leading surface and hemispherical at the trailing surface. The pressure drop across lamellae is therefore given by

$$\Delta p = \frac{2\gamma}{R_C} \tag{4.21}$$

It is not difficult to see that this is a small value but if n is the number of lamella per unit length, then

$$\frac{\Delta p_n}{L} = \frac{2n\gamma}{R_C} \tag{4.22}$$

This pressure drop over a length (L) can be compared to the viscous pressure drop over the same length as given by Darcy's law

$$\frac{\Delta p}{L} = \frac{\mu u}{k_r} \tag{4.23}$$

If Δp and Δp_n are of comparable size, then the pressure drop contributed by lamella is quite important. Of course, there may be cases for which $\Delta p_n \gg \Delta p$ and oil production will be severely restricted.

Example 4.6 The Number of Lamella and Their Significance

A porous medium has a relative permeability to oil, the nonwetting phase, of 10 mD. If the viscosity of the oil is 2×10^{-3} kg/m-sec and if the flux of oil is 2×10^{-2} m/sec, how many lamella per unit length must be created so that the oil flow into the wellbore is slowed significantly? Take the interfacial tension to be 2×10^{-2} N/m and the radius of the capillary to be 3 μm.

Solution The viscous pressure drop is given by Eq. (4.23) as

$$\frac{\Delta p}{L} = \frac{\left(2 \times 10^{-3} \dfrac{\text{kg}}{\text{m-sec}}\right)\left(2 \times 10^{-2} \dfrac{\text{m}}{\text{sec}}\right)}{(10 \text{ mD})\left(9.87 \times 10^{-16} \dfrac{\text{m}^2}{\text{mD}}\right)} = 4.05 \times 10^9 \text{ Pa}$$

Equating this to the $\Delta p_n/L$ created by the lamella and solving for n

$$n = \frac{(4.05 \times 10^9)(3 \times 10^{-6})}{(2)(2 \times 10^{-2})} = 3 \times 10^5 \text{ m}^{-1}$$

or the length of a drop should be about 3.3×10^{-6} μm. If many fewer lamella are created, the in-situ emulsification is an unimportant process.

For this example, the density of lamella must be the order of 10^6 m^{-1} to substantially increase the pressure drop. By comparing the results of Examples 4.5 and 4.6, several conclusions are evident. For in-situ emulsification to be significant, the following three conditions must be simultaneously satisfied:

1. Effective generating sites must be present in high density and scattered throughout the near wellbore region;
2. The wetting phase must be continuous; that is, thin films must be available to conduct the wetting phase into the pore throats; and
3. The lamella when produced must be stable.

Lamella are always produced when water and oil both flow, but they will normally rapidly thin and break under the action of van der Waals forces. If, however, surfactants are present, then stable lamella may form which may propagate for some distance through porous media.

It becomes evident that this type of damage should be rare since the three conditions must be simultaneously satisfied to produce a significant blockage. Most produced emulsions are formed upon entry of oil and water into the wellbore or attending their flow in the tubing. Thus, the presence of emulsions in surface storage tanks does not necessarily signal their existence in the near wellbore region. However, from time to time all three conditions will be simultaneously met. If the source of the stabilizing surfactant is the crude oil itself, then little can be done to prevent the formation of stable lamella. In carbonate formations, acidization can change the pore structure so as to prevent generation in the near wellbore region. It is also possible to destroy the integrity of the wetting films by neutralizing the surface charge, which is one of the primary stabilizing factors of thin films [49, 50]. This can be accomplished temporarily by injecting an aqueous solution containing an ionic surfactant having a charge opposite that of the surface charge. In limestones, this would generally be an anionic one whereas a cationic surfactant would generally be used in a sandstone (see Table 1.2). These surfactants will effectively alter the water wettability of a formation so that not only will the water films be disrupted and generating sites will not produce lamella, but the relative permeability to oil will also decrease. If the treatment does prevent in-situ emulsification, the production may increase despite the increase in oil wettability.

A surfactant treatment is temporary since the injected surfactant will desorb from the surface and be produced. Thus, the use of a surfactant to prevent in-situ emulsification is most effective when the surfactant responsible for stabilizing the emulsion has been inadvertently introduced into the formation.

The main point to be made here is that three special conditions must all be simultaneously satisfied before in-situ emulsification becomes a damaging process. It is suspected, but not known for certain, that this is rarely the source of formation damage. Certainly, it is less frequent than is imagined.

4.4 CHEMICAL PROCESSES THAT PRODUCE FORMATION DAMAGE

Most significant chemical processes that occur in porous media result from fluid/rock interaction. Generally, interactions between incompatible fluids are unimportant. This is because the mixing of two fluids in porous media is difficult to achieve. The exception to this rule is the region surrounding a production well. In this region, fluids moving along different streamlines are forcibly mixed together. Thus, although waterflooding with a fluid incompatible with the formation water may not create problems at the injection wells, it will certainly do so at the production wells.

The importance of fluid/rock interaction stems primarily from the ease with which these two phases mix. The interaction between two phases moving at different velocities is sometimes termed chromatography because fronts of various composition form and move through the porous medium. To appreciate the significance of these fronts to formation damage, it is first necessary to investigate their origin.

Frontal Movements in Porous Media

Linear systems. A fluid that enters a porous formation having a temperature or composition different from the resident fluid will, in general, have its composition or temperature modified as a result of the interaction with the reservoir rock. The modified fluid, which differs in some way from either the resident or the injected fluids, may then have a composition that will allow precipitation or it may have properties that will promote clay damage. Thus, a clear, nondamaging completion fluid, for example, may filter into the formation, contact the rock, and revert to a modified fluid that contains precipitate and is damaging. It is, therefore, important to understand how invading fluids are altered by the exchange of matter and heat with the rock and then to be able to properly assess the potential danger with regard to formation damage.

To quantify the exchange of heat and matter between the rock and the fluid, we shall ascribe to both the stationary and the mobile phases a capacity factor having as dimensions the amount of exchangeable quantity per unit of pore volume. The capacity of the rock will be designated by A and that of the fluid as c. Table 4.3 provides a listing of some of the capacity factors that will be of interest to us.

If, for example, we wish to calculate A for thermal energy, consider a cubic meter of formation (pores plus rock) containing $(1 - \phi)$ cubic meters of rock and ϕ cubic meters of pore volume. The heat contained in the *rock per unit of pore volume* is

$$A = \frac{c_{\mathrm{p}}\rho_{\mathrm{R}}T(1 - \phi)}{\phi} \tag{4.24}$$

which is the result given in Table 4.3.

It should be stressed that the capacity of the solid phase may depend upon the capacity of the fluid phase. For example, the amount of surfactant adsorbed onto the reservoir rock will vary with the solution concentration. An even simpler example is to note that the amount of heat stored in the rock depends on the temperature of the fluid.

One fundamental assumption invoked in all of the calculations described here is that the fluid at each point is in local equilibrium with the adjacent rock. This assumption does not preclude the variation of fluid composition and temperature (and hence, variation of the rock composition and temperature) from point to point within a porous medium. It simply assures that equilibrium prevails at each point. Several types of equilibria will be of interest to us. We will be concerned with ion exchange, surfactant and inhibitor adsorption, and many other modes of interaction between the fluid and the rock. To characterize the interaction, it is often necessary to have experimental data. For example, shown in

TABLE 4.3 Capacity Factors for Fluid/Solid Interactions (All Expressed per Unit of Pore Volume)

Process	Solid capacity A	Fluid capacity c
Ion exchange	$Q_v\theta_i$	$c_{\mathrm{an}}y_i$
Heat transfer	$\rho_{\mathrm{R}}c_{\mathrm{p_R}}\dfrac{1 - \phi}{\phi}T$	$\rho c_{\mathrm{p}}T$
Surfactant adsorption	$\rho_{\mathrm{R}}\dfrac{1 - \phi}{\phi}\Gamma$	c_{s}
Polymer adsorption	$\rho_{\mathrm{R}}\dfrac{1 - \phi}{\phi}\Gamma$	c_{p}
Precipitation-dissolution	$\overline{c}$	c

Notation:
Q_v = System cation-exchange capacity (kg eq/m^3)
Γ = Surfactant adsorption (kg/kg of rock)
c_{s} = Surfactant solution concentration (kg/m^3)
c_{p} = Polymer solution concentration (kg/m^3)
c_{aN} = Total counterion concentration (kg eq/m^3)
$\overline{c}$ = Solid concentration (kg moles/m^3)
c = Liquid concentration (kg moles/m^3)

Fig. 4.18 is a typical adsorption isotherm that has been determined experimentally. Note that the amount adsorbed per unit mass of rock, Γ, is a complex function of c_s, the solution concentration. The rock capacity is related to Γ as shown in Table 4.3.

Another example of interest is ion exchange. From Chapter 1 we have

$$K_{ij} = \frac{\theta_j^{z_i} c_i^{z_j}}{\theta_i^{z_j} c_j^{z_i}} \tag{1.10}$$

This equation defines ion-exchange equilibrium and, as noted in Chapter 1, the K_{ij} are experimentally determined constants. For a solution containing both calcium ($i \equiv Ca^{++}$, $z_i = +2$), and sodium ($j \equiv Na^+$, $z_j = 1$) cations, we have

$$K_{Ca,Na} = \frac{\theta_{Na}^2 c_{Ca}}{\theta_{Ca} c_{Na}^2} \tag{4.25}$$

The capacity for calcium ions of the rock is clearly

$$A_{Ca} = \theta_{Ca} Q_v \tag{4.26}$$

where Q_v is the system ion-exchange capacity and that of the solution is simply c_{Ca}.

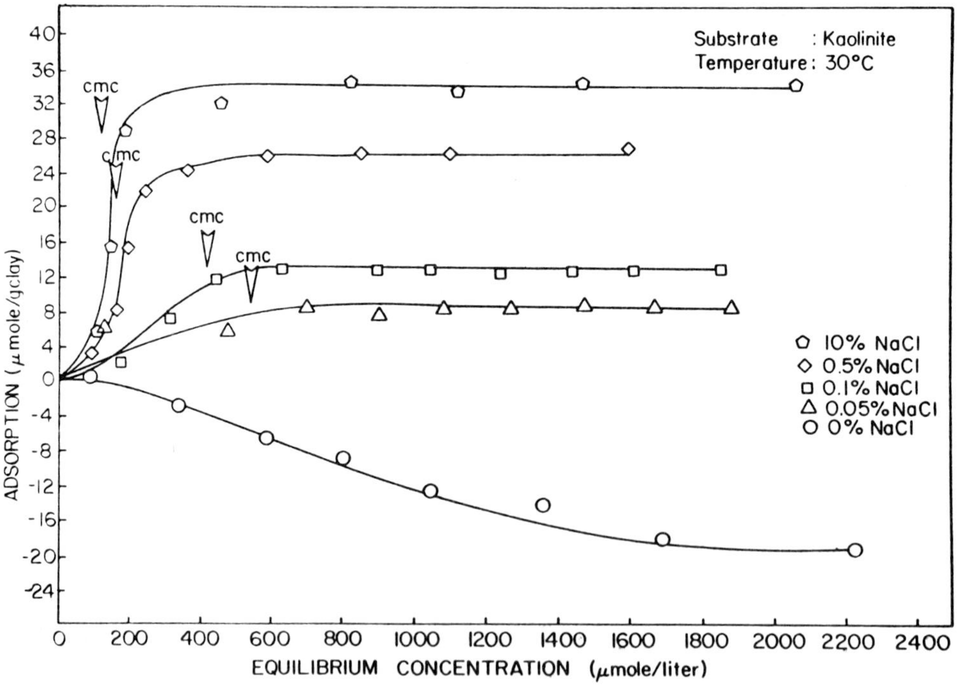

Figure 4.18 Adsorption 4-phenyl dodecyl benzene sulfonate on kaolinite at various salt concentrations [51].

Assuming that sodium and calcium are the only ions present, we note that

$$\theta_{Ca} + \theta_{Na} = 1 \tag{4.27}$$

When chemicals in the injected fluid interact with the rock, then zones having different compositions or temperatures are formed and these zones are propagated through the porous medium at a rate dictated by the capacities given in Table 4.3. Consider the zones shown in Fig. 4.19. One important feature is that the intermediate zone has a composition and/or a temperature that is different from either the injected or the resident fluid. It is within this zone that the composition is determined by the interaction of the fluid with the rock and that formation damage is likely to occur. For example, the injected fluid may be clear and free of suspended solids but the modified fluid may contain precipitates. Thus, it is important to assess the composition and temperature of the fluid in the intermediate zone.

The frontal velocities, v_f^I and v_f^{II}, shown by Fig. 4.19 are important variables. v_f^I is the velocity of the fluid. If the flux is u, then

$$v_f^I = \frac{u}{\phi} \tag{4.28}$$

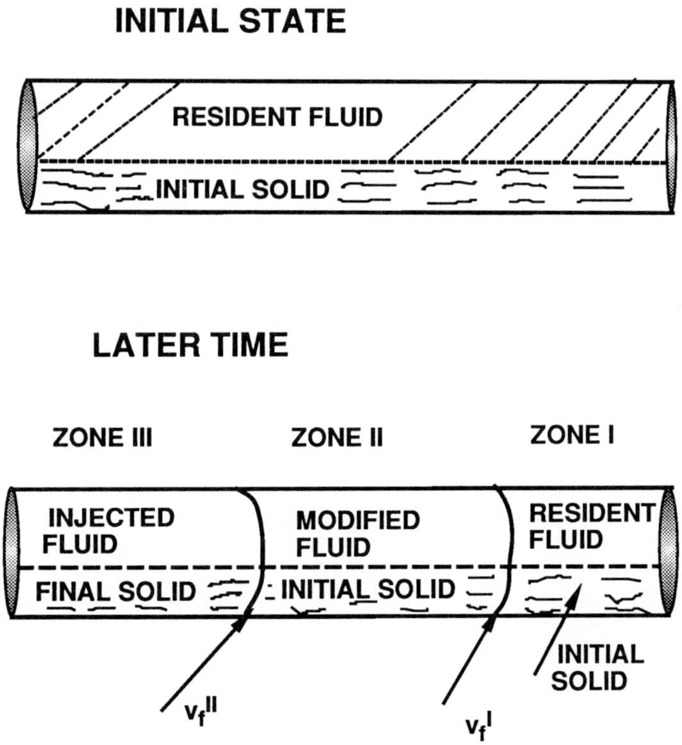

Figure 4.19 Development of zones as a result of fluid interaction with the formation rock.

Equation (4.28) shows that here the frontal velocity is equivalent to the fluid velocity. *This is only possible if the composition and temperature of the solid phase within Zone II is the same as that within Zone I.* Any interaction (exchange of energy or mass) between the stationary and flowing phases will retard the movement of that phase. Thus, we know that all frontal velocities must satisfy the inequality

$$\frac{v_f}{u/\phi} \leq 1 \tag{4.29}$$

with the equal sign applying if, and only if, there is *no* interaction between the stationary and mobile phases. This can be understood as follows. A molecule adsorbed on the rock does not progress through the porous medium. Hence, its average velocity is the fluid velocity multiplied by the fraction of the time the particle is flowing with the fluid. This fraction can obviously not exceed unity.

The frontal velocity (v_f^{II}) between Zones II and III is observed to be less than that of the fluid. Thus, Inequality (4.29) applies and there must as a consequence be some exchange between the flowing and stationary phases. To calculate velocity v_f^{II}, the capacities characterizing both the injected fluid/rock equilibrium (local equilibrium) and those for the modified fluid/rock interaction are required. If there are only three zones as shown in Fig. 4.19, then the rock composition and temperature in Zone II must be the same as that in Zone I. This is true because v_f^I is moving with the injected fluid velocity and this fact fixes the state of the stationary phase in Zone II. The composition of the fluid in Zone II can, therefore, be determined because the rock and fluid are in equilibrium. The problem then remains to calculate v_f^{II}. Let us consider this velocity by examining Fig. 4.20. A balance of any one of the components about the small interval Δx shown in this figure yields

Amount in during time Δt: $uc^i \,\Delta t$

Amount out during time Δt: $uc \,\Delta t$

Accumulation during time Δt: $v_f^{II}\,\Delta t\,\phi(A^i + c^i) + (\Delta x - v_f^{II}\,\Delta t)\phi(A + c)$
$- \Delta x\,\phi(A + c)$

Thus, since in–out = accumulation, we can write

$$\Delta t\,(uc^i - uc) = v_f^{II}\,\Delta t\,\phi\{A^i + c^i - A - c\} \tag{4.30}$$

Simplifying

$$v_f^{II} = \frac{u/\phi}{1 + \dfrac{\Delta A}{\Delta c}} \tag{4.31}$$

where $\Delta A = A^i - A$ and $\Delta c = c^i - c$

Equation (4.31) shows the case where the frontal velocity (v_f^{II}) is less than the fluid velocity (u/ϕ), because the capacity of the solid changes across the wave. Furthermore, we have $\Delta A/\Delta c \geq 0$ so that Inequality (4.29) is always valid.

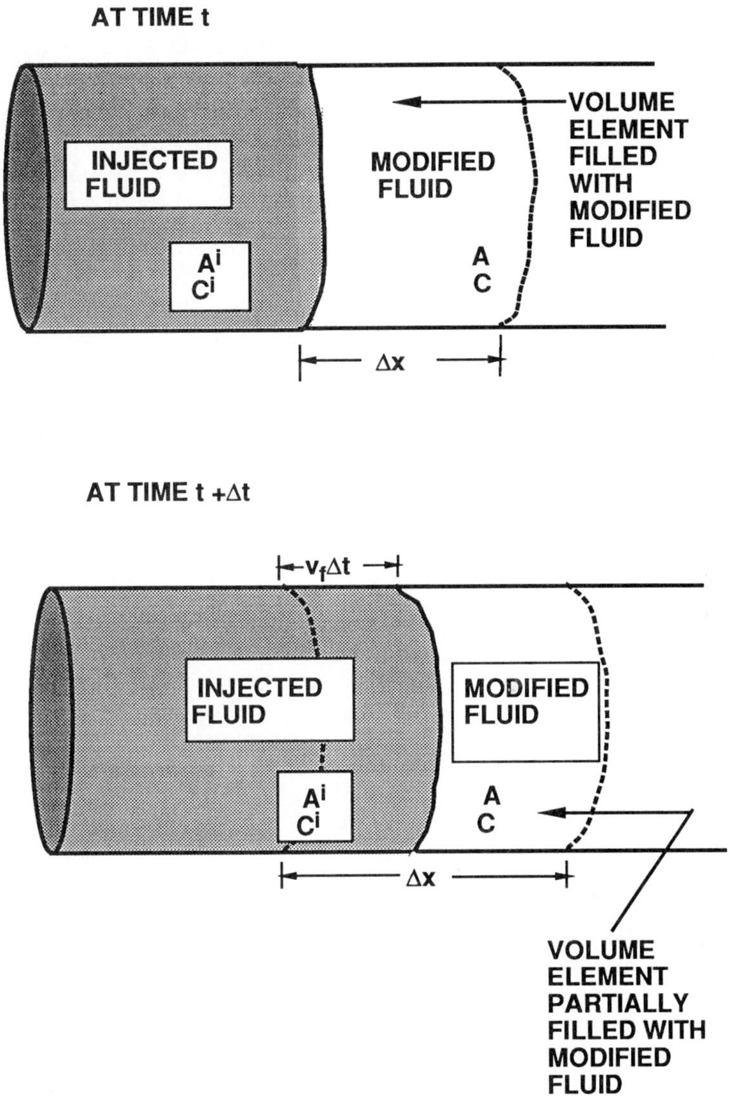

Figure 4.20 Injected fluid shown invading a region first occupied by modified fluid.

As the number of components interacting with the rock increases, the number of intermediate zones will also increase. For example, when an ion-exchange process includes three different cations rather than just two, then an additional intermediate zone may form. The same principles applied in solving for the composition and the frontal velocity when there is a single intermediate zone determine the movement of multiple zones, but the calculations are more complex [52, 53]. We will, therefore, restrict our attention here to problems having a single intermediate zone.

In practical applications, the existence of a single intermediate zone can be assumed, but this must be confirmed to assure that it is true. The approach is, as is often the case in many engineering problems, one of trial and error. Assuming a single intermediate zone, velocity v_f^{II} can be determined by applying Eq. (4.31) to any one of the interacting components. The entire calculation must then be repeated for each component present. If all the v_f^{II} so determined are the same, then the front is called *coherent* and the existence of a single intermediate zone is confirmed. If, however, the frontal velocities of all the interacting components are not the same, then other intermediate zones will appear.

The method for calculating the intermediate zone composition and frontal velocity assumes that the front is absolutely sharp. Such a frontal shape, called a *shock front*, is an idealization that will be applied here to simplify the calculations. In every case, diffusion and dispersion will tend to make the frontal zone diffuse. In some cases, the interaction with the solid will cause the front to become diffuse. Spreading waves are often found in ion exchange and other chromatographic processes [52, 53]. To obtain precise results, the influence of adsorptive spreading must be taken into account; however, to ascertain the potential for formation damage, the method of calculating fluid/rock interactions developed here is adequate.

Example 4.7 Cooling of a Formation

A linear core of rock 3 m long is heated together with its fluid contents to a temperature of 200°C. If a fluid having a temperature of 50°C is injected into the core so that the volumetric flux is 0.1 m/hr, how long will it take for fluid having a temperature equal to that of the injected fluid to emerge from the core? What volume of injected fluid per area core is heated to 200°C?

The porosity of the rock is 0.15, its density is 2700 kg/m^3, and its heat capacity is 0.922 kJ/kg-°C. The density of the injected fluid is 900 kg/m^3 and its heat capacity is 3.77 kJ/kg-°C.

Solution To calculate the frontal velocities, we note that three regions (at least) will develop. These are shown in Fig. 4.19. In this case, Zone I will contain formation fluid at its original condition, namely, 200°C. Since the frontal velocity between Zone II and Zone I is the fluid velocity, then the rock temperature *must* be the same in both zones. Thus, Zone II contains injected fluid at 200°C. This is a modified fluid in the sense that prior to injection, all of the injected fluid had a temperature of 50°C. As a result of interaction with the reservoir rock, some of it is heated to 200°C. This represents a modification of the injected fluid, which could lead to formation damage if any of the components in the injected fluid become less soluble with increasing temperature and tend to precipitate.

To determine the frontal velocity of the Zone III–Zone II interface, the capacities are required. In Zone II

$$A_{II} = \frac{c_{pR}\rho_R(T)(1-\phi)}{\phi} = \frac{(0.922)(2700)(200)(0.85)}{(0.15)} = 2.82 \times 10^6 \frac{\text{kJ}}{\text{m}^3}$$

$$c_{II} = \rho c_p T = 3.77(900)(200) = 6.79 \times 10^5 \frac{\text{kJ}}{\text{m}^3}$$

Since temperature differences will appear in the final calculation, it is not necessary to use absolute temperatures; but one can, of course, do so without changing the final result. For Zone III

$$A_{III} = \frac{0.922(2700)(50)(0.85)}{0.15} = 7.05 \times 10^5 \frac{kJ}{m^3}$$

$$c_{III} = 3.77(900)(50) = 1.7 \times 10^5 \frac{kJ}{m^3}$$

Then from Eq. (4.31)

$$v_f^{II} = \frac{\dfrac{u}{\phi}}{1 + \dfrac{A_{III} - A_{II}}{c_{III} - c_{II}}} = 0.129 \frac{m}{hr}$$

Time required for the cold fluid to emerge is, therefore,

$$t = \frac{L}{v_f^{II}} = \frac{3}{0.129} = 23.2 \text{ hr}$$

The time required for the Zone I–Zone II interface to emerge is

$$t = \frac{L}{v_f^{I}} = \frac{L}{\dfrac{u}{\phi}} = \frac{3}{\dfrac{0.1}{0.15}} = 4.5 \text{ hr}$$

Thus, all of the fluid which exits from the core between 4.5 and 23.2 hours is injected fluid heated to 200°C. The injected volume/area is

$$\frac{qt}{A} = ut = 0.1(23.2 - 4.5) = 1.87 \frac{m^3}{m^2}$$

This example shows that temperature waves travel at a retarded rate through porous media. The retardation is a result of heat exchange between the fluid and the rock. Here, 1.87 m^3 of injected fluid have been heated to 200°F and the rock has been cooled to 50°C.

Example 4.8 Ion-Exchange Waves in Porous Media

A rock core 3 m in length is filled with a formation water containing 0.5 kg eq/m^3 of sodium cation in the form of sodium chloride (NaCl) and 0.02 kg eq/m^3 of calcium cation as calcium chloride ($CaCl_2$). A water containing 1 kg eq/m^3 of sodium as sodium carbonate (Na_2CO_3) is injected into the core. If the volumetric flux is 0.1 m/hr, how many pore volumes of injected fluid will be required before water having the injected composition will emerge from the core? What is the composition of the modified zone? The porosity of the core is 15%.

The selectivity coefficient characteristic of the minerals composing the rock is known to be

$$K_{Ca,Na} = 0.2$$

and the system ion-exchange capacity is

$$Q_v = 1 \text{ kg eq/m}^3$$

Solution Assume that a single intermediate zone is created by the fluid rock interaction. The composition of cations adhering to the rock in Zone I and in Zone II must be the same because the frontal velocity between these two zones is the same as the fluid velocity. First, determine the adsorbed cation composition within Zone I. Zone I contains formation water. Thus,

$$c_{Na} = 0.5 \text{ kg eq/m}^3$$

$$c_{Ca} = 0.02 \text{ kg eq/m}^3$$

The corresponding adsorbed concentrations are given by Eq. (4.25)

$$\frac{\theta_{Na}^2 (0.2)}{(0.5)^2 \theta_{Ca}} = 0.2$$

Since there are only two cations present, $\theta_{Ca} + \theta_{Na} = 1$. Solving yields $\theta_{Na} = 0.766$ and $\theta_{Ca} = 0.234$. Although only 1% of solution cations are calcium, 23.4% of the surface is occupied by calcium. This shows the selectivity of many surfaces for divalent ions in preference to monovalent ones.

In Zone II the anion concentration must be the same as that of the injected fluid since anions do not interact with the rock. (*Note:* This is an assumption. In some cases anions may be adsorbed to some extent by reservoir rock.) Since there is 1 kg eq/m^3 of carbonate in the injected fluid, there will also be 1 kg eq/m^3 of carbonate in Zone II. Thus in Zone II, $c_{Ca} + c_{Na} = 1$ kg eq/m^3 to satisfy electrical neutrality, and the surface compositions are the same as Zone I. Equation (4.25) is written

$$\frac{(0.766)^2 c_{Ca}}{c_{Na}^2 (0.234)} = 0.2$$

Solving, one finds

$$c_{Na} = 0.93 \text{ kg eq/m}^3 \quad \text{and} \quad c_{Ca} = 0.07 \text{ kg eq/m}^3$$

The cations are present as carbonates in Zone II. Here, it is to be emphasized that the fluid in the intermediate zone is a modified injected fluid. It is different in composition from the injected fluid and in this case because calcium carbonate will precipitate, the modified fluid is damaging even though the formation or the injected fluids are not.

To complete the analysis, the frontal velocity v_f^{II} must be calculated. In Zone II, the sodium cation capacities are

$$A_{Na}^{II} = Q_v \theta_{Na} = (1)(0.766) = 0.766 \text{ kg eq/m}^3$$

$$c_{Na}^{III} = 0.93 \text{ kg eq/m}^3$$

In Zone III, the injected fluid only contains sodium cation; therefore, θ_{Na} is 1, and

$$A_{Na}^{III} = (1)(1) = 1 \text{ kg eq/m}^3$$

$$c_{Na}^{III} = 1 \text{ kg eq/m}^3$$

The frontal velocity is given by

$$v_f^{II} = \frac{\dfrac{u}{\phi}}{1 + \dfrac{A_{Na}^{III} - A_{Na}^{II}}{c_{Na}^{III} - c_{Na}^{III}}} = \frac{\dfrac{0.1}{0.15}}{1 + \dfrac{1 - 0.766}{1 - 0.93}} = 0.15 \text{ m/hr}$$

Since $$\frac{u}{\phi} = \frac{0.1}{0.15} = 0.667 \text{ m/hr}$$

The carbonate anion, which is essentially a tracer (it does not interact with the rock) will emerge from the core after 3/0.667 = 4.5 hr. This will be followed by the modified fluid for the next 3/0.15 = 20 hr. At this time the injected fluid will emerge. The reader must be impressed with the large volume of modified fluid created by the fluid/rock interaction, especially since this is a potentially damaging fluid.

Note v_f^{II} has been calculated using the sodium cation capacities. The same answer must result if calcium is considered. For calcium, $A_{Ca}^{II} = 0.234$ and $A_{Ca}^{III} = 0$. Also, $c_{Ca}^{II} = 0.07$ and $c_{Ca}^{III} = 0.0$; thus

$$v_f^{II} = \frac{\dfrac{0.1}{0.15}}{1 + \dfrac{0.234}{0.07}} = 0.15 \text{ m/hr}$$

which is the same as found before. In this case, the comparison is trivial and is always true if only two cations are present. If a third component is added, then it is likely that more than one intermediate zone will develop.

Example 4.8 is interesting in that it shows that sodium carbonate (Na_2CO_3), which is quite soluble, can be injected into a porous medium and produce a fluid that is a blend of sodium and calcium carbonate. Calcium carbonate ($CaCO_3$) may not be soluble to the extent shown in Example 4.8; thus, a precipitate may form which could clog the pores and thereby damage the formation. The solubility of $CaCO_3$ is shown as a function of temperature in Fig. 4.21. The $CaCO_3$ concentration in Zone II of Example 4.8 is 0.07 kg eq/m^3. This is equivalent to 3500 mg/liter, which exceeds the solubility limit at all temperatures shown. Thus, $CaCO_3$ will precipitate in the intermediate zone and tend to clog the pores.

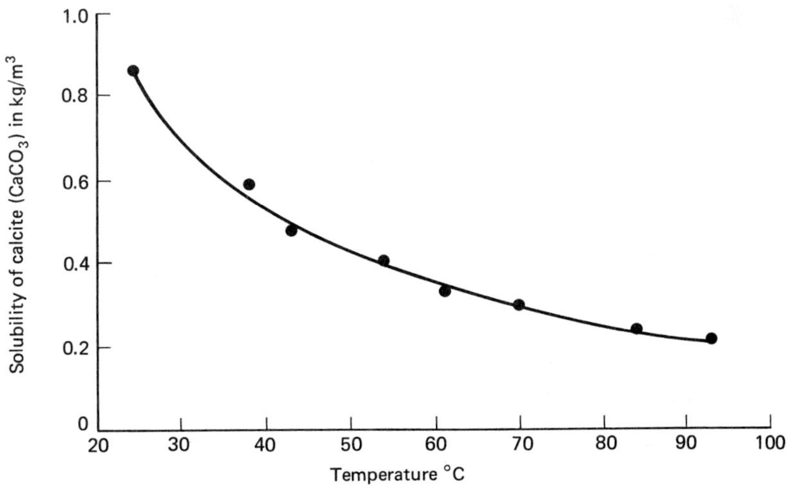

Figure 4.21 Solubility of $CaCO_3$ in pure water at 1 bar CO_2 partial pressure [55].

Radial systems. The injection of fluid into the formation radially from a wellbore will result in the formation of different zones that will propagate radially outward from the wellbore. If the fluids and rock all have compositions the same as a linear system, then the same zones will form in both the linear and radial cases and the ratio of the frontal velocity to the fluid velocity will be given by Eq. (4.31). For radial flow

$$i = 2\pi r h u$$

where i is the volumetric injection rate and h is the formation thickness. Thus,

$$v_f = \frac{i}{2\pi r h \phi}\left(\frac{1}{1 + \dfrac{\Delta A}{\Delta c}}\right) \tag{4.32}$$

The distance dr that the front advances during time dt is clearly $v_f\, dt$. Thus,

$$r_f\, dr = \frac{i\, dt}{2\pi h \phi}\left(\frac{1}{1 + \dfrac{\Delta A}{\Delta c}}\right)$$

at $t = 0$, $r_f = r_w$ (wellbore radius) and at some latter time, the front will have advanced a distance r_f from the wellbore. Integrating yields

$$\phi h \pi (r_f^2 - r_w^2) = it\left(\frac{1}{1 + \dfrac{\Delta A}{\Delta c}}\right) \tag{4.33}$$

This equation is easily understood if one considers the case of a tracer ($\Delta A = 0$). Then it shows that the volume injected is the pore volume a distance r_f from the wellbore.

Example 4.9 Surfactant Treatment in a Radial System

Given that adsorption of a certain surfactant is represented by the equation

$$\frac{\rho_R(1 - \phi)\Gamma}{\phi} = \frac{c_s}{0.1 + \dfrac{c_s}{c_0}} \tag{4.34}$$

where c_0 equals 0.05 kg/m^3. Calculate the volume per meter of formation thickness of fluid containing 0.1 kg/m^3 of surfactant that must be injected to treat a radius 0.5 m around the centerline of a wellbore having a radius of 0.1 m. The porosity of the formation is 10%.

Solution As shown in Fig. 4.19, three zones will develop upon injection of the surfactant solution. The same amount of surfactant must be adsorbed on the rock in Zone II as in Zone I. Since there is initially no surfactant in the formation waters, there can be no surfactant adsorbed on the rock in Zone I nor, therefore, in Zone II. Since there is no surfactant on the rock in Zone II ($\Gamma = 0$), Eq. (4.34) shows that $c_s = 0$. Hence, $A^{II} = 0$ and $c^{II} = 0$. From Eq. (4.34) we see that Γ can be found in Zone III using the fact that the solution concentration of the surfactant in this zone must be the same as that in the injected fluid. Thus,

$$\rho_R \frac{1 - \phi}{\phi} \Gamma = \frac{0.1}{0.1 + \dfrac{0.1}{0.05}} = 0.0476 \text{ kg/m}^3$$

Therefore, $A^{III} = 0.0476 \text{ kg/m}^3$ and $c^{III} = 0.1 \text{ kg/m}^3$

Then $$\frac{it}{h} = \left(1 + \frac{\Delta A}{\Delta c}\right)(\phi\pi)(r_f^2 - r_w^2) = \left(1 + \frac{0.0476}{0.1}\right)(0.1)(\pi)(0.5^2 - 0.1^2)$$

or $$\frac{it}{h} = 0.11 \frac{\text{m}^3 \text{ of solution}}{\text{m}}$$

In this section, we have tried to emphasize the importance of fluid/rock interactions by citing a series of examples where the interaction modifies the composition and the temperature of the injected fluid. Thus, to ensure that the injected fluid and the formation are compatible one has to examine the composition of the modified fluid. This fluid can be highly unstable even though the injected fluid is itself stable and is stable when mixed with formation fluids in all proportions.

Precipitation in the Intermediate Zone

The precipitates (e.g., $CaSO_4$, $BaSO_4$, $CaCO_3$) formed when injected fluid interacts with reservoir rock can migrate through the porous medium, become trapped in the pore throats, and impair the natural formation permeability [56]. Potter and Dibble [57] have shown that colloids other than clays can also cause formation damage. These may result from corrosion of tubular goods, originate when one mineral dissolves causing a second one to precipitate, or be formed by a process of ion exchange followed by precipitation.

Ion exchange and precipitation. This mechanism is illustrated by Example 4.8, showing that a perfectly harmless sodium carbonate solution can be modified to a carbonate solution containing substantial quantities of calcium which will then precipitate as calcium carbonate. This result reflects the importance of understanding the changes in the injected fluids which take place upon interaction with the formation rock. It is, for example, economically advantageous to waterflood offshore reservoirs with seawater. Table 4.4 gives a typical analysis of seawater showing the presence of substantial quantities of sulfate (SO_4^{--}). If the formation water contains traces of barium, then even though the seawater may be entirely compatible with it (no precipitation upon mixing) injectivity problems may rapidly develop because of the high concentrations of barium affixed to the

TABLE 4.4 Typical Seawater Analysis [58]

Solution cation	ppm
Sodium	11,000
Calcium	450
Magnesium	1400
Barium	0.1
Strontium	7
Iron	0.1
Chloride	19,800
Sulfate	2650
Bicarbonate	150
pH	7

rock surface. This barium will exchange with the cations in seawater and may very well precipitate as barium sulfate. Thus, it is not sufficient to consider the compatibility of the injected and the formation waters when determining the potential for formation damage.

It is perhaps worth noting that the injection of a small quantity of sulfate-free buffer fluid in the hope of separating the barium cations from the sulfate anions is not likely to be helpful. The barium travels through the formation at a velocity less than that of the fluid, whereas the sulfate essentially moves with the fluid velocity. Thus, the sulfate will overtake the barium.

Ion exchange is a very important mode of rock/fluid interaction. It can give rise to formation damage, but it can also be used advantageously in well stimulation. It is a process that one must be aware of.

Dissolution and precipitation. Ion exchange is one important type of rock/fluid interaction leading in some cases to formation damage. Another mechanism of importance entails the dissolution of one or more minerals and the consequent precipitation of others. Often the precipitation process is homogeneous (takes place in the solution rather than on solid surfaces), thereby creating suspended fine particles. This is, therefore, a process to be considered here because of the potential for formation damage.

The process of dissolution followed by precipitation leads to the formation of fronts propagating through the formation. It is a chromatographic process. A front is a position at which some minerals dissolve while others may precipitate. Since the waves are clearly formed as a result of a fluid/rock interaction, then the frontal velocities move at a speed slower than that of a tracer; and furthermore, the same principles describing other types of fluid/rock interactions still apply here [58, 59, 60, 61, 62].

There is, however, an additional principle needed to predict the compositions in all zones and speeds of all fronts. This new principle, applying strictly to dissolution/precipitation problems, is called the *downstream equilibrium condition* and can be stated as follows: The solution in a zone downstream of a solid must be in equilibrium with that solid even though it is not present in the zone. A proof of this condition has been established [57].

Using the downstream equilibrium condition and those which apply to moving fronts in general, most precipitation/dissolution problems can be solved.

Example 4.10 Precipitation on Waterflooding with Seawater

Seawater contains substantial quantities of sulfate anions. Evaluate the potential for formation damage when a water containing 0.0318 kg mole/m^3 SO_4^{--} (3055 ppm) and 0.0636 kg mole/m^3 of Na^+ is injected into the formation. There are many other components, but for the purposes of this problem these can be considered to be inert tracer material and not participate in the process. The formation contains primarily inert minerals, but does have 0.008 kg mole/m^3 of witherite ($BaCO_3$).

Solution The witherite may very well dissolve and barium sulfate may precipitate, since barium sulfate is generally less soluble than barium carbonate. If this is indeed the case, a possible sequence of composition zones is shown in Fig. 4.22. Thus, a solution to this problem will not only require the determination of both the fluid and

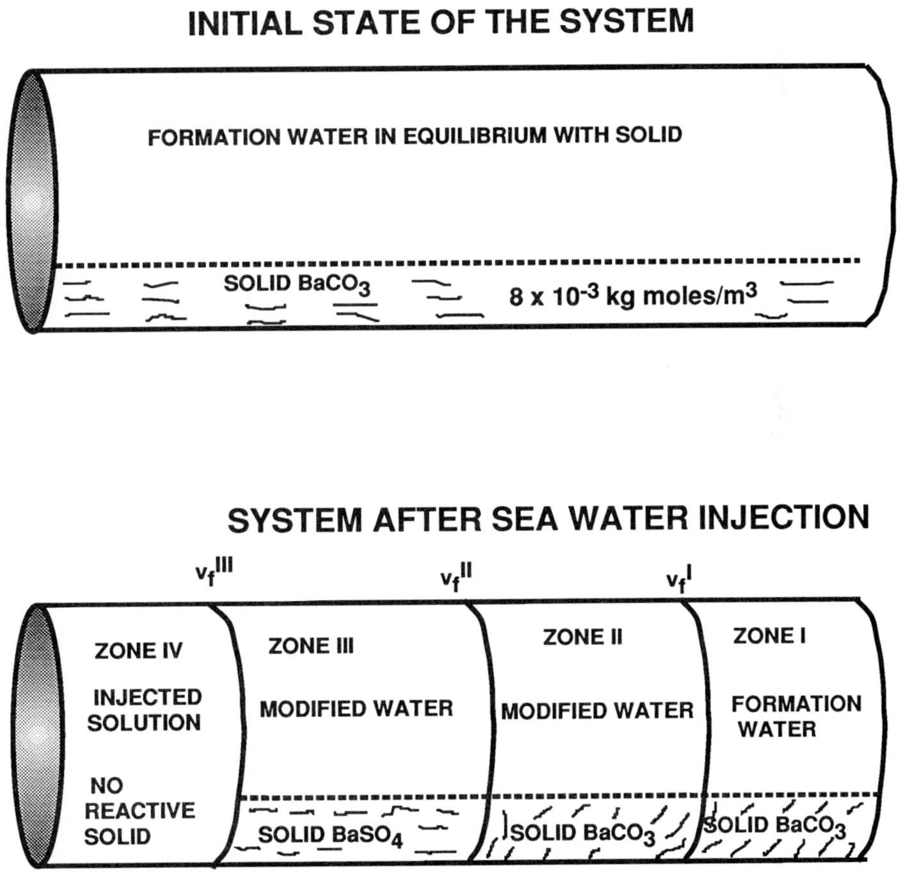

Figure 4.22 The injection of seawater results in the formation of four distinct zones. Each zone is moving with a different velocity.

mineral compositions in the different zones and the frontal velocities, but also an evaluation of the validity of the sequence shown.

As shown in Table 4.3, the capacity of the solid is $\bar{c}_i$ and the capacity of the solution is c_i. The solubility product determines the equilibrium between a solid and the ions in solution. Thus, for barium sulfate

$$(c_{Ba})(c_{SO_4}) = 1.26 \times 10^{-9}$$

and for barium carbonate

$$(c_{Ba})(c_{CO_3}) = 1.59 \times 10^{-9}$$

A second condition that applies in precipitation/dissolution problems is electrical neutrality. Thus, $\sum Z_i c_i = 0$ or in Zone III this equation is written as

$$2c_{Ba}^{III} + c_{Na}^{III} = 2c_{SO_4}^{III}$$

Note that the only reactive anion present in Zone III is SO_4^{--} since there is no CO_3^{--} in the injected seawater. Since Na^+ does not, in this example, interact with the rock, it is a tracer and its concentration is the same in the modified water as in the injected water. Thus,

$$c_{Na}^{IV} = c_{Na}^{III} = c_{Na}^{II} = 0.0636 \text{ kg mole/m}^3$$

Combining the condition of electrical neutrality with the condition for local equilibrium, the condition that applies in Zone III is

$$c_{Ba}^{III} c_{SO_4}^{III} = c_{SO_4}^{III}(c_{SO_4}^{III} - 0.0318) = 1.26 \times 10^{-9}$$

Solving $$c_{SO_4}^{III} = 0.0318 \text{ kg mole/m}^3$$

and $$c_{Ba}^{III} = \frac{1.26 \times 10^{-9}}{0.0318} = 3.96 \times 10^{-8} \text{ kg mole/m}^3$$

Finding the concentrations of the various ions in Zone II is interesting because to do so the downstream equilibrium condition must be applied. Thus,

$$c_{Ba}^{II} c_{SO_4}^{II} = 1.26 \times 10^{-9}$$

is to be satisfied even though there is no barium sulfate present. Also, equilibrium with barium carbonate requires that

$$c_{Ba}^{II} c_{CO_3}^{II} = 1.59 \times 10^{-9}$$

(Here, it is assumed that the solution pH is relatively high so that no bicarbonate is present. All carbonate exists as CO_3^{--}). Electrical neutrality requires that

$$2c_{Ba}^{II} + c_{Na}^{II} = 2c_{CO_3}^{II} + 2c_{SO_4}^{II}$$

There are three equations and three unknowns. These can be solved to yield:

$$c_{Ba}^{II} = 4.48 \times 10^{-8} \text{ kg moles/m}^3$$

$$c_{SO_4}^{II} = 2.81 \times 10^{-2} \text{ kg moles/m}^3$$

$$c_{CO_3}^{II} = 3.51 \times 10^{-2} \text{ kg moles/m}^3$$

Precipitation/dissolution solution phase compositions can be determined without knowing the solid concentrations. Only the solid identities in each zone are required. The solid concentrations are determined by applying the condition of coherence; that is, the frontal velocity must be the same irrespective of the component considered. In this case, $BaCO_3$ dissolves on the downstream side of the front and barium sulfate precipitates on the other side. This result provides two possible sources for formation damage. The precipitated mineral may clog the pore throats; and furthermore, the sand grains that were cemented together by $BaCO_3$ will now be free to migrate, also clogging pore throats.

Although this example is highly simplified, it serves to demonstrate that whenever a water is undersaturated with respect to one or more of the formation minerals, the potential for formation damage exists. Generally, there are so many possible aqueous species and solid precipitates that the evaluation of the danger requires a computer simulation [62].

Finally, it is worth noting that the frontal velocities indicate the position of the damaged zone. Knowledge of this position will be helpful in the selection of a stimulation method.

Paraffin Deposition

Depending on the paraffin-wax content, the cloud point and the pour point of a crude oil, a wax, which consists of straight and branched chain hydrocarbons usually ranging from $C_{18}H_{38}$ to about $C_{40}H_{82}$ mixed with other organic and inorganic materials [63], may deposit in the near-wellbore region [64], in perforation tunnels, on tubular goods, or in surface equipment [65, 66]. Paraffin solubility in crude oil depends on the chemical composition of the oil and the temperature. Paraffin deposits will often occur on surfaces cooler than the crude oil [64]. The viscosity of the crude increases in the presence of paraffin crystals and if the temperature is reduced sufficiently, the crude will become very viscous. The temperature at which this occurs is known as the *pour point*. Viscosity versus temperature curves are shown in Fig. 4.23 for two crude oils. As the temperature is decreased, a break in the curve is observed. This break is attributed to the presence of paraffin crystals and this occurs at a temperature less than the cloud point, which is when crystals first form. As the temperature is further decreased, the viscosity increases rapidly and a second break is observed at the temperature where the crystal concentration is exceedingly large. These high viscosities will give the appearance of formation damage if they develop within the porous medium.

The viscosity of both viscous and waxy crude oils can be significantly reduced by the presence of dissolved gases [63]. High-solubility gases such as butane lower the viscosity primarily by dilution, whereas low-solubility gases such as methane lower the viscosity by preventing the agglomeration of wax crystals.

The deposition of paraffin is therefore most sensitive to changes in temperature, but may also be triggered by the dissolution of gases from the crude oil. Both of these changes would be expected to take place in the near wellbore region or in the production tubing. To minimize this type of damage, it is desirable to maintain the temperature as near to the formation temperature as possible. When the crude oil has a high cloud point, it is often difficult to prevent paraffin deposition. The remedial treatment normally involves mechanical removal if the de-

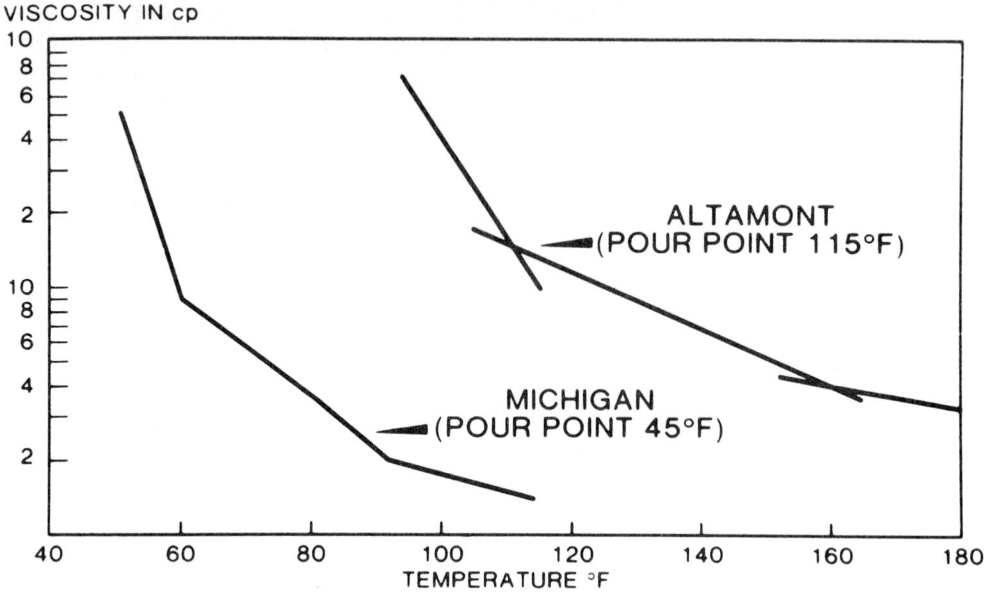

Figure 4.23 Viscosity versus temperature for Michigan and Altamont crude oils [From R. N. Tuttle, *J. Pet. Tech. 35* (1983), 1192.]

posits are primarily in the wellbore. This can be accomplished by wireline scrapers. An alternative is to attempt to dissolve the waxy deposit using a hot oil which has a high aromatic content [63, 66].

In some cases, it is possible to continuously inject chemicals that will retard paraffin deposition. These include polymers which inhibit crystal growth [66] or surfactants designed to render the paraffin particles water-wet so that they dissolve in that water produced together with the crude oil. This prevents the paraffin particles from agglomerating. The use of polymers to retard the crystal growth is apparently not widely practiced, because to be effective it must be tailored to match the individual crude oil [66].

Asphaltene Deposition

The asphaltene fraction of crude oil is defined as that part precipitated by the addition of a low boiling (low molecular weight) paraffin solvent such as *n*-pentane. The asphaltenes usually comprise condensed aromatic and napthenic molecules having molecular weights ranging from several hundred to several thousand [63, 68]. These normally contain significant quantities of oxygen, nitrogen, and sulfur. The asphaltic materials are generally thought to be colloidally dispersed rather than in solution [63, 69, 70]. The colloids are duplex in nature as they are dispersed by the adsorption of a resin on the surface of the small particles. The resins known as *maltenes* are structurally similar to the material composing the asphaltenes but of lower molecular weight [63].

The dispersed asphaltene particles are thought to be quite small in some cases (<6 nm in diameter) [73]. Such particles will readily pass through most pore throats and, in the dispersed state, asphaltenes are expected to be nondamaging. If, however, the resins are desorbed from the particle surface by the addition of a low molecular weight paraffin, water, acids, or CO_2 [63, 70], then the particles begin to flocculate [71] forming much larger entities. These larger particles can then become damaging [63, 72].

Furthermore, when some asphaltenic crude oils come into contact with acid, a rigid film forms at the oil/water interface. This film is thought to be a source of formation damage because it can stabilize emulsions [74, 75].

Difficulties with asphaltene flocculation normally occur either in or near the wellbore. Flocculation can result when natural gas and crude oil originating from different parts of the reservoir mix. The crude oil temperature may also change in the near wellbore region and this may result in flocculation [68].

It is also possible that CO_2 injection to enhance the recovery of oil will strip enough of the resins from the asphaltene to cause precipitation. This mechanism can give rise to damage remote from the wellbore.

Asphaltene flocculation can result in a severe loss in productivity [63]. Normally, it is removed from the near-wellbore region by dissolving the deposits in an aromatic solvent. The deposition process will continue, thus requiring frequent treatments.

4.5 BIOLOGICAL PROCESSES THAT PRODUCE FORMATION DAMAGE

Numerous investigations have found that bacteria are very effective in causing plugging of formation faces, particularly in water injection wells [77, 78, 79]. Bacteria can reproduce rapidly and they populate in extremely diverse conditions of pH and of temperature and even in the absence of oxygen. Patton [55] classifies bacteria found in oil field waters as follows:

1. Aerobic bacteria—growth dependent upon oxygen.
2. Anaerobic bacteria—growth best with no oxygen.
3. Facultative bacteria—growth independent of oxygen.

The primary bacterial problems found in oil field injection waters are caused by sulfate reducers, iron bacteria, and slime formers. The sulfate reducer is anaerobic and is a problem in injection waters. They reduce sulfate ions to sulfide ions, resulting in H_2S which causes corrosion. The end product, iron sulfide, causes plugging problems.

It has been shown that injection systems generally contain microorganisms of some type [80], including slime formers which are aerobic. These will also contribute to both formation plugging and to corrosion of the tubular goods.

These problems should be avoided as much as possible by the use of bacterocide where appropriate. Special treatments to remove formation damage created by bacterial growth have been reported [80], but these are not often applied and will not be described in this book.

REFERENCES

4.1. Rajagopalan, R., and Tien, C., *Amer. Inst. Chem. Eng. J., 22* (1976) 523.

4.2. Payatakes, A. C., Tien, C., and Turian, R. M., *Amer. Inst. Chem. Eng. J., 20* (1974) 889.

4.3. Einstein, A., *Investigations on the Theory of the Brownian Movement*, translated by A. D. Cowper, New York: Dover Publications (1905 publication date of original papers).

4.4. Dahneke, B., *J. Colloid and Interface Sci., 50* (1975) 89.

4.5. Ruckenstein, E., and Prieve, D. C., *Amer. Inst. Chem. Eng. J., 20* (1974) 1178.

4.6. Dullien, F. A. L., *Porous Media: Fluid Transport and Pore Structure*, New York: Academic Press, 1979.

4.7. Israelachvili, J. N., *Intermolecular and Surface Forces*, New York: Academic Press, 1985.

4.8. Mahanty, J., and Ninham, B. W., *Dispersion Forces*, New York: Academic Press, 1976.

4.9. Miller, C. A., and Neogi, P., *Interfacial Phenomena*, New York: Marcel Dekker, 1985.

4.10. Verwey, E. J. W., and Overbeek, J. Th. G., *Theory of the Stability of Lyophobic Colloids*, Amsterdam: Elsevier, 1948.

4.11. Kor, G., Chander, S., and Mika, T. S., *J. Colloid and Interface Sci., 44* (1973) 347.

4.12. Wilemski, G., *J. Colloid and Interface Sci., 88* (1982) 111.

4.13. Ninham, B. W., and Parsegian, V. A., *J. Theoret. Biol., 31* (1971) 405.

4.14. Healy, T. W., Chan, D., and White, L. R., *Pure Appl. Chem., 52* (1980) 1207.

4.15. Khilar, K. C., and Fogler, H. S., *J. Colloid and Interface Sci., 101* (1984) 214.

4.16. O'Niel, M. E., *Chem. Eng. Sci., 23* (1968) 1293.

4.17. Chamoun, H., "A Theoretical and Experimental Analysis of the Release of Non-Brownian Particles from Surfaces," PhD dissertation, The University of Texas at Austin, 1988.

4.18. Borchardt, J. K., Roll, D. L., and Payne, L. M., "Use of a Mineral Fines Stabilizer in Well Completions," SPE 12757, presented at the California Regional Meeting of the Society of Petroleum Engineers, Long Beach, California, 1984.

4.19. Derjaguin, B. V., Muller, V. M., and Tuporove, Yu. P., *J. Colloid and Interface Sci., 53* (1975) 314.

4.20. Gabriel, G. A., and Inamder, G. R., "An Experimental Investigation of Fines Migration in Porous Media," SPE 12168, presented at the 58th Fall Technical Conference and Exhibition of the Society of Petroleum Engineers, San Francisco, California, 1983.

4.21. Muecke, T. W., *J. Pet. Tech., 31* (1979) 144.

4.22. Gruesbeck, C., and Collins, R. E., *Soc. Pet. Eng. J., 22* (1982) 847.

4.23. Grey, D. H., and Rex, R. W., "Formation Damage in Sandstones Caused by Clay Dispersion and Migration," paper presented at the 14th National Conference on Clays and Clay Minerals, Chevron Company, London, 1966, 335.

4.24. Mungan, N., *J. Pet. Tech., 17* (1965) 1449.

4.25. Jones, F. O., *J. Pet. Tech., 16* (1964) 441.

4.26. Hewitt, C. H., *J. Pet. Tech., 15* (1963) 818.

4.27. Khilar, K. C., and Fogler, H. S., *Soc. Pet. Eng. J., 23* (1983) 55.

4.28. Khilar, K. C., and Fogler, H. S., in *Surface Phenomena in Enhanced Oil Recovery*, D. O. Shah (ed.), New York: Plenum Press, 1981.

4.29. Monaghan, P. H., Salathiel, R. A., Morgan, B. E., and Kaiser, A. D., *Trans. AIMME, 216* (1959) 209.
4.30. Atwood, D. K., *J. Pet. Tech., 16* (1964) 1405.
4.31. Land, C. S., and Baptist, O. C., *J. Pet. Tech., 17* (1965) 1213.
4.32. Riese, A. C., "Adsorption of Radium and Thorium onto Quartz and Kaolinite: A Comparison of Solutions/Surface Equilibria Models," PhD dissertation, Colorado School of Mines, Golden, Colorado, 1982.
4.33. Jenny, H., and Reitmeier, R. F., *J. Phys. Chem., 39* (1935) 593.
4.34. Sydansk, R. D., *J. Pet. Tech., 37* (1984) 1366.
4.35. Kia, S. F., Fogler, H. S., and Reed, M. E., *J. Colloid and Interface Sci., 118* (1987) 158.
4.36. Simon, D. E., McDaniel, B. W., and Coon, R. M., "Evaluation of Fluid pH Effects on Low Permeability Sandstones," SPE 6010, presented at the 51st Annual Fall Technical Conference and Exhibition, New Orleans, Louisiana, 1976.
4.37. Gdanski, R., and Peavy, M., "Well Return Analysis Causes Re-evaluation of HCl Theories," SPE 14825, presented at the Society of Petroleum Engineers Symposium on Formation Damage Control, Lafayette, Louisiana, 1986.
4.38. Bryant, S. L., and Buller, D. C., "Formation Damage from Acid Treatments," SPE 17597, presented at the International Meeting on Petroleum Engineering, China, 1988.
4.39. Roof, S. G., *Soc. Pet. Eng. J., 22* (1970), 85.
4.40. Mohanty, K. K., Davis, H. T., and Scriven, L. E., in *Surface Phenomena in Enhanced Oil Recovery*, D. O. Shah (eds.), New York: Plenum Press, 1981.
4.41. Falls, A. H., Gauglitz, P. A., Hirasaki, G. T., Miller, D. D., Patzek, T. W., and Ratulowski, J., "Development of a Mechanistic Foam Simulator: The Population Balance Approach and a Description of Generation by Capillary Snap-Off," SPE 14961, presented at the Fifth Symposium on Enhanced Oil Recovery of the Society of Petroleum Engineers, Tulsa, Oklahoma, 1986.
4.42. Radke, C. J., and Ranshoff, T. C., "Mechanism of Foam Generation in Glass Bead Packs," SPE 15441, presented at the 61st Annual Fall Technical Conference and Exhibition of the Society of Petroleum Engineers, New Orleans, Louisiana, 1986.
4.43. Ransohoff, T. C., Gauglitz, P. A., and Radke, C. J., *Amer. Inst. Chem. Eng. J., 33*, (1987) 753.
4.44. Sanchez, J. M., and Schechter, R. S., "Surfactant Effects on the Two-Phase Flow of Steam/Water and Nitrogen/Water through Permeable Media," *J. Pet. Sci. Eng., 3* (1989) 185.
4.45. Lake, L. W., *Fundamentals of Enhanced Oil Recovery*, Englewood Cliffs: Prentice Hall, 1988.
4.46. Hirasaki, G. J., and Lawson, J. B., "Mechanism of Foam Flow in Porous Media—Apparent Viscosity in Smooth Capillaries," SPE 12129, presented at the 58th Annual Fall Technical Conference and Exhibition, San Francisco, California, 1983.
4.47. Hoffman, R. L., *J. Colloid and Interface Sci., 50* (1975) 228.
4.48. Jiang, T., Oh, S., and Slattery, J. C., *J. Colloid and Interface Sci., 68* (1979) 74.
4.49. Melrose, J. C., Hall, A. C., and Collins, S. H., *Soc. Pet. Eng. J., 23* (1983) 249.
4.50. Hirasaki, G. J., "Wettability: Fundamentals and Surface Forces," SPE/DOE 17367, presented at the Society of Petroleum Engineers/Department of Energy Enhanced Oil Recovery Symposium, Tulsa, Oklahoma, 1988.
4.51. Fernandez, M., "Adsorption of Sulfonates from Aqueous Solutions onto Mineral Surfaces," MS thesis, The University of Texas at Austin, 1977.

4.52. Rhee, H. K., Aris, R., and Amundson, N. R., *Phil. Trans. Roy. Soc. Lond., A269* (1971) 187.
4.53. Helfferich, F. G., and Klein, G., *Multicomponent Chromatography*, New York: Marcel Dekker, 1970.
4.54. Rhee, H. K., Aris, R., and Amundson, N. R., *First-Order Partial Differential Equations: Volume I*, Englewood Cliffs: Prentice Hall, 1986.
4.55. Patton, C. C., *Oilfield Water Systems*, Campbell Petroleum Services, Norman, 1977.
4.56. Wu, G., Sharma, M. M., and Lake, L. W., *Amer. Inst. Chem. Eng. J., 35* (1989) 1385.
4.57. Potter, J. M., and Dibble, W. E., *J. Pet. Tech., 37* (1985), 1682.
4.58. Dria, M. A., "Chemical and Thermochemical Wave Behavior in Multiphase Fluid Flow Through Permeable Media: Wave-Wave Interactions," PhD dissertation, The University of Texas at Austin, 1988.
4.59. Walsh, M. P., Bryant, S. L., Lake, L. W., and Schechter, R. S., *Amer. Inst. Chem. Eng. J., 30* (1984) 317.
4.60. Lichtner, P. C., *Geochimica Cosmochimica Acta, 49* (1985) 779.
4.61. Bryant, S. L., Schechter, R. S., and Lake, L. W., *Amer. Inst. Chem. Eng. J., 32* (1986) 751.
4.62. Novak, C. R., Schechter, R. S., and Lake, L. W., *Amer. Inst. Chem. Eng. J., 34* (1988) 1607.
4.63. Tuttle, R. N., *J. Pet. Tech., 35* (1983) 1192.
4.64. Sutton, G. D., and Roberts, L. D., "Paraffin Precipitation During Fracture Stimulation," SPE 4411, presented at the Rocky Mountain Regional Meeting of the Society of Petroleum Engineers, Casper, Wyoming, 1973.
4.65. Shock, D. A., Sudbury, J. D., and Crockett, J. J., *J. Pet. Tech., 7* (1955) 23.
4.66. Patton, C. C., and Casad, M. M., *Soc. Pet. Eng. J., 10* (1970) 17.
4.67. McClafin, G. G., and Whitfil, D. L., *J. Pet. Tech., 36* (1984) 1965.
4.68. Hirschberg, A., deJong, L. N. J., Schipper, B. A., and Meijer, J. G., *Soc. Pet. Eng. J., 24* (1984) 283.
4.69. Dodd, C. G., Moore, J. W., and Denekao, M. O., *Ind. and Eng. Chem., 45* (1951) 1759.
4.70. Leontaris, K. J., and Mansoori, G. A., "Asphaltene Flocculation During Oil Production and Processing: A Thermodynamic Colloid Model," SPE 16258, presented at the International Symposium on Oilfield Chemistry, San Antonio, Texas, 1987.
4.71. Mitchell, D. L., and Speight, J. G., *Fuels, 52* (1973) 149.
4.72. Haskett, C. E., and Tartera, M., *J. Pet. Tech., 17* (1965) 387.
4.73. Witherspoon, P. A., and Munir, J. A., *Producers Monthly, 26* (October 1960) 20.
4.74. Jacobs, I. C., and Thorne, M. A., "Asphaltene Precipitation During Acid Stimulation Treatments," SPE 14823, presented at the Formation Damage Control Symposium, Lafayette, Louisiana, 1986.
4.75. Moore, E. W., Crowe, C. W., and Hendrickson, A. R., *J. Pet. Tech., 17* (1965) 1023.
4.76. Newberry, M. E., and Barker, K. M., "Formation Damage Prevention Through the Control of Paraffin and Asphaltene Deposition," SPE 13796, presented at the Production Operations Symposium, Oklahoma City, Oklahoma, 1985.
4.77. Plummer, F. B., Merkt, E. E., Power, H. H., Sawin, H. J., and Tapp, P., *Pet. Tech.* (1944) 1.
4.78. Rayleigh, J. T., and Floch, D. L., *J. Pet. Tech., 17* (1965) 201.
4.79. Kalish, P. J., Stewart, J. A., Rodgers, W. F., and Bennett, E. O., *J. Pet. Tech., 16* (1964) 805.

4.80. Myers, G. E., and Slabye, B. M., *Producers Monthly*. (May 1962) 12.
4.81. Crowe, C. W., *J. Pet. Tech.*, *20* (1968) 475.

PROBLEMS

4.1. The number of fine particles suspended in a fluid surrounding a sand grain that strike the sand grain per unit of time as a result of Brownian motion is essentially determined by the Einstein diffusion coefficient [Eq. (4.1)]. Derive an equation expressing this number of collisions in terms of the sand grain radius, R, the diffusion coefficient, D, and the concentration of fine particles, $c_A^{(\infty)}$. Assume that each particle striking the sand grain adheres to it so that the particle concentration vanishes at the surface of the sand grain. [*Hint:* Show that the particle concentration in the fluid near the sand grain is given by

$$\frac{\partial}{\partial r}\left(r^2 \frac{\partial c_A}{\partial r}\right) = 0$$

where r is the coordinate measuring distance from the center of the spherical sand grain. Recall that the particle flux in the radial direction is given by $-D\, \partial c_A/\partial r$].

If the concentration of fines having a radius of 1 μm is 10^8 particles/cm^3, how many of these particles will strike a 40-μm in radius sand grain each second? The fluid temperature is 60°C and the viscosity is 0.7 centipoise.

***4.2.** The force between a small spherical particle (radius = 0.1 μm and density = 2.6 g/cm^3) immersed in water (density = 1 g/cm^3) and a nearby sand grain (considered to be a flat slab) is characterized by a Hamaker constant whose value is 10^{-21} J. At what spacing between the particle and the sand grain will the van der Waals force and the net gravitational force be equal? The particle shape may be taken to be spherical.

***4.3.** If the electrostatic force between two similarly charged parallel plates separated by an aqueous solution containing 2×10^{-3} kg moles/m^3 of KCl at 20°C is reduced by a factor of 8 when the space between them is doubled, how far apart were the plates originally spaced?

***4.4.** What is the minimum concentration of Na_2SO_4 (expressed in wt%) required to eliminate the potential barrier between a particle (assumed to be a flat slab) having a surface charge of 3×10^{-3} C/m^2 and a similarly charged flat surface? The fluid temperature is 50°C.

The Hamaker constant characterizing the force between the two solids in the presence of the intervening water layer is 10^{-21} J.

****4.5.** The electrical interaction potential $\overline{U}_R^{(sw)}$ between a sphere (radius a) and a wall having surface potentials ψ_0^{sphere} and ψ_0^{wall} is approximately given by

$$U_R^{(sw)} = 16\epsilon\left(\frac{RT}{ZF}\right)^2 a \tanh\left(\frac{Z\psi_0^{\text{wall}} F}{4RT}\right) \tanh\left(\frac{Z\psi_0^{\text{sphere}} F}{4RT}\right) \exp\left(-\frac{h}{\lambda}\right)$$

where h is the spacing between the sphere and the wall. This equation applies to a symmetric electrolyte for which the cation has a valence Z and the anion a valence $-Z$ (see Miller and Neogi [9]).

Using the preceding expression for the repulsive energy and Eq. (4.7), which expresses the attractive van der Waals potential, determine the critical salt concentration when the aqueous phase contains sodium chloride and the following conditions apply:

$$\psi_0^{\text{wall}} = 60 \text{ mV}$$

$$\psi_0^{\text{sphere}} = 40 \text{ mV}$$

$$a = 4\ \mu\text{m}$$

$$T = 60°\text{C}$$

$$A_H = 10^{-21} \text{ J}$$

What will the critical concentration be if the temperature is reduced to 40°C, all other factors remaining the same?

***4.6.** Neglecting the repulsive force that might result from the overlapping of the electrical double layers, estimate the value of the deformation angle (α) [see Eq. (4.15)] when a silica spherical particle of radius 2 μm is positioned 0.2 nm from a wall. Take the Hamaker constant to be 10^{-21} J. Silica has a Young's modulus of 4×10^{10} Pa and Poisson ratio of 0.3.

****4.7.** Suppose a pore has a geometry represented by one cycle of a sine wave rotated about the axis as shown in Fig. P4.7 to form a surface of revolution. If twice the mean curvature of a surface of revolution is given by

$$2H = \frac{-\eta''}{(1 + \eta'^2)^{3/2}} + \frac{1}{\eta(1 + \eta'^2)^{1/2}}$$

and the capillary pressure is given by $p_c = 2\gamma H$, produce a map showing values of R/λ and δ that separates those regions of the map for which the pore is a generating site from those where it is not.

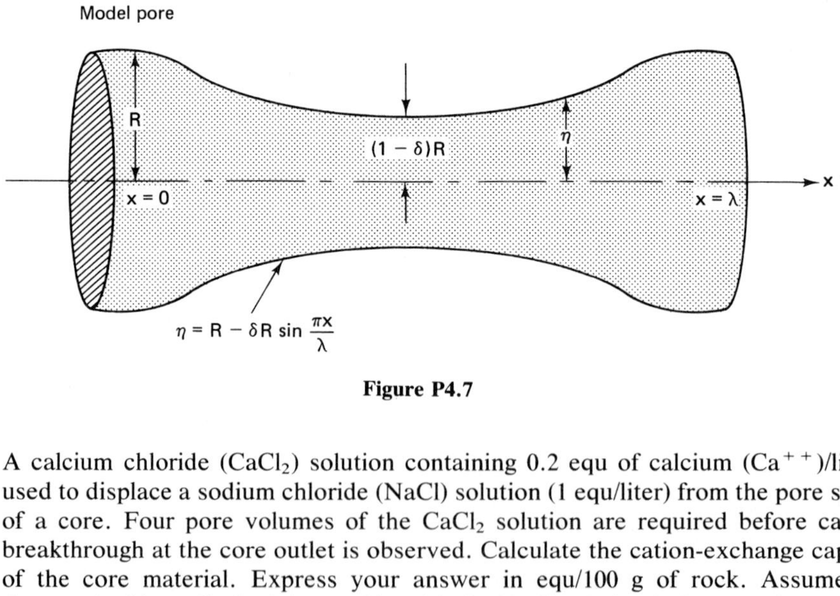

Figure P4.7

***4.8.** A calcium chloride ($CaCl_2$) solution containing 0.2 equ of calcium (Ca^{++})/liter is used to displace a sodium chloride (NaCl) solution (1 equ/liter) from the pore spaces of a core. Four pore volumes of the $CaCl_2$ solution are required before calcium breakthrough at the core outlet is observed. Calculate the cation-exchange capacity of the core material. Express your answer in equ/100 g of rock. Assume that the core had been flushed thoroughly with the NaCl solution before starting to inject the $CaCl_2$ solution. Thus all of the cation-exchange sites contain sodium ions at the moment the $CaCl_2$ solution is injected.

The density of the rock is 2.56 g/cm^3 and its porosity is 0.25.

***4.9.** A formation that is 10 m thick is treated with a 2 wt% surfactant solution (density 1.02 g/cm^3) designed to alter formation wettability. If 5 g of surfactant are adsorbed on each kg of formation rock, how many liters of surfactant solution are required to treat a radius of 1 m about a wellbore 0.2 m in radius? The density of rock is 2500 kg/m^3 and its porosity is 0.23. How deep into the formation does the modified solution penetrate?

***4.10.** If 2 pore volumes of the surfactant solution used in Problem 4.9 are injected into a linear core cut from the treated formation, how far will the surfactant front have penetrated? Express your answer as a fraction of the total length of the core.

***4.11.** An acid corrosion inhibitor can alter formation wettability if it adsorbs onto formation rock. Suppose in treating a well too much inhibitor was added to the fluid and a solution having a concentration of inhibitor equal to 2 kg/m^3 enters the formation at a rate of 10 liters/min/m of formation thickness. How deep into the formation will the corrosion inhibitor have penetrated after 10 min? The adsorption of corrosion in the inhibitor on reservoir rock is given by

$$\Gamma = \frac{0.001c}{1 + 0.02c}$$

where c is the concentration in the solution (kg/m^3) and Γ is the adsorption (kg/kg of rock). The wellbore radius is 0.2 m and the porosity of the formation is 0.2. Take the density of rock to be equal to 2500 kg/m^3.

***4.12.** A cylindrical formation core 50 cm in length and 4 cm in diameter (porosity = 0.18) is filled with formation fluid and heated to 100°C (the reservoir temperature). A laboratory experiment in which drilling fluid filtrate (heat capacity = 0.255 kJ/kg °C and density = 1030 kg/m^3) is injected at a temperature of 50°C into the core that is sealed and very well thermally insulated along its lateral sides shows that 238 cm^3 of the filtrate are required to cool the core from the 100°C temperature to 50°C.

(a) Estimate the heat capacity per unit volume of the reservoir rock.

(b) If during drilling operations 5 m^3 of drilling fluid having a temperature of 60°C are lost into the same formation rock as tested in the laboratory, how deep (to what radius) around the wellbore will the formation be cooled to 60°C? The wellbore radius is 0.3 m, the formation thickness is 10 m, and the reservoir temperature is 100°C.

***4.13.** A formation of water is primarily composed of sodium (molecular weight = 23) chloride (molecular weight = 35.5) and calcium (molecular weight = 40) chloride. The water contains 10 wt% NaCl and 1 wt% $CaCl_2$ and its density is 1031 kg/m^3. By analyzing core samples the system cation-exchange capacity is found to be 0.3 kg equ/m^3 of reservoir pore spaces and the selectivity coefficient for the sodium–calcium exchange is

$$K_{\text{Na,Ca}} = \left(\frac{\theta_{\text{Ca}}}{\theta_{\text{Na}}^2}\right)\left(\frac{c_{\text{Na}}^2}{c_{\text{Ca}}}\right) = 20$$

where c_{Na} and c_{Ca} are concentrations expressed in kg equ/m^3. Calculate the fraction of the formation surface occupied by calcium ions and determine the proportion of the calcium present in the reservoir that is in solution.

If a solution containing 0.2 kg mole/m^3 of Na_2SO_4 (sodium sulfate) is injected into the formation, precipitation of $CaSO_4$ will occur if the amount of $CaSO_4$ in water exceeds 2×10^{-3} kg equ/m^3. Will precipitation occur?

4.14. It has been suggested that formation damage may be in some way related to the rate at which the salinity of the water is changed. Show that the calcium ion concentration in the modified zone does depend on the sequence of fluids injected. Since the damage caused by fresh water injection is controlled by the presence of divalent ions, determine which of the following two modes of operation is likely to be most damaging.

Calculate the calcium composition within the modified zone for the case depicted in Table P4.14A. Take $K_{Ca,Na} = 5.0$.

TABLE P4.14A Case A: Data for Calculating Calcium Composition within One Modified Zone

Zone III (Injected fluid)	Zone II (Modified zone)	Zone I (Resident fluid)
		$Q_V = 100$ eq/m^3
$c_{An} = 2.5$ eq/m^3	$c_{An} = 2.5$ eq/m^3	$c_{An} = 250$ eq/m^3
$c_{Na} = 2.4$ eq/m^3	$c_{Na} = ?$	$c_{Na} = 240$ eq/m^3
$c_{Ca} = 0.1$ eq/m^3	$c_{Ca} = ?$	$c_{Ca} = 10$ eq/m^3

Now consider the same reduction in salinity to take place in two steps. Calculate the calcium ion concentrations in each of the modified zones shown in Table P4.14B.

TABLE P4.14B Case B: Data for Calculating Calcium Composition within Two Modified Zones

Step 1:

Zone III (Injected fluid)	Zone II (Modified zone)	Zone I (Resident fluid)
		$Q_v = 100$ eq/m^3
$c_{An} = 25$ eq/m^3	$c_{An} = 25$ eq/m^3	$c_{An} = 250$ eq/m^3
$c_{Na} = 24$ eq/m^3	$c_{Na} = ?$	$c_{Na} = 240$ eq/m^3
$c_{Ca} = 1$ eq/m^3	$c_{Ca} = ?$	$c_{Ca} = 10$ eq/m^3

Step 2:

Zone III (Injected fluid)	Zone II (Modified zone)	Zone I (Resident fluid)
		$Q_v = 100$ eq/m^3
$c_{An} = 2.5$ eq/m^3	$c_{An} = 2.5$ eq/m^3	$c_{An} = 25$ eq/m^3
$c_{Na} = 2.4$ eq/m^3	$c_{Na} = ?$	$c_{Na} = 24$ eq/m^3
$c_{Ca} = 0.1$ eq/m^3	$c_{Ca} = ?$	$c_{Ca} = 1$ eq/m^3

4.15. Suppose a solution consisting of 1 kg mole/m^3 of sodium carbonate (Na_2CO_3) dissolved in water is injected into a core which is equilibrated with 0.5 kg mole/m^3 of calcium chloride ($CaCl_2$). In this case two intermediate zones develop because of the precipitation of calcium carbonate ($CaCO_3$). These two intermediate zones (Zones II and III) are shown in Fig. P4.15.

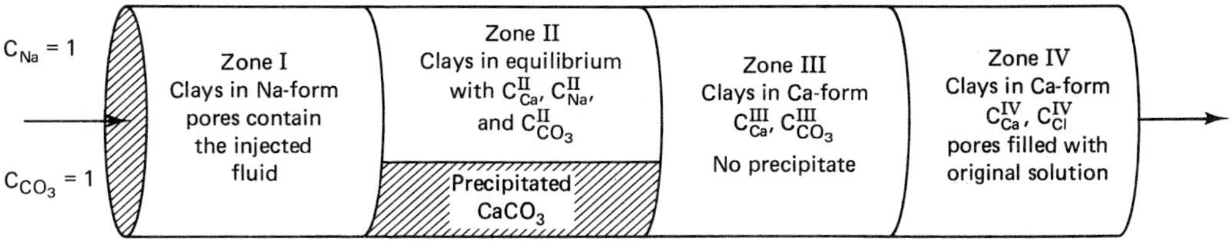

Figure P4.15

Given that the solubility product of $CaCO_3$ at the temperature of the system is represented by

$$c_{Ca} \cdot c_{CO_3} = 4 \times 10^{-8}$$

where both c_{Ca} and c_{CO_3} are expressed in kg equ/m^3, given $Q_v = 0.15$ kg equ/m^3 and $K_{Na,Ca} = 5$, calculate the fluid composition in each of the zones shown, the concentration of $CaCO_3$ precipitate in Zone II, the fraction of the clays in sodium form in Zone II, and the velocity of each of the wave fronts. In the calculation assume that $CaCO_3$ solid immediately attaches to the rock surface as it forms. [*Hint:* Be certain that the downstream equilibrium condition is satisfied within Zone III.]

5 Modeling Formation Damage

The mechanisms giving rise to formation damage are described in Chapter 4. Based on this discussion important guidelines emerge that will be helpful in avoiding or minimizing damage. The information developed in Chapter 4 is therefore useful; however, its impact would be far greater if the mechanisms were all integrated into a mathematical model capable of predicting the permeability as a function of both time and position. This model would greatly assist in the selection and design of well treatments to remove the damage. Furthermore, a model would help to assess the relative importance of the various mechanisms as well as to predict the permeability.

Unfortunately, a complete mathematical model does not now exist; but because of its potential importance, this is a subject of considerable current research. Furthermore, because of the similarity of this problem to other lattice or network problems, such as transport in homogeneous conductors [1, 2], important results have been recorded. Two concepts prove to be quite useful. One is the notion of *percolation* [1] and the other is the *conductivity* of an effective medium [2]. These concepts are both discussed in this chapter. Another perhaps more empirical approach considered in this chapter is the concept of *parallel pathways*. The essential feature of this model is the apportioning of particles between the large pores called *nonplugging pathways* and small pore throats called *plugging pathways*.

5.1 THE PARALLEL-PATHWAY MODEL

The movement of suspended particles through porous media is an important process with a variety of industrial applications, including deep bed filtration as well as the damaging of hydrocarbon-producing formations. If the particle sizes are much smaller than the vast majority of pore throats, then the process is called *deep bed filtration*. Due to its industrial significance, deep bed filtration has received considerable attention. A number of predictive models have been developed, essentially following the filtration equation proposed by Iwasaki [3]. An excellent description of the early work is presented in the 1970 review article by Herzig et al. [4]. Since the publication of this review, significant results have been reported by Payatakes et al. [5, 6] and Rajagopalan and Tien [7]. This work does not, however, directly relate to the process leading to formation damage since the crucial feature of damaging processes is the clogging of pore throats, whereas deep bed filtration is concerned primarily with the interception of small particles by large collector fibers or grains. The forces contributing to the interaction of particles with the pore walls are the same and are discussed in Chapter 4.

The feature of greatest significance is, therefore, the need to distinguish between those released particles which reattach to the pore walls and those which are mechanically trapped in pore throats.

Several approaches have been proposed. One of the most attractive has been proposed by Gruesbeck and Collins [8]. They introduced a concept which they called a *parallel-pathway model*. At any cross section in a porous medium there are two types of continuous parallel branches available for fluid flow: one of small pore size in which bridging by particles may occur and the other, of larger pore diameter, in which only nonplugging surface deposits occur (Fig. 5.1). From these surface deposits fines are entrained by the moving fluid. As the smaller pores become clogged, fluid is then diverted to the nonplugging pathways thereby ultimately attaining a steady-state permeability. Furthermore, if small fine particles are introduced into the fluid flowing into the core, this model, in correspondence with experimental evidence, predicts that ultimately a steady state will be reached in which the concentration of particles in the effluent equals that at the inlet.

Fundamental Concepts

For a fluid in linear flow, a mass balance on the fine particles can be written as follows:

$$\frac{\partial}{\partial t}(c + c_a) = -\frac{u}{\phi}\frac{\partial c}{\partial x} \tag{5.1}$$

where c is the concentration of fine particles in the flowing stream and c_a is the concentration of fine particles deposited on the surface, both concentrations most conveniently expressed as a volume per unit pore volume.

To complete the model, a law defining the rate at which particles are liberated and deposited must be postulated. For this model, f is defined to be the fraction of the pore pathways that are plugging. The value of f is a characteristic of both

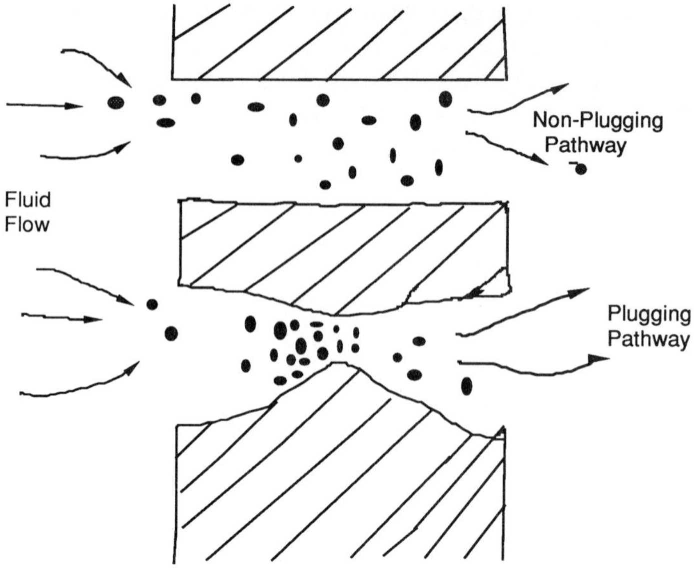

Figure 5.1 Parallel-pathway model of fines entrainment and deposition.

the pore sizes of the porous medium and the sizes of the fine particles in question. If the fines are very small, then f would be expected to be correspondingly small. Most of the pores are nonplugging. On the other hand if the particles are large, then they can be expected to be captured in the pore throats by the "log jam" mechanism (see Chap. 4) a very short distance from the point at which they are introduced or released from the pore walls. In this case f approaches unity. Thus, f should be predictable based on measurements characterizing the topology of the pore structure and the sizes of the particles. To date, no one has reported the steps required to carry out the determination of f and thus in terms of the modeling presented here, it is to be regarded as an empirical factor.

The factor f can also be considered to represent that fraction of the total pore area at any cross section which is composed of plugging pores. This dual meaning will be viable when the pores are randomly distributed so that the void volume fraction, ϕ, is equal to the void area fraction.

The flux (u) can be divided into two parts (u_p and u_{np}), the flux through the plugging and nonplugging pathways, respectively. Thus

$$u = fu_p + u_{np}(1 - f) \tag{5.2}$$

Similarly, we can divide attached or trapped fine particles into groups—those deposited in the plugging and those deposited in the nonplugging pathways. Thus

$$c_a = c_{an} + c_{ap} \tag{5.3}$$

where c_{an} is the concentration of fines adhering to the walls of the nonplugging pathways and c_{ap} is the concentration of fines that are plugging. The permeability will decrease as c_{ap} increases (see Fig. 5.1). In the nonplugging pathways, particles are removed from the surface by the action of the fluid flow (see Chap. 4) and some are redeposited under the influence of the van der Waals attractive forces. If, moreover, it is assumed that there exists a critical flux at which particle entrainment commences (see Chap. 4), then the simplest empirical law that can be postulated for the net rate of particle entrainment from nonplugging pore surfaces is

$$\frac{\partial c_{an}}{\partial t} = -\alpha(u_{np} - u_{npc})H(u_{np} - u_{npc})c_{an} + \beta c \tag{5.4}$$

where α and β are empirical rate constants and

$$H(u_{np} - u_{npc}) = \begin{cases} 1 & u_{np} > u_{npc} \\ 0 & u_{np} \leqslant u_{npc} \end{cases}$$

The role of the Heaviside function $H(x)$ is to cause the entrainment process to stop whenever u_{np} is less than the critical value, u_{npc}. Equation (5.4) is formulated assuming that the rate of particle entrainment is proportional to the flux and to the concentration of particles adhering to the surface. The recapture rate given by the last term on the right-hand side of Eq. (5.4) is taken to be proportional to the particle concentration in the flowing stream.

Equation (5.4) is strictly empirical. There is no experimental evidence suggesting that this particular equation is better than any other plausible form containing the same variables. The equation may, however, be useful. The key will be if α and β can be related to physical and chemical parameters known to be important. Such parameters include pH, electrolyte concentration, particle size and shape, etc., and are discussed in Chapter 4.

The rate that particles are trapped in plugging pores is assumed to be given by

$$\frac{\partial c_{ap}}{\partial t} = (\delta + \gamma c_{ap})u_p c \tag{5.5}$$

This equation is written assuming that once trapped, particles will not again be dislodged. Furthermore, the rate at which mechanical trapping takes place is proportional to the plugging flux, the suspension concentration, and the concentration of plugging particles already present. This model considers a process rather like the buildup of a filter cake. The more particles trapped, the easier it will be to trap others. Again this equation is empirical.

Equations (5.1)–(5.5) represent a set that can be solved if the parameters α, β, δ, and γ are all known. This solution will give c, c_{an}, and c_{ap} as a function of time and position given the initial and inlet conditions.

The missing quantity is the permeability, which must in some way depend on the concentration of plugging particles but should be relatively independent of the concentration of nonplugging ones. If it is assumed that each pathway has its own independent permeability, then

$$u_p = -\frac{k_p}{\mu}\frac{\partial p}{\partial x} \tag{5.6}$$

and

$$u_{np} = -\frac{k_{np}}{\mu}\frac{\partial p}{\partial x} \tag{5.7}$$

where k_p and k_{np} are the respective permeabilities. Substituting these equations into Eq. (5.2)

$$u = -\frac{k}{\mu}\frac{\partial p}{\partial x} = -\frac{k_p f}{\mu}\frac{\partial p}{\partial x} - \frac{k_{np}(1 - f)}{\mu}\frac{\partial p}{\partial x}$$

or

$$k = k_p f + (1 - f)k_{np} \tag{5.8}$$

This equation is still incomplete since the functional relationship between k_{np} and c_{ap} is unknown. Gruesbeck and Collins [8] take

$$k_p = k_{pf} \exp(-ac_p^q) \tag{5.9}$$

where again k_{pf}, a, and q are empirical constants which must in some way be a function of the pore topology. This will be discussed in greater detail in a subsequent section.

The parallel-pathway model is now completely stated. It has a number of parameters to be determined. That these equations can be used to match experimental data will become evident.

Example 5.1 Removal of Particles from Nonplugging Pathways

Consider a system composed of only nonplugging pathways containing an adhering particle concentration of 0.01 m^3/m^3 of pore volume (c_{an}) and suppose a fluid that is free of particles enters at a mean flux (u_{np}) of 5 cm/hr. The problem is to calculate the fluid phase particle concentration as a function of both time and position. To simplify the calculation, suppose that $\beta/\alpha^* = 3$ and that the fluid phase is in local equilibrium with the adhering phase. This implies that $c = \alpha^* c_{an}/\beta$. Here

$$\alpha^* = \alpha(u_{np} - u_{npc})H(u_{np} - u_{np})$$

[see Eq. (5.4)]. Take ϕ to be 0.15.

Solution Equation (5.1) reduces to the following form when c represents the suspension phase concentration:

$$\frac{\partial c}{\partial t} + \frac{\partial c_{an}}{\partial t} + \frac{u_{np}}{\phi}\frac{\partial c}{\partial x} = 0$$

Substituting $c_{an} = \beta c/\alpha^*$, this balance equation reduces to

$$\frac{\partial c}{\partial t} + \frac{u_{np}}{\phi(1 + \beta/\alpha^*)}\frac{\partial c}{\partial x} = 0$$

Make the change of variable so as to convert to a moving coordinate system. The new variables z and τ are expressed in terms of x and t as follows:

$$z = x - \frac{u_{np}}{\phi(1 + \beta/\alpha^*)}$$

$$\tau = t$$

The balance equation then reduces to

$$\left(\frac{\partial c}{\partial \tau}\right)_z = 0$$

This result implies

$$c = c(z)$$

This function of z can be evaluated by noting that for $z < 0$, the adhering particle concentration and the suspension concentration both vanish and for $z > 0$ the suspension concentration is $c = (0.01)(3) = 0.03\ m^3/m^3$ of pore volume. Thus

$$c = \begin{cases} 0 & \text{for } z < 0 \\ 0.03 & \text{for } z > 0 \end{cases}$$

This expression represents a wave moving at an effective wave velocity $u_{np}/(1 + \beta/\alpha^*)\ \phi = 8.3$ cm/hr. On the upstream side of the wave, the particle concentration in the flowing stream is zero and on the downstream side, it is $0.03\ m^3/m^3$ of pore volume.

This example indicates that when a particle-free fluid enters nonplugging pathways, the adhering particles will ultimately be stripped from the surface. These particles will either be produced with the effluent or will become trapped in pore throats along plugging pathways. According to this model the particles, once trapped in plugging pathways, remain trapped. Thus once all the particles are swept from the surface of the nonplugging pathways, no further permeability damage can be inflicted.

The number of pore volumes required to reach this state of ultimate damage is approximately $(1 + \beta/\alpha^*)$, as shown by Example 5.1. Actually, since this example assumes local equilibrium rather than considering all of the rate processes involved, $(1 + \beta/\alpha^*)$ is an underestimate. The number of pore volumes will depend on the absolute values of the rate constants α^* and β. If these rate constants are both small, the number of pore volumes required to reach the ultimate state may greatly exceed that predicted on the basis of local equilibrium.

The parallel-pathway model takes into consideration most of the mechanisms identified in Chapter 4. Some mechanisms have been addressed directly while others remain inherent in the values of the phenomenological constants. For example, the flux appearing in Eq. (5.4) is a measure of the shearing force (or torque) needed to mobilize fines, but factors known to be important such as ion concentration, pH, temperature, fluid viscosity, and particle geometry must be included in the values of α, β, and u_{npc}. Equation (5.5) describes the deposition of migrating fines in the pore throats. This process is proportional to the rate at which the fines are transported (u_p) and to the concentration of suspended fines (c). The rate of deposition, however, is also related to the particle size distribution as well as the pore size distribution. These factors must be in the phenomenological constants δ and γ.

One possible limitation of the model concerns the pore plugging deposition. Equation (5.5) does not allow for pore throat bridges to be broken. Once a fine particle is deposited in a plugging pore, it remains there. However, Muecke noticed that variations in fluid flux through the pore pathways cause pressure surges [9]. These pressure surges may be sufficient to break the bridges formed by fine particles in the pore throats. Therefore, plugging fines may be remobilized. In this case Eq. (5.5) is an incomplete description of the trapping process.

Some Experimental Results

It can be stated that there are few studies reported that have been conducted for the purpose of verifying a mathematical model of formation damage. Most have examined cores taken from productive formations to evaluate the injectivity decline that might be expected during a waterflood or to discover the mechanism of damage around producing wells. The goals are most often of a practical nature, not to find general rules or principles. Radke [10] has reported a study carried out using Berea cores. He found that for small flow rates (fluxes < 0.032 cm/sec) there was no measurable decline in permeability even after 1000 pore volumes of 2 wt% KCl solution had been injected. This finding is in full accord with the value reported by Gruesbeck and Collins [8]. As a word of caution, however, it should be noted that very large volumes in excess of 1000 pore volumes could result in damage even if the injection rate is less than critical. The notion of a critical velocity is not yet well established. What is certain, however, is that fluxes in excess of the critical cause a rapid decrease in the permeability. Thus in this sense there does exist a critical velocity. As noted in Chapter 4, it would be expected to depend on a number of factors and most especially the characteristics of the porous medium itself.

Figure 5.2 shows a graph of Radke's data and the comparison of the model predictions with the data. Because there are many parameters, the fit is excellent. The values of the parameters needed to fit the data are listed in Table 5.1 as Run 1. Also shown are the parameters used to fit data taken for four other runs.

All of the tests were conducted using Berea cores, but because there is some variability from sample to sample, it is perhaps reasonable to expect some differences in the parameters from run to run. These should not be large differences. The range of several of the parameters is reasonably narrow. For example, the initial volume of fines adhering to the wall per unit of pore volume, c_{an}^{i}, is observed to be almost constant. In addition the values of f, a and q fall within a relatively narrow interval. The constant γ is found to have little effect on the solution [10], whereas the parameters appearing in Eq. (5.9) and f have the most pronounced effect on the results. It appears that given a more fundamental method for evaluating the influence of pore plugging on the formation permeability, the other

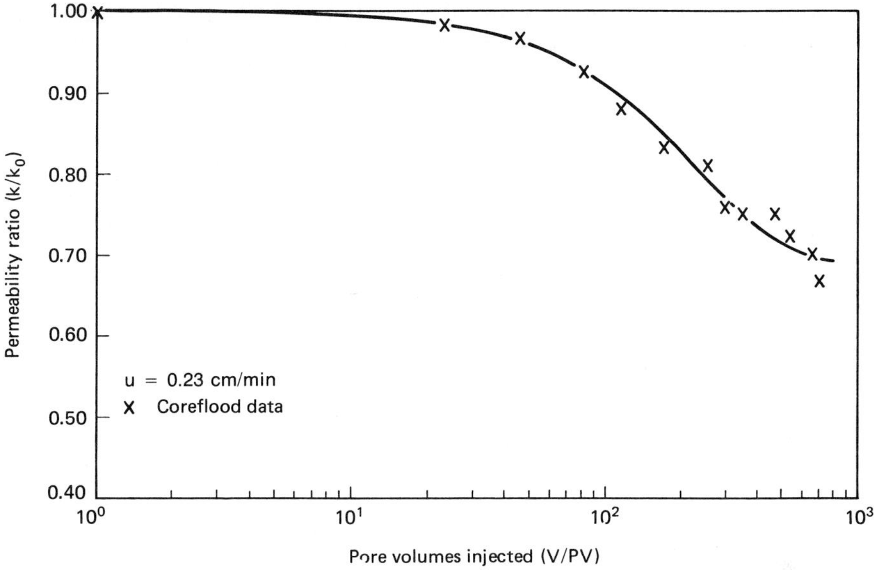

Figure 5.2 Graph of permeability ratio (k/k_0) versus pore volumes injected (V/PV).

aspects of the parallel-pathway model can be well defined based on a few experiments carried out with field cores. The parameters so obtained can apparently be used to predict the position of the formation damage.

The calculation of the permeability requires that both c_{an} and c_{ap} be determined. The plugging concentration essentially represents the formation damage.

TABLE 5.1 Parallel-Pathway Parameters Modeling Formation Damage in Berea Cores. Each Run Represents a Different Flow Rate [10].

Parameter	Units	Run 1	Run 2	Run 3	Run 4	Run 5
u	cm/sec	0.00383	0.00646	0.00653	0.00973	0.00984
u_c	cm/sec	0.00322	0.00322	0.00322	0.00322	0.0322
k_{pi}/k_0	—	0.760	0.710	0.760	0.630	0.820
k_{np}/k_0	—	1.633	2.321	2.760	1.951	2.184
c_{an}^{i}	—	0.230	0.230	0.250	0.220	0.240
f	—	0.725	0.82	0.880	0.720	0.820
α	1/sec	0.00280	0.00019	0.00014	0.0010	0.00021
β	1/sec	0.0170	0.120	0.150	0.10	0.120
δ	1/cm	2.1	19.0	29.0	3.80	15.50
γ	1/cm	1.0	2.0	12.0	1.0	1.0
a	—	10.0	16.5	15.0	12.0	12.0
q	—	1.040	0.90	0.950	0.950	1.0

Figure 5.3 shows the nonplugging concentration as a function of position. Note that this concentration decreases rather uniformly as a function of the pore volumes injected, just as we would anticipate. On the other hand the plugging concentration, shown by Fig. 5.4, increases monotonically from the inlet. The damage extends to the end of the core.

If one considers radial flow about the wellbore of an injection well, these results imply that the damage will extend into the formation, at least as far as the radial flux exceeds the critical value and perhaps somewhat further. In a production well the damage due to fines migration will, according to these results, be greatest in the region immediately around the wellbore.

Modeling Fresh Water Damage

It is known that the sudden injection of fresh water into a sandstone will generally produce a rapid decline in permeability [11, 12, 13]. The rate is much more pronounced than is observed when the damage is due to excessive fluid velocities in the pores. The reasons for this immediate response to the intrusion of fresh water have been discussed in Chapter 4. In terms of the parallel-pathway model,

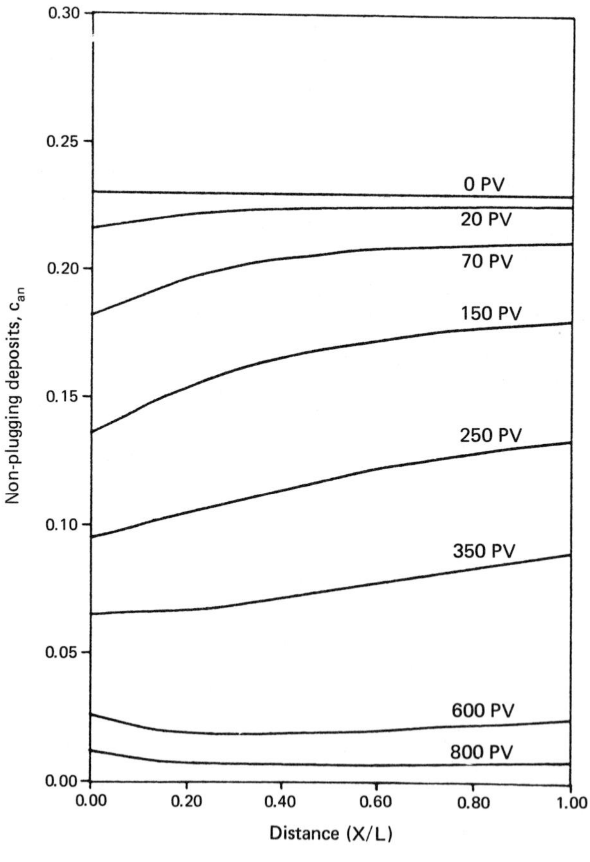

Figure 5.3 Graph showing the concentration of nonplugging deposits as a function of the fractional distance measured from the face of the core.

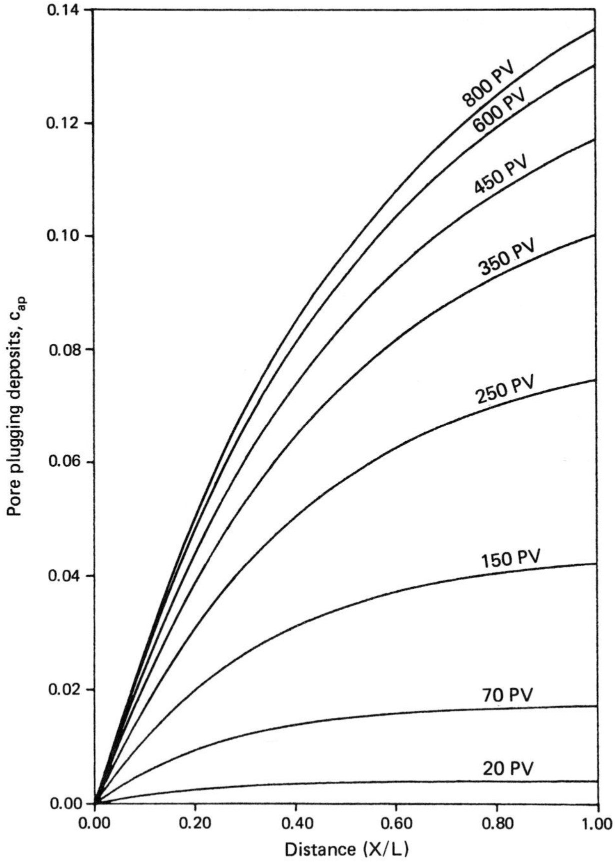

Figure 5.4 Graph showing the concentration of plugging deposits as a function of the fractional distance measured from the face of the core.

it seems evident that α, the release rate constant, may increase with declining salinity and certainly β, the rate constant for reattachment, will be greatly decreased. The particles have difficulty reattaching to the pore surface because the electrostatic repulsive forces between the pore well and the particles become very long range at low salinities (see Chap. 4). It is in fact reasonable to imagine that in fresh water $\beta = 0$. Thus the number of pore volumes required to remove all of the particles from nonplugging pathways, which is found in Example 5.1 to be approximately $(1 + \beta/\alpha^*)$, reduces to a single pore volume. The damage process in fresh water will essentially be completed when the first pore volume of fresh water has been injected. Contrast this with the results shown in Fig. 5.2. It becomes evident that even after hundreds of pore volumes have been injected, the permeability continues to decline. This difference can be primarily, although not entirely, ascribed to a marked decrease in β.

Since practically all particles released into fresh water are trapped mechanically at pore throats, then $f = 1$. All particles are essentially transported through plugging pathways. Convection of suspended particles can, therefore, be neglected. The particles are trapped almost as soon as they are released. For this case Eq. (5.1) reduces to

$$\frac{\partial c}{\partial t} + \frac{\partial c_a}{\partial t} = 0 \tag{5.10}$$

This simply means that at every point the fresh water has reached, the following equation applies:

$$c + c_a = \text{constant} \tag{5.11}$$

This result is true because convection of particles in fresh water has been neglected.

The constant in Eq. (5.11) can be evaluated at any time; in particular, it can be evaluated at $t = 0$, the time at which the fresh water front arrives at a certain position within the porous medium. At this time, there are no plugging deposits and few particles suspended in the flow stream. Essentially, all of the particles present are adhering to the pore walls. Thus if c_{an}^i represents the initial concentration of nonplugging deposits, then

$$c + c_{an} + c_{ap} = c_{an}^i \tag{5.12}$$

at initial conditions and according to Eq. (5.11), it must be true at all subsequent times.

Since in the presence of fresh water $\beta = 0$, then Eq. (5.4) reduces to the simple form

$$\frac{\partial c_{an}}{\partial t} = -\alpha(u_{np} - u_{npc})H(u_{np} - u_{npc})c_{an} = -\alpha^* c_{an} \tag{5.13}$$

Integrating Eq. (5.13) yields

$$c_{an} = c_{an}^i \exp(-\alpha^* t) \tag{5.14}$$

Similarly setting $\delta = 0$ (an approximation made for convenience), Eq. (5.5) reduces to

$$\frac{\partial c_{ap}}{\partial t} = (\gamma + \delta c_{ap})u_p c \cong \gamma^*(c_{an}^i - c_{an} - c_{ap}) \tag{5.15}$$

Integrating this expression yields

$$\frac{c_{ap}}{c_{an}^i} = [1 - \exp(-\gamma^* t)] + \frac{\gamma^*}{\alpha^* + \gamma^*}[\exp(-\alpha^* t) - \exp(-\gamma^* t)] \tag{5.16}$$

Equation (5.16) gives the concentration of plugging deposits at any point in the core as a function of time measured from the moment that the fresh water has reached that position within the porous medium.

Example 5.2 Rate of Damage in the Presence of Fresh Water

Determine the rate that plugging deposits build up once the fresh water front passes, assuming that $\alpha^* = 2.1 \text{ min}^{-1}$ and $\gamma^* = 1.1 \text{ min}^{-1}$. Plot c_{ap}/c_{an}^i as a function of time.

Solution Figure 5.5 shows a graph oı c_{ap}/c_{an}^{i} shown as a function of time. This is simply a graphical representation of Eq. (5.10). The important point is that in 3 or 4 minutes essentially all of the particles have been transferred from nonplugging to plugging pathways.

Khilar and Fogler [13] have modified Eq. (5.16) somewhat and used the following empirical equation to represent the permeability loss as a function of the concentration of plugging deposits:

$$\frac{k}{k_o} = \left[1 - A\frac{c_{ap}}{c_{an}^{i}}\right]^2 \tag{5.17}$$

where A is an empirical constant.

As fresh water is introduced into a core, nonplugging particles are transferred to plugging pathways at the core inlet, but in those zones further downstream which have not yet been contacted, the nonplugging particles essentially remain intact. Thus, the permeability at the front face of the core is reduced first and then in subsequent sections as the fresh water front progresses. The time during which each part of the core has been in contact with fresh water differs. If this difference is taken into account, then the permeability at each point of the core can be calculated. The average value permeability of the entire core can then be obtained. Figure 5.6 shows the calculated permeability decline as compared with experimental data [14].

Thus, when fresh water is suddenly introduced into a core, the response is immediate, giving rise to a sharp permeability decline. Generally, great care is taken to avoid using fresh water in well treatment operations.

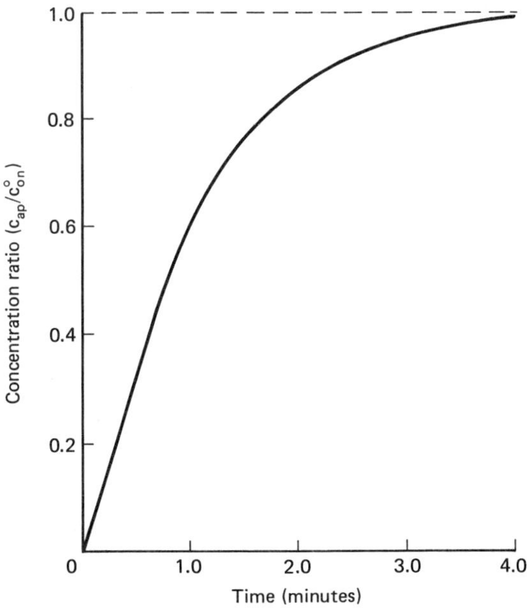

Figure 5.5 Graph showing rate at which plugging deposits build up.

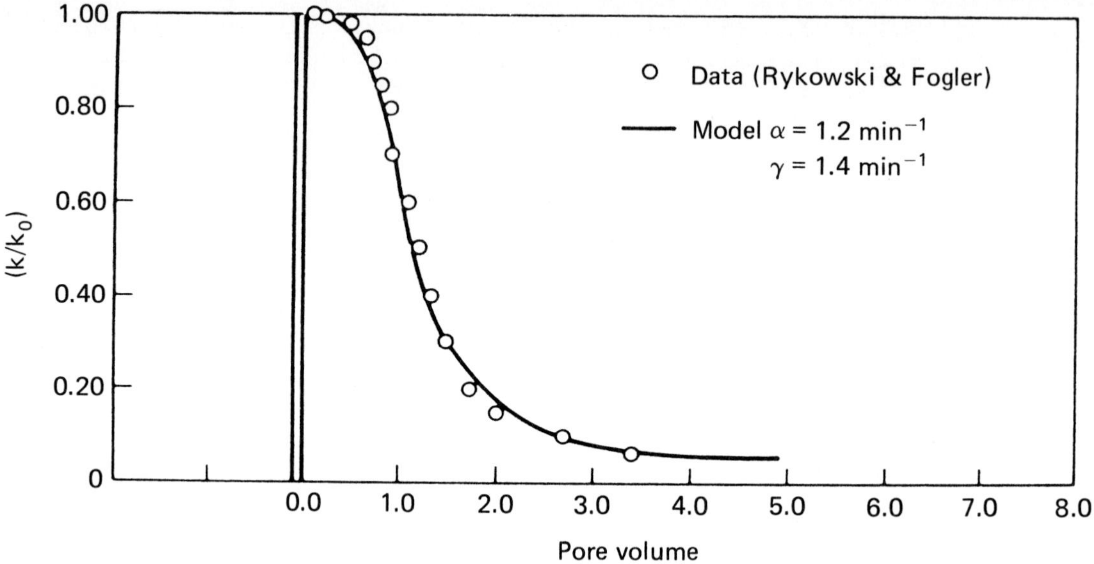

Figure 5.6 Graphical comparison of model predictions with experimental permeability decline. (With permission.)

5.2 NETWORK MODELS

Porous media can often be adequately modeled as a collection of larger void volumes connected to each other through a series of smaller channels. Such models may be thought of as a network composed of nodes and bonds. This model is similar to the dual-pathway concept, except a particle must flow through a bond to move to another node. These models have proven to be quite useful in many areas of science and engineering, including flow in porous media. Early work has been reported by Fatt [15, 16, 17] and by Hammersly [18].

Much of the work dealing with network or lattice models pertains to transport in inhomogenous conductors. One aspect of primary interest is that of percolation and percolation thresholds [2]. The concept is that in a lattice of voids, the average number of randomly placed bonds connecting the voids must exceed a certain critical value if there is to exist at least one continuous pathway extending from one end of the network to the other. For example, if the network is imagined to be a three-dimensional array of voids without any connection to one another, then fluid will not be able to flow from one void to another. If a few bonds are randomly introduced into the network, then fluid will be able to flow between the connected voids, but will not be able to flow from one end of the network to the other because an insufficient number of voids are connected. The percolation threshold is represented by that average number of bonds per void volume that must be present in the network so that the fluid can find at least one continuous connected path extending through the network. This critical coordination number is designated as Z_c and Kirkpatrick notes that the relationship

$$Z_c \approx \frac{d}{d-1} \tag{5.18}$$

is satisfied to within a few percent for all common two-dimensional and three-dimensional lattices [1]. Here d is the dimensionality of the network. For three-dimensional networks $Z_c \approx 3/2$. This number is of practical significance. In using a network model to study formation damage, once the number of channels clogged reduces the average coordination number so that $Z \approx Z_c$, then the network will no longer have a permeability. To the extent that the network actually models the behavior of porous sedimentary rock, then the permeability of the rock will also vanish.

Equation (5.18) can be used to help decide the point at which a porous medium will become completely clogged. This issue will be addressed, but first it is convenient to develop an approximate expression for the permeability in terms of the distribution of sizes of the small connecting channels (bonds) and the coordination number.

Effective Medium Theory

To quantify the transport of fluid through the network, it is necessary to assign a conductivity to each of the pore throats (bonds) connecting the large voids (nodes). This conductivity, when multiplied by the pressure difference between nodes, yields the volumetric flow rate through a bond. For simplicity, the channel or bond can be visualized as being a small cylindrical capillary of length l and radius r_{ij} (see Fig. 5.7). The subscripts i and j denote the two nodes (voids) that are linked together by the pore throat. Thus, the conductivity of a particular bond is given by

$$g_{ij} = \frac{\pi r_{ij}^4}{8\mu l} \tag{5.19}$$

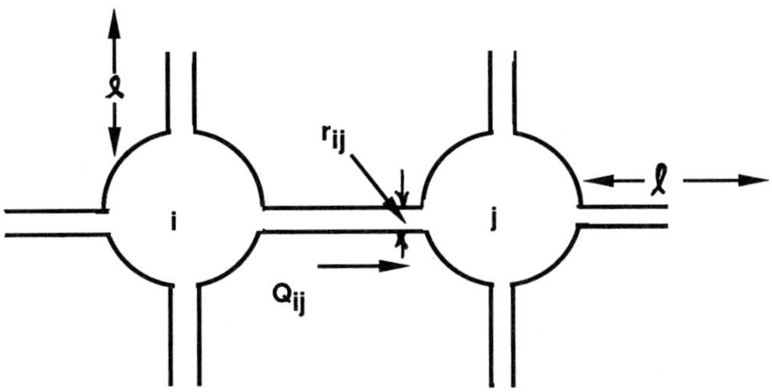

Figure 5.7 Sketch depicting a pore throat connecting pore body i to pore body j. The pore throat is considered to be a cylinder with radius r_{ij} and length l.

The fluid flow rate between i and j is therefore given by

$$Q_{ij} = \frac{\pi r_{ij}^4}{8\mu l} \Delta p = g_{ij} \Delta p \tag{5.20}$$

The network in question is depicted by Fig. 5.8, although the one considered here is a three-dimensional network rather than a two-dimensional network as shown in the figure and the average coordination number is Z. Note that the g_{ij} are randomly distributed; that is, there exists a distribution of conductivities and the value assigned to a particular g_{ij} is drawn from this distribution irrespective of its location within the lattice. Thus there is not a spatial bias which might, for example, require bonds near one side of the network to be less conductive on the average than bonds on the opposite side. Our goal is to replace this distribution of conductivities by a single value selected so that the average flow through the network caused by an applied pressure gradient will be the same as that for the actual network. This network where all of the pore throats have the same conductivity will be called an *effective medium*, and the conductivity of each of the pore throats will be denoted as g_m. Thus, there are two levels of approximation. First, the actual porous medium is replaced by a three-dimensional network having pore throats distributed according to a given probability function, and this network is in turn replaced by an effective network in which all pore throats have the same conductivity.

The flux from a column of nodes to the next column in the x-direction (as for example from the nodes in Column 4 of the network shown by Fig. 5.8 to the

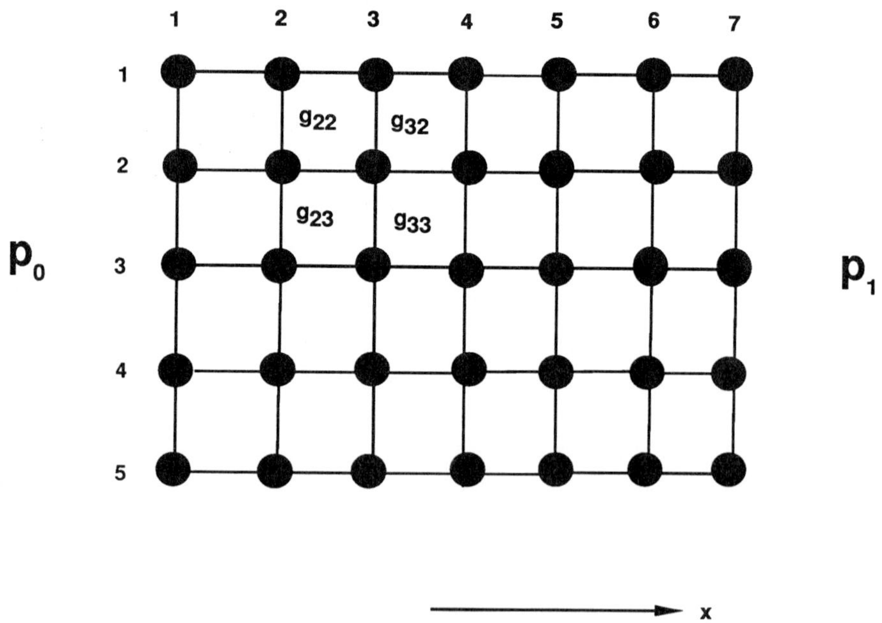

Figure 5.8 Sketch showing lattice of pore bodies connected by pore throats. The conductivities are randomly distributed.

nodes in Column 5) will depend not only on the mean conductivity of each of the bonds between the nodes, but also upon the total number of bonds per unit of area connecting nodes in Column 4 to nodes in Column 5. This number can be estimated as follows. Let N be the number of nodes per unit area in a plane perpendicular to the x-axis.

Each node is on the average connected by Z bonds to surrounding nodes. Of these roughly one third are parallel to the x-axis. The others on the average are parallel to the y- and z-axes. Thus, $NZ/3$ bonds are parallel to the x-axis but only one half of these extend in the forward x-direction from a given node. The flux is, therefore, found by multiplying the volumetric flow rate through one bond by $NZ/6$. Based on Eq. (5.20), the x-component of flux is given by

$$u_x = \left(\frac{NZ}{6} g_m\right) \Delta p \tag{5.21}$$

Comparing Eq. (5.21) with Darcy's law, the permeability of the mean effective network is

$$k = NZlg_m\mu/6 \tag{5.22}$$

The pressure at a point in the actual network can be thought to be composed of two parts. One part is the value that exists at the same point in the effective medium and can be thought of as that due to the externally imposed pressure. The second contribution can be considered to be a fluctuation arising from the fact that the g_{ij} are not constant and equal to g_m. The fluctuating pressure when averaged over a large enough region must vanish. As an approximation consider the network shown in Fig. 5.9. This shows a single pore throat having a conductivity g_0 surrounded by an effective medium in which all of the surrounding pore

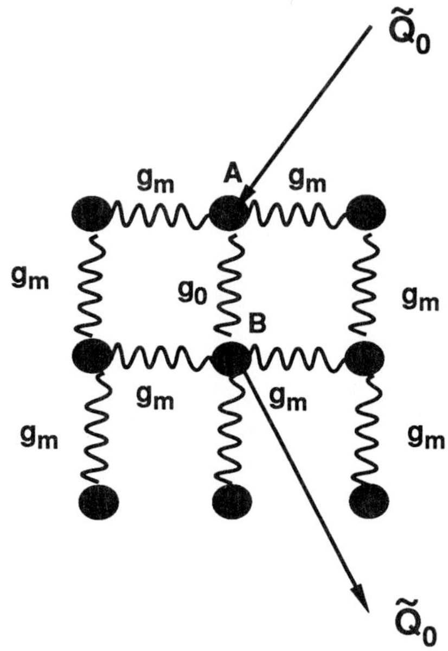

Figure 5.9 Sketch depicting one pore throat with a conductance g_0 surrounded by an effective medium.

throats have the effective conductivity, g_m. The fact that the conductance of one bond differs from all of the other surrounding bonds will cause a spurious volumetric flow from point A to point B. It is called *spurious* because when all conductances are the same and equal to the mean value, this extra volumetric flow vanishes. It is a volumetric flow which is over and above the average value. Since the flow rate is a linear function of the pressures, one would expect this extra volumetric flow to be proportional to $g_m - g_0$, the amount the conductance deviates from the mean value. Thus

$$\tilde{Q}_0 = A(g_m - g_0) \tag{5.23}$$

where $\tilde{Q}_0$ is the spurious or fluctuating volumetric flow rate and A is simply a proportionality constant.

It should be noted that not all of this flow which leaves node A and appears at node B actually goes through the bond having conductance g_0. Only a part of it does. Some flow may take more circuitous paths leaving A and arriving at B. This will always be a possibility in any highly interconnected network.

If G'_{AB} is the conductance of the network between pore body A and pore body B with g_0 removed, then the fluid flow $\tilde{Q}_0$ which goes both directly from A to B through g_0 and also through the other available paths (see Fig. 5.10) in the network causes a pressure fluctuation given by

$$\Delta\tilde{p}_0 = \frac{\tilde{Q}_0}{G'_{AB} + g_0} \tag{5.24}$$

The network conductance G_{AB} with g_0 removed is the conductance of the effective medium when all connections have a conductance g_m less g_m; that is

$$G'_{AB} = G_{AB} - g_m \tag{5.25}$$

where G_{AB} is the conductance of a network with all bonds having a conductance g_m.

The conductance G_{AB} can be found using the linearity of the network and

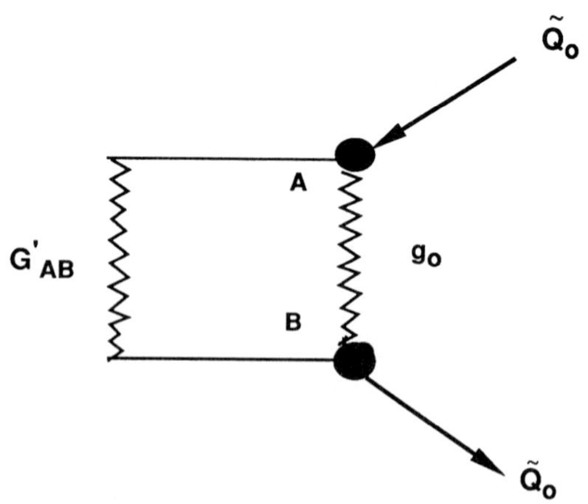

Figure 5.10 A simplified circuit that is equivalent to the one shown in Fig. 5.9.

a symmetry argument. Imagine that $\tilde{Q}_0$ is introduced at A in the effective medium with all conductances equal to g_m and removed at infinity. Since all connections are equally conductive, then a fraction $\tilde{Q}_0/Z$ of the fluid introduced into pore body A will flow directly from A to B on its route to infinity. Similarly if a volume of fluid per unit of time $\tilde{Q}_0$ is injected at infinity and removed at B, then on the average, volume per time flowing directly from A to B is also $\tilde{Q}_0/Z$. Since the sum of the two processes is equivalent to $\tilde{Q}_0$ added at A and $\tilde{Q}_0$ removed at B (the network is linear), then in the sum of the processes $2\tilde{Q}_0/Z$ flows through the pore throat connecting A to B when the total flow from A to B is $\tilde{Q}_0$. Since the flow through a bond is now known, the pressure difference can be calculated as follows:

$$2\tilde{Q}_0/Z = \Delta\tilde{p}_0 g_m \tag{5.26}$$

Clearly then, $G_{AB} = g_m Z/2$ or $G'_{AB} = (Z/2 - 1)g_m$. Substituting Eq. (5.24) into Eq. (5.23) for $\tilde{Q}_0$ yields

$$\Delta\tilde{p}_0 = A \frac{g_m - g_0}{g_0 + \left(\dfrac{Z}{2} - 1\right)g_m} \tag{5.27}$$

Equation (5.27) is an important result. It gives the pressure fluctuation or deviation in pressure caused by the conductance of one bond being different from the average. However, there must be other pressure fluctuations somewhere else in the network which on the average cancel this one. This must be so because by definition g_m is the conductance which when assigned to each bond will give a pressure gradient equal to the imposed one. Thus if g_{ij} of the actual network is distributed such that $f(g)\,dg$ is the fraction of the bonds having a conductance between g and $g + dg$, then

$$\int \Delta\tilde{p}_0 f(g)\,dg = 0 \tag{5.28}$$

From Eq. (5.27) this condition leads to

$$\int \frac{f(g)(g_m - g)\,dg}{\left[g + \left(\dfrac{Z}{2} - 1\right)g_m\right]} \tag{5.29}$$

This equation determines g_m. Given $f(g)$ and Z, there is a single value of g_m that will satisfy this equation.

Example 5.2 Mean Conductivity of a Binary Distribution

Suppose that a porous medium is composed of just two types of pore throats. A small fraction of the throats (ϵ) have a large conductivity g_L, whereas a fraction $(1 - \epsilon)$ have a small conductivity g_S. Calculate the mean conductivity for the case $\epsilon = 0.1$, $g_L = 10$, and $g_S = 0.1$. (The units associated with conductivity are discussed in Example 5.3.) Take $Z = 8$.

Solution Equation (5.29) reduces to a sum of two terms

$$\epsilon \left\{ \frac{g_m - g_L}{g_L + \left(\frac{Z}{2} - 1\right) g_m} \right\} + (1 - \epsilon) \left\{ \frac{g_m - g_S}{g_S + \left(\frac{Z}{2} - 1\right) g_m} \right\} = 0$$

This expression can be rearranged as follows:

$$\left(\frac{Z}{2} - 1\right) g_m^2 + g_m[\epsilon g_S + (1 - \epsilon) g_L] - g_m \left(\frac{Z}{2} - 1\right) [\epsilon g_L + (1 - \epsilon) g_S] - g_S g_L = 0$$

This is a quadratic equation, which can readily be solved giving

$$g_m = 0.161$$

It is easy to verify that g_m increases as Z increases, all other factors being held fixed. The interpretation is straightforward. The more connections a node has to other nodes, the higher is the probability that one of the bonds will be of the high conductivity variety. Thus in Example 5.2 for $Z = 12$, g_m is found to be 2.16 rather than 0.161.

Equation (5.29) is an approximation. Kirkpatrick [1] shows that this equation accurately fits the results of Monte Carlo simulations when the bonds satisfy the binary distribution considered in Example 5.2, provided the average number of bonds exceeds Z_c. When Z approaches Z_c, the predictions based on Eq. (5.29) become unreliable.

Even when Z exceeds Z_c, Eq. (5.29) must still be regarded as an approximation subject to some error that will depend on the pore size distribution. Despite these possible errors, Eq. (5.29) is generally applied when using networks to model formation damage.

Permeability of Damaged Formations

Given N, Z, and $f(g)$ the permeability of a network can be estimated. To the extent that the behavior of a network is representative of an actual porous medium, the permeability of the actual porous medium is also modeled. The processes leading to formation damage can be studied in terms of a network model, and this may also be an adequate representation of the processes going on within the porous medium. This is presumed here to be the case. The flow in the network is assumed to generate particles by dislodging them from the walls of the voids (nodes) causing them to be suspended in the flowing system. In this sense the network model is similar to the parallel-pathway model. Those suspended particles that are larger in size than the smallest bonds will ultimately become lodged in pore throats. It is assumed here that a clogged pore throat is no longer conducting.

If the damage process results in the clogging of all pores having a conductivity less than g^*, then the new $f^*(g)$ is given by

$$f^*(g) = f(g) \quad \text{for } g > g^* \qquad \text{and} \qquad f^*(g) = 2\epsilon\delta(g) \quad \text{for } g < g^* \tag{5.30}$$

Here $\delta(g)$ is a Dirac delta function which is a ''pathological'' function that vanishes everywhere except at the origin where it is infinite. The reader should regard this as a sort of a "spike" whose integral represents the fraction of the pores that are clogged. We really need to use only the following property of the delta function:

$$\int_0^g F(g)\delta(g)\,dg = \frac{1}{2}F(0) \tag{5.31}$$

where $F(g)$ is any function of g.

The quantity ϵ is the fraction of the pores that are clogged and have a zero conductivity. Since $f(g)\,dg$ represents the fraction of the pores having a conductivity between g and $g + dg$, then the fraction clogged is composed of all the pores having a conductivity less than g^*. Thus

$$\epsilon = \int_0^{g^*} f(g)\,dg \tag{5.32}$$

The mean conductivity of the undamaged formation is given by Eq. (5.29). Similarly, the mean conductivity of the damaged formation is given by substituting Eq. (5.30) into Eq. (5.29). This leads to the following equation for the mean conductivity of the damaged formation, g_m^*:

$$\frac{\epsilon}{\left(\frac{Z}{2} - 1\right)} + \int_{g^*}^{\infty} \frac{f(g')(g_m^* - g')\,dg'}{g' + \left(\frac{Z}{2} - 1\right)g_m^*} = 0 \tag{5.33}$$

The permeability ratio is then given by

$$\frac{k_D}{k} = \frac{g_m^*}{g_m} \tag{5.34}$$

These equations provide a model for the reduction of the permeability resulting from the clogging of pores having a conductivity less than g^*. The equations do not, however, define the rate at which the permeability declines. Sharma and Yortsos have formulated this rate by making population balances; that is, by maintaining an inventory accounting for the total number of particles of a particular size [18, 19]. This approach appears to be a fruitful one and may at some future time become suitable for modeling the dynamics of the processes resulting in formation damage. Their importance, however, remains to be experimentally verified. Although a number of network models have been applied to characterize porous media [20], it is still not clear just how one is to go about suitably defining N, Z, and $f(g)$ based on the available experiments. Thus, one can state that there remains a need for further research if the network models are to attain practical value.

Example 5.3 Permeability Loss

A certain network is defined by

$$f(g) = \begin{cases} 601.5\exp(-600g) & \text{for } g < 1 \times 10^{-2} \\ 0 & \text{for } g > 1 \times 10^{-2} \end{cases}$$

where g is the bond conductivity expressed as $(\mu\text{m})^4$-sec/kg.

(a) If the fluid flowing through the network has a viscosity of 5×10^{-3} kg/m-sec, estimate the radius of the largest pore throats. Take the length of the bonds to be 10^{-4} m.

(b) Find the mean conductivity for this network. Take $Z = 8$.

(c) When all bonds having a conductivity less than 10^{-3} $(\mu m)^4$-sec/kg become clogged, what is the mean conductivity of the damaged network?

(d) Find the permeability of the damaged formation relative to the undamaged permeability.

Solution

(a) Rearranging Eq. (5.19), yields

$$r = \sqrt[4]{\frac{8g\mu l}{\pi}}$$

Since the largest conductivity is 1×10^{-2} $(\mu m)^4$-sec/kg, then

$$r_{\max} = \sqrt[4]{\frac{8(1 \times 10^{-2})(5 \times 10^{-9})(100)}{\pi}} = 1.06 \times 10^{-2}\ \mu m$$

[*Note:* Viscosity $= (5 \times 10^{-3}$ kg/m-sec$)(10^{-6}$ m/μm$) = 5 \times 10^{-9}$ kg/μm-sec.]

This is a small maximum radius. Many particles adhering to the pore surface will be larger than this pore throat diameter, indicating that this network and the porous medium it represents will be easily damaged.

(b) The mean conductivity is found using Eq. (5.29). Thus

$$\int_0^{1\times 10^{-2}} \frac{(g - g_m)\exp(-600\,g)\,dg}{g + 3g_m} = 0$$

This cannot be integrated directly. The procedure entails assuming a value for g_m and performing the integration numerically. If the resultant integral is not zero, then a new value of g_m must be assumed. This process must be continued until the correct value of g_m is found. The solution to this trial-and-error problem is $g_m = 1.3 \times 10^{-3}$ $(\mu m)^4$-sec/kg.

(c) The fraction of pores that are clogged is determined based on Eq. (5.32) as follows:

$$\epsilon = \int_0^{1\times 10^{-3}} 601.5\exp(-600g)\,dg \qquad \text{or} \qquad \epsilon = \frac{601.5}{600}[1 - \exp(-0.6)] = 0.4523$$

The mean conductivity of the clogged network is then found by determining the g_m^* which satisfies the equation

$$\frac{0.4523}{3} + \int_{1\times 10^{-3}}^{1\times 10^{-2}} \left(\frac{g_m^* - g'}{g' + 3g_m^*}\right) 601.5\exp(-600\,g')\,dg' = 0$$

This equation cannot be solved explicitly for g_m^*, but it can again be solved by trial-and-error methods. This procedure yields

$$g_m^* = 3.2 \times 10^{-4}\ (\mu m)^4\text{-sec/kg}$$

(d) The permeability ratio is then given by Eq. (5.34):

$$\frac{k_D}{k_0} = \frac{3.2 \times 10^{-4}}{1.3 \times 10^{-3}} = 0.246$$

REFERENCES

5.1. Kirkpatrick, S., *Rev. Modern Phys., 45* (1973) 574.
5.2. Stauffer, D., *Introduction to Percolation Theory*, London: Taylor and Francis, 1985.
5.3. Iwasaki, T., *J. Amer. Water Works Assoc., 29* (1937) 1591.
5.4. Herzig, J. P., Leclerc, D. M., and LeGoff, P., *Ind. and Eng. Chem., 62* (1970) 8.
5.5. Payatakes, A. C., Rajagopalan, R., and Tien, C., *Canad. J. Chem. Eng., 52* (1974) 722.
5.6. Payatakes, A. C., Tien, C., and Turian, R. M., *Amer. Inst. Chem. Eng. J., 20* (1974) 889.
5.7. Rajagopalan, R., and Tien, C., in *Progress in Filtration and Separation I*, New York: North Holland (1979) 179.
5.8. Gruesbeck, C., and Collins, R. E., *Soc. Pet. Eng. J., 22* (1982) 847.
5.9. Muecke, T. W., *J. Pet. Tech., 31* (1979) 144.
5.10. Radke, K. T., "Fine Particle Migration," MS thesis, The University of Texas at Austin, 1984.
5.11. Atwood, D. K., *J. Pet. Tech., 16* (1964) 1405.
5.12. Mungan, N., *J. Pet. Tech., 17* (1965) 1449.
5.13. Khilar, C., and Fogler, H. S., *Soc. Pet. Eng. J., 23* (1983) 55.
5.14. Khilar, K. C., and Fogler, H. S., in *Surface Phenomena in Enhanced Oil Recovery*, D. O. Shah (ed.), New York: Plenum Press (1981) 721.
5.15. Fatt, I., *Trans. AIMME, 207* (1956) 144.
5.16. Fatt, I., *Trans. AIMME, 207* (1956) 160.
5.17. Fatt, I., *Trans. AIMME, 207* (1956) 164.
5.18. Sharma, M., and Yortsos, Y. C., *Amer. Inst. Chem. Eng. J., 33* (1987) 1636.
5.19. Sharma, M., and Yortsos, Y. C., *Amer. Inst. Chem. Eng. J., 33* (1987) 1654.
5.20. Dullien, F. A. L., *Porous Media*, New York: Academic Press, 1979.

PROBLEMS

****5.1.** A pore size distribution function for a collection of capillaries is defined as

$$p(r) = A \exp(-\lambda r^2) \quad \text{for } r > R_0$$

$$p(r) = 0 \quad \text{for } r < R_0$$

where $p(r)\,dr$ is the probability that a pore selected at random will have a radius between r and $r + dr$.

(a) For the case $\lambda = 25\ (\mu\text{m})^{-2}$ and $R_0 = 0.01\ \mu\text{m}$ show that $A = 5.979$.

(b) A suspension of particles having a diameter of 0.2 μm flows through the collection of capillaries defined in Part **(a)** If tubes having a diameter smaller than 3 times the particle diameter become clogged, what fraction of the capillaries can be described as providing nonplugging pathways?

(c) If the permeability of the system of capillaries is proportional to $\overline{r^2}$ [the average of the square of the radii which is $\int p(r) r^2\, dr$], what fraction of the permeability is lost when only the nonplugging pathways in Part **(b)** conduct fluid? The plugging pathways are clogged.

****5.2** The transport of suspended particles through a porous medium is described by Eq. (5.1). An interesting special case arises whenever particles that adhere are not remobilized. This mechanism is expressed by the rate law

$$\frac{\partial c_a}{\partial t} = \beta c$$

According to this equation, the concentration of adhering particles cannot decrease and it increases at a rate proportional to the suspension concentration, c.

(a) For this special case show that when $c_a = c = 0$ at $t = 0$, then Eq. (5.1) requires that

$$c = \begin{cases} 0 \quad \text{for } 0 < t < \dfrac{\phi x}{u} \\ \\ c_0 \exp\left(-\dfrac{\beta\phi x}{u}\right) \quad \text{for } t > \dfrac{\phi x}{u} \end{cases}$$

where c_0 is the suspension concentration of the fluid flowing into the porous medium (at $x = 0$; $c = c_0$).

(b) Using the result in Part **(a)**, develop an expression for $c_a(x, t)$.

(c) For $c_0 = 10^{-3}$ m^3/m^3 of solution and $\beta = 0.1$ sec^{-1}, plot c_a as a function of x at times 100, 1,000, and 10,000 sec, respectively, measured from the time flow is initiated. The Darcy flux through the core, which is 100 cm long and has a porosity of 10%, is 0.5 cm/sec.

****5.3.** The particle transport equation [Eq. (5.1)] applies to linear systems and is obtained by constructing a small volume element and noting that the number in less the number out must equal the accumulation of particles. A similar transport equation can be derived for radial transport of particles away from the wellbore. Construct a cylindrical shell within the porous medium and by making a balance conserving particles, show that for radial geometry

$$\frac{\partial c_a}{\partial t} + \frac{\partial c}{\partial t} + \frac{i}{2\pi rh}\frac{\partial c}{\partial r} = 0$$

where i is the volumetric injection rate, h is the formation thickness, and r is the distance measured from the center of the wellbore.

Show that this equation can be reduced to a form resembling Eq. (5.1) by defining a new measure of distance

$$\epsilon = r^2$$

Based on this similarity and the results obtained in Problem 5.2, find both $c(r, t)$ and $c_a(r, t)$ for the special case

$$\frac{\partial c_a}{\partial t} = \beta c$$

***5.4.** Assuming that there are no plugging pathways and that the particles in the fluid phase are in local equilibrium with those adhering to the wall, determine the time that will be required to displace half of the particles initially present in a 5-cm core ($\phi = 0.15$) entirely from the core. The following data apply:

$$c_a = 0.01 \text{ m}^3/\text{m} \quad \text{(initially)}$$

$$\beta/\alpha^* = 5$$

$$u_{np} = u = 3 \text{ cm/hr}$$

**5.5. When fresh water is injected into a sandstone, it is reasonable to neglect particle reattachment within nonplugging pathways. Thus, particles initially attached in nonplugging pathways will ultimately deposit in a plugging mode. This process is expressed by Eq. (5.16).

For the parameters given in Table P5.5, construct a graph showing the permeability ratio of a linear sandstone core sample as a function of pore volumes of fresh water injected. [*Hint:* Imagine the core sample to be divided into distinct sections and calculate the permeability ratio for each of the sections. Then determine the mean permeability for core sections that are connected in series. Also remember that the time the particles begin to be released is measured from the time the fresh water arrives at a particular section].

TABLE P5.5 Data for Graph of Permeability Ratio vs. Pore Volume

Parameter	Symbol	Value
Particle release parameter	α^*	1.5 min^{-1}
Particle capture parameter	γ^*	2.5 min^{-1}
Core length	L	1 m
Injection flux	u	1 cm/sec
Attached particle concentration	c_a	0.04 m^3/m^3 of pore volume
Core porosity	ϕ	0.15
Constant in permeability expression	q	1.0
Constant in permeability expression	a	70

**5.6. Suppose that the distribution of conductivities is given by

$$f(g) = Ag^2 \exp(-\lambda g)$$

where $\lambda = 200\ \text{kg}/(\mu\text{m})^4\text{-sec}$. Find the mean conductivity of the network, given that the coordination number is 11.

6
Pretreatment Well Tests

If a well is not producing at the anticipated volumetric rate, any one of a combination of factors may be responsible for the poor performance. Clearly, the success of any treatment designed to increase production will critically depend on the correct diagnosis of the problem.

Well performance curves (production shown as a function of time) are normally readily available and can provide clues to help define the problem. By comparing the actual well performance with what is considered normal for that type of operative reservoir drive mechanism, very often one can decide whether the formation pressure is depleted or whether the poor performance is instead due to formation damage or to mechanical problems in the wellbore such as sand production or ineffective artificial lift.

Production logging designed to determine flow profiles of oil, gas, and water entering the wellbore from different formations also provides very useful information. A substantial suite of production logs are now available to production engineers. Based on these logs he or she can decide which zones are producing what fluids and may even evaluate the extent to which individual perforations are flowing. Discussion of production logging, a subject suitable for a book in its own right [1], is beyond the scope of this text; however, it is important to keep the availability of these tools in mind.

A third source of information, which can be very useful, is flowing and static bottomhole pressure data, especially pressure drawdown and buildup tests. Such tests can be used to determine static bottomhole pressure, interwell kh, and the skin effect. We want to review briefly the use of pressure drawdown tests not only because the information so obtained is useful in designing well stimulations but also because postfracturing pressure drawdown tests are valuable in determining whether or not the designed treatment was actually realized. Postfracture pressure tests are always useful, but when the entire field is to be produced entirely through fractured wells, then such tests of early treatments become imperative. It will, therefore, be useful for the discussion in subsequent chapters to review prefacture pressure testing briefly here. This chapter is not intended to be comprehensive, as a number of other excellent sources of information on pressure testing are available [2, 3, 4].

6.1 PRETREATMENT PRESSURE DRAWDOWN TESTS OF OIL WELLS

To conduct a pressure drawdown test, the well must be shut in for a long period of time to permit the pressure to stabilize or the test must be performed on newly completed wells. Generally, a pressure buildup test is more accurate than a drawdown test since in a drawdown test one must maintain constant production. This may not be easy to do in all cases. Nevertheless, for the reasons cited previously, we shall consider only pressure drawdown tests. More extensive coverage of pressure testing is given by Earlougher [2].

In what follows, the reservoir is assumed to be homogeneous (except for a damaged zone surrounding the wellbore), isotropic, and drained by a fully penetrating well to ensure radial flow. The flow is assumed to be single phase and the fluid slightly compressible. The physical properties (viscosity and compressibility) are assumed constant. Furthermore, the test we consider here will assume that an infinite reservoir is being produced. This assumption will limit the time interval over which the equations apply to a time less than that required for the pressure wave to reach the boundaries of the system.

To be a good approximation, over the time period of interest

$$p_{\mathrm{wf}} = p_{\mathrm{i}} - \frac{Bq\mu}{4\pi kh}\left\{\ln\left(\frac{4kt}{\gamma\phi\mu\kappa r_{\mathrm{w}}^{2}}\right) + 2S\right\} \qquad (6.1)$$

where
p_{wf} = Bottomhole well flowing pressure
p_{i} = Initial reservoir pressure
q = Production rate measured at the well head
μ = Viscosity
h = Formation thickness
ϕ = Porosity
r_{w} = Wellbore radius
S = Skin factor
κ = Isothermal compressibility
γ = Constant = 1.781
B = Reservoir volume factor (reservoir volume/stocktank volume)

The skin factor is included in Eq. (6.1) to allow for formation damage or for wells that have been stimulated either by acidizing or hydraulic fracturing. The derivation of Eq. (6.1) is straightforward. The damaged or stimulated zone is considered equivalent to an altered zone of uniform permeability (k_S) and outer radius (r_S). The additional pressure drop across this zone (Δp_S) can be modeled by the steady-state radial flow equation. The geometry is depicted in Fig. 6.1. Thus

$$\Delta p_S = \frac{Bq\mu}{2\pi h} \ln \frac{r_S}{r_W} \left\{ \frac{1}{k_S} - \frac{1}{k} \right\} \tag{6.2}$$

From this equation it is easy to see that the first term on the right-hand side of Eq. (6.1) represents the actual pressure drop from the axial position r_S to the wellbore r_W and the second term represents the same pressure drop that would exist if the permeability in the zone would be k instead of k_S. Thus, as noted Δp_S is an additional pressure.

Equation (6.2) is generally written as

$$\Delta p_S = \frac{Bq\mu}{4\pi kh} S \tag{6.3}$$

where

$$S = \left(\frac{k}{k_S} - 1 \right) \ln \frac{r_S}{r_W} \tag{6.4}$$

and when this pressure drop Δp_S is added to that obtained assuming a uniform permeability, Eq. (6.1) results. The skin factor is determined by analyzing the pressure test, but there is no satisfying procedure for determining k_S and r_S sep-

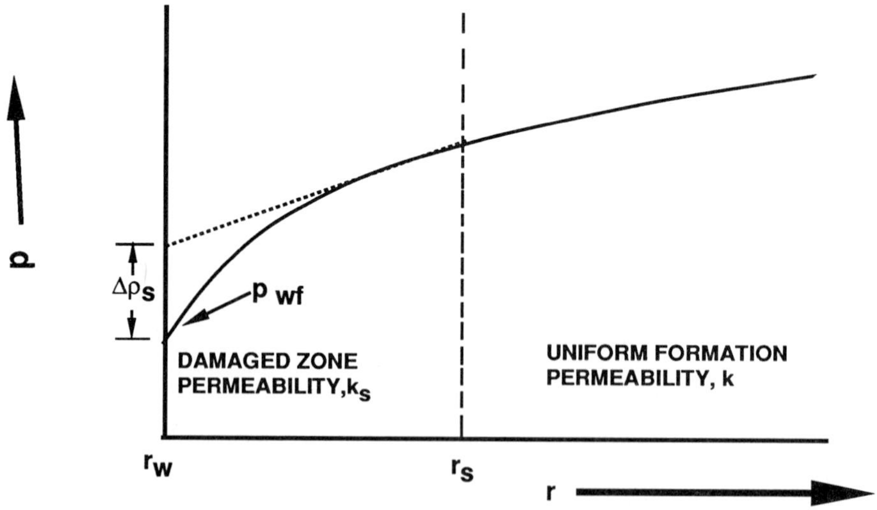

Figure 6.1 The solid line depicts the pressure loss as a function of position whereas the dashed line represents the pressure that would be observed in undamaged formation.

arately. It would be very helpful if there did exist a technique for estimating the depth of the damaged zone, as this would serve as a basis for choosing between matrix acidization or fracturing treatments and would also contribute to better designed matrix acid treatments.

The application of Eq. (6.1) to the analysis of pressure drawdown data has been discussed by Earlougher [2] and by Lee [3]. We will not detail the procedure here. We are, however, interested in showing that both k and S can be calculated given the well flowing pressures measured during the initial stages of production. A plot of typical pressure drawdown data is shown in Fig. 6.2. It is evident that a typical test can be divided into an early time response, a middle time response, and a late time response. It is the middle time response that is of interest here.

The deviation of the pressure at late times from the predicted straight line behavior is generally thought to be due to the finite extent of the reservoir. Equation (6.1) is derived assuming that the reservoir is infinite and therefore does not apply to the late times. The late time response can, however, be used to help define the size of the reservoir [2, 3].

The early time response deviates from the theoretical curve because the liquid levels may change in the well during the early stages of the test and because of the skin effect. Thus, all of the liquid entering the wellbore may not be initially produced. However, a steady state is ultimately reached in which the rate of fluid production is equal to the rate of fluid entry into the wellbore. At this point in time, the theoretical curve then applies. This is the middle time portion of the curve.

It is, therefore, the middle time response that interests us, for the data obtained during this time can be used to evaluate k and S.

Example 6.1 Analysis of Pressure Drawdown Data [3]

A pressure drawdown test of a newly completed oil well yields the data given in Table 6.1. Estimate the formation permeability and the skin factor.

Solution The first step is to determine the slope of the linear middle time response curve. This is most easily done by plotting $\Delta p = p_i - p_{wf}$ as a function of ln (t).

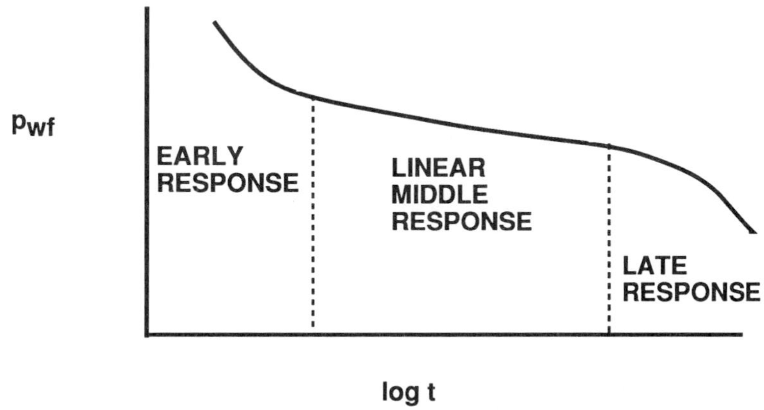

Figure 6.2 Plot of typical pressure response during a drawdown test.

TABLE 6.1 Well Test Data

Time (hr)	p_{wf} (kPa)	Δp (kPa)	ln (time)	Time (hr)	p_{wf} (kPa)	Δp (kPa)	ln (time)
0	30,421	0	—	35.8	24,436	5985	3.578
0.12	26,284	4137	−2.120	43.0	24,388	6033	3.761
1.94	25,505	4916	0.663	51.5	24,353	6068	3.942
2.79	25,187	5234	1.026	61.8	24,312	6109	4.124
4.01	25,070	5351	1.389	74.2	24,277	6144	4.307
4.82	24,932	5489	1.573	89.1	24,236	6185	4.490
5.78	24,870	5551	1.754	107	24,195	6226	4.673
6.94	24,822	5599	1.937	128	24,153	6268	4.852
8.32	24,774	5647	2.119	154	24,112	6309	5.037
9.99	24,725	5696	2.302	185	24,064	6357	5.220
14.4	24,636	5785	2.667	222	24,001	6420	5.403
17.3	24,594	5827	2.851	266	23,939	6482	5.583
20.7	24,533	5888	3.030	319	23,857	6564	5.765
24.9	24,512	5909	3.125	383	23,760	6661	5.948
29.8	24,470	5951	3.395	460	23,643	6778	6.131

Well Data:		
	Formation thickness (h)	21 m
	Formation porosity (ϕ)	0.04
	Wellbore radius (r_w)	0.06 m
	Oil production rate (q)	$4.6 \times 10^{-4} m^3/sec$
	Reservoir volume factor (B)	1.136
	Oil compressibility (κ)	2.45×10^{-9} Pa^{-1}
	Oil viscosity (μ)	8×10^{-4} kg/m-sec

Figure 6.3 shows the plot. The slope is 2.33×10^5 when Δp is expressed in pascals. This slope is obtained by a least squares fit of an equation of the form $\Delta p = m \ln (t) + b$ using only the data between $2 < \ln t < 5$. Another procedure is simply to read two values of Δp on the straight line portion of the curve. Then

$$\text{Slope} = \frac{(\Delta p_2) - (\Delta p_1)}{\ln t_2 - \ln t_1}$$

From Eq. (6.1) it is easily seen that

$$k = \frac{Bq\mu}{4\pi h(\text{slope})}$$

[*Note:* This slope is independent of the dimensions of time that are used.] Thus

$$k = \frac{(1.136)(4.6 \times 10^{-4})(8 \times 10^{-4})}{4\pi(21)(2.33 \times 10^5)} = 6.8 \times 10^{-15} \text{ m}^2$$

One way to determine the skin factor involves simply substituting into Eq. (6.1) and solving for S. Some authors recommend using an extrapolated value such as the one where $\ln t = 0$ when t is expressed in hours. For the given data, the

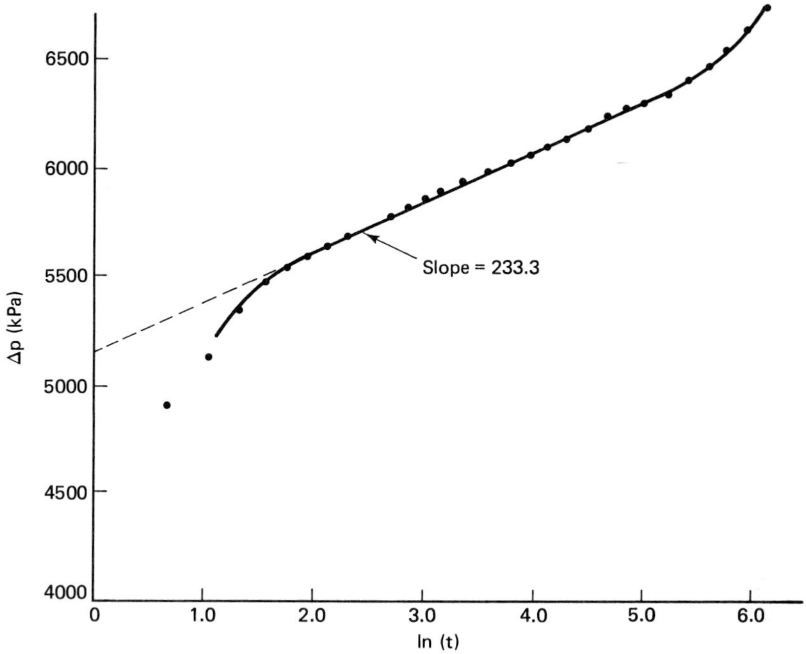

Figure 6.3 Plot of pressure response during a drawdown test. Data is from Table 6.1.

extrapolated Δp shown by the dashed line in Fig. 6.3 is 5.2×10^6 Pa. Those authors who recommend use of this value generally do so because all of the equations are expressed in field units and it is convenient to set $\ln t = 0$. However, in SI units, it is most convenient to express time in seconds. Thus the 1-hr extrapolated value can be used but it is no more convenient than simply taking the value of Δp at any given time and directly substituting into Eq. (6.1). Let us use the extrapolated value so that $\Delta p = 5.2 \times 10^6$ Pa at $t = 3600$ seconds. Thus

$$S = \frac{2\pi kh}{Bq\mu}\Delta p - \frac{1}{2}\ln\left(\frac{4kt}{\gamma\phi\mu\kappa r_w^2}\right)$$

or $$S = \frac{5.2 \times 10^6}{2(2.33 \times 10^5)} - \frac{1}{2}\ln\left(\frac{4(6.8 \times 10^{-15})(3.6 \times 10^3)}{1.781(0.04)(8 \times 10^{-4})(2.45 \times 10^{-9})(0.06)^2}\right)$$

Thus $$S = 5.1$$

This positive skin factor shows the well is damaged and production can be increased by removing this damage.

6.2 THE SIGNIFICANCE OF THE SKIN FACTOR

In Example 6.1, the value of S indicates some damage. Furthermore, the reservoir permeability is small. These data indicate that the well may be a candidate for a large fracture treatment. As will be evident, large S and large k may dictate small fracture or acid treatments.

The value of S can become negative if $k_S > k$. This situation indicates a stimulated well or perhaps the presence of natural fractures. S can range from $-\infty$ to $+\infty$, but generally about -10 is as small as S can be expected to be. If, however, the well does not flow at all, then S is $+\infty$. This is not a rare situation.

The significance of a particular value for S can be appreciated by noting that if the average reservoir pressure is $\bar{p}$, then the production during the semisteady-state inflow period which applies to a system having a finite drainage radius, r_e, is given by (see Dake [4])

$$\bar{p} - p_{wf} = \frac{Bq\mu}{2\pi kh}\left[\ln\left(\frac{r_e}{r_w}\right) - \frac{3}{4} + S\right] \tag{6.5}$$

From Eq. (6.5) we can deduce the following result:

$$\frac{q_{damaged}}{q_{undamaged}} = \frac{\ln\left(\frac{r_e}{r_w}\right) - \frac{3}{4}}{\ln\left(\frac{r_e}{r_w}\right) - \frac{3}{4} + S} \tag{6.6}$$

assuming that the same drawdown pressure will apply in both cases. From Eq. (6.6) we can develop some intuition as to the importance of a given skin effect and the intrinsic value of removing the damage by a well treatment.

Example 6.2 Restricted Production

The value of S found in Example 6.1 was about 5. Calculate the percentage reduction in production which can be expected because of this damage. The drainage radius is 200 m and the wellbore radius is 0.06 m.

Solution

$$\frac{Q_{damaged}}{Q_{undamaged}} = \frac{\ln\left(\frac{200}{0.06}\right) - \frac{3}{4}}{\ln\left(\frac{200}{0.06}\right) - \frac{3}{4} + 5} = 0.6$$

Therefore, the production can be roughly doubled if the damage to the well can be removed.

6.3 PRESSURE TESTING OF GAS WELLS

The equations used to describe the flowing pressure in the wellbore when oil is produced are, in general, not adequate for gases. The compressibility of gas is generally large, not small as is assumed for oil flow, and the viscosity of a gas may vary widely with pressure, whereas liquid viscosities are insensitive to changes in pressure over a wide range.

A number of investigators have considered this important problem [5, 6], but for our purposes, we shall assume simply that the compressibility factor (Z), the gas compressibility (κ), and the gas viscosity (μ) are all constants and can be evaluated therefore at the original reservoir pressure. We know these may be poor approximations and that better methods than those used here now exist for testing gas wells; however, the problems considered here are for illustrative purposes only. To actually interpret well tests, one should utilize the best procedures available.

The volume of gas produced at the surface measured at standard conditions of temperature (T_{STD}) and pressure (P_{STD}) is related to the volumetric flow into the wellbore by the equation

$$q = q_B\left(\frac{p_{wf}T_{STD}}{p_{STD}TZ}\right) \tag{6.7}$$

where q_B is the volumetric flow rate evaluated under bottomhole conditions. Based on this result and the assumptions cited previously, one can show that

$$p_{wf}^2 = p_i^2 - \left(\frac{Zqp_{STD}T\mu}{4\pi hkT_{STD}}\right)\left[\ln\left(\frac{4kt}{\gamma\phi\mu\kappa r_w^2}\right) + 2S\right] \tag{6.8}$$

Application of Eq. (6.8) to gas well testing is valid at low reservoir pressures generally less than 1.4×10^5 kPa [3]. At higher pressure, use of a pseudo pressure and a pseudo time appears to give better accuracy [6]. At very high pressures, the variation in gas compressibility may be slight and the same equations used for testing oil wells are applicable. In this text we shall consistently apply the same approximations used to derive Eq. (6.8) in pressure testing gas wells. It should be stressed that the viscosity (μ), the compressibility factor (Z) and compressibility (κ) are all evaluated at initial reservoir conditions. Also note that the reservoir temperature (T) and the standard temperature (T_{STD}) must both be expressed in absolute units. The pressures must also be absolute.

For tight formations containing gas, the initial response period depicted in Fig. 6.1 may be prolonged and use of type curves to obtain the skin effect and formation permeability may be the most convenient approach. The use of type curves is discussed in detail by Earlougher [2].

REFERENCES

6.1. Hill, A. D., *Production Logging—Theoretical and Interpretive Elements*, Mono. Ser., Society of Petroleum Engineers, Richardson, Texas, 1990.

6.2. Earlougher, R. C., *Advances in Well Test Analysis*, Mono. Ser., *5*, Society of Petroleum Engineers, Richardson, Texas, 1977.

6.3. Lee, J., *Well Testing*, Mono. Ser., *1*, Society of Petroleum Engineers, Richardson, Texas, 1982.

6.4. Dake, L. P., *Fundamentals of Reservoir Engineering*, Amsterdam: Elsevier, 1978.

6.5. Al-Hussainy, R., Ramey, H. J., and Crawford, P. B., *J. Pet. Tech., 18* (1966) 624.

6.6. Aziz, K., Mathur, L., Ko, S., and Brair, J. L. *J. Canad. Pet. Tech.* (April–June 1976) 58.

PROBLEMS

***6.1.** An oil well is producing at a constant rate of 2.1 stocktank m^3/hr from a reservoir having an initial pressure of 21 MPa. Other data are shown in Table P6.1.

TABLE P6.1 Additional Reservoir Data

Property	Symbol	Value
Permeability	k	5×10^{-14} m^2
Formation thickness	h	10 m
Wellbore radius	r_w	0.15 m
Fluid viscosity	μ	3 cp
Isothermal compressibility	κ	1.3×10^{-6} kPa^{-1}
Formation volume factor	B	1.25 reservoir m^3/m^3
Drainage radius	r_e	250 m
Porosity	ϕ	0.30

(a) Calculate the flowing bottomhole pressures at $t = 2, 5, 10$, and 15 hr, respectively, assuming that the reservoir is essentially infinite and that the well is undamaged ($S = 0$).

(b) Repeat Part **(a)**, except consider a well that is damaged so that the skin factor is 10.

***6.2.** A well test in a newly completed reservoir yields the data shown in Table P6.2A.

TABLE P6.2A Reservoir Data

t (hr)	Pressure (kPa)
0.25	17,728
0.50	17,475
0.75	17,327
1.0	17,222
3.0	16,821
4.0	16,716
6.0	16,588
8.0	16,463
10.0	16,381
15.0	16,233
20.0	16,128

What is the formation permeability and the skin factor? Additional data may be found in Table P6.2B.

TABLE P6.2B Additional Reservoir Data

Property	Symbol	Value
Formation thickness	h	15 m
Wellbore radius	r_w	0.15 m
Fluid viscosity	μ	5 cp
Isothermal compressibility	κ	2×10^{-6} kPa^{-1}
Oil production (reservoir volume)	q	2.5 m^3/hr
Porosity	ϕ	0.18

***6.3.** The results of a drawdown test of a gas well are tabulated in Tables P6.3A and P6.3B. The initial reservoir pressure is 27.58 MPa and the reservoir temperature is 65°C. The well is produced at a rate of 100 standard m^3/hr and the gas has a gravity of 0.65 compared to air. Using these data, estimate the reservoir permeability and the skin factor.

TABLE P6.3A Reservoir Data

Property	Symbol	Value
Formation thickness	h	25 m
Porosity	ϕ	0.07
Wellbore radius	r_w	0.15 m
Gas gravity	γ_g	0.65

TABLE P6.3B Pressure Data

Time (hr)	Pressure (MPa)	Time (hr)	Pressure (MPa)
0.5	26.05	10.0	24.78
1.0	25.76	20.0	24.48
2.0	25.47	30.0	24.30
3.0	25.30	40.0	24.17
4.0	25.17	50.0	24.07
5.0	25.08	100.0	23.76

PART 3
Perforating Methods

7
Perforating

Perforating is that process whereby the casing and the surrounding cement sheath are punctured in order to permit hydrocarbons contained in the zones surrounding the casing to flow into the wellbore and be subsequently produced. Acceptable perforations are crucial for optimizing production and for total recovery of hydrocarbons. By perforating at selected depths, individual zones can be isolated from one another when desired, water and gas zones may sometimes be separated from oil zones, damaged areas immediately surrounding the wellbore may be penetrated assuring better production, and better control of well stimulation treatments may be possible. Perforating is an important completion and stimulation function. A carefully designed perforating program will greatly improve the chances of achieving the desired objectives.

It is, therefore, necessary to understand just how the process is carried out and to appreciate how the results will differ under various conditions. These are the issues considered in this chapter. Primary attention is devoted to the jet gun. Other methods of perforating are used such as the bullet perforator, hydraulic perforators, or mechanical cutters [1], but the jet gun perforator is by far the most widely used. There are, however, problems associated with perforating using the jet gun. These problems will be pointed out here and, one hopes, their effect can be minimized by observing good engineering practices. Improved techniques for creating perforations that alleviate the difficulties associated with the jet gun may at some future date appear, but thus far, none has been developed.

7.1 THE JET GUN PERFORATOR

Since 1946 when the jet gun became commercially available, its usage has grown until nearly all perforations are created using it. The early method of firing a high-velocity projectile (bullet perforating) is rarely applied, except perhaps in soft formations (small rock embedment strength, see Chap. 2), and will not be described here.

The jet gun utilizes a shaped explosive charge to accelerate extremely fine metal particles into the form of a needle-like high-speed jet. These metal particles form a metal liner to protect the charge prior to detonation. The fine particles emanate from the tip of the cone shown in Fig. 7.1 at velocities approaching 10,000 m/sec [1]. The jet gun is composed of five main parts. The liner, which serves to protect the main explosive charge and which ultimately produces the jet action, the main explosive charge, a detonator, a primer charge to ignite the detonator, and the case. The liner is conically shaped. Figure 7.2 depicts the jet that forms

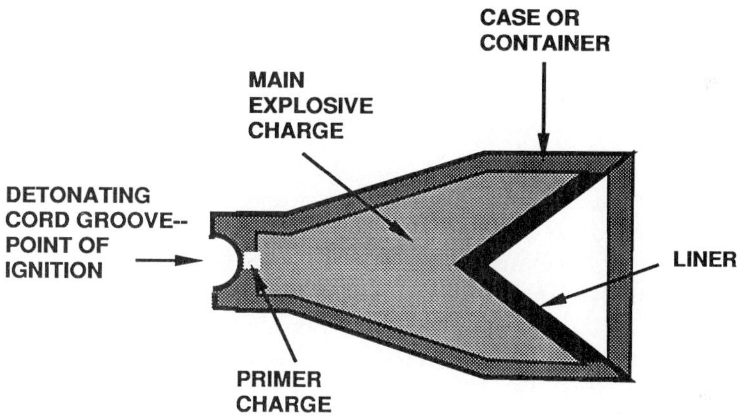

Figure 7.1 Components of a shaped charge.

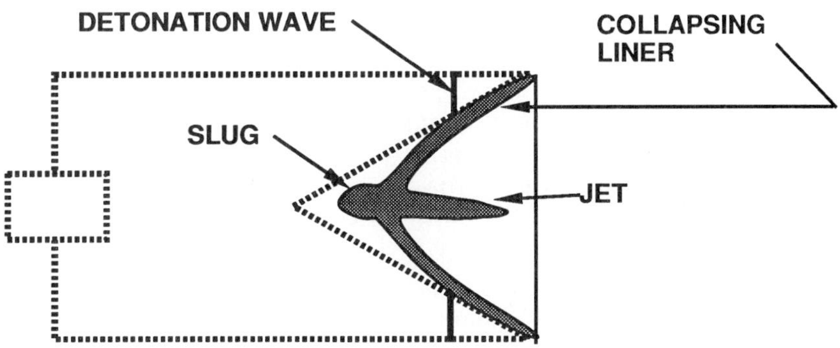

Figure 7.2 Mechanism of jet–slug formation.

and also a larger slower moving secondary slug of metal, which is sometimes called a *carrot*. This slug is propelled toward the perforation tunnel created by the fine jet, but does not contribute to the perforating process. Some practitioners believe that the slug may reduce the productivity of a perforation, and jet guns have been developed which essentially eliminate the slug [1].

There are three basic types of guns: casing guns, expendable guns, and through-tubing guns. The casing gun, which can be run into the well on a wireline or conveyed by tubing (Vann gun), is available in a variety of diameters ranging from approximately 3.5 to 18 cm. The charges are contained in a steel tube protected from impact and from the well fluids and are arranged so that they face radially outward from the vertical axis of the carrier. Some of the advantages of this type of gun is that it is retrievable, thereby enabling inspection to verify firing of all charges, it contains much of the debris, and it partially shields the casing from the force of the charge. This is a highly reliable device.

The expendable gun is composed of die-cast aluminum, ceramic, or glass charge cases with built-in links or arms hooked together. These materials are designed to disintegrate into small particles and drop to the bottom of the well when the gun is fired. The expendable gun is not used as much as the casing gun because there is no way to examine the gun and verify firing, most of the energy from the charge detonation must be adsorbed by the casing and fluid system which sometimes results in casing damage, and finally expendable guns are more difficult to position. These guns do have the advantage of containing a greater explosive charge for a fixed running diameter and are flexible enough to allow running through crooked casing or tubing.

Through-tubing guns are smaller in diameter than casing guns (~5.4 cm) and, therefore, they contain smaller charges, are usually fixed so that all of the guns fire in the same direction (0° phased), and may produce smaller entryholes and shorter perforations. However, because they can be run with tubing and packer in place, their poorer performance is sometimes overlooked. The main impetus for their use stems from the advantages of being able to perforate under conditions such that the pressure in the well is less than the reservoir pressure (underbalanced). This aspect will be discussed at some length in this chapter.

It is possible to perforate underbalanced using a tubing-conveyed casing gun. The gun is placed on the front end of a tubing string as it is run into the well. Using a sliding sleeve below a retrievable packer, the sleeve can be opened to equalize the pressure around the gun in the wellbore with the hydrostatic pressure of the fluid in the tubing. The tubing may be run in the well containing whatever fluid that is required to maintain the desired differential between the fluid in the casing (hydrostatic) and the formation. For example, the tubing-conveyed gun can be safely fired with high differential pressure into the wellbore (underbalanced) to help flush the debris from the perforations since a wellhead is in place at the surface. This, as will be evident in subsequent sections, is a highly desirable procedure which has resulted in increased use of tubing-conveyed or through-tubing jet guns.

7.2 PERFORMANCE OF JET GUNS

There are many parameters in the design of a jet gun that influence its performance. Some of the parameters affecting shaped-charge performance are indicated by Fig. 7.3. Included are the important liner design variables—geometry, material, and thickness. Charge configuration is of significance, including the distance from the liner apex to the primer. Explosive distribution and density (which determines the detonation velocity) may be more important than the overall amount of explosive. Shown in Fig. 7.3 are the factors that influence the entryhole diameter, the penetration depth, and the hole volume. These will be important factors in the selection of a jet gun. Note, however, that those factors which tend to increase the penetration depth also tend to decrease the entryhole diameter. Thus, these two factors are coupled and the final design may be a compromise.

Considering the other conditions in the well, gun clearance, which is the distance from the gun to the casing wall measured along the axis of the jet, can have an important effect [2]. Most guns are designed to fire in several directions.

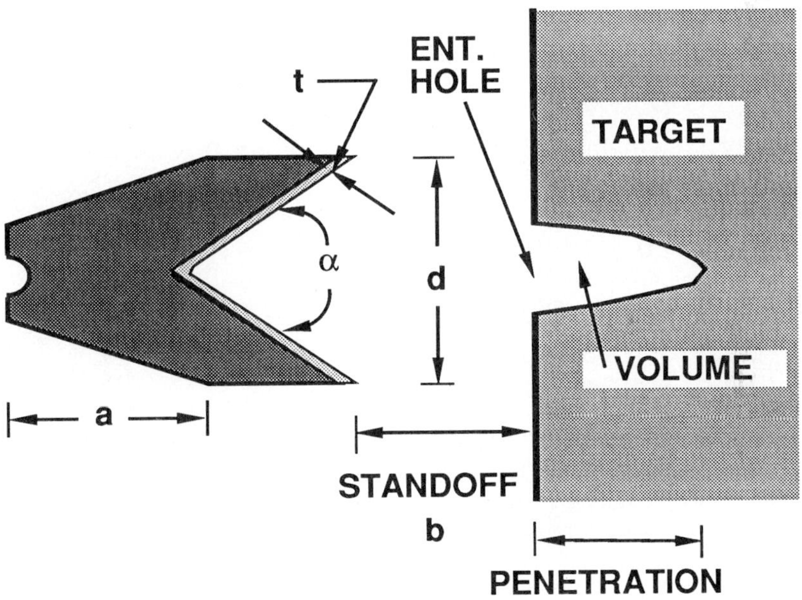

-To Increase

Penetration - reduce α,increase b,a,d,t

Entrance Hole -Increase α,d , reduce b,t

Hole Volume - Increase α,d,t, reduce b

Figure 7.3 Parameters affecting shaped-charge performance.

The phasing is defined as the angle between lines drawn through the axis of adjacent jet guns in a sequence.

An example of 90° phasing is shown in Fig. 7.4 [3]. In this figure, penetration and entryhole values are observed to vary with the larger values occurring at smaller clearances. Centralization of the gun has often been suggested as being desirable, but for the small through-tubing guns centralization would result in all four of the shots in a 90°-phased gun yielding small penetration depths and entryhole diameters. A better solution to the severe clearance problem presented by through-tubing guns is to press the gun against the wall of the casing by either mechanical or magnetic means and to use a 0° phase angle so that all shots are fired at a fixed low clearance.

The compressive strength or rock embedment strength of the formation is another factor influencing the penetration depth. Thompson [4] has shown that increasing the compressive strength of the formation tends to decrease the pen-

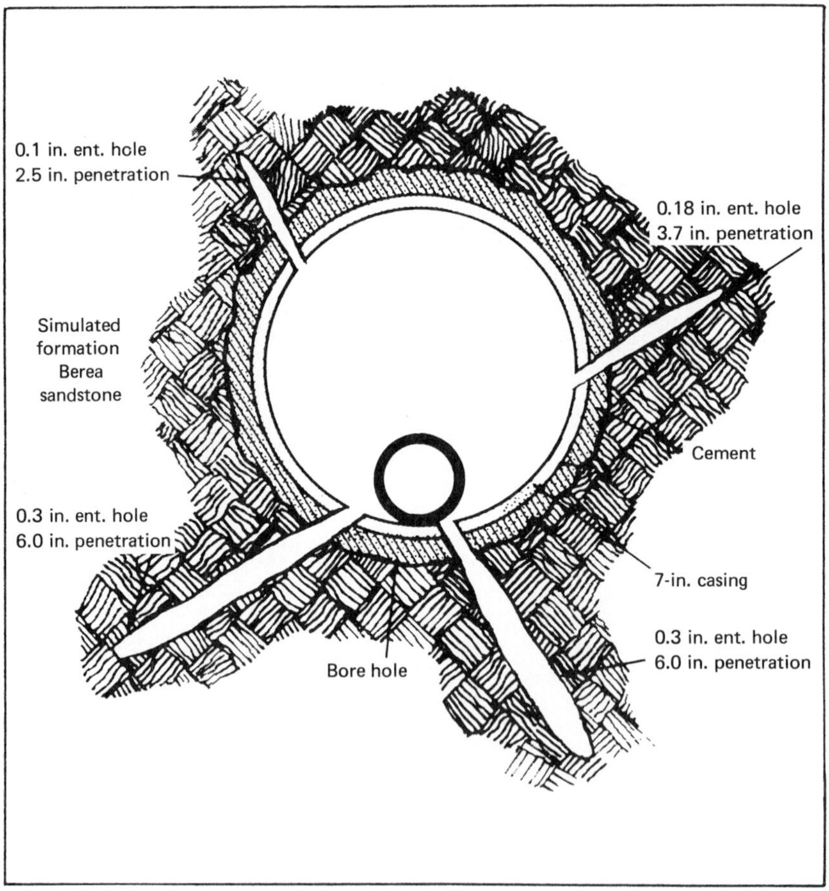

Figure 7.4 Random perforating pattern for conventional $1\frac{11}{16}''$, 90°-phased gun [4]. [With permission of the International Association of Drilling Contractors.]

etration depth. The decrease is dramatic. The penetration depth created by a 4″ jet gun fired into a formation having a compressive strength of 6 MPa is about 30 cm. If the compressive strength is increased to 60 MPa, the penetration depth is correspondingly reduced to 17 cm. Thus, penetration into hard formations may be expected to be small.

Core Flow Efficiency

It should not be imagined that the penetrating action of the jet gun is a result of the heating and subsequent melting of the rock. The entire action of the jet can be viewed as one of compression. Thus that portion of the casing, the cement sheath, and the surrounding formation which is penetrated by the force of the jet action has been displaced into the perforation tunnel and this debris is lodged along the face of the perforated formation as shown in Fig. 7.5. In addition to this debris, a portion of the formation rock is generally compacted by the force of the explosion. The extent of the compacted (or compressed) zone, which is depicted

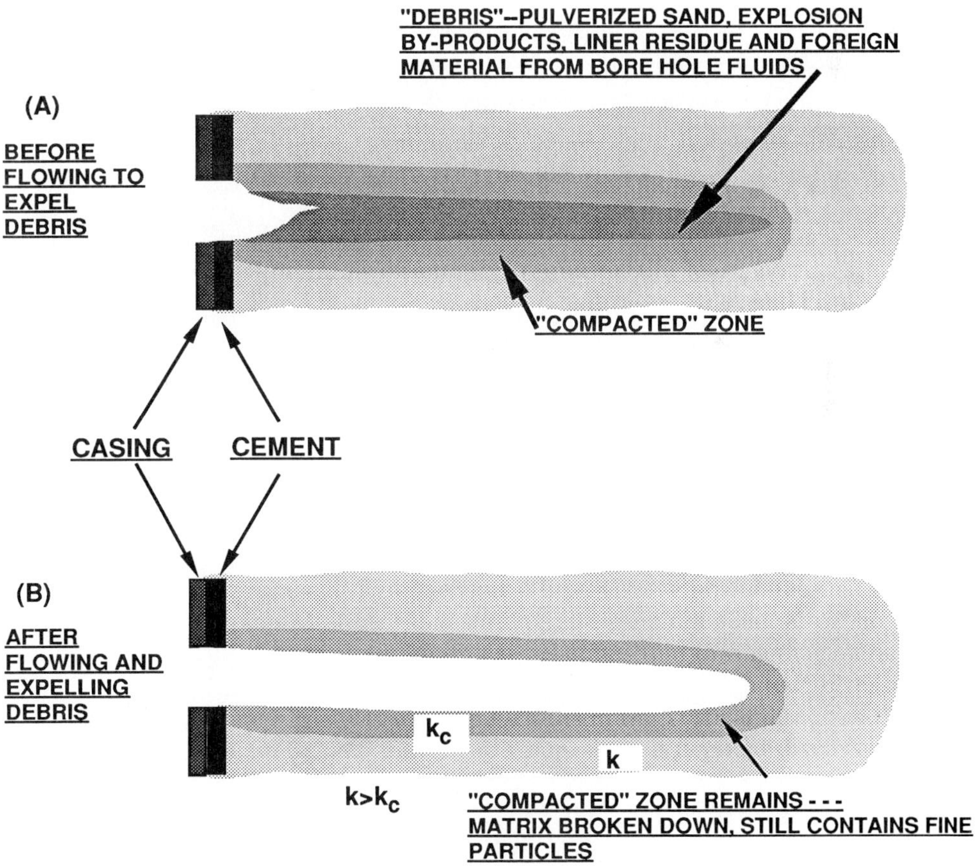

Figure 7.5 Characteristics of a perforated hole in a sandstone target.

in Fig. 7.5 as a crushed zone, is an important feature of the process since it may not readily be removed even when the well is put on production and the permeability of the crushed zone may be less, perhaps much less, than that of the original formation. Thus, this zone represents damage that can reduce the potential flow capacity of a perforation.

The possible implications of the damage resulting from the perforating action of the jet guns has been recognized. In 1962 a standard test procedure, API Recommended Practice No. 43 (RP43), for comparing the flow properties of perforations was established [5]. This test defined perforation effectiveness in terms the ratio of the apparent permeability, k_p (see Fig. 7.6), of a perforated Berea core to the permeability of the unperforated core, k, with both ends open. It was soon recognized that k_p/k is not really indicative of the degree of damage associated with the creation of the perforation. This is true because different perforating guns will yield different penetration depths, and, therefore, two different guns which create the same degree of damage may give quite different values of k_p. Without taking into account the actual depth to which the perforation penetrates into the Berea core target, a comparison of different values of k_p/k is meaningless.

In recognition of this problem, API RP43 was revised in 1971 [6] to provide a measure of the permeability damage surrounding a perforation. In the new procedure the apparent permeability (k_p) measured in a core with a damaged per-

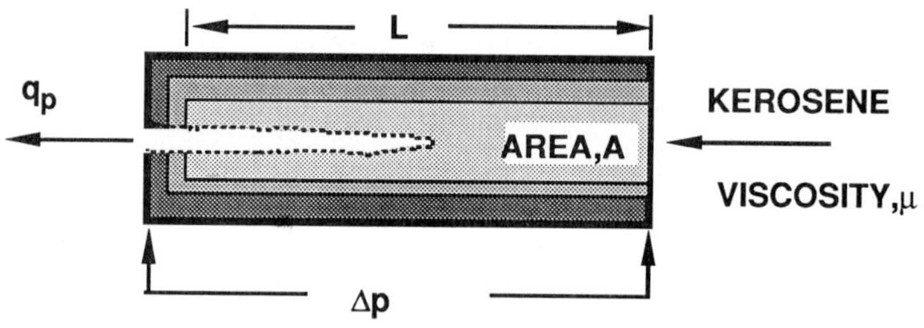

$$k_p = \frac{q_p \mu L}{A \Delta p}$$

$$CFE = k_p/k_i$$

k_i RepresentsThe Effective Permeability Of A Target Containing An Ideal (Drilled) Perforation Of The Same Depth And Diameter As TheActual Perforation

Figure 7.6 Core flow efficiency (CFE) determination.

foration is compared with an ideal permeability (k_i) that would be observed if an undamaged perforation (one that is carefully drilled to avoid creating a crushed zone) of the same depth as that actually created is subjected to the same flow conditions depicted by Fig. 7.6. A new measure of perforation performance called the *Core Flow Efficiency* (CFE) is then defined as

$$\text{CFE} = k_p/k_i \tag{7.1}$$

For a clean, undamaged perforation, k_i/k can be calculated using a numerical solution of the diffusivity equation applied to an ideal geometry consisting of a cylindrical perforation embedded into a cylindrical core (see Problem 7.4). The CFE does provide a measure of perforation quality. A CFE near 1.0 indicates a relatively undamaged perforation; a CFE $\ll 1.0$ indicates a highly damaged perforation. The reported values of CFE most often fall in the 0.65 to 0.85 range [7].

The CFE is an important measure of the damage caused by creating a perforation. It is generally available from the manufacturer of a jet gun. The CFE does not, however, always realistically represent the actual impairment caused by the existence of the crushed zone around the perforation under downhole conditions.

Perforation Efficiency

The perforation efficiency (PE) of a damaged perforation is defined as the ratio of the productivity under downhole conditions of an isolated damaged perforation to the productivity of an undamaged perforation of the same dimensions. The PE differs from the CFE because of the difference between flow patterns in downhole conditions versus those in the CFE test. This difference is illustrated by Fig. 7.7. Most of the fluid enters the tip of the perforation in the CFE test, since all of the fluid is introduced at one end of a cylindrical core sealed along its sides. On the other hand, the flow into a single isolated perforation in a cased hole comes from all surrounding parts of the reservoir. It is this difference in flow patterns which makes it difficult to interpret the CFE in terms of downhole performance. However, as the perforation density is increased so that there is a tendency for the perforations to interfere with each other, then the flow profile surrounding a perforation begins to approximate that of the CFE test. It is the similarity of the velocity profile between the CFE test and the production in a multiple perforated system that makes the CFE a valuable measure of perforation efficiency.

To understand just what influence the crushed zone thickness and permeability have on the performance of a perforation, let us consider a simple model first proposed by Bell et al. [7]. To a good approximation, a perforation can be considered to be one half of an ellipse that is rotated about its primary axis. Figure 7.8 shows this ellipsoidal geometry. Note that r_p is the radius of the entryhole and D_p is the penetration depth into the formation measured from the cement sheath. Assume that the ellipse will be productive over its entire surface. Far from the entrypoint, the fluid pressure is the reservoir pressure p_R. The influence that all the other perforations might have on the pressure distribution is neglected,

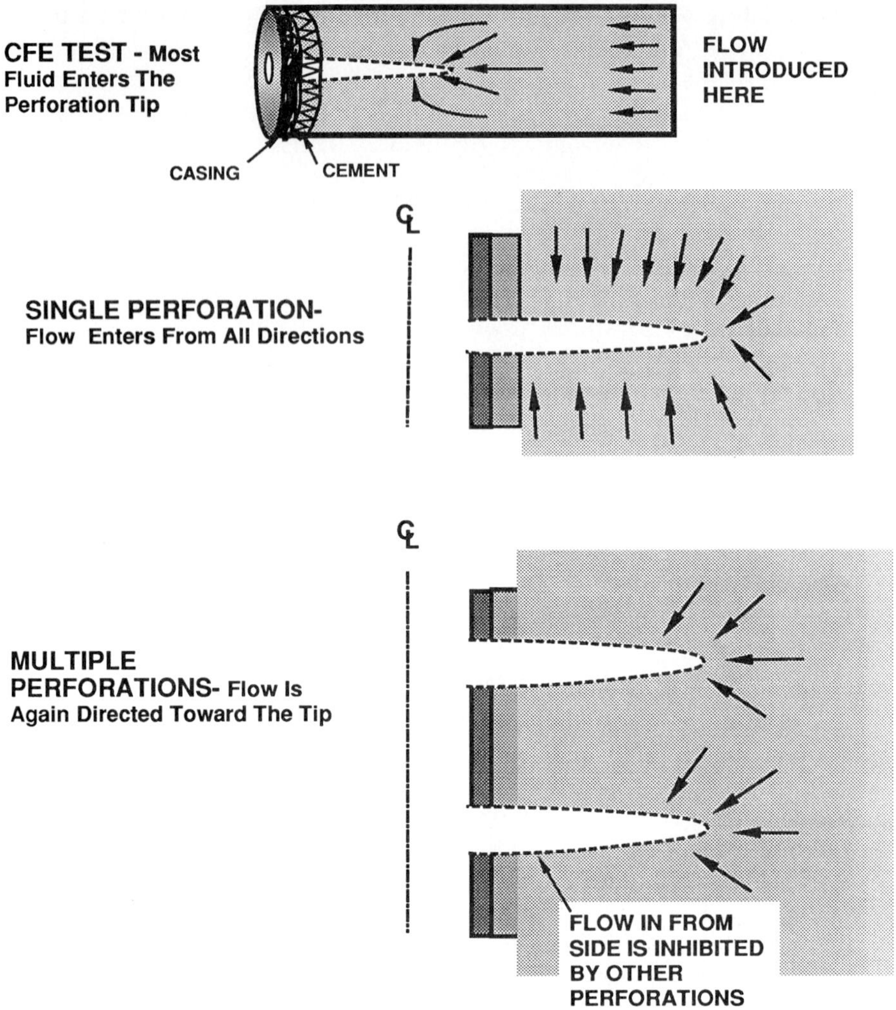

Figure 7.7 Comparison of fluid velocities in CFE test with those into a perforation.

or more precisely, our analysis is confined to the case of a single perforation in an infinite reservoir.

Within the interior of the perforation, the pressure is assumed to be constant and equal to p_w, the pressure in the wellbore or the drawdown pressure. Thus the pressure drop associated with flow within the perforation tunnel is neglected.

The pressure of any point within the formation is given by

$$\frac{\partial^2 p}{\partial x^2} + \frac{\partial^2 p}{\partial y^2} + \frac{\partial^2 p}{\partial z^2} = 0 \tag{7.2}$$

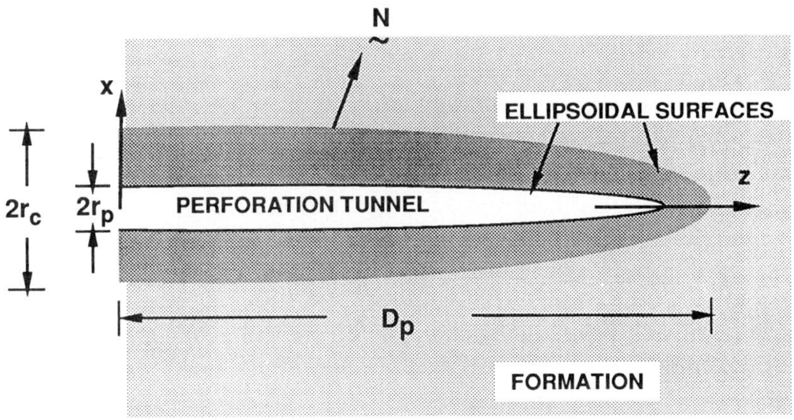

Figure 7.8 Idealized perforation geometry.

since in the near wellbore region the fluid compressibility is negligible. Equation (7.2) is difficult to solve in its present form because the pressure is fixed along the surface of the elliptical perforation and not at constant values of the coordinates x, y, or z. It is, therefore, convenient to make the following change of variables:

$$\begin{aligned} x &= H \sinh w \sin v \cos \gamma \\ y &= H \sinh w \sin v \sin \gamma \\ z &= H \cosh w \cos v \end{aligned} \tag{7.3}$$

Thus, we have a new coordinate system, (w, v, γ). It can be shown that surfaces of constant w are ellipses. Therefore, two constants (w_p and H) define the shape of a perforation. Their values can be expressed in terms of r_p and D_p, which are shown in Fig. 7.8. Eq. 7.3 demonstrates that $z = 0$ corresponds to $v = \pi/2$. For $z = 0$, the sum

$$x^2 + y^2 = r_p^2 = H^2 \sinh^2 w_p \cos^2 \gamma + H^2 \sinh^2 w_p \sin^2 \gamma \tag{7.4}$$

or

$$r_p^2 = H^2 \sinh^2 w_p \tag{7.5}$$

Furthermore $x = y = 0$, and $z = D_p$ represents a point on the tip of the perforation. Thus for $v = 0$, $x = y = 0$ and

$$D_p = H \cosh w_p \tag{7.6}$$

Example 7.1 Elliptical Perforation Parameters

Suppose that the entryhole radius and penetration depth of a perforation are estimated to be 2 cm and 20 cm, respectively. What is the equation for an ellipsoid of revolution having these two dimensions?

Solution To solve for H and w_p, note that from Eqs. (7.5) and (7.6)

$$r_p/D_p = \sinh w_p/\cosh w_p = 1/10$$

or solving for w_p, we find

$$w_p = 0.1007 \qquad \text{and} \qquad H = r_p/\sinh w_p = 2/0.1009 = 19.83$$

Substituting these values of w_p and H into Eq. (7.3) yields the desired solution.

The pressure is constant over the ellipsoidal perforation surface and it is also constant over all other ellipsoidal surfaces enveloping the perforation. For this reason, the pressure depends on the coordinate w but not on v and γ. Since the pressure depends only on the coordinate w, Eq. (7.2) takes the simple form

$$\frac{\partial}{\partial w}\left(\sinh w \frac{\partial p}{\partial w}\right) = 0 \tag{7.7}$$

which can be integrated to give

$$p(w) = p_R - (p_R - p_w)\frac{\ln\left(\dfrac{\xi + 1}{\xi - 1}\right)}{\ln\left(\dfrac{\xi_p + 1}{\xi_p - 1}\right)} \tag{7.8}$$

where $\xi = \exp w$ and $\xi_p = \exp w_p$. Equation (7.8) shows how the pressure surrounding the perforation depends on w.

Equation (7.8) can be used to calculate the rate of flow into the perforation using the general form of Darcy's law, namely

$$\mathbf{u} = -\frac{k}{\mu}\nabla p \tag{7.9}$$

To find the component of flux perpendicular to the ellipsoidal surface, the inner product of a unit normal to the surface $\mathbf{N}$ with the vector flux $\mathbf{u}$ is required. Thus

$$u_w = -\frac{k}{\mu}\mathbf{N} \cdot \nabla p \tag{7.10}$$

Since p depends on w, but not on v and γ,

$$\nabla p = \left(\mathbf{i}_x \frac{\partial w}{\partial x} + \mathbf{i}_y \frac{\partial w}{\partial y} + \mathbf{i}_z \frac{\partial w}{\partial z}\right)\frac{\partial p}{\partial w} \tag{7.11}$$

where $\mathbf{i}_x$, $\mathbf{i}_y$, and $\mathbf{i}_z$ are unit vectors in the x-, y-, and z-directions, respectively. Furthermore, a unit normal to the perforation surface is found to be

$$\mathbf{N} = \frac{\mathbf{i}_x(\cosh w \sin v \cos \gamma) + \mathbf{i}_y(\cosh w \sin v \sin \gamma) + \mathbf{i}_z(\sinh w \cos v)}{\sqrt{\sinh^2 w + \sin^2 v}} \tag{7.12}$$

After some algebra, it can be shown that

$$u_w = -\frac{k}{H\mu}\frac{1}{\sqrt{\sinh^2 w + \sin^2 v}}\frac{dp}{dw} \tag{7.13}$$

This is the flux perpendicular to the surfaces of constant w which are ellipsoids.

The production rate into a single perforation is determined by summing the flux multiplied by an element of area over the entire surface of the ellipse. Thus

$$q_p = -\iint u_w \big|_{w=w_p} dA \tag{7.14}$$

where $dA = H^2 \sinh w_p \sin v\sqrt{\sinh^2 w_p + \sin^2 v}\; dv\, d\gamma$ and the negative sign is selected so that q_p is positive when fluids are produced. Thus

$$q_p = -\int_0^{\pi/2}\int_0^{2\pi} \frac{kH}{\mu} \sinh w_p \sin v \left.\frac{dp}{dw}\right|_{w=w_p} dv\, d\gamma \tag{7.15}$$

or

$$q_p = \frac{2\pi kH}{\mu}\left[\frac{p_R - p_w}{\ln\left(\dfrac{\xi_p + 1}{\xi_p - 1}\right)}\right] \tag{7.16}$$

Example 7.2 Production of Oil Through a Single Undamaged Perforation

Assuming that the dimensions of the perforation are those given in Example 7.1, what pressure drawdown ($p_R - p_w$) is required to produce 10 m^3/day if the formation permeability is 9.87×10^{-15} m^2 and the fluid viscosity is 4×10^{-3} kg/m-sec (4 cp)?

Solution Since w_p is given as 0.1007 (dimensionless),

$$\xi_p = \exp(0.1007) = 1.1059$$

$$\frac{10}{24(3600)} = +\frac{2\pi(9.87 \times 10^{-15})}{(4 \times 10^{-3})}\left[\frac{0.1983}{\ln\left(\dfrac{1.1059 + 1}{1.1059 - 1}\right)}\right](p_R - p_w)$$

or

$$p_R - p_w = 1.13 \times 10^8 \text{ Pa}$$

This drawdown pressure is obviously unrealistic and to obtain adequate production, it is necessary to introduce multiple perforations throughout the productive zone.

It should be noted that p_w is the pressure assumed to be constant along the perforation surface, thus neglecting the pressure drop along the axis of the perforation tunnel and the friction loss associated with the sudden expansion of the fluid as it emerges from the perforation into the wellbore.

Equation (7.16) applies to an undamaged perforation. Let us consider the case when a crushed or compacted zone having a permeability (k_c) which is less than the formation permeability surrounds the perforation. As shown by Fig. 7.8 the compacted zone is taken to be an ellipsoid concentric with the ellipsoidal perforation. Thus, H is the same for both the compacted zone and the perforation. It should be noted that there are no experimental data supporting this shape for the crushed zone. Indeed, it seems more likely to have a shape that is thicker at the tip of the perforation than at the wellbore.

Example 7.3 The Compacted Zone

Suppose that the radius of the compacted zone measured at the cement sheath is 3 cm (see Fig. 7.8) and furthermore assume that the zone is concentric to the ellipsoidal perforation. What is the value of w_c that characterizes the position of the ellipsoidal boundary between the compacted zone and the undisturbed formation?

Solution To be concentric the same value of H must apply to both the perforation surface and the compacted zone/formation surface. Furthermore since $r_c = 3$ cm, then from Eq. (7.5)

$$\sinh w_c = \frac{r_c}{H} = 0.15$$

Solving gives $w_c = 0.1494$.

Having assumed an ellipsoidal shape for the compacted zone, it is possible to develop an equation for the pressure distribution. If we denote k_c as the permeability of the compacted zone, then the productivity of a damaged perforation is given by

$$q_p^{(D)} = +\frac{2\pi k_c H}{\mu}\left[\frac{p_R - p_w}{-\left(1 - \frac{k_c}{k}\right)\ln\left[\frac{(\xi_c + 1)}{(\xi_c - 1)}\right] + \ln\left[\frac{(\xi_p + 1)}{(\xi_p - 1)}\right]}\right] \tag{7.17}$$

where $\xi_c = \exp w_c$. For the same drawdown pressure in both the damaged and undamaged cases, the perforation efficiency, defined as $q_p^{(D)}/q_p$, is given by

$$\text{PE} = \frac{\frac{k_c}{k}\ln\left(\frac{\xi_p + 1}{\xi_p - 1}\right)}{-\left((1 - \frac{k_c}{k}\right)\ln\left(\frac{\xi_c + 1}{\xi_c - 1}\right) + \ln\left(\frac{\xi_p + 1}{\xi_p - 1}\right)} \tag{7.18}$$

Equation (7.18) gives the perforation efficiency in terms of the geometry of the compacted zone and the permeability contrast between the formation and the compacted zone.

As an approximation when $D_p \gg r_p$ and $D_p \gg r_c$, then

$$\xi_p \approx r_p/D_p + 1 \qquad \text{and} \qquad \xi_c \approx r_c/D_p + 1 \tag{7.19}$$

thus giving

$$\text{PE} \cong -\frac{\frac{k_c}{k}\ln\left(1 + \frac{2D_p}{r_p}\right)}{\left(1 - \frac{k_c}{k}\right)\ln\left(1 + \frac{2D_p}{r_c}\right) - \ln\left(1 + \frac{2D_p}{r_p}\right)} \tag{7.20}$$

Example 7.4 Perforation Efficiency

If k_c/k is 0.05, what is the perforation efficiency for the perforation and the compacted zone defined in Examples 7.1, 7.2, and 7.3?

Solution Using Eq. (7.20) and rearranging terms, we see that

$$\text{PE} = \frac{0.05 \ln\left(\frac{40}{2} + 1\right)}{\ln\left(1 + \frac{40}{2}\right) - (1 - 0.05)\ln\left(\frac{40}{3} + 1\right)} = 40\%$$

The perforation efficiency given by Eq. (7.20) is generally much smaller than that given by the CFE test because of the profound difference in the flow profiles depicted by Fig. 7.7. For example, Klotz et al. [8] have noted that the reported CFE values vary between 0.65 and 0.85. They calculated that for a perforation surrounded by a compacted zone having a thickness of 1.37 cm and permeability ratio $k_c/k = 0.06$, the CFE $= 0.65$. If these values are accepted and a perforation length of 30 cm with an entryhole radius of 0.3 cm is considered, then $r_p = 0.3$ and $r_c = 1.67$. Substituting these values into Eq. (7.20) yields a PE of 16%. Thus, a CFE of 65% corresponds to a PE of 16%. This result indicates that care must be exercised in applying CFE test results to obtain information about downhole response.

Bell et al. [7] have suggested that the flow into a perforation under downhole conditions might be better modeled by a radial flow arrangement rather than the linear flow used in the CFE test. In this arrangement fluid is permitted to enter along the lateral surfaces of the perforated core rather than just at one end. Deo et al. [9] have considered this proposal and found that under certain conditions, that is, no formation damage, low shot densities, and small crushed zone permeabilities, the radial flow model is a better one.

The CFE test is not a good measure of the expected field performance, especially under conditions of low shot densities in isotropic undamaged formations. However, any variation from these idealized conditions which reduces the flow into the perforation from the side and forces more of the fluid to enter from the tip will tend to cause the actual Well Flow Efficiency (WFE)* to approach the CFE. Shown in Fig. 7.9 are the three efficiencies plotted as a function of the permeability ratio. The values of CFE and WFE are taken from Klotz et al. [8], whereas the PE is computed using Eq. (7.20). It is important to recognize that all three efficiencies refer to precisely the same damaged perforation. Only the flow profiles have been changed by placing the perforation into the different flow settings illustrated by Fig. 7.7. For the particular case shown by Fig. 7.9, there is a moderate shot density and the interference between the adjacent perforation tends to divert flow from the side of the perforation into the tip. Thus, the WFE differs from both the CFE and the PE because the flow patterns differ.

* The WFE is the production through damaged perforations divided by the production that would be realized through same system of undamaged perforations. This differs from the PE which considers an individual perforation.

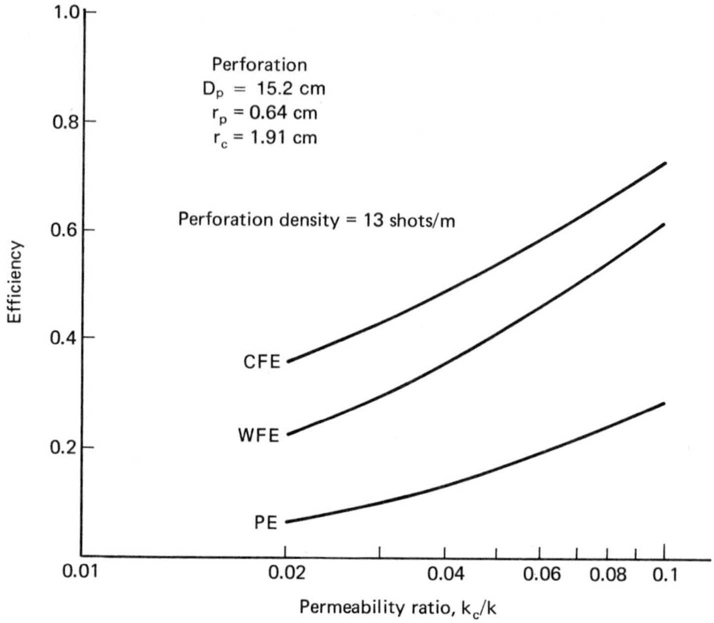

Figure 7.9 A comparison of different perforation efficiencies plotted as a function of the permeability ratio.

Figure 7.9 shows that the CFE is a better measure of the WFE than the PE is, although it is still a poor characterization. The CFE becomes more realistic whenever the formation is damaged, the vertical-to-horizontal permeability ratio is small, or the perforation density is high. These factors have been discussed by Deo et al. [9]. In some cases the CFE is, for all practical purposes, an adequate measure of perforation performance, whereas for others it is a gross overestimate.

7.3 FACTORS GOVERNING PERFORATION CLEANUP

Completion Fluids

There are steps that can be taken to reduce perforation damage. Clearly perforating with a clean, solid-free completion fluid is preferred to completion in a casing filled with drilling mud. If it is necessary to weight the fluid to control the well, then salts dissolved in an aqueous solid-free completion fluid can often yield the necessary density. In some cases, it may be desirable to add polymer to the completion fluid to reduce fluid loss into the formation.

Oil may be used as a completion fluid in water-sensitive formations. This may help to minimize damage to the perforation.

In carbonate formations perforations should, if possible, be made with hydrochloric acid in the wellbore (see Chap. 17) if the pressure in the wellbore exceeds that of the formation pressure (overbalanced case) or with a clean completion fluid if the pressure is underbalanced.

Flow of Produced Fluids

To clean the debris from a perforation and to partially restore the permeability in the compacted zone to an acceptable level, a flow from the formation into the wellbore is required. Figure 7.10 indicates that the CFE is initially quite low, but increases steadily with increased volume of produced fluid until a stable value is obtained [10]. The volume of fluid required to achieve a stable CFE is a complex function of the conditions under which the perforations were made, the size of the jet gun, the type of formation perforated, and the fluid velocity in the perforations. Bell et al. [7] have shown that at standard API RP43 conditions as much as 50 liters of fluid may be required to achieve a stable CFE. Because this is a considerable volume of fluid and because the debris appears to be removed with the first few liters of flow, it seems reasonable to think that the main factor which changes during the latter stages of the clean-up process is the permeability of the compacted zone.

Excessive flow of completion fluid into the formation before producing through the perforations will result in decreased final values of CFE [5].

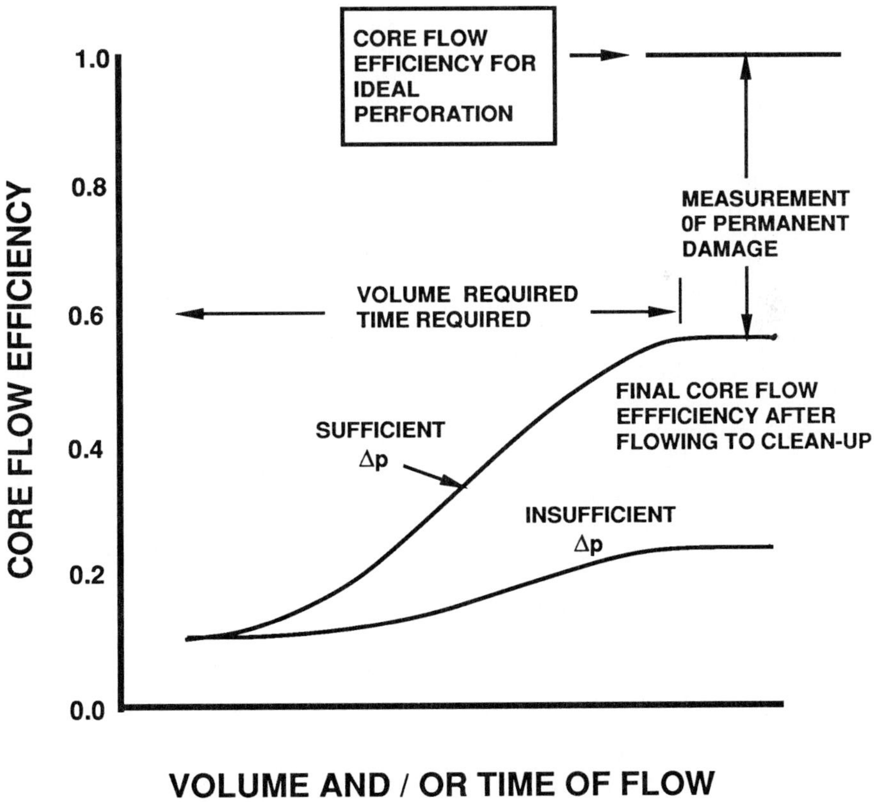

Figure 7.10 The CFE shown as a function of the volume of produced fluid.

Differential Pressure

The difference in pressure between the formation and the wellbore sharply affects perforation cleanup. At pressures less than about 1 MPa (150 psi), some experiments show that the CFE stabilize at values lower than those obtained at higher pressure differences [10, 11]. In other words, a minimum level of differential pressure is required to reach the plateau value of the CFE. Differential pressures greatly in excess of this minimum level do not substantially improve the CFE. Thus as indicated by Fig. 7.10, an insufficient level of pressure forcing fluid into the wellbore will result in a high level of permanent damage which will not be removed even though large volumes of fluid may be produced through the perforation.

Bell [12] suggests that the differential pressures given in Table 7.1 are adequate for cleanup. Presumably these differential pressures are sufficient to yield the plateau values of the CFE whether the perforations are created in underbalanced or in overbalanced conditions. Here overbalanced refers to wellbore pressures in excess of the formation pressure. The reverse is true for underbalanced conditions. Note that higher differential pressures are required to clean up in gas wells and in lower permeability formations.

In laboratory tests perforations tend to clean up better when liquid is not injected into the perforation before producing back through the perforation. Injected liquids apparently tend to further compress the compacted zone and perhaps tend to transport fines and debris further into the formation, making it more difficult to produce. Invasion of completion fluids, even those free of fines, into a gas well greatly increases the probability of plugging problems even when the differential pressures available for cleanup are large.

When a perforation is completed under overbalanced conditions, fluid will surge from the wellbore into the formation and, as noted previously, make perforation cleanup more difficult. Field experience has confirmed [10, 12, 13] that reverse pressure perforating (underbalanced conditions) is conducive to obtaining a more effective perforation system. Perforations are not exposed to a foreign fluid and the initial surge of formation fluids into the wellbore helps to maintain the maximum number of producing perforations. It should be noted that excessive underbalanced pressure differentials should be avoided since the sudden surge of fluid may cause fines to migrate and damage the near wellbore region.

What is remarkable is that reverse pressure perforating is often conducted through tubing with small-diameter perforators and even though the amount of

TABLE 7.1 Typical Ranges of Underbalanced Pressure for Perforated System Cleanup [12]

Permeability Range	Liquid	Gas
Higher Permeability (> 100 mD)	1,380 to 3,450 kPa	6,900 to 14,000 kPa
Lower Permeability (< 100 mD)	6,900 to 14,000 kPa	14,000 to 34,500 kPa

explosive charge used is generally smaller than would be used with casing guns, the results are frequently better, much better. Bell and Bell have termed this a paradox [10]. Nevertheless, the reason for the better performance of tubing guns used under reverse conditions is clear. Better cleanup is obtained. This procedure is now recommended for all cases in which neither fracturing nor gravel packing is contemplated following the completion of perforating. If the formation is to be fractured to achieve economic levels of production, then other factors (see Sec. 7.5) are of greater importance.

If more than one gun trip is required to perforate an interval, reverse pressure perforating is still preferred. During subsequent runs following the first, the well will be flowing to maintain the reverse pressure conditions. In this case it may be best to perforate low-permeability zones first, saving high-permeability zones for last.

7.4 PERFORATION DESIGN

Perforation cleanup is a crucial issue in obtaining performance. It probably overrides any other consideration; however, there are other factors that should also be taken into account since they can be controlled. For example, the density of perforations, the phasing, the entryhole diameter, and the depth to which a perforation penetrates will all determine to some extent the final performance of the system. In this section the influence of these factors is considered. To compare various perforation geometries and arrangements, the productivity of the cased, perforated system will be compared with the productivity of an equivalent undamaged open hole completion. Steady-state production of an incompressible fluid with constant pressure at a drainage radius (r_e) from the well will be considered. The ratio of the productivity of the perforated system to that of the open hole will be called the *productivity ratio*. A productivity ratio equal to unity will occur when both systems would produce at the same rate with the same drawdown pressure.

The comparisons shown in this section have been derived from the work of various researchers [8, 14, 15, 16, 17]. The results have all been obtained by computer simulation. The assumed geometry of a perforation may vary somewhat from investigator to investigator and the numerical techniques may differ; nevertheless, the results of the various calculations are in satisfactory agreement and are certainly more than adequate to permit an understanding of the trends.

It should be remarked that often the productivity ratio is not the primary concern. When perforating formations that are to be fractured [7] or gravel-packed [18] for sand control, other criteria are applied. Furthermore, the productivity ratio will depend on the extent to which either the perforations or the formation is damaged. In this section perforation damage will be simulated by imposing a different permeability for the crushed zone than for the formation permeability. Similarly, the formation damage resulting from drilling and completion operations will be modeled by imposing a low-permeability damaged zone around the wellbore. The depth of this damaged zone will vary. Of course, neither one of these models can actually be applied in selecting a perforator and perforation geometry because the extent of the damage is an unknown factor (see Chap. 6), but it is instructive to consider the trends associated with changing various parameters.

Depth of Penetration

The depth of penetration has a significant impact on the productivity ratio as shown for the case of 4 shots per foot, phased at 90° with no formation nor perforation damage. Figure 7.11, adapted from results presented by Harris [14], shows that the productivity ratio increases with penetration depth and reaches a value of 1.0 at about a 14-cm depth (5–6 in.). Doubling the penetration depth to 28 cm only increases the productivity ratio by about 10%.

If, however, the formation is damaged, it is desirable to penetrate a damaged zone surrounding the wellbore if at all possible. Figure 7.12 demonstrates the rapid decrease in the productivity ratio as the radius of the damaged zone surrounding the wellbore approaches the depth to which the perforation penetrates. In the case shown, the permeability of the damaged zone is 40% of the formation permeability. As the permeability of the damaged zone is further decreased, the decrease of productivity will be even more severe.

Shot Density

Figure 7.11 shows the productivity ratio for various shot densities. The densities shown are 1, 2, and 4 shots per foot (3, 6, and 13 shots/m). There is continued improvement as the shot density is increased. Note that 2 shots per foot (6 shots/m), 10 inches (25 cm) in depth are equivalent to 4 shots/foot (13 shots/m) only 4 inches (5 cm) in depth. Thus especially in hard formation where deep

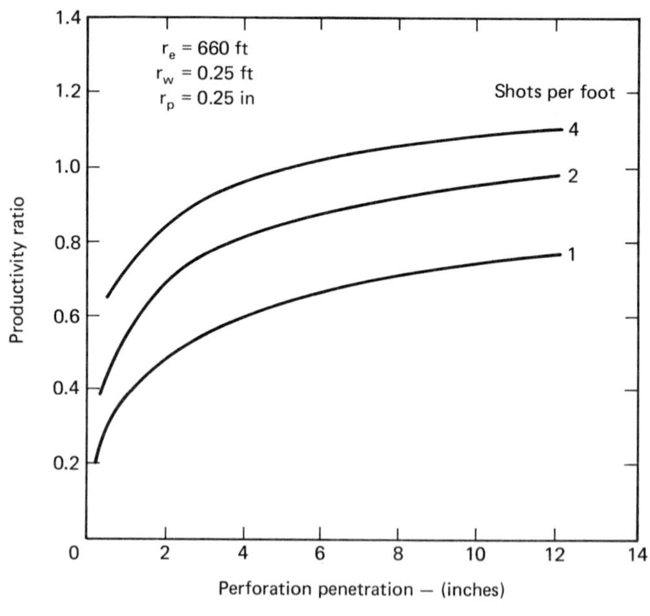

Figure 7.11 Productivity ratio versus penetration depth for various shot densities. (Adapted from Harris [14]).

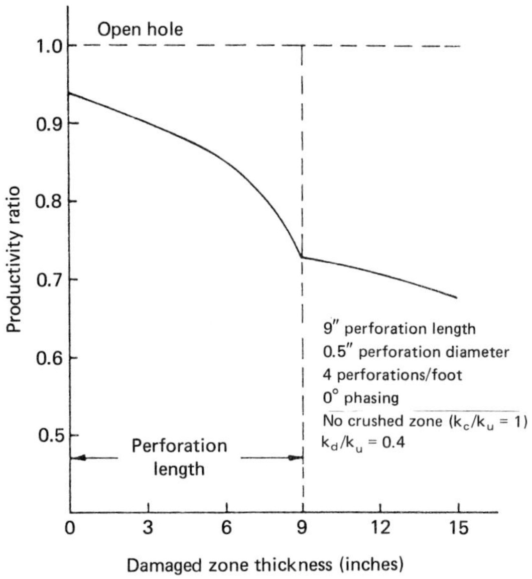

Figure 7.12 The effect of the damaged zone thickness on the productivity ratio [16]. [With permission of the Society of Petroleum Engineers.]

penetration is difficult to obtain, at least 4 and possibly more shots per foot (13 shots/m) are recommended.

The use of high shot densities is attractive to produce larger productivity ratios; however, the choice of shot density should also take into account difficulties that can arise because of casing damage [18].

Phasing

The angular phase between successive perforations is an important parameter. Figure 7.13 shows that 90° phasing is, as expected, optimum although many other phase angles except 0° give approximately the same result. This is true if all the perforations have the same penetration depth, which might be the case if a centered-casing jet gun is used. However, for underbalanced perforating where smaller through-tubing jet guns are used, it is necessary to position the guns facing and against the casing to achieve adequate penetration. For a phase angle of 0°, Fig. 7.13 shows there is a substantial loss in productivity due to the hydrodynamic interference between perforations. In fact, this interference may not be as severe in field operations as is indicated by Fig. 7.13 since vertical permeabilities are often smaller than horizontal permeabilities.

Entryhole Diameter

Figure 7.14 shows the influence of varying entryhole diameters with varying crushed zone permeabilities. The results show that diameters in excess of 0.25 in. (0.64 cm) give only marginal improvement. Thus in many cases, it would not be prudent to increase entryhole diameter at the expense of penetration depth; however, this is not always true. For example, when perforating for sand control

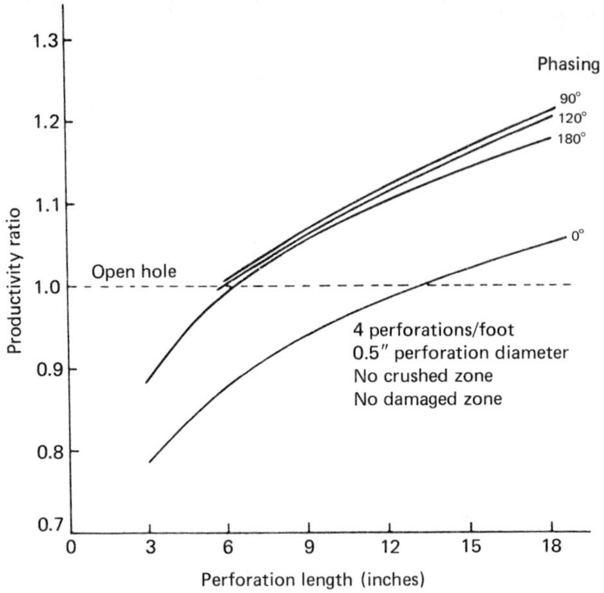

Figure 7.13 Phasing of perforations influences performance [16]. [With permission of the Society of Petroleum Engineers.]

by gravel packing larger perforation diameters will reduce the possibility of the gravel bridging the perforation rather than being forced into the perforation tunnel. Thus, larger entryholes are desired.

If the wells are extremely high flow, then the flow surrounding the perforation is turbulent and the results shown in Fig. 7.14 do not apply. Again larger diameter perforations may be desirable [19].

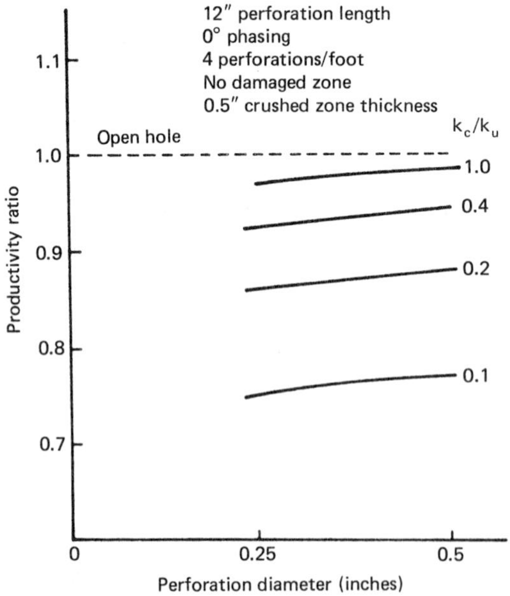

Figure 7.14 The influence of the crushed zone permeability on the productivity ratio [16]. [With permission of the Society of Petroleum Engineers.]

Generalized Design Nomograph

Locke [16] has constructed a nomograph, shown as Fig. 7.15, to allow the graphical solution of the productivity ratio for different geometrics and arrangements. This nomograph is quite useful.

Example 7.5 Productivity Ratio for a Perforated System

A jet gun produces a perforation depth of 12 in. (30.5 cm) and a perforation diameter of 0.5 in. (1.3 cm) as determined by API RP43. Assuming a damaged zone thickness of 6 in. (15 cm), a damaged zone to formation permeability, $k_s/k = 0.4$ (defined as k_D/k on nomograph), a crushed zone thickness of 0.5 in. (1.3 cm), and a crushed

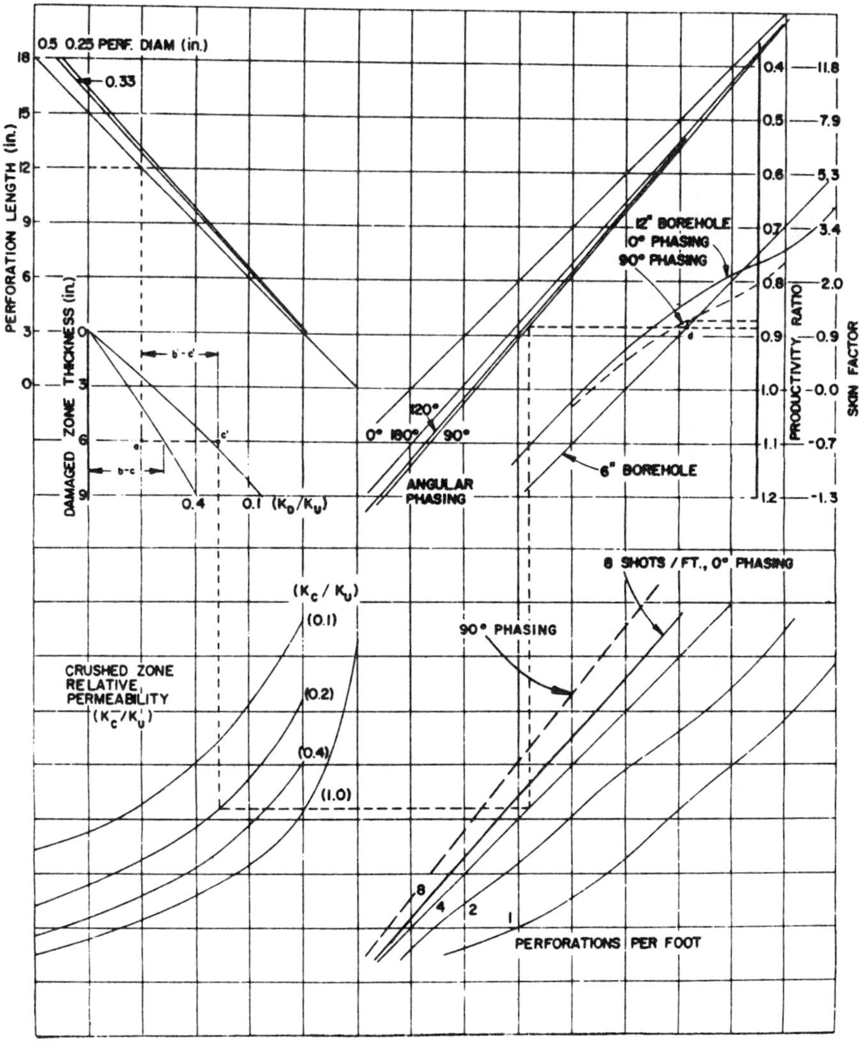

Figure 7.15 Nomograph for design of perforation system [16]. [With permission of the Society of Petroleum Engineers.]

zone to formation permeability of 0.2, determine the productivity ratio if a density of 4 shots per foot (13 shots/m) at 90° phasing is used.

Solution The solution can be obtained without calculation using the nomograph (Fig. 7.15). The dashed line refers to this example. Enter with the perforation length on the upper left. Proceed horizontally to intersect the appropriate entryhole diameter (perforation diameter). Go vertically downward to the 6-in. damaged zone thickness. This point is designated as *a* on Fig. 7.15. Now using a ruler measure along the 6-in. damaged zone thickness line from the vertical axis to the appropriate permeability damage ratio (0.4 in this example). A distance *b–c* shown in Fig. 7.15 is found. Transfer the distance *b–c* to *b′–c′* beginning at *a*. From *c′* go down to the crushed zone permeability ratio of 0.2 and across to the shot density of 4 per foot. Move up to the 90° phasing and read the productivity ratio of 0.88. This is the desired result.

The nomograph has been constructed for a wellbore diameter of 6 in. However, the effects of varying the wellbore diameter are partially compensating [15] and the nomograph is applicable over a considerable range of wellbore sizes. An additional correction for a 12 in. (30.5 cm) wellbore on 160-acre spacing is shown. At point *d*, simply move vertically to the 12 in. 90° phasing line and read horizontally the correct productivity ratio.

There are two other restrictions that are worth noting. The nomograph has been constructed assuming that perforations penetrate the damaged zone surrounding the wellbore and that the crushed zone thickness surrounding the perforation is in all cases 0.5 in. Furthermore, differences between horizontal and vertical formation permeabilities are not taken into account. Because of this, substantial differences between the calculated productivies and those actually observed may exist [17].

7.5 PERFORATING FOR HYDRAULIC FRACTURING

When perforating formations that are known to require a fracture treatment to achieve economic levels of production, the primary concern, other than the usual ones such as well control, is to obtain clean, open perforations in every zone to be hydraulically fractured.

Once perforations have been placed, the fracturing process is initiated. The surface pressures measured during the process will depend in part on the number and size of the perforations. It is of interest to study their contribution to the overall pressure drop.

Fracture Pressures

Figure 7.16 shows the various pressures that apply during fracture treatment. As discussed in Chapter 8, the pressure just inside the perforations, which is labelled p_{BISIP}, is the bottomhole instantaneous shut-in pressure. This is the pressure just required to propagate the fracture. It is generally measured during or at the start of a fracture treatment or can be estimated by the expression

$$p_{\mathrm{BISIP}} = (FG)(D) \tag{7.21}$$

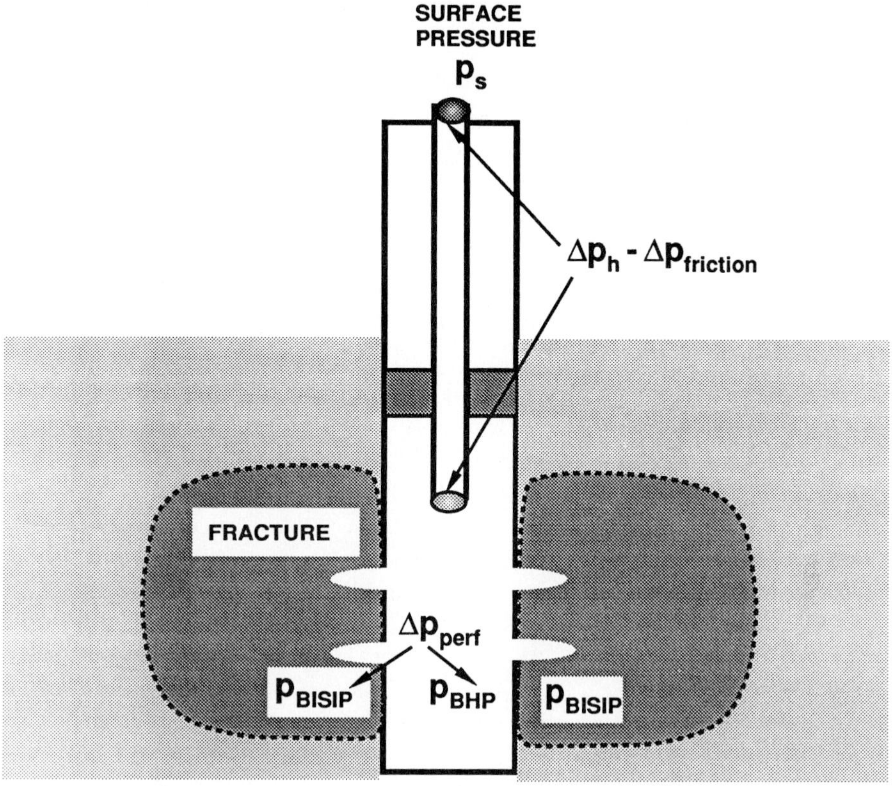

Figure 7.16 Various pressures that contribute to the surface pressure.

where *FG* is the fracture gradient and *D* is the formation depth. Methods for estimating the fracture gradient are described in Chapter 8. For the time being it is sufficient to note that each formation is characterized by a fracture propagation pressure which must be a complex function of the mechanical behavior of the rock, the overburden pressure, the degree of tectonic activity, and the fluid pressure of the reservoir (see Chapter 8). To propagate a fracture, the fluid pressure in the fracture must exceed this pressure. Smaller pressures will allow the fracture to close.

The hydrostatic pressure is the fluid head. Thus

$$\Delta p_{\mathrm{h}} = g\rho D \tag{7.22}$$

where ρ is the density of fluid in the wellbore and g is the acceleration due to gravity. If there is proppant suspended in the fracturing fluid, the density to be used in Eq. (7.22) is

$$\overline{\rho} = \frac{m_{\mathrm{i}} + \rho}{1 + m_{\mathrm{i}}/\rho_{\mathrm{p}}} \tag{7.23}$$

Here m_i is the proppant mass per unit volume of fluid and ρ_p is the density of the proppant.

In Fig. 7.16, p_{BHP} is the bottomhole pressure during the fracture treatment. The pressure drop across the perforation is given by

$$\Delta p_{perf} = p_{BHP} - p_{BISIP} \tag{7.24}$$

We may assume that the perforation behaves as an orifice restricting the flow. Thus

$$q_{perf} = \alpha \pi r_p^2 \sqrt{\frac{2\Delta p_{perf}}{\rho}} \tag{7.25}$$

where r_p is the radius of the perforation [21]. Squaring both sides of the equation and rearranging terms, we find

$$\Delta p_{perf} = \frac{q_{perf}^2 \rho}{2\alpha^2 \pi^2 r_p^4} \tag{7.26}$$

The orifice coefficient (α) can be taken to be approximately 0.90.

If the friction pressure drop attending flow of fluid down the tubing is given by $\Delta p_{friction}$, then the surface pressure is

$$p_{surface} = p_{BISIP} + \Delta p_{friction} - \Delta p_h + \Delta p_{perf} \tag{7.27}$$

Using Eq. (7.27) one can calculate the surface pressure. It should be noted that if there are N perforations, then assuming that the flow through each perforation is the same, we have

$$i = N q_{perf} \tag{7.28}$$

where i is the volumetric injection rate. Of course, Δp_{perf} is the same across each of the perforations.

The friction pressure loss during fracturing cannot be calculated using the classical friction-factor plot to describe turbulent pressure drop because most fracture fluids contain friction reducers. The pressure drop for these systems must be measured. Data are usually available from the service companies. Typical data are shown in Fig. 7.17. It should be noted that the curves shown in Fig. 7.17 are typical, but not general, and each fracture fluid will exhibit its own characteristic pressure drop.

Figure 7.17 shows the effect of the friction reducer on the friction pressure drop (see Section 3.3). Note that in accordance with Eq. (7.27), reducing the friction pressure drop directly reduces the pressure at the wellhead. Thus in many cases where the wellhead pressure is limiting, the addition of drag-reducing agents is required to achieve the desired injection rates.

Reduction of the surface pressure also reduces the hydraulic horsepower required to perform the fracture treatment. Hydraulic horsepower is defined as $ip_{surface}$. It is this factor which often determines the equipment needed to carry out the fracture treatment.

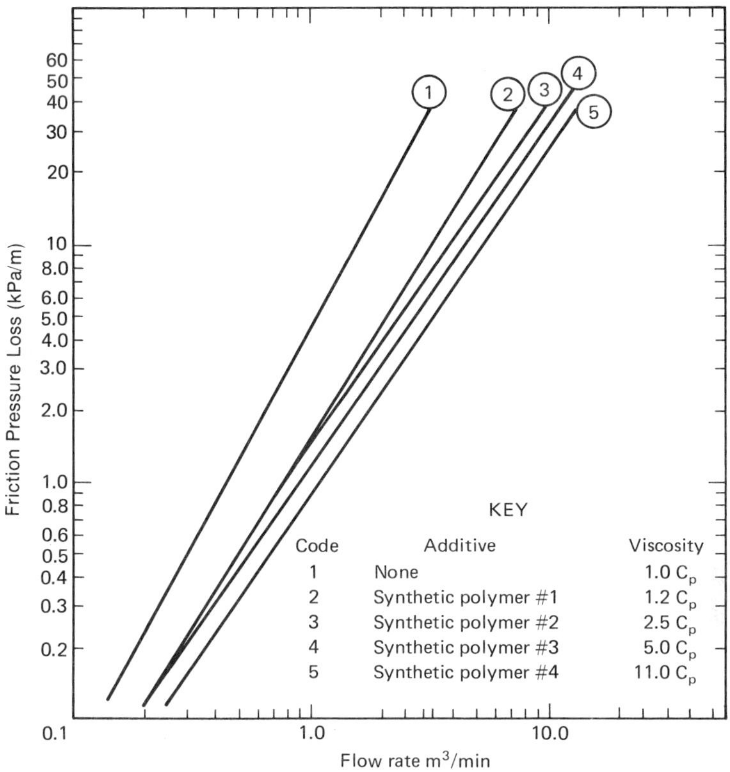

Figure 7.17 Friction pressure as a function of the fluid injection rate for flow of dilute polymer solution in $2\frac{1}{2}$-inch OD tubing. (Adapted from Howard and Fast [22].)

Limited Entry Perforating

Lagrone and Rasmussen [23] have proposed a means of ensuring the simultaneous treatment of multiple producing zones or the fracturing of thick pay zones by limiting the number of perforations and providing a sufficient fracture fluid injection rate to cause the diversion of the fluids to all or nearly all of the perforated interval(s) because of the restricted capacity of an individual perforation to accept the greater portion of the treatment fluid.

The normal procedure in perforating for hydraulic fracturing is to provide an ample number of perforations in each zone and, if necessary, isolate the zones being treated to ensure that the fracture fluid and proppant enter the desire perforations. An alternative procedure is to use diverting agents during the course of the treatment to try to treat all perforated zones.

The limited entry procedure restricts the number of perforations, thereby attempting to ensure that all will accept fluid at some point during the fracture treatment. Suppose, for example, two separate zones have been perforated as shown in Fig. 7.18. If Zone A breaks down first and a fracture is initiated, then

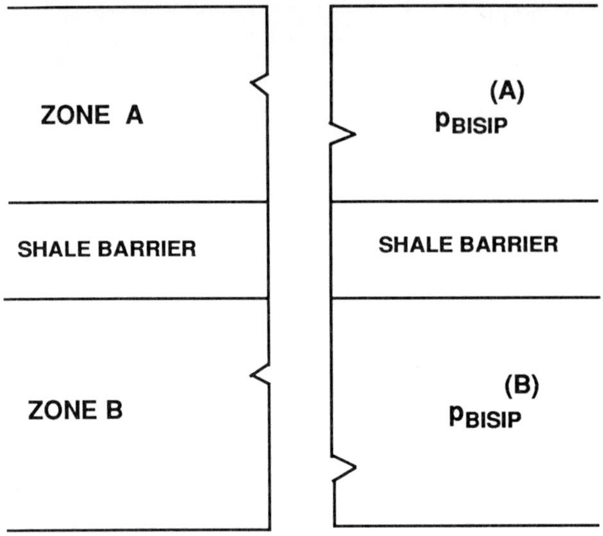

Figure 7.18 Sketch depicting two separate zones to be fractured. A few perforations have penetrated each zone.

all of the fluid is diverted through the few (limited number) perforations in that zone. This means that q_{perf} as calculated by Eq. (7.28) is large, yielding a correspondingly large Δp_{perf} [see Eq. (7.26)]. When the pressure drop through the perforations is large, this causes p_{BHP} to be large. This is true since p_{BISIP} is a property of the formation and does not vary substantially during the treatment. (This key point is discussed further in Chap. 8.) If p_{BHP} remains large and above the breakdown pressure of Zone B, it will fracture and thereby reduce q_{perf} and Δp_{perf}.

Thus, perforations play the role of a choke, causing the bottomhole treatment pressure to remain high until all zones have been fractured and are accepting fluid. Furthermore, by carefully monitoring the surface pressure, it is possible to estimate the number of perforations that are receiving fluid, provided of course, the number is limited. Once the number of perforations is increased beyond a value, it is then difficult to ascertain how many are accepting fluid. These points will become clear by considering Examples 7.6 and 7.7.

Example 7.6 Limited Entry Perforating

It has been decided that a low-permeability formation, consisting of three separate producing zones, will have to be fractured to produce at economic rates. Before perforating, reasonable injection rates for fracturing (4 m^3/min) and large pressure drops across each perforation (3.5 MPa) have been selected as being suitable. Calculate the surface pressure and the number of perforations required in each zone such that the proportion of fracture fluid entering each of the zones is proportional to the height of the zones. Additional data are provided in Table 7.2.

TABLE 7.2 Well Data for Example 7.6

Zone	Depth (m)	Net Pay (m)
A	2,130	9
B	2,225	7.5
C	2,255	14

Additional Data	
Fracture gradient	15.8 kPa/m
Tubing	6.5 lb/ft
Perforation ID	0.76 cm
Fracture fluid density	1042 kg/m³
Fracture fluid type	Water-based with synthetic polymer #3–5 cp (see Fig. 7.17)

Solution The radius of the perforation is given by

$$r_p = (7.6 \times 10^{-3}\ \text{m})/2.0 = 3.8 \times 10^{-3}\ \text{m}$$

$$q_{\text{perf}} = \alpha \pi r_p^2 \sqrt{\frac{2\Delta p_{\text{perf}}}{\rho}} = 0.9\pi(3.8 \times 10^{-3}\ \text{m})^2 \sqrt{\frac{2(3.5 \times 10^6)\ \text{Pa}}{\dfrac{1042\ \text{kg}}{\text{m}^3}}}$$

$$= 3.35 \times 10^{-3}\ \text{m}^3/\text{sec}$$

The number of perforations is given by [see also Eq. (7.28)]:

$$N = i/q_{\text{perf}} = \left(\frac{4}{60}\right)\left(\frac{1}{3.35 \times 10^{-3}}\right) = 19.9 \approx 20 \text{ perforations}$$

As a first approximation take the number of perforations in a zone in proportion to the height. Thus

$$\text{Zone A number of perforations} = \frac{9}{9 + 7.5 + 14}(20) = 6$$

$$\text{Zone B number of perforations} = \frac{7.5}{9 + 7.5 + 14}(20) = 5$$

$$\text{Zone C number of perforations} = \frac{14}{9 + 7.5 + 14}(20) = 9$$

The surface pressure is determined by Eq. (7.27) with

$$p_{\text{BISIP}} = (FG)D = \frac{15.8\ \text{kPa}}{\text{m}}\ \frac{(2130 + 2255)}{2}\ \text{m} = 3.46 \times 10^4\ \text{kPa}$$

$$\Delta p_{\text{friction}} = \left(\frac{8.2\ \text{kPa}}{\text{m}}\right) D$$

from data given in Fig. 7.17. Thus

$$\Delta p_{\text{friction}} = \left(\frac{8.2 \text{ kPa}}{\text{m}}\right)\left(\frac{2130 + 2255}{2}\right) \text{m} = 1.79 \times 10^4 \text{ kPa}$$

$$\Delta p_{\text{perf}} = 3.5 \times 10^3 \text{ kPa}$$

Furthermore,

$$\Delta p_{\text{h}} = \left(\frac{9.8 \text{ m}}{\text{sec}^2}\right)\left(\frac{1042 \text{ kg}}{\text{m}^3}\right)\left(\frac{2130 + 2255}{2}\right) \text{m} = 2.24 \times 10^7 \text{ Pa}$$

$$\Delta p_{\text{surface}} = 3.46 \times 10^4 \text{ kPa} + 1.79 \times 10^4 \text{ kPa} - 2.24 \times 10^4 \text{ kPa} + 3.5 \times 10^3 \text{ kPa}$$

$$= 3.36 \times 10^4 \text{ kPa}$$

Example 7.6 started with an assumed pressure drop across each perforation and a total injection rate. The example shows that these assumptions then fix the number of perforations and the surface pressure. If the surface pressure is too high, then the assumed conditions would have to be adjusted until a suitable value is obtained. One must also be certain that the number of perforations is not so large that it will not be possible to detect from surface pressure measurements whether or not all perforations are accepting fluid. Example 7.7 will illustrate this point.

Example 7.7 Number of Perforations Accepting Fluid

If a well is completed with ten perforations, estimate from the measurement of surface pressure obtained during the initial stages of a fracture treatment and the data given in Table 7.3 the number of perforations accepting fluid.

Solution We know [see also Eq. (7.27)]

$$p_{\text{surface}} = p_{\text{BISIP}} + \Delta p_{\text{perf}} + \Delta p_{\text{friction}} - \Delta p_{\text{h}}$$

TABLE 7.3 Well Data for Example 7.7

Well Data	Symbol	Value
Depth	D	1829 m
Fracture Gradient	FG	17 kPa/m
Perforation Radius	r_p	4.8×10^{-3} m
Treatment Data	**Symbol**	**Value**
Injection Rate	i	0.0318 m³/sec
Fluid Density	ρ_f	1041 kg/m³
Fluid Viscosity	μ	1.2 cp (synthetic #1)
Tubing Diameter		$2\frac{1}{2}$ in. OD (see Fig. 7.17)
Surface Pressure	p_{surface}	3.3×10^4 kPa

From Fig. 7.17

$$\Delta p_{\text{friction}} = \frac{(4.2 \text{ kPa})}{\text{m}} D \qquad \text{or} \qquad \Delta p_{\text{friction}} = 7.68 \times 10^3 \text{ kPa}$$

The hydrostatic head is given by

$$\Delta p_{\text{h}} = \left(\frac{9.8 \text{ m}}{\text{sec}^2}\right)\left(\frac{1041 \text{ kg}}{\text{m}^3}\right)(1829\text{m}) = 1.86 \times 10^7 \text{ Pa} = 1.86 \times 10^4 \text{ kPa}$$

The bottomhole instantaneous shut-in pressure can be easily calculated, based on the fracture gradient, as

$$p_{\text{BISIP}} = (FG)(D) = \left(\frac{17 \text{ kPa}}{\text{m}}\right)(1829 \text{ m}) = 3.11 \times 10^4 \text{ kPa}$$

Solving for Δp_{perf} we find

$$\Delta p_{\text{perf}} = 3.3 \times 10^{-4} \text{ kPa} - 3.11 \times 10^4 \text{ kPa} - 0.768 \times 10^4 \text{ kPa} + 1.86 \times 10^4 \text{ kPa}$$
$$= 1.28 \times 10^4 \text{ kPa}$$

From Eq. (7.25) we find

$$q_{\text{perf}} = (0.9)(\pi)(4.8 \times 10^{-3} \text{ m})^2 \sqrt{\frac{(2)(1.28 \times 10^7) \text{ Pa}}{\frac{1041 \text{ kg}}{\text{m}^3}}} = 9.38 \times 10^{-3} \text{ m}^3/\text{sec}$$

Therefore, the number of perforations accepting fluid is given by

$$N = \frac{1}{q_{\text{perf}}} = 3.4 \text{ perforations}$$

Example 7.7 is intended to illustrate the method by which one can estimate during the early stages of the fracture treatment the number of perforations that are accepting fluid. In the example, three of ten were flowing. This results in an excessively high surface pressure and also bottomhole flowing pressure, but despite these high pressures, the seven other perforations were not flowing. If the ten perforations had been distributed between two different zones, it might be presumed that only one of the zones fractured and at some time during the treatment it will be necessary to attempt to stop the flow through the open perforations (by using ball sealers dropped into the fracture fluid) and attempt to fracture the second zone.

Limited entry techniques have been used with some success and the approach should be considered when designing completions in areas where hydraulic fracturing is routinely practiced as a part of the completion program.

REFERENCES

7.1. Allen, T. O., and Roberts, A. P., *Production Operation*, Vol. 1, Chap. 7, Tulsa: Oil and Gas Consultants International, Inc., 1978.

7.2. Bell, W. T., Lebourg, M. P., and Bricaud, J., "Perforating Today—A Science," *Drilling and Production Practice*, Washington, D.C.: American Petroleum Institute (1959) 249.

7.3. Bell, W. T., and Auberlinder, G. A., *J. Pet. Tech., 12* (1960) 211.

7.4. Thompson, G. D., "Effects of Formation Compressive Strength on Perforator Performance," *Drilling and Production Practice*, Washington, D.C.: American Petroleum Institute (1962) 191.

7.5. Allan, T. O., and Warzel, H. C., "Productivity Method of Evaluating Gun Perforating," *Drilling and Production Practice*, Washington, D.C.: American Petroleum Institute (1956) 111.

7.6. "API Recommended Practice: Standard Procedure for Evaluation of Well Perforators," RP43, *Division of Production of American Petroleum Institute*, 2nd ed., Washington, D.C.: American Petroleum Institute, 1971.

7.7. Bell, W. T., Brieger, E. F., and Harrigan, J. W., *J. Pet. Tech., 24* (1972) 1095.

7.8. Klotz J. A., Krueger, R. F., and Pye, D. S., *J. Pet. Tech., 26* (1974) 1303.

7.9. Deo, M., Tarig, S., and Halleck, P. M., "Linear and Radial Flow Targets for Characterizing Downhole Flow in Perforations," SPE 16896, presented at the 62nd Fall Technical Conference and Exhibition of the Society of Petroleum Engineers, Dallas, Texas, 1987.

7.10. Bell, W. T., and Bell, R. M., "The Paradox of Gun Powder Versus Completion Efficiency," paper presented at the 1981 Explosives Conference on the International Association of Drilling Contractors, Houston, Texas, 1981.

7.11. Bell, W. T., "Perforating Techniques for Maximizing Well Productivity," SPE 10033, presented at the International Petroleum Exhibition and Technical Symposium of the Society of Petroleum Engineers, Beijing, China, 1982.

7.12. Bell, W. T., *J. Pet. Tech., 36* (1984) 1653.

7.13. Benedyczak, C., and Al-Towailib, A. A. A., *J. Pet. Tech., 36* (1984) 1511.

7.14. Harris, M. H., *J. Pet. Tech., 18* (1966) 518.

7.15. Hong, K. G., *J. Pet. Tech., 27* (1975) 1027.

7.16. Locke, S., *J. Pet. Tech., 33* (1981) 2481.

7.17. Tariq, S. M., Ichara, M. J., and Ayestaran, Y., "Performance of Perforated Completions in the Presence of Common Heterogeneities Anisotropy, Laminations, or Natural Fractures," SPE 14320, presented at the 60th Fall Technical Conference and Exhibition of the Society of Petroleum Engineers, Las Vegas, Nevada, 1985.

7.18. McLeod, H. O., and Crawford, H. R., "Gravel Packing for High Rate Completions," SPE 11008, paper presented at the 57th Fall Technical Conference and Exhibition of the Society of Petroleum Engineers, New Orleans, Louisiana, 1982.

7.19. Godfrey, W. K., and Methven, H. E., "Casing Damage Caused by Jet Perforating," SPE 3043, presented at the 1970 Society of Petroleum Engineers Annual Meeting, Houston, Texas, 1970.

7.20. McLeod, H. O., *J. Pet. Tech., 35* (1983) 31.

7.21. Bird, R. B., Steward, W. E., and Lightfoot, E. N., *Transport Phenomena*, New York: John Wiley, 1960.

7.22. Howard, G. C., and Fast, C. R., *Hydraulic Fracturing*, Mono. Ser., *2*. Society of Petroleum Engineers, Richardson, Texas, 1968.

7.23. Lagrone, K. W., and Rasmussen, J. W., *J. Pet. Tech., 15* (1963) 695.

PROBLEMS

***7.1.** A perforation is ellipsoidal in shape. If it extends 15 cm into the formation and is 1 cm in diameter at the wellbore, how much formation fluid (m^3/day) will be produced through this perforation if the formation permeability is 1×10^{-14} m^2 and the oil viscosity is 2 cp? The effective drawdown pressure is 2100 kPa. Construct a plot showing contours of constant pressure.

***7.2.** The API RP43 test of perforating guns involves creating a perforation in a confined Berea core, measuring the permeability and comparing that permeability to an ideal (undamaged) one. Suppose the flow through the ideal perforation is calculated to be 1×10^{-2} cm^3/sec when the applied pressure gradient across the test core is 15 kPa/cm. If the actual flow through the perforation created with a jet gun is 4×10^{-2} cm^3/sec when the applied pressure gradient is 42 kPa/cm, what is the CFE of the jet gun?

***7.3.** A damaged perforation produces at a rate equal to 20% of that of an undamaged one. If the penetration depth of the perforation is 25 cm and the radius of the perforation is 0.3 cm at the wellbore, calculate the permeability contrast (k_c/k) of the crushed zone relative to that of the undamaged formation permeability. Use the approximation that the perforation is independent of all other perforations. It is known that the outer radius of the crushed zone surrounding the perforation is 0.5 cm measured at the wellbore.

****7.4.** The CFE test is ultimately expressed in terms of the ideal permeability of an undamaged perforation. A numerical calculation is required to obtain the ideal permeability because the geometry shown by Fig. P7.4 is complex. The pressure within the perforation is constant (p_w) and the pressure applied at the open face of the cylindrical core is p_1. The heavy boundaries are sealed and cannot be penetrated by fluid.

(a) Write the diffusivity equation describing incompressible flow in cylindrical coordinates. Based on Darcy's law, eliminate the fluxes u_r and u_z to obtain a partial differential equation in which the pressure depends on r and z.

(b) Write the conditions that apply at each boundary (e.g., for all $L > z > L - D$ and $r = r_p$, then $p = p_w$).

(c) Using finite difference approximations express the partial differential equation as a set of algebraic equations suitable for computer calculations.

(d) Write a computer program that calculates k_i as a function of D/L, R/L, and r_p/L.

(e) Construct a graph showing k_i/k as a function of D/L for $R/L = 0.10$ and $r_p/L = 0.01$.

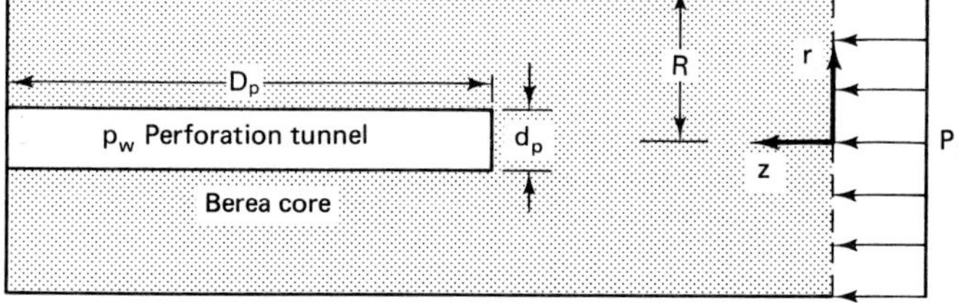

Figure P7.4

***7.5.** It is known that in some cases, part of the damage surrounding a perforation can be removed by increasing the drawdown pressure ($p_R - p_w$) when the well is first put on production. Suppose that a perforation having a depth of 30 cm and an entryhole radius of 0.25 cm is surrounded by a crushed zone that extends 33 cm into the formation measured along the axis of the perforation. Both the perforation and the crushed zone can be considered to be ellipsoidal and concentric (same H). When the drawdown pressure is increased from 7×10^6 Pa to 1.05 to 10^7 Pa, the production from one perforation increased from 1×10^{-5} m^3/sec to 1.85×10^{-5} m^2/sec. By what factor does the permeability of the crushed zone increase when the higher drawdown pressure is imposed? Note p_R is the reservoir pressure and p_w is the well flowing pressure. It is known that at a drawdown pressure of 7×10^6 Pa an undamaged performation (no crushed zone) will flow 2×10^{-5} m^3/sec.

***7.6.** Consider an oil well having the following characteristics:

$$\text{(Drainage radius)} \quad r_e = \text{(660 ft) 201 m}$$

$$\text{(Wellbore radius)} \quad r_w = \text{(3.5 in.) 8.9 cm}$$

$$\text{(Formation thickness)} \quad h = \text{(50 ft) 15.2 m}$$

$$\text{(Formation permeability)} \quad k = \text{10 mD}$$

$$\text{(Fluid viscosity)} \quad \mu = \text{5 cp}$$

Tests in the laboratory have indicated that perforations 1.27 cm (0.5 in.) in diameter and 25.4 cm (10 in.) deep may be expected in this formation.

If the perforation is ideal (nondamaging), how many perforations per meter and at what phasing are required to achieve the same production as with an openhole completion?

If a crushed zone having a permeability of 1 mD and a thickness of 1.27 cm (0.5 in.) at the wellbore is formed, what shot density and phasing will then be required?

***7.7.** A zone to be fractured has been perforated in ten equally spaced places throughout the 25 m productive zone. If the surface pressure is limited to 35 MPa, calculate the maximum injection rate (i) permissible and the hydraulic horsepower (expressed as kilowatts) required for the treatment. The data are given in Table P7.7.

TABLE P7.7 Data for Fractured Zone

Property	Symbol	Value
Formation thickness	h	25 m
Formation depth	D	2200 m
Fracture gradient	FG	18 kPa/m
Perforation radius	r_p	4×10^{-3} m
Coefficient of discharge	α	0.9
Fracture fluid density	ρ	1200 kg/m^3

Note: The friction pressure in the tubing is given by

$$\Delta p_f = \Delta p_f^0 \left(\frac{i}{i_0}\right)^2$$

where the constants Δp_f^0 and i_0 are

$$\Delta p_f^0 = 1200 \text{ kPa/1000 m of tubing}$$
$$i_0 = 5 \times 10^{-3} \text{ m}^3\text{/sec}$$

***7.8.** A 6.2 cm ID casing is 1000 m in length and inclined at an angle of 45° with respect to the horizontal. If water containing a drag-reducing agent is pumped through this tubing so that the pressure at the top of the tubing is 2 MPa and the pressure at the bottom is 67 MPa, how much fluid per unit of time is flowing (express your answer in m^3/min)?

The density of the fluid is same as that of water at 60°F and the *friction loss* for the drag-reducing fluid in the tube is expressed by

$$\frac{\Delta p_{\text{friction}}}{L} = 300i^2 + 740 \text{ (in Turbulent Flow)}$$

where $\Delta p_{\text{friction}}/L$ = Pressure drop due to friction (in kPa/1000 m)
i = Volumetric flow rate (in m^3/min)

How much water without drag-reducing agents will flow in the same tubing under the same pressure gradient?

PART 4
Hydraulic Fracturing

8
Dynamic Fracture Geometry

Hydraulic fracturing is the process of injecting a fluid into a well to create tensile stresses in a formation exposed to the fluid pressure, causing local stresses in the formation to exceed the tensile strength of the rock. This creates a crack, or fracture, propagating into the formation from the wellbore as fluid continues to be injected at a high rate. In some formations, acids may be used as the fracture fluid to etch the face of the crack, whereas in others, a proppant such as sand may be injected with the fluid so that upon cessation of pumping and crack propagation, the crack remains a conductivity pathway for fluids to flow from the reservoir into the wellbore. These two processes are referred to as *acid fracturing* and *proppant fracturing*. The dynamic geometry is the same regardless of which process is selected for use. Thus, this chapter will apply to both acid and proppant fracturing. Chapter 9 also applies to both since the mechanisms of heat transport are the same for both processes.

The mechanism by which fracture conductivity is created, however, differs between the two processes and these are considered in separate sections within Chapter 10.

There are two basic reasons for fracturing a well: to increase the rate or productivity and/or to improve the ultimate recovery. Additionally, other wells may also be fractured to aid in secondary recovery operations and to assist in the injection or disposal of waste water.

If a well has a large skin effect (see Chap. 6), then a relatively small fracture may increase productivity as much as a few hundred percent. Chapter 11 is devoted to the prediction of the increased well productivity that can result from fracturing. Fracturing is also attractive in very tight sands where the production rates may be normally too low to permit economic drainage of the formation hydrocarbon. In these cases long, conductive fractures may help to increase the ultimate recovery of hydrocarbon by extending the time before the economic limit of a well is reached.

The final chapters in Part 4 deal with optimum fracture design and pressure testing fractured wells. Both of these topics are of great importance to the practicing engineer. These chapters can be studied without reading the more fundamental aspects of fracturing presented in the initial chapters of Part 4.

8.1 FRACTURE ORIENTATION

Fracturing Pressure

In many applications of hydraulic fracturing it is essential to ascertain, if at all possible, the orientation of the fracture. The first question to be answered regards whether or not the fracture is vertical or horizontal. Generally, horizontal fractures are preferred, but in deeper formations they are, as will be evident, most often vertical. One of the important measurements that can help distinguish between horizontal and vertical fractures is the bottomhole pressure measured during the fracture treatment.

In hydraulic fracturing, enough pressure must be applied to initially break down or fracture the formation and enough pressure must continue to be imposed to allow the fracture to continue to grow or propagate. Normally, more pressure is required to initially break down a formation than is required to propagate a fracture. Once a fracture is formed, the fluid in the fracture acts as a wedge, forcing the fracture to grow. A fracture is more easily created using a low-viscosity, penetrating fluid than with a high-viscosity nonpenetrating fluid. A penetrating fluid pressurizes a larger area, and the total force on the formation is greater than if a nonpenetrating fluid, which acts only on the area near the wellbore, is used.

Pressure behavior during a fracturing treatment is illustrated by Fig. 8.1. The fluid injection rate is constant, except that at some time injection is stopped to obtain the instantaneous shut-in pressure. The bottomhole pressure is shown versus time from the initial injection of fluid until the treatment has been completed. The surface pressure is, of course, different from the bottomhole pressure because of the weight of the fluid and the friction losses in the wellbore. The critical portions of the pressure history shown in Fig. 8.1 are:

- *Breakdown Pressure:* the pressure required to break down the formation and initiate fracture.
- *Propagation pressure:* the pressure required to continually enlarge the fracture.
- *Instantaneous Shut-in Pressure:* the pressure that is required to just hold the fracture open.

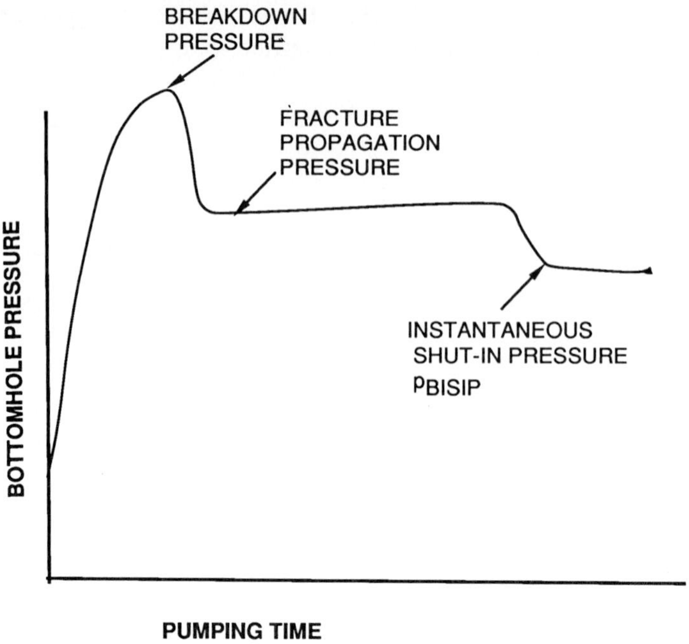

Figure 8.1 Idealized pressure behavior during fracturing.

The instantaneous shut-in pressure measured by stopping the flow will depend on the width of the fracture at this point and the pore pressure surrounding the fracture. If large quantities of fluid have been injected and the fracture width at the wellbore is large, then a larger shut-in pressure will be observed [1, 2]. If it is desired to measure the intrinsic tectonic stress, the shut-in pressure should be measured after only a small amount of low-viscosity fluid has been injected to create a fracture. At this stage the fracture width will be narrow and will have little effect. Even if larger quantities of fluid are injected, the effect of fracture width is normally less than 3000 kPa [1]. After shut-in, stresses in the earth squeeze the fluid in the fracture until the fracture walls close on the proppant or on the etched walls of an acid fracture. When the walls close, and support the earth's stresses, the pressure will decrease rapidly as more fluid leaks off into the formation.

The described pressure behavior is highly idealistic. Seldom will all the pressures described be observed during a fracture treatment. For example, if the reservoir had been previously fractured, there may not be any difference between breakdown pressure and fracture propagation pressure. If the reservoir pressure is very low, the well will go on vacuum when the fracture closes, and a static reservoir pressure will not be measured at the surface.

If p_{ISIP} is the instantaneous shut-in pressure measured at the surface, then the bottomhole shut-in pressure (p_{BISIP}) is given by

$$p_{\text{BISIP}} = p_{\text{ISIP}} + \rho g D \tag{8.1}$$

where D is the formation depth. This equation is precise because when flow is stopped, the friction pressure vanishes.

The bottomhole pressure required to maintain a fracture divided by the reservoir depth (D) is defined as fracture gradient (FG). Thus

$$FG = P_{\text{BISIP}}/D \tag{8.2}$$

Vertical or Horizontal Fractures

Figures 8.2 (a), (b), and (c) show data on fracturing pressures for both breakdown and propagation. Slightly more pressure is required to break down a formation than to propagate a fracture. In the Texas–Louisiana Gulf Coast area, breakdown pressure gradients are approximately 17.9 (0.8 psi/ft) to 19.0 kPa/m (0.84 psi/ft) on the average, whereas fracture gradients range from 15.6 to 16.7 kPa/m (0.69 to 0.74 psi/ft). The average fracture gradient in reservoirs having a depth greater than 1800 m is approximately 15.6 kPa/m. The spread in the data is approximately 2.2 kPa/m.

The overburden stress is normally 22.3–24.6 kPa/m (0.99–1.08 psi/ft). If the fracture gradient is less than this, it will be easier to part the earth than to lift it, and a vertical fracture will form. The curves for fracture gradient show that when reservoirs are deeper than about 600 m, the fracture gradient is less than 22.3 kPa/m, and vertical fractures should form. In shallow reservoirs, however, fracture gradients can be higher than 24.6 kPa/m and here horizontal fractures can form. There is quite a bit of scatter in the data, and these rules should not be considered to be precise.

The principle apparently governing the orientation of a fracture is that the crack opens and widens in an orientation requiring the least work, as depicted by Fig. 8.3. The crack is shown to be perpendicular to σ_x, the smaller of the three principal stresses.

Thus, it may be postulated that fractures should occur along planes normal to the least principal stress [3, 4]. Since, as shown by Fig. 8.2 the fracture gradient for deeper formations is often much smaller than the weight of the overburden, these fractures are vertical and the horizontal stresses are, therefore, less than the vertical ones.

The state of stress in the subsurface is a complex function of the depth and the tectonic activity in the region [5]. Several theories characterizing horizontal stresses have been proposed. Hubbert and Willis [4] have postulated that the rock is in a state of incipient faulting. This state of stress is independent of the elastic properties of the rock. A second hypothesis assumes that during diagenetic processes the movement of the earth in horizontal directions is limited. Stresses can only be accommodated by corresponding strains in the vertical direction [3]. The equations describing an elastic body (see Chap. 2 for a detailed discussion) are

$$\sigma_i - p = 2G\left(\epsilon_i + \frac{\nu}{1 - 2\nu}e\right) \qquad (i = x, y, z) \tag{8.3}$$

where p is the formation pressure, G is the shear modulus, and ν is Poisson's

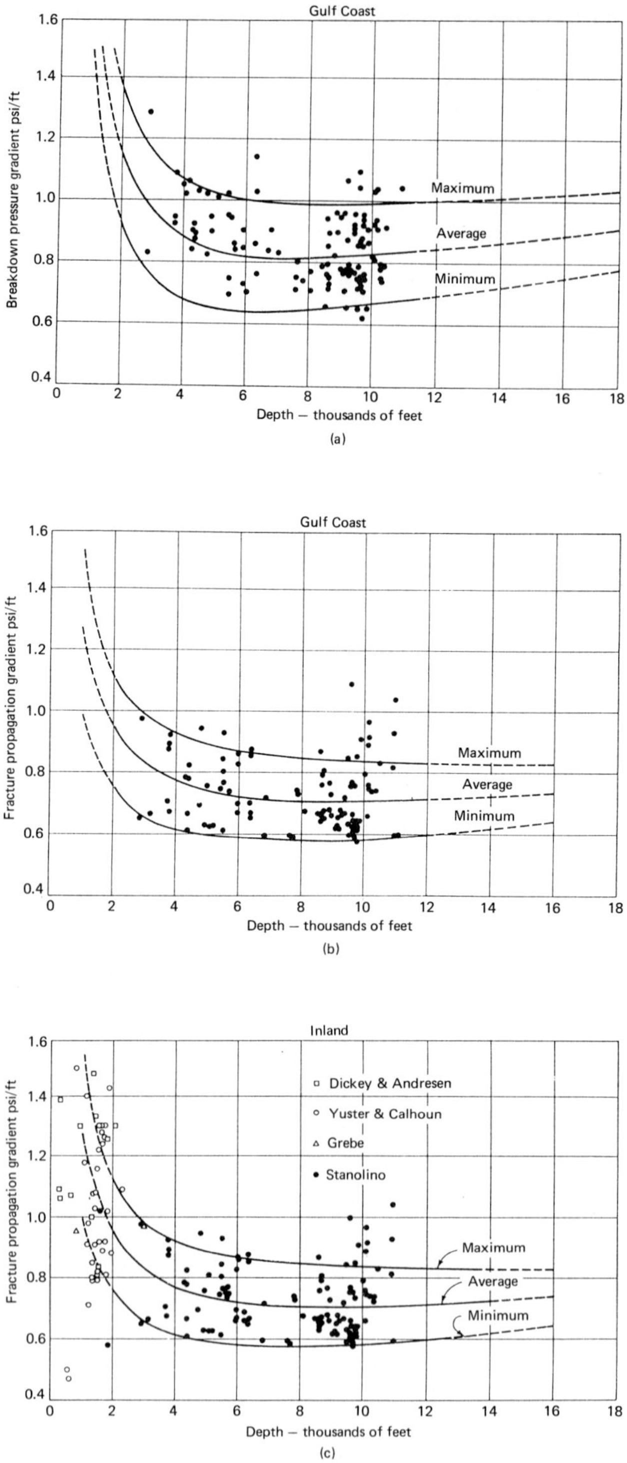

Figure 8.2 U.S. Gulf Coast and Inland fracture gradients [3].

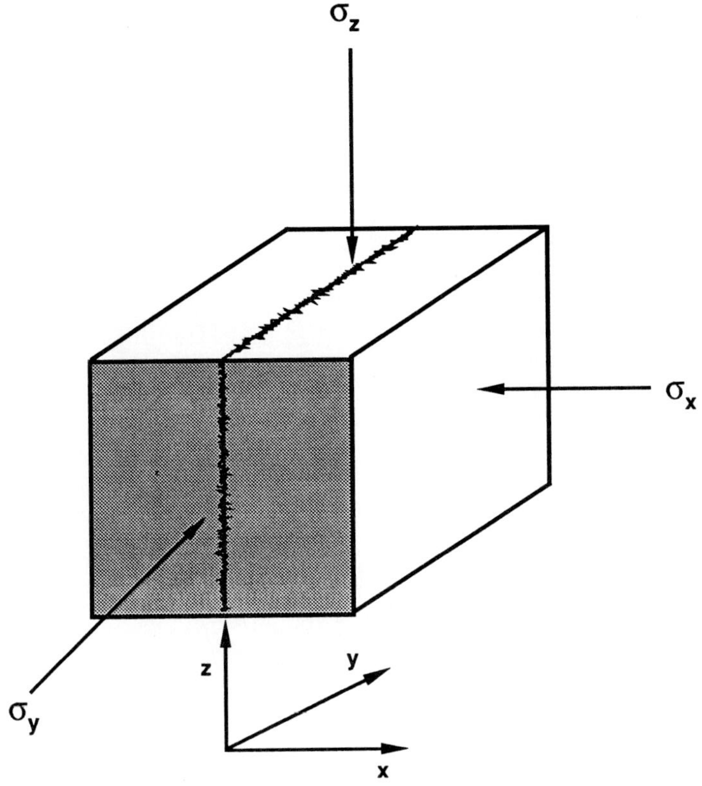

Figure 8.3 Rock fracture shown perpendicular to the direction of least stress.

ratio. If strain is restricted to the z (or axial) direction as shown in Fig. 8.4, then

$$\epsilon_x = \epsilon_y = 0 \qquad \text{and} \qquad e = \epsilon_z \tag{8.4}$$

The axial stress must equal the weight of the overburden. Therefore

$$\sigma_z = \rho_o D g \tag{8.5}$$

where ρ_o is the density of the overburden, which is approximately 2290 kg/m^3. The overburden stress then amounts to an increasing pressure with depth of about 22.4 kPa/m (1 psi/ft).

Substituting Eqs. (8.4) and (8.5) into Eq. (8.3), the strain in the z-direction is given by

$$\epsilon_z = \left[\frac{\rho_o D g - p}{2G}\right]\left(\frac{1 - 2\nu}{1 - \nu}\right) \tag{8.6}$$

We can now solve for either σ_x or σ_y. In the present case these two stresses are assumed equal; however, because of movements within the earth, it is usually not true that strains in both the x- and y-directions vanish nor are the two stresses necessarily equal. Thus, tectonic activity plays a very important role in deter-

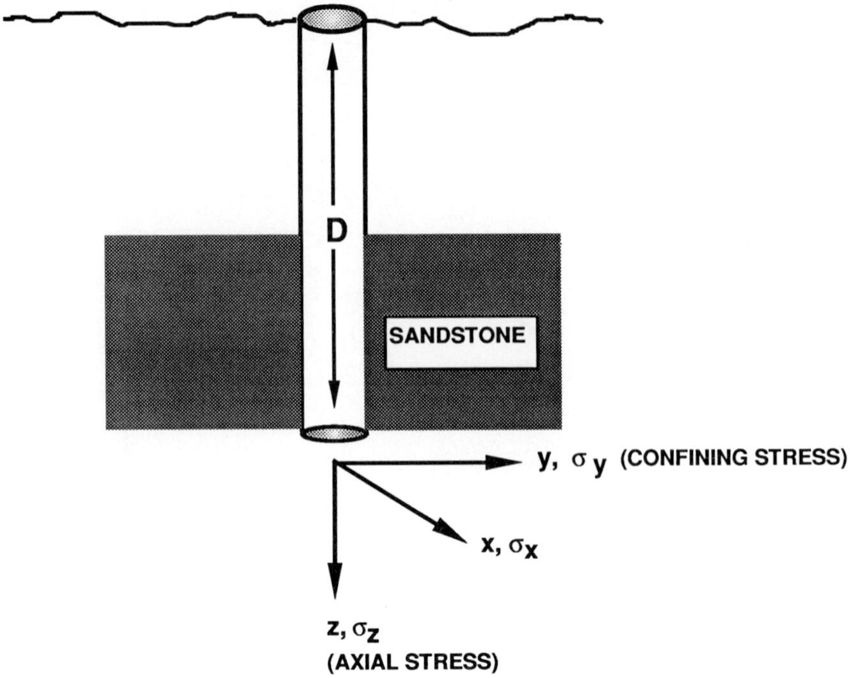

Figure 8.4 The stress acting in a subsurface formation. The strain is restricted to the axial direction.

mining the state of stress. Continuing with our present calculation by substituting Eq. (8.6) into Eq. (8.3) (with i set equal to x), we find the important result

$$\sigma_x = \left(\frac{\nu}{1-\nu}\right)\rho_o D g + p\left(1 - \frac{\nu}{1-\nu}\right) = \sigma_y \tag{8.7}$$

It is of interest to compare the magnitudes of σ_x or σ_y with that of σ_z, for if $\sigma_z > \sigma_x$, then one can expect vertical fractures. Examination of Table 2.1 will show that ν generally ranges between 0.25 to 0.4. This would mean that in general σ_z exceeds σ_x provided, of course, the assumptions made at the onset apply.

Since the fracture which opens vertically must act against the stress σ_x, it is possible to identify σ_x as p_{BISIP}. This is the pressure necessary to hold the fracture open. Thus

$$\frac{p_{\text{BISIP}}}{D} = FG = \left(\frac{\nu}{1-\nu}\right)\rho_o g + \frac{p}{D}\left(1 - \frac{\nu}{1-\nu}\right) \tag{8.8}$$

Equation (8.8) has been used in recent years to predict the fracture gradient in a new area where information regarding the fracture gradient is not available. Equation (8.8) shows that the fracture gradient decreases as the reservoir pressure declines. The equation can, therefore, be used to predict the fracture gradient for a partially depleted formation, given the fracture gradient measured at an earlier time when the reservoir pressure was higher. This application is illustrated by Example 8.1.

Example 8.1 Prediction of the Fracture Gradient

A formation 3,000 m in depth exhibited a fracture gradient of 16 kPa/m when the formation pressure was 24,000 kPa. What is the fracture gradient when the formation pressure has declined to 10,000 kPa?

Solution Equation (8.8) is useful in helping to solve this problem. The quantity $\nu/(1 - \nu)$ can be treated as an empirical constant, α, which is determined based on the conditions known to exist when the fracture gradient was measured. Thus

$$FG = \alpha(22.4) + (1 - \alpha)\frac{p}{D} \quad \text{or} \quad 16 = \alpha(22.4) + (1 - \alpha)\frac{24{,}000}{3{,}000}$$

Solving for α we find

$$\alpha = 0.56$$

Since α depends (in theory) on the Poisson ratio, it can be assumed to remain constant as the reservoir pressure decreases. Thus, the fracture gradient for the depleted reservoir may be estimated as follows:

$$FG = 0.56(22.4) + (1 - 0.56)\frac{10{,}000}{3{,}000} \quad \text{or} \quad FG = 14 \text{ kPa/m}$$

Equation (8.8) ignores both the influence of temperature and creep. Prats [5] has shown that these are important contributions. His analysis shows the horizontal in-situ stresses depend on the burial history and because of the differences in these properties, the tectonic stresses will differ in layers having different creep properties. This differential creep and tectonic activity are at least in part the origin of the stress contrasts which will be shown to limit the vertical growth of fractures into the under and over burdens.

8.2 VERTICAL FRACTURES

Although the final or static fracture dimensions are really the only ones that concern us since these determine the increased productivity that will result from the treatment, they cannot be calculated without first knowing the dynamic ones. The geometry of a vertical fracture is characterized by a length that is a function of time; a width that can depend on the distance from the wellbore, the vertical position and the time; and a fracture height that depends on the distance from the wellbore and the time. Figure 8.5 depicts one wing of a fracture extending outward from the wellbore. Normally a fracture propagates in a configuration that is symmetric about the wellbore. There will, therefore, be two wings which are assumed to be identical. The dimensions of the fracture change with time as fluid is injected. Let us consider the variation of each.

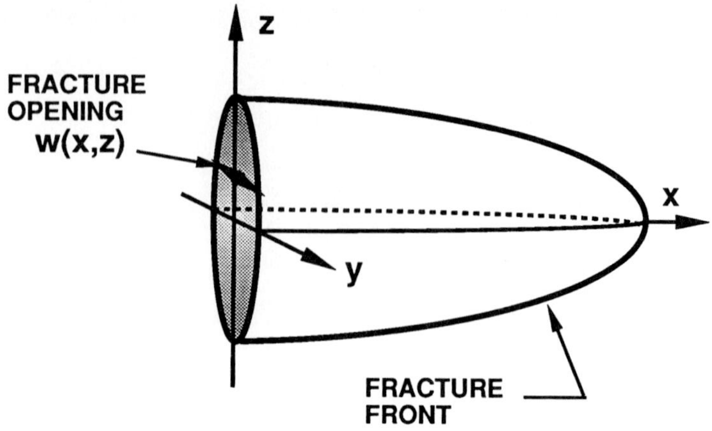

Figure 8.5 Geometry of a vertical hydraulic fracture.

Fracture Azimuth

In deep formations, fractures are generally vertical. It is important in many applications to be able to predict the direction or azimuth of the vertical fracture. This is especially true if fracturing is a planned part of the initial field development [6]. It is not difficult to imagine how certain well patterns will, in such a case, be much more efficient than others depending on the direction (i.e. North, South, etc.) that vertical fractures propagate. In the previous section we have assumed the principal horizontal stresses to be equal; however, movements of the earth during geological times have caused repeated deformation, faulting, and folding of rocks; and substantial variation in all three of the principal components of stress exist in subsurface formations. This point has been emphasized by Hubbert and Willis [4] using a simple example system, a box of sand divided into two parts by a partition as depicted in Fig. 8.6 (a).

A movement of the partition to the right causes the sand near the partition on the left side to slump, producing a series of normal faults. This is shown by Fig. 8.6(b) where the sand in the left-hand compartment dips at an angle of approximately 60°. It is not difficult to imagine that in this case the horizontal component of the principal stress perpendicular to the normal faults is reduced as a result of the movement of the partition. In this case σ_x will be the smaller of the three principal stresses and a vertical fracture perpendicular to this stress represents the least work. Thus, a vertical fracture parallel to the direction of the faults is to be expected.

The Gulf Coast region is normally faulted with faults running approximately parallel to the coast line, apparently due to the movement of sediments over the continental shelf. Fractures will, therefore, tend to be parallel to the coast line.

In Fig. 8.6, the movement of the barrier to the right (an uplift) produces a series of thrust faults and it seems reasonable to expect σ_x to be increased as a result. Thus, the fracture, if vertical, will be parallel to σ_x and perpendicular to the thrust faults. This is apparently the tectonic state that dominates in the central

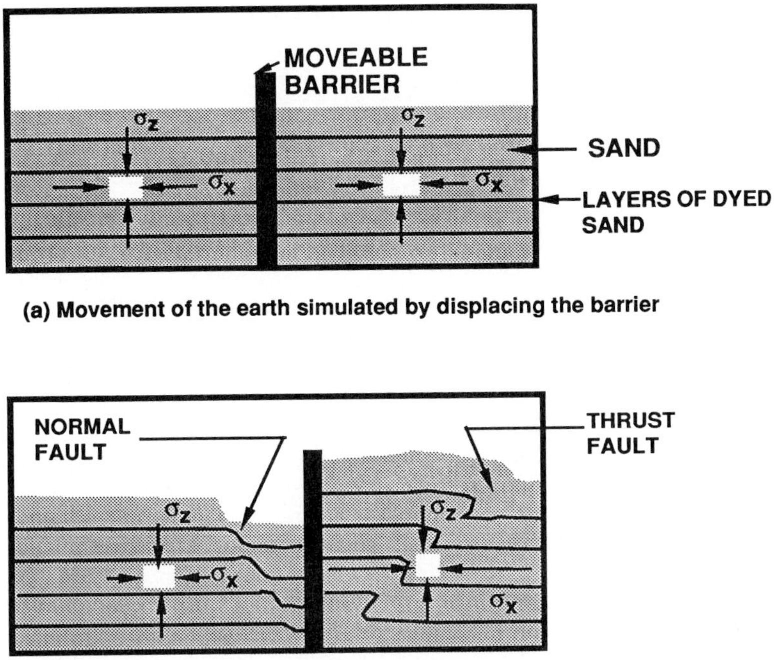

(a) Movement of the earth simulated by displacing the barrier

(b) In response to movement, sand in the right compartment slumps and is thrust upward in the left compartment

Figure 8.6 Sandbox model showing change in horizontal stresses as a result of simulated tectonic movement. **(a)** Approximately equal principal stresses. **(b)** Unequal horizontal stresses.

United States east of the Rocky Mountains. Fractures are expected to be roughly perpendicular to the thrust faults and in this region the fracture orientation tends to be N70°.

West of the Rocky Mountains, complex thrust-faulting systems have developed and no one feature can be said to dominate fracture orientation. Local tectonic activity is highly influential in determining the fracture azimuth and the only certain way to know the direction of propagation is to measure it. Fracture azimuths have been measured by using impression packers, caliper logs, tiltmeters, core analysis, and seismic methods.

A caliper log taken in a vertical open borehole can provide an indication of the direction of the minimum horizontal stress. The borehole will be deformed by the in-situ stresses [4, 7]. Therefore, unless excessive flow of fluids into and out of the formation has eroded the wellbore walls, the caliper log will reveal an elliptical wellbore with the major axis aligned in the direction of the least principal horizontal stress.

The availability of fresh oriented core samples can be used to determine the direction of the least horizontal stress. The method is based on the relaxation that a core sample undergoes when it is removed from its underground position [7, 8,

9, 10]. The strains are measured and the principal strain axes determined. The principal stresses are assumed to have axes coinciding with the principal strains.

Several authors [11, 12, 13] have reported the occurrence of acoustic emissions during and following hydraulic fracturing and have utilized these as a means of mapping the fracture plane. In the context of hydraulic fracturing, *acoustic emissions* may be thought of as extremely small microearth quake-like events occurring as a fracture propagates [13]. It is believed that these sounds emanate from the entire surface of the fracture, not just from the fracture tip as one might at first think. Given that the compressional (P-wave—see Chap. 2) and shear wave (S-wave) signals are spatially associated with the fracture plane, the fracture orientation can be determined if the source of the signals is known.

These signals are small and the best approach would seem to make these measurements in wells near the well being fractured, rather than attempting to make the measurements on the surface as is done by surface deployed geophysical techniques such as tiltmeters [14] or magnetometers [15]. These latter techniques are apparently limited to shallow wells [13].

Fracture Height

One of the most important issues to be addressed relates to fracture containment, or the tendency for a fracture to extend in length rather than increase vertically. Fractures that grow extensively in the vertical direction will ultimately extend beyond the pay zone and fracture into undesirable zones, such as aquifers, either above or below the pay zone. As will be observed, the in-situ stresses are the most important factors determining fracture containment, but first it is necessary to consider the mechanism of fracture propagation.

Stress intensity factor. In the linear theory of fractures, the stresses around a fracture are predicted to be singular with $r^{-1/2}$, where r is the distance to the crack tip as shown by Fig. 8.7. The strength of the singularity is measured

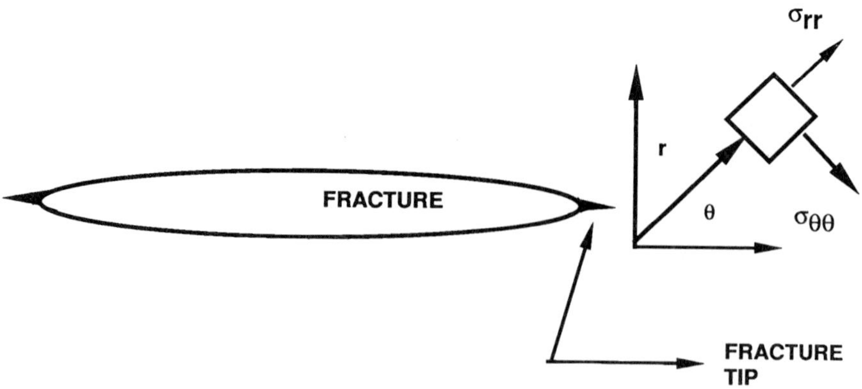

Figure 8.7 Stress components near the tip of a sharp crack.

by means of the stress intensity factor (K) which for a tensile crack is [16, 17]

$$K = \lim_{r \to 0} \sqrt{2\pi r}\, \sigma_{\theta\theta} \tag{8.9}$$

where $\sigma_{\theta\theta}$ is the tensile stress ahead of the fracture tip. The value of the stress intensity factor depends on the fracture geometry and on the applied load. For a long crack of height h in a homogeneous material [18]

$$K = 1.25\ \Delta p \sqrt{h} \tag{8.10}$$

where $\Delta p = p_{\mathrm{f}} - p_{\mathrm{BISIP}}$ and p_{f} is the pressure of the fluid in the fracture. Similarly for a penny-shaped crack of radius R, it is

$$K = 0.8\ \Delta p \sqrt{R} \tag{8.11}$$

A fracture will propagate whenever the stress intensity factor reaches a critical value (K_{c}), which is thought to be a material property called the *critical stress intensity factor, fracture toughness*, or *fracturability*. Measured values of K_{c} for sandstones range between 0.44 and 1.76×10^3 kPa/m$^{1/2}$ [18]. For carbonates the range is 0.44 to 1.04×10^3 kPa/m$^{1/2}$. Interestingly, the range of critical stress intensity factors in shales is similar. Thus the materials of interest in hydraulic fracturing all have similar critical stress intensity factors and it is not, therefore, anticipated that a fracture will be contained within the pay zone simply because of differences in this factor.

Field results do indicate that vertical fracture growth tends to be restricted by thick over- and underlying shale layers [19]. A number of factors other than differences in the critical stress intensity factor have been considered as the reason for this observation [18]. Variations of the shear modulus (G) have been investigated and rejected as the reason shales tend to terminate vertical fracture growth [20, 21].

Influence of horizontal in-situ stresses. Equations (8.10) and (8.11) suggest that since growth ceases when $K < K_{\mathrm{c}}$, then zones having smallest Δp will be the more difficult to fracture. Thus if one zone is subjected to a larger minimum horizontal stress than an adjacent zone, then growth in that zone will tend to be restricted.

Simonson et al. [22] postulated that the contrast in horizontal in-situ stresses is the most important factor inhibiting fracture growth. This has been borne out by laboratory experiments [20, 21, 22] and computer simulations of fracture growth. In the laboratory, stress contrasts of 2 to 3×10^3 kPa have been found to contain fractures [23].

Figure 8.8 shows the calculated growth of vertical fracture in a pay zone over- and underlain by zones in which the in-situ horizontal stresses are greater by 344 kPa (50 psi). For this case, the vertical growth is inhibited by the stress contrast but certainly not contained. The fracture is shown to have extended outward a distance of roughly 320 ft (97.5 m) and vertically a distance of 180 ft (54.9 m) after a treatment time of 31 min. Thus, even a small stress contrast does influence growth.

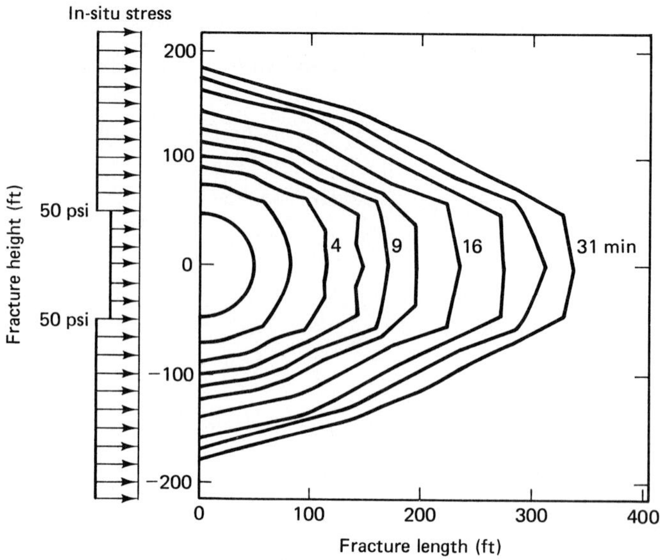

Figure 8.8 Fracture front propagation with an in-situ stress contrast of 344 kPa (50 psi). (E = 5.8 × 10^6 kPa. ν = 0.3, μ = 150 cp, i = 3.18 m^3/min, C = 1.83 × 10^{-4} m/$\sqrt{}$min) [24].

Figures 8.9 and 8.10 show the propagation of a fracture when the stress contrast is increased to 690 kPa and 1380 kPa, respectively, applying the same treatment conditions as used to obtain Figure 8.8. The difference is dramatic. Once the stress contrast has reached 1380 kPa (200 psia) the fracture is essentially contained.

The degree of containment depends primarily on the horizontal in-situ

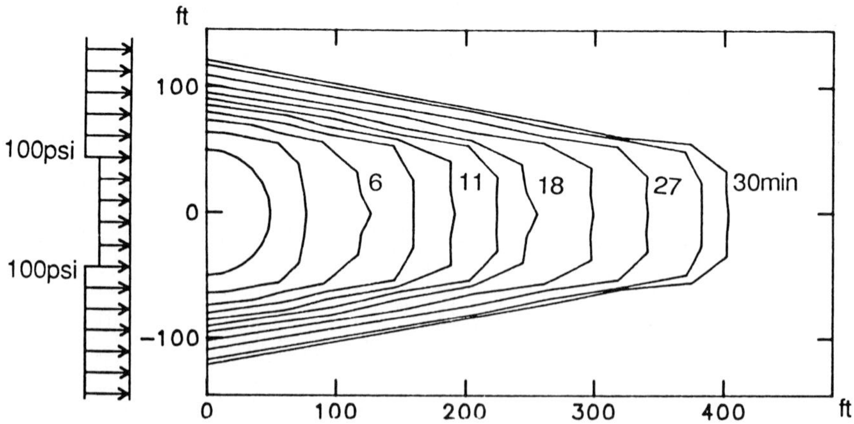

Figure 8.9 Fracture front propagation with an in-situ stress contrast of 690 kPa (100 psi) [24]. Conditions of the treatment are the same as those shown for Fig. 8.8.

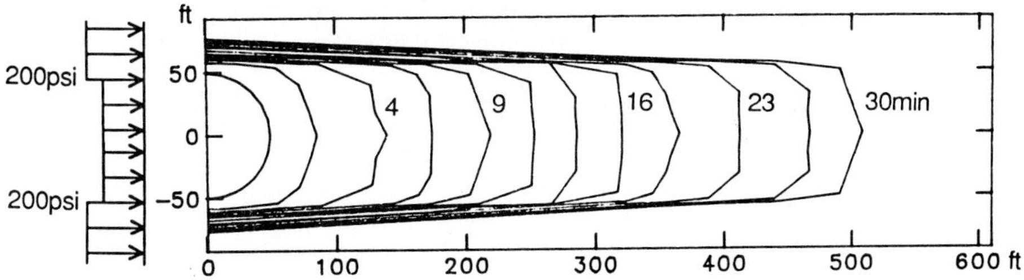

Figure 8.10 Fracture front propagation with an in-situ stress contrast of 1.38 × 10^3 kPa (200 psi) [24]. Conditions of the treatment are the same as those for Fig. 8.8.

stresses, but not entirely. Equations (8.10) and (8.11) show that when the fluid pressure within the fracture is increased, then K increases correspondingly and the possibility that K will exceed K_c is also increased. Figure 8.11 shows the same fracture treatment as that of Figure 8.10 except that the Young's modulus of the formation material has been increased substantially. In this case the fluid pressure in the fracture increases in order to propagate the fracture outward, but this increased pressure also results in K exceeding K_c along the vertical periphery of the fracture and vertical growth is observed. Gu [24] has calculated that an in-situ stress contrast of 5.5 × 10^3 kPa (800 psi) is required to contain the fracture in this case. Thus depending on the in-situ stress contrast and the formation prop-

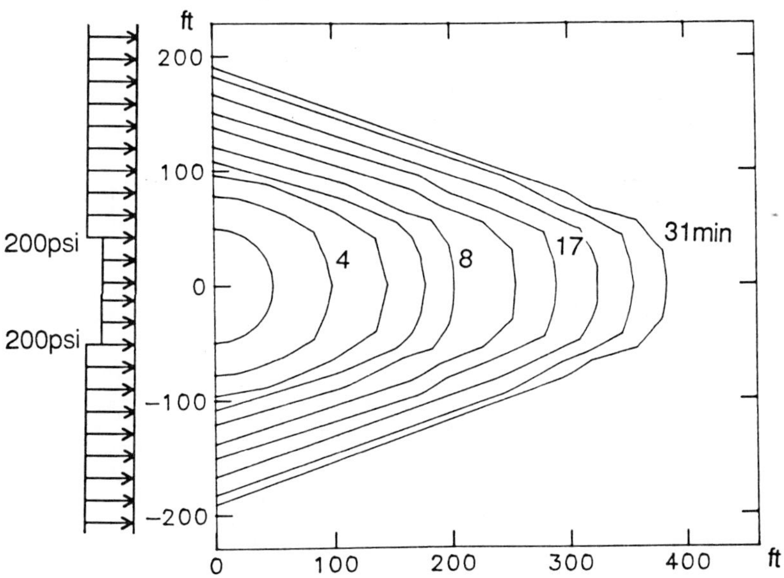

Figure 8.11 Fracture front propagation with an in-situ stress contrast of 1.38 × 10^3 kPa (200 psi) [24]. Treatment conditions same as those for Fig. 8.8, except that E is increased to 2.91 × 10^{10} kPa.

erties, a given fracture treatment may be contained within the pay zone. The most important variable is the contrast in the horizontal in-situ stresses between zones.

Measurement of in-situ stresses. Prats [5] has shown that the horizontal in-situ stresses in a given formation depend on the burial history of the sediments including the effects of temperature (thermal expansion), creep, and elasticity. This concept has been extended by Abou-Sayed [25]. Roughly it seems that soft, high-ductility rocks require a higher horizontal stress to confine them during plastic flow than do stronger low-ductility rocks [25]. Thus salt, which is a rock that continuously creeps, will tend to exhibit large in-situ stresses.

Higher Poisson ratio values have been proposed as an indicator of higher minimum horizontal stresses [26]. This indication has been found to be compatible with trends in the Lousiana Gulf Coast area where clean sandstone pay zones are prevalent [25]. This information provides a method for estimating horizontal stress contrasts based on sonic logs which measure both S-waves and P-waves. (see Chap. 2). This technique becomes less reliable as the clay content of the formation increases, thus diminishing the difference between shales and sandstones. Furthermore, as noted, the horizontal stresses depend on the burial history [5] and are not entirely determined by the current elastic properties. Thus the Poisson ratio may not be a reliable indicator of horizontal stress contrasts.

The most reliable method for determining the in-situ stresses is based on minifracture field tests [27–31]. The instantaneous shut-in pressures are measured for specific isolated zones. This method provides the best data for use in predicting the growth of fractures.

Dynamic Width

Introduction. Another critical design variable is the fracture width. The size proppant that can be transported into a fracture or the depth to which live acid penetrates is controlled primarily by the fracture width. The first fracture models were two-dimensional and these can be divided into two groups. The first group disregards strain in the z-direction and as shown by Case A of Fig. 8.12, the fracture is elliptical in the x-y plane. It is rectangular in the y-z plane. Khristianovich and Zheltov [32] were the first to apply this model. It has been improved by Geertsma and de Klerk [33] using an equation for plane strain developed by England and Green [34].

The second group disregards the x-component of strain. In this case the fracture is elliptical in the y-z plane, as shown by Case B of Fig. 8.12. Perkins and Kern [35] originally used this model and it was significantly extended by Nordgren [36].

Calculations based on the two different models reveal quite different predictions [37, 38]. Neither model is precisely correct, although it is perhaps true that Case B should be the better approximation for fractures that are long as compared to their height [39]. Fractures do not propagate with a constant height nor is the fluid flow one-dimensional, as is assumed in both models. Fluid enters the perforation and flows in all of the coordinate directions. The pressure in the

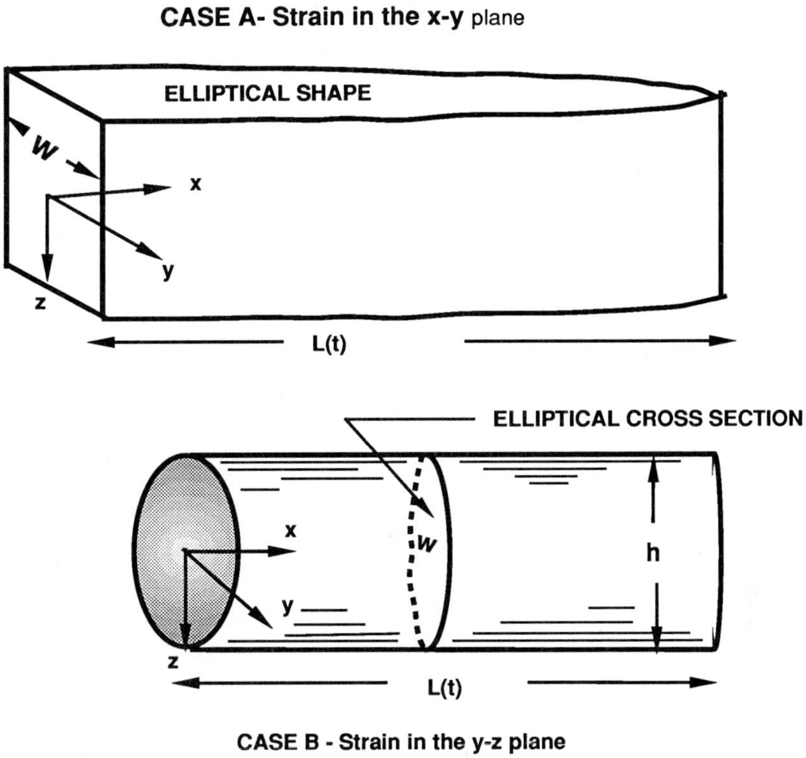

Figure 8.12 Two different simplified fracture geometries.

fracture is, therefore, a function of all coordinates and time although it is perhaps reasonable to disregard the flow normal to the fracture face caused by fluid loss. A few fracture models have been developed to better take into account these factors.

Settari and Cleary [40] developed a pseudo three-dimensional model in which the Nordgren and the Geertsma and de Klerk models were both applied. Advani and Lee [41] modified the Nordgren model by considering the stress intensity factors at the upper and lower fracture tips. Palmer and Carroll [42] assumed the fracture front to be elliptical. These models can all be considered to be pseudo three-dimensional because the fluid flow was one-dimensional and the strains were not treated with rigor.

True three-dimensional models have been reported [41, 43, 44, 45, 46, 47, 24]. These are all in the form of complex computer programs. Analytical solutions do not appear possible. For example, the results plotted in Figs. 8.8–8.11 have been obtained by Gu using his three-dimensional simulator [24]. We will draw heavily on his results in our discussion; however, for the purposes of understanding the fracturing process, a two-dimensional fracture will be considered. The theory presented will follow the development reported by Nordgren [36], which in certain limits leads to instructive analytical solutions that are useful in approximate design considerations. Certainly in the final design of a well treatment, especially large hydraulic fracture well stimulations, one should insist on using the more precise three-dimensional models that are now available and not apply the equations presented here.

The coupling with fracture fluid pressure. The width of a crack in an elastic body depends directly on the fluid pressure. If the strain is confined to the y-z plane, then the width is given by [48]:

$$w(x, z, t) = \frac{1 - \nu}{G}(h^2 - 4z^2)^{1/2}\,\Delta p(x, t) \qquad \text{for } |z| < (1/2)h \tag{8.12}$$

where h is the fracture height, G is the shear modulus, and ν is the Poisson ratio. In this model the fluid pressure less the in-situ stress depends on x and t but not on z. This implies that one-dimensional fluid flow is assumed. As noted, three-dimensional models allow Δp to depend on z as well as x and t.

Equation (8.12) shows that to create wide fractures, the fluid pressure must be large. This is one reason that viscous fracture fluids are preferred.

Example 8.2 Fracture Width

Assuming that the Δp in a fracture at a point near the wellbore is 1400 kPa, what fraction of the total fracture height will accept proppant particles having a diameter of 0.71 mm (25-mesh sand)? The shear modulus of the formation rock is 7×10^6 kPa and its Poisson ratio is 0.30. The height of the fracture at the wellbore is 15 m.

Solution

$$w(z) = \left(\frac{1 - 0.3}{7 \times 10^6}\right)(1.4 \times 10^3)(15)\left[1 - \frac{4z^2}{(15^2)}\right]^{1/2}$$

or

$$w(z) = (2.1 \text{ mm})\left[1 - \left(\frac{z}{7.5}\right)^2\right]^{1/2}$$

Suppose that proppant will enter for all $w(z) > 0.71$ mm. (Actually, experiments show that a fracture width of 4–5 times the proppant diameters is required to ensure entry.) Solving for z at $w = 0.71$, we find

$$z = 7.06 \text{ m}$$

or the fraction of height accepting proppant is $7.06/7.5 = 0.94$. Thus, 94% of the fracture is accepting proppant based on the criteria applied here. (We must again note that nowhere does the width of this fracture exceed the proppant diameter by more than a factor of three. The 25-mesh sand should be replaced by a smaller sand or steps must be taken to increase the fracture width.)

The dynamic pressure in the fracture is not constant. It depends on the volumetric flow rate (q) and the fluid viscosity (μ). The flow of a Newtonian fluid between parallel plates separated by a distance w is described by the equation

$$q = -\frac{w^3 h}{12\mu}\frac{d\,\Delta p}{dx} \tag{8.13}$$

Now the geometry of the fracture is an ellipse with a very small eccentricity [see Fig. 8.12(b)] rather than parallel plates. Lamb [49] has shown that the pressure gradient when a fluid flows in an elongated ellipse is $3\pi/16$ times as great as the pressure gradient when flowing the same fluid between parallel plates under the same conditions of height, maximum width, and flow rate. Thus, the flow in an elongated ellipse is

$$q = -\frac{\pi h w_{\max}^3}{64\mu}\frac{d\,\Delta p}{dx} \tag{8.14}$$

Equation (8.14) relates the fracture width to the pressure. This equation expresses a hydrodynamic balance of the forces acting within the fluid. Equation (8.12) is a different and independent relationship between the width and the pressure. It obtains from a consideration of the solid elasticity. Thus, there is a coupling between the equations of elasticity and the flow. Between the two equations, the fracture width and fluid pressure can be determined, but first it is necessary to consider the fracture length.

Fracture Length

The fracture length is determined by a volume balance. At any time the volume of fluid per unit time which enters each wing of the fracture is known and is a design variable. This taken together with the fracture width and height (assumed here to be constant, but precisely the same principle applies to variable-height calculations) would determine the fracture length, except for the fact that all of the fluid injected into the fracture does not remain in the fracture. Some of the fluid is lost to the formation since the fluid pressure in the fracture is higher than the pressure of the fluid in the pores. This fluid loss is sometimes called *leak-off*.

Thus to determine the fracture length, it is necessary to make a volume balance. Define $q(x, t)$ to be the volumetric rate through a cross section of the fracture perpendicular to x, $q_l(x, t)$ is the rate of fluid loss to the formation per unit length of fracture and $A(x, t)$ is the cross-sectional area of the fracture. Let us now make a volume balance about the small portion of the fracture shown in Fig. 8.13. The volume flowing in during the time interval Δt is

$$q(x, t)\,|_x \Delta t$$

and the volume of liquid flowing out during the same time interval is

$$q(x, t)\,|_{x+\Delta x}\,\Delta t + q_l \Delta x\,\Delta t$$

where the second term is the fluid loss. That amount of fluid which enters but does not leave is accumulated and serves to increase the volume of the segment of fracture shown in Fig. 8.13. This volume increase is given by

$$A(x, t)\,|_{t+\Delta t}\,\Delta x - A(x, t)\,|_t\,\Delta x$$

Since the volume of fluid entering less the volume leaving equals that accumulated

$$\frac{q(x, t)|_x - q(x, t)\,|_{x+\Delta x}}{\Delta x} - q_l = \frac{A(x, t)\,|_{t+\Delta t} - A(x, t)|_t}{\Delta t}$$

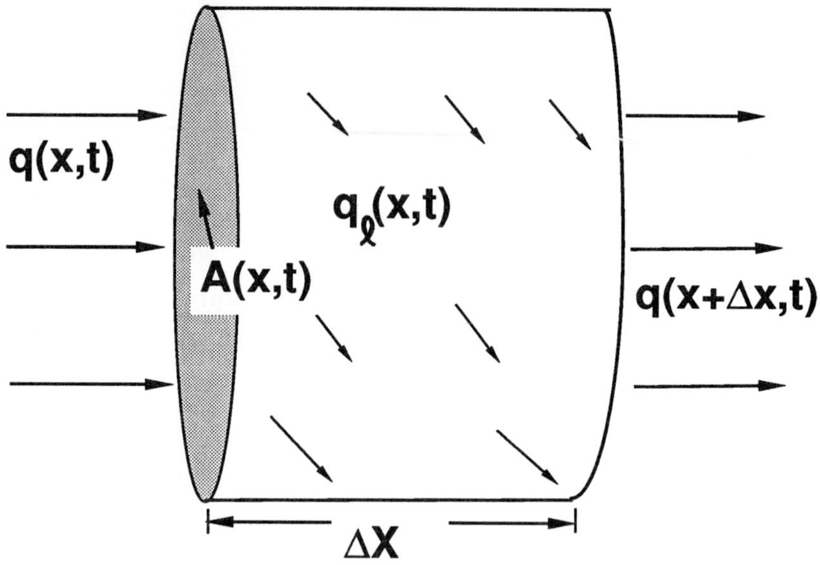

Figure 8.13 The volume rates of fluid into and out of a small part of the fracture. Note also that the cross-sectional area depends on time.

which in the limit as Δx and Δt go to zero yields

$$-\frac{\partial q}{\partial x} - q_l = \frac{\partial A}{\partial t} \tag{8.15}$$

In Section 8.4 it will be shown that the fluid loss is given by an equation of the form

$$q_l = \frac{2Ch}{\sqrt{t - \tau}} \tag{8.16}$$

where τ is the time at which the fracture at position x is opened and C is the overall fluid-loss coefficient. It should be noted that τ is a time measured from the time fracturing is initiated and depends on position. Clearly, positions near the wellbore will fracture at an early time and τ will be small. Thus, positions near the wellbore will have been exposed to fluid longer than are points more remote from the wellbore. For positions further from the wellbore τ will be larger and the quantity $t - \tau$ smaller. Equation (8.16) therefore implies that recently exposed areas lose fluid faster than areas that have been exposed for a longer time. The reason for this will be discussed in Section 8.4.

It is now convenient to relate the average width to the maximum one. An average is defined as

$$\overline{w} = \frac{1}{h}\int_{-h/2}^{h/2} w\,dz = \frac{1}{h} w_{\max} \int_{-h/2}^{h/2} \left(1 - \frac{4z^2}{h^2}\right)^{1/2} dz = \frac{\pi}{4} w_{\max} \tag{8.17}$$

Here Eq. (8.12) has been used with $w_{max} = w(0, x, t)$. In terms of the average fracture width $A = \overline{w}h$.

Substituting Eqs. (8.12) and (8.14), the volume balance equation becomes

$$-\frac{G}{\pi^3\mu(1-\nu)}\frac{\partial^2\overline{w}^4}{\partial x^2} + \frac{2hC}{\sqrt{t-\tau}} + h\frac{\partial\overline{w}}{\partial t} = 0 \tag{8.18}$$

This is a partial differential equation, which must be solved subject to the following boundary conditions:

$$\overline{w} = 0 \qquad (\text{at } x = L) \tag{8.19}$$

and

$$q = -\frac{G}{\pi^3(1-\nu)\mu}\frac{\partial\overline{w}^4}{\partial x} = \frac{i}{2} \qquad (\text{at } x = 0) \tag{8.20}$$

where i is the volumetric injection rate. Only one half of the injected fluid enters each wing of the fracture so that at the wellbore $q = i/2$. A final condition is

$$q = -\frac{G}{\pi^3\mu(1-\nu)}\frac{\partial\overline{w}^4}{\partial x} = 2V_{sp}h\frac{dL}{dt} \qquad (\text{at } x = L) \tag{8.21}$$

Boundary condition (8.21) takes into account the spurt loss, V_{sp}, which takes place at the tip of the fracture. This occurs when new rock is suddenly exposed to fluid and the fluid loss is rapid until a filter cake forms. The measurement of V_{sp} is discussed in Section 8.4.

We now, following Nordgren [36], introduce the following dimensionless variables:

$$t_D = \frac{16}{\pi^2}\left[\frac{2C^5hG}{(1-\nu)\mu i^2}\right]^{2/3} t \tag{8.22}$$

$$x_D = \frac{16}{\pi}\left[\frac{2C^8Gh^4}{(1-\nu)\mu i^5}\right]^{1/3} x \tag{8.23}$$

$$w_D = \left[\frac{C^2Gh}{4(1-\nu)\mu i^2}\right]^{1/3} w_{max} \tag{8.24}$$

Using these dimensionless groups, the volume balance Eq. (8.18) becomes

$$\frac{\partial^2 w_D^4}{\partial x_D^2} = \frac{1}{t_D - \tau_D} + \frac{\partial w_D}{\partial t} \tag{8.25}$$

Subject to the boundary conditions

$$\frac{\partial w_D}{\partial x_D} = -1 \qquad (\text{at } x_D = 0) \tag{8.26}$$

$$w_D = 0 \qquad (\text{at } x_D = L_D) \tag{8.27}$$

$$\frac{\partial w_D^4}{\partial x_D} = -V_D\frac{\partial L_D}{\partial t_D} \qquad (\text{at } x_D = L_D) \tag{8.28}$$

where V_D is the dimensionless spurt loss defined as

$$V_D = \frac{4}{\pi} V\text{sp} \left[\frac{2C^2Gh}{(1 - \nu)\mu i^2} \right]^{1/3}$$

and L_D and τ_D are dimensionless fracture length and time of rock exposure to fluid, respectively. Equation (8.25) subject to boundary conditions (8.26–8.28) represents a complete statement of the problem, and the dynamic fracture geometry can be obtained by solving these equations. Nordgren [36] has solved this set of equations numerically assuming that V_D is small (which is normally the case). These solutions are shown in Figs. 8.14 and 8.15. For a given t_D, the dimensionless fracture length can be obtained from Fig. 8.14 and the dimensionless maximum fracture width at the wellbore can be read from Fig. 8.15. The average fracture width is found using Eq. (8.17).

Note that there are two approximate solutions shown on these figures. The first disregards fluid loss so that the fracture volume is the volume of fluid injected. This is an interesting solution in that it represents an upper limit on the fracture

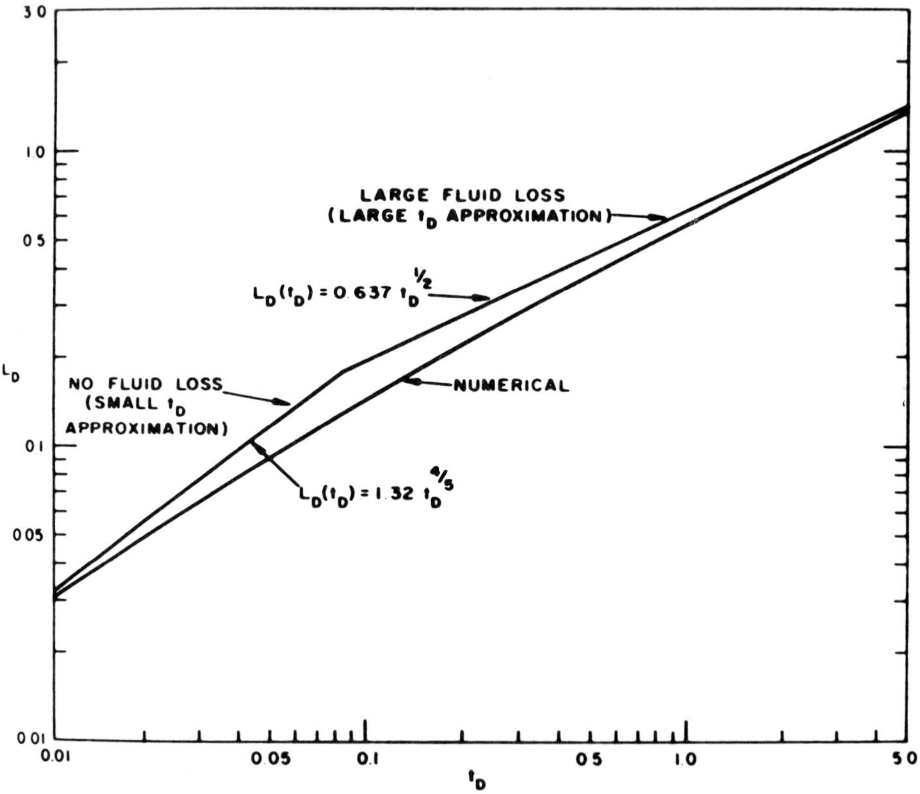

Figure 8.14 Dimensionless fracture length as a function of dimensionless time [36]. [With permission of the Society of Petroleum Engineers.]

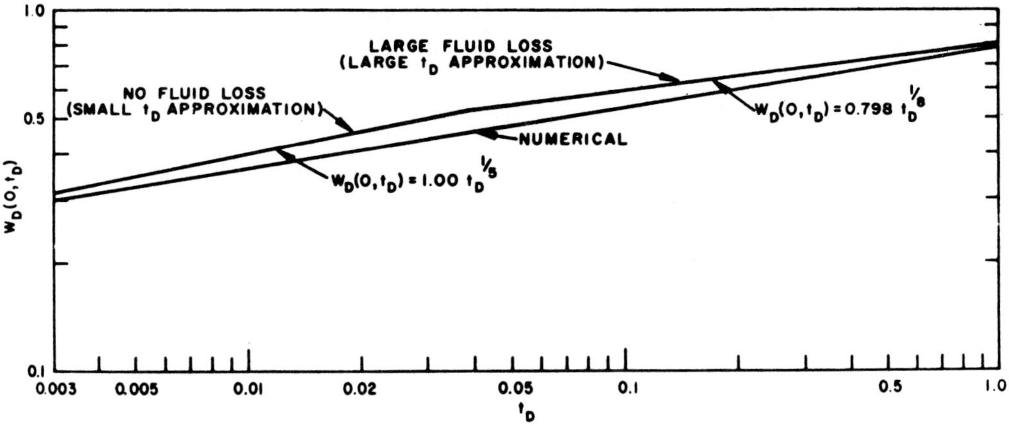

Figure 8.15 Dimensionless maximum fracture width at the wellbore as a function of dimensionless time [36]. [With permission of the Society of Petroleum Engineers.]

volume. If fluid losses are neglected, Eq. (8.25) reduces to

$$\frac{\partial^2 w_D^4}{\partial x_D^2} = \frac{\partial w_D}{\partial t_D} \tag{8.29}$$

subject to the boundary conditions

$$\begin{aligned} \frac{\partial w_D^4}{\partial x_D} &= -1 \qquad (\text{at } x_D = 0) \\ w_D &= 0 \qquad (\text{at } x_D = L_D) \\ \frac{\partial w_D^4}{\partial x_D} &= 0 \qquad (\text{at } x_D = L_D) \end{aligned} \tag{8.30}$$

The solution to this partial differential equation is obtained by similarity methods [36]:

$$w_D(0, t_D) = 1.00 t_D{}^{1/5} \tag{8.31}$$

$$L_D = 1.32 t_D{}^{4/5} \tag{8.32}$$

which gives

$$\overline{w}(0, t) = 1.50 \left[\frac{(1 - \nu)ui^2}{Gh}\right]^{1/5} t^{1/5} \tag{8.33}$$

$$L = 0.44 \left[\frac{Gi^3}{h^4 (1 - \nu)\mu}\right]^{1/5} t^{4/5} \tag{8.34}$$

Equations (8.33) and (8.34) disregard the fluid loss and neither the fluid-loss coefficient (C) nor the spurt loss (Vsp) appear. These equations would be expected to apply at early times. The approximation is actually only applicable for $t_D <<$ 1 (see Figs. 8.14 and 8.15). There is a second approximation that is applicable when the time is long, that is, $t_D \longrightarrow \infty$. In this limit the accumulation can be neglected and Eq. (8.25) becomes

$$\frac{\partial^2 w_D^4}{\partial x_D^2} = \frac{1}{\sqrt{t_D - \tau_D}} \tag{8.35}$$

where τ_D depends on x_D. Let us integrate with respect to x_D:

$$\frac{\partial w_D^4}{\partial x_D} - \left.\frac{\partial w_D^4}{\partial x_D}\right|_{x_D=0} = \int_0^{x_D} \frac{dx_D}{\sqrt{t_D - \tau_D}}$$

or using the boundary condition at $x_D = 0$, we find

$$\frac{\partial w_D^4}{\partial x_D} = \int_0^{x_D} \frac{dx_D}{\sqrt{t_D - \tau_D}} - 1 \tag{8.36}$$

Now at $x_D = L_D$

$$-V_D \frac{dL_D}{dt_D} = \int_0^{L_D} \frac{dx_D}{\sqrt{t_D - \tau_D}} - 1 \tag{8.37}$$

This leads to an equation for $L_D(t_D)$. Note it is an integrodifferential equation and is complete in the sense that given $L_D(t_D)$, the function $\tau_D(L_D)$ is determined by solving for t_D. We can clearly write

$$dx_D = \frac{dL_D}{d\tau_D}\, d\tau_D \tag{8.38}$$

so that now τ_D is the variable of integration and Eq. (8.38) has the form

$$-V_D \frac{dL_D}{dt_D} - \int_0^{t_D} \frac{\frac{dL_D}{d\tau_D}\, d\tau_D}{\sqrt{t_D - \tau_D}} - 1 \tag{8.39}$$

This equation can be solved in a straightforward way using Laplace transforms to give

$$L_D(t_D) = \frac{1}{\sqrt{\pi}}\left[\frac{2t_D^{1/2}}{\sqrt{\pi}} - \frac{V_D}{\sqrt{\pi}} + \exp\left(\frac{\pi t_D}{V_D}\right) \operatorname{erfc}\left(\sqrt{\frac{\pi t_D}{V_D}}\right)\right] \tag{8.40}$$

In the limit of long times, the spurt losses are unimportant and

$$L_D(t_D) = \frac{2}{\pi}\, t_D^{1/2} \tag{8.41}$$

or in dimensional terms

$$L(t) = \frac{1}{2\pi}\left(\frac{i}{Ch}\right) t^{1/2} \tag{8.42}$$

Equation (8.42) is valid in the limit of long times, but for our purposes it is the formulation that will be used here for many purposes since it is convenient to have an analytical solution to work with. Better results are obtained using Figs. 8.14 and 8.15 and for design of fracture treatments where containment and length are both critical variables, one should use a three-dimensional simulator.

Thus, based on Eq. (8.42) $\tau_D = (\pi/2)^2 x_D^2$ and we can write $t_D = (\pi/2)^2 L_D^2$. Thus from Eq. (8.36)

$$\frac{\partial w_D^4}{\partial x_D} = \frac{2}{\pi} \sin^{-1}\left(\frac{x_D}{L_D}\right) - 1 \tag{8.43}$$

Integrating a second time yields

$$w_D^4 = \int_{L_D}^{x_D} \frac{2}{\pi} \sin^{-1}\left(\frac{x_D}{L_D}\right) dx_D + (L_D - x_D) \tag{8.44}$$

or

$$w_D^4 = x_D \left[\frac{2}{\pi} \sin^{-1}\left(\frac{x_D}{L_D}\right) - 1\right] + \frac{2}{\pi} L_D \sqrt{1 - (x_D/L_D)^2} \tag{8.45}$$

Equation (8.45) represents the fracture width as a function of position x_D at a time when the fracture length is L_D. At the wellbore ($x_D = 0$), the dimensionless maximum width is given by

$$w_D^4(0, t_D) = \frac{2}{\pi} L_D \tag{8.46}$$

Expressed in terms of the dimensionless time

$$w_D(0, t_D) = 0.798 t_D^{1/8} \tag{8.47}$$

Thus, in dimensional terms, the average width ($= \pi/4\ w_{max}$) is given by

$$\overline{w}(0, t) = 1.12 \left[\frac{(1 - \nu)\mu i^2}{GCh}\right]^{1/4} t^{1/8} \tag{8.48}$$

Example 8.3 Comparison of Two-Dimensional and Three-Dimensional Predictions

The results shown by Fig. 8.10 are those predicted by using a three-dimensional simulator for the conditions listed in Table 8.1. Compare the fracture length shown in Fig. 8.10 at $t = 16$ min with that predicted by using Figs. 8.14 and 8.15.

TABLE 8.1 Treatment Conditions

Property	Symbol	Value
Injection rate	i	3.18 m^3/min or 5.3 × 10^{-2} m^3/sec
Young's modulus	E	5.8 × 10^9 Pa
Poisson's ratio	ν	0.3
Fluid-loss coefficient	C	1.83 × 10^{-4} m/$\sqrt{\text{min}^{1/2}}$ 2.36 × 10^{-5} m/$\sqrt{\text{sec}^{1/2}}$
Viscosity	μ	150 cp or 0.15 kg/m-sec

Solution The first issue to be resolved when applying a two-dimensional model is the selection of a fracture height. In the absence of any other information, it is usual to assume that the fracture will penetrate a distance of 2 to 3 m into the over- and underlying shales. In Fig. 8.10, the shales are presumably at $z = \pm 50$ ft, which is the point at which the in-situ stress increases. Thus for our fracture height, we take

$$h = (100 \text{ ft}) \left(\frac{0.3048 \text{ m}}{\text{ft}} \right) + 2(3 \text{ m}) \cong 36 \text{ m}$$

This is an approximation that assumes the in-situ stress contrast in the over- and underlying zones is sufficient to contain the fracture.

From Eq. (8.22) for $t = 16$ min or 960 sec

$$t_D = \frac{16}{\pi^2} \left[\frac{2(2.36 \times 10^{-5})^5 \, 36 \left(\dfrac{5.8 \times 10^9}{2(1 + 0.3)} \right)}{(1 - 0.3)(0.15)(5.3 \times 10^{-2})^2} \right]^{2/3} 960$$

or

$$t_D = 3.91 \times 10^{-3}$$

Since t_D is small, Eq. (8.32) applies. Thus

$$L_D = 1.32(3.91 \times 10^{-3})^{4/5} = 1.57 \times 10^{-2}$$

Converting this dimensionless length to the dimensional one

$$L = (0.0157) \frac{\pi}{16} \left[\frac{(1 - 0.3)(0.15)(5.3 \times 10^{-2})^5}{2(2.36 \times 10^{-5})^8 \left(\dfrac{5.8 \times 10^9}{2(1.0 + 0.3)} \right) (36)^4} \right]^{1/3}$$

This gives

$$L = 121 \text{ m} = 398 \text{ ft}$$

The three-dimensional calculations presented in Fig. 8.10 show the fracture length to be about 350 ft. The major reason for this discrepancy is our underestimation of the fracture height, which was obtained using a rule of thumb that recommends adding about 6 m to the thickness of the pay zone in the absence of any other information. Figure 8.10 shows the fracture height to be about 150 ft or 46 m, rather than the 36 m used in the calculation. Substituting this value into the calculation, one finds $L = 347$ ft predicted by the two-dimensional calculation.

Wellbore Pressure

The net pressure in the wellbore (fluid pressure less the instantaneous shut-in pressure) is thought to be an important diagnostic tool revealing, for example, the point in time that a fracture may extend vertically suddenly penetrating through a bounding layer of high in-situ stress [26, 27] into a zone of low in-situ stress. The pressure also provides information regarding the degree of containment. For example, Fig. 8-16 shows the net wellbore pressure corresponding to three different levels of stress contrast. The net pressure is plotted as a function of time corresponding to the three different levels of in-situ stress contrast shown in Figs. 8.8, 8.9, and 8.10 as well as for a penny-shaped fracture, which develops when there is no stress contrast.

As long as the fracture height is the same or nearly the same as the fracture length, the net pressure declines during the treatment. For limited vertical growth the pressure increases once the length exceeds the height. This latter trend is predicted by the two-dimensional model presented here.

The net pressure is determined from the fracture width using Eq. (8.12). Thus for the short-time, no-fluid-loss case, Eq. (8.33) can be combined with Eq. (8.12) to give the pressure in excess of p_{BISIP} measured at the wellbore for small t_{D} as follows:

$$\Delta p(0, t) = 1.89 \left[\frac{G^4 \mu i^2}{(1 - \nu)^4 h^6} \right]^{1/5} t^{1/5} \tag{8.49}$$

For the other extreme, where the growth of the fracture width can be disregarded, then for large t_{D}

$$\Delta p(0, t) = 1.43 \left[\frac{\mu i^2 G^3}{C h^5 (1 - \nu)^3} \right]^{1/4} t^{1/8} \tag{8.50}$$

Thus Δp increases with time, but not rapidly. This may make Δp difficult to use as a diagnostic tool. However, as shown by Figure 8.16, there is a distinct difference in the pressure behavior depending on the containment and clearly accurate pressure measurements can help define the extent of vertical fracture growth.

Example 8.4 Pressure Predicted by Two-Dimensional Model

For the conditions stated in Example 8.3, compare the pressure shown in Fig. 8.16 at 16 min with that predicted by Eq. (8.49) or (8.50).

Solution Since t_{D} is small for the conditions relating to Example 8.3, Eq. (8.49) should apply. Thus

$$\Delta p = 1.89 \left[\frac{\left(\frac{5.8 \times 10^9}{2(1 + 0.3)} \right)^4 (0.15)(5.3 \times 10^{-2})^2}{(1 - 0.3)^4 \, 36^6} \right]^{1/5} (960)^{1/5}$$

or $$\Delta p = 8.6 \times 10^5 \text{ Pa} = 124 \text{ psi}$$

This value compares favorably with that presented in Fig. 8.16. The comparison is somewhat deceiving since a fracture height of 36 m was used in the calculation. If the more accurate value of 45 m is used, then $\Delta p = 6.4 \times 10^5$ Pa or 93 psi. This value does not compare as well.

Effective Non-Newtonian Viscosity

In subsequent chapters fracture design will involve consideration of realistic, non-Newtonian fluids, but the design equations developed in this chapter assume that the fluids are Newtonian. This simplification is necessary to enable us to develop

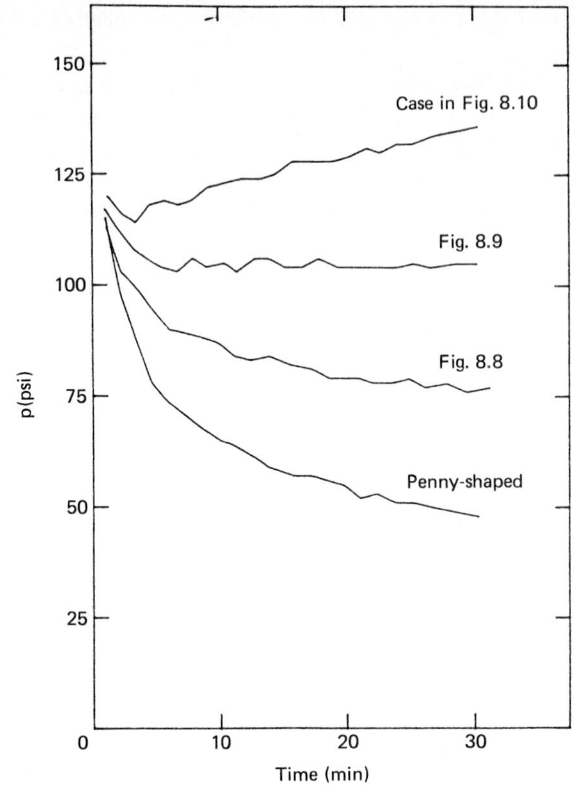

Figure 8.16 Time variation of net borehole pressure under different in-situ stress contrasts. Treatment conditions correspond to those given in the caption of Fig. 8.8.

design equations that do not require extensive computer computations to apply. In this section a method for obtaining an effective viscosity is proposed, but this approach is not necessarily recommended for use in actual practice.

The following equation describes the flow of a power-law fluid between parallel plates spaced a distance w apart (see Problem 3.10).

$$-\frac{d\,\Delta p}{dx} = 2m\left(\frac{i}{h}\right)^n\left(\frac{2n+1}{n}\right)\frac{1}{w^{2n+1}} \tag{8.51}$$

Replacing i by $2q$ and setting $n = 1$ and $m = \mu$, this equation reduces to Eq. (8.13). As noted, the pressure drop for a Newtonian fluid flowing between parallel plates is smaller by a factor of $3\pi/16$ than for one flowing through an ellipse for which $w_{max}/h << 1$. Assuming that same factor is applicable when non-Newtonian fluids flow (an unlikely result), Eq. (8.51) for an elliptical cross section is

$$-\frac{d\,\Delta p}{dx} = \frac{3\pi}{8}\,m\left(\frac{i}{h}\right)^n\left(\frac{2n+1}{n}\right)\frac{1}{w_{max}^{2n+1}} \tag{8.52}$$

Comparing Eqs. (8.52) and (8.14), it is evident that if an effective viscosity is defined as

$$\mu_{\text{EFF}} = \frac{m}{3}\left(\frac{i}{h}\right)^{n-1}\left(\frac{2n+1}{n}\right) w_{\max}^{2-2n} \tag{8.53}$$

and substituted into Eq. (8.14), then the correct pressure gradient at the fracture inlet is predicted. This is the sole merit of the effective viscosity defined by Eq. (8.53). Clearly, different effective viscosities can be defined which satisfy other desirable conditions. Moreover, because of fluid loss this effective viscosity will not be representative of other positions along the fracture at some distance from the wellbore.

Despite these evident shortcomings, Eq. (8.53) will be used here to obtain an estimated effective viscosity that can be used instead of the Newtonian viscosity (μ) in fracture design discussed in Chapter 12.

8.3 HORIZONTAL AND PENNY-SHAPED FRACTURES

At shallow depths horizontal fractures may be created. These will be circular in shape. In fact, their appearance will be similar to vertical fractures created in massive formations where there is little or no in-situ stress contrast. Apart from some difference in the inlet conditions at the wellbore, the two fractures are identical. The equations describing horizontal or penny-shaped fractures differ from those relating to contained vertical ones, but the same principles apply. The pressure distribution along the fracture, assumed to be symmetric about the wellbore, determines the fracture width through an equation of ideal elasticity, but the width must be known to obtain the fluid pressure drop. Thus, these two coupled variables must be solved together. The width and the pressure cannot be uncoupled. We will, however, consider the pressure loss first to find the mathematical relationship between them.

Laminar Flow in Radial Fractures

In this case, the fluid is forced to flow radially outward from the wellbore. Figure 8.17 depicts a horizontal fracture. If we ignore the effect of fluid loss, assume the fracture width to be a constant, consider only Newtonian fluids, disregard the fluid inertia which exists because the radial component of velocity depends on the radial position, and assume the flow to be laminar, then the velocity distribution and the pressure as a function of radial position can all be calculated. It is this latter factor, the pressure as a function of position, which we wish to know since its value as a function of position determines the fracture width.

We start by noting $v_r = v_r(r, z)$ and $v_\theta = v_z = 0$. The equation of motion is [50]

$$-\frac{\partial p}{\partial r} + \mu\left[\frac{\partial}{\partial r}\left(\frac{1}{r}\frac{\partial}{\partial r}(rv_r)\right) + \frac{\partial^2 v_r}{\partial z^2}\right] = 0 \tag{8.54}$$

and the continuity equation is simply

$$\frac{\partial}{\partial r}(r v_r) = 0 \tag{8.55}$$

Here z is a coordinate direction measured from the center of the fracture (see Fig. 8.17). The continuity equation simply insists that the same volume of fluid crosses each ring of radius r surrounding the wellbore per unit of time. Equation (8.55) can be written in the equivalent form

$$r v_r = \phi(z) \tag{8.56}$$

where $\phi(z)$ is a function of z, but not of r.

Substituting Eq. (8.56) into (8.54) yields

$$-r\frac{\partial p}{\partial r} + \mu\frac{d^2\phi}{dz^2} = 0 \tag{8.57}$$

We note that $r(\partial p/\partial r)$ does not depend on z. Equation (8.57) can be integrated to yield

$$\phi(z) = -\frac{r}{\mu}\frac{dp}{dr}\frac{w^2}{8}\left[1 - \frac{4z^2}{w^2}\right] \tag{8.58}$$

The integration of Eq. (8.57) has used $v_r(r, z) = 0$ for $z = \pm w/2$.

The fluid injection rate is related to v_r as follows:

$$i = \int_{-w/2}^{w/2} 2\pi r v_r \, dz \tag{8.59}$$

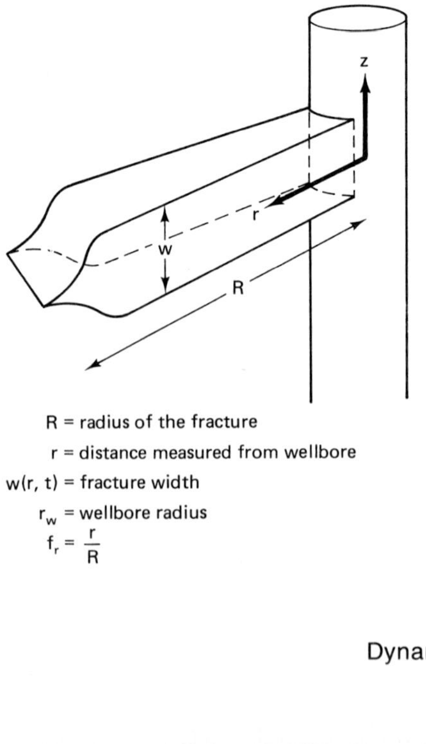

Figure 8.17 Coordinate system for radial flow calculation.

where any value of r can, of course, be used. Thus

$$i = -\frac{\pi}{6}\frac{r}{\mu} w^3 \frac{dp}{dr} \tag{8.60}$$

This formula can be rearranged to give

$$p_w - p = \frac{6\mu i}{\pi w^3} \int_{r_w}^{r} \frac{dr}{r} = \frac{6\mu i}{\pi w^3} \ln\left(\frac{r}{r_w}\right) \tag{8.61}$$

assuming the width to be constant. Equation (8.61) thus determines the pressure distribution in a horizontal fracture. It is convenient to define the fractional distance down the fracture as

$$f_r = \frac{r}{R}$$

where R is the radius of the fracture

$$p_w - p = \frac{6\mu i}{\pi w^3} \ln\left(\frac{f_r}{f_{rw}}\right) \tag{8.62}$$

Volume Balance for Horizontal Fractures

If we let $q(r, t)$ be the volume of fluid flowing outward in the horizontal fracture shown in Fig. 8.17 per unit of time, then this quantity will vary in both time and position in accordance with the volume balance

$$-\frac{\partial q}{\partial r} - \frac{4\pi r C}{\sqrt{t - \tau(r)}} = 2\pi r \frac{\partial w}{\partial t} \tag{8.63}$$

This volume balance is similar to that written for a linear, vertical fracture in the preceding section [see Eq. (8.15)]. Just as in the case of vertical fractures, after the fracture area becomes large ($R \gg r_w$), the fluid loss dominates and $\overline{w}/\partial t$ can be neglected. In this case, it can be shown that (see Problem 8.4)

$$R(t) = \frac{1}{\pi}\sqrt[4]{\frac{i^2 t}{C^2}} \tag{8.64}$$

Thus the radius of a horizontal (or penny-shaped) fracture grows as $t^{1/4}$, whereas the length of linear fractures increases as $t^{1/2}$. We shall use this asymptotic equation since numerical calculations including the accumulation term $[2\pi r(\partial\overline{w}/\partial t)]$ have not been reported.

It is perhaps worth noting that spurt losses are neglected in deriving Eq. (8.64). For large $R(t)$ this is an excellent approximation.

Fracture Width

The pressure varies with distance from the wellbore in accordance with Eq. (8.62) and is dependent on the fracture width. The theory of elasticity provides a general relationship between the fracture width, $w(r, t)$, and the fluid pressure within the fracture. For a radially expanding fracture [33]

$$w(f_r, t) = \frac{4(1-\nu)R}{\pi G}\left\{\int_{f_r}^{1} \frac{df_2}{\sqrt{f_2^2 - f_r^2}} \int_{f_{rw}}^{f_2} \frac{f_1 p(f_1)\, df_1}{\sqrt{f_2^2 - f_1^2}} - p_{BISIP}\frac{\pi}{2}\sqrt{1 - f_r^2}\right\} \tag{8.65}$$

Equation (8.65) makes explicit the relationship between the fracture width and the pressure distribution within the fracture. The bottomhole instantaneous shut-in pressure is the pressure at the tip of the fracture which is just necessary to maintain the fracture open.

To obtain a precise solution for $R(t)$, $w(f_R, t)$ and $p(f_R, t)$, Eqs. (8.61), (8.63), and (8.65) must all be solved simultaneously. To date, a computer solution for horizontal fractures has not been reported. It is, however, clear that the solution would necessarily be numerical, Geertsma and de Klerk [33] have reported that for a radially propagating horizontal fracture, the width at the wellbore is approximately given by

$$w_w = 2\sqrt[4]{\frac{i\mu R}{G}} \tag{8.66}$$

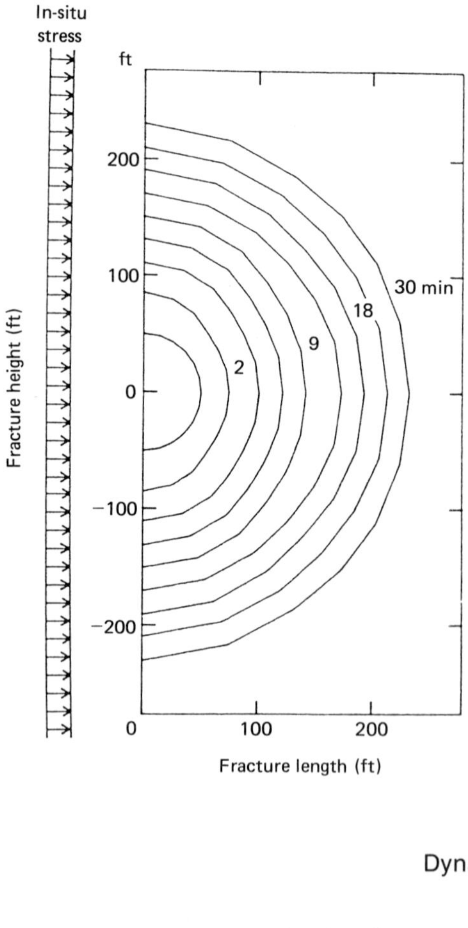

Figure 8.18 Fracture front propagation with a uniform in-situ stress [24]. The treatment conditions are the same as those shown for Fig. 8.8.

We will accept this equation (actually restricted to long fractures and $\nu = 0.25$) for design purposes. Thus to this same approximation,

$$\Delta p = \frac{5}{4\pi} \frac{G w_w}{R} \ln \left(\frac{R}{r_w} \right) \tag{8.67}$$

As R increases, Eq. (8.67) predicts that the pressure will decline. This is the trend shown in Fig. 8.16 for a penny-shaped fracture. The radial growth of that fracture is shown in Fig. 8.18. This growth shown has been calculated using an accurate three-dimensional model. Let us compare these results with the approximations advocated here.

Example 8.5 Growth of a Penny-Shape Fracture

Calculate the radial extent of a penny-shaped fracture at a time 16 min from the start of fluid injection assuming the conditions of the treatment are the same as the ones specified in Example 8.3.

Solution From Eq. (8.64), at 960 sec

$$R = \frac{1}{\pi} \sqrt[4]{\frac{(5.3 \times 10^{-2})^2(960)}{(2.36 \times 10^{-5})^2}} \qquad \text{or} \qquad R = 84 \text{ m} = 275 \text{ ft}$$

This compares with the more precise value of 210 ft shown by Fig. 8.18. The long-time approximation overestimates the fracture radius.

8.4 FLUID LOSS

The overall fluid-loss coefficient (C) has been seen to be an important, perhaps the most important, factor determining the effectiveness of a given fracture treatment. It is therefore necessary to estimate C as accurately as is possible if reasonable approximations to fracture geometry are to be obtained. As depicted by Fig. 8.19 the total fluid loss is controlled by three mechanisms: (1) the compression of the reservoir fluids, (2) the thickness of the invaded zone which is filled with viscous fracture fluid, and (3) the filter cake which may or may not be present depending on the additives contained in the fracture fluid. We shall at first consider each of these mechanisms separately and then combine them to obtain the overall fluid-loss coefficient.

Compression of Formation Fluids

As fluid is forced into the formation by the higher pressures generally found within the hydraulic fracture, the thickness of the invaded zone shown in Fig. 8.19 increases steadily and the formation fluid is, as a consequence, compressed. If p_I is defined as the pressure within the reservoir at the fracture fluid/formation fluid interface and if p_R is the formation fluid pressure, then the process of compression

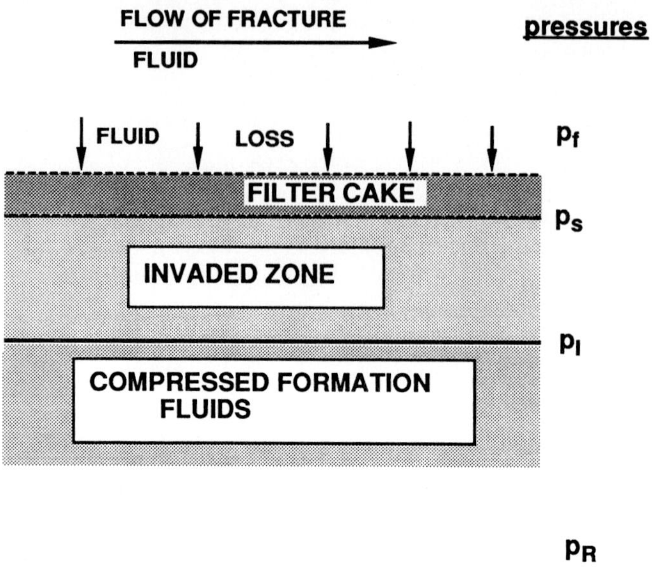

Figure 8.19 Fluid lost from fracture is shown to result in a series of processes.

can be defined by the following differential equation and boundary conditions:

$$\frac{\phi \mu_f \kappa_{fl}}{k} \frac{\partial p}{\partial t} = \frac{\partial^2 p}{\partial y^2} \tag{8.68}$$

$$\begin{aligned} p &= p_R \qquad (\text{at } t = 0) \\ P &\longrightarrow p_R \qquad (\text{as } y \longrightarrow \infty) \\ p &= p_I \qquad (\text{at } y = 0) \end{aligned} \tag{8.69}$$

This equation is written for a slightly compressible fluid being compressed in a linear system. κ_{fl} is the isothermal compressibility of the formation fluid, k is the formation permeability, μ_f is the viscosity of the formation fluid, and ϕ is the porosity.

The solution to this partial differential equation is easily shown to be

$$p_I - p = \frac{p_I - p_R}{\sqrt{\pi}} \int_0^{\eta} \frac{1}{\lambda^{1/2}} e^{-\lambda}\, d\lambda \tag{8.70}$$

where

$$\eta = \frac{\phi \mu_f \kappa_{fl} y^2}{4kt}$$

The Darcy flux (u_N) at the interface may be found as follows:

$$u_N = -\frac{k}{\mu_f} \frac{\partial p}{\partial y} \bigg|_{y=0} \tag{8.71}$$

Differentiating Eq. (8.70) yields

$$u_N = \sqrt{\frac{\phi \kappa_{fl} k}{\pi \mu_f}} \frac{(p_I - p_R)}{\sqrt{t}} = \alpha_c \frac{p_I - p_R}{\sqrt{t}} \tag{8.72}$$

The fluid loss can be calculated using this equation provided, of course, both p_I and p_R are known. The reservoir pressure is generally given; however, p_I is undetermined. Therefore, the unknown pressure must be eliminated before we can hope to predict the fluid loss.

Viscous Invaded Zone

In Region 2 of Fig. 8.19, Darcy's law applies so that

$$u_N = -\frac{k}{\mu}\frac{dp}{dy} \tag{8.73}$$

The pressure difference is constant so that

$$u_N = \frac{k}{\mu}\left(\frac{p_s - p_I + \Delta p_{CAP}}{D(t)}\right) \tag{8.74}$$

In this equation Δp_{CAP} is the capillary pressure, which is the difference in the pressure between the oil or gas phase and the water phase. This capillary pressure depends on saturation, but for the purposes of the present calculation we shall assume a sharp interface exists between the invading fluid and the formation fluid so that the capillary pressure can be evaluated at the water saturation of the reservoir. This assumes that the invading fracture fluid is water and the reservoir fluid is oil or gas. If the fracture fluid is miscible with the formation fluid, there is no interfacial tension and the capillary pressure vanishes. Inclusion of the capillary pressure is generally important only in the case of low-permeability gas wells fractured with a water-based fluid. If the formation is water-wet, Δp_{CAP} is positive and fracture fluid tends to be drawn into the formation by the capillary forces. As fracture fluid enters the invaded zone, the distance $D(t)$ increases. Clearly,

$$\frac{dD}{dt} = \frac{u_N}{\phi} \tag{8.75}$$

Substituting into Eq. (8.74) we find

$$D(t)\frac{dD}{dt} = \frac{k(p_s - p_I + \Delta p_{CAP})}{\mu\phi} \tag{8.76}$$

Integrating

$$D(t) = \sqrt{\frac{2k(p_s - p_I + \Delta p_{CAP})t}{\phi\mu}} \tag{8.77}$$

since $D(t) = 0$ at $t = 0$.

Finally since

$$u_N = \phi \frac{dD}{dt}$$

then $$u_N = \sqrt{\frac{\phi k(p_s - p_I + \Delta p_{CAP})}{2\mu t}} = \frac{\alpha_v \sqrt{(p_s - p_I + \Delta p_{CAP})}}{\sqrt{t}} \tag{8.78}$$

Since the pressure driving force in Eq. (8.78) contains unknown pressures, it cannot be used directly to evaluate the fluid-loss flux, u_N.

Wall-Building Fracture Fluids

The fluid-loss characteristics of a filter cake can be determined by plotting the cumulative filtrate volume that passes through a core wafer per unit area at constant pressure drop versus the square root of time. This is a static test.

Hall and Dollarhide [51] have reported that fluid-loss rates through filter cakes increase if the fracture fluid circulates across the cake rather than being contained in a static cell. These dynamic tests have been repeated a number of times [52, 53, 54] and the conclusion seems to be that the flow tends to erode the cake, causing the buildup to occur at a slower rate. Indeed, the square root of time dependence normally associated with static test results may not be found in dynamic tests. If at all possible, the fluid-loss characteristics of fracture fluid should be evaluated under realistic conditions [55, 56].

To eliminate relative permeability effects, the core wafers are saturated under a vacuum with mineral oil or water, depending on the fracture fluid, prior to the fluid-loss tests. The fluid-loss tests should be conducted at the estimated mean fracture temperature and the differential pressure between the fracture fluid and the pore fluid.

The data shown in Fig. 8.20 are typical and can be represented by an equation of the form

$$V = m_w \sqrt{t} + V_{sp} \tag{8.79}$$

where V is the filtrate volume per unit area (cumulative filtrate volume divided by the wafer area), m_w and V_{sp} are the slopes and the intercept, respectively.

The straight lines shown in Fig. 8.20 do not intercept the origin when extrapolated to $\sqrt{t} = 0$. Of course, carefully taken data must begin at the origin; however, to an approximation the early initial rapid fluid loss can be represented by a spurt loss, V_{sp}. This is the fluid loss that occurs rapidly as new fracture area is created and continues until a filter cake is built up on the fracture surface.

The slope m_w depends on a number of factors. Certainly the fluid-loss additive must play a major role (see Chap. 10), but also the applied pressure is an important factor. Obviously, increasing the differential pressure must increase m_w. If the cake formed is not compressible, then the filtrate volume found after a given time will be increased in direct proportion with applied pressure. Ex-

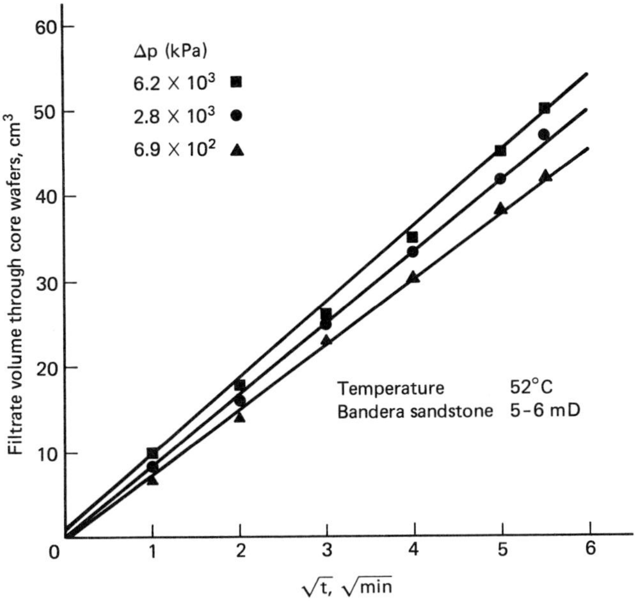

Figure 8.20 Fluid-loss characteristics of paraffin base crude. [Adapted from data given in reference [3].]

amining the few data reported comparing values of m_w at different differential pressures given by Howard and Fast [3] shows that most often the wall cake is compressible and the increase in filtrate volume is not proportional to the pressure increase. An examination of the few data suggests a relationship with the pressure difference of the form

$$m_w = 2\alpha_w \sqrt{p_f - p_s} \tag{8.80}$$

The assertion that m_w varies with the square root of the pressure difference is based on very few data. Equation (8.80) may be subject to revision when more data are available or when dynamic effects are taken into account.

Differentiating Eq. (8.79) we find

$$u_N = \frac{dV}{dt} = \frac{m_w}{2\sqrt{t}} = \frac{\alpha_w \sqrt{p_f - p_s}}{\sqrt{t}} \tag{8.81}$$

Again if $p_s - p_f$ is known during the fracturing treatment, the flux (u_N) can be calculated based on Eq. (8.81); however, this pressure drop is generally an unknown.

Combining of Fluid-Loss Coefficients

The overall pressure difference $p_f - p_R$ can be estimated as can the capillary pressure, p_{CAP} but intermediate pressures are unknown. Starting with the identity

$$p_f - p_R + \Delta p_{CAP} = (p_f - p_s) + (p_s - p_I + p_{CAP}) + (p_I - p_R)$$

and Eqs. (8.72), (8.78), and (8.81), we can write

$$\Delta p = \frac{u_N \sqrt{t}}{\alpha_c} + \frac{u_N^2 t}{\alpha_v^2} + \frac{u_N^2 t}{\alpha_w^2}$$

where $$\Delta p = p_f - p_R + p_{CAP}$$

This equation can be written in the following simpler form

$$1 = \frac{u_N \sqrt{t}}{C_c} + \frac{u_N^2 t}{C_v^2} + \frac{u_N^2 t}{C_w^2} \tag{8.82}$$

where $C_c = \alpha_c \, \Delta p = \sqrt{\dfrac{\phi \kappa_{fl} k}{\pi \mu_f}} \, \Delta p$ (compressibility fluid-loss coefficient)

and $C_v = \alpha_v \, (\Delta p)^{1/2} = \sqrt{\dfrac{\phi k}{2\mu}} \, (\Delta p)^{1/2}$ (viscous fluid-loss coefficient)

Also $C_w = \alpha_w \, (\Delta p)^{1/2}$ (wall-building, fluid-loss coefficients)

or $$C_w = \frac{m_w}{2}$$

where m_w is the slope of the curve shown in Fig. 8.20 evaluated at Δp. Note that all of these fluid-loss coefficients are based on the same pressure difference, the overall pressure difference including capillary pressure. These definitions correspond to those given by Howard and Fast [3].

Equation (8.82) is a quadratic equation in the unknown $u_N \sqrt{t}$. Solving for $u_N \sqrt{t}$, yields

$$u_N \sqrt{t} = \frac{-\dfrac{1}{C_c} + \sqrt{\dfrac{1}{C_c^2} + 4\left(\dfrac{1}{C_v^2} + \dfrac{1}{C_w^2}\right)}}{2\left(\dfrac{1}{C_v^2} + \dfrac{1}{C_w^2}\right)} = C \tag{8.83}$$

where C is called the *overall fluid-loss coefficient.*

Thus the fluid loss is given by

$$u_N = \frac{C}{\sqrt{t}} \tag{8.84}$$

or $$q_l = 2hu_N = \frac{2hC}{\sqrt{t}}$$

as was assumed in calculating the fracture geometry. The overall fluid-loss coefficient is then determined by a combination of the individual coefficients as shown by Eq. (8.83). Each individual coefficient is based on a pressure difference between the fracture fluid and the reservoir fluid.

Average Fluid Loss for Vertical Fractures

In some cases it is convenient to use an average fluid-loss flux instead of the local value. As shown, the fluid-loss flux depends on both time and position in accordance with the equation

$$u_N = \frac{C}{\sqrt{t - \tau(x)}} \tag{8.85}$$

As an approximation for vertical fractures after a long time, Eq. (8.42) gives

$$L = \frac{1}{2\pi}\left(\frac{i}{Ch}\right) t^{1/2} = \frac{t^{1/2}}{a} \tag{8.86}$$

From this equation, the time at which a position x is first exposed to fracture fluid is clearly

$$x = \frac{\tau^{1/2}}{a} \tag{8.87}$$

Substituting Eqs. (8.86) and (8.87) into Eq. (8.85) yields

$$u_N = \frac{C}{aL} \frac{1}{\sqrt{1 - \dfrac{x^2}{L^2}}} \tag{8.88}$$

From this equation it is evident that for practical purposes u_N is nearly independent of x/L, at least for the first 70% to 80% of the fracture. Thus in many cases, an average value of u_N will suffice. Define

$$\overline{u}_N = \frac{1}{L}\int_0^L u_N \, dx \tag{8.89}$$

Substituting Eq. (8.88) into Eq. (8.89) yields

$$\overline{u}_N = \frac{\pi C}{2\sqrt{t}} \tag{8.90}$$

Equation (8.90) shows that the average fluid-loss flux can be calculated by using an overall fluid-loss coefficient multiplied by the factor $\pi/2$. Equation (8.90) is useful when an average fluid-loss flux is required at a certain time t after the fracturing process has been initiated.

If an average is required for the entire period of the fracture treatment, then

$$\langle \overline{u}_N \rangle = \frac{1}{t_f}\int_0^{t_f} \overline{u}_n \, dt \tag{8.91}$$

or

$$\langle \overline{u}_N \rangle = \frac{\pi C}{t_f^{1/2}} \tag{8.92}$$

where t_f is the total treatment time and $\langle \underline{u}_N \rangle$ is the average fluid-loss flux which applies over the entire period of the treatment. It is independent of either time or position.

Example 8.6 Overall Fluid-Loss Coefficient

Suppose a fracture treatment requires an overall fluid-loss coefficient $C = 6.10 \times 10^{-4}$m/min$^{1/2}$ to be effective. It is necessary to specify the value of m_w required in laboratory measurements which are conducted using a differential pressure of 3.44×10^6 Pa to assure that the desired value of C will be realized. The relevant data are presented in Table 8.2.

Solution The pressure in the fracture is approximately that required to hold the fracture open (p_{BISIP}). Of course, there is an additional propagation pressure varying from a maximum in the wellbore to zero at the tip of the fracture. This additional pressure can, however, generally be disregarded so that

$$\Delta p = FG(D) + p_{CAP} - p_R$$

or

$$\Delta p = (2.26 \times 10^4)(2.5 \times 10^3) + 2.06 \times 10^6 - 2.07 \times 10^7$$

Thus

$$\Delta p = 3.79 \times 10^7 \text{ Pa}$$

The fluid-loss coefficient for compressibility is found as follows:

$$C_c = \sqrt{\frac{\phi \kappa_{fl} k}{\pi \mu}}\,\Delta p = \sqrt{\frac{(0.15)(9.674 \times 10^{-10})(2.96 \times 10^{-15})}{\pi(3 \times 10^{-3})}}\,(3.79 \times 10^7)$$

or

$$C_c = 2.56 \times 10^{-4} \text{ m/sec}^{1/2} = 1.98 \times 10^{-3} \text{ m/min}^{1/2}$$

Similarly,

$$C_v = \sqrt{\frac{\phi k}{2\mu}}\,(\Delta p)^{1/2} = \sqrt{\frac{(0.15)(2.96 \times 10^{-15})(3.79 \times 10^7)}{2(0.1)}}$$

or

$$C_v = 2.90 \times 10^{-4} \text{ m/sec}^{1/2} = 2.25 \times 10^{-3} \text{ m/min}^{1/2}$$

TABLE 8.2 Reservoir and Fluid Properties for Example 8.6

Property	Symbol	Value
Porosity	ϕ	0.15
Permeability	k	2.96×10^{-15} m^2
Fracture gradient	FG	2.26×10^4 Pa/m
Depth	D	2500 m
Reservoir pressure	p_R	2.07×10^7 Pa
Oil compressibility	κ_{fl}	9.674×10^{-10} Pa^{-1}
Oil viscosity	μ_f	3×10^{-3} kg/m-sec
Capillary pressure	p_{CAP}	2.06×10^6 Pa
Fracture-fluid viscosity	μ	0.1 kg/m-sec

Equation (8.83) can be written as

$$\frac{C}{C_c} = \frac{-1 + \sqrt{1 + 4\left(\dfrac{C_c^2}{C_v^2} + \dfrac{C_c^2}{C_w^2}\right)}}{2\left(\dfrac{C_c^2}{C_v^2} + \dfrac{C_c^2}{C_w^2}\right)}$$

which can be rearranged in the form

$$\frac{C}{C_c}\left(\frac{\psi^2 - 1}{2}\right) = \psi - 1$$

where $$\psi = \sqrt{1 + 4\left(\frac{C_c^2}{C_v^2} + \frac{C_c^2}{C_w^2}\right)}$$

Since $$\frac{C}{C_c} = \frac{6.1 \times 10^{-4}}{1.98 \times 10^{-3}} = 0.308$$

Then $$\psi = 5.494$$

(*Note:* The root $\psi = 1$ is unacceptable.) We can now solve for C_w:

$$C_w = 1 \times 10^{-4} \text{ m/sec}^{1/2} = 7.75 \times 10^{-4} \text{ m/min}^{1/2}$$

Thus under field conditions, we need

$$(m_w)_{\text{field}} = 2C_w = 1.55 \times 10^{-3} \text{ m/min}^{1/2}$$

This value of the slope, which is defined by Eq. (8.79), is designated as a field value since it applies at the Δp between the fracture and the reservoir. If the Δp used in the laboratory is the same as that in the field, then the $(m_w)_{\text{field}}$ will equal $(m_w)_{\text{laboratory}}$. Generally, however, these two pressure differences are not equal. Thus to obtain $(m_w)_{\text{laboratory}}$, one must correct for difference in applied pressure.

Thus $$\frac{(m_w)_{\text{laboratory}}}{(m_w)_{\text{field}}} = \sqrt{\frac{\Delta p_{\text{laboratory}}}{\Delta p_{\text{field}}}}$$

If, for example the laboratory value of m_w is obtained by applying a pressure of 3.44 MPa, then,

$$(m_w)_{\text{laboratory}} = 1.55 \times 10^{-3}\sqrt{\frac{3.44 \times 10^6}{3.79 \times 10^7}} = 4.66 \times 10^{-4} \text{ m/min}^{1/2}$$

If the filter cake builds at a rate in laboratory experiments such that $m_w = 4.66 \times 10^{-4}$ m/min$^{1/2}$, the overall fluid-loss coefficient required for a successful treatment will be obtained.

REFERENCES

8.1. Nierode, D. E., *J. Pet. Tech., 37* (1985) 1831.

8.2. McLennan, J. D., and Roegiers, J.-C., "How Instantaneous Are Instantaneous Shut-In Pressures," SPE 11064, presented at the 57th Fall Technical Conference and Exhibition of the Society of Petroleum Engineers, New Orleans, Louisiana, 1982.

8.3. Howard, G. C., and Fast, C. R., *Hydraulic Fracturing*, Mono. Ser., *2*, Society of Petroleum Engineers, Richardson, Texas, 1970.
8.4. Hubbert, M. K., and Willis, D. G., *Trans. AIMME, 210* (1957) 153.
8.5. Prats, M., *Soc. Pet. Eng. J., 21* (1981) 658.
8.6. Smith, M. B., "Effect of Fracture Azimuth on Production with Application to the Wattenburg Gas Field," SPE 8298, presented at the 54th Fall Technical Conference and Exhibition of the Society of Petroleum Engineers, Las Vegas, Nevada, 1979.
8.7. Roegiers, J. C., in *Reservoir Stimulated*, M. J. Economides and K. G. Nolte, (eds.), Houston: Schlumberger Education Services, 1987.
8.8. Blanton, T. L., "The Relation Between Recovery Deformation and the In-Situ Stress Magnitude," SPE 11624, presented at the 1983 Society of Petroleum Engineers/Department of Energy Symposium on Low Permeability Gas Reservoirs, Denver, Colorado, 1983.
8.9. Clark, J. A., "The Prediction of Hydraulic Fracture Azimuth through Geological, Core, and Analytical Studies," SPE 11611, presented at the 1983 Society of Petroleum Engineers/Department of Energy Symposium on Low Permeability Gas Reservoirs, Denver, Colorado, 1983.
8.10. Teufel, L. W., "Determination of In-Situ Stress from Anelastic Strain Recovery Measurements of Oriented Core," SPE 11649, presented at the 1983 Society of Petroleum Engineers/Department of Energy Symposium on Low Permeability Gas Reservoirs, Denver, Colorado, 1983.
8.11. Schuster, C. L., "Detection within the Wellbore of Seismic Signals Created by Hydraulic Fracturing," SPE 7448, presented the 53rd Fall Technical Conference and Exhibition of the Society of Petroleum Engineers, Houston, Texas, 1978.
8.12. Albright, J. N., and Pearson, C. F., *Soc. Pet. Eng. J., 22* (1982) 523.
8.13. Dobecki, T. L., "Hydraulic Fracture Orientation Using Passive Borehole Seismics," SPE 12110, presented at the 58th Fall Technical Conference and Exhibition of the Society of Petroleum Engineers, San Francisco, California, 1983.
8.14. Evans, K., "The Growth and Consolidation Characteristics of Shallow Hydraulic Fractures as Viewed Through the Surface Deformation Field," SPE 10851, presented at the Society of Petroleum Unconventional Gas Recovery Symposium, Pittsburgh, Pennsylvania, 1982.
8.15. Wood, M. D., Parkin, C. W., Yotam, R., Hanson, M. E., Smith, M. B., Abbott, R. L., Cox, D., and O'Shea, P., "Fracture Proppant Mapping by Use of Surface Superconducting Magnetometers," SPE 11612, presented at the Society of Petroleum Engineers Symposium on Low Permeability Gas Reservoirs, Denver, Colorado, 1983.
8.16. Shah, R. C., and Kobayashi, A. S., *Eng. Fracture Mech., 3* (1971) 71.
8.17. Mastrojannis, E. N., Keer, L. M., and Mura, T., *Internat. J. Fracture, 15* (1979) 247.
8.18. Van Eckelen, H. A. M., *Soc. Pet. Eng. J., 22* (1982) 341.
8.19. Williams, B. B., Gidley, J. L., and Schechter, R. S., *Acidizing Fundamentals*, Mono. Ser., *6*, Society of Petroleum Engineers, Richardson, Texas, 1979.
8.20. Schmidt, R., Northrop, D., and Warpinski, N., "Hydraulic Fracturing Near an Interface: Observations and Calculations Regarding Geometry and Containment," *Proceedings of the 20th United States Symposium on Rock Mechanics, 2* (1979).
8.21. Warpinski, N. R., Schmidt, R. A., and Northrop, D. A., *J. Pet. Tech., 34* (1982) 653.
8.22. Simonson, E. R., Abou-Sayed, A. S., and Clifton, R. J., *Soc. Pet. Eng. J., 18* (1978) 27.

8.23. Warpinski, N. R., Clark, J. A., Schmidt, R. A., and Huddle, C. W., *Soc. Pet. Eng. J., 22* (1982) 333.
8.24. Gu, H., "A Study of Propagation of Hydraulically Induced Fractures," PhD dissertation, The University of Texas at Austin, 1987.
8.25. Abou-Sayed, A. S., "Laboratory Evaluation of In-Situ Stress Contrast in Deeply Buried Sediments," SPE 11069, presented at the 57th Fall Technical Conference and Exhibition of the Society of Petroleum Engineers, New Orleans, Louisiana, 1982.
8.26. Labudovic, V., *J. Pet. Tech., 37* (1984) 287.
8.27. Nolte, K. G., "Determination of Fracture Parameters from Fracture Pressure Decline," SPE 8341, presented at the 54th Fall Technical Conference and Exhibition of the Society of Petroleum Engineers, Las Vegas, Nevada, 1979.
8.28. Nolte, K. G., "Fracture Design Considerations Based on Pressure Analysis," SPE 10911, presented at the Society of Petroleum Engineers Cotton Valley Symposium, Tyler, Texas, 1982.
8.29. Kehle, R. O., *J. Geophys. Res., 69* (1964) 259.
8.30. Warpinski, N. R., Branagan, P., and Wilmer, R., "In-Situ Stress Measurements at DOE's Multiwell Experiment Site, Mesaverde Group, Rifle, Colorado," SPE 12142, presented at the 58th Fall Technical Conference and Exhibition of the Society of Petroleum Engineers Annual Meeting, San Francisco, California, 1983.
8.31. Cleary, M. P., *J. Pet. Tech., 40* (1988) 13.
8.32. Khristianovich, S. A., and Zheltov, P. A., "Formation of Vertical Fractures by Means of Highly Viscous Fluids," *Proceedings of the Fourth World Petroleum Congress. Rome, II* (1955) 579.
8.33. Geertsma, J., and de Klerk, F., *J. Pet. Tech., 21* (1969) 1571.
8.34. England, A. H., and Green, A. E., *Proceedings of the Cambridge Philosophical Society, 59* (1963) 498.
8.35. Perkins, T. K., and Kern, L. R., *J. Pet. Tech., 13* (1961) 937.
8.36. Nordgren, R. P., *Soc. Pet. Eng. J., 12* (1972) 306.
8.37. Geertsma, J., and Haafkens, R., *Trans. ASME 101* (1979) 8.
8.38. Daneshy, A. A., *J. Pet. Tech., 25* (1973) 83.
8.39. Perkins, T. K., *J. Pet. Tech., 25* (1973) 93.
8.40. Settari, A., and Cleary, M. P., *J. Pet. Tech., 36* (1984) 1177.
8.41. Advani, S. H., and Lee, J. K., *Soc. Pet. Eng. J., 22* (1982) 209.
8.42. Palmer, D., and Carroll, H. B., *Soc. Pet. Eng. J., 23* (1983) 870.
8.43. Mendelsohn, D. A., *J. Energy Res. Tech., 106* (1984) 369.
8.44. Cleary, M. P., Kavvades, M., and Lam, K. Y., "Development of a Fully Three-Dimensional Simulator for Analysis and Design of Hydraulic Fracturing," SPE 11631, presented at the Society of Petroleum Engineers Symposium on Low Permeability Gas Reservoirs, Denver, Colorado, 1983.
8.45. Clifton, R. J., and Abou-Sayed, A. S., "On the Computation of Three-Dimensional Geometry of Hydraulic Fractures," SPE 7943, presented at the Society of Petroleum Engineers Symposium on Low Permeability Gas Reservoirs, Denver, Colorado, 1979.
8.46. Abou-Sayed, A. S., Sinha, K. P., and Clifton, R. J., "Evaluation of the Influence of In-Situ Reservoir Conditions on the Geometry of Hydraulic Fractures Using a Three-Dimensional Simulator: Part I. Technical Approach," SPE 12877, presented at the Society of Petroleum Engineers Unconventional Gas Recovery Symposium, Pittsburgh, Pennsylvania, 1984.
8.47. Lam, K. Y., Cleary, M. P., and Barr, D. T., "A Complete Three-Dimensional Simulator for Analysis and Design of Hydraulic Fracturing," SPE 15264, presented at

the Society of Petroleum Engineers Unconventional Gas Recovery Symposium, Louisville, Kentucky, 1986.
8.48. Sneddon, I. N., and Elliot, H. A., *Quart. Appl. Math., 4* (1946) 262.
8.49. Lamb, H., *Hydrodynamics*, 6th ed., New York: Dover Publications, 1932.
8.50. Bird, R. B., Stewart, W. E., and Lightfoot, E. N., *Transport Phenomena*, New York: John Wiley, 1960.
8.51. Hall, C. D., and Dollarhide, F. E., *J. Pet. Tech., 20* (1968) 763.
8.52. Sinha, B. K., "Fluid Leakoff Under Dynamic and Static Conditions Utilizing the Same Equipment," SPE 6126, presented at the 51st Fall Technical Conference and Exhibition of the Society of Petroleum Engineers, New Orleans, Louisiana, 1976.
8.53. McDaniel, R. R., Deysarker, A. K., Callanan, M. J., and Kohlhaas, C. A., "An Improved Method for Measuring Fluid Loss at Simulated Fracture Conditions," SPE 10259, presented at the 56th Fall Technical Conference and Exhibition of the Society of Petroleum Engineers, San Francisco, California, 1981.
8.54. Gulbis, J., "Dynamic Fluid Loss of Fracture Fluids," SPE 12154, presented at the 56th Fall Technical Conference and Exhibition of the Society of Petroleum Engineers, San Francisco, California, 1983.
8.55. Penny, G. S., Conway, M. W., and Lee, W., *J. Pet. Tech., 37* (1985) 1071.
8.56. Roodhart, L. P., *Soc. Pet. Eng. J., 25* (1985) 629.

PROBLEMS

***8.1.** The fracture gradient for a reservoir 3,000 m deep was measured to be 17 kPa/m when the reservoir pressure was 28 MPa.

(a) Estimate the Poisson ratio (ν) of the formation rock.

(b) Predict the fracture gradient when the reservoir has been depleted to the extent that the pressure is 18 MPa.

(c) What bottomhole instantaneous shut-in pressure would you anticipate for in the depleted reservoir?

***8.2.** Both the overall fluid-loss coefficient (C) and the fracture height (h) are difficult quantities to predict in advance. If it would be possible to measure the average width at the wellbore, then based on the long-time approximation one could predict the fracture length without knowing C or h. For example, suppose at a time 30 min after initiating fracture fluid injection, the average width at the wellbore is found to be 0.25 cm. How long is the fracture at this time?

$$i = 2 \times 10^{-3}\ \text{m}^3/\text{sec}$$

$$G = 8 \times 10^6\ \text{kPa}$$

$$\nu = 0.37$$

$$\mu = 250\ \text{cp}$$

***8.3.** A fracture treatment consists of injecting a fluid ($\mu = 0.05$ kg/m-sec, $\rho = 1250$ kg/m^3) at a rate of 0.03 m^3/sec. The fracture length of the vertical fracture is calculated to be 100 m and the average fracture width measured at the wellbore is calculated to be 8×10^{-3} m at a time 2200 sec after starting the treatment. These calculations are all based on the assumption that the fracture height is 30 m. Suppose that the actual fracture height created by the fracture was 40 m rather than the 30 m used in the calculation. Assuming the long-time approximation to be valid, answer the following:

(a) What is the actual fracture length and average fracture width measured at the wellbore after 2200 sec?

(b) Estimate the bottomhole pressure after 2200 sec. The instantaneous shut-in pressure is 25,000 kPa.

****8.4.** Prove, starting with the volume balance expressed by Eq. (8.63) for horizontal or penny-shaped fractures, that the long-time approximation for the fracture radius expressed by Eq. (8.64) is valid. Note that in the derivation $R >> r_w$, the fracture width is taken to be a constant and the following mathematical approximation is applied

$$\sqrt{\pi} Z \exp Z^2 \text{erfc} Z \cong 1$$

for large Z.

****8.5.** To evaluate the fracturing program in a particular area, careful measurements of both the bottomhole pressure and the fracture length (by well testing) gives the data listed in Table P8.5A and final fracture length is 270 m. The treatment data are listed in Table P8.5B.

TABLE P8.5A Well Test Data

Time (min)	Bottomhole Pressure During Fracture Treatment (MPa)
5	28.42
10	28.66
15	28.80
20	28.90

TABLE 8.5B Treatment Data

Property	Symbol	Value
Fluid viscosity	μ	250 cp
Injection rate	i	4.77 m^3/min
Bulk modulus	G	6.9 × 10^9 Pa
Poisson ratio	ν	0.25
Fracture gradient	FG	17 kPa/m
Depth	D	1,524 m

The injection time was 20 min and it can be assumed that the dynamic length is the same as the final length. Estimate from this information the fluid-loss coefficient and the fracture height. These values will be used in future fracture designs.

***8.6.** A static fluid-loss test of a fracture fluid yields the filtrate volumes as a function of time shown in Table P8.6. The test conditions include an imposed pressure difference of 8 × 10^6 Pa. If the spurt loss is independent of the imposed pressure difference, what would the filtrate volume, V/A, be after 13 min if the imposed pressure is 1.5 × 10^7 Pa? What value of C_w would you use in calculating the overall fluid-loss coefficient for a fracture treatment in a well at 3000 m? The fracture gradient is estimated to be 16.5 kPa/m and the reservoir pressure is 2.4 × 10^4 kPa.

TABLE P8.6 Filtrate Volumes as a Function of Time

t (min)	V/A (cm^3/cm^2)
10	1.34
15	1.60
25	2.00

***8.7.** A fracture fluid tested by a service company in a static test ($\Delta p = 6.9 \times 10^3$ kPa) yields the data shown in Table P8.7A.

TABLE P8.7A Filtration Data

Time (min)	Volume (m^3/m^2 of area)
1	5.66×10^{-4}
4	1.032×10^{-3}
16	1.964×10^{-3}
25	2.43×10^{-3}

Calculate the overall fluid-loss coefficient that might be expected if this fluid is used to fracture the well defined by the data in the Table P8.7B.

TABLE P8.7B Reservoir and Fluid Properties for Well to be Fractured

Property	Symbol	Value
Porosity	ϕ	0.15
Permeability	k	6×10^{-15} m^2
Fracture gradient	FG	2.26×10^4 Pa/m
Depth	D	3000 m
Reservoir pressure	p_R	2.70×10^7 Pa
Oil compressibility	κ_{fl}	1×10^{-9} Pa^{-1}
Oil viscosity	μ_f	2×10^{-3} kg/m-sec
Capillary pressure	p_{CAP}	2×10^6 Pa
Fracture fluid viscosity	μ	0.2 kg/m-sec

***8.8.** A gas well is to be fractured. Since the compressibility of the gas is large, $C_c \gg C_w$ and $C_c \gg C_v$ where C_c, C_v, and C_w are the fluid-loss coefficients for compression, fluid invasion (viscous zone), and filter cake (wall), respectively. If the reservoir pressure is 35 MPa, the fracture gradient is 17 kPa/m, and the reservoir depth is 4000 m, what overall fluid-loss coefficient would you use to design the fracture? The fracture fluid has a viscosity of 2 kg/m-sec. The reservoir permeability is 1 mD (9.87×10^{-16} m^2) and its porosity is 0.1.

The service company reports C_w to be 7.5×10^{-5} $m/sec^{1/2}$ when a differential pressure of 7000 kPa is used in laboratory tests to evaluate the buildup of a filter cake.

****8.9.** Two bids have been received for a fracture treatment of a tight gas sand. One bid includes a fluid-loss additive, which in laboratory experiments gives a wall-building coefficient $C_w = 2 \times 10^{-6}$ $m/sec^{1/2}$ at an imposed pressure differential pressure of 3100 kPa. The second bid does not include any fluid-loss additive. All other factors in the two bids are identical. The first bid claims at least a 15% increase in fracture length can be obtained by using the additive. Is this a valid claim? Refer to Table P8.9 for data.

TABLE P8.9 Treatment Data

Properties of the Fracture Fluid (same for both bids)		
Viscosity		0.02 kg/m-sec
Density		1100 kg/m^3
Formation Properties		
Fracture Gradient	FG	17 kPa/m
Permeability	k	0.05 mD
Depth	D	3000 m
Porosity	ϕ	0.08
Reservoir pressure	p_R	20.0 MPa
Gas compressibility	κ_{fl}	5×10^{-5} kPa^{-1}
Gas viscosity	μ_f	1.5×10^{-5} kg/m-sec

****8.10.** A service company indicates that by adding methanol to the fracture fluid the fracture length can be increased by 20% at any given treatment time assuming that the only effect of the methanol is to decrease the surface tension between the fracture fluid between the reservoir gas (essentially methane) from 70 dynes/cm to 31 dynes/cm. All other factors can be assumed to be unchanged. Based on the data presented in Table P8.10, determine whether or not the claim is valid.

TABLE P8.10 Fracture Fluid Data

Treatment Data		
Fracture fluid viscosity	μ	300 cp
Capillary pressure (without methanol)	p_{CAP}	1×10^7 Pa
Wall cake fluid-loss coefficient (reported by service company for $\Delta P = 7 \times 10^3$ kPa)	C_w	6×10^{-4} m/min$^{1/2}$
Reservoir Data		
Permeability	k	0.1 mD
Reservoir gas pressure	p_R	2.8×10^4 kPa
Bottomhole fracture pressure	p_{BISIP}	5.6×10^4 kPA
Temperature	T	210°F
Gas isothermal compressibility	κ_{fl}	3.57×10^{-5} kPa^{-1}
Gas viscosity	μ_f	1.5×10^{-2} cp
Porosity	ϕ	0.1

9 Fracture Fluid Temperature

In this chapter we want to consider the temperature experienced by a fracture fluid. This will be a consideration when selecting a fluid for a particular application. All polymer gels, crosslinked or not, tend to degrade at elevated temperatures and to obtain a satisfactory final effective viscosity, one may be required to substantially increase the amount of polymer added to the solution when the fracture fluid is subjected to elevated temperatures for long periods of time. This will become a particularly important consideration in massive hydraulic fracturing of very tight gas sands. In this case, the fractures are long and tend to close slowly (see Sec. 10.1). The fracture fluid must be stable at reservoir temperatures over long periods of time. This represents an extreme situation.

For smaller fracture treatments in more permeable formations, the fractures close rapidly and the fluid in the fracture may not be exposed to the reservoir temperature during the entire time period.

The discussion of fracture length and fracture closure time is aimed primarily at hydraulic fracturing with proppants. Temperature also plays an important role in acid fracturing, since the rate of heterogeneous reaction is temperature sensitive. Thus to properly evaluate the reaction rate and acid penetration distance, knowledge of the fluid temperature is needed. In Sec. 10.2 we assume the rate to be infinite, but in subsequent chapters the rate of this reaction as a function of temperature will be defined. The validity of the assumption of finite rate can be ascertained.

The temperature within the fracture is considered first in this chapter, followed by a calculation of the fluid temperature at the perforation. The solution of both problems follows the same line of attack. The temperature in the surrounding solid is calculated independently from the fluid temperature, and then the two are brought together by making an energy balance on the flowing fluid. This uncoupling of the two problems simplifies the calculation. The results are, however, compared to more comprehensive equations (but more complex) and found to be in good agreement. The equations developed in this section should provide an adequate representation of the fluid temperatures for all practical purposes.

9.1 TEMPERATURE OF FRACTURE FLUID IN VERTICAL FRACTURES

The temperature of the fluid varies both with time and position during the fracturing process. An exact calculation of this temperature is not needed for most purposes; however, it is of interest to have an approximate knowledge of the fluid temperature to help in the selection of the fracture fluid, to be able to evaluate the effective viscosity of the fracturing fluid since this viscosity is quite sensitive to temperature and in the case of acid fracturing, to be able to obtain an estimate of the surface reaction rate. Wheeler [1] has developed analytical solutions, which take into account both conduction and convection of heat into the formation as well as convection of heat along the fracture. Wheeler's solutions are in the form of integral equations, which cannot be readily solved except by numerical integration. Since our needs do not generally require a knowledge of the temperature with the accuracy that Wheeler's model yields, we shall adopt a less rigorous approach which has the advantage of giving a simple, easily used relationship for the temperature within the fracture as a function of both time and position and which compares favorably with the temperature profiles given by Wheeler.

Volume and Energy Balances

The model we will use is depicted in Fig. 9.1. The average width ($\overline{w}$) is assumed to be a constant over the entire length of the fracture, the fluid-loss flux ($\overline{u}_N$) depends on time but not on position along the fracture and as shown, the inlet or bottomhole temperature T_I is taken as constant independent of time.

The mathematical model is developed in differential equation form by making energy and volume balances about the element of fracture length shown in Fig. 9.2. We will, for simplicity, neglect the accumulation of energy within the small volume element shown (i.e., $\partial T/\partial t \approx 0$) and assume that the fluid density and heat capacity are constant. The quantity $Q(x, t)$ is defined to be the rate of heat conduction from the formation to the fracture fluid expressed per unit area of fracture surface. It is this heat flow which results in the heating of the fracture fluid as it moves from the fracture inlet toward the tip of the fracture. We will evaluate this

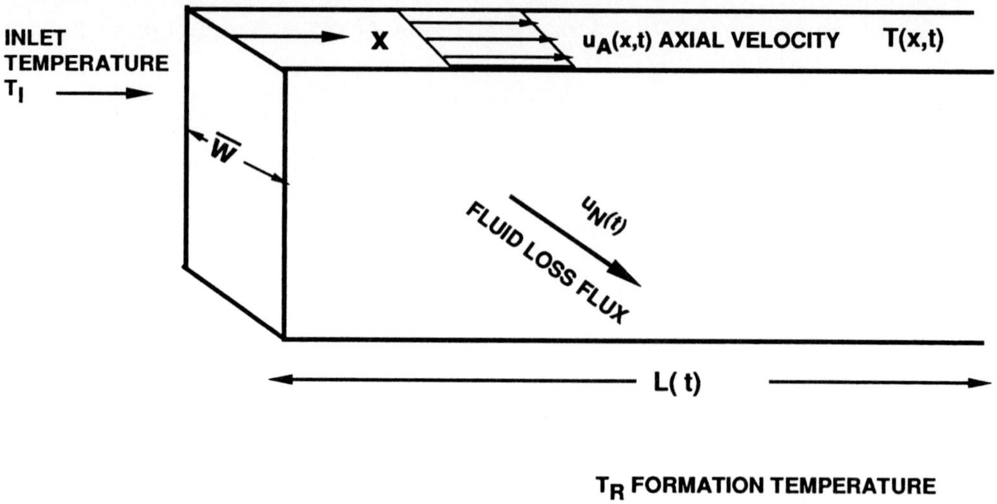

Figure 9.1 Diagram of simplified system used to model fracture fluid temperatures.

quantity by studying separately the process of heat conduction within the rock. Strictly speaking the two processes are coupled and they should, as Wheeler has done, be considered together; however, this leads to complex mathematical solutions which we aim to avoid.

An energy balance taken over a unit of time states that in the absence of energy accumulation the energy flow in must equal that flowing out. Thus

$$\underset{\text{(convected in)}}{u_A h\overline{w}\rho C_p T\,|_x} + \underset{\text{(conducted in)}}{2h\,\Delta x\,Q} = \underset{\text{(convected out)}}{u_A h\overline{w}\rho C_p T\,|_{x+\Delta x}} + \underset{\text{(convected out)}}{2\overline{u}_N h\rho C_p T\,\Delta x}$$

There are three convection terms. One, the last term on the right-hand side of the equation, accounts for the rate at which energy is convected into the formation

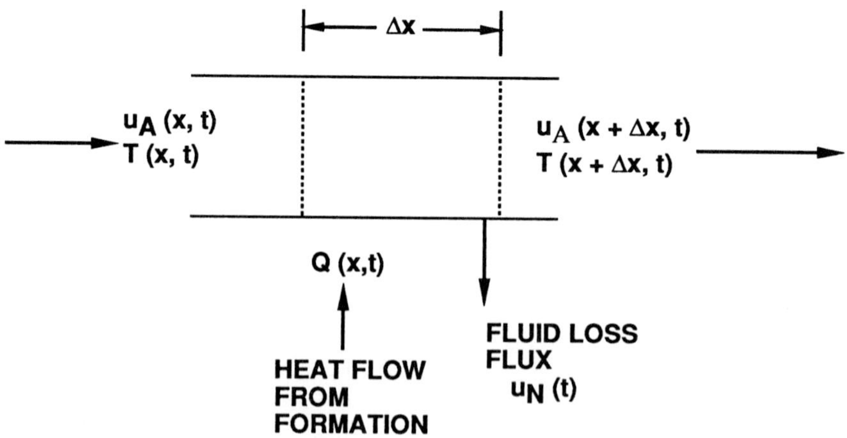

Figure 9.2 Sketch showing volume element fixed at a position x through which a fluid is flowing.

because of fluid loss. The other convection terms arise because of flow along the axis of the fracture.

Dividing by Δx and taking the limit as Δx approaches zero, we obtain

$$\overline{w}\rho C_p \frac{d}{dx}(u_A T) = 2Q - 2\overline{u}_N \rho C_p T \tag{9.1}$$

Similarly, a volume balance yields

$$\overline{w}\frac{d}{dx}u_A = -2\overline{u}_N \tag{9.2}$$

Equation (9.2) states quite simply that the axial velocity within the fracture varies because a portion of the fluid is lost to the formation.

Heat Conducted into Fracture

Before $T(x, t)$ can be obtained by integrating Eqs. (9.1) and (9.2) simultaneously, the function $Q(x, t)$ must be evaluated, or in this case it is more appropriate to say approximated, since as noted previously we will uncouple the two problems. Figure 9.3 depicts how this will be accomplished. For the moment, the temperature of the fracture fluid will be imagined to be constant so that we can readily calculate the rate at which heat is conducted from the formation to the fracture without having to know the temperature T in advance. This procedure used to uncouple the process of heat conduction into the fracture from the convective processes occurring within the fracture has been proposed by Dysart and Whitsitt [2].

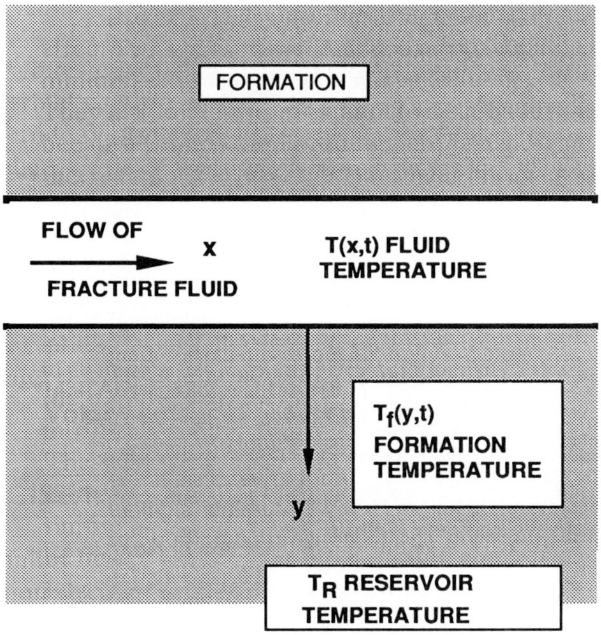

Figure 9.3 Sketch depicting formation temperature distribution near the fracture.

Referring again to Fig. 9.3, we note that the heat conduction equation applied to the system depicted yields [3]

$$\frac{\partial T_f}{\partial t} = \frac{K_f}{\rho_f C_{pf}} \frac{\partial^2 T_f}{\partial y^2} \tag{9.3}$$

The boundary and initial conditions are

$$T_f = T \qquad (\text{at } y = 0)$$

$$T_f = T_R \qquad (\text{at } y \to \infty)$$

and

$$T_f = T_R \qquad (\text{at } t = 0)$$

The solution to this problem is given by [3]

$$\frac{T_f - T_R}{T - T_R} = 1 - \text{erf}\left(\frac{y}{\sqrt{4K_f t/\rho_f C_{pf}}}\right) \tag{9.4}$$

The erf(x) represents the error function of x and it should perhaps be stressed that K_f, ρ_f, and C_{pf} are the thermal conductivity, density and heat capacity, respectively, of the fluid-filled formation rock. These will differ substantially from the properties of the formation fluid or of air-filled rocks [4] (see Chap. 2).

One observation worth stressing is that Eq. (9.3) neglects the convective effect of the fluid loss into the formation. Flow into the formation from the fracture will tend to reduce the rate at which formation heat arrives at the fracture surface. The value of $Q(x, t)$ obtained neglecting the convective effect will clearly be larger than a value found by taking it into account. Thus, we would expect that the temperatures calculated based on Eq. (9.1) will be too large. We will find, for example, that the temperatures we predict will tend to be larger than those given by Wheeler's equations, although not necessarily so since other assumptions have also been invoked in the analysis. Neglecting the convection resulting from fluid loss in Eq. (9.3) is thought to introduce the most serious error in our analysis. The analysis is, however, apparently quite accurate, as will be seen, and there appears to be little incentive to correct for convection in Eq. (9.3).

To obtain $Q(x, t)$ from Eq. (9.4), we note that [3]

$$Q(x, t) = K_f \left.\frac{\partial T_f}{\partial y}\right|_{y=0} \tag{9.5}$$

Upon differentiating Eq. (9.4), we find

$$Q(x, t) = \sqrt{\frac{K_f \rho_f C_{pf}}{\pi t}}\,(T_R - T) \tag{9.6}$$

Temperature Distribution

Substituting Eqs. (9.2) and (9.6) into Eq. (9.1), the following differential equation is found:

$$\overline{w} \rho C_p u_A \frac{dT}{dx} = 2\sqrt{\frac{K_f \rho_f C_{pf}}{\pi t}}\,(T_R - T) \tag{9.7}$$

To solve this equation for T, an expression for $u_A(x, t)$ is required. The most convenient form is obtained by recognizing that the fracture length is determined by a volume balance which expresses that the fluid injected is lost to the formation. Thus

$$2\bar{u}_N hL = u_A(0, t)\overline{w}h = \frac{i}{2}$$

Integrating Eq. (9.2) and imposing $u_A = 0$ at $x = L$ and $u_A = i/2\overline{w}h$ at $x = 0$, one can readily show that

$$u_A(x, t) = \frac{i}{2\overline{w}h}\left(1 - \frac{x}{L}\right) \tag{9.8}$$

Using this result to eliminate the axial velocity from Eq. (9.7), the temperature can be shown to satisfy the following differential equation:

$$\frac{(1 - x/L)}{(T_R - T)}\frac{dT}{dx} = \frac{4h}{\rho C_p i}\sqrt{\frac{K_f \rho_f C_{pf}}{\pi t}}$$

Noting (see Fig. 9.1) that $T = T_I$ at $x = 0$ and that the right-hand side of this equation does not depend on x, then the equation can readily be integrated to yield

$$\frac{T_R - T}{T_R - T_I} = \left(1 - \frac{x}{L}\right)^{\alpha_v} \tag{9.9}$$

where

$$\alpha_v = \frac{4hL}{\rho C_p i}\sqrt{\frac{K_f C_{pf} \rho_f}{\pi t}} \tag{9.10}$$

If we use the approximation, valid for long times,

$$L = \frac{it^{1/2}}{2\pi Ch}$$

then we obtain the result

$$\alpha_v = \frac{2}{C\rho C_P}\sqrt{\frac{K_f \rho_f C_{pf}}{\pi^3}} \tag{9.11}$$

Equation (9.11) is significant because α_v is seen to be independent of time. Thus, the temperature expressed as a function of the fractional distance along the fracture is a constant. A curve such as one of those shown in Fig. 9.4 can be prepared and applied during the entire course of the process.

If α_v is small, which occurs when there is substantial fluid loss [C large in Eq. (9.11)], then the fracture fluid retains essentially its inlet temperature for a large portion of the travel down the fracture. In fact, C is the only parameter which can readily be adjusted (by the addition of fluid-loss additives) by the designer. The other properties are fixed, although it should be noted that for foams, α_v will generally be quite large since ρ is small and foam will rapidly be heated

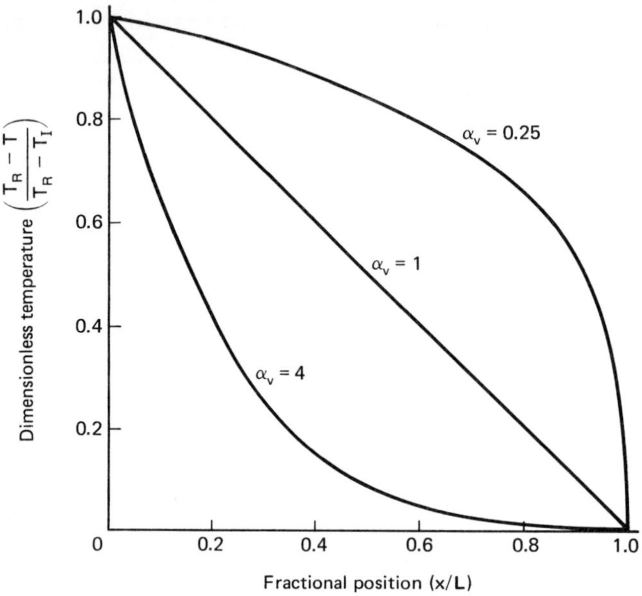

Figure 9.4 Fluid temperature shown as a function of α_v.

to the formation temperature as it enters the fracture. Appropriate design measures should be taken to account for this rapid heating.

Another feature suggested by Eq. (9.11) is that the use of a pad fluid which is rapidly lost to the formation will (that is one which has a large fluid-loss coefficient) help to cool the fluid which follows. This procedure may be especially valuable in acid fracturing where the fluid viscosity tends to rapidly decay in presence of acid at the higher temperatures and where the reaction rates are very sensitive to temperatures.

Example 9.1 Fluid Temperature

When treating a formation having a temperature of 150°C, what is the fluid temperature a distance 60 m from the wellbore at time 10 min after initiation of the fracture treatment? The fluid enters the perforations at 70°C. Refer to Table 9.1 for pertinent data.

TABLE 9.1 Treatment Data

Property	Symbol	Value
Injection rate	i	3×10^{-2} m³/sec
Fluid-loss coefficient	C	4×10^{-5} m/sec$^{1/2}$
Fluid density	ρ	1×10^3 kg/m³
Fluid heat capacity	C_p	4.18 kJ/kg-°C
Formation thermal conductivity	K_f	2.51×10^{-3} kJ/m-sec-°C
Formation density	ρ_f	2.1×10^3 kg/m³
Formation heat capacity	C_{pf}	0.879 kJ/kg-°C
Fracture height	h	20 m

Solution At the end of 10 min using the long-time approximation, Eq. (8.42), the fracture length is given by

$$L(t) = \frac{1}{2\pi}\left(\frac{3 \times 10^{-2}}{(20)(4 \times 10^{-5})}\right)(600)^{1/2}$$

Thus

$$\frac{x}{L} = \frac{60}{146} = 0.41$$

Also

$$\alpha_v = \frac{2}{(4 \times 10^{-5})(1 \times 10^3)(4.18)} \sqrt{\frac{(2.51 \times 10^{-3})(2.1 \times 10^3)(0.897)}{\pi^3}} = 4.6$$

Therefore

$$\frac{150 - T}{150 - 70} = (1 - 0.41)^{4.6}$$

which yields

$$T = 143°\text{C}$$

Sinclair [5] has used Wheeler's formulation to calculate temperature profiles in a fracture. Shown in Fig. 9.5 is a comparison of the more precise calculations given by Sinclair with those calculated using Eq. (9.9). It is seen that the approximation is quite satisfactory for all practical purposes.

Sinclair compared his result with those obtained using the Dysart and Whitsitt approximation, which is essentially the approach used here, and found a much larger discrepancy than is shown in Fig. 9.5. However, the comparison shown by Sinclair is not valid since Dysart and Whitsitt made an error in their derivation. That error has been corrected here.

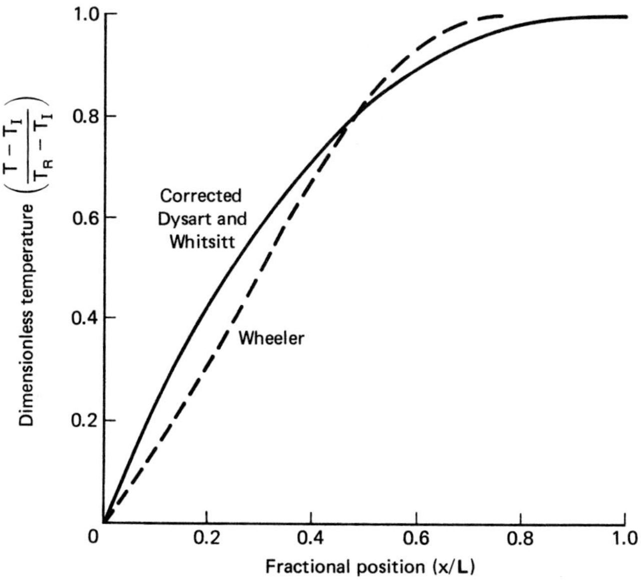

Figure 9.5 Comparison of Wheeler's solution with the approximation developed here, a corrected and modified version of Dysart and Whitsitt's solution.

Average Fluid Temperature

The average fluid temperature in the fracture will be of interest for design purposes. In some cases it may be necessary to develop complex computer programs which track precisely the time/temperature history of the fracture fluid so that its viscosity can be predicted. Given the fluid viscosity, the fluid pressures in the fracture can then be computed so as to accurately obtain values for the fracture width. In most cases, however, a knowledge of the average fluid temperature is sufficient. If the average temperature is defined as

$$\langle T \rangle = \frac{1}{L} \int_0^L T \, dx \tag{9.12}$$

Then substituting for T from Eq. (9.10) and integrating, we find the following expression for the average fluid temperature:

$$\langle T \rangle = \frac{\alpha_v T_R + T_I}{1 + \alpha_v} \tag{9.13}$$

If, for example, $\alpha_v = 1$, then as indicated by Fig. 9.4, the average temperature is the arithmetic average of the fluid entrance temperature and the formation temperature. As α_v increases, the average temperature closely approaches that of the reservoir.

9.2 FLUID TEMPERATURE IN HORIZONTAL OR PENNY-SHAPED FRACTURES

Since horizontal fractures usually occur in shallow formations, the temperatures are not normally a problem; however, the same equation developed here will also apply when the fracture is vertical, but penny-shaped. The temperature distribution is obtained based on the same approximations as applied in the previous section. The fracture geometry is depicted in Fig. 9.6. An energy balance yields

$$\overline{w}\rho C_p \frac{d}{dr}(ru_r T) = 2rQ(k, t) - 2\overline{u}_N r\rho C_p T \tag{9.14}$$

assuming, as before, that $\overline{w}$ is a constant and that the fluid-loss flux ($\overline{u}_N$) depends on time but not on position. Similarly, a volume balance is

$$\overline{w} \frac{d}{dr}(ru_r) = -2r\overline{u}_N \tag{9.15}$$

Substituting Eq. (9.15) into Eq. (9.14) and using Eq. (9.6) for Q, we find

$$\overline{w}\rho C_p u_r \frac{dT}{dr} = 2\sqrt{\frac{K_f \rho_f C_{pf}}{\pi t}}\,(T_R - T) \tag{9.16}$$

Let us now consider a volume balance by integrating Eq. (9.15) and noting that

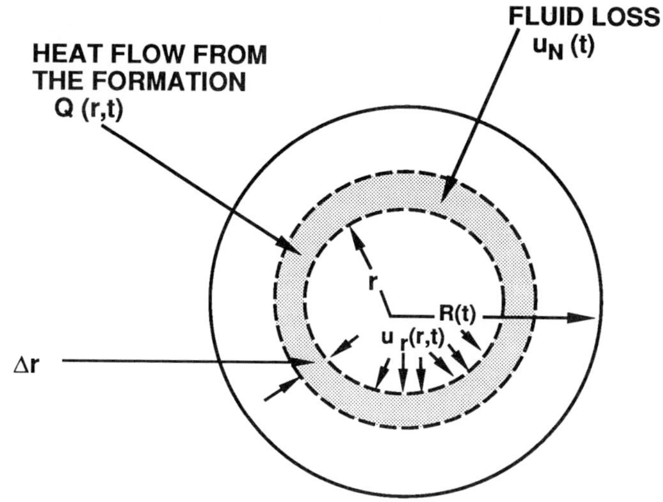

Figure 9.6 Sketch showing geometry of a horizontal fracture.

$u_r = 0$ at $r = R$. This gives

$$\overline{w}u_A = \frac{\overline{u}_N R^2}{r}\left(1 - \frac{r^2}{R^2}\right)$$

Also, the volumetric injection must be equivalent to the volumetric loss to the formation. Thus

$$2\pi R^2 \overline{u}_N = i$$

provided the fracture length is much larger than the wellbore radius. The radial velocity is represented by

$$u_r = \frac{i}{2\pi \overline{w} r}\left(1 - \frac{r^2}{R^2}\right) \tag{9.17}$$

Substituting this equation into Eq. (9.16) and integrating the fluid temperature within the fracture yields

$$\frac{T_R - T}{T_R - T_I} = \left(1 - \frac{r^2}{R^2}\right)^{\alpha_H} \tag{9.18}$$

where

$$\alpha_H = \frac{R^2}{\rho C_p i}\sqrt{\frac{\pi K_f \rho_f C_{pf}}{t}}$$

Equation (9.18) is similar to Eq. (9.9) and may be applied in the same way. We note that for horizontal or for penny-shaped fractures, the approximate relationship between the fracture radius as a function of time, Eq. (8.64) is expressed as follows:

$$R = \frac{1}{\pi}\sqrt[4]{\frac{i^2 t}{C^2}}$$

and α_H becomes
$$\alpha_H = \frac{1}{C\rho C_p}\sqrt{\frac{K_f\rho_f C_{pf}}{\pi^3}} \tag{9.19}$$

The exponent α_H is like α_v, dependent primarily upon the fluid-loss coefficient. Large values of C tend to maintain a cooler fluid within the fracture.

9.3 TEMPERATURE ENTERING PERFORATIONS

Calculation of fluid temperatures in a hydraulic fracture has assumed that the fluid temperature entering the perforation is known and is constant. In practical field situations, the surface temperature is generally known or in some cases perhaps even carefully controlled. As the fluid flows through the tubing, its temperature is modified by heat exchange with the formation rock. It is of interest here to estimate the effect of this heat exchange. Several numerical models have been developed, which are primarily intended to predict mud and circulating cement temperatures that with slight modification can be used to predict injection temperatures [6–10]. For cases where precise prediction of bottomhole temperatures is required, many effects believed to be of secondary importance can be taken into account using one of these numerical approaches. We shall in this section consider a simplified problem—one which can be solved analytically and which can be used to estimate the bottomhole temperature taking into account the most important effect; namely, the heat exchange with the surrounding formation. Other factors, such as heat conduction across the casing wall or the cement sheath, will be neglected.

Squire et al. [11] and Ramey [12] have both obtained analytical solutions to the problem depicted by Fig. 9.7. Their analyses neglect heat conduction in the vertical direction both within the formation and within the fluid flowing in the tubing. Also neglected are variations in the fluid injection rate, horizontal temperature gradients in the flowing fluid, and any variations of either the heat capacities or the densities of the formation materials or the injected fluids.

The mathematical analysis considers that the fluid initially present in the wellbore and the reservoir are at a temperature given by the surface temperature plus the product of the depth times the geothermal gradient. At large distances from the wellbore ($r \to \infty$), the formation is assumed to remain at this temperature throughout the workover operation.

The energy balance over a differential section of tubing shown in Fig. 9.7 is expressed by the following equation:

$$2\pi r_w Q(z, t) = \rho C_p i \frac{dT}{dz} + \rho C_p \pi r_w^2 \frac{\partial T}{\partial t} \tag{9.20}$$

where r_w is the tubing radius, $Q(z, t)$ is the rate at which heat is conducted from the formation to the fluid in the tubing per unit surface area, and z is the depth measured downward from the surface. In general, $Q(z, t)$ would have to take into

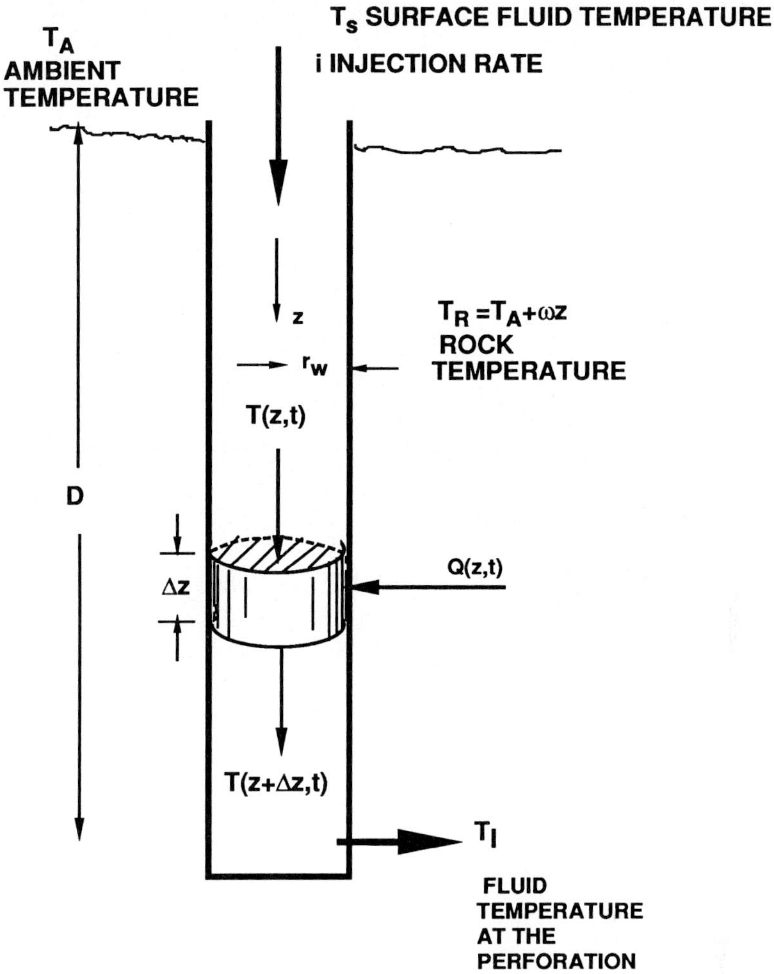

Figure 9.7 Sketch showing fluid flow down tubing and heat exchange with rock at geothermal temperature.

account the cylindrical geometry of the system by solving the heat conduction equation in cylindrical coordinates [3]

$$\frac{\partial^2 T_f}{\partial r^2} + \frac{1}{r}\frac{\partial T_f}{\partial r} = \frac{\rho_f C_{pf}}{K_f}\frac{\partial T_f}{\partial t} \tag{9.21}$$

Subject to the boundary conditions

$$\begin{aligned} T &= T_f = T_A + \omega z && (\text{at } t = 0) \\ T &= T_f && (\text{at } r = r_w) \\ T_f &\rightarrow T_A + \omega z && (\text{at } r \rightarrow \infty) \end{aligned} \tag{9.22}$$

Here ω is the geothermal gradient, T_A is the surface temperature of the earth (yearly mean average), so that $T_A + \omega z$ is the temperature of the earth at depth z prior to fluid injection.

Squire et al. [11] solved Eqs. (9.20) and (9.21) simultaneously subject to the conditions (9.22), but the solution applied to time scales, which are of interest to well treatment, are complex and difficult to use without performing numerical integrations. One simplification consistent with those previously made is to neglect the accumulation of heat. Therefore we neglect, $\partial T/\partial t$ and for short times note, that the cylindrical and the linear conduction of heat are identical. Thus, to an excellent approximation, Eq. (9.6) applies and

$$Q(z, t) = \sqrt{\frac{\rho_f C_{pf} K_f}{\pi t}}\,(T_R - T) \tag{9.6}$$

where $T_R = T_A + \omega z$. Substituting into Eq. (9.20) and integrating we obtain

$$T = T_s e^{-\alpha_f z} + \left(T_A - \frac{\omega}{\alpha_f}\right)(1 - e^{-\alpha_f z}) + \omega z \tag{9.23}$$

This expression has a form very similar to that derived by Ramey [12] as would be expected since the assumptions made in the development of this simple model are similar to those imposed by Ramey. Here

$$\alpha_f = \frac{2\pi r_w}{\rho C_p i}\sqrt{\frac{K_f \rho_f C_{pf}}{\pi t}}$$

Note that in this case, as contrasted to the fracturing calculation, α_f depends on time.

Example 9.2 Bottomhole Fluid Temperature

Let the fluid injection rate be 3×10^{-2} m^3/sec through tubing with a radius of 2.5×10^{-2} m. What will the temperature of the fluid entering the perforation be after 10 min if the surface injection temperature is 25°C and the yearly mean earth surface temperature is also 25°C? The well is 1000 m deep and the geothermal gradient is 0.032°C/m. Take the physical properties to be those given in Example 9.1.

Solution First calculate α_f.

$$\alpha_f = \frac{2\pi(0.025)}{(1 \times 10^3)(4.18)(3 \times 10^{-2})}\sqrt{\frac{(2.51 \times 10^{-3})(2.1 \times 10^3)(0.879)}{600\pi}} = 6.2 \times 10^{-5}\,\text{m}^{-1}$$

Substituting into Eq. (9.23) we find

$$T = 25e^{-0.062} + (1 - e^{-0.062})\left(25 - \frac{0.032}{6.2 \times 10^{-5}}\right) + 1000(0.032) = 26°\text{C}$$

This compares with the formation temperature

$$T_R = 25 + 0.032(1000) = 57°\text{C}$$

Thus, the fluid temperature at the perforation is essentially the same as the injected temperature. Lower injection rates or shorter times will increase this value.

It should be stressed that Eq. (9.23) applies only for short times. For long injection periods, the equations developed by Squire et al. [11] are recommended for use.

REFERENCES

9.1. Wheeler, J. A., "Analytical Calculations for Heat Transfer from Fractures," SPE 2494, 1969.
9.2. Dysart, R. G., and Whitsitt, N. F., "Fluid Temperature in Fractures," SPE 1902, presented at the 1967 Society of Petroleum Engineers Annual Meeting, Houston, Texas, 1967.
9.3. Bird, R. B., Stewart, W. E., and Lightfoot, E. N., *Transport Phenomena*, New York: John Wiley, 1960.
9.4. Somerton, W. H., *Trans. AIMME, 213* (1958) 375.
9.5. Sinclair, A. E., *J. Pet. Tech., 23* (1971) 1484.
9.6. Leutwyler, K., *J. Pet. Tech., 21* (1969) 1157.
9.7. Holmes, C. S., and Swift, S. C., *J. Pet. Tech., 22* (1970) 670.
9.8. Raymond, L. R., *J. Pet. Tech., 21* (1969) 333.
9.9 Keller, H. H., Couch, E. J., and Berry, P. M., *Soc. Pet. Eng. J., 13* (1973) 23.
9.10. Sump, G. E., and Williams, B. B., *Trans. ASME, 958* (1973) 1083.
9.11. Squire, D. P., Smith, D. A., and Dougherty, E. L., *J. Pet. Tech., 14* (1962) 436.
9.12. Ramey, H. J., Jr., *J. Pet. Tech., 14* (1962) 427.

PROBLEMS

***9.1.** Graph the temperature along the length of a fracture at times 5 min, 10 min, 15 min, and 20 min after initiation of fracturing. The data are given in Table P9.1.

TABLE P9.1 Treatment Data

Property	Symbol	Value
Reservoir temperature	T_R	138°C
Fracture fluid temperature at perforations	T_I	52°C
Fracture height	h	24 m
Injection rate	i	1.6 m^3/min
Fluid-loss coefficient	C	0.0015 m/$min^{1/2}$
Density of fracture fluid	ρ	1050 kg/m^3
Heat capacity of fracture fluid	C_p	4.18 kJ/kg-°C
Heat capacity of formation	C_{pf}	0.879 kJ/kg-°C
Thermal conductivity of formation	K_f	9.28 kJ/hr-m-°C
Density of the formation	ρ_f	2700 kg/m^3

Show how this temperature profile is changed if the fluid-loss coefficient is reduced to 0.00015 m/$min^{1/2}$.

***9.2.** The temperature at a point x_p, which is one half the length of the fracture at time t, is 200°F (reservoir temperature = 250°F and the fluid temperature measured at the perforations is 150°F). What will the temperature be measured at the same point (a distance x_p from the wellbore) if the fluid fracturing injection rate is doubled, all other factors remaining the same? Assume that the long-time fluid-loss dominated approximation applies.

***9.3.** The temperature measured at a distance x from the wellbore along a vertical fracture was 200°F 10 min after the fracture treatment had been initiated. The temperature measured at the same point 15 min after beginning the treatment was 177.28°F. What will it be 20 min after beginning the treatment?

***9.4.** If the fluid injection rate is 4×10^{-2} m³/sec through tubing having a radius of 2.5×10^{-2} m, how long will be required before the fluid temperature at the perforations reaches 28°C? The yearly average ambient temperature is 18°C and the fluid surface temperature is 25°C. Other pertinent data are shown in Table P9.4.

TABLE P9.4 Treatment Data

Property	Symbol	Value
Geothermal gradient	ω	0.032°C/m
Fluid density	ρ	1030 kg/m³
Fluid heat capacity	C_p	4.18 kJ/kg-°C
Formation thermal conductivity	K_f	2.5×10^{-3} kJ/m-sec-°C
Formation density	ρ_f	2.3×10^3 kg/m³
Formation heat capacity	C_{pf}	0.96 kJ/kg-°C
Formation depth	D	1500 m

****9.5.** Two asymptotic limits were considered in Chapter 8. The long-time approximation is the one used extensively in this chapter. The second extreme is the zero fluid-loss limit valid at early times. For this case [Eq. (8.34)] applies

$$L = 0.44 \left[\frac{Gi^3}{h^4(1 - \nu)\mu} \right]^{1/5} t^{4/5} \qquad (8.34)$$

Using this result develop an equation that expresses the fracturing fluid temperature as a function of both time and the fractional distance, x/L.

10
Final Fracture Conductivity

In Chapter 8 we developed the equations for the dynamic fracture geometry, but these dimensions are temporary. They only apply so long as fluid is being injected to propagate the fracture; however, once the treatment ends and the well is shut in, the fluid contained in the fracture is forced by the overburden pressure into the formation and the fracture begins to close. To ensure that the created fracture remains a conductive path for fluid to flow from the formation into the wellbore, two approaches are now commonly practiced.

Proppant fracturing requires that small particles be pumped with the fluid into the fracture. An effort is made to pack the fracture with a bed of these particles in order to support the walls of the fracture. Thus "propped" apart, the bed of particles will form a conductive path to the wellbore.

Acid fracturing requires that an acid capable of reacting rapidly with the fracture walls be pumped into the fracture. The acid will react unevenly with different parts of the fracture surface so that when the fluid pressure is released, the etched portions of the two walls will not fit together perfectly leaving gaps or spaces that form a conductive channel leading into the wellbore.

These two methods for creating fracture conductivity are very different, relying on entirely different physical principles. They are, therefore, considered separately in this chapter. Our purpose, however, will be to obtain a means of estimating final fracture conductivity. This goal will permit us in Chapter 11 to predict the production rates that can be realized from a fractured well and ultimately (in Chap. 12) to optimize the result.

10.1 PROPPANT FRACTURING

The objective of proppant fracturing is to pack the dynamic fracture with proppant (small particles) so that when the fracture treatment has terminated and production commences, the fracture will remain conductive. The need to pack the fracture with some form of propping agent was recognized early in the development of the process [1]. Production from fractured wells not propped declined rapidly.

The first fracture fluids used in the initial stages of development of the fracturing process were primarily low-viscosity oils or in a few cases, water. Proppant transported by these fluids tended to settle rapidly. In such cases, an interesting study reported by Kern and Perkins [2] showed that the mechanism of proppant (sand) transport and packing involved the formation of a sand dune near the entrance of the fracture and the subsequent movement of sand across the top of the dune once an equilibrium height was attained.

This mode of proppant transport applies when the apparent viscosity of the transporting fluid is small. Viscosities of 100–400 cp are now attainable using cross-linked polymer systems, and the settling of the proppant can be controlled to achieve the desired proppant distribution in the fracture. Thus, there is no need to consider further the mechanism of proppant transport by low-viscosity fluids. One of the important design considerations will be to select a fluid capable of transporting and holding the proppant particles in suspension until the fracture has closed.

Proppant Types

Many materials including glass beads, walnut shells, plastic beads, and aluminum pellets have at one time or another been used to prop hydraulic fractures [1]. As the fracture fluids have improved, the number of different materials used has steadily been reduced. Presently, sand and bauxite are in common use.

Sand. Sand has proven to be successful as a proppant for all types of reservoirs, and it is less expensive than other types of proppant. In the United States, quality well-graded sand is available in most oil-producing areas. Sand for use as proppant should not contain more than 5 wt% fines which, if present in excessive quantities, reduce the fracture conductivity. Sand has the additional advantage that when crushed, it breaks into smaller fragments, rather than being powdered. This particular advantage helps to maintain high fracture conductivities even when the closure stresses supported by the proppant are large.

As the closure stresses increase, sand has a tendency to crush and, at some point, the decrease in fracture conductivity as a result of the crushed particles becomes severe. In long fractures, significant reductions cannot be tolerated.

Sintered bauxite [3]. A high-strength proppant (compressive strength in excess of 1×10^5 kPa), which does not crush as readily as sand under high closure stresses, is bauxite. Bauxite is denser ($\rho_p = 3400$–3800 kg/m^3) than sand (2650 kg/m^3), and thus the fracture fluid designed to transport bauxite will have to be more viscous and hence more expensive than a fluid that will transport sand.

Ceramics. Other high-strength proppants have been developed [4, 5] which appear to have advantages with respect to sintered bauxite; however, these are not yet widely applied.

Propped Fracture Conductivity

In the previous discussion the term *fracture conductivity* was used extensively, but not defined. It is given as follows:

$$FC = \overline{w}_f k_f \tag{10.1}$$

where $\overline{w}_f$ is the final average fracture width and k_f is the permeability of proppant-packed fracture. FC has the dimensions of length cubed, although in much of the literature, it may be reported as darcy-feet, darcy-inches, or even millidarcy-feet. The reader should be aware of the different units used and be alert to the possible errors that can be introduced unless they are properly converted.

The fracture conductivity defined by Eq. (10.1) is a product of the final fracture width (not the dynamic width) and the fracture permeability. This equation is only valid if the proppant forms a multilayer, as shown in Fig. 10.1. Interestingly, partial monolayers (see Fig. 10.1) can have higher fracture conductivities than multilayer packing and there was some effort by the industry to develop methods of achieving partial monolayer packing in the field. The results of these attempts were mixed, and partial monolayer packing was found difficult to achieve. The concept we will apply and the one which most often dictates fracture design considerations at the present time is to get the most proppant possible into the fracture. We will strive for multilayer packing, and Eq. (10.1) will serve as the definition of fracture conductivity.

Fracture permeability. Final fracture permeability is strictly a function of the diameter of the proppant particles used in the treatment. According to the Blake–Kozeny equation [6]

$$k_f = \frac{d_p^2 \phi_f^3}{150(1 - \phi_f)^2} \tag{10.2}$$

where d_p is the diameter of the proppant particle and ϕ_f is the porosity of the packed, multilayer bed of proppant particles. As a fair approximation and in the absence of experimental data, $\phi_f \approx 0.32$–0.38.

Equation (10.2) is not often used in fracture design as will be observed shortly, but the equation does provide a very important message; namely, the fracture permeability increases with the square of the proppant particle diameter.

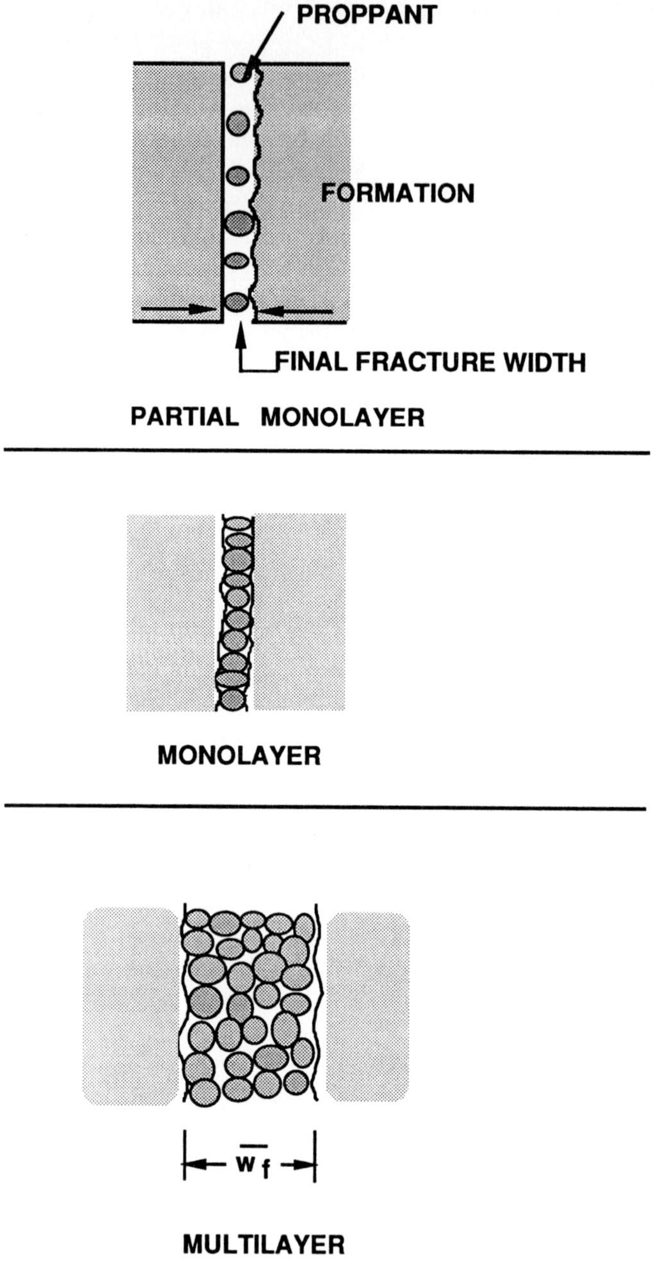

Figure 10.1 Sketch depicting various modes of proppant arrangement.

Therefore, it is desirable to use large proppant particles. Actually, the size of the proppant is an optimization problem that must always be settled on economic grounds. Larger particles will require more expensive fluids to transport them. The optimum will depend on a large number of factors, all of which will be discussed in Chapter 12.

For the purposes of the present discussion, we note that larger proppant diameters are preferred. Proppant sizes are specified by U.S. Mesh Standard Sieves. The sieve openings are provided in Table 10.1. Also given are the permeabilities and porosities of graded sands. The table indicates that sands are available as 8/12, 10/20, 10/30, 20/40, and 40/60 mesh ranges. (*Note:* 8/12 means that the sand particles will pass through a U.S. #8-mesh screen, but be retained on a U.S. #12-mesh screen. This, in accordance with Table 10.1, fixes the particle size range between 1.68–2.38 mm in diameter.)

TABLE 10.1 Properties of Commercially Available Graded Sands

Mesh size range (U.S. Series)	8–12	10–20	10–30	20–40	40–60
Sieve openings (mm)	2.38–1.68	2.00–0.84	2.00–0.589	0.84–0.42	0.42–0.250
Approximate permeability $(\mu m)^2$ [7]	1722	321	188	119	44
Porosity [7]	0.36	0.32	0.32	0.35	0.32

Example 10.1 Application of the Blake–Kozeny Equation

It is instructive to compare the fracture permeability given by Eq. (10.1) with the values reported in Table 10.1. Determine the fracture permeability of a fracture system packed with 10/20 U.S. Mesh graded sand, assuming the bed porosity to be 0.32.

Solution The average particle diameter is found from Table 10.1 as

$$d_p = \frac{(2.0 + 0.84)}{2} = 1.42 \text{ mm} = 0.142 \text{ cm}$$

Therefore

$$k_f = \frac{(0.142)^2}{150} \frac{(0.32)^3}{(1 - 0.32)^2} = 9.53 \times 10^{-6} \text{ cm}^2 = 953(\mu m)^2$$

This does not compare well with the value given in Table 10.1. The agreement is essentially qualitative.

Fracture width. The final fracture width is strictly related to the concentration of proppant in the fracture when it closes. This final concentration is related to, but is not equal to, the injected proppant concentration. The final concentration of proppant in the fracture depends on the injected concentration if there is little or no proppant settling. For a well-designed fracture fluid, proppant settling is minimal.

Consider the fracture shown in Fig. 10.2. $\overline{w}(t_f)$ is the average dynamic fracture width at the end of pumping. This width is a function of the length of the fracture treatment and the distance from the wellbore. If m_i is the mass of proppant per unit volume of fluid (the common method of reporting proppant concentra-

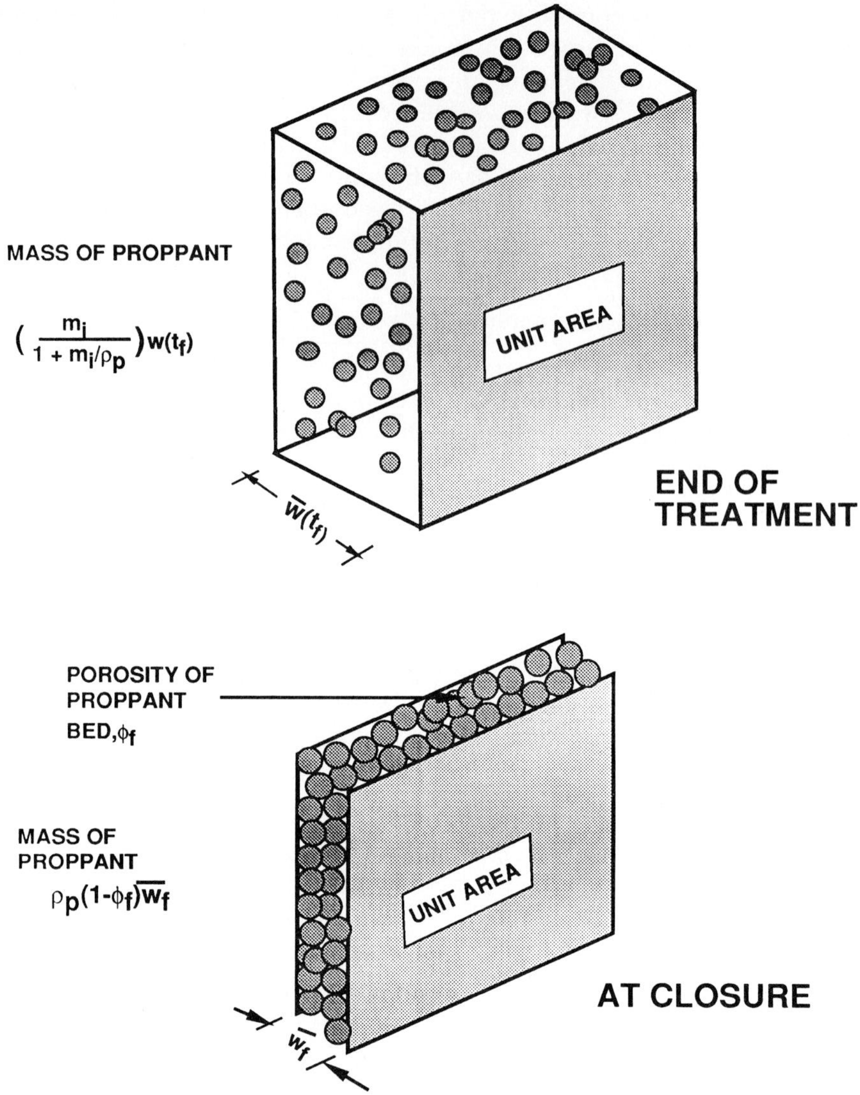

Figure 10.2 Diagram showing the relationship between dynamic and final fracture widths.

tions) in the fracture when fluid injection has ceased and ρ_p is the density of proppant, then clearly by mass balance

$$\overline{w}_f(1 - \phi_f)\rho_p = \overline{w}(t_f)\left(\frac{m_i}{1 + m_i/\rho_p}\right)$$

where $m_i/(1 + m_i/\rho_p)$ is the mass of proppant per total volume, including both proppant and fluid.

$$\text{Alternatively, } \overline{w}_f = \frac{\overline{w}(t_f)}{\rho_p(1 - \phi_f)} \left(\frac{m_i}{1 + m_i/\rho_p} \right) \tag{10.3}$$

This equation shows that in the absence of particle settling, the final fracture width is proportional to the dynamic width at the time all of the fluid has been injected and to the proppant concentration. Thus to have wide final fractures, one must start with a wide dynamic fracture. In addition, high proppant concentrations are desirable.

Normally, the injected concentration of proppant will range from 0.12 kg of proppant/ liter of fluid to 0.80 kg of proppant/liter of fluid (1 lb_m/gal to 6 lb_m/gal) although even larger concentrations have been reported when cross-linked fluids are used. Proppant concentrations in excess of 1 kg/liter of fluid should be used with care, because it may be difficult to get all of the proppant into the fracture if the fluid loss is somewhat more than anticipated.

The effect of closure stresses. The fracture conductivity can, in principle, be calculated by the multiple of $\overline{w}_f$ times k_f obtained from Eqs. (10.2) and (10.3). This calculated conductivity will, however, exceed the field value when the closure stresses exerted by the overburden become large. In this case, the proppant will embed into the formation causing the actual fracture width to be less than that calculated and also proppant crushing may cause the effective proppant radius to be reduced, thereby reducing the permeability of the fracture. Both effects will contribute to a decreased fracture conductivity.

The loss of permeability due to proppant crushing can be estimated using an apparatus such as that shown in Fig. 10.3. The proppant is packed between two horizontal plates mounted on a hydraulic press. In the experiment, a known pressure is exerted on the two plates to simulate those forces exerted by the

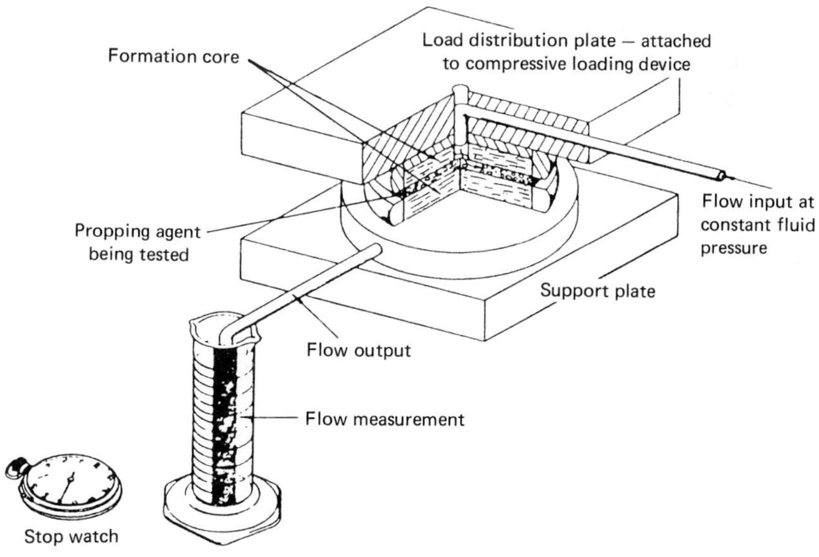

Figure 10.3 Diagram of a fracture flow capacity test apparatus.

overburden. The pressure will tend to crush the proppant. The conductivity of the crushed bed is measured by forcing fluid into the center of the horizontal plates outward and measuring the pressure required to attain a given flow. From this measurement the fracture conductivity can be obtained.

The experiment should be repeated at several different hydraulic pressures to simulate different overburden pressures. Figure 10.4 shows the permeability of various sand sizes as a function of closure stress, and Figs. 10.5 and 10.6 show the results of the laboratory fracture conductivity experiments as a function of the proppant concentration. These curves will be used in Chapter 11 for design purposes. It should be noted, however, that the values of the fracture conductivity shown are probably optimistic. The experiment does not include embedment. If the rock embedment strength, S_{RES} (see Chap. 2 for the definition of this quantity and a listing of some typical values) is low (the formation is "soft"), then considerable embedment may be expected and the fracture conductivity may be correspondingly reduced.

Despite their potential shortcomings, data such as those shown in Figs. 10.5 and 10.6 are used for fracture design. The closure stress exerted on the proppant particles is obtained by recalling that p_{BISIP} (see Chap. 8) is the pressure required to maintain the fracture open. When the fracture closes, this pressure is supported

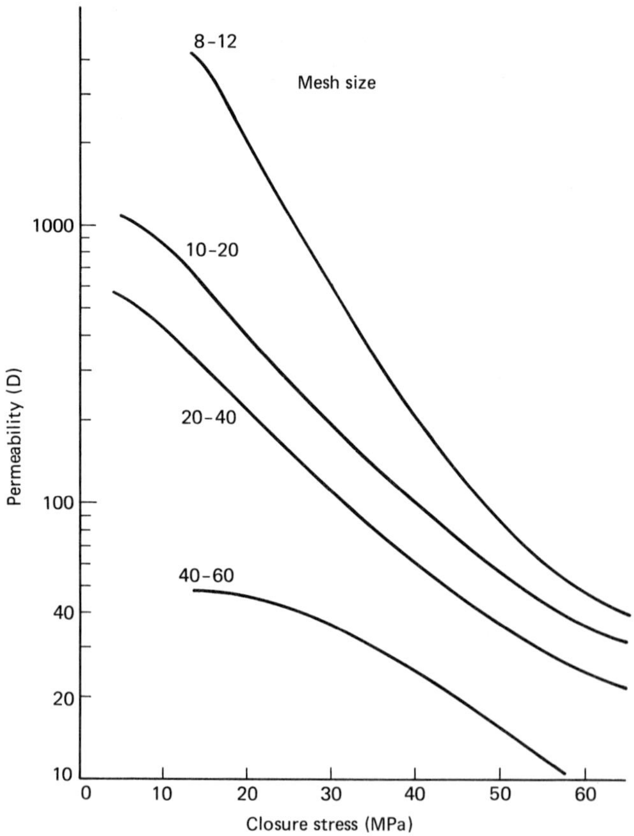

Figure 10.4 Graph showing the permeability of a propped fracture as a function of the closure stress. (Adapted from data supplied by the Norton Company.)

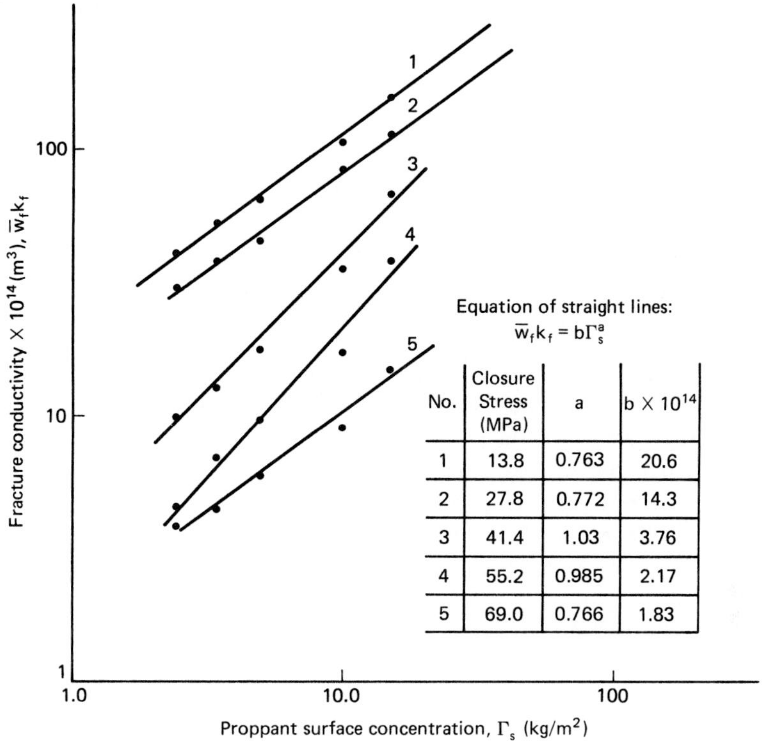

No.	Closure Stress (MPa)	a	b × 10^{14}
1	13.8	0.763	20.6
2	27.8	0.772	14.3
3	41.4	1.03	3.76
4	55.2	0.985	2.17
5	69.0	0.766	1.83

Figure 10.5 Graph showing conductivity of 20/40 sand as a function of proppant concentration. (Adapted from data supplied by the Norton Company.)

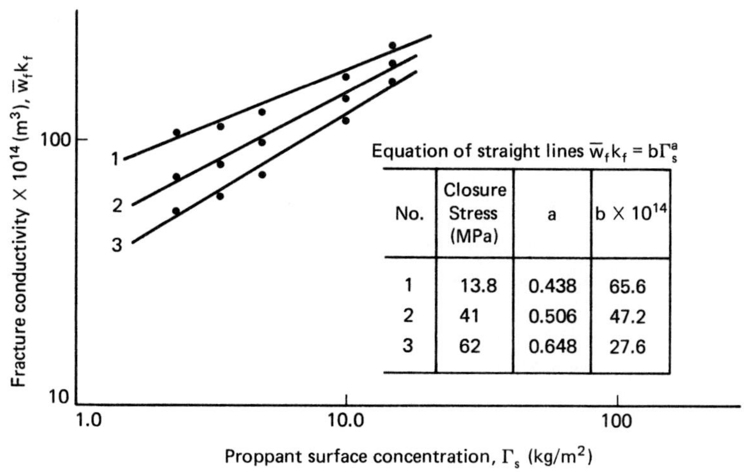

No.	Closure Stress (MPa)	a	b × 10^{14}
1	13.8	0.438	65.6
2	41	0.506	47.2
3	62	0.648	27.6

Figure 10.6 Graph showing fracture conductivity for 20/40 mesh bauxite as a function of proppant concentration. (Adapted from data supplied by the Norton Company.)

by the pressure of the fluid in the fracture and the proppant. Thus, the closure stress on the proppant is given by

$$\text{Closure stress} = p_{\text{BISIP}} - p_{\text{wf}} \tag{10.4}$$

Equation (10.4) shows that by maintaining the bottomhole well flowing pressure at a high level, part of the overburden stresses can be supported by the fluid in the fracture. Generally, however, to produce the well at an economic rate, p_{wf} is much less than the reservoir pressure (large drawdown) and the proppant must support nearly the entire overburden.

Proppant Settling Velocities

The selection of a fluid is one of the critical steps in the design of a fracture treatment. One of the important properties required of the fluid is an ability to transport and hold the proppant in suspension. Thus, it is important to be able to calculate the rate at which particles settle under the influence of gravity. Two different types of fluids will be considered here—non-Newtonian polymer solutions and foams—since these are the two primary types of fracture fluids now in use.

Non-Newtonian fluids. When a particle settles in a fluid under the influence of gravity, it reaches a constant velocity so that the frictional forces are in balance with the gravitational forces. For Reynold's numbers less than about 2, that is, for

$$N_{\text{Re}} = \frac{d_{\text{p}} v_{\text{s}} \rho}{\mu_{\text{APP}}} < 2 \tag{10.5}$$

Novotny [8] has found that the settling velocity (v_{s}) takes on the following very simple form as predicted theoretically by Slattery and Bird [9]:

$$v_{\text{s}} = \frac{g(\rho_{\text{p}} - \rho) d_{\text{p}}^2}{18\mu_{\text{APP}}} \tag{10.6}$$

which is Stokes's law for Newtonian fluids, but represents a modification when the fluid is non-Newtonian since the apparent viscosity is used rather than an actual viscosity of a Newtonian fluid. Equation (10.6) is then an empirically derived equation since the apparent viscosity depends on shear rate (see Chap. 3) and is therefore not a constant. Slattery and Bird [9] have shown that for particles settling in a quiescent non-Newtonian fluid, μ_{APP} can be evaluated at a shear rate

$$\dot{\gamma} = \frac{v_{\text{s}}}{d_{\text{p}}} \tag{10.7}$$

and the proppant settling velocity can be predicted using Eq. (10.6); the modified Stokes's law. This has been confirmed to be true in cross-linked polymer solutions by Novotny [8].

Thus for a power-law fluid (see Chap. 3)

$$\mu_{\text{APP}} = m \, | \dot{\gamma} |^{n-1} \tag{3.3}$$

the appropriate viscosity to be used in Eq. (10.6) is

$$\mu_{\text{APP}} = m\left(\frac{v_s}{d_p}\right)^{n-1} \tag{10.8}$$

Substituting this expression into Eq. (10.6), one finds

$$v_s = \left[\frac{g(\rho_p - \rho)d_p}{18m}\right]^{1/n} d_p \tag{10.9}$$

Example 10.2 Settling Velocities in Non-Newtonian Fluids

Compare the settling velocities of 20/40-mesh sand particles ($\rho_p = 2650$ kg/m^3) with 20/40-mesh bauxite particles ($\rho_p = 3500$ kg/m^3) in a power-law fluid ($m = 0.07$ kg/m-sec$^{1.5}$ and $n = 0.5$). The density of the power-law fluid is 1100 kg/m^3.

Solution The average particle diameter as observed from Table 10.1 is

$$d_p = \frac{(0.84 + 0.42)}{2} = 0.63 \text{ mm} = 6.3 \times 10^{-4} \text{ m}$$

For sand

$$v_s = \left[\frac{(9.8)(2650 - 1100)(6.3 \times 10^{-4})}{18(0.07)}\right]^2 (6.3 \times 10^{-4})$$

or

$$v_s = 0.036 \text{ m/sec}$$

For bauxite

$$v_s = \left[\frac{9.8(3500 - 1100)(6.3 \times 10^{-4})}{18(0.07)}\right]^2 (6.3 \times 10^{-4})$$

or

$$v_s = 0.087 \text{ m/sec}$$

Thus, the bauxite particle because of its greater density settles more than twice as fast as sand in the same fluid.

Equation (10.9) applies when the proppant concentration is small. As the concentration increases, the settling velocity becomes a complex function of the fluid rheology which is not characterized by the power-law indices m and n. In some fluids, clusters of proppant particles are observed to form and these entities settle collectively at a faster rate than that predicted by Eq. (10.9). In other fluids the particles are observed to settle in a "rope-like" fashion, where again the settling velocity may exceed that predicted by Eq. (10.9) because the particles are moving in a fluid that has been sheared by the particles preceding them. Clustering or roping seems to be prevalent at modest volume fractions of proppant ranging between 10 to 20% [10].

At higher volume fractions, the settling velocities are greatly hindered by the presence of the other particles and by the upward flow of fluid caused by the downward movement of the particles. If the proppant concentration in the fluid is inhomogeneous so that there is a relatively particle-free path for the fluid to take in its movement upward, then the settling will be more rapid than in the case of a uniformly distributed concentration [11]. Thus, it is known that settling is more rapid in an inclined tube rather than in a vertical one. This is called the *Bowcott effect* [12, 13].

All of these complex phenomena contribute to the settling velocity of proppant particles in a fracture and should be accounted for. Unfortunately, there do not now exist theories or models that will permit us to anticipate their contribution. Generally, one would require experimental data for design of large fracture treatment.

Hindered settling has been considered by Roodhart [14]. As proppant is added to the slurry, the apparent viscosity of the fluid increases. Evidently m changes, but the power-law index (n) is independent of the proppant volume fraction. Roodhart has found that for a slurry, an effective value of m is given by

$$m_s = m\left(1 + \frac{1.25\phi_p}{1 - 1.8n\phi_p}\right) \tag{10.10}$$

Here ϕ_p is the volume fraction of the proppant particles. Thus as the volume fraction is increased, the slurry consistency index (m_s) increases. To estimate hindered settling, this average value for the slurry should replace m in Eq. (10.9) in the calculation of settling velocity.

Equation (10.9) considers settling in a stagnant fluid. This equation applies during the time interval between cessation of flow and fracture closure. As we shall see, this time period can be the same order of magnitude as the length of the fracture treatment and therefore, it will be important to be able to calculate the distance the particles settle during this time.

During the fracture treatment, the fracture fluid is flowing and is, therefore, subjected to shear because of its motion through the fracture. This motion will tend to reduce the viscosity of shear-thinning fracture fluids, and thus particles will settle faster than is predicted if the shear rate is estimated using Eq. (10.7). To resolve this difficulty Novotny [8] has proposed that Eq. (10.7) be replaced by the following average rate:

$$\dot{\gamma}_{ave} = \sqrt{\dot{\gamma}^2_{settling} + \dot{\gamma}^2_{flow}} \tag{10.11}$$

Here $\dot{\gamma}_{settling}$ is given by Eq. (10.7) and represents that part of the fluid shear rate resulting from the downward movement of the particle and $\dot{\gamma}_{flow}$ is that part of the shear rate due to the flow of the fracture fluid down the channel. It should be emphasized that the concept of dividing the shear rate into parts, as shown by Eq. (10.11), is without any firm theoretical foundation. The only theoretical significance that can be attributed to Eq. (10.11) is that it works in the limits of no flow and small settling rates. Novotny [8] has tested Eq. (10.11) by measuring the settling of particles between coaxial rotating cylinders. Calculations based on the apparent viscosity evaluated at differing shear rates showed good comparison with experiment. Thus, Eq. (10.11) must be viewed as a reasonable means of evaluating the average shear rate, although there is considerable evidence that proppant settling cannot in any event be determined independent of the proppant concentration [10, 11, 15].

There is one difficulty which becomes evident when applying Eq. (10.11) to the fracturing process. The shear rate due to flow between the fracture walls is not constant. Indeed, for a power-law fluid flowing between parallel plates, we derived (see Problem 3.5)

$$\tau_{yx} = m \left| \frac{dv_x}{dy} \right|^{n-1} \frac{dv_x}{dy} = y \frac{dp}{dx} \tag{10.12}$$

This equation shows that the shear rate $\dot{\gamma}$ (recall $\dot{\gamma} = dv_x/dy$) is a function of position within the channel of the fracture. The shear rate vanishes when $y = 0$ and is maximum when $y = \overline{w}/2$. Thus, particles near the centerline will settle more slowly than particles near the wall. Furthermore, another complicating feature is the tendency of particles to migrate from the wall to the centerline [16]. It does not seem reasonable to attempt to account for all of these factors in view of the uncertainties in the settling mechanism already described. A reasonable approach will be to use an average shear rate over the entire width of the channel. For a Newtonian fluid, Eq. (10.12) shows $\dot{\gamma}$ to be a linear function of y. Since the maximum velocity is $3u_A/2$, then for a Newtonian fluid,

$$\dot{\gamma}_{\text{flow}} \cong \frac{3u_A}{2\overline{w}} = \frac{3i}{2h\overline{w}^2} \tag{10.13}$$

Of course, because of fluid loss to the formation both the flow rate and the width vary with distance measured from the wellbore.

Solution for the settling velocity based on Eq. (10.6) requires a trial-and-error approach, as illustrated by Example 10.3.

Example 10.3 Proppant Settling During Fracturing

Calculate the settling velocity of individual sand proppant particles in the fluid described in Example 10.2 when the fluid is being injected into a fracture at the rate of 1.5 m^3/min. The fracture height is 17.8 m and the average width at the wellbore is 5×10^{-3} m.

Solution From Eq. (10.13) we can calculate the average shear rate due to flow as

$$\dot{\gamma}_{\text{flow}} = \frac{3\,(1.5)}{2\,(60)(17.8)(5.0 \times 10^{-3})^2} = 84 \text{ sec}^{-1}$$

Then

$$\dot{\gamma}_{\text{ave}} = \sqrt{\frac{v_s^2}{(6.3 \times 10^{-4})^2} + 84^2}$$

and the modified Stokes's Eq. (10.6) gives

$$v_s = \frac{9.8(2650 - 1100)(6.3 \times 10^{-4})^2}{18(0.07)\dot{\gamma}_{\text{ave}}^{-0.5}}$$

Solving for v_s gives

$$v_s = 0.051 \text{ m/sec}$$

This can be compared to the 0.036 m/sec found in Example 10.2, which is the settling velocity of the same particle in a stagnant fluid. The effect of the fluid motion is to increase the rate of proppant settling.

Foams. The settling of proppant particles in foams must be a complex function of the wettability of the particles, the quality of the foam, and its stability. No general theory has been presented which shows the relationship of these factors. Blauer and Kohlhaas [17] have reported very slow settling rates in an 83% quality foam. The excellent carrying capacity of foams appears to be one of its advantages as a fracture fluid.

Fracture Closure Time

A fracture does not immediately close once fluid injection has stopped and the well is shut-in. The fluid contained in the fracture will be forced into the formation since the fluid pressure will be approximately p_{BISIP}, which is in excess of the reservoir pressure. Under the influence of this driving pressure, the fluid in the fracture will, on the average, flow into the formation in accordance with the equation

$$\overline{u}_{\mathrm{N}} = \frac{\pi}{2} \frac{C}{\sqrt{t_{\mathrm{f}} + \Delta t}} \tag{10.14}$$

where C is the overall fluid-loss coefficient and t_{f} is the time required for the fracture treatment. The time elapsed since the fracture treatment is designated as Δt. This equation gives the average fluid loss along the entire length of the fracture, a concept developed in Chapter 8 and given by Eq. (8.90).

The average flux into the formation is $\overline{u}_{\mathrm{N}}$. As fluid is lost into the formation, the fracture width decreases in accordance with the volume balance

$$\frac{d\overline{w}}{dt} = -2\overline{u}_{\mathrm{N}} \tag{10.15}$$

This rate of closure will continue until the fracture reaches its final propped width $\overline{w}_{\mathrm{f}}$ starting with the dynamic width which existed at the time pumping ceased, $\overline{w}(t_{\mathrm{f}})$.

By substituting Eq. (10.14) into Eq. (10.15) and integrating, one can show

$$\overline{w}(t_{\mathrm{f}}) - \overline{w}_{\mathrm{f}} = 2\pi C[(t_{\mathrm{f}} + \Delta t)^{1/2} - t_{\mathrm{f}}^{1/2}] \tag{10.16}$$

This equation is important because it defines the time (Δt) required to reach the final width. During this time, proppant will continue to settle and thus Δt is an important factor in fracture design.

Example 10.4 Fracture Closure Time

Given a fluid-loss coefficient, $C = 3 \times 10^{-4}$ m/min$^{1/2}$, how long will be required for a fracture which required 20 mins to create to close to its final width of 1.5×10^{-3} m? The average dynamic width at the end of 20 min is 5×10^{-3} m.

Solution Substituting into Eq. (10.16)

$$5 \times 10^{-3} - 1.5 \times 10^{-3} = 2\pi(3 \times 10^{-4})[(20 + \Delta t)^{1/2} - (20)^{1/2}]$$

Solving for Δt we find

$$\Delta t = 20 \text{ min}$$

Thus, a closure time equal to the treatment time is required. The proppant will have to remain suspended for a total of 40 min.

10.2 ACID FRACTURING

The mechanism by which permanent conductivity is achieved by acid fracturing differs entirely from that of proppant fracturing; therefore, we are compelled to distinguish between the two types of fracture treatments in this chapter.

The effective fracture length for acid fracturing may not be at all related to the dynamic fracture length created during the fracturing process. The important length is the distance that acid moves along the fracture before it has been completely reacted (spent). This distance is a function of many factors including the acid fluid-loss characteristics, the rate of acid reaction with the rock, the fracture width, and the acid injection rate. These variables are all included in the detailed model of acid penetration distance presented in this chapter, but before considering the detailed model it may be useful to consider qualitatively the trends one might expect.

Increasing the fracture width can significantly increase the acid penetration distance. This is true because live acid must diffuse from the center of the fracture to the fracture wall before it can react. The mechanism of this reaction is discussed in Chapter 15 in some detail, but here we simply note that the process of molecular diffusion in liquids is a slow one as compared to the reaction rate of carbonates with hydrochloric acid and by widening the fracture, the rate of overall reaction is slowed. This means that wide acid fractures are required to obtain deep penetration distances.

The penetration distance also increases with acid injection rate as a result of the shorter residence time for reaction. However, in practice, increasing the rate will increase the fracture width and thus change the residence time in a complex way. However, increasing the flow will generally lead to increased acid penetration distances.

The acid penetration distance is almost independent of temperature in limestones but depends on the temperature in dolomites [18]. The reaction rate of hydrochloric acid with limestone is an extremely rapid one even at low temperatures; thus, increasing the temperature only serves to increase an already fast reaction and does not alter the penetration distance.

On the other hand, the reaction of dolomite with hydrochloric acid is slower than that of limestone (see Chap. 14 for a discussion of this reaction rate) and at low temperatures the finite reaction rate slows acid spending, permitting deeper penetration. As the temperature is increased, the rate of reaction increases and at sufficiently high temperatures ($>70°C$) there is little, if any, difference in the acid penetration distance betwen limestone and dolomite.

One of the important variables is the fluid loss. When an acid enters the fracture it will react with the fracture walls and can eliminate the filter cake created by fluid-loss agents added to the pad fluid. Once this occurs the fluid loss will be determined by the properties of the acid. The mechanism whereby acid bypasses the filter cake is known as *wormholing*, which occurs when the acid enters the matrix pores and reacts rapidly therein [18]. The process of wormholing, which results in the formation of large passages through which acid flows, is described in detail in Chapter 17. The change in fluid-loss characteristics associated with acid injection into a fracture following a viscous pad cannot be accurately predicted. If, however, the fluid loss is greatly increased in the presence of acid, it may be expected that the fracture will begin to close. This will be discussed further in Chapter 12.

To be effective the acid must react with the walls of the fracture to form a channel that will stay open after the fracture treatment. Flow channels can be formed as a result of uneven reaction with the rock surface or preferential reaction with minerals heterogeneously distributed throughout the fracture surface. The fracture conductivity resulting from acid reaction is modeled in detail in this chapter.

Acid Penetration Distance

The distance to which live acid penetrates down a fracture determines the effective length of the fracture just as it is the propped length that is productive after hydraulic fracturing. In this section, we want to understand those factors which contribute to the acid spending rate and we want to devise a method for calculating the concentration as a function of length. A quasi-stationary procedure will be used in which terms involving the partial derivatives with respect to time holding the spatial coordinate constant will be set equal to zero. This assumption implies that the concentration at a point along the fracture changes at a much slower rate with time than with position. Thus, convection and diffusion will be the important features of the process. Accumulation can be disregarded. This assumption is extremely accurate for the acid fracturing process. The length and width of the fracture are slowly increasing as a function of time and relative to the reaction rate measured following a fluid element down the fracture, they are negligibly slow.

Also slow is the rate of molecular diffusion. This will be emphasized in Section 14.2. The rate of acid reaction will, therefore, be governed by the rate at which acid is convected to the walls by the small fluid-loss generated velocities that are perpendicular to walls. Thus to accurately model the acid reaction and the attendent penetration distance, we must first understand the fluid dynamics.

Laminar fluid velocity in a fracture. We would like to know the magnitude and direction of the velocity at any point in the fracture channel assuming that the flow is steady state, the fluid is incompressible, the flow is laminar, and the fluid-loss flux $\langle \overline{u}_n \rangle$ is a constant. It is an average value taken over the entire area of the fracture during the time of treatment [see Eq. (8.92) for its value in terms of the fluid-loss coefficient].

The solution to the flow problem with a fully developed laminar velocity profile within a fracture was first obtained by Berman [19]. Let x and y again be the distances measured parallel and perpendicular to the fracture wall, respectively, and let $u(x, y)$ and $v(x, y)$ be the velocity components in the increasing x- and y-directions, respectively. Berman showed that the Navier–Stokes equation and the equation of continuity can be satisfied by equations of the form

$$u(x, y) = [u_A^0 - \langle \bar{u}_N \rangle x] f'(\eta) \tag{10.17}$$

and

$$v(x, y) = \langle \bar{u}_N \rangle f(\eta) \tag{10.18}$$

where $u_A^0 = i/2\overline{w}h$ and $\langle \bar{u}_N \rangle$ is the fluid-loss flux. The variable η $(= 2y/\overline{w})$ is defined so that the walls are located at $\eta = \pm 1$. The prime denotes the derivative of $f(n)$ with respect to η.

Define a Reynolds number based on the fluid-loss velocity as follows:

$$N_{\mathrm{Re}^*} = \frac{\langle \bar{u}_N \rangle \overline{w} \rho}{2\mu} \tag{10.19}$$

Then for small values of N_{Re^*} Berman showed that

$$f(\eta) = \frac{\eta}{2}(3 - \eta^2) + \frac{N_{\mathrm{Re}^*}}{280}\eta(-\eta^6 + 3\eta^2 - 2) + \frac{3N_{\mathrm{Re}^*}^2}{280}\eta\left(\frac{\eta^{10}}{990} - \frac{\eta^8}{36} + \frac{\eta^6}{70} + \frac{146}{2310}\eta^2 - \frac{703}{13860}\right) + O(N_{\mathrm{Re}^*}^3) \tag{10.20}$$

where O denotes order of magnitude Terrill [20] found this solution to be accurate in the range $|N_{\mathrm{Re}^*}| < 7$.

Calculation of the axial velocity profile will show that the usual laminar velocity is flattened near the centerline and increased near the wall by the fluid loss. The small rates of fluid loss encountered in fracturing operations do not greatly alter the flow pressure drop at a point along the fracture. The hydrodynamic flow induced by the fluid loss does, however, have a profound influence on the acid penetration distance. In calculating the acid concentration profile, the fluid loss cannot be disregarded.

Finally, it should be noted that the solution obtained here is for a Newtonian fluid. Many of the gelled acid solutions, acid emulsions, or acid foams behave as non-Newtonian fluids. We will use the equations developed here to describe these complex situations but will replace the Newtonian viscosity by an effective viscosity given by Eq. (8.53). While there have been no studies reported that support or deny the validity of this approach, it is clear that because the overall reaction rate is sensitive to the flow large errors may be expected.

Fast acid reactions in laminar flow. Denote the acid concentration as c and write an acid material balance equation to obtain (see Sec. 14.2)

$$u(x, y)\frac{\partial c}{\partial x} + v(x, y)\frac{\partial c}{\partial y} = D_A \frac{\partial^2 c}{\partial y^2} \tag{10.21}$$

where D_A is the acid molecular diffusion coefficient. Let

$$\Gamma = c/c_0$$

$$\xi = 2x/\overline{w}$$

Substituting Eqs. (10.17) and (10.18) for the velocity components u and v, respectively, into the acid balance Eq. (10.21) we obtain

$$\left(1 - \frac{N_{Re^*}}{N_{Re}}\right) f'(\eta) \frac{\partial \Gamma}{\partial \xi} + \frac{N_{Re^*}}{N_{Re}} f(\eta) \frac{\partial \Gamma}{\partial \eta} = \frac{N_{Re^*}}{N_{Pe} N_{Re}} \frac{\partial^2 \Gamma}{\partial \eta^2} \tag{10.22}$$

where N_{Re} is the Reynolds number defined as

$$N_{Re} = \frac{u_A^0 \overline{w} \rho}{2\mu} = \frac{i\rho}{4h\mu} \tag{10.23}$$

and

$$N_{Pe} = \frac{\langle \overline{u}_N \rangle \overline{w}}{2D_A} \tag{10.24}$$

is called the *Peclet Number.* N_{Re^*}, N_{Re}, and N_{Pe} are all dimensionless groups. Since these are the parameters that appear in Eq. (10.22), we would, of course, expect the solution to depend on them.

The boundary conditions to be imposed are

$$\begin{aligned} \Gamma &= 0 \qquad (\text{at } \eta = \pm 1) \\ \Gamma &= 1 \qquad (\text{at } \xi = 0) \end{aligned} \tag{10.25}$$

The boundary condition $\Gamma = 0$ at $\eta = \pm 1$ implies that the reaction at the wall is fast. It is, in fact, so fast that each time an acid molecule arrives at the solid boundary, reaction is instantaneous and the acid concentration near the boundaries is vanishingly small. In reality, of course, the rate of reaction is never infinite and assuming it to be so will yield a lower limit for the acid penetration distance. It should be noted, however, that the rates of hydrochloric reaction with carbonates are generally quite fast and the approximation of a fast reaction is expected to be realistic.

Terrill [21] obtained a formal solution of Eq. (10.22) by separation of variables. The final solution for the mean acid concentration as a function of x, that is, the acid concentration $c(x, y)$ averaged over y, is

$$\frac{\overline{c}}{c_0} = \sum_{n=0}^{\infty} G_n \left(1 - \frac{N_{Re^*}}{N_{Re}} \xi\right)^{2\lambda_n^2/3N_{Pe}} \tag{10.26}$$

where the λ_n are eigenvalues and the G_n are constants.

If this solution for the mean acid concentration is restricted to $0.001 < N_{Re^*} < 1$ and $N_{PE} < 8$, then only about five terms in the series expressed by Eq. (10.26) are required. Thus, the first five λ_n and G_n are needed and these can be well represented as follows:

$$\underset{n}{\lambda} = \sum_{i=0}^{3} g_{i,n} N_{\mathrm{Pe}}^{i} + \sum_{i=1}^{2} h_{i,n} N_{\mathrm{Re}^*}^{i} \tag{10.27}$$

and

$$G_n = \sum_{i=0}^{3} \overline{g}_{i,n} N_{\mathrm{Pe}}^{i} + \sum_{i=1}^{2} \overline{h}_{i,n} N_{\mathrm{Re}^*}^{i} \tag{10.28}$$

The constants $g_{i,n}$, $h_{i,n}$, $\overline{g}_{i,n}$, and $\overline{h}_{i,n}$ are given in Table 10.2. These values can be substituted into Eq. (10.26) and the average concentration may be determined as a function of axial position.

TABLE 10.2A Coefficients for Eq. (10.27) to Calculate $\underset{n}{\lambda}$

n	$g_{0,n}$	$g_{1,n} \times 10$	$g_{2,n} \times 10^3$	$g_{3,n} \times 10^4$	$h_{1,n} \times 10^3$	$h_{2,n} \times 10^3$
0	1.68231	−2.26693	6.7544	−1.8408	6.7593	−4.6274
1	5.67053	−0.69600	17.2931	−2.9304	1.0032	−3.4376
2	9.66842	−0.39587	10.7745	−0.5564	−5.7028	−0.4705
3	13.66772	−0.27662	7.9375	−0.1358	−9.1500	−0.5668
4	17.66740	−0.21305	6.3431	−0.0373	−12.4496	−0.71169

TABLE 10.2B Coefficients for Eq. (10.28) to Calculate G_n

n	$\overline{g}_{0,n} \times 10$	$\overline{g}_{1,n} \times 10^4$	$\overline{g}_{2,n} \times 10^4$	$\overline{g}_{3,n} \times 10^5$	$\overline{h}_{1,n} \times 10^4$	$\overline{h}_{2,n} \times 10^4$
0	9.10378	−2.38279	14.9298	−8.97017	−7.08188	−1.18392
1	0.53126	1.88909	−12.5375	8.13482	4.01538	0.35148
2	0.15272	0.39035	−1.6607	0.680785	1.03940	0.51540
3	0.06807	0.07330	−0.4172	0.111312	0.58639	0.141225
4	0.03739	0.01901	−0.1503	0.027559	0.35277	0.056322

Although the solution to Eq. (10.26) was derived without explicit reference to a fracture length, a tacit assumption about length is inherent in the steady-state flow assumption. In particular the length is fixed by an overall material balance between the constant flow into the fracture, u_{A}^{0}, and the constant flow out of the fracture, $\langle \overline{u}_{\mathrm{N}} \rangle$. Thus the flow into one wing of a fracture must, at steady state, equal the fluid loss. This result implies

$$\frac{i}{2} = u_{\mathrm{A}}^{0} h \overline{w} = 2\langle \overline{u}_{\mathrm{N}} \rangle hL \tag{10.29}$$

Furthermore

$$\frac{N_{\mathrm{Re}^*}}{N_{\mathrm{Re}}} \xi = \frac{\dfrac{\langle \overline{u}_{\mathrm{N}} \rangle \overline{w} \rho}{2\mu}}{\dfrac{u_{\mathrm{A}}^{0} \overline{w} \rho}{2\mu}} \left(\frac{2x}{\overline{w}} \right) = \frac{4h \langle \overline{u}_{\mathrm{N}} \rangle}{i} x$$

Combining this result with Eq. (10.29) shows that

$$\frac{N_{\mathrm{Re}^*}}{N_{\mathrm{Re}}}\xi = \frac{x}{L}$$

The mean acid concentration is now seen to be a function of the fractional distance along the fracture. Equation (10.26) in its most useful form is

$$\frac{\bar{c}}{c_0} = \sum_{n=0}^{\infty} G_n \left(1 - \frac{x}{L}\right)^{2\lambda_n^2/3N_{\mathrm{Pe}}} \tag{10.30}$$

A second interesting result of the solution is that λ_n and G_n are relatively independent of N_{Re^*} for $N_{\mathrm{Re}^*} < 1$. This is usually the condition that applies in acid fracturing treatments when viscous acids are used. If acids without gelling agents are used, their viscosity is small and large fluid losses can be expected. The equations presented here may not apply for two reasons: $N_{\mathrm{Pe}} > 8$ and $N_{\mathrm{Re}} > 1000$ (turbulent flow). Turbulent flow in acid fracturing is treated by Roberts and Guin [22] and is reviewed by Williams et al. [18]. It is not considered here because turbulent conditions will rarely exist within the fracture.

Figure 10.7 shows $\bar{c}/c_0$ represented as a function of x/L for various values of N_{Pe}. The curves shown apply when N_{Re^*} is small. These curves show an interesting result. For $N_{\mathrm{Pe}} < 1$, the acid concentration is reduced to small values at small fractional values of the total fracture length. This result indicates that the effective fracture length will be shorter than the dynamic fracture length. The effective fracture length is said to be *reaction-rate limited*. The acid is spent prior to reaching the end of the fracture.

On the other hand, if N_{Pe} is greater than unity, live acid penetrates to the end of the fracture. In this case, the effective fracture length will correspond to the dynamic length. The fracture is said to be *fluid-loss limited* if the acid concentration is high at the end of the fracture.

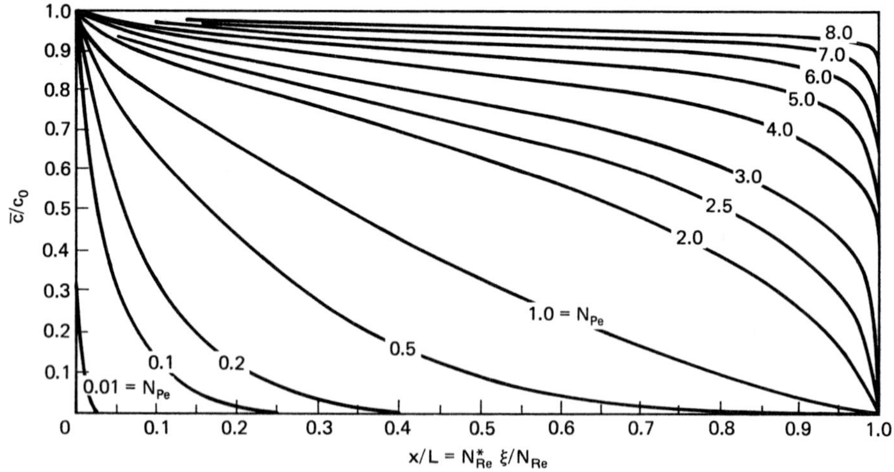

Figure 10.7 Graph of mean concentration versus distance along channel.

The Peclet number is defined based on the fluid-loss velocity, and large Peclet numbers correspond to short fractures. Thus if all other factors are held constant, live acid penetrates to the end of the fracture because it is short. If this is the case, then improved performance from a given treatment is possible by decreasing the fluid loss.

The diffusion coefficient, which appears in Eq. (10.24), is the molecular diffusion coefficient; however, in some cases more accurate results are obtained if a value of the diffusion coefficient measured under experimental conditions closely corresponding to those encountered in fracturing is used. It has been found by both Nierode and Williams [23, 24] and Roberts and Guin [22, 25] that the effective diffusion coefficient measured under realistic conditions is larger than the molecular diffusion coefficient of acid in aqueous solutions. This fact has been attributed to surface roughness and to secondary flows induced by density differences [26].

Nierode and Williams [23] took special precautions to assure the fracture walls used in their laboratory studies were of a surface roughness typical of surfaces formed when a formation is fractured. Cores that were used in the study were prepared from cylindrical specimens that were fractured in tension. This technique allowed the walls of the test system to have very rough surfaces with a uniform fracture width. During the experiments the fracture was oriented vertically to allow gravity forces to influence acid mixing. For low Reynolds numbers $N_{Re} < 1000$ (laminar flow), the effective diffusion coefficient was found to be approximately 10^{-8} m^2/sec, which is larger than the corresponding molecular diffusivity of hydrogen ions in acid solutions (see Sec. 14.2). Values of the effective diffusivity as a function of the Reynolds number may be found in the *Acidizing Monograph* [18].

Roberts and Guin [25] have noted that although the average value of the effective diffusion coefficient is approximately 10^{-8} m^2/sec, there exists a definite minimum in the plot of effective diffusivity as a function of the Reynolds number. They attributed this minimum to a secondary convective motion which is triggered because the reaction products are, as shown in Table 10.3, of higher density than the reactants. Thus, the fluid near the wall of a vertical fracture experiences a net downward force because of its greater density. Chang and Guin [26] have examined the contribution of free convection to the overall reaction rate. They

TABLE 10.3 Properties of Hydrochloric Acid Solutions

	Acid Density (38°C)		Acid Viscosity (25°C)	
Acid Concentration (wt%)	Live Acid (kg/m^3)	Spent Acid (kg/m^3)	Live Acid (kg/m-sec)	Spent Acid (kg/m-sec)
15	$1.07 \times 10^{+3}$	$1.18 \times 10^{+3}$	1.15×10^{-3}	1.7×10^{-3}
20	$1.10 \times 10^{+3}$	$1.25 \times 10^{+3}$	1.27×10^{-3}	2.4×10^{-3}
25	$1.12 \times 10^{+3}$	$1.31 \times 10^{+3}$	1.41×10^{-3}	3.8×10^{-3}
30	$1.15 \times 10^{+3}$	$1.56 \times 10^{+3}$	1.57×10^{-3}	5.9×10^{-3}

show that the secondary flows do significantly increase the rate of mass transfer to the wall when compared to systems reacting without fluid loss. It is not clear that the effects would be as profound in the presence of fluid loss. Experiments or theoretical calculations to ascertain the importance of the secondary flows in the presence of fluid loss have not been reported.

Thus we see that the proposed method of calculating the acid penetration distance in laminar flows represents an approximation because surface roughness, secondary flows, and finite reaction rates are all neglected. To somewhat compensate for the first two difficulties, a diffusivity larger than the molecular coefficient can be used. In the absence of any other information, take $D_A = 10^{-8}$ m^2/sec as a first approximation.

Example 10.5 Calculation of Acid Penetration Distance

If the fluid-loss flux is 3×10^{-5} m/sec and the acid injection rate is 2×10^{-3} m^3/sec per meter of fracture height, then determine to what depth acid having a concentration of 10% of its inlet concentration will penetrate into a fracture having an average width of 2.5×10^{-3} m. Assume the reaction rate to be infinite and take the diffusion coefficient of the acid to be 10^{-8} m^2/sec.

Solution The first step in determining the penetration distance is to calculate N_{Re^*}, N_{Re}, and N_{Pe}. Take the viscosity from Table 10.3 to be 1.15×10^{-3} kg/m-sec and the density to be 1.07×10^3 kg/m^3. The average acid concentration is not sensitive to these numbers and any reasonable value will yield approximately the same answer.

$$N_{Re^*} = \frac{(3 \times 10^{-5})(1.07 \times 10^3)(2.5 \times 10^{-3})}{2(1.15 \times 10^{-3})} = 0.035$$

For most cases, we expect N_{Re^*} to be small.

$$N_{Re} = \frac{u_A^0 \rho \overline{w}}{2\mu} = \frac{(0.4)(1.07 \times 10^3)(2.5 \times 10^{-3})}{2(1.15 \times 10^{-3})} = 465$$

The Reynolds number does not enter the calculation at all; however, it should be calculated to be certain that the flow is laminar. If $N_{Re} < 1000$ the procedure used here applies.

$$N_{Pe} = \frac{\langle \overline{u}_N \rangle \overline{w}}{2D_A} = \frac{(3 \times 10^{-5})(2.5 \times 10^{-3})}{2(1 \times 10^{-8})} = 3.75$$

The values of $\underset{n}{\lambda}$ and G_n are now required.

From Eq. (10.27) and the values given in Table 10.2

$$\underset{0}{\lambda} = 1.68231 - 0.226693 N_{Pe} + 0.00675 N_{Pe}^2 - 0.00018 N_{Pe}^3$$
$$+ 0.00676 N_{Re^*} - 0.00463 N_{Re^*}^2 = 0.9179$$

and from Eq. (10.28)

$$G_0 = 0.910378 - 0.000238 N_{Pe} + 0.001493 N_{Pe^*}^2 - 0.0000897 N_{Pe}^3$$
$$- 0.000708 N_{Re^*} - 0.000118 N_{Re^*}^2 = 0.9258$$

Other values of $\underset{n}{\lambda}$ and G_n can be similarly calculated; however, the contribution of these terms to the acid concentration is small except for small values of x/L (near the wellbore). Thus, Eq. (10.30) becomes

$$\frac{\bar{c}}{c_0} \cong 0.9258 \left(1 - \frac{x}{L}\right)^{2(0.9179)^2/11.25} \quad \text{or} \quad \frac{\bar{c}}{c_0} = 0.9258 \left(1 - \frac{x}{L}\right)^{0.1498}$$

Solving for x/L which yields $\bar{c}/c_0 = 0.1$, we find

$$\left(\frac{x}{L}\right)_{\text{penetration}} \cong 1.0$$

Live acid penetrates to the end of the fracture. By volume balance

$$L = \frac{i}{4h\langle \bar{u}_N \rangle} = \frac{2 \times 10^{-3}}{4(1)(3 \times 10^{-5})} = 17 \text{ m}$$

Thus, a fracture having an acidized length about 17 m will be produced. To extend this fracture, the fluid-loss should be reduced.

Ideal Fracture Width

In this section we calculate an ideal fracture width, which is defined as that width a fracture would have based on the volume of rock dissolved by the acid, all other factors being neglected. Clearly, this ideal fracture width never exists at any time during the treatment process. While acid is being injected, the width of the fracture is the dynamic one, determined primarily by the pressure forces in the fracture and the mechanical properties of the reservoir. Once pumping ceases the closure stresses act on the fracture and tend to reseal it. The fracture remains conductive for two reasons despite the forces exerted by the closure stress. First, the fluid that is lost to the formation dissolves rock, creating a highly permeable zone along the boundaries of the fracture. It is also visualized that the surface of the fracture is not eroded uniformly and even after closure, the fracture is open to some extent owing to a ''pillar-and-post'' arrangement such as is depicted in Fig. 10.8.

The actual conductivity of a fracture will be discussed in the next section. Its calculation is empirical, but requires knowledge of the ideal fracture width.

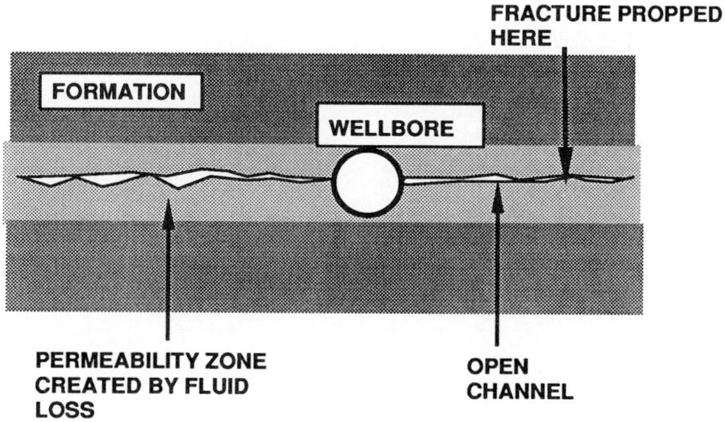

Figure 10.8 Sketch depicting final acidized fracture.

Thus, it is important to develop equations that permit calculation of the ideal fracture width.

Consider the element of fracture volume depicted in Fig. 10.9. The mass of acid reacted in the fracture contained between x and $x + \Delta x$ is clearly given by

$$\overline{w}h\overline{c}u_A t_A |_x - \overline{w}h\overline{c}u_A t_A |_{x+\Delta x}$$

where t_A is the time during which acid is injected. It is not generally the total treatment time, since an inert pad fluid may be used to create the fracture before initiating the injection of the acid. Now, if we multiply this by β, which is the mass of rock dissolved per unit mass of acid reacted (see Sec. 3.6), then

$$\text{Mass of rock dissolved} = \beta h\overline{w}t_A(\overline{c}u_A |_x - \overline{c}u_A |_{x+\Delta x})$$

$$\text{Volume of rock dissolved} = \frac{\beta h\overline{w}t_A}{\rho_f}(\overline{c}u_A |_x - \overline{c}u_A |_{x+\Delta x})$$

where ρ_f is the density of the rock. This volume of rock would create a fracture volume given by

$$w_i(1 - \phi)h\Delta x$$

where w_i is the ideal fracture width. Equating the ideal with the actual amount dissolved, we find in the limit as $\Delta x \to 0$

$$w_i(1 - \phi) = -\frac{\beta\overline{w}t_A}{\rho_f}\frac{d}{dx}(\overline{c}u_A) \tag{10.31}$$

In this section the fluid loss is taken to be a constant over the entire fracture surface. Thus

$$u_A = u_A^0\left(1 - \frac{x}{L}\right) \qquad \text{and} \qquad i = 2u_A^0\overline{w}h$$

Thus we write

$$w_i = -\frac{\beta i t_A c_0}{2\rho_f h(1 - \phi)}\frac{d}{dx}\left[\left(1 - \frac{x}{L}\right)\frac{\overline{c}}{c_0}\right] \tag{10.32}$$

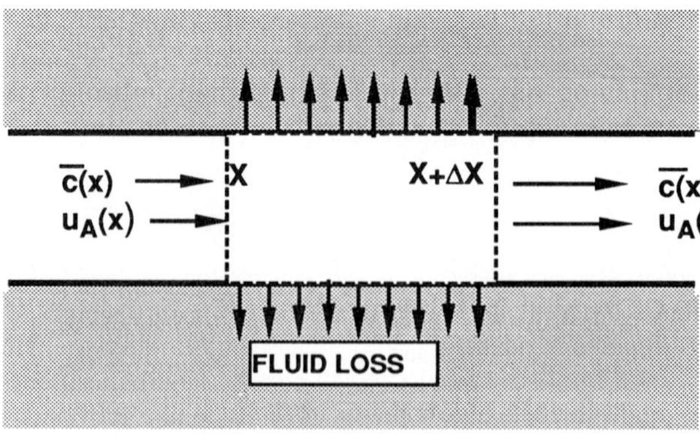

Figure 10.9 Sketch showing volume element of fracture.

Note that it_A/h is the total acid volume injected per unit of fracture height and that w_i depends on x/L.

Substituting Eq. (10.30) into Eq. (10.32) and differentiating, we find

$$w_i = \frac{\beta c_0 i t_A}{2\rho_f(1 - \phi)hL} \sum_{n=0}^{\infty} G_n \left(1 + \frac{2\lambda_n^2}{3N_{Pe}}\right)\left(1 - \frac{x}{L}\right)^{2\lambda_n^2/3N_{Pe}} \tag{10.33}$$

The ideal fracture width depends strongly on position as seen in Fig. 10.10. The quantity $\langle w_i \rangle$ shown on this graph represents the ideal fracture width that would be obtained if the acid reacted uniformly along the entire length of the fracture. This average fracture width is thus given by

$$\langle w_i \rangle = \frac{\beta c_0 i t_A}{2\rho_f(1 - \phi)hL} = \frac{XV}{2(1 - \phi)hL} \tag{10.34}$$

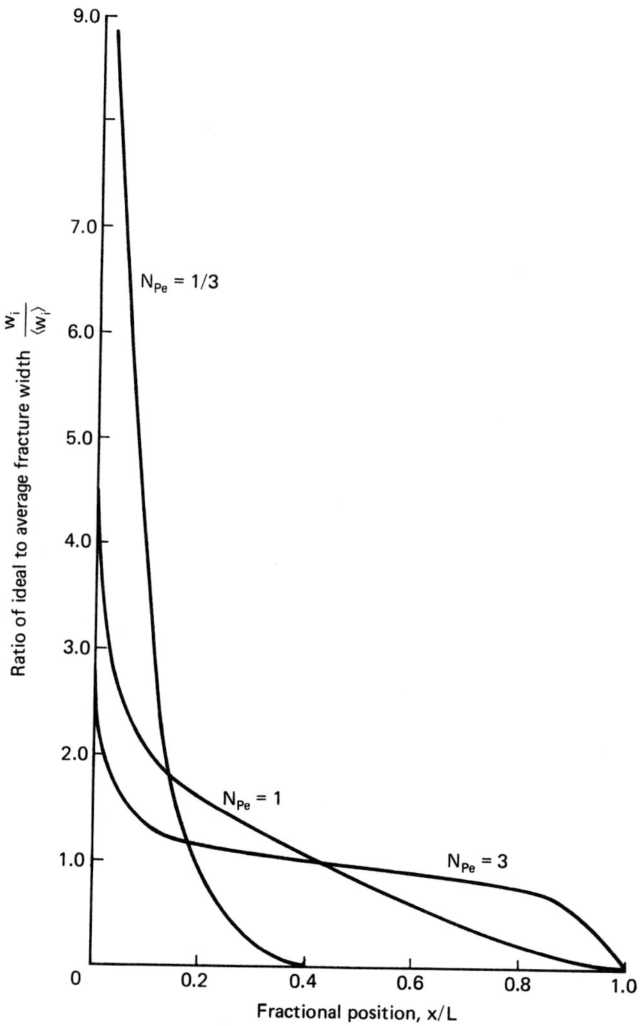

Figure 10.10 Graph of ideal fracture width as a function of position.

where X is the acid dissolving power defined in Sec. 3.6 and V is the acid volume used in the treatment. The area under any one of the curves shown in Fig. 10.10 must equal unity, since a given volume acid will dissolve a fixed volume of rock in accordance with its dissolving power.

Despite the fact that the same volume of rock is dissolved in each of the three cases shown in Fig. 10.10, the position that it is dissolved from depends strongly on the Peclet number. Thus for small Peclet numbers, most of the acid is spent in the first 20% of the dynamic fracture length. The remaining part of the fracture would not be contacted by live acid and would not be expected to be conductive following the acid fracture treatment.

For larger Peclet numbers the acid produces a more uniform ideal width along the entire length of the fracture. It should be noted that in some cases, as will be evident in Chapter 11, it is desirable to restrict the acid penetration distance so as to create conductive fractures near the wellbore.

The design of acid fractures is considered in some detail in Sec. 12.2. However, it must be noted at this point that Fig. 10.10 may be somewhat misleading in that a small fractional penetration distance may in fact yield a longer fracture than a case for which live acid penetrates practically to the end of the fracture. This is because the penetration distance is a fraction of the fracture length. Clearly, a small fraction of a very long fracture may represent a greater distance than a large fraction of a short fracture.

The curves shown in Fig. 10.10 have been calculated assuming N_{Re^*} is a small number. In general, one would calculate $w_i(x)$ using Eq. (10.33) together with values of $\underset{n}{\lambda}$ and G_n found from Eqs. (10.27) and (10.28). If N_{Re^*} is large, results substantially different from those shown in Fig. 10.10 may arise.

Example 10.6 Calculation of the Ideal Fracture Width

If the fluid-loss flux can be considered to be constant and equal to 3×10^{-5} m/sec and the acid (15wt% HCl) injection rate is 2×10^{-3} m^3/sec for each meter of formation thickness, calculate the ideal fracture width as a function of position after acid has been injected for 600 sec. Assume the flow in the fracture to be laminar and the reaction rate to be infinite. The effective diffusivity of the acid can be taken to be 1.25×10^{-8} m^2/sec. The dynamic fracture width is 2.5×10^{-3} m. The formation porosity is 0.1. The acid density and viscosity are found in Table 10.3.

Solution The N_{Pe} and N_{Re^*} are calculated as follows:

$$N_{Re^*} = \frac{\langle \bar{u}_N \rangle \rho \bar{w}}{2\mu} = \frac{(3 \times 10^{-5})(1.07 \times 10^3)(2.5 \times 10^{-3})}{2(1.15 \times 10^{-3})} = 0.035$$

$$N_{Pe} = \frac{\langle \bar{u}_N \rangle \bar{w}}{2D_A} = \frac{3 \times 10^{-5}(2.5 \times 10^{-3})}{2(1.25 \times 10^{-8})} = 3$$

Using Eqs. (10.27) and (10.28), calculate $\underset{n}{\lambda}$ and G_n. The results are shown in Table 10.4. The values of $w_i/\langle w_i \rangle$ are calculated using Eq. (10.33). The results as a function of x/L are given in Table 10.5.

The fracture length is calculated as follows:

$$L = \frac{i}{4h\langle \bar{u}_N \rangle} = 17 \text{ m}$$

TABLE 10.4 Parameters for Evaluation of w_i

n	$^{*}\lambda_n$	$^{**}G_n$	$2\lambda_n^2/3N_{Pe}$	$G_n(2\lambda_n^2/3N_{Pe} + 1)$
0	1.058	0.9207	0.249	1.15
1	5.609	0.04461	6.991	0.36
2	9.645	0.01408	20.672	0.31
3	13.66	0.00648	41.47	0.28
4	17.66	0.00362	69.31	0.25

* Calculated using Eq. (10.27).
** Calculated using Eq. (10.28).

TABLE 10.5 The Ideal Fracture Width for Example 10.6

x/L	$w_i/\langle w_i \rangle$	$w_i \times 10^3$ (m)	x/L	$w_i/\langle w_i \rangle$	$w_i \times 10^3$ (m)
0.05	1.53	4.9	0.5	0.97	3.1
0.1	1.33	4.3	0.6	0.92	3.0
0.2	1.17	3.8	0.7	0.85	2.7
0.3	1.09	3.5	0.8	0.77	2.5
0.4	1.02	3.3	0.9	0.65	2.1

Note $X = 0.082$ (from Sec. 3.5) and $V/h = (600)(1 \times 10^{-3}) = 1.2\ m^3/m$. Thus from Eq. (10.34)

$$\langle w_i \rangle = \frac{(0.082)(1.2)}{2(1 - 0.1)(17)} = 3.22 \times 10^{-3}\ m$$

Finally, w_i is shown in Table 10.5.

Acidized Fracture Conductivity

To be effective, the acid must react with the walls of the fracture to form a channel that will stay open after the fracturing treatment. Flow channels can be formed as a result of uneven reaction with the rock surface or because of the reaction of the live acid that flows from the fracture surface into the formation. The reaction of this acid with the formation will increase the permeability along the fracture boundary. It is not possible to separate the contribution of these mechanisms, and an empirical approach is required. The method to be considered has been developed by Nierode and Kruk [27]. Data that were correlated represented conductivities measured under realistic conditions, which included fluid loss. The experiments are complex and have not been reproduced. The correlation is based on fewer than 40 experiments, so that the potential for error is significant. Most data have been obtained using techniques similar to those reported by Knox et al. [28] and Broaddus and Knox [29]. An apparatus such as that shown by Fig.

10.4 is used. Fluid loss is not permitted and in the author's opinion, the fracture conductivity results are therefore not as realistic as those reported by Nierode and Kruk [30].

The Nierode–Kruk correlation begins by defining an ideal permeability as follows:

$$k_{fi} = \frac{w_i^2}{12} \tag{10.35}$$

This is the permeability of two parallel plates spaced a distance w_i apart. Thus, the ideal fracture conductivity is

$$(wk_f)_i = \frac{w_i^3}{12} \tag{10.36}$$

This ideal fracture conductivity is generally considerably larger than that observed in laboratory experiments, primarily because of the closure stress. Nierode and Kruk have proposed the following empirical equation, which in effect corrects the ideal fracture conductivity for the closure stress σ_c and the rock embedment strength S_{RES}.

$$wk_f = c_1 \exp(-c_2\sigma_c) \tag{10.37}$$

Here wk_f is in m^3, where

$$c_1 = 2.94 \times 10^{-4}\,(wk_f)_i^{0.822}$$

and $c_2 = [36.82 - 1.885 \ln S_{RES}] \times 10^{-7}$ for $0 < S_{RES} < 1.38 \times 10^8$ Pa

$c_2 = [9.1 - 0.406 \ln S_{RES}] \times 10^{-7}$ for $1.38 \times 10^8 < S_{RES} < 3.45 \times 10^9$ Pa

Here S_{RES} *must* be expressed in pascals (Pa). Equation (10.37) is used to calculate the actual fracture conductivity (in m^3) as a function of the ideal fracture conductivity (in m^3), the closure stress, and the rock embedment strength.

Example 10.7 Acid Fracture Conductivity

Suppose the rock embedment strength is 3.1×10^8 Pa, the fracture gradient is 18 kPa/m, and the drawdown pressure is 6.8×10^6 Pa. Calculate the fracture conductivity at $x/L = 0.05$ for the case considered in Example 10.6. The reservoir is 3000 m deep.

Solution The ideal fracture conductivity is calculated using Eq. (10.33) for $w_i = 4.9 \times 10^{-3}$ m (see first entry in Table 10.5).

$$(wk_f)_i = \frac{1}{12}(4.9 \times 10^{-3})^3 = 9.80 \times 10^{-9}\ m^3$$

The constant c_1 is given by

$$c_1 = 2.94 \times 10^{-4}\,(9.8 \times 10^{-9})^{0.822} = 7.68 \times 10^{-11}\ m^3$$

Also since $S_{Re} > 1.37 \times 10^8$ Pa, then

$$c_2 = [9.1 - 0.406 \ln (3.1 \times 10^8)] \times 10^{-7} = 1.162 \times 10^{-7}$$

The actual fracture conductivity is calculated as follows [Eq. (10.37)].

$$(wk_f) = 7.68 \times 10^{-11} \exp - \{1.162 \times 10^{-7} [(18000)(3000) - 6.8 \times 10^6]\}$$

Finally

$$wk_f = 3.19 \times 10^{-13} \text{ m}^3$$

This can also be expressed in terms of mD-in:

$$wk_f = \frac{(3.19 \times 10^{-13} \text{ m}^3)(12 \text{ in/ft})}{(0.3048 \text{ m/ft})(9.869 \times 10^{-16} \text{ m}^2/\text{mD})} = 1.27 \times 10^4 \text{ mD-in}$$

REFERENCES

10. 1. Howard, G. C., and Fast, C. R., *Hydraulic Fracturing*, Mono. Ser., *2*, Society of Petroleum Engineers, Richardson, Texas, 1970.

10.2. Kern, L. R., and Perkins, T. K., *Trans. AIMME, 216* (1959) 403.

10.3. Cooke, C. E., *J. Pet. Tech., 29* (1977) 1222.

10.4. Neel, N. K., Parmley, A. S., and Colpoys, J. R., "Oxide Ceramic Proppants for Treatment of Deep Well Fractures," SPE 6816, presented at the 52nd Fall Technical Conference and Exhibition of the Society of Petroleum Engineers, Denver, Colorado, 1977.

10.5. Sarda, W. P., "Use of High Strength Ceramic Beads for Propping Deed Hydraulic Fractures," SPE 7572, presented at the 53rd Fall Technical Conference and Exhibition of the Society of Petroleum Engineers, Houston, Texas, 1978.

10.6. Bird, R. B., Stewart, W. E., and Lightfoot, E. N., *Transport Phenomena*, New York: John Wiley, 1960.

10.7. B. J. Hughes Engineering Bulletin, (1979).

10.8. Novotny, E. J., "Proppant Transport," SPE 6813, presented at the 52nd Fall Technical Conference and Exhibition of the Society of Petroleum Engineers, Denver, Colorado, 1977.

10.9. Slattery, J. C., and Bird, R. B., *Chem. Eng. Sci., 16* (1961) 231.

10.10. Thiele, M. R., personal communication, Sept. (1987).

10.11. Kirby, L. L., and Rockefeller, H. A., "Proppant Settling Velocities in Non-flowing Slurries," SPE 13906, presented at the Society of Petroleum Engineers Symposium on Low Permeability Gas Reservoirs, Denver, Colorado, 1985.

10.12. Acrivos, A., and Herbolzheimer, E., *J. Fluid Mech., 92* (1979) 435.

10.13. Herbolzheimer, E., and Acrivos, A., *J. Fluid Mech., 108* (1981) 485.

10.14. Roodhart, L. P., "Proppant Settling in Non-Newtonian Fracturing Fluids," SPE 13905, presented at the society of Petroleum Engineers Symposium on Low Permeability Gas Reservoirs, Denver, Colorado, 1985.

10.15. Clark, P. E., and Güler, N., "Proppant Transport in Vertical Fractures: Settling Velocity," SPE 11636, presented at the Society of Petroleum Engineers Symposium on Low Permeability Gas Reservoirs, Denver, Colorado, 1983.

10.16. Gauthier, F., Goldsmith, H. L., and Mason, S. G., *Trans. Soc. Rheology, 15* (1971) 297.

10.17. Blauer, R. E., and Kohlhaas, C. A., "Formation Fracturing with Foam," SPE 5003, presented at the 1974 Society of Petroleum Engineers Annual Meeting, Houston, Texas, 1974.

10.18. Williams, B. B., Gidley, J. L., and Schechter, R. S., *Acidizing Fundamentals*, Mono. Ser., *6*, Society of Petroleum Engineers, Richardson, Texas, 1979.

10.19. Berman, A. S., *J. Appl. Phys., 24* (1953) 1232.
10.20. Terrill, R. M., *Aeronautical Quarterly, 15* (1964) 299.
10.21. Terrill, R. M., *Internat. J. Heat and Mass Transfer, 8* (1965) 1491.
10.22. Roberts, L. D., and Guin, J. A., *Soc. Pet. Eng. J., 14* (1974) 385.
10.23. Nierode, D. E., and Williams, B. B., *Soc. Pet. Eng. J., 11* (1971) 408.
10.24. Williams, B. B., and Nierode, D. E., *J. Pet. Tech., 24* (1972) 849.
10.25. Roberts, L. D., and Guin, J. A., *Soc. Pet. Eng. J., 15* (1975) 277.
10.26. Chang, C. V., and Guin, J. A., *Amer. Ins. Chem. Eng. J., 22* (1976) 252.
10.27. Nierode, D. E., and Kruk, K. F., "An Evaluation of Acid Fluid Loss Additives, Retarded Acids and Acidized Fracture Conductivities," SPE 4549, presented at the 1973 Society of Petroleum Engineers Annual Meeting, 1973.
10.28. Knox, J. A., Pollock, R. W., and Beechcroft, W. H., *J. Canad. Pet. Tech., 4* (1965) 5.
10.29. Broaddus, G. C., and Knox, J. A., "Influence of Acid Type and Quantity in Limestone Etching," API 851-39-1, presented at the American Petroleum Institute Mid-Continent Meeting, Wichita, Kansas, 1965.
10.30. Flach, T. A., "Creating and Measuring Acid Fracture Conductivity in the Laboratory, MS thesis, The University of Texas at Austin, 1987.

PROBLEMS

***10.1.** A fracturing fluid ($\rho = 0.95$ g/cm^3) exhibits the stress/strain rate response given in Table P10.1. These measurements are at reservoir temperature and pressure. How long will it take a proppant particle ($d_p = 0.035$ cm, $\rho_p = 2.6$ g/cm^3) to settle 20 m in a fracture containing this fluid?

TABLE P10.1 Stress/Strain Rate Response Data

Stress (Pa)	Shear Rate (sec^{-1})
0.05	1
0.18	5
0.32	10
0.55	20
0.96	40

***10.2.** A fracture fluid (density 1140 kg/m^3) is known to behave as a power-law fluid and has properties such that 3% of a 20/40 U.S. Mesh sand (density 2540 kg/m^3) will settle out near the fracture entrance until the fracture closes following a treatment, and similarly, it is estimated that 18.9% of a 12/20 U.S. Mesh sand would settle if used in the same treatment. Based on these data, what percentage of a 20/40 U.S.

Mesh bauxite (density 3980 kg/m^3) would you expect to settle out during the same treatment?

***10.3.** If a fracture fluid containing 500 kg of proppant (10/20 U.S. Mesh sand) per m^3 of fracture fluid is injected into a fracture having an average width of 2×10^{-3} m, estimate the final fracture conductivity. The final porosity of the proppant packed in the fracture is 0.35. Express your answer in D-cm^2. Neglect proppant settling and assume the proppant to be uniformly distributed.

***10.4.** A fracture treatment required 90 min to complete. The average dynamic fracture width at the end of the treatment is estimated to be 1 cm and if the proppant concentration is 1000 g per liter of fracture fluid (*not* per liter of total solution volume), estimate the final fracture conductivity assuming the final porosity of the proppant in the fracture is 0.42. The average proppant diameter is 0.05 cm and the proppant density is 3 g/cm^3. Neglect proppant settling.

Also estimate the time required for the fracture to close once pumping has ceased and the well is shut in. The overall fluid-loss coefficient is 0.02 cm/min$^{1/2}$.

***10.5.** Estimate the surface concentrations of 20/40-mesh particles of both sand and bauxite which constitute a proppant monolayer. Take the proppant particles to be spheres of uniform size (the mean radius), and assume that they are close packed. Mark these monolayer concentrations on both Fig. 10.4 and Fig. 10.5 for future reference.

***10.6.** A fracture treatment is to be carried out in a calcite ($CaCO_3$) formation using 15 wt% hydrochloric acid. The relevant treatment data are given in Table P10.6. To what distance, measured from the wellbore, will acid penetrate along the fracture before the concentration has decreased to 1.5 wt% (assume density remains constant as acid reacts)? How can you increase this penetration distance?

TABLE P10.6 Treatment Data

Property	Symbol	Value
Fluid density	ρ	1050 kg/m^3
Density of formation rock	ρ_f	2700 kg/m^3
Acid viscosity	μ	50 cp
Fracture gradient	FG	17 kPa/m
Injection rate	i	6×10^{-2} m^3/sec
Fracture height	h	10 m
Average fluid-loss flux	$\langle \overline{u}_n \rangle$	5×10^{-5} m/sec
Effective diffusivity	D_A	10^{-8} m^2/sec
Formation porosity	ϕ	0.1
Formation permeability	k	1×10^{-15} m^2
Average dynamic fracture width	$\overline{w}$	4×10^{-3} m

***10.7.** A limestone formation is to be fractured using a total of 12 m^3 of 15 wt% HCl. Make a graph showing the final fracture conductivity as a function of position measured from the wellbore. Consider two cases. In the first case, the rock embedment strength is 7×10^5 kPa. In the second case, it is 2×10^5 kPa. The data in Table P10.7 apply.

TABLE P10.7 Treatment Data

Property	Symbol	Value
Average fluid-loss flux	$\langle \overline{u}_N \rangle$	1×10^{-6} m/sec
Acid injection rate	i	1×10^{-3} m^3/sec
Average dynamic fracture width	$\overline{w}$	2×10^{-3} m
Acid diffusion coefficient	D_A	10^{-8} m^2/sec
Fracture height	h	10 m
Acid viscosity	μ	3×10^{-2} kg/m-sec
Porosity	ϕ	0.10
Fracture gradient	FG	17 kPa/m
Formation depth	D	3×10^3 m
Bottomhole well flowing pressure	p_{wf}	3.5×10^3 kPa

11
Well Productivity of Fractured Systems

Final fracture dimensions and the ultimate fracture conductivity determine the hydrocarbon production rate resulting from treatment. This topic is the subject of the present chapter. Useful methods for well testing and fracture treatment design are described. The chapter is concerned with vertical fractures because of their pervasiveness.

11.1 FLUSH PRODUCTION

When a fractured well is first put on production, the flow is essentially linear as illustrated in Fig. 11.1. After an initial period of linear flow, pressure profiles having a somewhat radial character will begin to develop and fluid from the recesses of the drainage area will begin to flow.

Finally, the flow will tend to be more or less radial and the final steady-state production phase will have been reached. Steady-state production will be achieved in a few days in some cases where the reservoir is permeable and the fluid compressibility is small. Oil wells will most likely reach steady-state production in a relatively short time (less than 1 or 2 months).

On the other hand, in tight formations containing highly compressible hydrocarbons (gas), it may take years before steady-state production is achieved.

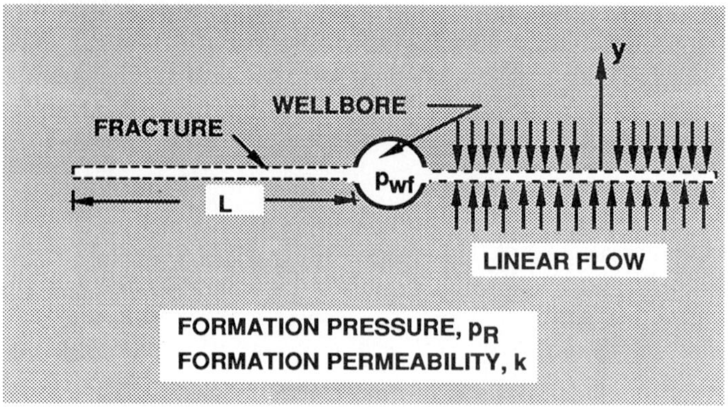

Figure 11.1 Sketch depicting initial linear flow into a fracture.

Economic considerations will, in this case, be determined by the unsteady-state or flush production.

Before discussing the stimulation ratio for the steady-state and semisteady-state period when it is effectively constant, it is prudent to determine the time that must elapse before the period of initial flush production is over.

The semisteady-state period arises because the reservoir or drainage area is finite. Ramey and Cobb [1] have considered a well situated in the center of a regular-shaped drainage area—for instance a circle, square or hexagon—and concluded that, approximately, the steady-state or semisteady-state period will be observed after the dimensionless time (t_{DA}) reaches a value of approximately 0.1. Thus, the period of flush production is given by

$$t_{DA} = \frac{kt}{A\phi\mu_f\kappa_f} \approx 0.1 \tag{11.1}$$

where A is the drainage area.

Morse and Von Gonten [2] found for fractured systems $t_{DA} \approx 0.2$ was required to attain stabilized flow in a square drainage area. We shall use this value since our interest is in fractured systems.

Example 11.1 Flush Production Period

A fractured gas well ($\mu_f = 0.01$ cp and $\kappa_f = 1.5 \times 10^{-4}$ kPa^{-1}) in a region having a drainage area of 4×10^4 m^2, a porosity of 0.15, and a permeability of 0.987×10^{-16} m^2 (0.1 mD) is put on production. How long will be required before the period of steady-state production is reached with $t_{DA} \approx 0.2$?

Solution

$$t = 0.2\left(\frac{A\phi\mu_f\kappa_f}{k}\right) = 0.2\left[\frac{(4 \times 10^4)(0.15)(1 \times 10^{-5})(1.5 \times 10^{-7})}{0.987 \times 10^{-16}}\right]$$

$$= 1.8 \times 10^7 \text{ sec} = 211 \text{ days}$$

Thus, 211 days of production will be required to reach the stabilized period. During this time, production will be larger than the steady-state or semisteady-state values.

If this production persists for a very long time, as would be the case for Example 11.1 if the reservoir permeability is reduced by a factor of 10, then economic considerations must take into account the amount produced during the initial transient period.

11.2 SEMISTEADY-STATE OR STEADY-STATE PRODUCTION

After the initial period of flush production, a period of semisteady-state or steady-state production, depending on the boundary conditions imposed at the drainage radius, will ensue. These two flows are described in detail by Dake [3]. Semi-steady-state production corresponds to an isolated well, whereas steady-state production results from a constant pressure imposed at the drainage radius.

To discuss the effectiveness of a well stimulation treatment, it is most convenient to consider ratios of a productivity index defined as follows:

$$J = \frac{q}{p_{\mathrm{R}} - p_{\mathrm{wf}}} \quad \text{(oil wells)} \qquad (11.2)$$

or

$$J = \frac{q}{p_{\mathrm{R}}^2 - p_{\mathrm{wf}}^2} \quad \text{(gas wells)} \qquad (11.3)$$

These production indices vary with time for semisteady-state production; however, ratios taken of them before and after treatment are nearly constant. Let J_{s} be the productivity index following stimulation and J_0 be the productivity index of an undamaged and unstimulated well. If the well is damaged prior to stimulation, the appropriate ratio is $J_{\mathrm{s}}/J_{\mathrm{d}}$.

In Chapter 6 we saw that if the well is damaged prior to stimulation, then

$$\frac{J_{\mathrm{d}}}{J_0} = \frac{\ln\left(\frac{r_{\mathrm{e}}}{r_{\mathrm{w}}}\right) - \frac{3}{4}}{\ln\left(\frac{r_{\mathrm{e}}}{r_{\mathrm{w}}}\right) - \frac{3}{4} + S} \quad \text{(semisteady-state production)} \qquad (11.4)$$

where S is the skin factor.

Stimulation Ratios for Undamaged Wells

McGuire and Sikora [4] have developed a graphical method for determining J_{s}/J_0 for rectangular vertical fractures having a uniform fracture conductivity along their entire length. Figure 11.2 represents their results. These curves are not restricted to a given drainage area. A correction that depends on the drainage area (A) must be applied to both the abscissa and the ordinate.

Example 11.2 Application of McGuire–Sikora Chart

An undamaged well completed in a formation having a permeability of 5×10^{-15} m^2 (5.07 mD) is to be stimulated using hydraulic fracturing. The proppant has been selected so that the fracture conductivity will be 2.5×10^{-12} m^3 (1×10^5 mD-in).

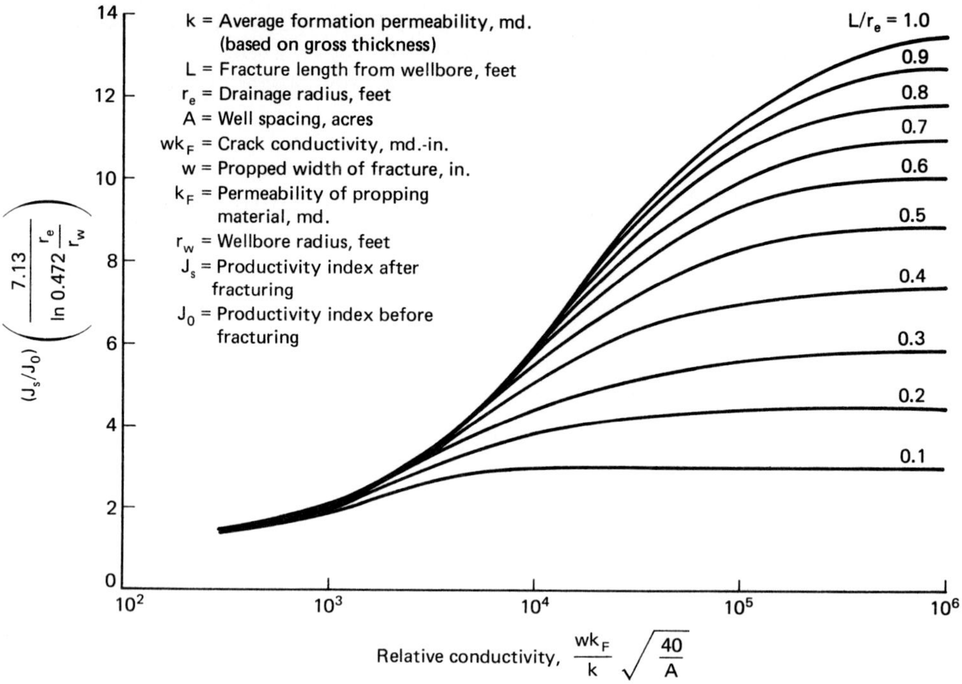

Figure 11.2 Graph showing increase in productivity from fracturing [4]. (With permission from the Society of Petroleum Engineers.)

What length propped fracture would you recommend? What will be the stimulation ratio? How can this stimulation ratio be improved?

The drainage area may be taken to be 40 acres and the wellbore radius to be 0.25 ft.

Solution The relative conductivity must be expressed in inches to use Fig. 11.2. Thus

$$\frac{\overline{w}k_f}{k}\sqrt{\frac{40}{A}} = \frac{1 \times 10^5 \text{ mD-in}}{5.07 \text{ mD}} = 2 \times 10^4 \text{ in}$$

From Fig. 11.2 we observe the following data for this relative conductivity:

$\frac{L}{r_e}$	$\frac{J_s}{J_0}$
0.1	3.0
0.2	4.1
0.3	4.9
0.4	6.0
0.5	6.6
0.6	7.3
0.7	7.6
0.8	7.8

The selection of a fracture length is an economic issue; however, because the stimulation ratio reaches a plateau, it would seem that values of L/r_e greater than 0.6 would not be cost effective. Increasing the fracture conductivity will permit longer fractures to be useful.

The previous example makes an important point. If the relative conductivity is less than 10^{+4} ins, there is no need to create fractures having a length greater than one half of the drainage radius ($L/r_e < 0.5$). For smaller values of the relative conductivity, even shorter fractures will yield the plateau stimulation ratio. When this situation exists, the stimulation is fracture conductivity limited; the way to increase the stimulation ratio is to increase the conductivity of the fracture.

For short fractures, such as $L/r_e = 0.2$, Fig. 11.2 shows that the stimulation ratio reaches a plateau independent of the conductivity ratio. On this plateau, the stimulation ratio is length limited; longer fractures are required to increase productivity.

Damaged Wells

Figure 11.2 gives the stimulation ratio for a fractured well that was initially undamaged. In practice, much larger stimulation ratios are observed since damaged wells are often treated. In this case, the stimulation ratio is given by

$$\frac{J_s}{J_d} = \left(\frac{J_s}{J_0}\right)\left(\frac{J_0}{J_d}\right) \tag{11.5}$$

where J_d/J_0 can be calculated using Eq. (11.4) if the skin factor prior to treatment is known.

Turbulent Flow (Gas Wells)

The McGuire–Sikora chart may not give the correct stimulation ratio for gas wells. The chart is developed assuming that the flow into the wellbore of the unstimulated well follows Darcy's law and that the flow of fluid through the fracture into the wellbore is laminar. Both of these assumptions may be incorrect when applied to gas wells [5]. Although corrections for turbulence may become significant, they are not considered in this text. These corrections are generally best handled using a reservoir simulator so that the nonlinear, turbulent flow behavior can be accurately modeled. The same general behavior shown by the McGuire–Sikora chart will apply to gas wells, although if the reservoir pressure is low, significant deviations in the actual stimulation value will be observed.

Stimulation Ratios for Variable Fracture Conductivities

In this section we consider an alternative procedure that will permit the calculation of stimulation ratios when the fracture is nonuniform; that is, when the fracture conductivity varies along the fracture length. As we have seen, acidized fractures (see Chap. 10) exhibit considerable variability in fracture conductivity because of the uneven degree of acid attack along the fracture.

Propped fractures can also have variable fracture conductivities when the proppant concentration and/or the proppant diameter are varied during the course of the treatment. As will be described in Chapter 12, variation of these two factors is common practice.

The development starts by assuming that the flow is radial but that the radial permeability is a function of position. Thus, the radial flux given by Darcy's law is

$$u_r = -\frac{k(r)}{\mu}\frac{\partial p}{\partial r} \tag{11.6}$$

Multiplying both sides by $2\pi rh$, we find

$$q = \frac{2\pi rhk(r)}{\mu}\frac{\partial p}{\partial r} \tag{11.7}$$

where q is production rate ($q = -2\pi rhu_r$). Assuming a constant pressure p_e at the drainage radius r_e, Eq. (11.7) can be integrated to give

$$J = \frac{q}{p_e - p_{wf}} = \frac{2\pi h}{\mu \displaystyle\int_{r_w}^{r_e} \frac{dr}{rk(r)}} \tag{11.8}$$

Now, following Raymond and Binder [6], we consider J_d as the productivity index of a well damaged as shown in Fig. 11.3. Thus

$$J_d = \frac{2\pi kh/\mu}{\dfrac{k}{k_s}\ln\left(\dfrac{r_s}{r_w}\right) + \ln\left(\dfrac{r_e}{r_s}\right)} \tag{11.9}$$

The stimulated productivity index is calculated for the geometry shown in Fig. 11.4. Shown is a top view. The average permeability for any position $r > r_s$ is given as

$$k(r) = k + \frac{w_i}{\pi r}(k_{fi} - k) \tag{11.10}$$

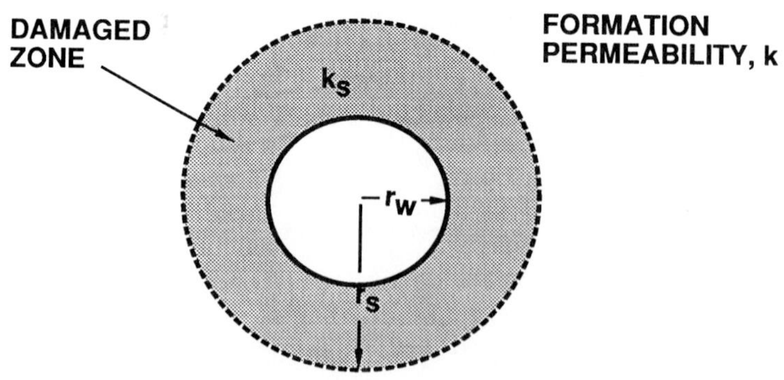

Figure 11.3 Sketch depicting the geometry of a damaged well.

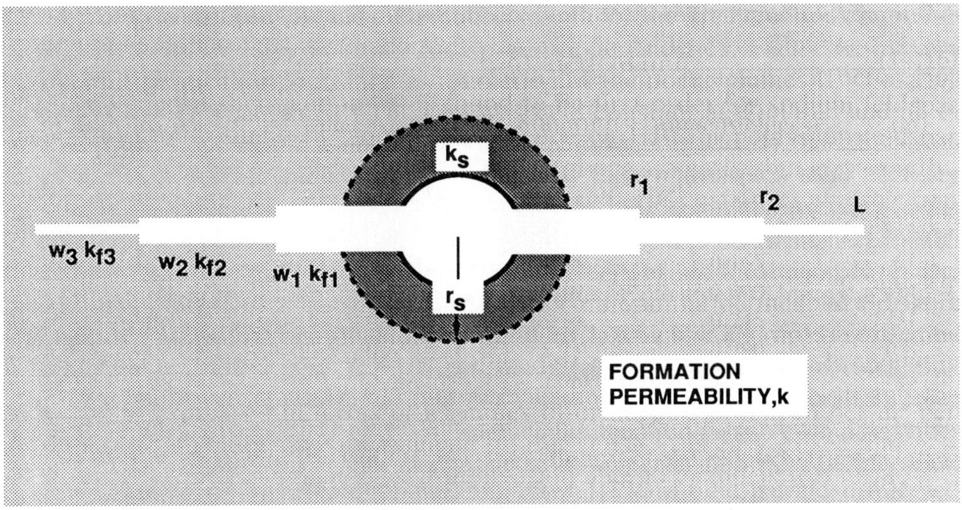

Figure 11.4 Sketch of top view of vertically fractured well with variable fracture conductivity.

This equation is derived by noting that over an arc of length $2\pi r - 2w_i$ the permeability is the formation permeability and over the very small length $2w_i$ the permeability is the fracture permeability k_{fi}. The index i refers to that section of the fracture position at the radial distance r. If $r < r_s$ (inside the damaged zone), then

$$k(r) = k_s + \frac{w_i}{\pi r}(k_{fi} - k_s) \tag{11.11}$$

Here we assume that the first segment of the fracture penetrates the damaged zone as shown by Fig. 11.4. This is the only case we will consider here; however, Raymond and Binder have derived the appropriate equations for other cases.

Substituting Eqs. (11.10) and (11.11) into Eq. (11.7), and then forming the stimuation ratio, one finds

$$\frac{J_s}{J_d} = \frac{\dfrac{k}{k_s}\ln\left(\dfrac{r_s}{r_w}\right) + \ln\left(\dfrac{r_e}{r_s}\right)}{\dfrac{k}{k_s}\ln\left[\dfrac{1 + \dfrac{w_1}{\pi r_s}\left(\dfrac{k_{f1}}{k_s} - 1\right)}{\dfrac{r_w}{r_s} + \dfrac{w_1}{\pi r_s}\left(\dfrac{k_{f1}}{k_s} - 1\right)}\right] + \ln\left(\dfrac{1 + CR_1}{\dfrac{r_s}{r_1} + CR_1}\right) + \sum_{i=2}\ln\left(\dfrac{1 + CR_i}{\dfrac{r_{i-1}}{r_i} + CR_i}\right) + \ln\left(\dfrac{r_e}{L}\right)} \tag{11.12}$$

where the fracture conductivity ratio is defined as

$$CR_i = \frac{w_i}{\pi r_i}\left(\frac{k_{fi}}{k} - 1\right)$$

It should be understood that Eq. (11.12) is an approximation; however, for an initially undamaged well vertically fractured with a uniform conductivity fracture, it predicts stimulation ratios that compare very well with the curves shown in Fig. 11.2 for $L/r_e < 0.5$. For longer fractures, the assumption of radial flow even during the steady-state period is never valid and Eq. (11.12) is in error. Its use should, therefore, be restricted to short fractures.

The stimulation ratio also applies to gases where J is defined by Eq. (11.3), provided the flow in both the damaged and stimulated wells is everywhere laminar.

Equation (11.12) can be used to draw some interesting and important conclusions. Example 11.3 illustrates the use of small fracture treatments to obtain large stimulation ratios when the well is damaged.

Example 11.3 Stimulation of a Damaged Well

A well ($r_w = 0.05$ m) completed in a formation having a permeability of 4.93×10^{-14} m^2 and a drainage radius of 215 m is stimulated by hydraulic fracturing. If the well has been damaged by the drilling process such that the damaged zone encompasses a radius of 1 m about the wellbore ($r_s = 1$ m, see Fig. 11.3) and has a permeability of 4.93×10^{-15} m^2, what stimulation ratio would you expect from the following fracture geometries?

(a) $L = 2$ m; $k_f = 1.97 \times 10^{-12}$ m^2; $w = 5 \times 10^{-3}$ m

(b) $L = 2$ m; $k_f = 1.97 \times 10^{-11}$ m^2; $w = 5 \times 10^{-3}$ m

(c) $L = 10$ m; $k_f = 1.97 \times 10^{-12}$ m^2; $w = 5 \times 10^{-3}$ m

Solution For this example, Eq. (11.12) reduces to

$$\frac{J_s}{J_d} = \frac{\dfrac{k}{k_s}\ln\left(\dfrac{r_s}{r_w}\right) + \ln\left(\dfrac{r_e}{r_s}\right)}{\dfrac{k}{k_s}\ln\left[\dfrac{1 + \dfrac{w}{\pi r_s}\left(\dfrac{k_f}{k_s} - 1\right)}{\dfrac{r_w}{r_s} + \dfrac{w}{\pi r_s}\left(\dfrac{k_f}{k_s} - 1\right)}\right] + \ln\left[\dfrac{1 + \dfrac{w}{\pi L}\left(\dfrac{k_f}{k} - 1\right)}{\dfrac{r_s}{L} + \dfrac{w}{\pi L}\left(\dfrac{k_f}{k} - 1\right)}\right] + \ln\left(\dfrac{r_e}{L}\right)}$$

For all of the cases to be considered

$$\frac{J_s}{J_d} = \frac{10\ln\left(\dfrac{1}{0.05}\right) + \ln(215)}{10\ln\left[\dfrac{1 + \dfrac{5 \times 10^{-3}}{\pi}\left(\dfrac{k_f}{k_s} - 1\right)}{0.05 + \dfrac{5 \times 10^{-3}}{\pi}\left(\dfrac{k_f}{k_s} - 1\right)}\right] + \ln\left[\dfrac{1 + \dfrac{5 \times 10^{-3}}{\pi L}\left(\dfrac{k_f}{k} - 1\right)}{\dfrac{1}{L} + \dfrac{5 \times 10^{-3}}{\pi L}\left(\dfrac{k_f}{k} - 1\right)}\right] + \ln\left(\dfrac{215}{L}\right)}$$

For Case (a): $J_s/J_d = 2.6$

For Case (b): $J_s/J_d = 5.6$

For Case (c): $J_s/J_d = 3.1$

We see that if a fracture penetrates through a damaged zone, then significant stimulation ratios are possible even if the fracture is short and the formation permeable.

Equation (11.12) can also be used to demonstrate the importance of maintaining proppant in the fracture near the wellbore. This is illustrated by Example 11.4.

Example 11.4 Unpropped Fracture Near the Wellbore

What will be the resultant reduction in the productivity index caused by failing to prop the first 1% of the fracture length? This means that the portion of the fracture near the wellbore is unfilled and will close following the treatment, restoring the initial formation permeability. Take $r_e/L = 2$ and $r_e/r_w = 5000$. Also assume that the fracture conductivity ratio is large; that is, $CR \rightarrow \infty$.

Solution The well is originally undamaged and if the entire fracture from the wellbore to the tip is packed, then Eq. (11.12) reduces to

$$\frac{J_s}{J_0} = \frac{\ln\left(\frac{r_e}{r_w}\right)}{\ln\left(\frac{r_e}{L}\right)} \tag{11.13}$$

since $CR \rightarrow \infty$. Call this stimulation ratio Case (a).

Next consider the case for which the radius r_1 around the wellbore has $CR = 0$ (no fracture conductivity). Then

$$\frac{J_s}{J_0} = \frac{\ln\left(\frac{r_e}{r_w}\right)}{\ln\left(\frac{r_1}{r_w}\right) + \ln\left(\frac{r_e}{L}\right)} \tag{11.14}$$

call this stimulation ratio Case (b) and consider the ratio

$$\frac{(J_s/J_0)_b}{(J_s/J_0)_a} = \frac{(J_s)_b}{(J_s)_a} \quad \text{when } r_1 = 0.01L$$

Thus, taking the ratio of Eq. (11.14) to Eq. (11.13) we obtain

$$\frac{(J_s)_b}{(J_s)_a} = \frac{\ln\left(\frac{r_e}{L}\right)}{\ln\left(\frac{0.01L}{r_w}\right) + \ln\left(\frac{r_e}{L}\right)} = \frac{\ln(2)}{\ln[(0.01)(5000)]} = 0.18$$

Therefore, leaving the first 1% of the fracture unpacked reduces the stimulation to 18% of its fully packed valued.

Example 11.4 shows that there is a drastic reduction in stimulation if only a very small percentage of the fracture near the wellbore is left unpropped. It is therefore important to avoid overflushing following the injection of proppant because of the danger of flushing proppant from the near wellbore region. Also, one should avoid returning the well to production too soon after completing the fracture treatment. A finite time is required for the fracture walls to close onto the proppant (see fracture closure time in Chap. 10), and initiating production before the fracture is closed can result in proppant production from the near wellbore region of the fracture.

For acidized fractures, overflushing and immediate return to production do not appear to represent a problem.

As noted, when discussing Fig. 11.2, there are cases for which the stimulation is fracture conductivity limited. It will be shown that if the conductivity ratio

$$\frac{wk_f}{\pi kL} > 10 \tag{11.15}$$

then the fracture is not conductivity limited; that is, further increases in the fracture conductivity will not lead to significant increases in the stimulation ratio. Thus Inequality (11.15) serves as a guideline to identify conductivity limited fractures. The consequences of this inequality can be illustrated using Eq. (11.12). If we consider the stimulation ratio of an undamaged well with a uniform fracture conductivity, then Eq. (11.12) reduces to the following very simple form:

$$\frac{J_s}{J_0} = \frac{\ln\left(\frac{r_e}{r_w}\right)}{\ln\left(\frac{1 + CR}{r_w/L + CR}\right) + \ln\left(\frac{r_e}{L}\right)} \tag{11.16}$$

If $J_s(\infty)$ is designated as the productivity index for a fracture with an infinite conductivity ratio and $J_s(CR)$ is defined as the productivity index of a fracture having a conductivity ratio CR, then the results given in Table 11.1 can be obtained using Eq. (11.16).

It is evident that for long fractures, a fracture conductivity of 10 yields about 80% of the stimulation possible. Chapter 12 will illustrate that the desired final fracture conductivity is an economic issue, one that must be addressed by the designer. Often, fracture conductivity ratios less than 10 may prove to be optimum. Inequality (11.15) simply serves as a guideline, indicating that fracture conductivity ratios greater than 10 will not greatly improve the stimulation ratio. Thus,

TABLE 11.1 Productivity Index Decrease in Finite Conductivity Fractures

$J_s(CR)/J_s(\infty)$

	CR				
L/r_e	1	5	10	20	100
0.3	0.635	0.863	0.927	0.961	0.992
0.5	0.500	0.792	0.879	0.934	0.986
0.6	0.424	0.737	0.843	0.913	0.981
0.7	0.340	0.662	0.789	0.880	0.973

$CR = 10$ may be considered to be a target value if economic limitations are not important.

Equation (11.12) requires that both w_i and k_{fi} be known separately; but we have observed in Chapter 10 that most often they are reported as a product, the fracture conductivity, and are often not measured separately. It should be noted that very little error may be incurred if the following approximation is used:

$$CR_i = \frac{w_i}{\pi r_i}\left(\frac{k_{fi}}{k} - 1\right) \cong \frac{w_i k_{fi}}{\pi k r_i} \tag{11.17}$$

Since w_i and k_{fi} are not known separately, this form, out of necessity, will be used in fracture design. It has also been applied in developing Eq. (11.16).

11.3 INITIAL PRODUCTION RATES

When a fractured well is first put on production, the flow is essentially linear into the fracture (see Fig. 11.1). If the well flowing pressure is constant, the production rate will initially be large and as the flow field approaches that of steady-state or semisteady-state production, the production rate will decline. Conversely, if the production rate is held fixed, the well flowing pressure will decline. This initial linear period is most often not of great significance with regard to the optimization of a fracture treatment; however, the very early stages of production can be used as a well test to evaluate both fracture length and fracture conductivity. Such a test is essential if the size of the treatment is large and if similar treatments are to be carried out in the same area.

To understand the interpretation of pressure drawdown data, it is convenient to first consider fractures having essentially infinite fracture conductivity.

Infinite Fracture Conductivity

Gas wells. For fractures with $CR > 100$, it is reasonable to assume the fracture conductivity ratio to be infinite and to take the pressure everywhere in the fracture to be the well flowing pressure, p_{wf}, as is depicted in Fig. 11.1. When the well is first put into production, the flow will be linear into the fracture. This is an excellent approximation, as can be verified from the Gringarten et al. [7] type curves where the early portion of the type curve is a straight line with a slope of one half on a log–log graph.

Take the gas density to be

$$\rho = \frac{M_w p}{Z_i RT}$$

where Z is the compressibility factor and the subscript i denotes a property evaluated at the reservoir pressure and taken to be constant; then the diffusivity equation in one dimension can be written as [12]

$$\frac{\partial^2 p^2}{\partial y^2} = \frac{\phi \mu_i \kappa_i}{k} \frac{\partial p^2}{\partial t} \tag{11.18}$$

It is further assumed that the initial pressure is everywhere the reservoir pressure; that is

$$p^2(y, t) = p_R^2 \qquad (\text{at } t = 0) \tag{11.19}$$

Also

$$p^2(y, t) = p_R^2 \qquad (\text{as } y \rightarrow \infty) \tag{11.20}$$

Equation (11.20) implies that the fluid presure approaches the reservoir pressure as y increases. This also implies that the reservoir is infinite in extent, an approximation valid for short times.

The gas production rate is the same over all of the fracture area so that the volumetric gas rate, evaluated at bottomhole conditions is

$$-4Lhu_y \big|_{y=0}$$

where $u_y \big|_{y=0}$ is the Darcy flux evaluated at the fracture surface. The volumetric rate at the surface measured at standard pressure, p_{STD}, and standard temperature, T_{STD} (remember to use absolute units) is

$$q = -4Lhu_y \big|_{y=0} \left(\frac{T_{\text{STD}}}{T}\right)\left(\frac{p_{\text{wf}}}{p_{\text{STD}}}\right)\left(\frac{1}{Z_i}\right) \tag{11.21}$$

where q is the constant volumetric rate of gas production.

Using Darcy's law to evaluate flux into the fracture, we can write the third boundary condition as

$$q = \frac{4Lhp_{\text{wf}}k}{\mu_{\text{i}}} \frac{\partial p}{\partial y}\bigg|_{y=0} \left(\frac{T_{\text{STD}}}{T}\right)\left(\frac{1}{p_{\text{STD}}Z_i}\right) \tag{11.22}$$

or rearranging

$$\frac{\partial p^2}{\partial y} = \frac{qTp_{\text{STD}}Z_i}{2hLT_{\text{STD}}} \frac{\mu_{\text{i}}}{k} \quad (\text{at } y = 0) \tag{11.23}$$

Solving Eq. (11.18) subject to Initial Condition (11.19) and Boundary Conditions (11.20) and (11.23), we find for the pressure in the wellbore as a function of time the following simple result:

$$p_{\text{wf}}^2 = p_{\text{R}}^2 - M_{\text{L}}\sqrt{t} \tag{11.24}$$

where

$$M_{\text{L}} = \frac{qTZ_{\text{i}}}{2\sqrt{\pi}\, hL}\left(\frac{p_{\text{STD}}}{T_{\text{STD}}}\right)\sqrt{\frac{\mu_{\text{i}}}{\phi k \kappa_{\text{i}}}}$$

Thus a plot of well flowing pressure squared (p_{wf}^2) versus the square root of time ($\sqrt{t}$) will, at early times, be a straight line. The slope of this straight line can be used to calculate the fracture length, provided the formation permeability is known from a prefracture well test. If not, then it may be necessary to continue the drawdown test until the steady-state period is reached. This yields a second equation relating the productivity, formation permeability, and the fracture length. This second equation can be solved together with the known value of M_{L} to give both L and k. This procedure is, however, not recommended [8, 9].

Example 11.5 Fracture Length from Pressure Drawdown Data

Based on the pressure drawdown data given in Table 11.2A, estimate the length of the hydraulic fracture having the characteristics listed in Table 11.2B.

Solution The first step is to recalculate the given data so that p_{wf}^2 is tabulated as a function of $\sqrt{t}$. The result is shown in Table 11.2C. Figure 11.5 shows a plot of p_{wf}^2 versus $\sqrt{t}$. The straight line portion of the curve is seen to have a slope $M_L = 4.6 \times 10^6$ kPa^2/hr$^{1/2}$. Solving for the length of the fracture we see that

$$L = \frac{qTZ_i}{2\sqrt{\pi}\,h}\left(\frac{p_{STD}}{T_{STD}}\right)\sqrt{\frac{\mu_i}{\phi k \kappa_i}}\,\frac{1}{M_L}$$

Converting the slope to the proper units

$$M_L = 4.6 \times 10^6 \frac{\text{kPa}^2}{\text{hr}^{1/2}}\left(\frac{10^3\,\text{Pa}}{\text{kPa}}\right)^2 \frac{\text{hr}^{1/2}}{(3600\ \text{sec})^{1/2}} = 7.66 \times 10^{10}\frac{\text{Pa}^2}{\text{sec}^{1/2}}$$

Therefore

$$L = \frac{(0.328)(134 + 273)(1.151)(101 \times 10^3)}{2\sqrt{\pi}\,(50)(7.66 \times 10^{10})(14.7 + 273)}$$

$$\times \sqrt{\frac{2.7 \times 10^{-5}}{(0.018)(6.2 \times 10^{-17})(1.294 \times 10^{-8})}}$$

or

$$L = 172\ \text{m}$$

Oil wells. The previous analysis for gas wells is the more useful since the period for which the linear approximation applies is more likely to be sufficient to permit a successful well test free of cleanup and well storage problems; whereas for permeable oil wells stimulated with short fractures penetrating a damaged zone, the time period for linear flow will not be long enough (see Example 11.1).

The time for which the linear flow approximation applies can be estimated from the type curves presented by Gringarten et al. [7]. These curves show [10] that the initial period ends approximately at a time such that

$$\frac{kt}{\phi \mu_f \kappa_f L^2} \cong 10^{-2} \tag{11.25}$$

This equation applies to both oil and gas wells. For gas wells the viscosity and compressibility are again to be taken at the initial reservoir pressure.

Equation (11.25) shows that for short fractures in permeable formations, the initial period of linear flow will be over in a matter of hours making its measurement difficult. However, for longer fractures in less permeable formations, the determination of the fracture length by a drawdown test is not only feasible, but advisable since if the well performance is unexpectedly poor, one can attempt to ascertain whether the source of the difficulty is an intrinsic reservoir problem or one of simply not achieving the designed treatment.

An analysis similar to that shown for gas wells gives for oil wells

$$p_{wf} = p_R - M_L\sqrt{t} \tag{11.26}$$

where

$$M_L = \frac{Bq}{2\sqrt{\pi}\,hL}\sqrt{\frac{\mu_i}{\phi k \kappa_i}}$$

TABLE 11.2A Pressure Drawdown Data

Time (hr)	Pressure (kPa)	Time (hr)	Pressure (kPa)
0	20,634	78.72	19,187
0.024	20,568	198.7	18,653
0.0576	20,488	318.7	18,179
0.312	20,411	438.7	17,785
0.960	20,317	558.7	17,441
2.904	20,195	678.7	17,131
8.736	20,019	720	17,026
26.23	19,742	—	—

TABLE 11.2B Treatment Data

Property	Symbol	Value
Total net pay	h	50m
Initial reservoir pressure	p_R	20,684 kPa
Reservoir temperature	T	134°C
Dry gas flow rate	q	1180 m^3/hr
Porosity	ϕ	0.018 (total porosity × water saturation)
Separator gas gravity	—	0.65
Gas viscosity	μ_i	2.7×10^{-5} kg/m-sec
Gas compressibility	κ_i	1.294×10^{-5} kPa^{-1}
Compressibility factor	Z_i	1.151
Formation gas permeability	k	$6.2 \times 10^{-17} m^2$
Standard temperature	T_{STD}	14.7°C
Standard pressure	p_{STD}	101 kPa

TABLE 11.2C Recalculated Data

$\sqrt{t}$ ($hr^{1/2}$)	p^2 (kPa^2)	$\sqrt{t}$ ($hr^{1/2}$)	p^2 (kPa^2)
0	4.278×10^8	8.872	3.720×10^8
0.1549	4.230×10^8	14.10	3.479×10^8
0.2400	4.198×10^8	17.85	3.305×10^8
0.5586	4.666×10^8	20.95	3.163×10^8
0.9800	4.128×10^8	23.64	3.042×10^8
1.704	4.078×10^8	26.05	2.935×10^8
2.956	4.008×10^8	26.83	2.899×10^8
5.122	3.897×10^8	—	—

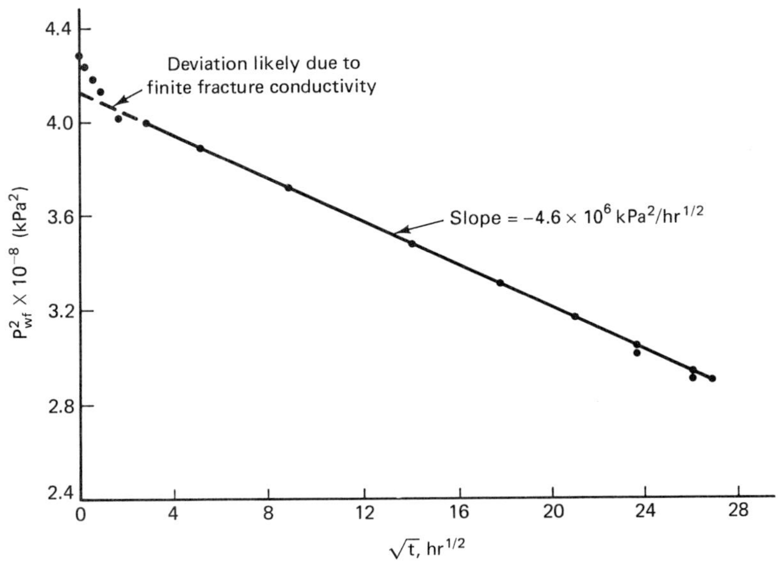

Figure 11.5 Graph of p_{wf}^2 versus $\sqrt{t}$. Fracture length can be determined from the slope.

Thus, a plot of p_{wf} versus $\sqrt{t}$ will initially be linear. From the slope of this line the length of the fracture can be determined.

The analysis detailed in this section is for drawdown tests. Lee and Holditch [9] recommended buildup tests as being more precise. We shall, however, consider only drawdown tests to illustrate the need and application of well testing to good stimulation practices. To present more detail would expand this text beyond its intended scope.

Finite Fracture Conductivity

Cinco et al. [11] have presented type curves that apply to finite conductivity fractures and Aggarwal et al. [8] have extended their analysis. Aggarwal et al. show that finite conductivity fractures can seriously alter the initial pressure response predicted by Eqs. (11.24) and (11.26) and have suggested the use of early time pressure drawdown (or buildup) data to provide an estimate of both fracture conductivity and fracture length. Their analysis assumed that the fracture conductivity is uniform, thus neglecting the possibility that conductivity may vary as the distance from the wellbore increases.

Figure 11.6 shows $\Delta(p^2)$ plotted as a function of the $\sqrt{t}$. The fracture conductivity expressed in mD-ft is shown as a parameter. The early time pressure data for higher fracture conductivities fall on a straight line, as predicted by Eq. (11.24). The slope of the line (M_L) is inversely proportional to the fracture length. The early time pressure data for finite capacity fractures exhibit a curved portion before approaching the expected straight line behavior. The duration of this curved

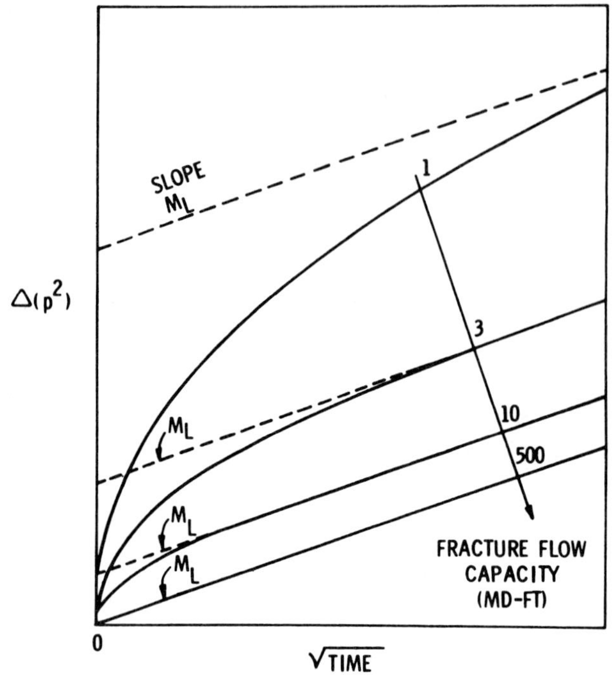

Figure 11.6 Graph of Δp^2 versus $\sqrt{t}$ for finite capacity vertical fractures [8]. (With permission of the Society of Petroleum Engineers.)

portion depends on the fracture capacity. For very low-capacity fractures, it may be almost impossible to determine the correct straight line slope. In other cases, the time to approach straight line behavior may be so brief that it is not practical to run the test. In any case, if the early time slope for a finite capacity fracture is used in Eq. (11.24) the estimated slope (M_L) will be too high and the computed fracture length will be less than the one actually created. Thus for long fractures or even poorly conductive short ones, neglecting the fracture conductivity can lead to serious errors.

Figures 11.7 and 11.8 show generalized plots as presented by Aggarwal et al. Figure 11.7 represents the dimensionless pressure as a function of dimensionless time for a constant rate drawdown test. Various fracture conductivity ratios are shown. It is important to note that the fracture conductivity ratios shown as parameters in Fig. 11.7 are a factor of π greater than those defined in the previous section. Thus

$$CR = \frac{1}{\pi} F_{CD} \tag{11.27}$$

where F_{CD} is the fracture conductivity ratio displayed as a parameter in Figs. 11.7 and 11.8. This difference, which is simply a matter of definition, should be recognized and taken into account.

Figure 11.8 is a plot of the intercept obtained by extrapolating the linear portion of the pressure versus square root of time curve shown in Fig. 11.7 back to the origin. Based on these two curves it is possible to estimate both fracture

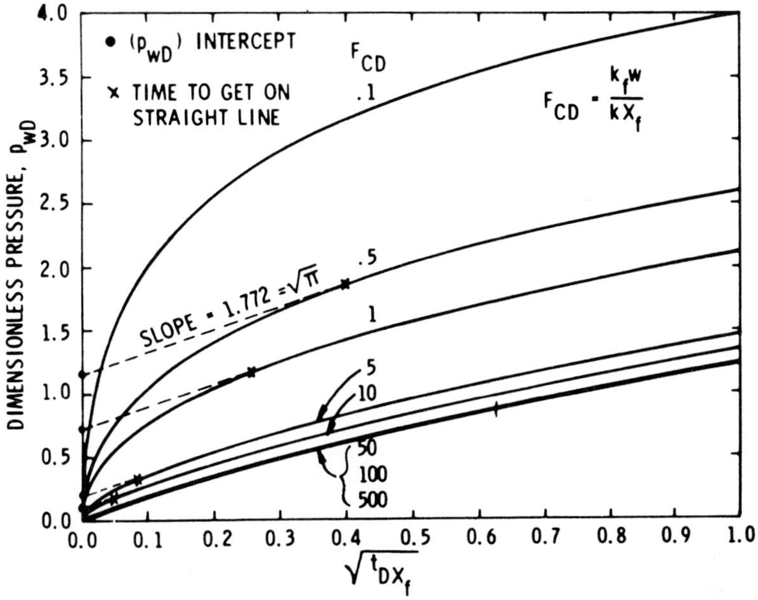

Figure 11.7 Generalized plot of constant rate type curves for finite capacity vertical fractures p_{wD} intercept versus $\sqrt{t_{DXf}}$ [8]. (With permission of the Society of Petroleum Engineers.)

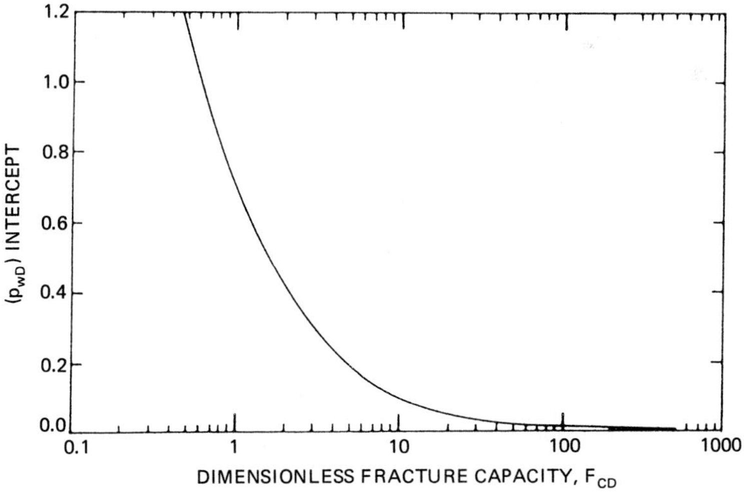

Figure 11.8 Generalized plot of (p_{wD}) intercept versus F_{CD} for finite capacity vertical fractures [8]. (With permission of the Society of Petroleum Engineers.)

length and fracture conductivity, although for very small conductivities the straight line portion of the curve may not be seen. Deviations from linear flow may take place before a straight line portion of the curve is established.

The dimensionless time (t_{DX}) is defined as

$$t_{DX} = \frac{kt}{\phi\mu_i\kappa_i L^2} \tag{11.28}$$

and the dimensionless pressure as

$$p_{wD} = \frac{p_R^2 - p_{wf}^2}{\dfrac{q\mu_i T Z_i p_{STD}}{2\pi hkT_{STD}}} \tag{11.29}$$

for gas wells or

$$p_{wD} = \frac{p_R - p_{wf}}{\dfrac{Bq\mu}{2\pi hk}} \tag{11.30}$$

for oil wells. In either case Eqs. (11.24) and (11.26) reduce to

$$p_{wD} = \sqrt{\pi t_{DX}} \tag{11.31}$$

for infinitely conductive fractures (*Note:* The 1.772 slope shown in Fig. 11.7 is $\sqrt{\pi}$.)

Example 11.6 Estimation of Fracture Conductivity

The extrapolated p^2 versus $\sqrt{t}$ curve for the pressure drawdown data presented in Example 11.5 does not intersect the $t = 0$ axis at the initial reservoir pressure, as shown in Fig. 11.5. If we interpret this deviation as being due to a finite fracture conductivity, what is the fracture conductivity? The data and results from Example 11.5 may all be used to help answer this question.

Solution The dimensionless pressure for a gas well is defined by Eq. (11.29). The pressure squared extrapolated from the straight line is 4.12×10^8 kPa^2, as shown in Fig. 11.5. The square of the reservoir pressure is 4.278×10^8 kPa^2. Thus

$$\Delta p^2 = 4.278 \times 10^8 - 4.12 \times 10^8 = 0.158 \times 10^8 \text{ kPa}^2 = 1.58 \times 10^{13} \text{ Pa}^2$$

so

$$(p_{wD})_{INTERCEPT} = \frac{1.58 \times 10^{13}}{\left[\dfrac{(0.328)(2.7 \times 10^{-5})(407)(1.151)(101 \times 10^3)}{(2\pi)(50)(6.2 \times 10^{-17})(287.7)}\right]}$$

$$= 0.21$$

Reading from Fig. 11.8, we find

$$F_{CD} = 5$$

or in terms of the conductivity ratio defined in this text

$$CR = \frac{F_{CD}}{\pi} = \frac{5}{\pi} = 1.6$$

Thus, improvement could be obtained by either extending the fracture by about a factor of 2 or increasing the fracture conductivity by about a factor of 3 to 4. It should be noted, however, that any change in the treatment will have to be justified on economic grounds.

11.4 FLUSH PRODUCTION

There are essentially three periods that can be identified during the productive lifetime of a fractured well. The initial period is characterized by a linear flow into the fracture and the semisteady-state or steady-state production is achieved when the flow is almost radial. There is an intermediate transition period which can, according to Morse and Von Gonten [3], be long in duration. To calculate realistic production rates during both the initial and transition period, use of a type curve such as that shown in Fig. 11.9 is expedient although reservoir simulators may have considerable merit [9].

Figure 11.9 is a type curve developed by Aggarwal et al. [8] and it has been calculated assuming the well flow pressure is constant and the production rates vary. Thus the quantity $1/q_D$ is defined by Eq. (11.29), except that $p_R^2 - p_{wf}^2$ is a constant and q is a function of time. This is a convenient representation especially for gas wells, since the wellhead pressure is often held fixed during the lifetime of the well and the flow is allowed to vary. Aggarwal et al. [8] and Cinco et al. [11] have presented similar curves for constant flow rate production, but they are not needed here.

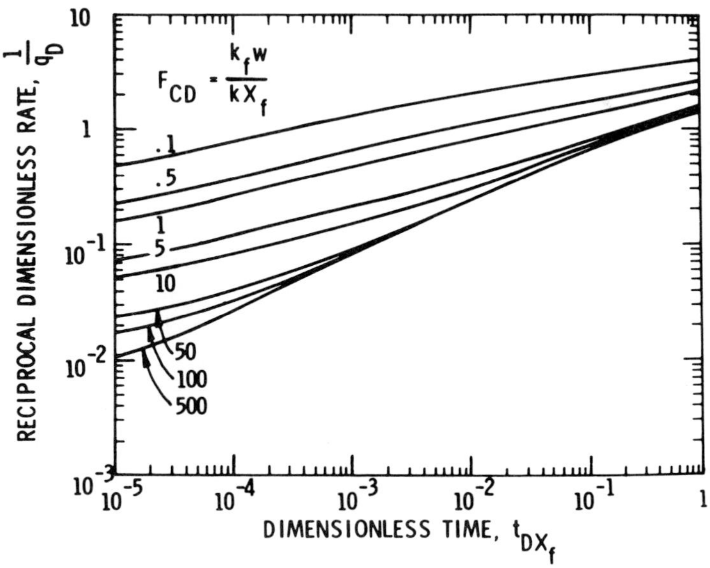

Figure 11.9 Graph of constant pressure log–log type curves for finite capacity vertical fractures: $1/q_D$ versus t_{DXf} [8]. (With permission of the Society of Petroleum Engineers.)

The results plotted in Fig. 11.9 can be used to predict the production rate for any combination of fracture length and fracture conductivity. After long times, the flow will become semisteady and Fig. 11.2 then applies.

Example 11.7 Productivity as a Function of Fracture Conductivity

Estimate the volumetric gas production rate (measured at 60°F and 1 atm) during a 5-year period for Cases (A) and (B) having the following specifications:

Case (A): Fracture length (one wing) = 400 m
Fracture conductivity ratio (CR) = 10 (bauxite)
Infinite reservoir

Case (B): Fracture length = 400 m
Fracture conductivity ratio (CR) = 2 (sand)
Infinite reservoir

The pertinent treatment data are given in Table 11.4.

Solution Based on Fig. 11.9, prepare a table of $1/q_D$ versus t_D for Cases (A) and (B). This should range from very small values to

$$t_D = \frac{(5)(365)(24)(3600)(1.974 \times 10^{-17})}{(0.06)(4.6 \times 10^{-5})(2.6 \times 10^{-9})(400)^2} = 2.71$$

for a time corresponding to five years.

Case (A): $CR = 10$

t_D	$\frac{1}{q_D}$	q (m^3/sec)	t (days)
0.01	0.24	1.97	6.7
0.10	0.62	0.76	67.0
0.50	1.1	0.43	337
0.75	1.25	0.38	505
1.00	1.35	0.35	673
1.5	*1.56	0.32	1010
2.5	*1.83	0.26	1684
2.71	*1.90	0.25	1825

* Extrapolated values.

$$\frac{1}{q_D} = \frac{p_R^2 - p_{wf}^2}{\dfrac{q\mu TZp_{STD}}{2\pi hkT_{STD}}} = \frac{(35 \times 10^3)^2 - (10.5 \times 10^3)^2}{q\,\dfrac{(4.6 \times 10^{-5})(138 + 273)(1.10)(101 \times 10^3)}{2\pi(25)(1.974 \times 10^{-17})(14.7 + 273)}}$$

or

$$q = q_D[0.474]$$

[*Note:* This equation applies to both Cases (A) and (B).]

Case (B): $CR = 2$

t_D	$\frac{1}{q_D}$	q (m^3/sec)	t (days)
0.01	0.36	1.32	6.7
0.10	0.86	0.55	67
0.50	1.35	0.35	337
0.75	1.52	0.31	505
1.00	1.62	0.29	673
1.50	1.90	0.25	1010
2.50	2.20	0.22	1684
2.71	2.26	0.21	1825

From this example two things can be observed. Firstly, the more conductive well will have produced more gas by the time the well reaches its economic lower limit of gas productivity. Thus, its ultimate yield will be higher. Example 11.7 applies to an essentially infinite reservoir, but the conclusion is likely to be true also for finite systems. Secondly, the production rates are substantially higher for the more conductive fracture. To achieve this higher productivity, the treatment will be more costly. Thus, economic issues will come into play. Each individual or company will have its own objectives and each will settle on a different treatment. Nevertheless, the route to a decision is to examine the costs and the incremental benefits. These concepts are discussed in Chapter 12.

It should perhaps be stressed that the same type curves apply to both gas and oil wells, but the definition of q_D or p_D differs. The reason that this is true is, of course, that the differential equation solved for p^2 in the case of a gas is the same as the differential equation that applies to p for the case of a liquid. The boundary conditions have the same form. Thus the solutions, namely, the type curves are the same. The reader should bear this in mind because many type curves that apply to slightly compressible fluids can also be used for gases.

TABLE 11.4 Treatment Data

Property	Symbol	Value
Formation depth	D	3048 m
Formation gas permeability	k	1.974×10^{-17} m^2
Formation gas porosity	ϕ	0.06
Net gas pay	h	25 m
Reservoir pressure	p_R	35×10^3 kPa
Flowing bottomhole pressure	p_{wf}	10.5×10^3 kPa
Reservoir temperature	T	138°C
Compressibility factor	Z_i	1.10
Gas viscosity	μ_i	4.6×10^{-5} kg/m-sec
Gas compressibility	κ_i	2.6×10^{-6} kPa^{-1}

REFERENCES

11.1. Ramey, H. J., and Cobb, W. M., *J. Pet. Tech., 23* (1971) 1493.
11.2. Morse, R. A., and Von Gonten, W. D., *J. Pet. Tech., 24* (1972) 807.
11.3. Dake, L. P., *Fundamentals of Reservoir Engineering*, Amsterdam: Elsevier, 1978.
11.4. McGuire, W. J., and Sikora, V. J., *Trans. AIMME, 219* (1960) 401.
11.5. Katz, D. L., *Handbook of Natural Gas Engineering*, New York: McGraw-Hill, 1959.
11.6. Raymond, L. R., and Binder, G. G., *J. Pet. Tech., 19* (1967) 120.
11.7. Gringarten, A. C., Ramey, H. J., and Raghaven, R., *Soc. Pet. Eng. J., 14* (1974) 347.
11.8. Aggarwal, R., Carter, R. D., and Pollock, C. B., *J. Pet. Tech., 31* (1979) 362.
11.9. Lee, W. J., and Holditch, S. A., *J. Pet. Tech., 33* (1981) 1776.
11.10. Earlougher, R. C., *Advances in Well Test Analysis*, Mono. Ser., Society of Petroleum Engineers, Richardson, Texas, 1977.
11.11. Cinco, H., Samaniego, F., and Dominguez, N., *Soc. Pet. Eng. J., 18* (1978) 253.
11.12. Millheim, K. K., and Cichowicz, L. *J. Pet. Tech., 20*. (1968) 193.

PROBLEMS

****11.1.** It is important in hydraulic fracturing to be certain that the fracture near the wellbore is fully packed with proppant. If, for example, the proppant is totally flushed from the region near the wellbore for a distance of 4 m along the first part of the fracture, what stimulation ratio would be expected if the fracture treatment was originally designed to yield a stimulation ratio equal to 6.0. The fracture conductivity ratio used in the design is 5. It can be assumed that the fracture conductivity over the entire length of the fracture (except the first 4 m) is equal to the design value. Furthermore, the following data apply:

Wellbore radius: 0.2 m
Drainage radius: 250 m

***11.2.** A fracture treatment has been designed so that a vertical fracture 70 m long is expected to yield a stimulation ratio of 5.0 in an undamaged formation. Unfortunately, in attempting to flush the tubing clean of proppant following the treatment, too much overflush fluid was used decreasing the fracture conductivity in that portion of the fracture nearest the wellbore to a level only one eighth ($\frac{1}{8}$) as large as the design value. At distances along the 70-m fracture further removed from the wellbore, the fracture conductivity is equal to the desired value. The removal of some of the proppant from the region near the wellbore of the fracture resulted in an actual stimulation ratio of 4.0 rather than the anticipated value of 5.0. Calculate the distance that the zone of reduced fracture conductivity extends out from the wellbore. Assume the flow is steady state and that the drainage and wellbore radii are 250 m and 0.2 m, respectively.

**11.3. Assume, for the purpose of this problem, that the reaction of the acid with the fracture wall during an acid treatment takes place uniformly, that is, the volume of rock dissolved is the same at each point along the fracture. If the volume of acid is fixed at 32 m^3 of 15 wt% HCl and if the penetration distance is fixed by fluid loss to be 50 m, what is the greatest closure stress that can be tolerated and still maintain a stimulation ratio in excess of $J/J_0 > 3$? The following data are given:

Fracture height:	20 m
Fracture gradient:	17 kPa/m
Rock embedment strength:	50 MPa
Formation composition:	dolomite
Drainage radius of well:	300 m
Wellbore radius:	0.15 m
Formation permeability:	4 mD
Formation porosity:	0.1

Assume that the well produces oil at bottomhole conditions and that the treated well is undamaged.

**11.4. An acid fracture treatment of an undamaged limestone formation is defined by the data given in Table P11.5. The fluid-loss flux that is given, 1.3×10^{-5} m/sec, is an average taken over the entire fracture surface during the entire course of the treatment. This fluid-loss flux should be used in all of the calculations relating to the questions that follow. Based on the data given in Table P11.5, answer these questions:

(a) Is the treatment fluid-loss limited or is it limited by the acid penetration distance?
(b) What steady-state increase in production would be realized by this treatment?
(c) Assuming that the volume of acid is fixed, how can the treatment be improved? Increase the injection rate? Reduce the average fluid-loss flux? Increase acid viscosity? You must substantiate your answer with hard evidence (numbers, calculations, equations, etc.).

TABLE P11.5 Treatment Data

Formation Property	Symbol	Value
Rock embedment strength	S_{RES}	3×10^8 Pa
Fracture gradient	FG	17 kPa/m
Depth	D	2500 m
Permeability	k	5 mD
Drainage radius	r_e	250 m
Wellbore radius	r_w	0.05 m
Bottomhole well flowing pressure	p_w	6×10^3 kPa
Porosity	ϕ	0.12

Treatment Property	Symbol	Value
Fluid-loss flux	$\langle \bar{u}_n \rangle$	1.3×10^{-5} m/sec
Fracture height	h	20 m
Acid injection rate	i	3×10^{-2} m^3/sec
Dynamic fracture width (average)	$\bar{w}$	2×10^{-3} m
Acid diffusion coefficient	D_A	5×10^{-8} m^2/sec
Acid strength	c_0	15 wt% HCl
Acid injection time	t_A	1800 sec
Acid viscosity	μ	0.02 kg/m-sec
Acid density	ρ	1040 kg/m^3

***11.5.** A fracture treatment of an oil well using 3000 kg of 20/40-mesh bauxite (density 3900 kg/m^3) yielded a stimulation ratio of 5.5. The fracture length by well test was 30 m. If it is shown by laboratory experiment that the fracture permeability of bauxite proppant under the closure stresses expected in this case is 2150 D and its porosity in the fracture is 0.36, determine the extent to which the well was damaged prior to stimulation by finding the skin factor. Use the following data:

Formation permeability:	7 mD
Wellbore radius:	0.1 m
Drainage radius:	220 m
Fracture height:	15 m

****11.6.** Given the data shown in Tables P11.7A and P11.7B, estimate the length of the fracture created in a low permeability gas reservoir.

TABLE P11.7A Treatment Data

Property	Value
Depth to top of formation	1700 m
Total net pay	15 m
Wellbore radius	0.1 m
Reservoir pressure	20.69 MPa
Bottomhole temperature	81°C
Dry-gas flow rate	0.4 m^3/sec (at 10^5Pa and 16°C)
Total porosity	0.036
Water saturation	0.5
Separator gas gravity	0.65
Gas viscosity	2.7×10^{-5} kg/m-s
Gas compressibility	1.31×10^{-5} kPa^{-1}
Compressibility factor	1.15
Formation gas permeability	0.013 mD

TABLE P11.7B Pressure Data

Bottomhole Pressure (MPa)	Time (hr)
20.690	0
20.568	0.0240
20.488	0.0576
20.411	0.312
20.318	0.960
20.196	2.904
20.020	8.736
19.743	26.23
19.288	78.72
18.654	198.7
18.180	318.7
17.786	438.7
17.442	558.7
17.132	678.7
17.027	720

***11.7.** *Estimation of Gas Production* (Refer to Table P11.8 for given data.)
Find the gas production from an essentially infinite reservoir over a 15-year period for the following two cases:

Case (A)

Fracture length	520 m
Proppant	Infinite fracture conductivity
Infinite reservoir	

Case (B)

Fracture length	520 m
Proppant	20/40-mesh sand @ 3kg/m^2
Infinite reservoir	

Evaluate all fluid properties at the initial reservoir pressure and plot the gas production for both cases as a function of time.

TABLE P11.8 Treatment Data

Property	Symbol	Value
Formation depth	D	3000 m
Formation gas permeability	k	0.01 mD
Formation gas porosity	ϕ	0.045
Net gas pay	h	18 m
Initial reservoir pressure	p_R	34.48 MPa
Flowing bottomhole pressure	p_{wf}	10.34 MPa
Bottomhole temperature	T_R	127°C
Gas gravity	—	0.62
Fracture gradient	FG	17.2 kPa/m

12 Design and Optimization of Fracture Processes

In this chapter the task of designing a fracture treatment is considered. All of the concepts needed have been developed, but here the complex interrelationship between those parameters which can be manipulated is considered. Most often, extensive computer modeling is required before specifying the final process. This modeling includes sensitivity studies and economic analyses to identify both the optimum treatment and those variables which require careful control. This latter factor, quality control, is a crucial, but often neglected, aspect.

The final design will be best in some economic sense. However, because economic methods and criteria vary from company to company, no attempt to optimize will be made here. Instead a procedure for optimization will be proposed, discussed, and illustrated. Implementation will require rather lengthy, repetitive, and complex calculations. These are best carried out with the aid of a computer.

The design of acid and proppant fractures requires different considerations. It is, therefore convenient to discuss them separately. Proppant fracturing will be considered first.

12.1 PROPPANT FRACTURING

Selection of Fracture Fluid and Additives

Howard and Fast [1] note that the fracture fluid selected for use in a treatment should possess the following properties:

- Low fluid loss
- Ability to carry and suspend the proppant
- Low friction loss
- Easy to recover from the formation
- Compatible with formation fluids and nondamaging
- Reasonable cost

It is now possible to achieve most or all of these objectives by properly selecting the chemical composition of a fracture fluid. There are very few situations for which fracturing is technically impossible because of fluid limitations. This is, of course, not to say that all cases which are technically possible will also be economically attractive.

Both oil-based and water-based fluids are used. In recent years, the overwhelming majority of fracture jobs have been carried out with water-based fluids. Oil-based fluids would only be selected for stimulating oil wells in water-sensitive sandstones. Condensate-based fluids have occasionally been used to treat gas wells. In the past, refined oils, kerosene, lease crude, or condensate have all been used for hydraulic fracturing, but these fluids present a fire hazard and increase the cost of the treatment without, in most cases, contributing to the stimulated productivity.

Water-based fluids are discussed here primarily because of their widespread usage. Water has many advantages with respect to oil in addition to costing less and not being flammable. Water has a higher density than oil, thereby reducing the surface pressures (see Chap. 7). Furthermore, a number of additives are now available for use in water so that the properties of the water-based systems can be varied to accommodate a wide range of reservoir conditions.

Fluid loss. Those factors determining fluid loss have been formulated in Chapter 8. While C_c and C_v can be calculated, the spurt-loss and the wall coefficient for a cross-linked hydroxypropylguar are shown in Figs. 12.1 and 12.2. These data are typical of those provided by service companies. It is important to remember that the fluid-loss coefficient must be corrected to account for the difference between the standard pressure difference applied across the filter cake in the laboratory, typically 7×10^3 kPa, and the pressure difference corresponding to that expected during the fracture treatment (see Eq. 8.81).

Figure 12.2 shows that the fluid-loss coefficient of this particular fluid is not changed much with increasing polymer concentration; however, this is somewhat misleading since the viscous coefficient, C_v, depends on the viscosity of the fracture fluid. Increasing the polymer concentration will in turn increase the viscosity and decrease C_v assuming, of course, that the polymer is not filtered from the fracture fluid as it invades the reservoir. Thus the overall fluid loss-coefficient will generally be decreased as the polymer concentration is increased.

The effective viscosity in calculating C_v using Eq. (8.78) can be estimated by noting that the average shear rate for flow in a permeable medium is roughly [2, 3]

$$\dot{\gamma} \cong \frac{u_N}{\phi d_g} = \frac{C}{\phi d_g \sqrt{t}} \tag{12.1}$$

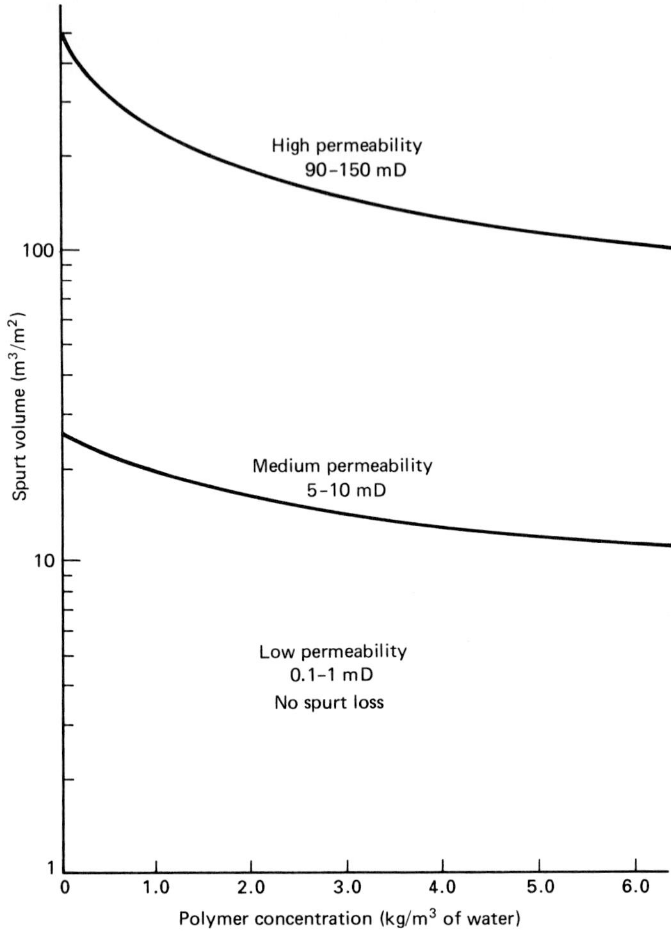

Figure 12.1 Graph showing typical values of spurt volume as a function of polymer concentration. The data correspond to a cross-linked hydroxypropylguar.

where d_g is the effective size of a sand grain and C is the overall fluid-loss coefficient. For a power-law fluid, the apparent viscosity is given by

$$\mu_{APP} = m \left(\frac{C}{\phi d_g \sqrt{t}} \right)^{n-1} \tag{3.3}$$

The effective sand grain size can be estimated by using the Blake–Kozeny [Eq. (10.12)] and the formation permeability. Thus

$$d_g = \sqrt{\frac{150k (1 - \phi)^2}{\phi^3}} \tag{12.2}$$

These equations yield a value of C_v which depends on time since u_N depends on time. Fracture calculations all assume constant C_v. Realistically, it is unnecessary

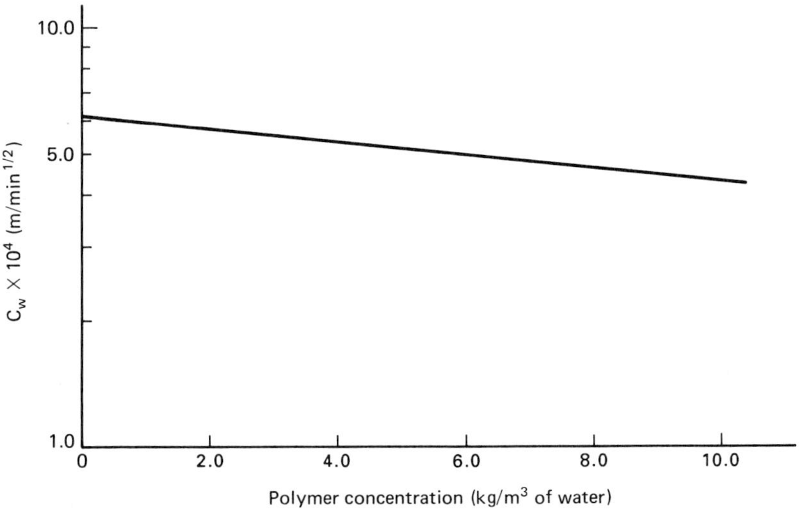

Figure 12.2 Graph showing typical values of fluid-loss coefficient (C_w) as a function of polymer concentration.

to consider a time-dependent value since many of the parameters used in the calculation of fracture geometry are only imperfectly known. This is not to say that one should not consider all factors in the greatest detail possible, for this is the only approach that will lead to precise designs which is our ultimate goal.

One approach that can be used to control fluid loss is to increase the effective viscosity of the fracture fluid. Other additives that can be introduced to reduce the fluid loss are 100-mesh sand, silica flour, hydrocarbon soluble resins, or small quantities of diesel oil. In formations where no natural fractures are expected to occur, either hydrocarbon-soluble resins or diesel oil are recommended over silica flour. Upon producing the well, the resins or the diesel oil will dissolve while the silica flour will not. These small particles can then clog up the fracture and reduce the fracture conductivity. In areas where natural fractures do occur, 100-mesh sand used in conjunction with hydrocarbon-soluble resins is often effective. The 100-mesh sand tends to bridge off in the fractures and the resins can then build a wall cake on the sand. In gas wells an overflush of solvent can be used to dissolve the resin.

Acid-soluble calcium carbonate coated with oil-soluble resin is another fluid-loss additive that is effective in oil wells. An acid overflush is used to dissolve the calcium carbonate.

Low fluid loss is claimed as one significant advantage of foams [4]. However, as the formation permeability increases, the difference between the fluid loss of foam and that of gelled water decreases. Thus foams would appear to flow readily through larger pore spaces, which suggests that the mechanism responsible for the increased resistance to foam flow through porous media is one of gas blockage by soap films bridging the pore space. It would be interesting to have additional data obtained in the presence of oil and at elevated temperature and absolute pressures to see how these factors influence the fluid loss. The available data do, however, indicate a considerable potential for foams.

Proppant transport. The fracture height used in production calculations is not the created height, but the created height less the distance that the proppant settles. Also, it must be remembered that the height lost due to proppant settling must be subtracted from the top of the fracture. Since the fracture will extend a distance into the overburden depending on the insitu-stress contrast, some particle settling can be tolerated. Proppant settling has been discussed in Chapter 10. The settling velocity depends upon the rheological properties of the fracture fluid. Table 12.1 presents some representative data for a popular fracture fluid. The data show that both n and m change with time and temperature. The trends shown are typical of all gelled waters whether crosslinked or not. Effective viscosity decreases with time. After 6 hr at 135°C, the effective viscosity at a shear rate of 170 sec^{-1} of a solution containing 3.6 kg of polymer/m^3 decreases from 137 cp to 39 cp. As the temperature is increased, the rate of decrease is more dramatic.

To combat the detrimental effect of temperature, the amount of polymer added to the solution must be increased. Ultimately, the polymer concentration required to ensure adequate proppant transport capability is determined by both temperature and process time.

If the fluid is selected so that proppant settling is minimal (less than 8–10 m during the injection and closure period) then essentially "perfect" suspension is achieved.

Recoverability. The high viscosity of each fracture fluid must be temporary. The use of enzymes or oxidizing agents to "break" the polymer is, as noted in Chapter 3, common practice. In deep, hot fractures agents to break the polymer may not be required. The rate of thermal degradation may very well be sufficient to permit the efficient recovery of the fracture fluid. It is important to leave the well shut in long enough to permit the polymer to break completely. If this is not done, a residue may remain [5].

Foams have the advantage of permitting easier cleanup. Thus, the use of foam should be considered in shallow, water-sensitive formations. Because foam has a relatively small density, the hydrostatic head is smaller than when water-based fluids are used. The surface pressure during a foam treatment will therefore be higher.

Compatibility with formation fluids. This is not a critical issue. Very little mixing between the resident water and the fracture fluid is anticipated; nevertheless, by carefully examining the water analyses of both the reservoir fluid and the fracture fluid, one can anticipate precipitation. As noted in Chapter 4, it is the modified fluid not the fracture fluid that mixes with the formation water; however, if one water is rich in divalent anion and the other in divalent cation, then some difficulty with compatibility may be expected. Steps should then be taken to improve the quality of the fracture fluid. There is no need to risk the

TABLE 12.1 Rheology of Cross-linked Hydroxypropyl Guar*

Fluid (kg of polymer/m^3)	Temp (°C)	Time (hr)	n (index)	m (kg/m-sec^{2-n})	Apparent Viscosity (cp @ 170 sec^{-1})
3.6	110	1	0.570	3.10	341
3.6	110	2	0.588	2.14	258
3.6	110	4	0.630	1.00	150
3.6	110	6	0.672	0.52	96
3.6	120	1	0.656	1.29	220
3.6	120	2	0.674	0.90	169
3.6	120	4	0.712	0.45	103
3.6	120	6	0.752	0.22	62
3.6	135	1	0.740	0.52	137
3.6	135	2	0.757	0.36	103
3.6	135	4	0.792	0.19	65
3.6	135	6	0.835	0.09	39
4.8	110	1	0.65	3.48	607
4.8	110	2	0.685	2.33	462
4.8	110	4	0.718	1.29	303
4.8	110	6	0.755	0.86	244
4.8	120	1	0.689	1.48	300
4.8	120	2	0.712	1.00	228
4.8	120	4	0.745	0.62	167
4.8	120	6	0.781	0.45	146
4.8	135	1	0.718	0.67	157
4.8	135	2	0.740	0.48	126
4.8	135	4	0.771	0.29	89
4.8	135	6	0.805	0.23	84

* These data correspond approximately to values provided in several service company brochures; however, these values are not recommended for use in fracture design.

consequences of using incompatible water even though the probability of damage is small.

Formation damage. The fluid loss into the formation adjacent to the fracture will result in formation damage. Some of the mechanisms responsible for this damage have been detailed in Chapter 4. The damage must be severe, however, if it is to have a significant effect on the rate of hydrocarbon production. The situation is quite different for damage occurring near the face of a fracture than it is for damage surrounding a wellbore. The flow into the fracture is approximately linear, so that substantial permeability damage is needed to restrict the flow, whereas all of the fluid must flow through a decreasing area in order to enter the wellbore and thus small decreases in permeability can greatly increase the pressure drop. Prats [6] has shown that even if the permeability within a small zone around the fracture surface is reduced by as much as a factor of 100 or more, the productivity is not decreased markedly. One would seldom expect damage to be that severe.

Tannich [7] has shown that for gas wells, the time required to obtain cleanup following fracturing will depend on the viscosity of the fracture fluid. Viscous fluids, having viscosities of several hundred centipoise, will take hundreds of days to be produced. If the capillary pressure is large as compared to the reservoir gas pressure, severe permanent damage is possible [8].

Thus, in most cases, the damage to the formation adjacent to the fracture surfaces will not be severe enough to influence production. Nevertheless, precautions should be taken to minimize the damage.

It is recommended that the fracture fluid contain at least 2 wt% KCl and higher concentrations may be helpful. Fresh water without added salts should be avoided (see Chap. 4). Surfactants, on the other hand, should not be added to fracture fluid for oil well application without supporting laboratory work to demonstrate that oil/water emulsions are made less stable if the surfactant is added. For gas wells, addition of surface tension reducing surfactants is recommended. Lowering the surface tension will reduce capillary pressure effects and be beneficial. Surfactants used in this application should be quite water soluble at the bottomhole condition and should be applied at concentrations well above the critical micelle concentration (refer to Chap. 3). Anonic surfactants are preferred. If the reservoir temperature is high (>80°C), then the surface activity of surfactants is greatly decreased and their contribution to decreasing the surface tension may not justify their application.

The compatibility of the surfactant with the polymer gelling agent, the friction reducer (if one is used), and the ions in the water should be considered. Cationic surfactants used with anionic polymers should, for example, be avoided. The interaction between the two will reduce the effectiveness of both chemicals.

Clay stabilizers can be added to the prepad or pad fluid if clay migration is thought to represent a severe problem. Discussion of clay entrainment and pore clogging mechanisms are presented in Chapter 4. This information should be reviewed.

Design of Proppant Fracturing Treatments

The design strategy for optimizing the fracture treatment once it has been decided to fracture, must evidently include economic considerations. We will not carry the procedure out to the extent that the return on investment is calculated, since all of the factors including interest rates, oil or gas prices, taxes, treatment costs, etc., will vary, making obsolete any results based on assumed values. The technical problem of optimizing fracture design can, however, be separated from the economic aspects if the procedure recommended here is followed.

The first step is to select the amount and type of proppant to be used. This is equivalent to specifying the "size" of the treatment. Once the amount of a certain proppant is selected, the fracture length is fixed, because, as will be demonstrated, there is an optimum length corresponding to each total amount of proppant. Given the fracture length, the next task is to select a fluid that can transport and suspend the proppant to the extent necessary as well as create the desired fracture geometry.

For the amount of proppant selected the return on the investment or some other economic measure can then be determined, a different amount of proppant may be assumed, and the return can again be computed. A plot of return on investment versus mass of proppant will reveal the optimum.

Optimum fracture length. Figures 10.5 and 10.6 showed the fracture conductivity as a function of proppant surface concentration. If M_0 is the selected amount of proppant, then for uniform coverage the proppant surface concentration is given by

$$M_0 = 2hL\Gamma_s \tag{12.3}$$

Given the total amount of proppant, there exists an optimum fracture length which maximizes the stimulation ratio. To understand just why an optimum length does exist, it is helpful to note that for a given closure stress, the fracture conductivity can over a reasonable range of Γ_s be represented by an empirical equation of the form

$$w_f k_f = b\Gamma_s^a \tag{12.4}$$

Values for constants a and b were given by Figs. 10.5 and 10.6. Referring to the McGuire–Sikora chart (Fig. 11.2) the relationship between fracture conductivity, fracture length, and stimulation ratio is seen. If a long fracture is selected, then since M_0 is fixed, Γ_s is small. This, according to Eq. (12.4), yields a small fracture conductivity. The stimulation is, therefore, limited by fracture conductivity. By decreasing the fracture length, both the fracture conductivity and the stimulation ratio will increase until it is no longer limited by fracture conductivity. Further decreases in length will ultimately result in a decrease of the stimulation ratio. The length which maximizes the stimulation ratio is the optimum one.

Figure 11.2 applies when the fracture conductivity is uniform from the wellbore to the fracture tip. However, it seems clear that productivity can be improved if more of the proppant is placed near the wellbore and less in the more distant parts of the fracture. In fact, one can seek to find that distribution of proppant which maximizes the stimulation ratio. This problem has been considered by Sevougian [9]. He found that placing the proppant along the length of the fracture in the best possible way produces only a marginally better result than distributing it uniformly. This is fortunate because even if the best distribution is calculated, there is really no means for depositing the proppant in the preferred surface concentration which depends on position. Thus for practical reasons, our attention is restricted to uniform conductivities.

For an undamaged well, Eq. (11.12) gives the stimulation ratio as

$$\frac{J_s}{J_0} = \frac{\ln\left(\frac{r_e}{r_w}\right)}{\left(\ln \frac{L + \frac{\overline{w}_f k_f}{\pi k}}{r_w + \frac{\overline{w}_f k_f}{\pi k}}\right) + \ln\left(\frac{r_e}{L}\right)} \tag{12.5}$$

To further simplify the calculation, neglect the term r_w/L as compared to the fracture conductivity ratio, CR. This is generally an excellent approximation.

To maximize J_s/J_0, the fracture length should be selected to make the denominator of Eq. (12.5) as small as possible. Thus, our optimization problem reduces to the following:

$$\min_L \left[\ln\left(\frac{1 + CR}{CR}\right) - \ln L\right] \tag{12.6}$$

where $CR = w_f k_f/\pi k L$. A necessary condition for the existence of an optimum

$$\frac{d}{dL}\left[\ln\left(\frac{1 + CR}{CR}\right) - \ln L\right] = 0$$

It can be shown that when Eq. (12.4) applies

$$\frac{1}{CR}\frac{dCR}{dL} = -\frac{1}{L}(1 + a) \tag{12.7}$$

Using this result, one finds from Eq. (12.6) the remarkably simple, yet instructive, result

$$CR^* = a \tag{12.8}$$

The asterisk is appended to denote this value of CR as the optimum one. CR^* is the value of CR that satisfies Eq. (12.6).

Combining Eq. (12.8) with Eq. (12.3), the optimum length is found to be

$$L^* = \left(\frac{M_0}{2h}\right)^{a/1+a}\left(\frac{b}{\pi k a}\right)^{1/1+a} \tag{12.9}$$

Example 12.1 Optimum Fraction Length

A deep gas well is to be fractured. The well data are given in Table 12.2. If 22,680 kg of 20/40-mesh sand is to be used as proppant, what is the optimum fracture length and fracture conductivity? What steady-state stimulation ratio will be obtained?

Solution At this point, an estimate of the final propped fracture height is required. This is the gross fracture height less the distance proppant has settled. Until a fracture fluid is selected, this distance is not known. Initially we will allow the proppant to settle about 5 m, because in this case the entire producing zone (net fracture height) will be open to a conductive fracture. This leaves a propped height of 28.5 m.

TABLE 12.2 Well and Formation Data for Example 12.1

Well Data	Value
Young's modulus (E)	1.724×10^{10} Pa
Poisson's ratio (ν)	0.3
Permeability (k)	3.95×10^{-16} m^2
Reservoir fluid compressibility (κ_{fl})	9.8×10^{-8} Pa^{-1}
Reservoir fluid viscosity (μ_f)	2×10^{-5} kg/m-sec
Reservoir pressure (p_R)	3.103×10^7 Pa
Fracture gradient (FG)	18.1 kPa/m
Closure pressure (σ_c)	5.5×10^7 Pa
Formation depth (D)	3658 m
Reservoir temperature (T_R)	135°C
Drainage radius (r_e)	400 m
Wellbore radius (r_w)	0.1 m
Porosity (ϕ)	0.12
Treatment Data	**Value**
Gross fracture height (h)	33.5 m
Net fracture height (h_N)	24.4 m
Amount proppant 20/40-mesh sand (M_0)	22,680 kg
Type of gel (hydroxypropyl guar)	$n = 0.74$ and $m = 0.52$ kg/m-sec$^{1.26}$
Overall fluid-loss coefficient (C)	7.9×10^{-5} m/sec$^{1/2}$
Maximum injection rate (i)	0.0375 m^3/sec

From Figure 11.4 for 20/40 sand at a closure pressure of 55 MPa we have

$$w_f k_f = b\Gamma_s^a = (2.17 \times 10^{-14})\Gamma_s^{0.9846}$$

The optimum dimensionless fracture conductivity is

$$CR^* = a = 0.9846$$

and from Eq. (12.9)

$$L^* = \left[\frac{22{,}680}{2(28.5)}\right]^{0.9846/1.9846} \left[\frac{2.17 \times 10^{-14}}{\pi(3.95 \times 10^{-16})(0.9846)}\right]^{1/1.9846}$$

or

$$L^* = 83 \text{ m}$$

The proppant surface concentration corresponding to the final fracture geometry is

$$\Gamma_s = \frac{22{,}680}{2(28.5)(83)} = 4.8 \text{ kg/m}^2$$

The steady-state stimulation ratio expected from this treatment is

$$\frac{J_s}{J_0} = \frac{\ln(400/0.1)}{\ln(1.9846/0.9846) + \ln(400/83)} = 3.6$$

There are several important points to be made before moving on to the next phase of the design process. First, one may wonder whether or not a proppant not susceptible to the degree of crushing characteristic of sand will improve performance. There are several possible choices available, including resin coated sand and bauxite (see Chap. 10). If, for example, bauxite rather than sand is used as a proppant, then based on the results shown by Fig. 10.6, $CR^* = 0.6478$ at a closure pressure of 60 MPa. The optimum fracture length found by applying Eq. (12.9) is 439 m, or a distance in excess of the drainage radius. If we take L^* to be equal to the drainage radius, the stimulation ratio is found to be 8.9 which is considerably larger than that obtained using an equivalent mass of sand.

This result obtained using bauxite must, however, be viewed with some caution because the proppant surface concentration is 1 kg/m^2 rather than the 4.8 kg/m^2 found for sand. The final fracture width is related to the surface concentration by the equation

$$\Gamma_s = \overline{w}_f(1 - \phi_f)\rho_p \tag{12.10}$$

where ϕ_f is the final porosity of the propped fracture. For sand, $\rho_p = 2500$ kg/m^3 and for a surface concentration of 4.8 kg/m^2 (see Example 12.1), $w_f \approx 3$ mm. For 20/40-mesh sand, this represents about 5 layers of sand particles which except for soft formations where extensive embedment is expected, should be adequate. For bauxite ($\rho_p = 3800$ kg/m^3) a similar calculation shows there is just about a monolayer of proppant at optimum (see Prob. 10.5). This amount is probably insufficient. One should reduce the fracture length and increase CR until an adequate number of layers is obtained. This would be a suboptimum solution and J_s/J_0 would be decreased from the optimum value.

It should be noted that some practitioners recommend $CR = 10$ [10]. This choice is not, however, optimum. In some cases it may be a prudent criterion as, for example, when the number of proppant layers at optimum appears to be too small, but for the case considered in Example 12.1—namely, sand under high closure pressures—arbitrarily choosing $CR = 10$ cannot be justified.

Another issue which must be considered is the proposition that optimizing the steady-state stimulation ratio is sufficient. If flush production occurs over an extended period, then this added production occurs over an extended period and will contribute to the economic evaluation perhaps changing the optimum fracture geometry. For example 12.1, 200 days of flush production is predicted based on Eq. (11.1). This period will probably not significantly change the optimum fracture geometry; however, a detailed analysis to ascertain the change must be based on economic considerations.

Thus, the optimum length corresponding to the choice of proppant amount is now fixed. The increased production rate is predicted and can be assigned an economic value. To calculate the return on investment, the fracture process must be defined and its cost estimated.

Selection of a fracture fluid. Generally, this is a trial-and-error process. If the fracture length is long, then fluids which maintain their viscosity at the reservoir temperature for several hours may be required. For relatively short fractures, such as that found to be optimum in Example 12.1, the entire process may require less than one hour and therefore, the polymer concentration can be reduced. Having selected a fluid for consideration, one must ensure that both the desired fracture geometry can be created and the proppant transporting capabilities are satisfactory. If one or both of these conditions are not satisfied, the entire calculation must be repeated until a fluid is found which just satisfies them.

Example 12.2 Apparent Viscosity in Fractures

For the treatment conditions given in Example 12.1, determine the effective viscosity of a fracture fluid consisting of 3.6 kg of polymer/m^3. According to the values given in Table 12.1, $n = 0.74$ and $m = 0.52$ kg/sec$^{1.26}$-m at a temperature of 135°C.

Solution Let us first estimate an effective viscosity. This is a task preliminary to the determination of the dynamic fracture geometry. It is required here because the design equations are restricted to Newtonian fluids. However, in practice, computer programs simulating the fracturing process including the effect of non-Newtonian behavior are used, making the estimation of an effective viscosity unnecessary.

The effective viscosity is defined as follows:

$$\mu_{EFF} = \frac{m}{3}\left(\frac{i}{h}\right)^{n-1}\left(\frac{2n+1}{n}\right)(w_{max})^{2-2n} \tag{8.53}$$

where $w_{max} = 4\overline{w}/\pi$. Furthermore, the fracture width at the wellbore in the long-time limit is found by substituting μ_{EFF} for the Newtonian viscosity as follows:

$$\overline{w}(0, t) = 1.12\left[\frac{(1-\nu)\mu_{EFF}\, i^2}{GCh}\right]^{1/4} t^{1/8} \tag{8.48}$$

These two equations can be solved simultaneously to yield an effective viscosity evaluated at the wellbore. This effective value will be used here in our fracture design.

The bulk modulus G is related to the Young's modulus by Eq. (2.4):

$$G = \frac{E}{2(1+\nu)} = \frac{1.724 \times 10^{10}}{2(1+0.3)} = 6.63 \times 10^9 \text{ Pa}$$

The value of C reported in Example 12.1 is taken from Fig. 12.2. This is actually C_w and in practice the viscous and compressible fluid-loss coefficients must be determined and C calculated. These steps have been described in Chapter 8 and for conciseness are omitted here. There is, however, an additional correction which must be made in each case. The gross fracture height is 33.5 m, but the productive or permeable zone height is 24.4 m (see Table 12.2). Since fluid is lost only in the permeable zone, then the effective fluid-loss coefficient is

$$C = (7.9 \times 10^{-5})\left(\frac{24.4}{33.5}\right) = 5.75 \times 10^{-5} \text{ m/sec}^{1/2}$$

TABLE 12.3 Effective Viscosity as a Function of Time

Time (hr)	$\overline{w}(0,t)$ (mm)	μ_{EFF} (kg/m-sec)
0.1	4.86	0.242
0.3	5.69	0.262
0.5	6.12	0.272

From Example 12.1, $i = 0.0375$ m^3/sec and $\nu = 0.3$. Solving for μ_{EFF}, the results given in Table 12.3 are found.

As expected, the effective viscosity increases as the fracture width increases. Since these calculations are based on the long-term approximation, the width here will be larger than that calculated using more precise equations. Thus, a reasonable effective viscosity is roughly 0.262 kg/m-s.

Example 12.3 The Dynamic Fracture Geometry

Having established in Example 12.2 an effective viscosity for a fracture fluid containing 3.6 kg/m^3 of polymer, determine the dynamic geometry of the fracture. The process variables are given in Table 12.2.

Solution The following dimensionless variables are defined in Chapter 8:

$$t_D = \frac{16}{\pi^2}\left[\frac{2C^5hG}{(1-\nu)\mu i^2}\right]^{2/3} t \qquad \text{(time)} \qquad (8.22)$$

$$L_D = \frac{16}{\pi}\left[\frac{2C^8Gh^4}{(1-\nu)\mu i^5}\right]^{1/3} L \qquad \text{(length)} \qquad (8.23)$$

$$w_D = \frac{4}{\pi}\left[\frac{C^2Gh}{4(1-\nu)\mu i^2}\right]^{1/3} \overline{w} \qquad \text{(width)} \qquad (8.24)$$

Thus

$$t_D = \frac{16}{\pi^2}\left[\frac{2(5.75\times 10^{-5})^5(33.5)(6.63\times 10^9)}{0.7(0.262)(0.0375)^2}\right]^{2/3} t$$

or

$$t_D = 1.709 \times 10^{-4} t$$

Figures 8.14 and 8.15 provided the dimensionless fracture width (w_D) and length (L_D) as a function of dimensionless time (t_D). The results obtained are listed in Table 12.4.

The dynamic fracture width created by this fracture fluid is satisfactory. It is about 7 times the proppant diameter after $t = 0.1$ hr. A value of 3 or less would be unacceptable. A value of at least 5 is desirable.

Another test of adequacy that a fracture fluid must satisfy is to ensure that the proppant settling is limited. In this case we have decided to restrict the level to 5 m. The time to create a fracture having the optimum length of 83 m is about 36 min (see Table 12.4). As discussed in Chapter 10, once pumping has ceased, an additional time (Δt) is required for the fracture to close. This time interval was shown in Chapter 10 to be

$$\overline{w}(t_f) - \overline{w}_f = 2\pi C\,[(t_f + \Delta t)^{1/2} - t_f^{1/2}] \qquad (10.16)$$

TABLE 12.4 Fracture Geometry as a Function of Time for the Treatment Considered in Examples 12.1, 12.2, and 12.3

Time (hr)	$t_D \times 10^1$ (dimensionless)	w_D (dimensionless)	$\overline{w}$ (mm)	L_D (dimensionless)	L (m)
0.05	0.308	0.47	4.13	0.066	24.6
0.10	0.615	0.49	4.31	0.110	40.9
0.15	0.923	0.51	4.49	0.140	52.1
0.20	1.23	0.55	4.84	0.165	61.4
0.30	1.85	0.59	5.19	0.200	74.5
0.40	2.46	0.63	5.54	0.245	91.2
0.50	3.08	0.65	5.71	0.290	108
0.60	3.69	0.68	5.98	0.323	120

Based on the sum of the treatment time and the closure time, the distance of proppant settling can be estimated.

Example 12.4 Proppant Settling Distance

Estimate the settling distance of the 20/40 sand during the fracture treatment prescribed by Examples 12.1, 12.2, and 12.3.

Solution The settling velocity was discussed in Chapter 10. There are two periods to be considered. The first is the period during which fluid is injected. This period is, according to Table 12.4, about 30 min. When pumping ceases, the fracture must decrease in width from 5.71 mm to its final width of 3 mm (the thickness for which $\Gamma_s = 4.8$ kg/m^2, see Example 12.1). This time interval is from Eq. (10.16):

$$5.71 \times 10^{-3} - 3 \times 10^{-3} = 2\pi(5.75 \times 10^{-5})[(2.16 \times 10^3 + \Delta t)^{1/2} - (2.16 \times 10^3)^{1/2}]$$

or

$$\Delta t = 753 \text{ sec} = 0.21 \text{ hr}$$

In the first stage, Eq. (10.11) applies and the mean shear rate due to both flow and settling is given by

$$\dot{\gamma}_{\text{ave}} = \sqrt{\dot{\gamma}^2_{\text{settling}} + \dot{\gamma}^2_{\text{flow}}}$$

where the average shear rate due to flow is estimated using Eq. (10.13). Thus

$$\dot{\gamma}_{\text{flow}} \cong \frac{3i}{2\overline{w}^2 h} = \frac{3(0.0375)}{2(4.13 \times 10^{-3})^2(33.5)} = 98 \text{ sec}^{-1}$$

This value has been calculated using the smaller of the widths listed in Table 12.4. By taking a lower estimate of the width, the average shear rate is then overestimated and thus the distance the proppant settles will also be overestimated. This provides a conservative design. $\dot{\gamma}_{\text{settling}}$ is, according to the discussion in Chapter 10, approximately v_s/d_p.

The settling velocity will be estimated here using Eq. (10.6)

$$v_s = \frac{g(\rho_p - \rho)d_p^2}{18\mu_{\text{APP}}}$$

where

$$\mu_{\text{APP}} = m\dot{\gamma}_{ave}^{n-1} \qquad (3.13)$$

Thus we solve for v_s given $d_p = 0.66$ mm, $n = 0.74$, $m = 0.52$, $\rho_p = 2500$ kg/m^3, and $\rho = 1000$ kg/m^3. The value can be found to be

$$v_s = 2.46 \times 10^{-3} \text{ m/sec}$$

While pumping, the maximum distance a proppant particle injected at $t = 0$ will settle is equal to $v_s t_f$ or for this case 4 m in 30 minutes. Once pumping stops, the shear rate is entirely due to settling. In this case, the settling velocity is found to be

$$v_s = 3.1 \times 10^{-4} \text{ m/sec}$$

or during time Δt

$$\text{settling distance} = (0.21)(3600)(3.1 \times 10^{-4}) = 0.23 \text{ m}$$

Thus, the total settling distance is 4.2 m. The fracture fluid, therefore, provides satisfactory proppant carrying capacity. In fact, since hindered settling occurs in a slurry (see Chap. 10), this amount of settling is an overestimate and a fluid containing smaller amounts of polymer should be considered.

In addition to creating the desired fracture geometry and being able to suspend proppant, a fracture fluid must be pumpable, that is, the friction loss in the tubing must not produce an excessive surface pressure (Chap. 7). It will also be assumed that various fluid-loss additives have been investigated to determine whether they are cost effective and need, therefore, be included. Longer fractures will require fluid-loss additives. The selection of an appropriate fracture fluid together with its additives determines the cost of materials. From knowledge of surface pressure and the injection rate, the fluid horsepower and hence equipment costs are known. It is now possible to evaluate the cost of the treatment. This information taken together with the stimulation ratio determines the return on investment or some other economic measure that will be used to judge the process. This calculation should be repeated until the optimum size treatment is found.

Injection schedule. A fracture treatment is generally initiated by first injecting water containing small quantities of polymer selected so as to reduce the friction pressure (see Chap. 3). This fluid is sometimes called *slick water*. Its viscosity is essentially that of water and it readily invades the formation surrounding the wellbore, thereby increasing the pore pressure. This situation is helpful in initiating a fracture; that is, in "breaking down" the formation. If in this initial stage of the operation some unexpected mechanical difficulty should arise, then the tubing is filled with a water-like fluid and not a highly non-Newtonian fracture fluid.

Following the slick water, the polymer solution is injected but proppant is not immediately added. This fluid which contains polymer but not proppant is called a *pad fluid* and the volume of this fluid that is injected is called the *pad volume*. The purpose of the pad volume is to create a fracture of sufficient width and length so that when proppant is introduced, it can be freely transported along the fracture. It is not desirable for proppant to reach the end of the fracture because the fracture narrows sharply at the end and proppant particles could conceiveably bridge across the width of the fracture, thereby prematurely terminating proppant transport down the fracture. Thus, the pad volume must be sufficient to prevent proppant from reaching the end of the fracture. On the other hand, the pad volume cannot be so large as to leave a significant fraction of the fracture length devoid of proppant. Thus, the pad volume should be selected with some care.

The injection schedule is simply a listing of the total volumes and compositions of each of the stages of a fracture treatment. To resolve this issue it is useful to be able to track the movement of a particle as it progresses down the length of a fracture. In particular, we would like to know the time t^* at which the fluid element occupying the position x at time t was injected.

Our calculation will assume that the mean fracture width is a constant. This result is a reasonable approximation toward the end of the treatment when proppant is to be injected (see Table 12.4). The approach will be to calculate the axial velocity u_A at the point x, because the distance a particle moves during the time interval dt is given by

$$\frac{dx}{dt} = u_A \tag{12.11}$$

A volume balance gives

$$\overline{w}hu_A = \frac{1}{2}i - 2hC\int_0^x \frac{dx'}{\sqrt{t - \tau(x')}} \tag{12.12}$$

The second term on the right-hand side is the rate at which fluid is being lost to the formation through that fracture surface area contained between the wellbore and the point x measured along the fracture (see Chap. 8). When the fracture width becomes constant, then

$$L(t) = \frac{1}{2\pi}\left(\frac{i}{Ch}\right)t^{1/2}$$

(see Eq. 8.42). Therefore

$$\tau(x) = \left(\frac{2\pi Chx}{i}\right)^2 = x^2/a^2 \tag{12.13}$$

Substituting this result into Eq. (12.12) yields

or

$$u_A = \frac{dx}{dt} = \frac{i}{2\overline{w}h}\left[1 - \frac{2}{\pi}\sin^{-1}\left(\frac{x}{L}\right)\right] \tag{12.14}$$

We would like to integrate this equation to obtain $x(t, t^*)$ where t^* is the time at which the fluid element that is at the position x at time t was introduced into the fracture. However, the integration is complicated by the fact that L depends on time. We can eliminate the time by noting $t = L^2/a^2$ and $t^* = L^{*2}/a^2$. The differential Eq. (12.14) can be written as

$$\frac{dX}{d\psi} + \frac{X}{\psi} = \beta\left(1 - \frac{2}{\pi}\sin^{-1}X\right) \tag{12.15}$$

where

$$\psi = \frac{L}{L^*} = \frac{L}{at^{*1/2}}$$

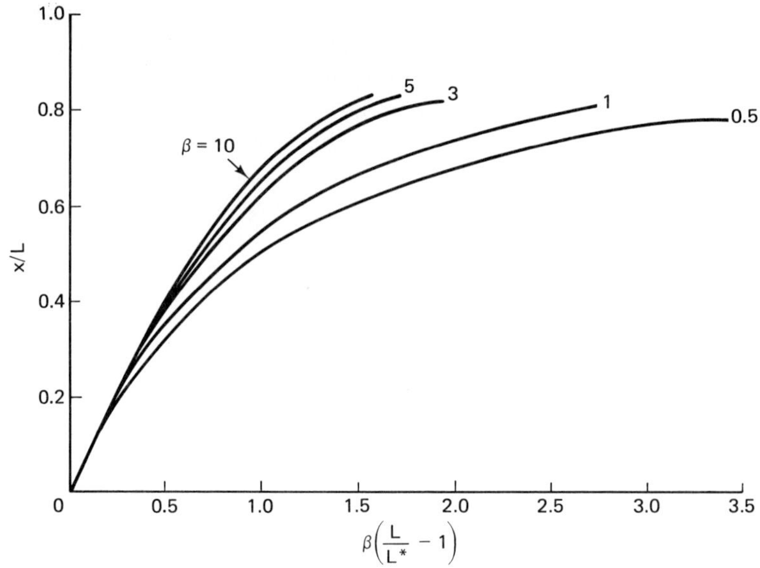

Figure 12.3 Graph showing dimensionless distance traveled down a fracture by a fluid element introduced into the fracture at a time when the length is L^*.

$$X = x/L$$

$$\beta = \frac{2\pi C t^{*1/2}}{\overline{w}}$$

Equation (12.15) is an ordinary differential equation that can be integrated to yield X as a function of ψ where β is a parameter. The values of X so obtained are represented in Fig. 12.3. This graph can be used to calculate the time at which a fluid element arrives at a particular $X = x/L$ along a vertical fracture.

Nolte [11, 12] has developed a different approach for determining the pad volume. He considers both the long-time and the short-time rates at which fluid is lost, whereas the method developed here is conservative. Only the long-time behavior is considered and a lower limit on the fluid-loss rate taken. Thus, the distance traveled by a proppant particle is overestimated. Nolte did not, however, attempt to follow the movement of particular particles along the fracture. Thus, this method may have some advantages.

Example 12.5 Determination of the Pad Volume

Continuing our design of the fracture treatment studied in Example 12.1, determine a reasonable fluid and proppant injection schedule.

Solution The optimum fracture length is 83 m. To be safe in avoiding premature clogging of the fracture by proppant, a fracture of about 100 m will be created; however, only 83 m of the created fracture will be propped. Thus according to the results given in Table 12.4 to achieve a fracture length of 100 m, fracture fluid should be injected for about 0.45 hr. This gives a total volume of fracture fluid of it_f, which for this case is 61 m^3. Proppant injection should be initiated at a time such that the proppant has advanced 83 m down the fracture when fluid injection is terminated. Thus, we desire to know L^* for which $X/L = 0.83$ when $L = 100$ m. One should try various values of L^*, to see when these conditions are satisfied. For example, take $L^* = 41$ m, then according to the results tabulated in Table 12.4, $t^* = 0.1$ hr. For this value of t^*

$$\beta = \frac{2\pi(5.75 \times 10^{-5})(360)^{1/2}}{4.31 \times 10^{-3}} = 1.6$$

Thus

$$\beta\left(\frac{L}{L^*} - 1\right) = 1.6\left(\frac{100}{41} - 1\right) = 2.3$$

Reading from Fig. 12.3, we find $x/L = 0.84$. Thus, proppant injected at t = 0.1 hr will reach 84 m at t = 0.45 hr. This is the desired propped length.

A typical treatment procedure appears thusly:

- 10 m^3 of slick water injected at a rate such that the surface pressure does not exceed the maximum limit
- 13.5 m^3 of pad volume composed of cross-linked hydroxypropyl guar 3.6 kg guar/m^3 injected at 0.0375 m^3/sec
- 47.3 m^3 of cross-linked hydroxypropyl guar solution (3.6 kg guar/m^3) blended with 480 kg of 20/40 sand/m^3 of fracture fluid injected at 0.0375 m^3/sec

The proppant concentration found in the previous example is not excessive. Typical ranges of proppant concentration are given in Table 12.5. The injection schedule, as defined in Example 12.5, is a relatively simple one because the process is to be completed in a half hour. Since this fracture fluid is capable of carrying sand/concentrations higher than 480 kg/m^3, it might be wise to initially inject a somewhat smaller concentration of proppant and finally to increase it to a level sufficient to transport all of the proppant into the fracture. Care should be taken so as not to flush the proppant away from the region near the wellbore (see Example 11.4).

Practical Considerations in Designing Fracture Treatment

Pumping rate. The fluid injection rate is an important design parameter that should be as large as possible. It is, however, limited by the strength of wellhead and tubular goods. Safety factors should be incorporated into the design if the fracture gradient is not precisely known, if the pressure drop through the

TABLE 12.5 Typical Sand Concentrations for Various Types of Fluid

Fluid	Proppant (kg/m^3)
Gelled water	120
Foam	360
Cross-linked gelled water	840
Polymer emulsion	840

perforation is significant, or if friction pressures are not well established. Small errors can result in substantial increases in surface pressures.

Fracture height. The design is based on an assumed knowledge of the fracture height and the net sand pay thickness. Uncertaintities in these values can lead to uncertainties in the results. Sensitivity analyses should be carried out. It is often better to overdesign a treatment until the fracture height normally created in a particular formation can be established or until measurements of the in-situ stresses as a function of depth are available. If the in-situ stresses are known, then the design can be carried out using a three-dimensional fracture simulator.

Fluid diversion. In thick horizons it may be necessary to fracture isolated sections or to attempt to divert the fluid from one zone to another by plugging perforations during the course of a treatment. Diverting of fracture fluid during proppant fracturing is not desirable and can obviously lead to difficulties. An alternate plan is limited entry perforating, as described in Chapter 7.

Fluid loss. Knowledge of the fluid-loss coefficient is critical. Excessive fluid loss can lead to premature bridging of the proppant across the fracture and ultimate "sand out." A buildup of proppant within the fracture or at the wellbore will be signaled by a sudden increase in surface pressure, forcing premature termination of the treatment. "Sand out" is most often a result of poor fluid-loss control. In new formations the design should include a safety factor (large pad volume) to ensure that the proppant is placed.

12.2 ACID FRACTURING

Design considerations for acid fracturing of carbonate formations are not greatly different from those applied to proppant fracturing. In the latter case an amount of proppant is selected to characterize the treatment size and the particular size of treatment optimized. In this section the same approach will be used. The amount of acid to be used in the treatment will be fixed. This fixed amount of acid will in turn fix the optimum fracture length. The properties of the available fluids will be studied to determine whether or not the desired fracture length can be created. Once the composition of the fracture fluid has been specified, its cost is known and the return on investment can be calculated. This design can be repeated for different acid volumes and the best one may be selected using some economic criterion.

Selection of Fracture Fluid and Additives

In an acid fracturing treatment, either an acid alone is injected into the formation or an acid preceded by a viscous pad fluid to form a wide, deep fracture. The viscous pad will generally contain suitable flow-loss control agents such as 100-mesh sand when fracturing carbonate formations that tend to be naturally fractured. If a viscous pad is used and the acid contains neither fluid-loss control agents nor viscosifiers, then the fracture created by the pad fluid will begin to close as the acid is injected and will approach a geometry expected if acid alone had been used [13]. This decrease in fracture volume occurs because the acid wormholes through the region invaded by the viscous fluid (see Chap. 17) and thereby increases the rate of fluid loss. In recognition of this problem some treatments are performed so that the acid and viscous pad fluid are injected in alternate batches and thus even though the fracture would tend to close as acid injection begins, it would not become too small before the following viscous solution is injected to reestablish the long wide fractures.

An alternative solution is to viscosify the acid and to use appropriate fluid-loss control agents. One system which has met with some success is an acid external polyemulsion that is 33% acid by volume and generally contains silica flour to control fluid losses. In preparing this system, polymer, often guar, is added to the acid. This system is effective but obviously has only one third the dissolving power of a plain acid solution.

Gelled acids (14, 15) using xanthan gums are also used for acid fracture treatments. Crowe et al. [15] have reported extensive tests of various polymer thickening agents in acid solutions at elevated temperatures and concluded that the xanthan gums showed the best overall potential. Gelled 15 wt% hydrochloric acids exhibited good stability up to temperatures of 150°F. These acids can be used in hotter wells if care is taken to precool the formation. Precooling the rock surrounding a fracture was discussed in Chapter 9.

Crowe et al. [15] report that gelled acids exhibit both lower fluid loss and lower reaction rates than plain HCl. This is true perhaps because the increased viscosity may reduce the effective diffusion of acid to the fracture surface.

Cross-linked gelled acids are also available [16]. Substantially larger quantities of polymer (9.6 kg/m^3) are required to achieve acceptable viscosities in concentrated acid solutions. Thus both gelled and cross-linked acids will tend to be more costly than processes using viscous prepads; however, it is clear that such fluids do have advantages and can potentially extend the applicability of acid fracturing.

Fluid-loss control additives such as oil-soluble resins, silica flour, 100-mesh sand, and diesel oil are all used in acid solutions. In plain acid, their effectiveness has been questioned [17]. The fluid-loss coefficient of acid systems is difficult to predict and is, as will be evident, important in determining the effectiveness of the acid treatment.

Other additives that are blended with the acid are:

- *Corrosion Inhibitors:* For all acid treatments, these inhibitors should be included to protect the metal goods from acid attack. Hydrochloric acid is particularly corrosive because of its pitting action; that is, its reactivity tends to be localized on the metal surfaces. The types and effectiveness of corrosion inhibitors are discussed in Chapter 18.
- Oil-Wetting Surfactants: Oil-wetting surfactants are added to reduce or retard the rate of acid attack on carbonate rocks so that the acid can penetrate

further down the fracture. The value of these additives has been shown to be only marginal and their use for retarding the rate of acid reaction is not recommended.

- *Emulsion-Breaking Surfactants:* These surfactants are useful for avoiding emulsions that tend to form when the spent acid and formation fine material mix with formation oil. If these emulsions reduce the effective permeability along the fracture surface drastically, surfactants will be useful provided laboratory tests, carried out at reservoir conditions, support their applicability. Surfactants should be compatible with the corrosion inhibitors, polymers, clay stabilizers, and sequestering agents and therefore, care should be taken in the choice of surfactant (see Chap. 3 for further discussion).
- *Friction Reducers:* Just as drag-reducing agents (see Chap. 3) can reduce the friction pressure drop for water, small amounts of polymer (usually polyacrylamides) can also reduce the friction losses of plain HCl.
- *Iron-Sequestering Agents:* In some formations, when acid reacts, an iron precipitate can form. To prevent this precipitation, which can damage the formation permeability or reduce the fracture conductivity, a small amount of organic acid is normally added to keep the pH of the reacted acid low. These sequestering agents are also called *complexing agents* or *stabilizers*. This type of additive should only be used when it is known that an iron-precipitation problem exists in a given well or reservoir (see Chap. 18 for further details).
- *Clay Stabilizers:* Since acids are used in treating carbonate formations, clay-stabilization problems generally do not arise or are not severe. In rare cases, clay stabilizers may be used; however, it should be remembered that in acid solution, the charge on clay particles becomes positive (below pH 3, see Chap. 1) and stabilizers that work at higher pH will not be effective.

The design of acid fracturing fluids is more difficult than proppant fracturing fluids, because of the complex chemical reactions that occur. However, acid fracturing is generally a more forgiving process. Errors in design do not result in "sand outs" as is the case for proppant fracturing.

Design of Acid Fractures

The basics of acid fracturing treatment design are similar to proppant fracturing treatment design in that the size of the treatment is dictated by economics. In a certain sense acid fracturing treatments are easier to design because of the limited choice of fluids and because of the limited control over the fracture conductivity. Given an acid volume, there is, however, an optimum length. The reason for this is illustrated in Fig. 12.4. The short fracture has a high fracture conductivity since, as shown in Chap. 10, the more rock dissolved the greater will be the fracture conductivity. The long fracture has a smaller fracture conductivity since less rock is dissolved within a given fracture area. Thus if the volume of acid is specified, the volume of rock which can be dissolved is fixed. However, just as in proppant

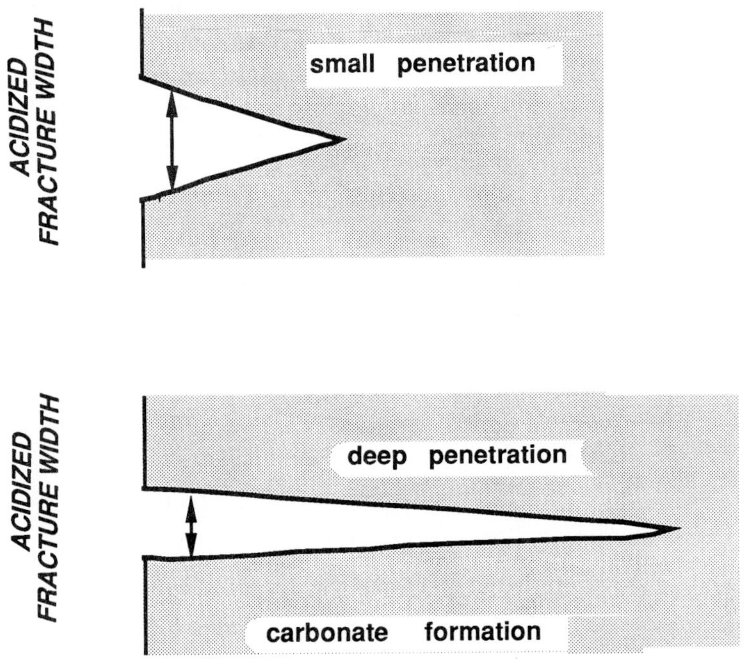

Figure 12.4 Sketch depicting two different fractures, both created using the same volume of acid.

fracturing, long poorly conductive or short highly conductive fractures can be created. Between these two extremes, there is an optimum length.

Optimum fracture length. Given a volume of acid and the ability to dictate where along the length of a fracture this acid reacts, a maximum stimulation ratio can be found. However, we cannot direct the precise point at which the acid reacts and therefore, this maximum stimulation ratio is unattainable. It can be shown that a uniform fracture conductivity yields very nearly the same stimulation ratio as the maximum one [9]. This is an important result since it is possible to select the process conditions so that a nearly uniform fracture conductivity is created. In this section the optimum fracture length is selected so that a fracture of uniform conductivity maximizes the stimulation ratio. It will then be necessary to determine just how to achieve this optimum fracture geometry.

A uniform conductivity implies that rock is dissolved uniformly over the entire fracture surface. Thus the ideal fracture width is given by

$$\langle w_i \rangle = \frac{X V_{\text{acid}}}{2\,(1 - \phi) h L} \tag{10.34}$$

Furthermore, assuming the same ideal fracture width is created over the entire fracture surface, the final fracture conductivity is found from the equation

$$\overline{w}_{\text{f}} k_{\text{f}} = (2.9 \times 10^{-4}) \left[\frac{\langle w_i \rangle^3}{12} \right]^{0.822} \exp(-c_2 \sigma_{\text{c}}) \tag{10.37}$$

where c_2 depends on the rock embedment strength and $\overline{w}_f$ is used to denote this conductivity is to apply over the entire fracture surface.

It is both interesting and instructive to note the term $V_{acid}/2hL$, where $V_{acid} = it_A$, is precisely analogous to the proppant surface concentration. To emphasize this comparison, substitute Eq. (10.34) into Eq. (10.37) so that the fracture conductivity can be put into the form

$$\overline{w}_f k_f = b\left(\frac{V_{acid}}{2hL}\right)^{2.47} \tag{12.16}$$

where $$b = (2.9 \times 10^{-4})\left[\frac{X}{2.29\,(1 - \phi)}\right]^{2.47} \exp(-c_2\sigma_c)$$

Since this relationship has the same form as Eq. (12.4), it follows that the optimum conductivity and fracture length are related by the same result. Thus, by analogy, Eqs. (12.8) and (12.9) apply to acid fracturing as follows:

$$CR^* = a = 2.47 \tag{12.17}$$

and $$L^* = \left(\frac{b}{2.47\pi k}\right)^{1/3.47}\left(\frac{V_{acid}}{2h}\right)^{2.47/3.47} \tag{12.18}$$

Example 12.6 Optimum Acid Fracture Length

Consider the oil well described by the data given in Table 12.6. Determine the optimum fracture length and the corresponding stimulation ratio if the acid treatment is to use 40 m^3 of 28 wt% HCl.

Solution One of the important parameters is the closure stress that the fracture will be required to sustain. If the well flowing pressure is estimated to be about 3000 kPa, then

$$\sigma_c = (FG)D - p_{wf} = (15.8)(1524) - 3000 = 21{,}100 \text{ kPa} = 2.11 \times 10^7 \text{ Pa}$$

To obtain c_2 refer to Eq. (10.37),

$$c_2 = [9.1 - 0.406 \ln(3.79 \times 10^8)] \times 10^{-7} = 1.08 \times 10^{-7}$$

The value of b is then given by

$$b = (2.9 \times 10^{-4})\left[\frac{0.17}{2.29(1 - 0.1)}\right]^{2.47} \exp[-(2.11 \times 10^7)(1.08 \times 10^{-7})] = 6.25 \times 10^{-8}$$

Thus $$L^* = \left[\frac{6.25 \times 10^{-8}}{\pi(2.47)(9.87 \times 10^{-15})}\right]^{1/3.47}\left(\frac{40}{2(30.5)}\right)^{2.47/3.47} = 37.4 \text{ m}$$

The stimulation ratio for the undamaged formation is found by substituting into Eq. (11.12)

$$\frac{J_s}{J_0} = \frac{\ln(200/0.05)}{\ln[(1 + 2.47)/2.47] + \ln(200/37.4)} = 4.11$$

TABLE 12.6 Well and Treatment Data Required for Design of Acid Fractures

Formation Properties	Values
Fracture gradient (FG)	15.8 kPa/m
Porosity (ϕ)	0.1
Reservoir temperature (T_R)	104°C
Young's modulus (E)	6.07×10^7 kPa
Depth (D)	1524 m
Permeability (k)	9.87×10^{-15} m^2
Rock embedment strength (S_{RES})	3.79×10^8 Pa
Drainage radius (r_e)	200 m
Wellbore radius (r_w)	0.05 m
Poisson ratio (ν)	0.25
Reservoir Fluid Properties	**Values**
Viscosity (μ_f)	1×10^{-3} kg/m-sec
Density (ρ_{rock})	2801 kg/m^3
Compressibility (κ_{fl})	2.9×10^{-5} kPa^{-1}
Pressure (p_R)	1.72×10^4 kPa
Treatment Parameters	**Values**
Gross fracture height (h)	30.5 m
Net thickness (h_n)	15.3 m
Maximum injection rate (i)	0.056 m^3/sec
Injection temperature (T_I)	60°C
Acid concentration (c_0)	28% HCl
Dissolving power (X)	0.17 m^3 of rock/m^3 of acid
Acid density (ρ)	1140 kg/m^3

For future reference, the ideal width is

$$w_i = \frac{(0.17)(40)}{2(0.9)(30.5)(37.4)} = 3.31 \times 10^{-3} \text{ m}$$

where the dissolving power is 0.17.

The fracture process will have to be designed to create as nearly as possible this ideal width uniformly along the fracture.

Example 12.6 shows that for a given acid volume, the optimum fracture length will decrease as the rock embedment strength decreases or as the closure stress increases. In fact, serious questions have been raised concerning the stability of a conductive acid fracture when formations of low embedment strength are subjected to high closure stresses [18]. Under these conditions experiments show that rock tends to creep and the fracture tends to heal.

Once the optimum fracture length for a given acid volume has been found, it is necessary to define a process that will create the desired fracture. This problem is difficult because it is difficult to control the distance live acid penetrates down a fracture. Acid will react as soon as it reaches the fracture surface and is displaced into the formation.

Process design. The optimum fracture length has been calculated assuming that a uniform conductivity exists. With acids it is not possible to achieve a perfectly uniform conductivity. Because acid concentration is highest at the wellbore, the amount of rock dissolved at that point will always exceed the amount dissolved at positions remote from the wellbore. Thus, the fracture conductivity will be greater near the wellbore and decrease with increasing distance (see Chap. 10).

The fact that a uniform fracture conductivity can never be achieved is not a real disadvantage. In fact because the fracture conductivity is largest near the wellbore, actual stimulation ratios slightly greater than the optimum value found based on uniform conductivities are possible. It has been found that the most important factor is the fracture length [9]. The designer should make every effort to ensure that the acid is spent over the optimum length even if the conductivity is not precisely uniform. If this priority is followed, then stimulation ratios approaching or perhaps somewhat exceeding the maximum value for uniform fracture conductivities will be found. If, however, the acidized fracture length differs markedly from the optimum one, the stimulation ratio will fall short of that predicted for the optimum fracture length and uniform conductivities.

As has been noted, there are several possible processes that can be used to create the desired fracture geometry. One process that should be considered because of its simplicity is the use of the aqueous hydrochloric acid solution together with fluid-loss and anticorrosion additives but without viscosifying agents. This process is limited to very short fractures because acid tends to form wormholes in carbonate formations (see Chap. 17) and acid flowing through these wormholes will tend to bypass filter cake that may form on the fracture surface. The primary restraint limiting fluid-loss, therefore, is the compression of the formation fluids. The average fracture length for purposes of acid fracture design is defined by the simple volume balance

$$i = 4hL\langle \overline{u}_{\mathrm{N}} \rangle \tag{12.19}$$

where the average fluid-loss flux is determined from Eq. (8.92) as

$$\langle \overline{u}_{\mathrm{N}} \rangle = \frac{\pi C}{t_{\mathrm{f}}^{1/2}} \tag{8.92}$$

Generally the acid injection rate (i) is limited and therefore if the average fluid-loss flux is high, L will be correspondingly limited.

Example 12.7 Inadequacy of Plain Acid Solutions

The optimum fracture length found in Example 12.6 is 37.4 m. Is it possible to create an acidized fracture of this length for the conditions stated in that example using an acid without gelling agents?

Solution If the fluid loss is limited strictly by the compressibility of the formation fluids, then

$$C = C_c = \sqrt{\frac{\phi \kappa_{fl} k}{\pi \mu_f}}\, \Delta p \tag{8.72}$$

This result represents the largest possible fluid loss since C_v and C_w are neglected. Based on the information provided in Table 12.6,

$$C_c = \sqrt{\frac{0.1(2.9 \times 10^{-8})(9.87 \times 10^{-15})}{\pi(1 \times 10^{-3})}}\,[(1.58 \times 10^4)(1.524 \times 10^3) - 1.72 \times 10^7]$$

or

$$C_c = 6.6 \times 10^{-4}\ \text{m/sec}^{1/2}$$

This fluid-loss coefficient must be corrected to take into account the proportion of the gross fracture height which is permeable. In this case half of the gross is permeable so that the corrected value is 3.3×10^{-4}. Also given in Table 12.6 is a maximum injection rate of 0.056 m^3/sec. If the 40 m^3 of acid are injected at this rate, then

$$t_f = \frac{V_{\text{acid}}}{i} = \frac{40}{0.056} = 714\ \text{sec}$$

and

$$\langle \bar{u}_N \rangle = \frac{\pi C}{\sqrt{t_f}} = \frac{\pi(3.3 \times 10^{-4})}{(714)^{1/2}} = 3.88 \times 10^{-5}\ \text{m/sec}$$

The average acid fracture length is

$$L = \frac{i}{4h\langle \bar{u}_N \rangle} = \frac{0.056}{4(30.5)(3.88 \times 10^{-5})} = 11.8\ \text{m}$$

This is far short of the desired length.

Example 12.7 illustrates the limited fracture lengths that can be obtained using plain acid solution. This length can be extended somewhat by using oil-soluble resins to further limit the fluid loss [17]. However, because the acid solution viscosity is small only narrow fractures can be created and this will also severely limit the distance that live acid can penetrate along the fracture. Thus, plain acids are satisfactory fracture fluids whenever the optimum fracture length is relatively short, perhaps less than 10–15 m. Longer fractures corresponding to larger acid volumes require more complex processes to attain.

It should be noted that many highly successful fracture treatments have been realized using plain acids. These undoubtably result when damaged wells are stimulated.

The process recommended here will use viscosified acid preceded by a small pad volume of viscous polymer solution designed to initiate a wide fracture. To simplify the calculations, the viscosity of the acid and the pad volumes will be taken to be equal and both fluids considered to be Newtonian.

The process design entails specifying the acid viscosity and the injection rate, since the acid volume and concentration are already specified. These two variables are intimately related. Increasing the acid viscosity reduces the fluid-loss rate and increases the fracture width. The injection rate can then be adjusted so that the desired fracture length is attained. Thus, there are a range of acid viscosities and corresponding injection rates that will result in a satisfactory fracture length. This being the case, which of the many possible choices is best? The conditions yielding the more nearly uniform fracture conductivity will be selected. As shown in Chap. 10, uniformity is achieved at high Peclet numbers. Thus, the set of conditions yielding the proper fracture length and simultaneously maximizing

$$N_{Pe} = \frac{\langle \overline{u}_N \rangle \overline{w}}{2D_A} \tag{10.24}$$

are to be preferred.

Example 12.8 Selection of an Acid Viscosity

For the well and treatment conditions defined in Example 12.6, what is the recommended acid viscosity and corresponding acid injection rate? The pad fluid is to be injected for five minutes. This time has been arbitrarily selected and one should check to determine if a fracture of suitable size is created prior to acid injection.

Solution The use of viscosified acid decreases the fluid-loss rate. The formation of an invaded zone filled with a viscous fluid results in an additional fluid-loss coefficient.

$$C_v = \sqrt{\frac{\phi k \,\Delta p}{2\mu}} \tag{8.82}$$

As a first try, take the acid viscosity to be 5×10^{-2} kg/m-sec. Then for the data given in Table 12.6, one finds $C_v = 2.61 \times 10^{-4}$ m/sec$^{1/2}$. Correcting this value for the fraction of the gross fracture height which is permeable, yields $C_v = 1.31 \times 10^{-4}$ m/sec$^{1/2}$. An estimate for C_c is found in Example 12.7. If C_w ∞ (neglecting the resistance of the filter cake), then Eq. (8.83) reduces to

$$C = \frac{1}{2}\left[-\frac{C_v^2}{C_c} + \sqrt{\frac{C_v^4}{C_c^2} + 4C_v^2}\right]$$

Substituting for C_v and C_c, we find the overall fluid-loss coefficient

$$C = 1.07 \times 10^{-4} \text{ m/sec}^{1/2}$$

Furthermore,

$$V_{acid} = i(t_f - t_{pad})$$

where t_{pad} is the time the viscous pad fluid is injected. This equation together with Eqs. (12.19) and (8.92) can be used to eliminate $\langle \overline{u}_N \rangle$ and t_f to give the following expression for i:

$$t_{pad} i^2 + V_{acid} i - (4\pi h L^* C)^2 = 0$$

Thus with $h = 30.5$ m, $L = 37.4$ m, $V_{acid} = 40$ m^3, and $t_{pad} = 300$ sec,

$$i = 0.044 \text{ m}^3\text{/sec}$$

is found. This gives

$$t_f = \frac{V_{acid}}{i} + t_{pad} = 1200 \text{ sec}$$

and

$$\langle \overline{u}_N \rangle = \frac{\pi(1.07 \times 10^{-4})}{\sqrt{1200}} = 9.7 \times 10^{-6} \text{ m/sec}$$

Thus, corresponding to an acid viscosity of 0.05 kg/m-sec is an injection rate of 0.049 m^3/sec. This combination will give the desired fracture length of 40 m. To decide if this combination is acceptable, the Peclet number is evaluated.

The dynamic average fracture width is estimated here using the long-term approximation, Eq. (8.48),

$$\overline{w}(0, \tfrac{1}{2}t_f) = 1.12 \left[\frac{0.75(0.05)(0.044)^2}{(2.428 \times 10^{10})(1.07 \times 10^{-4})(30.5)} \right]^{1/4} \left(\frac{1200}{2} \right)^{1/8} = 2.44 \times 10^{-3} \text{ m}$$

The fracture width is a slowly changing function of time, and over the period of acid injection it is relatively constant. Thus

$$N_{Pe} = \frac{(9.7 \times 10^{-6})(2.44 \times 10^{-3})}{2(1.5 \times 10^{-8})} = 0.79$$

The acid diffusion coefficient used here is that for aqueous HCl solutions at reservoir temperature (see Fig. 14.2).

The Peclet number is less than unity, indicating that the fracture conductivity will not be uniform. It is of interest to repeat this calculation for several different acid viscosities. The results shown in Table 12.7 are found.

The results presented in Table 12.7 show that it is not, in this case, possible to find conditions which both give the proper fracture length and at the same time also give a large Peclet number. Thus, it will not be possible to obtain a uniform fracture. The acid viscosity cannot be reduced to a value less than 40 cp, because then injection rates higher than 0.056 m^3/sec will be required. This, however, is not permitted under the conditions of the problem (see Table 12.6). It should be remarked that acid fracturing is generally best carried out at high injection rates with the fluid viscosity adjusted so that the proper length is obtained.

Since the Peclet number is less than about 3, the fracture will be nonuniform. Thus to complete the design, it is necessary to evaluate the stimulation ratio under the best possible conditions. If the result is unsatisfactory, then perhaps one would want to consider using a proppant rather than acid or attempt to allow higher injection rates by whatever means is available.

TABLE 12.7 Different Acid Viscosities and the Corresponding Injection Rates and Peclet Numbers

μ (kg/m-sec)	C (m/sec$^{1/2}$)	i (m^3/sec)	$\langle \overline{u}_N \rangle$ (m/sec)	t_f (sec)	$\overline{w}$ (m)	N_{Pe} (Peclet No.)
0.035	1.24×10^{-4}	0.056	1.23×10^{-5}	1014	2.37×10^{-3}	0.97
0.050	1.07×10^{-4}	0.044	0.97×10^{-5}	1200	2.44×10^{-3}	0.79
0.100	0.80×10^{-4}	0.027	0.60×10^{-5}	1760	2.57×10^{-3}	0.51

Example 12.9 Final Stimulation Ratio Resulting from Acid Fracture

What stimulation ratio is expected from an acid treatment using 40 m^3 of 28% HCl having a viscosity of 40 cp and injected at a rate of 0.056 m^3/sec? The well and treatment data given in Table 12.6 all apply here to this example.

Solution From Chapter 10, the ideal width along a fracture is given by

$$w_i = \frac{XV_{\text{acid}}}{2(1 - \phi)hL} \sum_n G_n \left(\frac{2\lambda_n^2}{3N_{\text{Pe}}} + 1\right)\left(1 - \frac{x}{L}\right)^{2\lambda_n^2/3N_{Pe}} \tag{10.33}$$

Using Tables 10.4 and 10.5, it can be shown that for $N_{\text{Pe}} = 0.971$, $G_0 = 0.9115$, and $\lambda_0 = 1.468$. For small Peclet numbers only a single term of the series is required. Thus

$$w_i = 7 \times 10^{-3}\left(1 - \frac{x}{L}\right)^{1.44}$$

Using this equation together with Eq. (10.37), which gives the fracture conductivity as a function of w_i, Table 12.8 can be constructed.

The stimulation ratio for an undamaged fractured well is

$$\frac{J_s}{J_0} = \frac{\ln\left(\dfrac{r_e}{r_w}\right)}{\sum \ln\left(\dfrac{1 + CR_i}{\dfrac{r_{i-1}}{r_i} + CR_i}\right) + \ln\left(\dfrac{r_e}{L}\right)} \tag{11.12}$$

Substituting the values given in Table 12.8 yields

$$\frac{J_s}{J_0} = 3.72$$

TABLE 12.8 Calculation of the Fracture Conductivity for a Nonuniform Case

$\frac{x}{L}$	w_i (mm)	$w_f k_f$ ($m^3 \times 10^{12}$)	Interpolated $(w_f k_f)_j$ ($m^3 \times 10^{12}$)	r_j (m^2)	CR_j
0.0	7.0	18.4	15.4	4	124.0
0.1	6.0	12.5	10.3	8	42.0
0.2	5.0	8.1	6.2	12	16.60
0.3	4.1	4.2	3.4	16	7.10
0.4	3.3	2.6	2.2	20	3.50
0.5	2.5	1.5	1.10	24	1.40
0.6	1.8	0.65	0.44	28	0.50
0.7	1.2	0.23	0.14	32	0.14
0.8	0.65	0.052	0.03	38	0.03
0.9	0.23	0.004	0.002	40	0.00
1.0	0	0			

This result is compared with the value 4.1, which would be obtained if a uniform fracture conductivity could somehow be created. In this case, however, it is not possible. Thus, either the lower stimulation ratio must be accepted or the limit on the injection rate somehow increased. Hydraulic rather than acid fracturing is another possibility.

There are a number of difficulties still to be resolved associated with the design of acid fractures. The fluid-loss rates of viscosified acid solutions have not really been adequately studied. Here the fluid-loss coefficient (C_v) is calculated as though the acid solution is nonreactive. In fact, because of acid fluid loss, the formation permeability in the region near the fracture surface will increase substantially. Thus, the formation permeability used in calculating C_v should be modified in some way to account for this increase. Unfortunately, because the interaction of the acid solution with the rock is complex, the influence of the acid on C_v is not now known. Thus, the actual fracture length may fall short of the optimum value. If the difference is significant, then the observed stimulation ratio will be smaller than that predicted.

The optimization process requires that the calculation be repeated for a series of acid volumes and the volume giving the best economic return is the one selected. As the acid volume increases, so does the optimum fracture length. This will in turn require higher and higher acid fracture fluid viscosity to obtain the desired fracture. The amount of polymer required to fulfill the viscosity requirement may become excessive in longer fractures. This means that acid fracturing is in general restricted to smaller treatments of damaged wells. Furthermore, acid fracturing is a process restricted to carbonate formations.

REFERENCES

12.1. Howard, G. C., and Fast, C. R., *Hydraulic Fracturing*, Mono. Ser., *2*, Society of Petroleum Engineers, Richardson, Texas, 1970.

12.2. Yuan, M. K., "A Rheological Study of Polymer and Microemulsion in Porous Media," MS thesis, The University of Texas at Austin, 1981.

12.3. Maerker, J. M., *Soc. Pet. Eng. J.*, *15* (1975) 311.

12.4. Blauer, R. E., and Kohlhaas, C. A., "Formation Fracturing with Foam," SPE 5003, presented at the 1974 Society of Petroleum Engineers Annual Meeting, Houston, Texas, 1974.

12.5. Almond, S. W., "Factors Affecting Gelling Agent Residue under Low Temperature," SPE 10658, presented at the Society of Petroleum Engineers Formation Damage Control Symposium, Lafayette, Louisiana, 1982.

12.6. Prats, M., "Effect of Vertical Fractures on Reservoir Behavior Incompressible Fluid Case," *Soc. Pet. Eng. J.*, *2* (1961) 105.

12.7. Tannich, J. D., *J. Pet. Tech.*, *27* (1975) 1309.

12.8. Holditch, S. A., *J. Pet. Tech.*, *31* (1979) 1515.

12.9. Sevougian, S. D., "Optimization of Vertical Acid Fractures in Steady-State Flow," MS thesis, The University of Texas at Austin, 1986.

12.10. Veatch, R. W., *J. Pet. Tech.*, *35* (1983) 677.

12.11. Nolte, K. G., "Determination of Fracture Parameters from Fracturing Pressure Decline," SPE 8341, presented at the 54th Fall Technical Conference and Exhibition of the Society of Petroleum Engineers, Las Vegas, Nevada, 1979.
12.12. Nolte, K. G., *Soc. Pet. Eng. Prod. Eng., 1* (1986) 255.
12.13. Williams, B. B., Gidley, J. L., and Schechter, R. S., *Acidizing Fundamentals*, Mono. Ser., *6*, Society of Petroleum Engineers, Richardson, Texas, 1979.
12.14. Church, D. C., Quisenberry, J. L., and Fox, K. B., *J. Pet. Tech., 33* (1981) 2471.
12.15. Crowe, C. W., Martin, R. C., and Michaelis, A. M., *Soc. Pet. Eng. J., 21* (1981) 415.
12.16. Pabley, A. S., and Holcomb, D. L., "A New Stimulation Technique: High Strength Crosslinked Acid," SPE 9241, presented at the 55th Fall Technical Conference and Exhibition of the Society of Petroleum Engineers, Dallas, Texas, 1980.
12.17. Nierode, D. E., and Kruk, K. F., "An Evaluation of Acid Fluid Loss Additives, Retarded Acids and Acidized Fracture Conductivity," SPE 4549, presented at the 1973 Society of Petroleum Engineers Annual Meeting, Las Vegas, Nevada, 1973.

PROBLEMS

12.1. An undamaged well in a formation having a permeability of 1 mD is to be stimulated using 20,000 kg of proppant (ρ_p = 2500 kg/m^3) which is structured so that the final fracture permeability (k_f) will be 120 D and the final bed porosity (ϕ_f) will be 35% under the reservoir closure pressure. If the drainage radius of the well is 500 m and the wellbore radius is 0.1 m, determine the *maximum* steady-state stimulation possible. The fracture height is 17 m. [*Note:* Assume that the proppant is packed uniformly in its entire final length, (L_f).]

(** preceding 12.1 as printed: **12.1.)

***12.2.** At what time measured from the start of a fracture treatment was an element of fluid which at a time 15 min is at a distance 72 m from the wellbore injected into the fracture? The data in Table P12.2 apply.
(*Note:* This problem uses the long-time approximation for the fracture length and assumes $\overline{w}$ to be constant along the length of the fracture.)

TABLE P12.2 Treatment Data

Property	Symbol	Value
Overall fluid-loss coefficient	C	8×10^{-5} m/sec$^{1/2}$
Injection rate	i	2×10^{-2} m^3/sec
Fracture height	h	11 m
Fracture width	$\overline{w}$	5×10^{-3} m

***12.3.** A well is to be acid fractured using 20m^3 of 28 wt% HCl. The process is controlled so that the volume of rock dissolved is uniform over the entire surface of the fracture and the acid penetration distance can be varied. For the conditions given in Table P12.3 calculate the optimum fracture length (this is the length that maximizes J_s/J_0). Assume the treated well is undamaged.

Repeat your calculation for a well that is 3500 m deep.

TABLE P12.3 Reservoir Characteristics

Property	Data
Fracture height	25 m
Porosity	0.12
Composition	Dolomite
Rock embedment strength	2.1×10^5 kPa
Depth	1500 m
Fracture gradient	18 kPa/m
Producing pressure (bottomhole)	7000 kPa
Formation permeability	6 mD
Drainage radius	200 m

PART 5
Acid Reaction Rates and Matrix Acidizing

13
Acidizing Methods

13.1 DESCRIPTION OF ACID TREATMENT

In this section matrix acidizing of both sandstone and carbonate formations is described. This treatment method is defined as the injection of acid into the formation porosity at a pressure less than the pressure at which a fracture can be opened. The goal of matrix acidizing is to achieve, more or less, radial penetration into a formation. Stimulation is usually accomplished by removing the effect of a formation permeability reduction near the wellbore by enlarging the pore spaces and dissolving particles plugging these spaces. Matrix acidizing is most useful in treating highly permeable, but damaged formations or in less advantageous cases where acid fracturing cannot be risked because a shale break or other natural flow boundaries must be maintained to minimize or prevent water or gas production. When performed successfully, matrix acidizing often will increase oil production without increasing the percentage of either water or gas produced.

We also will consider acid washing when it is applied to open a perforation. This procedure involves injecting a small quantity of acid into the perforations to remove the damage created by perforating (see Chap. 7 for a full discussion). Acid washing is also used to remove scale which may have formed on the tubular goods by spotting a small quantity of acid in the wellbore and allowing it to react without agitation. If this type of treatment is to be successful, it is necessary to be absolutely certain that the problem scale is soluble in the acid used.

In addition to the uses described, acids are also used for acid fracturing (see Chaps. 10 and 11) and for the following purposes:

- As a spearhead when fracturing to clean out perforations, thereby assuring that the fracture fluid enters all perforations
- To break acid-sensitive gels used in a fracture treatment if the gel fails to break following the treatment
- As a preflush before squeeze cementing
- Acidizing a completed gravel pack positioned for sand control when the permeability of the gravel is impaired

Which Acid to Apply

In Chapter 3 the various acid types available for well treatment and their properties were defined. For acid fracturing of carbonate formations, hydrochloric acid is invariably the acid of choice, but in matrix acidizing various other acids may be appropriate depending on the particular situation. Table 13.1 has been prepared by McLeod [1] as a guide for acid selection. This compilation is based on his own experience as well as that of others (e.g., see [2–5]). The choices stated in Table 13.1 should be considered as guides rather than rules. Any choice of treatment acid should be derived from consideration of the specific problem and should utilize whatever information one has available. One purpose of this section is to explore those fundamentals which influence the choice of an acid, the concentration best suited for the particular job, the total volume of acid to be applied, and the optimum injection rate. Furthermore, an acid treatment in sandstones often includes a preflush and/or an afterflush and these fluids must also be selected. The purpose of these stages in an acid treatment will also be considered in this section. We note that Table 13.1 shows that a preflush is recommended whenever a sandstone formation is to be treated using hydrofluoric acid. As will be evident, this appears desirable; however, an afterflush, not considered by McLeod, is in some cases also advantageous.

Another factor to be considered in selecting an appropriate formulation for sandstone acidizing is the mineralogy and petrology of the formation rock. We will see that there are excellent reasons to support the general recommendations given in Table 13.1, but when reliable information about the formation to be treated is available more precise designs are possible. It is hoped that the practitioner will recognize the need for reliable petrographic information and will demand it before embarking on an extensive treatment program.

Table 13.1 shows that any one of three acids are used in the matrix acidizing of carbonates. Often, corrosion rates are one of the significant factors to be considered in selecting an acid. For deep, hot wells it is easier to protect the tubular goods if acetic rather than hydrochloric acid is used. Thus in some cases where Table 13.1 expresses a preference for hydrochloric acid, acetic may be substituted.

The mechanisms underlying the process of carbonate acidizing are still not completely understood and past experience in a particular reservoir is still an important factor guiding the selection of an acid. We will see that when the reaction

TABLE 13.1 Acid Use Guidelines [1]

SITUATION	
Carbonate Acidizing:	
Perforating fluid	5% acetic acid
Damaged perforations	(a) 9% formic acid (b) 10% acetic acid (c) 15% HCl
Deep wellbore damage	(a) 15% HCl (b) 28% HCl (c) Emulsified HCl
Sandstone Acidizing:	
HCl soluibility > 20%	Use HCl only
High permeability (100 mD plus)	
High quartz (80%), low clay (< 5%)	10% HCl–3% HF[a]
High feldspar (>20%)	13.5% HCl–1.5 HF[a]
High clay (>10%)	6.5% HCl–1% HF[b]
High iron chlorite clay	3% HCl–0.5% HF[b]
Low permeability (10 mD or less)	
Low clay (<5%)	6% HCl–1.5% HF[c]
High chlorite	3% HCl–0.5% HF[d]

[a] Preflush with 15% HCl
[b] Preflush with sequestered 5% HCl
[c] Preflush with 7.5% HCl or 10% acetic acid
[d] Preflush with 5% acetic acid

rate is rapid, the larger pores tend to grow much more rapidly than do the smaller ones. This gives rise to a phenomenon known as *wormholing* in which a few large holes form and conduct all, or nearly all, of the acid. In laboratory tests with short cores or disks, often a single wormhole forms. The growth of wormholes is considered in Chapter 17; however, it is not yet possible to predict just how many wormholes will form at a single perforation or from an uncased wellbore. This gap in our present knowledge makes precise design of carbonate matrix acid treatments difficult.

Calculation will show that wormholes may range in length from a few centimeters to a few meters. Because of this limited penetration, matrix treatments in carbonates can only bypass near wellbore flow restrictions and do not create significant stimulation above that achieved from damage removal. Only in exceptional cases can matrix treatments give significant stimulation in undamaged wells. This can occur, for example, in naturally fractured formations where acid can flow along existing fractures; however, this welcome result is often unexpected.

Well Preparation

It is advisable to remember that before a fluid enters the formation, it is pumped from surface containers and through tubing. Unless all of the materials that come into contact with the acid are thoroughly cleaned, acid-insoluble solids deposited on the tubing walls (oxide scale, pipe dope, paraffin, asphaltenes, etc.) will all be carried into the formation creating further damage. It is highly recommended that to wash the tubing, acid be pumped down the tubing near the bottom and subsequently produced back into the waste pit before injecting into the formation. If the reservoir pressure is too low to lift the acid out of the tubing, a foamed acid should be used [1]. Furthermore, the proper spotting of the acid is crucial. To be successful, the acid must be directed into the damaged zone.

Additives

To ensure that the acid is diverted into the damaged zones, a diverting agent consisting of particulate matter to be subsequently removed is often added to the acid. These agents may consist of benzoic acid powder or oil-soluble resins. Both of these components are soluble in crude oils and will presumably dissolve when the well is put in production. The use of diverting agents or other methods for channeling the acid, such as ball sealers, are described in Chapter 18.

Other additives to inhibit the precipitation of iron, to reduce interfacial tension, to prevent corrosion, or to restore water wettability are all used and described in Chapter 18. It is now clear that an acid formulation is a chemically complex "cocktail" designed to simultaneously carry out several functions in such a way that a chemical added for one purpose does not inhibit the effectiveness of chemicals added for other purposes. Thus, considerable skill is required to obtain an effective fluid.

13.2 THEORETICAL PRODUCTIVITY IMPROVEMENT FROM ACIDIZATION

A matrix acid treatment will be effective primarily in a well with a near wellbore flow restriction; that is, a damaged well. In this type of system a zone of reduced permeability (k_s) extends from the wellbore radius (r_w) to a radius r_s beyond which the formation has a constant permeability to the drainage radius (r_e). It is not difficult to show that the fluid production for this system compared with that of a similar system of uniform permeability is given by [6]

$$\frac{J_d}{J_0} = \frac{(k_s/k)\ln(r_e/r_w)}{\ln(r_s/r_w) + (k_s/k)\ln(r_e/r_s)} \tag{13.1}$$

where J_0 is the productivity index of an undamaged formation [see Eq. (11.2)] and J_d is the productivity index of the damaged well at the same well flowing pressure.

Example 13.1 Productivity of a Damaged Wall

Suppose a well with a wellbore radius of 0.1 m is known to be damaged. The skin factor (S) is 15, as determined by a pressure buildup test. Based on knowledge of the mechanism causing damage, it is estimated that the radius of damage is $r_s = 0.4$ m. What increase in productivity is available by removing the damage if the formation permeability is 10 mD and r_e is 200 m.

Solution Based on Eq. (6.4), the damaged zone permeability is found from

$$15 = \left(\frac{10}{k_s} - 1\right) \ln\left(\frac{0.4}{0.1}\right) \quad \text{or} \quad k_s = 0.85 \text{ mD}$$

Then
$$\frac{J_d}{J_0} = \frac{\dfrac{0.85}{10} \ln\left(\dfrac{200}{0.1}\right)}{\ln\left(\dfrac{0.4}{0.1}\right) + \left(\dfrac{0.85}{10}\right) \ln\left(\dfrac{200}{0.4}\right)} \quad \text{or} \quad J_0 = 2.96\, J_d$$

Thus, the productivity of a damaged well can be increased by a factor of 3 if the damage is removed and the original productivity restored.

Depending on the degree of damage, it is clear the the productivity increase can be substantial. If, on the other hand, the well is undamaged, the increased productivity achieved by increasing the permeability in a zone around the wellbore will not be large.

Example 13.2 Acidizing Undamaged Wells

What increase in productivity can be expected if the well described in Example 13.1 ($r_w = 0.1$ m, k $= 10$ mD, $r_e = 200$ m) is initially undamaged and stimulated so that the permeability is increased by a factor of 10 in a zone 0.4 m around the wellbore?

Solution Eq. (13.1) also applies when the well is stimulated. Thus

$$\frac{J_s}{J_0} = \frac{10 \ln\left(\dfrac{200}{0.1}\right)}{\ln\left(\dfrac{0.4}{0.1}\right) + 10 \ln\left(\dfrac{200}{0.4}\right)} = 1.20$$

Thus the stimulation of an undamaged well is not substantial. This is a general conclusion.

13.3 MAXIMUM INJECTION RATE

It is not difficult to show that the well pressure p_{wf} to inject fluid into a damaged well having a drainage radius r_e and reservoir pressure p_R is given by [7]

$$p_{wf} - p_R = \frac{i\mu}{2\pi kh}\left[\ln\left(\frac{r_e}{r'_w}\right) - \frac{3}{4}\right] \tag{13.2}$$

where r'_w is an effective radius in the presence of a skin (damage).

$$r'_w = r_w \exp(-S) \tag{13.3}$$

Eq. (13.2) shows that as i increases, the pressure in the wellbore correspondingly increases. Since for matrix treatments, it is necessary to maintain p_{wf} under the fracture limit, the maximum injection rate, i_{max}, is found when

$$p_{wf} = \text{FG}D \tag{13.4}$$

where FG is the fracture gradient and D is the formation depth. This effectively limits the acid injection rate, especially the initial rate, to small levels.

Example 13.3 Maximum Injection Rate

Consider a well ($r_w = 0.1$, $S = 15$) completed in a formation having a permeability of 20 mD and an average reservoir pressure of 7000 kPa. If the formation depth is 2000 m and the fracture gradient is 18 kPa/m and the drainage radius is 200 m, what is the maximum acid injection rate per unit of formation thickness permitted in a matrix treatment? The acid has a viscosity 0.9 cp at the treatment temperature.

Solution It should be emphasized that since the acid is immiscible with oil, the permeability in Eq. (13.2) should be the permeability to an aqueous phase in the presence of oil, that is, k should be multiplied by the relative permeability to water. Suppose k is 10 mD or 9.87×10^{-15} m^2. Furthermore

$$r'_w = (0.1)\exp(-15) = 3.06 \times 10^{-8} \text{ m}$$

$$\mu = 0.9 \text{ cp} = 9 \times 10^{-4} \text{ kg/m-sec}$$

$$p_R = 7 \times 10^6 \text{ Pa}$$

$$p_{wf} = (18)(2000)(1000) = 3.6 \times 10^7 \text{ Pa}$$

Substituting these values into Eq. (13.2), we find

$$\frac{i_{max}}{h} = 9.1 \times 10^{-5} \text{ m}^3\text{/m-sec}$$

This is an exceedingly small injection rate. This value will be observed to increase with formation depth; however, it is the factor which most often limits the depth to which live acid can penetrate into the matrix of a formation.

If the initial surge of acid into the matrix removes some damage, then S will decrease and the injection rate can be correspondingly increased without fracturing.

REFERENCES

13.1. McLeod, H. O., *J. Pet. Tech., 36* (1984) 2055.

13.2. McLeod, H. O., Ledlow, L. B., and Till, M. V., "The Planning, Execution and Evaluation of Acid Treatments in Sandstone Formations," SPE 11931, presented at the 58th Fall Technical Conference and Exhibition of the Society of Petroleum Engineers, San Francisco, October 1983.

13.3. Williams, B. B., Gidley, J. L., and Schechter, R. S., *Acidizing Fundamentals, 6,* Mono. Ser., Society of Petroleum Engineers, Richardson, Texas, 1979.

13.4. Fitzgerald, P. E., Martinez, S. J., and Staadt, H., "Acidizing," Chap. 46, *Petroleum*

Production Handbook, Vol II, Reservoir Engineering, T. C. Frick (ed.) New York: McGraw-Hill, 1962.

13.5. Muecke, T. W., "Principles of Acid Stimulation," SPE 10038, presented at the International Petroleum Exhibition and Technical Symposium of the Society of Petroleum Engineers, Beijing, China, March 1982.

13.6. Muskat, M., *Physical Principles of Oil Production*, New York: McGraw-Hill, (1949) 242.

13.7. Dake, L. P., *Fundamentals of Reservoir Engineering*, Amsterdam: Elsevier Publishing Co. (1978) 143.

PROBLEMS

***13.1.** Consider a calcite reservoir that is damaged within a 1-m radius about the centerline of the wellbore. The damaged zone permeability is 1% of the undamaged formation permeability. If the permeability in the damaged zone can be increased using a 15 wt% HCl acid solution in accordance with the equation

$$\frac{k_A}{k_S} = \left(\frac{\phi_A}{\phi_0}\right)^m$$

where k_A is the permeability following the treatment , ϕ is the porosity, and m = 10, how many m^3 of acid per m of interval treated are required to achieve a stimulation ratio of 15 compared to the productivity of the damaged well?

Assume that the acid reacts uniformly in the damaged zone and that only the damaged zone is treated. Use the following data:

Drainage radius: 200 m

Wellbore radius: 0.15 m

$\phi_{initial}$: 0.15

****13.2.** A deep carbonate well is severely damaged to a radius of 3 m and in this zone the permeability is smaller than the formation permeability by a factor of 100. Laboratory experiments show that by using 8 wt% formic acid, the region 2.5 m around the wellbore can be uniformly (assume) acidized even though the formation temperature is high (135°C). Since 2.5 m represents a considerable distance, it is possible that the cost of acidizing may be prohibitive. If more than 50 m^3 of 8 wt% formic acid per m of formation are required to increase the productivity ratio by a factor of 10, then fracturing, not matrix acidizing, is the method of choice. Should we fracture or should we matrix acidize?

Assume that the permeability increases in the damaged zone according to the following equation:

$$\left(\frac{k_A}{k_s}\right) = \left(\frac{\phi_A}{\phi_0}\right)^{10}$$

where k_s is the damaged zone permeability and k_A is the acidized zone permeability when the porosity is increased from ϕ_0 to ϕ_A. The value of ϕ_0 is 0.16.

The drainage radius is 250 m and the wellbore radius is 0.1 m.

(*Note:* Remember that all of the acid will not react.)

14
Rates of Acid Reactions with Reservoir Minerals

14.1 ACID REACTIONS WITH RESERVOIR MINERALS

To design a matrix or acid fracture treatment, it is necessary to establish the distance that live acid will penetrate into the formation. This distance can be obtained either by laboratory tests when formation samples are available or by attempting to model each of the individual steps in the process and to then combine them in an appropriate manner so as to simulate the actual process. Acid reactions with minerals are termed *heterogeneous reactions* because they occur at a boundary between the solid and the liquid rather than in the bulk phases. Those reactions which occur in bulk phases are called *homogeneous reactions*.

For a given treatment process, the observed overall reaction rate represented by the disappearance of acid from the solution is composed of various mechanisms occurring in series. It is important to understand these individual steps, for in some situations (very often, in fact) there is one slow step—a bottleneck—and the rate at which this slow step progresses determines the overall reaction rate. Upon changing the chemical or thermal environment, the rate-controlling step at one set of conditions may not be rate controlling at the second set. Thus to appreciate the importance of each of the individual steps in the design of an acid treatment, it is necessary to model each of them if at all possible.

The observed reaction rate is the time rate of change of the concentration of one component in the bulk liquid phase. This overall, or observed rate, can be controlled by one or a combination of steps illustrated schematically by Fig. 14.1.

The rate of acid transfer to the reactive surface by diffusion, flow-induced mixing (forced convection), mixing resulting from density gradients (free convection), or fluid loss into the formation is the first step. The second step is the reaction rate determined by the kinetics of the surface reaction once the acid reaches the rock surface. Finally, the reaction products must be transported away from the surface. Often, those mechanisms which dictate the rate of this transport are the same as those which bring live acid to the surface.

To understand the overall reaction process, it is necessary to develop an understanding of each individual step. In this chapter acid mass transfer and surface reaction kinetics are discussed, the important rate-controlling parameters are identified, and models for the processes are developed. Procedures for evaluating the reaction kinetics are described.

Rates of Surface Reactions

The rate of heterogeneous reaction may be thought of as being a function of the chemical composition of the liquid phase evaluated near the mineral surface [1]. Such concentrations are called *surface concentrations* or *boundary concentrations* to distinguish them from the concentrations in the bulk phase. In this section the rate of chemical reaction will be discussed in terms of surface concentrations, and in subsequent sections these surface concentrations will be related to the bulk concentrations. By this strategy, it is possible to separate mass transfer rates from reaction rates. Even if the chemical composition of the solution varies from point

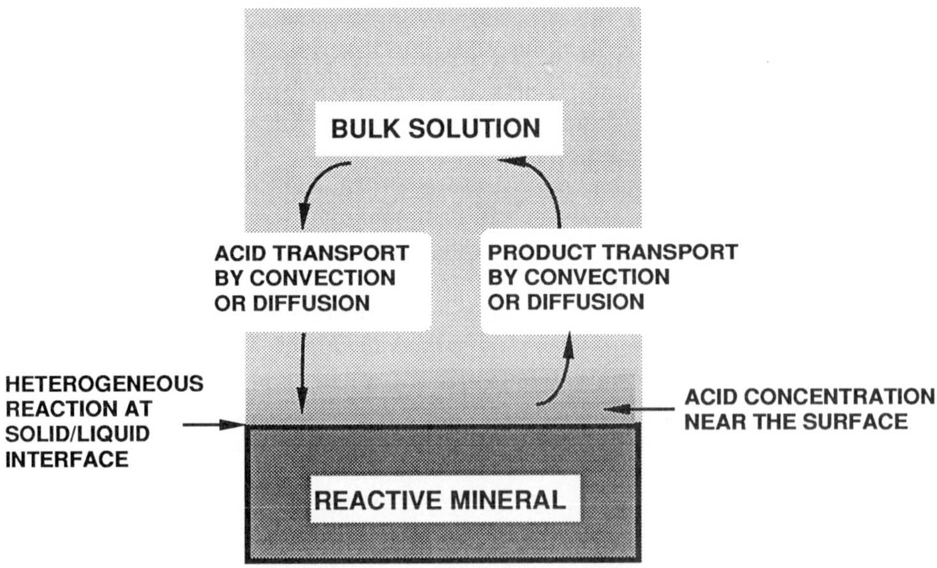

Figure 14.1 Schematic diagram of system in which acid reaction occurs.

to point, the rate of heterogeneous reaction is a function of the composition in contact with the surface.

The rate of heterogeneous reaction is generally expressed as an amount of chemical species produced (entering the solution) per unit of surface area per unit of time. The surface area refers to the mineral-solution wetted area. Thus the reaction rate of hydrochloric acid (HCl) with limestone ($CaCO_3$) can be expressed either as kg moles of HCl per m^2 per sec which react or kg moles of CO_2 per m^2 per sec produced by the reaction. By convention, when referring to a species consumed at the surface, the reaction rate will be negative.

Consider the heterogeneous reaction of an acid with calcium carbonate (limestone):

$$2H^+ + CaCO_3 \underset{E_R}{\overset{E_f}{\rightleftharpoons}} CO_2 + Ca^{++} + H_2O \tag{14.1}$$

or in symbolic terms

$$2A + B \underset{E_R}{\overset{E_f}{\rightleftharpoons}} C + D + E \tag{14.2}$$

For this reaction a general rate expression is [1]

$$-r_A = E_f c_A^\alpha c_B^\beta - E_R c_C^\gamma c_D^\delta c_E^\epsilon \tag{14.3}$$

where c_i is the surface concentration of component i, E_f and E_R are the forward and backward reaction rate constants, respectively, and α, β, γ, δ, and ϵ are exponents which, in general, are related by the expression

$$\frac{\alpha}{2} = \beta = \gamma = \delta = \epsilon \tag{14.4}$$

The term $\alpha/2$ arises because the stoichiometric coefficient of A in Eq. (14.2) is 2.

The r_A denotes the production of component A (here H^+) per area per time and since A (acid) is actually consumed r_A as calculated by Eq. (14.3) will be negative.

Strong acids. Highly dissociated acids such as HCl are often used for well stimulation in carbonate formations or to remove acid-soluble precipitates from perforations. For a reaction that is α-order in hydrogen ion concentration, the general rate expression is

$$-r_A = E_f c_A^\alpha - E_R c_C^{\alpha/2} c_D^{\alpha/2} \tag{14.5}$$

Experimental data show that this reaction goes essentially to completion. Therefore, the backward rate will be negligible in most applications of strong acids in carbonates and the rate expression thus reduces to

$$-r_A = E_f c_A^\alpha \tag{14.6}$$

where α is a constant and is called the *order* of the reaction.

Weak acids. Acids that are only slightly dissociated in aqueous solution are also used to acidize formations (see Chap. 3). Examples are formic, acetic, propionic, and other organic acids. The kinetics of these acids is complicated by the dissociation process. In Chapter 3 we have discussed the dissociation equilibria of weak organic acids in aqueous solution. As shown, an acid (HA) dissociates according to the relation

$$HA \rightleftharpoons H^+ + A^- \tag{14.7}$$

with equilibrium described by a dissociation constant (K_d) defined as

$$K_d = \frac{c_H c_A}{c_{HA}} = \frac{c_{H^+}^2}{c_{HA}} \tag{14.8}$$

Assuming α-order kinetics with respect to the hydrogen ion concentration and introducing acid dissociation gives the rate expression

$$-r_A = E_f c_H^\alpha - E_R c_{CO_2}^{\alpha/2} c_{Ca}^{\alpha/2} \tag{14.9}$$

Some weak acids reach reaction equilibrium at high acid concentration, indicating the reverse rate is high. For these acids the reverse reaction term must be retained in the rate equation.

Equation (14.9) is written to emphasize that the forward reaction rate is determined by the surface concentration of hydrogen ions and not by the acid concentration. For weak acids these two concentrations are different. If local equilibrium of undissociated acid and the ionic species is assumed (this is not always a valid assumption), then the dissociation equilibrium expressed by Eq. (14.8) can be used to relate the hydrogen ion concentration to that of the undissociated acid.

For formic acid, which reacts very nearly to completion, the reverse rate can be assumed to be small. If both HCl and formic acid react as α-order in hydrogen ion concentrations, then

$$-r_{HCl} = E_f c_{HCl}^\alpha \tag{14.10}$$

and

$$-r_{formic} = E_f K_d^{\alpha/2} c_{formic}^{\alpha/2} \tag{14.11}$$

If the rate constant E_f is independent of acid type, and this is a reasonable assumption implying that it is the hydrogen ion which is reactive, then the ratio of reaction rates at equivalent concentrations is

$$\frac{r_{HCl}}{r_{formic}} = \left(\frac{c_{HCl}}{K_d^{1/2} c_{formic}^{1/2}}\right)^\alpha \tag{14.12}$$

and for typical values of K_d (see Chap. 3) the surface reaction rate of HCl will be about one hundred times the rate for formic acid. The reaction rate of acetic acid will be even smaller than that of formic acid because it is less dissociated and because the reverse reaction is significant. Thus, acetic acid may be termed a retarded acid since its rate of reaction with carbonates is several orders of magnitude less than HCl.

Hydrochloric Acid–Limestone Reactions

Determining the rate of reaction with limestone is difficult since the reaction rate is high. Nierode and Williams [2] and Lund et al. [3] have both considered the reaction using different types of experiments and both have found the reaction order (α) to be fractional. Lund et al. [3] give the reaction rate expressed in kg moles of HCl/m^2-sec as

$$r_{\text{HCl}} = -E_{\text{f}} c_{\text{HCl}}^{0.63} \tag{14.13}$$

for experiments carried out at relatively low temperatures. If c_{HCl} is expressed as kg moles/m^3, then the forward rate constant is related to the temperature as follows:

$$E_{\text{f}} = 7.314 \times 10^7 \exp\left(-\frac{\Delta E}{RT}\right) \tag{14.14}$$

where the activation energy ΔE is 15×10^3 kcal/kg mole and R is the gas constant (1.987 kcal/kg mole-°K). Because high reaction rates are difficult to measure, this rate law which is extrapolated from measurements taken at subzero temperatures may not be accurate under reservoir conditions. In many cases of interest, it will suffice to ascertain that the surface reaction rate is fast as compared to the rates of mass transport of live acid to the mineral surface. The rate of mass transport then becomes the limiting step.

Example 14.1 Rate of Chemical Reaction

Consider a rod of $CaCO_3$, which is rapidly (very rapidly) rotated in an acid solution so that diffusion and other mass transfer rates noted in Fig. 14.1 can be neglected with respect to the reaction rate. The reaction rate is then the slow step. Suppose the $CaCO_3$ rod is immersed in a bath containing 15 wt% HCl. If the initial rod radius is 1 cm, what is its radius 15 sec after it has been immersed in the bath? The bath temperature is 40°C.

Solution If the mass transfer rate is high because the rod is rapidly rotated, the surface acid concentration will approximately equal the bulk acid concentration, 15 wt%. Since the density of 15 wt% HCl is about 1.07×10^3 kg/m^3 (see Table 3.8), then

$$c_{\text{HCl}} = \left(\frac{1.07 \times 10^3\ \text{kg}}{\text{m}^3}\right)\left(\frac{0.15\ \text{kg HCl}}{\text{kg}}\right)\left(\frac{\text{kg mole HCl}}{36.5\ \text{kg HCl}}\right) = \frac{4.40\ \text{kg moles HCl}}{\text{m}^3}$$

Substituting into Eq. (14.13) yields

$$r_{\text{HCl}} = -7.314 \times 10^7 \exp\left[-\frac{15{,}000}{(1.987)(273 + 40)}\right](4.40)^{0.63}$$

$$= -6.24 \times 10^{-3}\ \frac{\text{kg moles HCl}}{\text{m}^2\text{-sec}} = -0.023\ \frac{\text{g HCl}}{\text{cm}^2\text{-sec}}$$

If a = rod radius, ρ_{CaCO_3} = density of $CaCO_3$, and β = grams of calcite dissolved per gram of HCl reacted (β is found in Table 3.9)

$$\pi \frac{d}{dt}(a^2) = \frac{2\pi a r_{HCl}\beta}{\rho_{CaCO_3}} \quad \text{or} \quad \Delta a = \frac{r_{HCl}\beta\,\Delta t}{\rho_{CaCO_3}}$$

$$\Delta a = -\frac{\left(0.023\,\dfrac{\text{g HCl}}{\text{cm}^2\text{-sec}}\right)\left(1.37\,\dfrac{\text{g CaCO}_3}{\text{g HCl}}\right)(15\text{ sec})}{2.72\,\dfrac{\text{g CaCO}_3}{\text{cm}^3}} = -0.17\text{ cm}$$

The rod radius after 15 sec has been reduced to 0.83 cm.

The rate of chemical reaction as given by Eq. (14.14) is observed to increase exponentially with increasing temperature. This is typical of most chemical reactions. It is useful to keep this in mind and to remember that reactions classified as "slow" at room temperature may become rapid under reservoir conditions.

Dolomite–Hydrochloric Acid Reactions

The reaction of dolomite with HCl is not as fast as that between acid and calcite. Lund et al. [4] give the reaction as follows:

$$r_{HCl} = -E_f c_{HCl}^{\alpha} \tag{14.15}$$

where if c_{HCl} is expressed in kg moles/m^3 and r_{HCl} has the units kg moles/m^2-sec, then

$$E_f = 4.48 \times 10^5 \exp\left(-\frac{7900}{T}\right)$$

$$\alpha = \frac{6.32 \times 10^{-4}\,T}{1 - 1.92 \times 10^{-3}\,T}$$

Feldspar–Hydrofluoric Acid Reactions

The reactions of carbonate with HCl are important even in sandstone acidizing since mixtures of HF and HCl are used (refer to Chap. 3) and the HCl will react rapidly and completely with carbonates if they are present. The silicates, primarily feldspars, clays (with the possible exception of chlorite), and silica do not react with HCl (actually there is a slow reaction between HCl and these minerals, often resulting in the preferential leaching of aluminum, but the rate is so slow that it is of no practical importance). These minerals do, however, react with HF.

Feldspars may be divided into two major groups: the orthoclase group ($KAlSi_3O_8$) and the plagioclase group, consisting of a class of solid solutions ranging from albite ($NaAlSi_3O_8$) to anorthite ($CaAl_2Si_2O_8$). Feldspars are solid solutions having a crystalline lattice structure and the cations K^+, Na^+, Ca^{++}, and Al^{+++} can be replaced by similar cations in varying amounts. Typical compositions are given in Chapter 1. Fogler et al. [5] have investigated the dissolution kinetics of feldspars in HF–HCl mixtures and found that the dissolution rates obey the following rate law:

$$+r_{\text{feldspar}} = E_f[1 + K(c_{\text{HCl}})^{\beta}](c_{\text{HF}})^{\alpha} \tag{14.16}$$

The order of the reaction (α) is found to be first order for sodium feldspar (plagioclase) indicating a first-order dependence on HF; for potassium feldspar (microcline), however, α was found to be 1.2 giving a nearly, but not exactly, first-order dependence. This equation indicates that HCl catalyzes the reaction with HF. This interesting mechanism has been explored in some detail by Fogler and Kline [6].

For a potassium feldspar (orthoclase), the kinetic parameters have the values [5]

$$\beta = 0.4; \qquad \alpha = 1.2$$

If the rate is expressed in terms of kg moles of orthoclase produced per m^2 per sec, the rate constants have the values

$$E_f = 1.27 \times 10^{-1} \exp\left(-\frac{4680}{T}\right) \quad \text{and} \quad K = 5.66 \times 10^{-2} \exp\left(\frac{956}{T}\right)$$

For a sodium feldspar (albite), the kinetic parameters representing the kg moles of albite per m^2 per sec have the values

$$\beta = 1.0; \qquad \alpha = 1.0$$

and $E_f = 9.50 \times 10^{-3} \exp\left(-\frac{3930}{T}\right)$; $\quad K = 6.24 \times 10^{-2} \exp\left(\frac{554}{T}\right)$

These rate expressions are expressed in terms of the amount of orthoclase or albite which enters the solution. The rate expressions given by Lund et al. [5] have been corrected so that the reaction rates given here approximately apply to pure minerals rather than those for which either Na^+ or K^+ has been partially replaced by other cations.

In many cases it is convenient to know the rate at which HF is consumed. To obtain this value from the given rate expression, one must define the stoichiometry of the reaction; that is, the ratio of HF consumed to the orthoclase or albite produced must be specified. As noted in Chapter 3, this ratio changes as the HF concentration decreases [7]. It is not a constant. However, to an approximation 20 moles of HF will be required per mole of mineral produced. This number is obtained by assuming $(AlF_2)^{+1}$ and SiF_6^{-2} to be the reaction products. As the concentration of aluminum increases, the silicon will tend to form SiF_4 first and then $Si(OH)_4$ finally, which will precipitate from solution. However, to an approximation adequate for most design purposes:

$$-r_{\text{HF}} = 20 r_{\text{feldspar}}$$

Quartz–Hydrofluoric Acid Reactions

The reaction rate of α-quartz (SiO_2) with HF has been investigated by Bergman [8]. At 25°C

$$r_{\text{HF}} = -E_f c_{\text{HF}} \tag{14.17}$$

where E_f ranges from 2.1×10^{-9} to 3.4×10^{-9}. This rate is expressed in terms of kg moles of HF reacted per m^2 per sec. Note c_{HF} is the concentration in kg moles of HF per m^3 evaluated, of course, at a point near the surface. These same units are used throughout.

Hill et al. [9] have found that in acidization of Berea sandstone $E_f = 2.9 \times 10^{-9}$ fits the effluent acid concentration at 25°C and at 52°C $E_f = 4 \times 10^{-9}$ best fits the data. Assuming an Arrhenius equation, then

$$E_f = 1.39 \times 10^{-7} \exp\left(-\frac{1150}{T}\right) \tag{14.18}$$

where T is in °K.

It should be noted that this value of reaction rate is at least one order of magnitude smaller than that found for amorphous glass [10, 11, 12]. This result is consistent with solubility studies [13, 14] which indicate a similar ratio of α-quartz solubility to amorphous silica solubility in water.

The reaction rates reported here are also smaller than those reported by Fogler et al. [15], who fitted data given by Gatewood et al. [16]. These data were obtained using α-quartz, but the experiments were run at higher HF concentrations than are normally used in HF acidizing perhaps indicating that the reaction is not first order in HF concentration over a wide range of concentrations. The values given by Hill et al. [9] will be used in subsequent calculations.

Example 14.2 Comparison of HF Reaction Rates

A small particle of albite (sodium feldspar) 1 mm in radius is dropped into a large volume of acid solution composed of 3 wt% HF and 12 wt% HCl ($\rho_{acid} \cong 1075$ kg/m^3 at 25°C) which is stirred vigorously so that the acid concentration near the reactive surface is the same as that in the bulk phase; that is, diffusion is rapid. If the bath temperature is 25°C, how long will it take for the albite particle to dissolve in the acid solution? Compare this time with the time required to dissolve the same size particle of α-quartz.

Solution The concentrations are

$$c_{HF} = \left(\frac{3 \text{ kg HF}}{100 \text{ kg solution}}\right)\left(\frac{1075 \text{ kg solution}}{\text{m}^3}\right)\left(\frac{\text{kg mole HF}}{20 \text{ kg HF}}\right) = 1.61 \frac{\text{kg mole HF}}{\text{m}^3}$$

and

$$c_{HCl} = \left(\frac{12 \text{ kg HF}}{100 \text{ kg solution}}\right)\left(\frac{1075 \text{ kg solution}}{\text{m}^3}\right)\left(\frac{\text{kg mole HCl}}{36.5 \text{ kg HCl}}\right) = 3.53 \frac{\text{kg mole HCl}}{\text{m}^3}$$

Based on Eq. (14.16) and using the parameters for sodium feldspar, the enhancement factor due to the presence of HCl yields

$$1 + K(c_{HCl})^\beta = 1 + \left[6.24 \times 10^{-2} \exp\left(\frac{554}{298}\right)\right](3.53) = 2.41$$

In this case the rate of reaction is enhanced considerably owing to the catalytic influence of HCl. For sodium feldspar (albite)

$$E_f = 9.5 \times 10^{-3} \exp\left(-\frac{3930}{298}\right) = 1.78 \times 10^{-8}$$

or the rate law

$$r_{\text{albite}} = (1.78 \times 10^{-8})(2.41)(1.61) = (6.9 \times 10^{-8}) \frac{\text{kg moles albite}}{\text{m}^2\text{-sec}}$$

The theoretical formula of albite is $NaAlSi_3O_8$ (molecular weight 262 and density 2630 kg/m^3). By material balance, the rate of change of the particle volume must be related to the reaction rate multiplied by the surface area. Thus

$$-\frac{d}{dt}\left(\frac{4\pi}{3} R(t)^3\right) = r_{\text{feldspar}} 4\pi R^2 (M_{\text{w albite}})/(\rho_{\text{albite}})$$

or

$$-\frac{dR}{dt} = \left[(6.9 \times 10^{-8}) \frac{\text{kg moles albite}}{\text{m}^2\text{-sec}}\right] \times \left(262 \frac{\text{kg albite}}{\text{kg mole albite}}\right)\left(\frac{\text{m}^3}{2630 \text{ kg albite}}\right)$$

Thus

$$-\frac{dR}{dt} = 6.88 \times 10^{-9} \text{ m/sec}$$

Integrating from $R = 10^{-3}$ m at $t = 0$ to $R = 0$ at $t = t_f$ we find the time for a 1 mm particle of albite to completely dissolve

$$t_f = 1.45 \times 10^5 \text{ sec} = 40.4 \text{ hr}$$

For α-quartz, Eqs. (14.17) and (14.18) yield

$$r_{\text{HF}} = -\left[1.39 \times 10^{-7} \exp\left(-\frac{1150}{298}\right)\right](1.61)$$

or

$$r_{\text{HF}} = -4.72 \times 10^{-9} \frac{\text{kg moles HF}}{\text{m}^2\text{-sec}}$$

This result refers to the rate of disappearance of HF and it is necessary to relate this rate to the rate of appearance of α-quartz. Here we assume that H_2SiF_6 is the reaction product so 6 moles of HF are required to dissolve 1 mole of SiO_2; Thus,

$$r_{\text{SiO}_2} = -r_{\text{HF}}/6 = 7.87 \times 10^{-10}$$

Since ρ_{SiO_2} is 2650 kg/m^3 and the molecular weight is 60, we find

$$t_f = 1.27 \times 10^6 \text{ sec} = 353 \text{ hr}$$

This time is considerably longer than that for the feldspar particle, thus providing a comparison of the relative reaction rates of similarly sized particles. We shall use this result in Chapter 15.

Clay–Hydrofluoric Acid Reactions

The reactions of the clay minerals with HF have been studied by Kline and Fogler [17]. The results of this study were quite unexpected. The clay reactivity was found not to be proportional to the surface area of the clays in contact with the acid solution (wetted surface area) as had been previously assumed. The wetted surface areas for various clay minerals are shown in Table 14.1. Based on these

TABLE 14.1 Rate Constants (E_f) for the Dissolution of Clay Minerals in Hydrofluoric Acid. (Adapted from Kline and Fogler [17].)

Clay Mineral	Wetted Surface Area (m^2/kg)	Reactive Surface Area (m^2/kg)	Temperature (°C)	*Rate Constant (E_f) $\dfrac{\text{kg moles mineral}}{m^2\text{-sec-}\dfrac{\text{kg moles HF}}{m^3}}$
Kaolinite: $Al_4Si_4O_{10}(OH)_8$	16×10^3	8×10^3	0	1.3×10^{-11}
			25	9.8×10^{-11}
			38	2.6×10^{-10}
			49	5.4×10^{-10}
Sodium montmorillonite: $Al_4Si_8O_{20}(OH)_4 * nH_2O$	500×10^3	8.7×10^3	0	5.2×10^{-11}
			25	2.6×10^{-10}
Illite: $K_{O-2}Al_4(Al, Si)_8O_{20}(OH)_4$	79×10^3	21×10^3	25	8.1×10^{-12}
			35	1.7×10^{-11}
			45	3.0×10^{-11}
Muscovite: $KAl_3Si_3O_{10}(OH)_2$	30×10^3	0.25×10^3	38	3.6×10^{-10}
			49	8.4×10^{-10}
			60	1.5×10^{-9}
			70	2.7×10^{-9}

* Based on reactive surface area.

areas, one would expect a unit mass of montmorillonite to react about 60 times as fast as a similar mass of kaolinite since the surface area exposed to acid is 60 times greater for montmorillonite than for kaolinite. This, as noted, turns out to be incorrect. In fact, montmorillonite reacts at a rate just about equal to that of kaolinite.

This surprising difference in the reaction rates per unit area found between kaolinite and montmorillonite can be understood in terms their respective crystalline structures. It will be recalled (Fig. 1.4) that kaolinite is a two-layer clay. It consists of an alumina sheet and silica sheet bound together. On the other hand, montmorillonite is a three-layer clay (Fig. 1.5) in which one alumina sheet is sandwiched between two silica sheets. The alumina sheet is much more reactive than the silica one [17]. Therefore, kaolinite tends to be attacked faster than montmorillonite because for this clay an alumina sheet is exposed. The acid reacts with the alumina sheet producing holes that expose other layers to the acid. Montmorillonites, on the other hand, tend to be dissolved from the edges first. This difference in the mode of acid attack has been proposed by Kline and Fogler to account for the fact that montmorillonites dissolve more slowly than kaolinites per unit of wetted area. The actual reactive area of montmorillonites is, therefore, only a small fraction of the total wetted area. For kaolinite the reactive surface is about one half of the wetted area since alumina sheets are exposed on about half of the surfaces.

Onc other important feature of clay dissolution in HF–HCl mixtures reported by Kline and Fogler is that clays for which magnesium has been substituted extensively into the alumina layer react more slowly. Thus, chlorite may then be expected to react more slowly and perhaps release larger quantities of fines on initial contact with acid because, as shown by Fig. 1.6, it is a mixed clay—being three layered with a single layer sandwiched between. This potential difficulty has not, however, been demonstrated to be true in laboratory experiments.

Despite the fact that the alumina layers and the magnesia layers, when they exist as in chlorite, react with HF more rapidly than does the silica layer, the reaction is congruent; that is, aluminum and silicon appear in solution in stoichiometric ratios. It must be concluded therefore that once a silica layer is exposed by virtue of an alumina layer being partially dissolved, then the exposed silica layer rapidly reacts. This means, for example, that when 1 mole of aluminum appears in solution upon dissolving montmorillonite in HF, that, according to the ideal chemical formula given in Table 14.1, 2 moles of silicon can be expected to appear simultaneously. Thus the rate constants given in Table 14.1 are expressed in terms of moles of mineral dissolved with the calculations being based on the ideal compositions given in the table.

The rate constants can also be expressed in terms of aluminum produced. For example, the dissolution of one mole of kaolinite will result in 4 moles of aluminum appearing in solution. The rate of dissolution of clay is given by

$$r_{\text{clay}} = -E_f c_{\text{HF}} \tag{14.19}$$

where E_f is the forward rate constant given in Table 14.1 for the various clays; however, as noted the values are not widely different and an average value can be used. The activation energy is approximately 13 kcal/g mole and [17] therefore E_f can be found at any temperature given E_f^0 at temperature T_0 by the equation

$$E_f = E_f^0 \exp\left[\left(\frac{1}{T_0} - \frac{1}{T}\right) 7550\right] \tag{14.20}$$

where T_0 and T are in °K.

It is interesting to note that the widely held view that the clays react with HF much faster than other minerals present in the formation may not be true. Indeed it is observed that at 25°C, the reaction constant for the feldspars is about two orders of magnitude larger than for clay minerals both based on reactive area. Thus, feldspars have the highest reaction rate followed by clays and silica. The effective reaction rates in the pore spaces will, however, depend on the surface area of the particular mineral in contact with the acid. This in turn will be a function of the nature of the depositional process, as for example, whether the clays are clastic or not.

In the acidizing of Berea sandstone cores Hill et al. [9] found that feldspars were dissolved first and that clays, which were present primarily as fragments, react at a rate comparable to silica. These observations are consistent with the results presented here.

14.2 MASS TRANSFER IN ACID SOLUTIONS

It is as important to understand and be able to model those mechanisms which control the rate at which acid is transported to the surface as it is to understand the rate at which acid reacts when it arrives there. Mass transfer can be brought about by two different processes. Diffusion is that process which occurs because of the random movement of the molecules resulting in a net flow of molecules whenever a concentration gradient exists. If there are regions in which the concentration of acid is higher than in others, these regions of high concentration will tend to become diluted because more molecules will leave these regions than will enter. Thus, diffusion is one mechanism by which acid will be transported from the bulk of the solution to the mineral surfaces where the acid has become depleted because of reaction.

A second mechanism for acid transport is convection (see Fig. 14.1). Any movement of the fluid will result in acid being transported with the flow. Generally, fluid motion resulting from an applied pressure differential is called *forced convection*, whereas a motion resulting from a density difference is termed *natural convection*. Both types of fluid movement are of importance in acid reactions.

Diffusion of Acid

We shall represent diffusion by Fick's law:

$$J_y^{(A)} = -D_A \frac{\partial c_A}{\partial y} \tag{14.21}$$

Bird et al. [18] have given a review of the proper formulation of diffusional fluxes, $J_y^{(A)}$, and discuss the conditions under which this flux (expressed in moles, mass, or molecules per unit area per unit time across a plane perpendicular to y) can be adequately represented by Eq. (14.21). The coefficient D_A is called the *diffusion coefficient* and y is a coordinate direction.

Strong acids. In Chapter 3 we distinguished between strong and weak acids to predict the extent of their reaction with carbonates. The ionic character of the acid solution is also important in determining the diffusion rate of acids. In aqueous ionic solutions, such as HCl, electrical forces are long range and positive and negative charges must move together more or less collectively to maintain a local balance of charges [19]. This is different from diffusion in solutions of nonionized species or weakly ionized electrolytes, where the rate is essentially independent of other components in the solution except for second-order coupling of the diffusional fluxes [20, 21, 22]. Sherwood and Wei [23], Gilliland et al. [24], Vinograd and McBain [25], and others [21, 22] have studied the problems of diffusion in complex ionic systems and have presented techniques for predicting ionic diffusion rates. In the diffusion problems considered here, the effect of the electrical forces acting on the acid ions will be disregarded and in the calculations requiring the diffusion of the hydrogen ion, the HCl diffusivity shown in Fig. 14.2 will be applied even though this can lead to serious errors as the ionic reaction products (e.g., $CaCl_2$) begin to accumulate in the bulk solution [2].

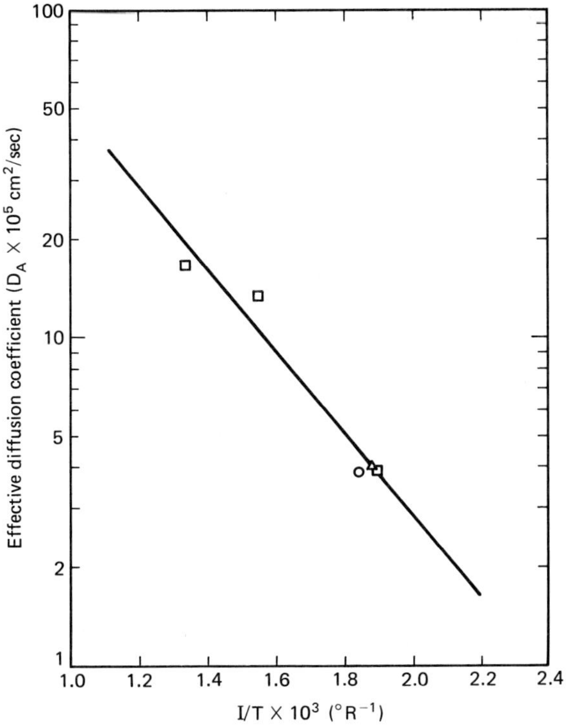

Figure 14.2 Graph showing effective diffusivity for 1.4*N* HCl [26].

Thus for strong acids the c_A to be used in Eq. (14.21) represents the hydrogen ion concentration and the surface reaction rates will depend on this concentration evaluated at a point immediately adjacent to the surface.

Weak acids. Weak acids are slightly dissociated and these acids diffuse primarily as the associated neutral species. Thus for acetic, formic, or propionic acids the concentration c_A in the diffusion flux expression, Eq. (14.21), will represent the neutral acid. Then, as shown by Eq. (14.11), the rate of reaction will depend both on the acid concentration immediately adjacent to the carbonate surface and also on the degree of acid dissociation.

Representative diffusion coefficients for weak acids are presented in Fig. 14.3.

Convective Mass Transfer

The rate of acid transport is determined by the combination of diffusion and convection. If the flow is laminar, which is the case for many problems of interest here, the total mass transfer is the sum of $J_y^{(A)}$, the diffusional flux and the convection flux. Thus

$$N_y^{(A)} = -D_A \frac{\partial c_A}{\partial y} + c_A v_y \tag{14.22}$$

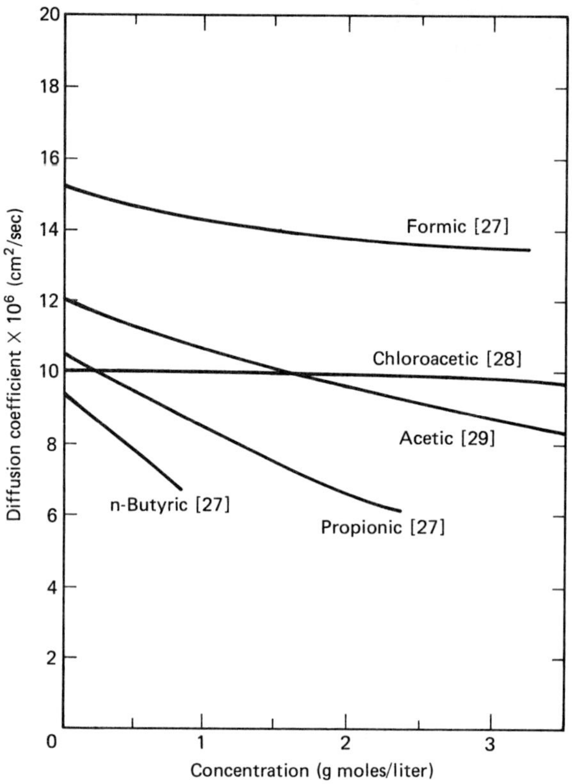

Figure 14.3 Graph showing diffusion coefficients of weakly ionized acids at 25°C [27, 28, 29].

Here v_y is the component of fluid velocity in the y-direction. Similar expressions apply for $N_x^{(A)}$ and $N_z^{(A)}$, the fluxes of mass in the x- and z-directions, respectively.

The important point to be gleaned from Eq. (14.22) is that to establish the rate of mass transport one must also determine the velocity field. The hydrodynamic equations which apply to each individual flow geometry must therefore be solved. In some cases, such as when the density gradients created by the reaction drive the flow (natural convection), the hydrodynamic equations and the acid balance equations must be solved simultaneously. The effects cannot be separated. Reaction creates flow and flow moderates the reaction [30]. The problems studied in this chapter will be simpler than those requiring solution of the coupled hydrodynamic and acid balance equations. We will assume that the hydrodynamic equations have been solved and that the velocity field is therefore established independent of the reaction rate. The flux as expressed by Eq. (14.22) then contains a single unknown, the acid concentration.

Modeling heterogeneous reactions. Heterogeneous chemical reactions take place at the solution–mineral surfaces; thus an acid balance taken about a small volume entirely contained within the bulk solution will not include any terms related to the chemical reaction. The acid balance will therefore simply be the convection–diffusion equation and chemical reactions will enter the analysis as

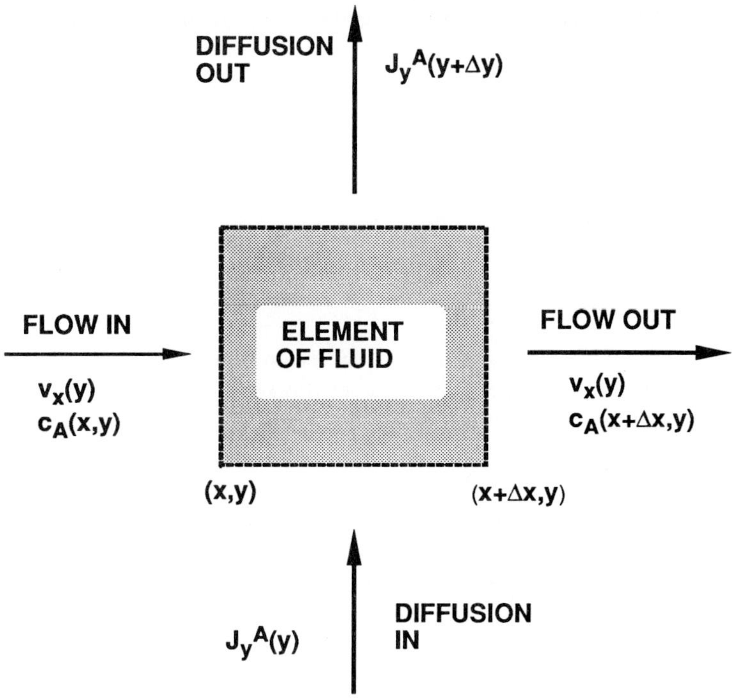

Figure 14.4 Sketch depicting a small volume element wholly contained within a fluid.

boundary conditions. Figure 14.4 depicts a volume element drawn in a fluid and as shown, there is a flow in the x-direction as well as diffusive transport in both the x- and y-directions. Diffusion in the z-direction perpendicular to the plane of the figure is disregarded. A material balance taken about the small element is as follows:

$$\underbrace{v_x c_A\Big|_x \Delta y - v_x c_A\Big|_{x+\Delta x} \Delta y}_{\text{convection}}$$

$$\underbrace{- D_A \frac{\partial c_A}{\partial y}\Big|_y \Delta x + D_A \frac{\partial c_A}{\partial y}\Big|_{y+\Delta y} \Delta x - D_A \frac{\partial c_A}{\partial x}\Big|_x \Delta y + D_A \frac{\partial c_A}{\partial x}\Big|_{x+\Delta x} \Delta y}_{\text{diffusion}} = 0 \tag{14.23}$$

Dividing by Δx and Δy and taking the limit as Δx and Δy both approach zero, we find

$$-\frac{\partial}{\partial x}(v_x c_A) + D_A \frac{\partial^2 c_A}{\partial y^2} + D_A \frac{\partial^2 c_A}{\partial x^2} = 0 \tag{14.24}$$

This is a special form of the convection–diffusion equation. In general the convection–diffusion equation is

$$\frac{\partial c_A}{\partial t} + \boldsymbol{v}\cdot\nabla c_A = D_A \nabla\cdot\nabla c_A \tag{14.25}$$

These equations have a unique solution, provided an appropriate number of boundary and initial conditions are imposed. This can best be demonstrated by an example.

Parallel Plate Reactor

Figure 14.5 depicts a parallel plate reactor composed of closely spaced mineral slabs with acid flowing in the axial direction. A similar system has been used by Barron et al. [31], Nierode and Williams [2], and Roberts and Guin [32] to study acid reaction rates. This arrangement is not the current preferred approach for obtaining acid–mineral reaction rates. The rotating disk system, which is discussed in the next section, represents a better approach; however, the parallel plate system is an interesting one in that the roles of heterogeneous reactions and diffusive transport to the surface are clearly delineated.

The flow system to be considered is shown in Fig. 14.5. The flow is laminar and the walls are solid and smooth. There is, therefore, no fluid loss and the laminar velocity is given by

$$v_x = \frac{3}{2}\bar{v}_A\left[1 - \left(\frac{2y}{w}\right)^2\right] \tag{14.26}$$

where w is the spacing between plates and $\bar{v}_A$ is average axial velocity. Substituting Eq. (14.26) into Eq. (14.24), which is the appropriate form to model the physical system depicted in Fig. 14.5, yields

$$\frac{3}{2}\bar{v}_A\left[1 - \left(\frac{2y}{w}\right)^2\right]\frac{\partial c_A}{\partial x} = D_A\frac{\partial^2 c_A}{\partial x^2} + D_A\frac{\partial^2 c_A}{\partial y^2} \tag{14.27}$$

We note that the diffusive flux directed toward the wall (the y-direction) is given by $-D_A\, \partial c_A/\partial y$. When acid arrives at the surface, it reacts. This equality

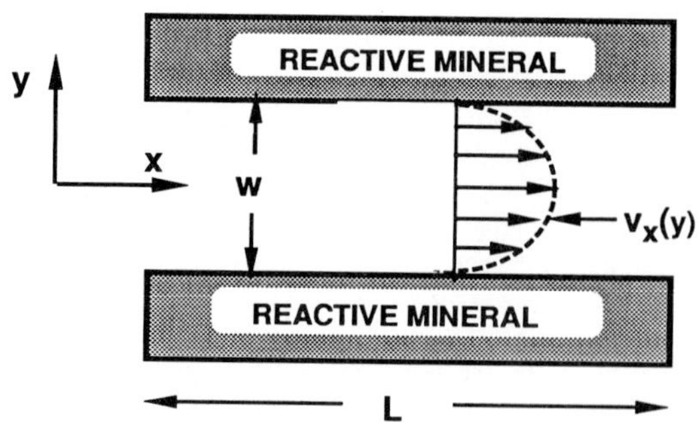

Figure 14.5 Sketch showing flow between parallel plates.

is expressed by

$$D_A \frac{\partial c_A}{\partial y} = r_A \qquad (\text{at } y = w/2) \tag{14.28}$$

where r_A is the rate of production of acid by chemical reaction and

$$-D_A \frac{\partial c_A}{\partial y} = r_A \qquad (\text{at } y = -w/2) \tag{14.29}$$

Thus, we see that the reaction rates enter as boundary conditions. If we take the reaction to be represented by Eq. (14.6) with $\alpha = 0.6$, then the differential equation can be integrated using numerical techniques. The results are shown in Fig. 14.6 [33]. The quantities plotted are defined as follows:

$$P_f = \frac{wE_f c_0^{\alpha - 1}}{2D_A} \tag{14.30}$$

$$L_{fD} = \frac{8D_A L}{3\overline{v}_A w^2} \tag{14.31}$$

$$R_f = 1 - \frac{\overline{c}_A}{c_0} \tag{14.32}$$

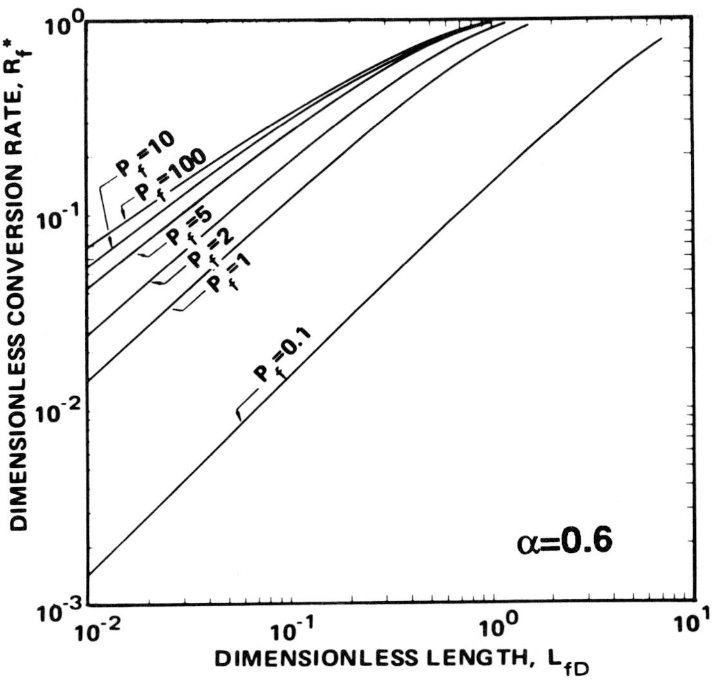

Figure 14.6 Graph of fractional order surface chemical reactions associated with laminar flow between parallel plates [33].

where $\bar{c}_A$ is the average acid concentration at length L and c_0 is the inlet acid concentration. Thus based on these definitions R_f represents the degree of acid reaction, L_{fD} is a dimensionless distance or length along the parallel plates, and P_f is a quantity representing the rate of chemical reaction divided by the rate of diffusion. Thus small values of P_f signal a small reaction rate as compared to the diffusion rate, whereas a large P_f relates to very rapid reactions. Larger P_f are said to represent diffusion-limited reactions, and a calculation will show that the acid concentration near the surface is quite small. This is the case for most reactions involving HCl and carbonates.

Similarly, graphs for other reaction orders and for laminar flow in circular tubes have been presented by Williams et al. [33].

Example 14.3 Measurement of Reaction Rate

Suppose that a 15 wt% HCl solution is made to flow at an average velocity of 0.5 cm/sec between parallel slabs of $CaCO_3$ that are spaced 0.5 cm apart. The slabs are 100 cm long. If the outlet acid concentration is 13 wt% and $\alpha = 0.6$, what is the forward reaction rate constant, E_f? The temperature is 25°C.

Solution For this case $D_A = 5.1 \times 10^{-5}$ cm²/sec (from Fig. 14.2)

and

$$R_f = 1 - 0.13/0.15 = 0.133$$

Also

$$L_{fD} = \frac{8(5.1 \times 10^{-5}\ \text{cm}^2/\text{sec})(100\ \text{cm})}{(3)(0.5\ \text{cm/sec})(0.5\ \text{cm})^2} = 0.109$$

Reading from Fig. 14.6 we find $P_f = 1.0$. Based on this result

$$E_f = \frac{2P_fD_A}{wc_0^{\alpha-1}}$$

Generally, we express the acid concentration in terms of kg moles/m³. As shown in Example 14.1, 15 wt% HCl corresponds to a concentration of 4.4 kg moles/m³.

Thus $$E_f = \frac{(2)(5.1 \times 10^{-9}\ \text{m}^2/\text{sec})}{(5 \times 10^{-3}\ \text{m})(4.4\ \text{kg mole/m}^3)^{-0.4}} \text{ or } E_f = 3.7 \times 10^{-6}\ \frac{(\text{kg mole})^{0.4}}{\text{sec-m}^{0.2}}$$

Note that for fractional order reactions, the dimensions of the forward rate constant look strange. The important point to remember is that the product $E_fc_A^\alpha$ must have the units of heterogeneous reaction rates, in this case kg moles/m²-sec. The units of E_f found in Example 14.2 fulfill this requirement.

Rotating Disk System

As previously noted, a preferred procedure for determining surface kinetics is to use a system where the rate of acid transfer to the reactive surface greatly exceeds the reaction rate at the surface. One system for which these conditions can often be satisfied is the rotating disk system shown in Fig. 14.7. This system consists of a large disk of the mineral to be studied immersed in the fluid and rotated around its center. Because of the rotation, there is a centrifugal effect that will cause the fluid to flow in a radial direction along the disk and reduce the thickness of the stagnant fluid layer near the surface (often called the *diffusion boundary layer*). According to Levich [34] each point on the disk surface is then equally accessible.

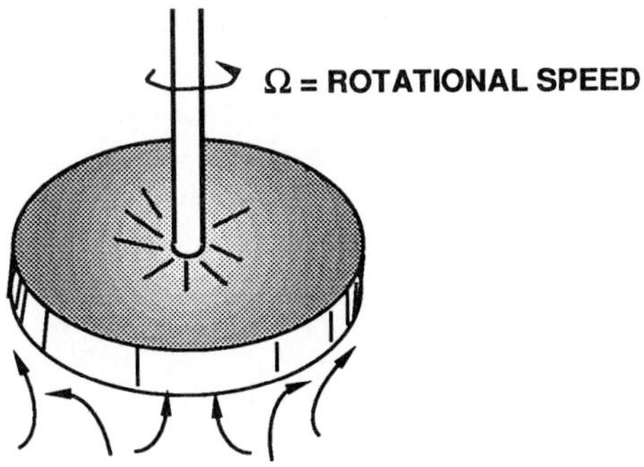

Figure 14.7 Diagram of rotating disk reactor.

In designing a rotating disk reactor system, the disk radius must be much larger than the boundary-layer thickness (often about 0.01 cm) and flow in the system must be laminar. In theory, the system is designed so that the disk spins in an infinite volume; however, Gregory and Riddiford [35] demonstrated that, in a liquid system, if the vessel diameter is at least twice the disk diameter, the observed rate is independent of the vessel diameter. Boomer et al. [36] have reported the details of a system suitable for studying acid reaction kinetics.

Since each point on the surface is equally accessible, the following equality between reaction rate and mass transfer rate must apply [37]:

$$\frac{D_A}{\delta}(c_A - c_A^s) = -r_j \tag{14.33}$$

c_A^s is the acid concentration near the surface and δ is the diffusion boundary-layer thickness. At small, angular velocities δ is large, the reaction is diffusion controlled, and the difference between the bulk and surface concentrations is large. As rotational speed is increased, the boundary-layer thickness decreases, thereby decreasing the mass transfer resistance.

A typical plot of reaction rate versus rotational speed is shown schematically in Fig. 14.8. When the reaction rate no longer increases with an increase in ro-

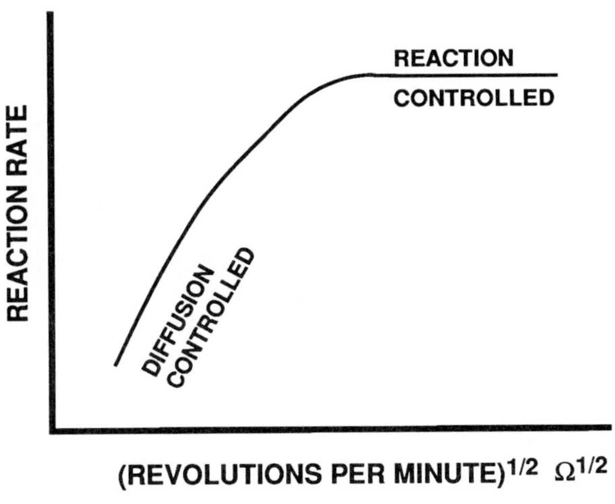

Figure 14.8 Graph showing typical reaction rate measurement as a function of rotation speed of reactive disk.

tational speed, the observed reaction rate is controlled entirely by the surface kinetics. Under these conditions, the surface kinetics are obtained directly, and diffusion rates do not have to be taken into account as was done for parallel plate reactors. This system has been used by several authors to measure the surface reaction kinetics of acid–mineral systems [36] and is the method now used by several service companies to determine reaction rates for use in designing fracture treatments.

REFERENCES

14.1. Laidler, K. J., *Chemical Kinetics*, 2nd ed., New York: McGraw-Hill (1965) 286.
14.2. Nierode, D. E., and Williams, B. B., *Soc. Pet. Eng. J., 11* (1971) 406.
14.3. Lund, K., Fogler, H. S., McCune, C. C., and Ault, J. W., *Chem. Eng. Sci., 30* (1975) 825.
14.4. Lund, K., Fogler, H. S., and McCune, C. C., *Chem. Eng. Sci., 28* (1973) 691.
14.5. Fogler, H. S., Lund, K., and McCune, C. C., *Chem. Eng. Sci., 30* (1975) 1325.
14.6. Fogler, H. S., and Kline, W. E., *J. Colloid and Interface Sci., 82* (1981) 93.
14.7. Hekim, Y., and Fogler, H. S., *Chem. Eng. Sci., 32* (1977) 1.
14.8. Bergman, I., *J. Appl. Chem., 3* (1963) 356.
14.9. Hill, A. D., Lindsay, D. M., Silberberg, I. H., and Schechter, R. S., *Soc. Pet. Eng. J., 21* (1981) 30.
14.10. Blumberg, A. A., *J. Phys. Chem., 63* (1959) 1129.
14.11. Blumberg, A. A., and Stavrinon, S. G., *J. Phys. Chem., 64* (1960) 1438.
14.12. Glover, M. C., and Guin, J. A., *Amer. Inst. Chem. Eng. J., 19* (1973) 1190.
14.13. Wollast, R., *Sea, 5,* M. N. Hill et al. (eds.), New York: John Wiley, 1974.
14.14. Owen, L. B., *Precipitation of Amorphous Silica from High Temperature Hypersaline Geothermal Brine*, Lawrence Livermore Laboratory, Publication UCRL-51866, June 1975.
14.15. Fogler, H. S., Lund, K., McCune, C. C., and Ault, J. W., "Dissolution of Selected Minerals in Mud Acid," paper 52c, presented at the American Institute of Chemical Engineers 74th National Meeting, New Orleans, Louisiana, March 1973.
14.16. Gatewood, J. R., Hall, B. E., Roberts, L. D., and Lasater, R. M., *J. Pet. Tech., 22* (1970) 693.
14.17. Kline, W. E., and Fogler, H. S., *Chem. Eng. Sci., 36* (1981) 871.
14.18. Bird, R. B., Stewart, W. E., and Lightfoot, E. N., *Transport Phenomena*, New York: John Wiley, 1960, 554.
14.19. Resibois, P. M. V., *Electrolyte Theory*, Chap. 3, New York: Harper and Row, 1968.
14.20. De Groot, S. R., and Mazur, P., *Non-Equilibrium Thermodynamics,* Amsterdam: North-Holland Publishing Co. (1962) 367.
14.21. Miller, D. G., *J. Phys. Chem., 70* (1966) 2639.
14.22. Miller, D. G., *J. Phys. Chem., 71* (1967) 66.
14.23. Sherwood, T. K., and Wei, J. C., *Amer. Inst. Chem. Eng. J., 4* (1955) 1.
14.24. Gilliland, E. R., Baddour, R. F., and Goldstein, D. J., *Canad. J. Chem. Eng., 37* (1957) 10.
14.25. Vinograd, J. R., and McBain, J. W., *J. Amer. Chem. Soc., 63* (1941) 2009.
14.26. Roberts, L. D., and Guin, J. A., *Soc. Pet. Eng. J., 15* (1975) 277.
14.27. Dunn, L. A., and Stokes, R. H., *Austral. J. Chem., 18* (1965) 285.
14.28. Garland, C. W., Tong, S., and Stockmayer, W. H., *J. Phys. Chem., 69* (1965) 2469.
14.29. Vitagliano, V., *J. Amer. Chem. Soc., 78* (1955) 4538.

14.30. Chang, C. Y., Guin, J. A., and Roberts, L. D., *Amer. Inst. Chem. Eng. J., 22* (1976) 252.
14.31. Barron, A. N., Hendrickson, A. R., and Wieland, D. R., *J. Pet. Tech., 14* (1962) 409; *Trans. AIMME, 225* (1962).
14.32. Roberts, L. D., and Guin, J. A., *Soc. Pet. Eng. J., 14* (1974) 385; *Trans. AIMME, 257* (1974).
14.33. Williams, B. B., Gidley, J. L., and Schechter, R. S., *Acidizing Fundamentals, 6,* Mono. Ser., Society of Petroleum Engineers, Richardson, Texas, 1979.
14.34. Levich, V. G., *Physicochemical Hydrodynamics,* Englewood Cliffs, N.J.: Prentice-Hall (1962) 60.
14.35. Gregory, D. P., and Riddiford, A. C., *J. Chem. Soc.,* (1956) 3756.
14.36. Boomer, D. R., McCune, C. C., and Fogler, H. S., *Rev. Sci. Instruments, 43* (1972) 225.
14.37. Fogler, H. S., Lund, K., McCune, C. C., and Ault, J. W., "Kinetic Rate Expressions for Reactions of Selected Minerals with HCl and HF Mixtures," SPE 4348, presented at the Society of Petroleum Engineers-American Institute of Mining Engineers Oilfield Chemistry Symposium, Denver, Colorado, May 1973.

PROBLEMS

***14.1.** Two rods, one composed of calcite ($CaCO_3$) and the other of dolomite [$CaMg(CO_3)_2$], are coated along the lateral surfaces so that only one end is exposed. The exposed ends are immersed in a 5 wt% HCl solution and the rods rotated rapidly for a period of 30 sec. If the overall reaction rate is controlled by the surface reaction, calculate the amount by which the length of each rod is reduced. The temperature of the reaction is 120°F.

The concentration of acid is approximately constant during the process.

***14.2.** Formic acid is thought to react with carbonates (both calcite and dolomite) at a reduced rate because formic acid is incompletely dissociated. Thus, formic acid is sometimes regarded as a retarded acid. To evaluate the magnitude of this effect, compare the rate of reaction of a 1 wt% HCl solution with $CaCO_3$ at a temperature of 65°C to the rate that a formic acid solution of equal dissolving power reacts with the same mineral.

****14.3.** A quartz rod (0.4 cm in diameter) is immersed into a 1 wt% HF solution contained in a cylindrical plastic (1.0 cm in diameter) vessel. The length of both the rod and the plastic container is the same. If the reaction temperature is maintained at 70°C, what will the acid concentration be after being contacted with the quartz rod for 40 hr? Neglect the variation in the diameter of the rod and note that the rate quartz reacts with HF is the controlling step. Why is the cylinder containing the HF solution made of plastic rather than glass?

****14.4.** The reaction between HCl and a formation is known to satisfy the equation

$$r_{HCl} = -E_f c_{HCl}^{\alpha}$$

If this acid flows through a small irregular pore of perimeter Γ and area A, develop an equation that will describe the acid concentration as a function of distance down the pore. The reaction rate is slow and the pore is small so that you may assume that the system is reaction rate controlled. [*Hint:* Begin by making an acid balance over a small differential axial distance.] Define all terms. Do not assume that the reader will know the meaning of any of your symbols.

**14.5. A small spherical $CaCO_3$ particle (0.5 cm in diameter) is suspended in a 3 wt% solution of HCl and if it is assumed that the natural convection currents generated by the difference in density between the reacting solution and the reaction products do not contribute to acid transport, then calculate the time for the particle diameter to be reduced by a factor of 2 when the acid temperature is 40°C. The reaction rate can be considered to be *infinite* but since the acid must diffuse from the surrounding solution to the particle surface, the overall rate of reaction is finite and is diffusion controlled. Diffusion in liquids is a slow process and the process can be considered to be quasi-steady state.

**14.6. The reaction rate of acid with a mineral can be represented by the equation

$$r_{HCl} = -E_f c_{HCl}$$

(first-order reaction) where c_{HCl} is the acid concentration at the mineral surface. Consider the case for which acid must diffuse through a stagnant layer of thickness δ to reach the reactive surface. Assuming that there is no flow within this stagnant film and that the acid concentration on one side of the film is constant, develop an expression for c_{HCl} and based on this expression determine the conditions characterizing a diffusion controlled reaction.

*14.7. Acid is caused to flow between parallel plates constructed of a reactive solid material. The acid reaction rate with the solid is given by

$$r_{acid} = -E_f c_A^{0.6}$$

where $E_f = 10^{-6}$ in units such that the rate is in kg moles of acid/m²-sec when c_A is expressed in kg moles/m³. Develop a curve showing the acid penetration distance as a function of the spacing between the plates for a fixed average fluid velocity of 0.1 cm/sec. Define the penetration distance as that distance required for 90% of the acid to be reacted. The initial acid concentration is 5 wt% HCl, and the acid diffusion coefficient is 10^{-5} cm²/sec.

15 Sandstone Matrix Acidizing Theory

Matrix acidizing is that process whereby acid is injected into a formation at pressures less than the fracture pressure. The acid is forced into the formation pores in the region surrounding the wellbore. Rock will consequently be dissolved in the invaded region, leaving permeable channels leading into the wellbore to assist the production of oil. This well treatment is normally applied to remove near-wellbore damage. In this chapter, we discuss the mechanisms underlying the matrix acidizing process.

Three important mechanisms that contribute to the final productivity of an acidized well include: erosion of the pore structure as the acid flows through these pores, consumption of the acid, and selective dissolution of certain minerals. Erosion of the pore structure leads to both increased porosity and permeability. Rate of acid consumption is an important feature, since this limits the depth to which acid penetrates. Finally, the mineral composition in the acidized region depends both on time and distance from the wellbore face because minerals do react with acid at differing rates (see Chap. 14). These three mechanisms are coupled together. It is not possible to consider them separately. The first step is to try to express in mathematical terms the full structure of this coupling. We accomplish this by making an acid balance.

15.1 ACID BALANCE EQUATIONS

Pore Structure

To quantify the changes that occur when an acid reacts with a porous medium, it appears to be necessary to adopt a simplified model to describe the pore structure because the actual pore geometry is so complex that any detailed description of it is likely to be mathematically unmanageable. Schechter and Gidley [1] have used a capillary model in which the pores are assumed to be a collection of cylinders all having an equal length (l) but differing cross-sectional areas (A). Capillary models have been used to represent porous media with only mixed success [2, 3, 4]. The capillary model does have the advantage that the distribution of capillary cross-sectional areas can be estimated based on mercury capillary pressure curves [5]. The reader should be cautioned, however, to remember that capillary models are nothing more than a greatly simplified picture of a complex structure and results derived from the model may not be representative of the real system.

In the treatment given here, model results will only be used in ratios relative to the initial state of the porous medium and absolute values will not be of interest. By employing this tactic, errors inherent in model calculations will hopefully be minimized. This means, of course, that additional data such as initial permeability, porosity, surface area, etc., will be required to supplement the mercury capillary pressure curves.

The relative abundance of capillaries in a unit volume of porous medium having a cross-sectional area between A and $A + dA$ is represented by $\eta(A, x, t)\, dA$. The pore density function η depends on position and time, represented here respectively by the variables x and t. Thus $w^2\, dx\, \eta\, dA$ represents the number of pores within the volume $w^2\, dx$ shown in Fig. 15.1 having an area between A and $A + dA$. A typical pore density function is shown in Fig. 15.2. The data shown were obtained based on capillary pressure measurements. The pore density function shown in Fig. 15.2 represents several sandstone samples and indicates that there are very few pores with cross-sectional areas exceeding 2500 $(\mu m)^2$. Most are pores having a small cross-sectional area. Interaction with acid will tend to change the pore density function [1].

The porosity at a position x and a time t can be calculated as follows. If l is the average length of a pore, then lA is the volume of a single pore having a cross-sectional area A. The total number of such pores in a unit volume is $\eta\, dA$, and the porosity is the sum of all such void volumes per unit volume. Thus

$$\phi(x, t) = l \int_0^\infty A\eta(A, x, t)\, dA \tag{15.1}$$

We will not attempt to assign a value to l. Instead if the initial (before acidizing)

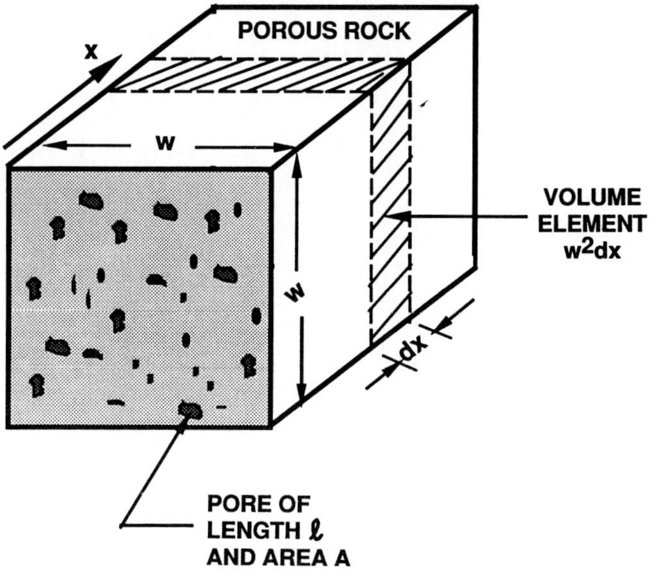

Figure 15.1 Schematic diagram of a model parallel capillary system.

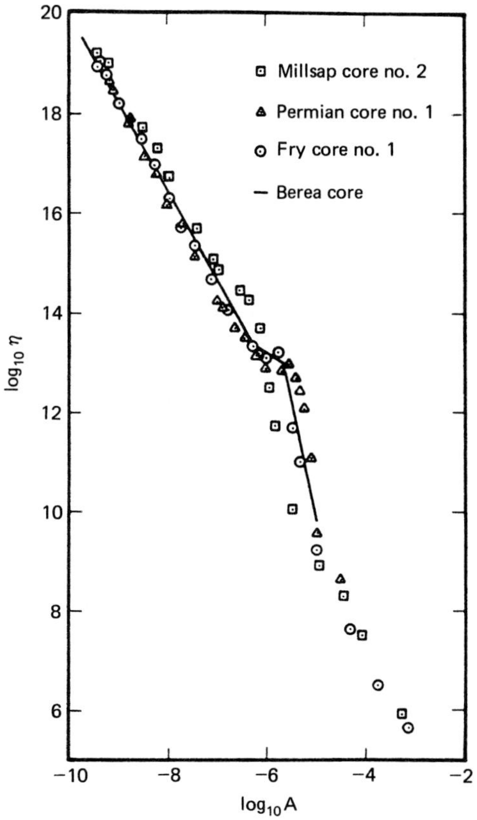

Figure 15.2 Graph of a typical pore density function, representing a comparison of pore size distributions from various sandstone samples [6]. (The area is expressed in units of cm^2.)

porosity ϕ_0 is known, then

$$\phi/\phi_0 = \frac{\int_0^\infty A\eta(A, x, t)\, dA}{\int_0^\infty A\eta(A, x, 0)\, dA} \tag{15.2}$$

where $\eta(A, x, 0)$ is the initial pore size density.

The formation permeability can be obtained by a similar argument. Note that the pressure drop across each capillary is the same regardless of its cross-sectional area since they are connected in parallel. Furthermore, the velocity in laminar flow is proportional to the area of a capillary multiplied by the pressure drop and divided by the viscosity. Thus if γ is a proportionality constant that depends on the shape of the capillary, then the velocity in a capillary is

$$v = -\frac{\gamma}{\mu}\frac{dp}{dx} A \tag{15.3}$$

The flux is the sum of all of the volumetric flow rates (Av) expressed per unit of volume. Thus

$$u = -\frac{\gamma}{\mu}\frac{dp}{dx}\int_0^\infty A^2\eta(A, x, t)\, dA \tag{15.4}$$

or taking the ratio between the initial permeability and the permeability at any time t, the unknown proportionality constant is cancelled as follows:

$$\frac{k}{k_0} = \frac{\int_0^\infty A^2\eta(A, x, t)\, dA}{\int_0^\infty A^2\eta(A, x, 0)\, dA} \tag{15.5}$$

Finally, the surface area of a single pore is proportional to $\sqrt{A}l$. Thus, the surface area per unit volume is

$$\frac{S}{S_0} = \frac{\int_0^\infty \sqrt{A}\eta(A, x, t)\, dA}{\int_0^\infty \sqrt{A}\eta(A, x, 0)\, dA} \tag{15.6}$$

where S_0 is the initial surface area per unit volume.

Schechter and Gidley [1] have developed an equation describing the change in the pore density function resulting from acidization of porous rock. Guin et al. [7] and Glover and Guin [8] have confirmed the validity of this theory for the acidization of porous sintered glass disks. Swift and Fogler [9] have studied the change in the pore size distribution during acidization of a Phacoides sandstone with hydrofluoric acid (HF). They found that the larger pores are chiefly responsible for the increase in permeability resulting from acidization, whereas the smaller pores play a minor role. The importance of the larger pores in determining the permeability is emphasized by the fact that the pore size distribution is weighted by the square of the area, as shown by Eq. (15.5). Furthermore, their role is even more crucial because theoretical considerations show that the distribution of the smaller pores is almost unaffected by the acidization process since the larger pores receive practically all of the acid [1] (see Chap. 17).

Example 15.1 Porosity and Permeability Changes

As a simple, but admittedly unrealistic, example take

$$\eta(A) = \exp(-\beta A) \tag{15.7}$$

where β is a parameter that is decreased by acidization (a smaller value of β corresponds to a greater number of large pores, which is the main effect of acidization). If β_0 is the value prior to acidization and β_A is the value following acidization, derive a relationship between the porosity change created by the process and the resultant permeability.

Solution Let us note the following results:

$$\int_0^\infty A \exp(-\beta A)\, dA = \frac{1}{\beta} \tag{15.8}$$

and

$$\int_0^\infty A^2 \exp(-\beta A)\, dA = \frac{2}{\beta^2} \tag{15.9}$$

Thus, based on Eq. (15.2)

$$\frac{\phi}{\phi_0} = \frac{\beta_0}{\beta_A} \tag{15.10}$$

Similarly, from Eq. (15.5)

$$\frac{k}{k_0} = \left(\frac{\beta_0}{\beta_A}\right)^2 \tag{15.11}$$

and the relationship sought is

$$\frac{k}{k_0} = \left(\frac{\phi}{\phi_0}\right)^2 \tag{15.12}$$

The pore size distribution used in this example is unrealistic because pores of zero area have a finite probability of existence. More realistic distributions are used in Problem 15.1.

The extreme sensitivity of the permeability to the density of larger pores implies that k depends primarily on $\eta(A, x, t)$ for large A. It is, however, precisely these pores whose density is most difficult to determine by mercury injection since the pressure needed to force mercury into them is normally quite small. For these pores even small variations in the level of mercury above the rock sample become significant. Because the larger pores are difficult to accurately characterize, it has proven difficult to use the theoretical approach proposed by Schechter and Gidley [1] for predicting the permeability change resulting from acidization. Experimentally, very small changes in porosity lead to large increases in permeability. Calculations reported by Guin and Schechter [10] based on the

Schechter and Gidley [1] theory are not in correspondence with most existing data. It is therefore necessary to resort to empiricism if a realistic relationship between the amount of rock dissolved and the resultant permeability increase is to be obtained.

Labrid [11] proposed an equation

$$\frac{k}{k_0} = M\left(\frac{\phi}{\phi_0}\right)^n \tag{15.13}$$

where the constants M and n were determined to be 1 and 3, respectively, for a Fontainbleau sandstone. These values are comparable to those found in theory by Guin and Schechter [10]. However, in sandstones that are less homogeneous, n is found to be larger than 3.

Lund and Fogler [12] have fitted data reported by McCune et al. [13] obtained by acidizing a Phacoides sandstone and found the empirical equation

$$k/k_0 = \exp\left[M\left(\frac{\phi - \phi_0}{\Delta\phi_{max}}\right)\right] \tag{15.14}$$

where $M = 7.5$ and $\Delta\phi_{max} = 0.08$ gave the best fit for that particular sandstone. A similar expression has been used by Walsh et al. [14] based on extensive core studies reported by Lambert [15]. Walsh et al. found $M/\Delta\phi_{max} = 45.7$.

It is important to recognize that the relationship between k/k_0 and $\Delta\phi(= \phi - \phi_0)$ strongly depends on the change in the density of the larger pores and this must vary from sandstone to sandstone. The more homogeneous sandstones composed almost entirely of well-sorted sand grains will not yield values of $M/\Delta\phi_{max}$ as large as the more heterogeneous ones. Thus, it is normally necessary to have experimental data from core samples of the formation to be treated if an accurate expression for the permeability is required. It will, however, be observed that generally an accurate relationship may not be required. The important fact is that $M/\Delta\phi_{max}$ is often a large number. Small increases in the porosity will lead to large increases in permeability. Thus it becomes most crucial to evaluate the porosity increase, or equivalently, to estimate the depth to which live acid penetrates.

The fact that the many smaller capillaries do not change perceptibly upon acidization leads us to another interesting and important conclusion. The surface area is, as shown by Eq. (15.6), dependent on the $\sqrt{A}$ and is, therefore, most sensitive to the number of smaller pores. Thus the initial surface area (S_0) will not change greatly during the acidization process and can, for some practical purposes, be taken to be a constant.

Flow and Reaction in Porous Media

To model both the rate of acid consumption and the local change in mineral composition, a balance for each of the reactive acid types present must be written. Consider the small volume element $w\ \Delta x^2$ shown in Fig. 15.1. Writing an acid balance about this small element over a small period of time (Δt) yields

$$\underbrace{\Delta t\, w^2 u c_i|_x}_{\text{acid in}} - \underbrace{\Delta t\, w^2 u c_i|_{x+\Delta x}}_{\text{acid out}} + \underbrace{\Delta x\, \Delta t w^2 l \sum_j \int_0^\infty \sqrt{A} f_j r_{i,j} \eta(A,x,t)\, dA}_{\text{acid generated by reaction}}$$
$$= \underbrace{[c_i\phi|_{t+\Delta t} - c_i\phi|_t] w^2 \Delta x}_{\text{acid accumulated}} \tag{15.15}$$

where u is the acid flux and c_i is the concentration of acid i. Dividing by $w^2\, \Delta x\, \Delta t$ and taking the limit as Δx and Δt both approach zero, the acid balance for a linear system is obtained:

$$-u \frac{\partial c_i}{\partial x} + \sum_j l \int_0^\infty f_j r_{i,j} \sqrt{A} \eta(A, x, t)\, dA = \frac{\partial}{\partial t} (\phi c_i) \tag{15.16}$$

Here f_j is the fraction of the total surface area occupied by mineral j and $r_{i,j}$ is the rate of reaction of acid i with mineral j expressed as an amount of acid produced per unit of area per unit of time. (Recall that $r_{i,j}$ is positive, provided acid is created by the reaction. Thus, $r_{i,j}$ will generally be a negative quantity.)

It is difficult to assign values for f_j. In acidizing limestones with hydrochloric acid (HCl) the surface in contact with the acid solution (at least the reactive surface) is primarily carbonate, although both clays and quartz may be present in large quantities. Since HCl does not react with these minerals, only reactions with carbonates will be important and f_j will be the fraction of the total surface area which is carbonate.

For sandstones where mixtures of HF and HCl are used in matrix treatments, there are many reactive minerals and the situation is more complex than for limestones. Generally, the chemical composition of the core material is known but not its morphology. Those minerals which are authogenic in origin will generally occupy a position atop the detrital components and will, therefore, tend to shield them from direct contact with the reactive solutions. Thus while quartz sand grains may be present in large quantities according to the chemical analysis, the fraction of the surface area f_{quartz} actually in contact with acid may in fact be initially small. As the reaction proceeds, the fractional area coverage occupied by a particular mineral changes. For example, once those clays covering the surface of the sand grains have dissolved, the underlying quartz surface will be exposed and f_{quartz} will then increase. Thus, f_j is a function of time and depends on the depositional environment.

For the purpose of classifying reactions with HF, it will be convenient to divide the acid-soluble minerals present in sandstones into two different categories: fast-reacting minerals and slow-reacting minerals. The first group will include the authogenic clays, feldspars, and amorphous silica. These minerals will react more rapidly with HF than will clay fragments and detrital quartz. The feldspars will dissolve more rapidly because their reaction rates are high. In Chapter 14 data are presented showing that feldspars react at a rate which is, in some cases, two orders of magnitude faster than the clays or silica. The second group, the slow-reacting minerals, includes quartz, clay fragments, and pyrites.

The authogenic clays are included in the group that reacts rapidly. Actually, the clay rate of reaction with HF is about the same as that of silica (see Chap. 14) but these clays are exposed to the acid solution (f_i is large). Thus, the apparent reaction rate as measured by the rate of disappearance of these clays will be high.

The precipitated silica should be included together with authogenic clays and detrital feldspars. This type of silica generally appears as small spherical deposits atop detrital sand grains. The spheres are thus exposed to the acid solution and since they are often amorphorous, they react with HF about one order of magnitude more rapidly than does α-quartz (see Chap. 14).

Detrital clays (fragments) will react at the same rate as α-quartz (sand grains). Chapter 14 shows that the reaction rates between clays and silica are in fact similar on a unit area basis. Hill et al. [16] reported that clay fragments found in Berea sandstone cores appeared to react at about the same rate as silica, but that the feldspars reacted more rapidly. This conclusion was obtained by examining a sequence of thin sections from a Berea core that had been exposed to HF for increasing times and is entirely consistent with the rates of reaction reported in Chapter 14.

Based on this discussion, Eq. (15.16) applied to HF can be reduced to the following equation by "lumping" all of the acid-soluble minerals into one of two categories—fast reacting and slow reacting:

$$-u\frac{\partial c_{\mathrm{HF}}}{\partial x} + f_{\mathrm{F}} r_{\mathrm{F}} S + f_{\mathrm{S}} r_{\mathrm{S}} S = \frac{\partial}{\partial t}(\phi c_{\mathrm{HF}}) \tag{15.17}$$

where f_{F} and f_{S} are the fraction of the total surface area in contact with the acid composed of the fast- and slow-reacting minerals, respectively. Clearly $f_{\mathrm{F}} + f_{\mathrm{S}} = 1$. The rates of the fast and slow reactions, r_{F} and r_{S}, are average values of the reaction rates.

Example 15.2 The Lumped Reaction Rate Parameters

Suppose that a sandstone is composed of 88 wt% α-quartz, 5 wt% kaolinite fragments, and 7 wt% orthoclase. Furthermore, petrographic analysis shows all of these minerals to be a detrital with the average size of the quartz grains being 10 μm, the feldspar grains 3 μm, and the kaolinite fragments 0.8 μm. Estimate the fraction f_{F} appropriate for use in Eq. (15.17).

Solution The emphasis here must be on the word *estimate*. It is not really possible to precisely determine the reaction rate parameters without the aid of acidization experiments carried out in core samples of the formation to be treated. For most cases, core experiments are not available and the procedures recommended here do provide a means of estimation.

Since the weight fraction and the volume fractions of formation minerals are about the same (they have approximately the same densities), the volume fraction of the fast-reacting mineral is 7% (feldspar). Furthermore, the surface area of a particle compared to its volume is certainly proportional to the reciprocal of the particle radius. Thus

$$f_F = \frac{0.07\left(\dfrac{1}{3 \times 10^{-6}\text{ m}}\right)}{0.88\left(\dfrac{1}{10 \times 10^{-6}\text{ m}}\right) + 0.05\left(\dfrac{1}{0.8 \times 10^{-6}\text{ m}}\right) + 0.07\left(\dfrac{1}{3 \times 10^{-6}\text{ m}}\right)}$$

or

$$f_F = 0.134 \qquad \text{and} \qquad f_S = 0.866$$

This calculation is based on the assumption that all formation minerals are present roughly as spherical particles and are equally available for reaction with the acid. Furthermore, as the feldspar reacts faster than the other minerals present, f_F will then change as the acid treatment proceeds.

All of the design procedures recommended for sandstone matrix treatments in this text assume that both the fast and slow reactions are first order in the concentration of HF. A review of the rate expressions given in Chapter 14 will show that the potassium feldspar reaction exhibits a slight deviation from first-order behavior but that α-quartz, albite, and the most abundant clays exhibit first-order kinetics. Thus the mineral reactions with HF are in general first order.

An assumption that is difficult to justify is now made. However, there is really no published information available to assist us in obtaining a better approximation or even cause us to reject the hypothesis that is applied here. It is assumed that

$$Sf_F = \alpha_F V_F(1 - \phi) \tag{15.18}$$

and

$$Sf_S = \alpha_S V_S(1 - \phi) \tag{15.19}$$

where V_F and V_S are the volume fractions (or weight fractions) of the fast- and slow-reacting minerals, respectively, and α_F and α_S are constants that are assumed not to change during the acidizing process. Equations (15.18) and (15.19) assert that the surface area available for reaction decreases in proportion to the volume of mineral remaining. This hypothesis has been used extensively by Fogler and Rege [17] with some success. It has also been applied by Hill et al. [16] to fit the acid effluent concentrations from a Berea sandstone core during acidization. These data will be presented in a subsequent section.

The application of Eqs. (15.18) and (15.19) permit us to complete the specification of the problem. One can relate the rate of acid reaction, and hence the rate of acid disappearance, to the rate at which the mineral volume decreases since there is a fixed stoichiometric ratio between the kg moles of mineral dissolved to the kg moles of acid reacted. This ratio is defined to be β, and approximate values for β are given in Table 15.1.

Note that $(1 - \phi)V_F$ represents the volume of fast-reacting mineral per unit volume of formation so that

$$\frac{\partial}{\partial t}[(1 - \phi)V_F]$$

TABLE 15.1 Approximate Stoichiometric Coefficients for Reactions Occurring in Acidization*

Reaction	β $\left(\frac{\text{kg moles of mineral}}{\text{kg moles of acid reacted}}\right)$
HF–muscovite [$KAl_2Si_3AlO_{10}(OH)_2$]	0.042
HF–montmorillonite [$Al_4Si_8O_{20}(OH)_4 \cdot 10H_2O$]	0.018
HF–illite [$Al_4Si_8O_{20}(OH)_4 \cdot 10H_2O$]	0.018
HF–kaolinite [$Al_4Si_4O_{10}(OH)_8$]	0.031
HF–K–feldspar [$KAlSi_3O_8$]	0.050
HF–Na–feldspar [$NaAlSi_3O_8$]	0.050
HF–α-quartz [SiO_2]	0.167

* The coefficients are estimates calculated by assuming that the reaction products for any aluminosilicate are AlF_2^+ and SiF_6^{--}. These are the products when the HF concentration is high.

is the rate of mineral volume increase per unit of volume. On the other hand if r_F is the rate of acid appearance per unit of area, then the rate of mineral *dissolution* per unit area is $-\beta_F r_F$. Furthermore, the area per unit of formation volume is Sf_F. Thus

$$\frac{\partial}{\partial t}[(1 - \phi)V_F] = \frac{M_F S f_F \beta_F r_F}{\rho_F} \tag{15.20}$$

where ρ_F is the average density of the fast-reacting minerals and the M_F is its average molecular weight. This equation is not an approximation—it is an identity, provided the area per volume of the fast reacting-mineral is the correct one. Now introduce Eq. (15.18) and obtain

$$\frac{\partial}{\partial t}[(1 - \phi)V_F] = \frac{M_F \alpha_F V_F \beta_F r_F (1 - \phi)}{\rho_F} \tag{15.21}$$

Similarly for the slow-reacting minerals

$$\frac{\partial}{\partial t}[(1 - \phi)V_S] = \frac{M_S \alpha_S V_S \beta_S r_S (1 - \phi)}{\rho_S} \tag{15.22}$$

Equation (15.17) can then be written as follows, provided it is assumed that the porosity ϕ is nearly a constant and equal to the initial value ϕ_0:

$$\phi_0 \frac{\partial c_{HF}}{\partial t} + u \frac{\partial c_{HF}}{\partial x} = [\alpha_F V_F r_F + \alpha_S V_S r_S](1 - \phi_0) \tag{15.23}$$

where α_F and α_S will be considered to be constant. Equations (15.21), (15.22), and (15.23) represent a complete set of partial differential equations in the unknowns c_{HF}, V_S, V_F. All other terms are parameters that must be independently determined. Furthermore, from Chapter 14 the following rate expressions were shown to apply:

$$r_F = -E_f^{(F)} c_{HF} \tag{15.24}$$

and

$$r_S = -E_f^{(S)} c_{HF} \tag{15.25}$$

where $E_f^{(F)}$ and $E_f^{(S)}$ are average forward reaction rate constants for the fast and slow reactions, respectively. These reaction rates refer to the rate of acid appearance.

Example 15.3 The Validity of Surface Area Models

Show by contriving a contradictory example that Eqs. (15.18) and (15.19) can not always be true.

Solution Figure 15.3 depicts one mineral, say the fast-reacting one, as coating in a uniform layer the slow-reacting one. Thus until the outer layer is partially dissolved, α_S must obviously be zero. The coated mineral cannot react until the surrounding shell is at least partially removed. At that point, as depicted by Fig. 15.3, the value of α_S must increase from zero to a finite value. Thus for the geometry depicted in Fig. 15.3, α_F and α_S cannot be constant during the entire acidizing process. It should

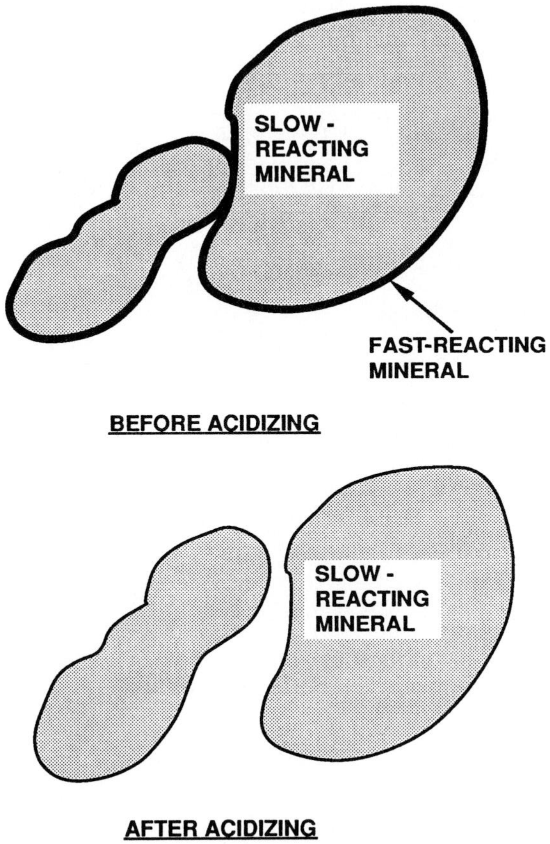

Figure 15.3 Sketch showing one mineral coated by a second mineral, which precipitated during diagenesis.

be stated, however, that a complete coating of any one mineral by another seems highly unlikely.

The lumped reaction model. An approximate analytical solution of Eqs. (15.21) and (15.22) is possible if the rate of the slow reaction is much less than that of the fast reaction so that only the fast reaction need be considered. This analysis is taken from Fogler and Rege [17] but it is apparently due to Amundson [18]. If $E_f^{(S)}$ is set equal to zero, then Eqs. (15.23) and (15.24) reduce to

$$-u\frac{\partial c_{HF}}{\partial x} - (1 - \phi_0)V_F\alpha_F E_f^{(F)}c_{HF} = \phi_0\frac{\partial c_{HF}}{\partial t} \tag{15.26}$$

The following dimensionless variables can now be defined:

$$\begin{aligned}\psi &= c_{HF}/c_{HF}^0\\ \Lambda &= V_F/V_F^0\\ \epsilon &= x/L\\ \theta &= t/\tau = tu/\phi_0 L\end{aligned} \tag{15.27}$$

where L is the length of a linear core and θ is the number of pore volumes injected into the core. Here c_{HF}^0 and V_F^0 are the initial values of c_{HF} and V_F, respectively. The balance equation (15.26) can be written in terms of these new variables as follows:

$$\frac{\partial\psi}{\partial\theta} + \frac{\partial\psi}{\partial\epsilon} + \left[\frac{(1 - \phi_0)V_F^0 E_f^{(F)}\alpha_F L}{u}\right]\psi\Lambda = 0 \tag{15.28}$$

and Eq. (15.21) becomes

$$\frac{\partial\Lambda}{\partial\theta} = -\left(\frac{\phi_0\beta_F c_{HF}^0 M_F}{V_F^0\rho_F}\right)\left(\frac{V_F^0 E_f^{(F)}\alpha_F L}{u}\right)\psi\Lambda \tag{15.29}$$

There are two dimensionless parameters appearing in these equations. The Damkohler number ($D_a^{(F)}$) is defined as

$$D_a^{(F)} = \frac{(1 - \phi_0)V_F^0 E_f^F \alpha_F L}{u}$$

$$= \frac{\text{rate at which acid is consumed by fast-reacting minerals}}{\text{rate at which acid is transported by convection}}$$

and the acid capacity number for the fast-reacting minerals is

$$A_C^{(F)} = \frac{\phi_0\beta_F c_{HF}^0 M_F}{(1 - \phi_0)V_F^0\rho_F} = \frac{\text{mass of minerals dissolved per unit volume}}{\text{mass of minerals present per unit volume}}$$

Example 15.4 Evaluation of Acid Capacity and Damkohler Numbers

If a core composed of the minerals considered in Example 15.2 is acidized with a mixture of 3 wt% HF and 12 wt% HCl, calculate the acid capacity and the fast-reaction Damkohler number for the following conditions:

$$u = 10^{-2} \text{ m/sec}$$
$$\phi_0 = 0.23$$
$$L = 0.40 \text{ m}$$
$$T = 50°\text{C } (323°\text{K})$$
$$S_0 = 1.8 \times 10^6 \text{ m}^2/\text{m}^3 \text{ of formation}$$
$$\rho_{\text{rock}} = 2650 \text{ kg of rock/m}^3$$

Solution The fast-reacting mineral is the 7 vol% orthoclase with $f_F = 0.134$ (see Example 15.2). Thus, $V_F^0 = 0.07$ and based on Eq. (15.18)

$$\alpha_F = \frac{(0.134)(1.8 \times 10^6)}{0.07(1 - 0.23)} = 4.47 \times 10^6 \text{ m}^2/\text{m}^3 \text{ of feldspar}$$

Since the acid concentrations given correspond to $c_{HF}^0 = 1.61$ kg moles of HF/m^3 and $c_{HCl}^0 = 3.53$ kg moles of HCl/m^3 (see Example 14.2), then the effective forward reaction rate as given by Eq. (14.16) for orthoclase is

$$E_f^{(F)} = 0.127 \exp\left(\frac{-4680}{323}\right)\left[1 + 5.66 \times 10^{-2} \exp\left(\frac{956}{323}\right)(3.53)^{0.4}\right]$$

or

$$E_f^{(F)} = 1.82 \times 10^{-7}$$

This forward reaction rate constant corresponds to the rate of mineral reaction. The value of $E_f^{(F)}$ must be divided by the appropriate β_F given in Table 15.1 for the feldspar to obtain the value corresponding to the rate of acid disappearance. Thus, the value of $D_a^{(F)}$ is

$$D_a^{(F)} = \frac{\left(\frac{0.77 \text{ m}^3 \text{ min.}}{\text{m}^3 \text{ form.}}\right)\left(0.07 \frac{\text{m}^3 \text{ feld.}}{\text{m}^3 \text{ min.}}\right)\left(\frac{1.82 \times 10^{-7} \text{ m}^3 \text{ soln.}}{0.05 \text{ m}^2 \text{ surf.-sec}}\right)\left(\frac{4.47 \times 10^6 \text{ m}^2 \text{ surf.}}{\text{m}^3 \text{ feld.}}\right)(0.4 \text{ m form.})}{10^{-2} \frac{\text{m}^3 \text{ soln.}}{\text{m}^2 \text{ form.-sec}}}$$

or $D_a^{(F)} = 35$

It will be observed that this large Damkohler number signals the existence of a sharp front separating the upstream region from which all of the fast-reacting mineral has been removed from the downstream region where the fast-reacting mineral still exists at its original concentration. The acid capacity is given by

$$A_C^{(F)} = \frac{\left(0.23 \frac{\text{m}^3 \text{ soln.}}{\text{m}^3 \text{ form.}}\right)\left(0.05 \frac{\text{kg mol feld.}}{\text{kg mol HF}}\right)\left(1.61 \frac{\text{kg mol HF}}{\text{m}^3 \text{ soln.}}\right)\left(278 \frac{\text{kg feld.}}{\text{kg mol feld.}}\right)}{\left(0.77 \frac{\text{m}^3 \text{ min.}}{\text{m}^3 \text{ form.}}\right)\left(0.07 \frac{\text{m}^3 \text{ feld.}}{\text{m}^3 \text{ min.}}\right)\left(2650 \frac{\text{kg feld.}}{\text{m}^3 \text{ feld.}}\right)}$$

or $A_C^F = 3.60 \times 10^{-2}$

It will be observed that this small value of $A_C^{(F)}$ means that an acid front moves slowly through the rock.

Solution of the equations will show that a reaction front develops and is propagated through the linear porous system. The speed at which this front moves is determined by the acid capacity number and its shape by the Damkohler number. Let us further investigate the solution to the lumped parameter system.

The dimensionless balance equations now become

$$\frac{\partial \psi}{\partial \theta} + \frac{\partial \psi}{\partial \epsilon} + D_a^{(F)} \psi \Lambda = 0 \tag{15.30}$$

$$\frac{\partial \Lambda}{\partial \theta} = -D_a^{(F)} A_C^{(F)} \psi \Lambda \tag{15.31}$$

Since there is no acid present in the system initially,

$$\psi = 0 \qquad (\text{at } \theta = 0) \tag{15.32}$$

At the reactor inlet

$$\psi = 1 \qquad (\text{at } \epsilon = 0) \tag{15.33}$$

and at any point within the reactor not yet contacted with acid

$$\Lambda = 1 \qquad (\text{for } \epsilon > \theta) \tag{15.34}$$

Combining Eqs. (15.30) and (15.31) yields

$$\frac{\partial \psi}{\partial \theta} + \frac{\partial \psi}{\partial \epsilon} - \frac{1}{A_C^{(F)}} \frac{\partial \Lambda}{\partial \theta} = 0 \tag{15.35}$$

This equation can be solved using the method of characteristics. Let us define new independent variables

$$\theta = \frac{1}{A_C^{(F)}} Y + Z \tag{15.36}$$

and

$$\epsilon = Z \tag{15.37}$$

Using the chain rule for partial differentiation, one can obtain

$$\left(\frac{\partial \psi}{\partial Z}\right)_Y = \left(\frac{\partial \psi}{\partial \epsilon}\right)_\theta + \left(\frac{\partial \psi}{\partial \theta}\right)_\epsilon \tag{15.38}$$

and

$$\left(\frac{\partial \Lambda}{\partial Y}\right)_Z = \frac{1}{A_C^{(F)}} \left(\frac{\partial \Lambda}{\partial \theta}\right)_\epsilon \tag{15.39}$$

Combining Eqs. (15.35), (15.38), and (15.39), yields

$$\frac{\partial \psi}{\partial Z} = \frac{\partial \Lambda}{\partial Y} \tag{15.40}$$

which is the condition for an exact differential. Consequently, a function F exists such that

$$dF = \Lambda \, dZ + \psi \, dY \tag{15.41}$$

where
$$\frac{\partial F}{\partial Z} = \Lambda\,(Z, Y) \tag{15.42}$$

and
$$\frac{\partial F}{\partial Y} = \psi(Z, Y) \tag{15.43}$$

Substituting Eq. (15.39) into Eq. (15.31), the mineral balance equation becomes

$$\frac{\partial \psi}{\partial Z} + D_{\rm a}^{\rm (F)}\,\Lambda\psi = 0 \tag{15.44}$$

Substitute Eqs. (15.42) and (15.43) into Eq. (15.44) to obtain

$$\frac{\partial^2 F}{\partial Y\,\partial Z} + D_{\rm a}^{\rm (F)}\,\frac{\partial F}{\partial Z}\frac{\partial F}{\partial Y} = 0 \tag{15.45}$$

This hyperbolic nonlinear partial differential equation may be readily solved by making the substitution

$$F(Z, Y) = (1/D_{\rm a}^{\rm (F)})\ln\,(U(Z, Y)) \tag{15.46}$$

in Eq. (15.45), which yields

$$\frac{\partial^2 U}{\partial Z\,\partial Y} = 0 \tag{15.47}$$

A general solution is of the form

$$U(Z, Y) = N(Z) + M(Y) \tag{15.48}$$

The original variables when expressed in terms of U are

$$\Lambda = \frac{1}{D_{\rm a}^{\rm (F)}}\frac{\partial \ln U}{\partial Z} \tag{15.49}$$

and
$$\psi = \frac{1}{D_{\rm a}^{\rm (F)}}\frac{\partial \ln U}{\partial Y} \tag{15.50}$$

The boundary conditions, Eqs. (15.33) and (15.34), are satisfied if we make the following choices:

$$U(0, Y) = \exp(D_{\rm a}^{\rm (F)}Y) \tag{15.51}$$

and
$$U(Z, 0) = \exp(D_{\rm a}^{\rm (F)}Z) \tag{15.52}$$

Combining Eqs. (15.47), (15.51), and (15.52)

$$U(Z, \mathrm{Y}) = \exp(D_{\rm a}^{\rm (F)}Z) + \exp(D_{\rm a}^{\rm (F)}Y) - 1 \tag{15.53}$$

or in terms of the original variables

$$\psi = \frac{1}{1 + \exp[D_{\rm a}^{\rm (F)}A_{\rm C}^{\rm (F)}(\epsilon + \epsilon/A_{\rm C}^{\rm (F)} - \theta)] - \exp[D_{\rm a}^{\rm (F)}A_{\rm C}^{\rm (F)}(\epsilon - \theta)]} \tag{15.54}$$

$$\Lambda = \frac{1}{1 + \exp[D_{\rm a}^{\rm (F)}A_{\rm C}^{\rm (F)}(\theta - \epsilon - \epsilon/A_{\rm c}^{\rm (F)})] - \exp[-D_{\rm a}^{\rm (F)}\epsilon)} \tag{15.55}$$

There are two fronts associated with the dissolution process:

1. The front between the reacting acid and the fluid initially present in the core. The equation representing this front is

$$\epsilon = \theta \tag{15.56}$$

This equation can be written in the form of a velocity as follows. Since $\epsilon = x/L$ and $\theta = tu/\phi_0 L$, one can write an expression for the frontal velocity v_f^I by noting that Eq. (15.56) implies that

$$v_f^I = \frac{u}{\phi_0} \tag{15.57}$$

which is simply the interstitial velocity of the fluid in the porous medium, i.e., the velocity of a tracer (see Chapter 4). The fluid downstream of this front is that originally present in the formation, whereas just upstream of the front is found the spent acid solution. The composition of this spent acid solution will turn out to be quite important in determining whether or not reprecipitation of silica will occur. This matter is taken up in some detail in the next chapter. Here this front is essentially ignored.

2. The front between the unreacted acid and the unreacted solid

$$\epsilon = \frac{A_C^F}{1 + A_C^F}\theta \tag{15.58}$$

The velocity of this front can be expressed as

$$v_f^{II} = \frac{x_f^{II}}{t} = \frac{A_C^{(F)}}{1 + A_C^{(F)}}\left(\frac{u}{\phi_0}\right) \tag{15.59}$$

The time for significant acid concentrations to emerge from a core is clearly

$$t_B = \frac{L}{v_f^{II}} = \frac{(1 + A_C^{(F)})\phi_0 L}{A_C^{(F)} u} \tag{15.60}$$

This frontal velocity will turn out to be an important factor in deciding how much acid to use in a given treatment. If the acid has a small capacity, then the movement of the front will also be slow.

The Damkohler number determines the steepness of the profile (the larger $D_a^{(F)}$, the steeper the profile).

Example 15.5 Frontal Movements

How many pore volumes (PV) of acid are required to dissolve the fast-reacting minerals from the core sample considered in Example 15.4 to the extent that no more than 20% of these minerals remain at any point in the sample? Take the conditions of the acidization to be those defined in Example 15.4.

Solution The highest concentration of fast-reacting minerals remaining will evidently be at the outlet of the core, i.e., at $\epsilon = 1$. Thus to obtain an answer to the question posed, Λ must be set equal to 0.2 at this point. Furthermore, from Example

15.4 we recall that $D_a^{(F)} = 35$ and $A_C^{(F)} = 3.6 \times 10^{-2}$. Substituting all of these values into Eq. (15.55), we find

$$0.2 = \frac{1}{1 + \exp[1.26(-28.8 + \theta)] - \exp(-35)}$$

Solving

$$\theta = 29.9$$

Thus, about 30 PV of acid are required to reduce the concentration of the fast-reacting minerals to 20% of its original value at the outlet end of the core. It should be noted that the reaction front is sharp and only a few more PV of acid are required to remove essentially all of the fast-reacting minerals.

The shape of the reaction front and its dependence on position is shown in Fig. 15.4. In this case the Damkohler number has been fixed to be 10 and the front, although quite sharp, is definitely not a shock. Also the acid capacity number is unity and according to Eq. (15.59), the velocity of the reaction front relative to the fluid velocity ($v_f^{II}\phi_0/u$) is 0.5. Thus after 1 PV of acid has been injected ($\theta = 1$), the reaction front, denoted as the point for which $\Lambda = 0.5$, has moved halfway through a linear core. The position of the front is denoted by ϵ_f. To cause the mineral concentration to be reduced to 50% of its initial value at the end of the core, 2 PV of acid are required.

The reaction front as seen in Fig. 15.4 is contained within that zone of width $\Delta\epsilon$ where both the acid concentration and the solid concentration vary substantially. The actual position of the front can be associated with the point at which the fast-reacting solid concentration is reduced to half of its initial value, i.e., to $\Lambda = 0.5$. This is, as shown in Fig. 15.4, the same point at which the acid concentration is reduced to one half of its initial value. This front moves through the

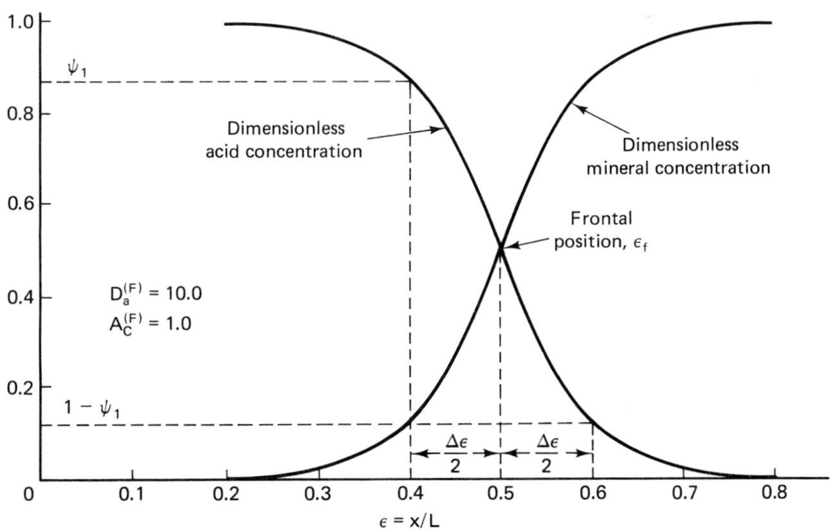

Figure 15.4 Detailed structure of the reaction zone shown for a single fast reaction.

porous medium as more acid is injected (θ increases). This movement is depicted by Figs. 15.5(a) and (b). It is shown in Fig. 15.5(a) for $A_C^{(F)} = 0.2$ that $v_f^{II}\phi_0/u = 0.167$ and 6 PV of acid are required to move the $\Lambda = 0.5$ fast-reacting solid concentration the full length of the core. Figure 15.5(b) shows the frontal position on a distance–time diagram. This plot provides the essential information in those cases for which the width of the reaction zone is much smaller than that of the core length, i.e., when the reaction front is sharp.

Figure 15.6 shows a case for which the reaction front is large compared to the length of the core sample. This occurs when the Damkohler number is small. For the example shown, $D_a^{(F)} = 0.5$ and it is observed that both the acid and the mineral concentrations vary continuously from one end of the core sample to the other.

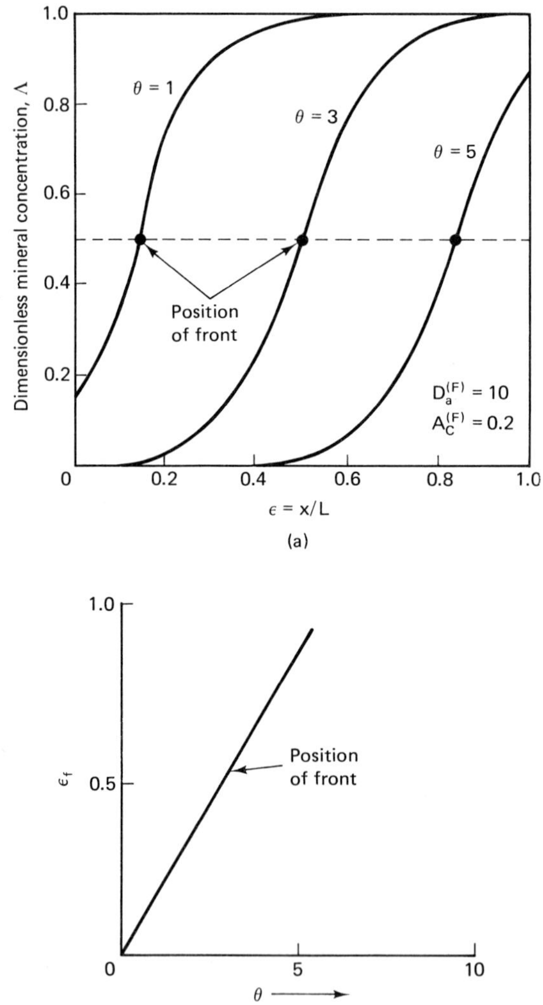

Figure 15.5 Graphs showing movement of a reaction front. The movement of the frontal zone is measured here by the point at which the mineral concentration is reduced to 50% of its original value.

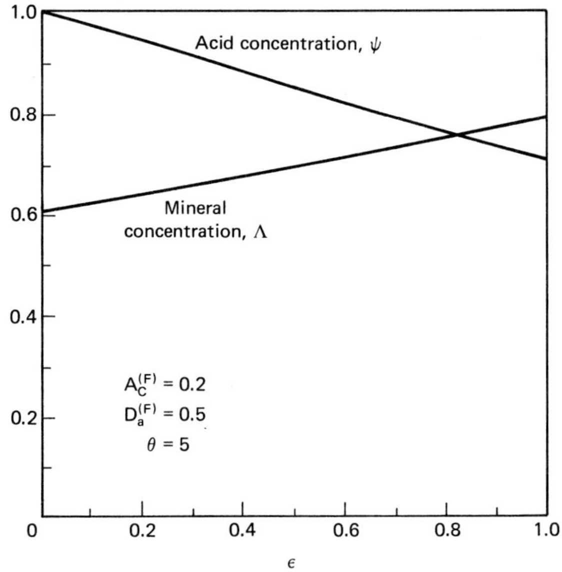

Figure 15.6 Plotted data for a broad reaction zone that is larger than the length of the core sample.

The results presented here apply only if the slow reaction rate can be completely neglected. It has not been included in any of the calculations. It will become evident that in considering acidizing of sandstone formations, one cannot neglect the slow reaction. The reasons for this being true are discussed in the next section.

Effluent concentrations. Normally a laboratory core test entails observation of the effluent composition, the flow rate of the acid, the temperature, and the core permeability. A primary variable is the effluent acid concentration as a function of time. For the case of a single fast-reacting mineral, the effluent acid concentration is given by Eq. (15.54) when ψ is evaluated at $\epsilon = 1$. The equation can be rearranged to yield

$$\ln\left(\frac{1}{\psi_e} - 1\right) = -D_a^{(F)}A_C^{(F)}\theta + \ln\{\exp[D_a^{(F)}A_C^{(F)}(1 + 1/A_C^{(F)})] - \exp D_a^{(F)}A_C^{(F)}\} \tag{15.61}$$

where ψ_e is the ratio of the exit concentration of the acid to the initial concentration. A plot of $\ln[(1/\psi_e) - 1]$ versus θ should, according to Eq. (15.61), be a straight line with an intercept equal to $\{\exp[D_a^{(F)}A_C^{(F)}(1 + 1/A_C^{(F)})] - \exp D_a^{(F)}A_C^{(F)}$ and negative slope $D_a^{(F)}A_C^{(F)}$. Thus from the slope and the intercept, $D_a^{(F)}$ and $A_C^{(F)}$ can be determined individually.

Figure 15.7 shows the effluent HF acid concentration when a Berea sandstone core 2.54 cm in diameter by 10 cm in length is acidized [19]. This figure shows that the first volumes of fluid to emerge from the core sample contain little unreacted acid, i.e., ψ_e is small. As the acidization proceeds, the effluent acid concentration increases. Figure 15.7 shows that the shapes of the various effluent acid concentration curves have little apparent character. It is difficult from this

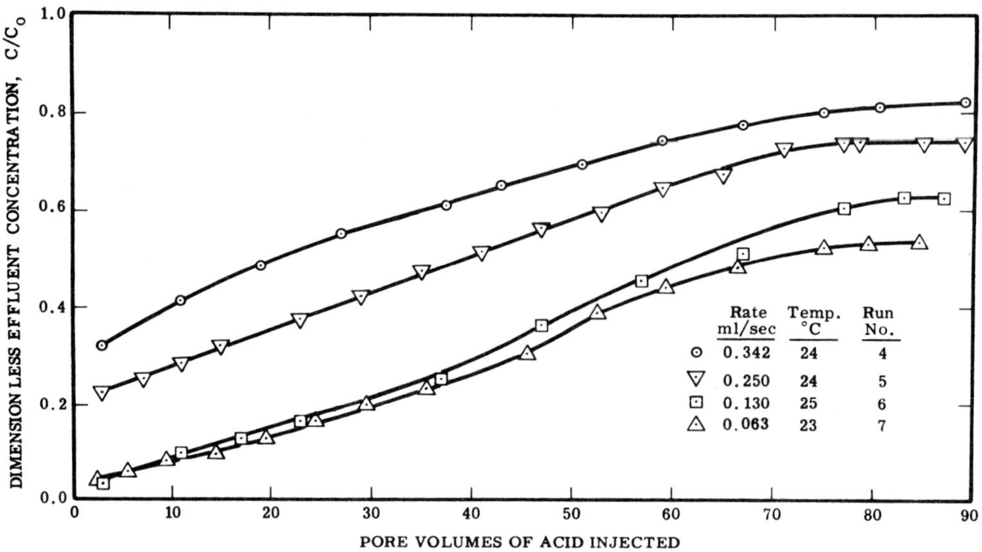

Figure 15.7 Plotted data of effluent acid concentration as a function of PV injected for Berea Sandstone [19]. The various curves show the effect of the flow rate.

plot to determine much about the reactions that are occurring. There do not appear to be any particular features that help us to decide when the fast-reacting minerals have been removed.

Hekim and Fogler [20] have replotted these data for the slow injection rate, as shown in Fig. 15.8. Two very distinct linear segments appear. The steeper portion corresponds to the dissolution of the fast-reacting minerals, such as feldspars and authogenic clays, whereas the shallower almost horizontal portion is attributed to the reaction with quartz and clay fragments, such as those often found in Berea core samples [16]. The two linear segments join together as a smooth curve, indicating the transition region where the dissolution rates become comparable. This aspect of the acidizing process will be discussed in the next section.

The important point to be made here in this section is that the effluent acid concentration is determined partly by the presence of slow-reacting substances. It does not fit Eq. (15.61). In other words both reactions must, therefore, be considered. Study of the acidizing process in the laboratory and also any design of acidizing treatment must, therefore, take into account the slow reaction rate as well as the fast reaction rate. One cannot be studied without understanding the other. Thus, we must now turn our attention to the more complex cases.

Multiple reactions. If two or more minerals are dissolving at different rates, then two or more slopes when $\ln[(1/\psi_e) - 1]$ is plotted as a function of θ will be observed. This implies that two or more reaction fronts are progressing through the porous medium. When these fronts are all well established, they can all move at well-defined frontal velocities. The propagation of these fronts has

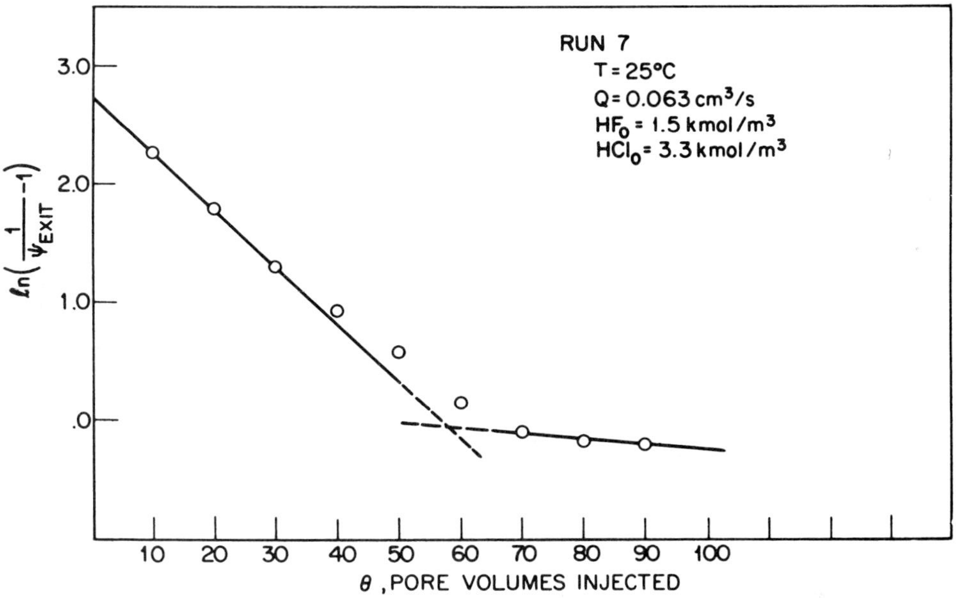

Figure 15.8 The effluent HF acid concentration for an injection rate of 0.063 cm^3/s replotted from Fig. 15.7 to show the two reaction zones [20]. (With permission of American Institute of Chemical Engineers.)

been described by Hekim and Fogler [20]. They show that a particular front is well defined only if the associated Damkohler number is >5. For small values of the Damkohler number, there is no sharp mineral front. As shown in Fig. 15.6, the reaction zone may extend through the entire length of the core sample. This is the case for the quartz reaction with HF. The rate of reaction is small, and furthermore the volume of mineral to be dissolved is large. In general, the volume of acid injected is not sufficient to dissolve all of the quartz even in a small region near the wellbore nor in a small region near the inlet of a core sample. The reason for this is evident. Too much acid would be required to do so and furthermore, as we shall see, such an extravagant application of acid will not yield a corresponding improvement in the result of a matrix treatment. Because it is uneconomical to dissolve all of the quartz or other slow-reacting minerals, some generalizations can be made to assist us in our design of a matrix treatment.

Characteristics of the Slow Reaction Zone

The general features governing the design of a matrix acidizing treatment can best be visualized by first considering the penetration of acid into a core whose solid matrix is composed entirely of a slow-reacting substance, say, quartz. Since the volume of the slow-reacting solid is large, then the acid capacity number of the slow-reacting quartz ($A_C^{(S)}$) is small; that is $A_C^{(S)} \rightarrow 0$. From Eq. (15.54) it is observed that in the limit of small acid capacity numbers

$$\psi = \exp(-D_a^{(S)}\epsilon) \tag{15.62}$$

This equation is interesting since it suggests that the acid concentration depends on the distance measured from the core inlet, but does not depend on θ, the number of PV injected. This implies that the effluent concentration from a core should be constant once the fast-reacting minerals have been removed. Referring back to Figs. 15.7 and 15.8, this is shown to be essentially the case.

Figure 15.8, e.g., shows that the second segment is practically horizontal indicating that for this particular Berea sample the effluent concentration becomes essentially constant, independent of θ once the fast-reacting materials have been removed. A number of experiments on Berea all show similar behavior [19].

We expect tha' acidization carried out in the field will exhibit similar behavior. Thus, it is worthwhile to investigate the implications of Eq. (15.62) in somewhat greater detail.

Example 15.6 Acid Penetration Distance

If the acid is defined to be spent whenever $\psi = 0.05$, then to what distance would live acid be expected to penetrate into a long Berea sandstone core given the results of a laboratory test shown by Fig. 15.8?

Solution The data taken from Fig. 15.8 can be summarized as follows:

$$\psi_e = 0.54 \text{ after } 70 \text{ PV}$$

$$Q = 0.063 \text{ cm}^3/\text{sec}$$

$$A = \frac{\pi}{4}(2.54 \text{ cm})^2 = 5.07 \text{ cm}^2$$

$$L = 10 \text{ cm}$$

Thus from Eq. (15.62) with $\epsilon = 1$, we find

$$D_a^{(s)} = -\ln \psi_e = 0.62$$

It is now important to note that the Damkohler number is proportional to the core length L. Thus, for a longer core sample of length L', the Damkohler number for the longer core sample is given by

$$D_a^{(S)} = (0.62)\frac{(L')}{(L)} \qquad (\text{with } L = 10 \text{ cm})$$

Furthermore,

$$\epsilon = \frac{x}{L'}$$

thus the dimensionless acid concentration as a function of position is

$$\psi_e = \exp\left(-\frac{0.62}{10}x\right)$$

where x is the distance measured from the core inlet in cm. The acid penetration distance is determined from this equation. The x for which $\psi_e = 0.05$ is

$$x = -\frac{10}{0.62}\ln(0.05) = 48.3 \text{ cm}$$

This is the distance into a core sample that live acid will penetrate for the conditions of the experiment.

If it is desired to increase the acid penetration distance, the Damkohler number for the slow-reacting minerals must be reduced. The only variables that are available to the experimenter are the temperature, the acid concentration, and the acid flux.

Reducing the formation or core temperature will substantially decrease $E_f^{(s)}$ (see Chap. 14) so that it is possible to reduce the Damkohler number in this way. As will be discussed, it may be possible to decrease the reaction temperature in field treatments by using an adequate preflush to precool the formation (see Example 16.2).

The Damkohler number is independent of the inlet acid concentration, so that increasing it will not change the position at which ψ takes on a particular value. Of course for fixed ψ, the acid concentration increases as the inlet acid concentration increases. We shall, however, find that for a variety of reasons increasing the concentration of HF in excess of some minimum level may be undesirable.

The Damkohler number can also be decreased by increasing the acid flux. In core tests u can be varied over a wide range; however, in matrix acid treatments the acid flux is limited by the fracture gradient.

Equation (15.62) is a key result for the design of matrix acid treatments. It shows that there is a limit to the depth surrounding the wellbore from which damage can be removed by acid treatment.

Speed of the Fast-Reacting Mineral Front

We have described the effluent acid concentration in two extreme cases—one for which the system contains only a small quantity of fast-reacting mineral with the remainder of the solid matrix being composed of an inert solid, and at the other extreme, one which is composed entirely of a single slow-reacting mineral. In practice, of course, both types of minerals will be present and one must, at a minimum, take into account the presence of both since they interact with one another. This interaction is illustrated by the curves shown in Fig. 15.9. The fast-reacting mineral is shown to exhibit a reaction front shaped much the same as that shown in Fig. 15.4, where there is no slow-reacting mineral. The difference between the reacting zones shown in Figs. 15.4 and 15.9 is in the acid concentration. Because of the presence of a large quantity of the slow-reacting mineral, the acid concentration decreases continuously from the core inlet until it reaches the reaction zone where quantities of fast-reacting mineral remain. At this point, the acid concentration decreases rapidly causing the fast-reacting front to advance.

Let us define ϵ_f as the position of the reaction front and consider the rate at which this front moves. We know, of course, that as ϵ_f increases, the concentration of acid arriving at the front will decrease in accordance with Eq. (15.62). Since the acid concentration decreases at the front, the velocity of the front will

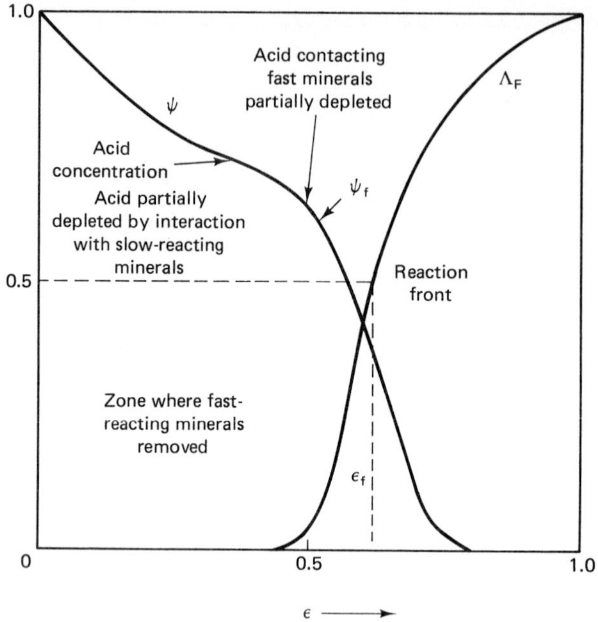

Figure 15.9 Graph showing the interaction of two mineral fronts. The acid concentration is plotted as a function of position for the case of a fast-reacting mineral and a slow-reacting mineral.

also decrease since the dissolving power of the acid is directly proportional to the acid concentration. Thus because the acid must flow through a porous medium of slow-reacting mineral to reach the reaction front, a portion of the acid dissolving power of the acid is lost. Clearly, a fast-reacting mineral cannot be removed for a distance beyond the acid penetration distance. This is a key concept that will be applied in the design of an acid treatment.

To complete those concepts needed for a rational design, the frontal position (ϵ_f) as a function of the PV injected must now be calculated.

To do this calculation it is noted that according to Eq. (15.59) the front moves at a rate dictated by the acid capacity number. To account for the reduced acid concentration as the acid arrives at the front as shown by Fig. 15.9, the acid capacity number at the front becomes a function of the frontal position. Thus, Eq. (15.59) becomes

$$\frac{\phi_0 v_f^{II}}{u} = \frac{d\epsilon_f}{d\theta} = \frac{a_C^{(F)}(\epsilon_f)}{1 + a_C^{(F)}(\epsilon_f)} \tag{15.63}$$

where $a_C^{(F)}$ is the acid capacity number evaluated at the reaction front. It is clearly less than the acid capacity number of the injected fluid, $A_C^{(F)}$ since, as observed in Fig. 15.9, the acid concentration at the front is less than that injected. From Eq. (15.62)

$$a_C^{(F)}(\epsilon_f) = A_C^{(F)} \cdot \psi = A_C^{(F)} \exp(-D_a^{(S)} \epsilon_f) \tag{15.64}$$

Substituting Eq. (15.64) into Eq. (15.63) and integrating, the following relationship is found:

$$\theta = \frac{\exp(D_a^{(S)}\epsilon_f) - 1}{A_C^{(F)} D_a^{(S)}} + \epsilon_f \tag{15.65}$$

This expression is the desired one. It relates the number of PV injected to the position of the front. There are two parameters that appear, the Damkohler number for the slow-reacting system and the acid capacity number for the fast-reacting mineral. The Damkohler number for the fast reaction is a third important variable. If it is sufficiently large, it is possible to assume the front to be sharp. The acid capacity number for a slow-reacting mineral is small ($A_C^{(S)} \rightarrow 0$) and is not needed in the design of an acid treatment.

Equation (15.65) can be used to calculate $A_C^{(F)}$, given $D_a^{(S)}$ and the point of intersection between the two line segments in Fig. 15.8. At this intersection it can be assumed that $\epsilon_f \cong 1$, i.e., this is the number of PV, θ, at which the front emerges from the core.

Example 15.7 Determination of the Acid Capacity Number

Consider the experiment described in Example 15.6. Based on the Damkohler number for the slow-reacting minerals and the data plotted in Fig. 15.8 estimate the acid capacity number for the fast-reacting minerals.

Solution The intersection of the two line segments shown in Fig. 15.8 appears to be at about 58 PV and from the work presented in Example 15.6, we know that $D_a^{(S)} = 0.62$. Thus, from Eq. (15.65)

$$58 = \left(\frac{\exp(0.62) - 1}{0.62 A_C^{(F)}}\right) + 1$$

Solving for $A_C^{(F)}$ we find

$$A_C^{(F)} = 2.43 \times 10^{-2}$$

Thus, from a single coreflood experiment at least two of the three essential design quantities can be determined. Actually, all three can be determined.

The intersection of the two line segments shown in Fig. 15.8, therefore, provides one measure of $A_C^{(F)}$. The values of both the acid capacity and the Damkohler numbers for the fast reaction can be estimated by applying Eq. (15.61) to the effluent acid concentration, which emerges in early times. Evidently, the slow-reacting minerals do not influence the rate of reaction as the acid first contacts the core.

Example 15.8 Alternate Method for the Determination of Acid Capacity and Damkohler Numbers

Consider the experiment described in Example 15.6 and plotted in Fig. 15.8. Using these data for small θ, estimate the acid capacity and Damkohler numbers characteristic of the fast-reacting minerals.

Solution The intercept of the first line segment shown in Fig. 15.8 can be estimated to be 2.75 and its slope is -0.5. Thus, from Eq. (15.61)

$$D_a^{(F)}A_C^{(F)} = 0.5 \quad \text{and} \quad \ln\{\exp[D_a^{(F)}A_C^{(F)}(1 + 1/A_C^{(F)})] - \exp(D_a^{(F)}A_C^{(F)})\} = 2.75$$

Solving these two equations simultaneously, gives

$$A_C^{(F)} = 1.81 \times 10^{-2} \quad \text{and} \quad D_a^{(F)} = 2.76$$

The value for $A_C^{(F)}$ obtained here should be compared with the value 2.41×10^{-2} obtained in Example 15.7. This difference between the two is significant and can only really be resolved by accurate solutions to the acid balance equation [Eq. (15.31)] and a careful fit of all of the data rather than only early time results as is done here.

15.2 PERMEABILITY OF THE ACIDIZED ZONE

As has been noted, it has proven quite difficult to predict permeability on purely theoretical grounds except for very simple porous systems such as sintered glass [7] or copper [21]. The approach that has been applied in the design of acid treatment [13] is the use of empirical equations such as Eqs. (15.13) or (15.14) or perhaps some modified version of one of these equations. It will be evident that the form of the equation used is not particularly critical, since matrix acidizing is used primarily to remove damage and restore the original formation permeability in a zone surrounding the wellbore. Once the damage is removed, then there is little gained by further increases in permeability. Thus, it is important to establish the zone in which damage in the form of particles clogging the pore throats has been cleared. Here we assume that damage removal is accomplished for all $\epsilon \leq \epsilon_f$, where ϵ_f is the frontal position described in the previous section and defined by Eq. (15.65). This view is summarized by the sketch shown in Fig. 15.10. The zone behind the front has a permeability k_A (the A denotes *acidized*), whereas downstream of the front the permeability is k_S. Since the average permeability is

$$\frac{\bar{k}(\theta)}{k_S} = \frac{1}{\displaystyle\int_0^1 \frac{k_S\, d\epsilon}{k}} = \frac{k_A}{\epsilon_f k_S + (1 - \epsilon_f)k_A} \qquad (\text{for } \epsilon_f < 1) \tag{15.66}$$

The mean permeability ($\bar{k}$), which is the average value obtained by measuring the pressure drop across the entire length of the core, depends on the position of the front as shown in Fig. 15.10. Upstream from the front, the fast-reacting minerals have been removed. The porosity has been increased by virtue of their removal. Since V_F^0 is the volume of fast-reacting minerals per unit volume of minerals, the increase in porosity is given by

$$\Delta\phi = \phi - \phi_0 = V_F^0(1 - \phi_0) \tag{15.67}$$

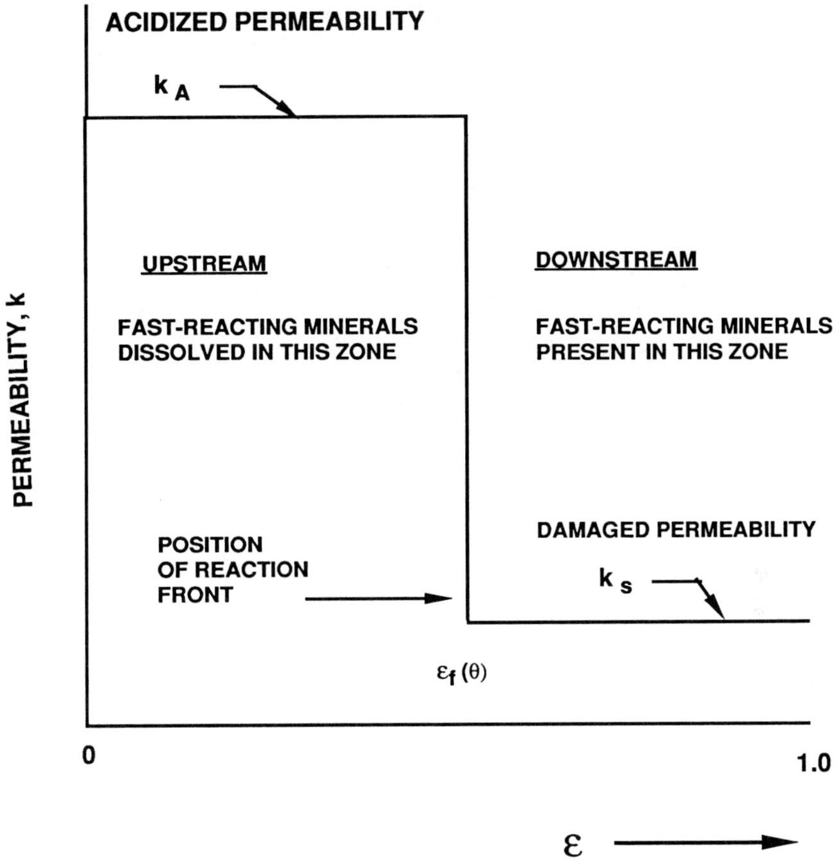

Figure 15.10 Graph showing permeability as a function of position in a linear system.

Example 15.9 Permeability of Acidized Cores

Determine the permeability after 30 PV of acid have been injected into a Berea core, which is 2.54 cm in diameter and 10 cm in length, assuming that the experimental conditions are the same as those stated in Example 15.6. Assume that the permeability of the acidized zone is given by [14]

$$k_A/k_S = \exp(48.5\,\Delta\phi)$$

Solution Based on the results obtained in Examples 15.6 and 15.7, we have

$$D_a^{(S)} = 0.62 \qquad \text{and} \qquad A_C^{(F)} = 2.43 \times 10^{-2}$$

We recall that the acid capacity number is defined as

$$A_C^{(F)} = \frac{\phi_0 \beta_F c_{HF}^0 M_F}{(1 - \phi_0) V_F^0 \rho_F}$$

For the purposes of estimating V_F^0 we can take the fast-reacting material to be a feldspar [16], and thus we have approximately

$$\beta_F = 0.05 \text{ kg moles of solid/kg mole of acid} \qquad \text{(from Table 15.1)}$$

$$\rho_F = 2650 \text{ kg of solid/m}^3$$

$$M_F = 278 \text{ kg of solid/kg mole of solid}$$

Furthermore, 3 wt% HF is equivalent to 1.61 kg moles of HF/m^3 and $\phi_0 = 0.22$ for this core sample. Using these numbers and the acid capacity, we obtain

$$V_F^0 = 9.8 \times 10^{-2}$$

(*Note:* This value exceeds the amount of feldspar reported to be present in the samples of Berea tested. Other minerals such as illites, which are also present in the samples [19] must also behave as fast-reacting minerals).

The porosity increase is therefore

$$\Delta\phi = (9.8 \times 10^{-2})(1 - 0.22) = 7.6 \times 10^{-2}$$

and

$$\frac{k_A}{k_S} = \exp[(48.5)(7.6 \times 10^{-2})] = 40$$

The permeability is increased 40-fold in the acidized zone.

After 30 PV have been injected, then from Eq. (15.65) we have

$$1 + (30)(0.62)(2.43 \times 10^{-2}) = \exp(0.62\epsilon_f) + (0.62)(2.43 \times 10^{-2})\epsilon_f$$

and solving for ϵ_f we find

$$\epsilon_f = 0.591$$

that is, the front has progressed more than halfway through the core. At this time, the average permeability is found from Eq. (15.66) to be

$$\frac{\bar{k}}{k_S} = \frac{40}{0.591 + (1 - 0.591)40} = 2.36$$

The observed value for this experiment is reported to be about 1.8 [19].

15.3 RADIAL FLOW OF ACIDS THROUGH POROUS MEDIA

Comparison with Linear Flows

The arguments presented in the previous sections suggest that the acidization of a core sample can be characterized by three parameters—the acid capacity number for the fast-reacting minerals, the Damkohler number associated with the slow-reacting minerals, and the Damkohler number for the fast-reacting minerals. The acid capacity number of the slow-reacting minerals is not on this list because for the purposes of acidization it is essentially zero. This implies that the treatment will be carried out in such a way that a large fraction of the slow-reacting minerals will still remain undissolved at the termination of the treatment. The strategy described in Chapter 16 states that a well-designed treatment will minimize the amount of acid expended in dissolving the slow-reacting minerals. Most of it should be used to dissolve fast-reacting minerals. Thus the acid capacity of the slow-reacting minerals is not an important factor to be considered in the design of an acid treatment.

The Damkohler number for the fast-reacting minerals is only important to the extent that it is required to assure that the reaction front is sharp enough so that essentially all of the fast-reacting minerals are removed behind the reaction front.

Considering the radial flow of acid around the wellbore, however, in linear flow the flux (u) is a constant; for radial flow the injection rate (i) is generally a constant, and the flux depends on radial position as follows:

$$u = \frac{i}{2\pi h r} \tag{15.68}$$

Equation (15.23) applied to the zone behind the reaction front (see Fig. 15.10), where all of the fast-reacting mineral has been removed, reduces to

$$\phi_0 \frac{\partial c_{\mathrm{HF}}}{\partial t} + \frac{i}{2\pi h r}\frac{\partial c_{\mathrm{HF}}}{\partial r} = -\alpha_{\mathrm{S}} V_{\mathrm{S}} E_{\mathrm{f}}^{(\mathrm{S})}(1 - \phi_0) c_{\mathrm{HF}} \tag{15.69}$$

This acid balance equation must be solved subject to the boundary and initial conditions

$$\begin{aligned} c_{\mathrm{HF}} &= c_{\mathrm{HF}}^{0} \qquad (\text{at } r = r_{\mathrm{w}}) \\ c_{\mathrm{HF}} &= 0 \qquad \left(\text{for } r > \sqrt{\frac{it}{\pi h \phi_0} + r_{\mathrm{w}}^2}\right) \end{aligned} \tag{15.70}$$

where r_{w} is the wellbore radius.

The solution to this partial differential equation is considered in the next section.

Acid Concentration at the Reaction Front

As shown in Fig. 15.11, the acid enters the formation at the wellbore and must flow through the zone containing only slow-reacting minerals to reach the reaction front. To calculate the speed at which the reaction front moves, it is necessary to determine the concentration of acid at this front. Start by introducing the following dimensionless quantities into Eq. (15.69):

$$\begin{aligned} \psi &= c_{\mathrm{HF}}/c_{\mathrm{HF}}^{0} \\ \epsilon^{(\mathrm{R})} &= \frac{r^2}{r_{\mathrm{w}}^2} - 1 \\ \theta^{(\mathrm{R})} &= \frac{it}{\pi r_{\mathrm{w}}^2 h \phi_0} \\ D_{\mathrm{a}}^{(\mathrm{R,S})} &= \frac{\alpha_{\mathrm{S}} V_{\mathrm{S}}^{0}\, \pi r_{\mathrm{w}}^2 h E_{\mathrm{f}}^{(\mathrm{S})}(1 - \phi_0)}{i} \end{aligned} \tag{15.71}$$

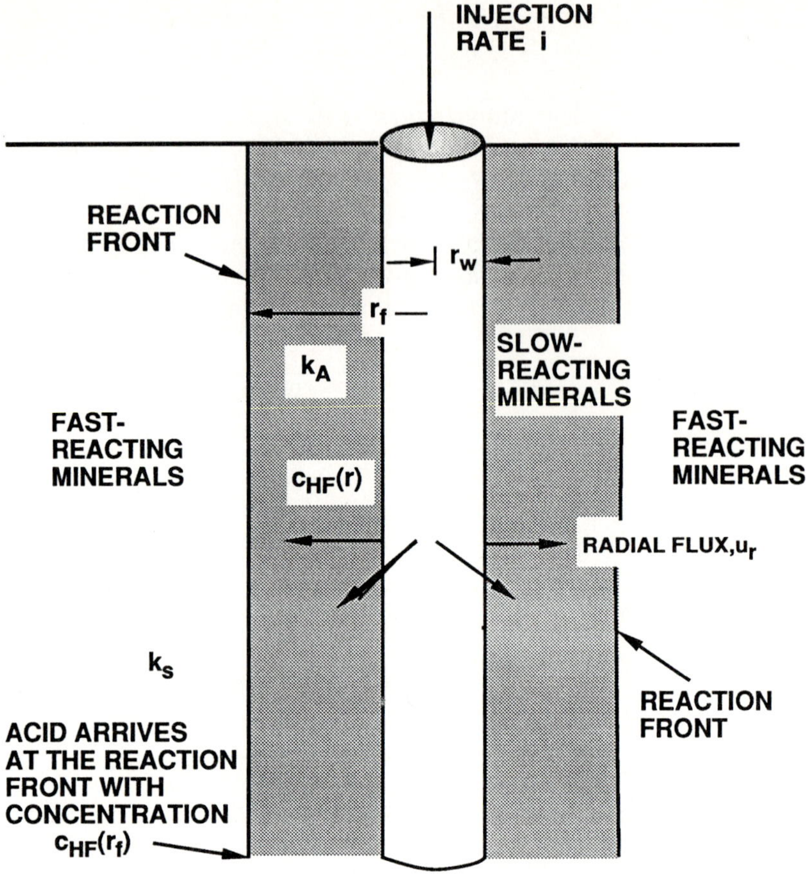

Figure 15.11 Sketch depicting acidizing in an uncased wellbore.

The superscript (R) is intended to denote a radial system, and it should be noted that the dimensionless radial variables differ somewhat from their linear counterparts. The difference stems from the variability of the radial flux, as shown by Eq. (15.68). Using Eq. (15.71), Eq. (15.68) can be rewritten as

$$\frac{\partial \psi}{\partial \theta^{(R)}} + \frac{\partial \psi}{\partial \epsilon^{(R)}} = -D_a^{(R,S)}\psi \tag{15.72}$$

since V_S is essentially constant. The solution to this equation is for a sharp front readily shown to be

$$\psi = \exp(-D_a^{(R,S)}\,\epsilon^{(R)}) \qquad (\text{for } \epsilon^{(R)} \leq \epsilon_f^{(R)}) \tag{15.73}$$

This expression permits us to calculate the acid concentration at the acid front. The acid capacity of the acid arriving at the front is reduced so that

$$a_C^{(F)}(\epsilon_f^{(R)}) = A_C^{(F)} \exp(-D_a^{(R,S)}\epsilon_f^{(R)}) \tag{15.74}$$

Movement of the Reaction Front

Equation (15.59) applies even in radial flow. Thus assuming that the Damkohler number in radial flow is sufficiently large so the front is sharp (see Example 15.11), one has

$$\frac{dr_f}{dt} = \frac{a_C^{(F)}(\epsilon_f^{(R)})}{1 + a_C^{(F)}(\epsilon_f^{(R)})} \frac{u}{\phi_0} \tag{15.75}$$

Substituting the dimensionless quantities defined by Eq. (15.71) yields

$$\frac{d\epsilon_f^{(R)}}{d\theta^{(R)}} = \frac{a_C^{(F)}(\epsilon_f^{(R)})}{1 + a_C^{(F)}(\epsilon_f^{(R)})} \tag{15.76}$$

which is identical to Eq. (15.63) for a linear system. Substituting Eq. (15.74) and integrating, we obtain a result that is identical to the result obtained for a linear system [see Eq. (15.65)]:

$$\theta^{(R)} = \frac{\exp(D_a^{(R,S)}\epsilon_f^{(R)}) - 1}{A_C^{(F)}D_a^{(R,S)}} + \epsilon_f^{(R)} \tag{15.77}$$

This equation applies to a radial geometry, provided the radial definitions of the dimensionless variables are applied. Equation (15.77) will play an essential role in our design of matrix treatments.

Example 15.10 Acid Penetration Into Uncased Wellbores

Suppose that a sandstone formation composed of Berea is to be acidized through an uncased wellbore that is 0.1 m in radius. Let us further suppose that the results found in Examples 15.6 and 15.8 apply to this sandstone. If the acid (3 wt% HF and 12 wt% HCl) injection rate is to be 10^{-3} m^3/sec-m of formation thickness, how far will the reaction front have progressed after 1 min? What penetration depth is attained when a total of 5 m^3/m of formation thickness has been injected (this is a relatively large volume of acid).

Solution The radial Damkohler number and the one for a linear system can be observed to be related as follows:

$$D_a^{(R,S)} = D_a^{(S)}\left(\frac{\pi u r_w^2 h}{iL}\right)$$

From Example 15.6

$$u = \frac{Q}{\pi r^2} = \frac{0.063 \text{ cm}^3/\text{sec}}{\pi(2.54 \text{ cm}/2)^2} = 1.24 \times 10^{-2} \text{ cm/sec} = 1.24 \times 10^{-4} \text{ m/sec}$$

$$L = 10 \text{ cm} = 0.1 \text{ m}$$

and

$$D_a^{(S)} = 0.62$$

$$D_a^{(R,S)} = (0.62)\left(\frac{1.24 \times 10^{-4}}{0.1}\right)\left(\frac{\pi(0.1)^2}{10^{-3}}\right) = 2.42 \times 10^{-2}$$

$D_a^{(R,S)}$ is the radial Damkohler number for the slow-reacting mineral. Since the core sample considered in Example 15.8 is the same as the formation to be acidized, then the acid capacity is the same (assuming, of course, that the same concentration acid is used), and $A_C^{(F)} = 1.81 \times 10^{-2}$.

After 1 min

$$\theta^{(R)} = \frac{(60 \text{ sec})(10^{-3} \text{ m}^3/\text{sec-m})}{\pi(0.1 \text{ m})^2(0.23)} = 8.3$$

Substituting into Equation (15.77)

$$8.3 = \frac{\exp[(2.42 \times 10^{-2})\epsilon_f^{(R)}] - 1}{(1.81 \times 10^{-2})(2.42 \times 10^{-2})} + \epsilon_f^{(R)}$$

Solving for $\epsilon_f^{(R)}$ we find

$$\epsilon_f^{(R)} = 0.147$$

Thus

$$r_f = r_w(1 + \epsilon_f^{(R)})^{1/2} = 0.1071 \text{ m}$$

or the front has progressed a distance

$$r_f - r_w = 0.0071 \text{ m} = 0.71 \text{ cm}$$

This is a very short penetration distance; however, depending on the depth to which the formation is damaged, this could represent a very effective treatment.

For $it/h = 5 \text{ m}^3/\text{m}$,

$$\theta = 692 \qquad \text{and} \qquad \epsilon_f^{(R)} = 13.9$$

which gives

$$r_f - r_w = 0.39 \text{ m} = 39 \text{ cm}$$

In the previous example the position of the reaction front has been fixed assuming the front to be sharp. This depends on $D_a^{(R,F)}$, the radial Damkohler number for the fast-reacting minerals. Referring to Fig. 15.4 and Eq. (15.55), the width of the reaction zone, $\Delta\epsilon^{(R)}$, can be shown to be given as

$$\Delta\epsilon^{(R)} = -\frac{2 \ln\left[\left(\frac{1}{\psi_1}\right) - 1\right]}{D_a^{(R,F)}} \tag{15.78}$$

where ψ_1, is the acid concentration ratio on the upstream side of the reaction front as shown by Figure 15.4. The derivation of this equation requires that $A_C^{(F)} \ll 1$ and neglects the slow reaction altogether. For a radial system the thickness of the front is Δr and in the design strategy to be adopted in the next chapter, it will be necessary to minimize $\Delta r/r_f$ subject to other restrictions. If $\Delta r/r_f = \delta$, then since

$$\frac{\Delta\epsilon^R}{2} = \left(\frac{r_f + \frac{\Delta r}{2}}{r_w}\right)^2 - \left(\frac{r_f}{r_w}\right)^2 \tag{15.79}$$

it can be shown by neglecting terms of second order in Δr that $\Delta\epsilon^{(R)} = r_f \Delta r/r_w^2$ or

$$\Delta\epsilon^{(R)} = 2\delta(\epsilon_f^{(R)} + 1) \tag{15.80}$$

If it is desired to maintain a certain dimensionless reaction zone width (δ), then

$$D_A^{(R,F)} = -\frac{\ln\left[\left(\frac{1}{\psi_1}\right) - 1\right]}{\delta(\epsilon_f^{(R)} + 1)} \tag{15.81}$$

Example 15.11 The Reaction Zone Width

Suppose that a sandstone formation composed of Berea is acidized through an uncased wellbore having a radius of 0.1 m, and assume that the formation is the same one considered in Examples 15.6 and 15.8. The acid composition is 3 wt% HF and 12 wt% HCl, and the formation temperature is 25°C. If a total of 5 m^3 of acid/m of formation thickness is injected at a rate of 10^{-3} m^3/sec-m of formation thickness, the penetration distance of the reaction front is 0.49 m (see Example 15.10). Estimate the width of the reaction zone.

Solution The radial Damkohler number for the fast-reacting minerals, $D_a^{(R,F)}$, is required to estimate the width of the reaction zone. For the linear system, $D_a^{(F)}$ = 2.76 and the scaling used in Example 15.10 also applies here. Thus

$$D_a^{(R,F)} = (2.76)(1.24 \times 10^{-3}) \frac{\pi(0.1)^2}{1 \times 10^{-3}} = 0.108$$

Let us take $\psi_1 = 0.75$ and $1 - \psi_1 = 0.25$ (see Fig. 15.4). Thus, the reaction zone is defined as the distance over which the acid concentration drops from 75% of its initial value to 25% of its initial value. Therefore, from Eq. (15.78)

$$\Delta\epsilon^{(R)} = -\frac{2 \ln(1/0.75 - 1)}{0.108} = 20.3$$

To compare this value in terms of the fraction of the actual penetration distance, the dimensionless distance is required. Thus

$$\epsilon_f^{(R)} = \left(\frac{0.49}{0.1}\right)^2 - 1 = 23$$

Thus, from Eq. (15.80)

$$\frac{\Delta r}{r_f} = \delta = \frac{20.3}{(2)(24)} = 0.42$$

This is a diffuse reaction front. The thickness of the zone is 42% of the depth of penetration. To sharpen the front, it will be necessary to reduce the injection rate.

The equations developed in this section will be the foundation for our design procedure to be described in Chapter 16. The radial flow model is appropriate for uncased holes and also for perforated completions if relatively large acid volumes, such as 1–4 m^3/m of formation thickness are applied. For smaller acid volumes, it may be necessary to take the geometry of the perforation into account.

15.4 ACIDIZING THROUGH PERFORATIONS

In this section we consider the acidization of the formation surrounding a single perforation using the same geometric model of a perforation that is applied in Chapter 7 for studying the damaged zone around perforations. Referring to Chapter 7, recall that the following transformation proved useful:

$$\begin{aligned} x &= H \sinh w \sin v \cos \gamma \\ y &= H \sinh w \sin v \sin \gamma \\ z &= H \cosh w \cos v \end{aligned} \tag{15.82}$$

The geometry of a perforation is defined by the penetration depth D and a radius of the perforation at the wellbore r_p. These two parameters can be related to H and w_p as follows:

$$\begin{aligned} r_p^2 &= H^2 \sinh^2 w_p \\ D_p^2 &= H^2 \cosh^2 w_p \end{aligned} \tag{15.83}$$

For an indepth discussion of this geometry and the origin of these equations the reader should refer to Chapter 7. Figure 15.12 depicts the process to be considered. Surfaces of constant w are also surfaces of constant pressure. The flux u_w is, therefore, perpendicular to surfaces of constant w. It can be shown that

$$u_w = -\frac{k}{H\mu}\frac{dp}{dw}\left[\frac{1}{\cosh^2 w - \cos^2 v}\right]^{1/2} \tag{15.84}$$

This is an important equation since u_w is the flux we need to incorporate into the acid balance. To make the equation more convenient for use in acidizing, we want to eliminate the pressure gradient and substitute in its place the acid volumetric

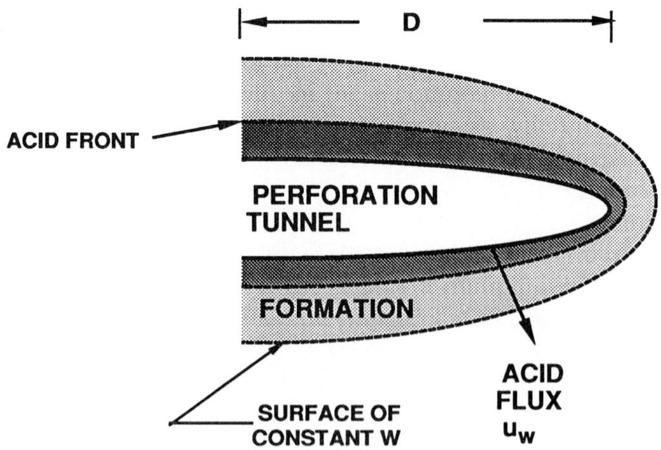

Figure 15.12 Sketch depicting the acidization of a formation through a single perforation.

injection rate, i, which is a variable that is controllable. From Eq. (7.10) we see that

$$i_p = -\frac{k}{\mu} 2\pi H \sinh w \frac{dp}{dw} \tag{15.85}$$

where i_p is the injection rate into a single perforation.

Substituting this expression into Eq. (15.84) yields

$$u_w = \frac{i_p}{2\pi H^2 \sinh w}\left[\frac{1}{\cosh^2 w - \cos^2 v}\right]^{1/2} \tag{15.86}$$

This shows that although the flux is perpendicular to surfaces of constant w, its magnitude, on the other hand, depends on its position on the perforation surface. Thus, some portions of the perforation surface will receive larger amounts of acid than others. This will cause some difficulty in evaluating the penetration distance, since it will depend on the angle v as well as w.

Slow Reactions in the Matrix Around Perforations

The acid balance Eq. (15.23) can be applied in the zone behind the reaction front where all of the fast-reacting minerals have been removed so that in this region $V_F = 0$ and for the geometry shown in Fig. 15.12, the acid balance can be written as:

$$\phi_0 \frac{\partial c_{HF}}{\partial t} + \frac{i_p}{2\pi H^3 \sinh w}\left(\frac{1}{\cosh^2 w - \cos^2 v}\right)\frac{\partial c_{HF}}{\partial w} = -(1 - \phi_0)\alpha_S V_S^0 E_f^{(S)} c_{HF} \tag{15.87}$$

To solve this equation, it is convenient to define the following dimensionless variables:

$$\epsilon^{(P)}(w, v) = (\tfrac{1}{3}\cosh^3 w - \cos^2 v \cosh w) - (\tfrac{1}{3}\cosh^3 w_p - \cos^2 v \cosh w_p)$$

$$\theta^{(P)} = \frac{t i_p}{2\pi H^3 \phi_0} \tag{15.88}$$

$$\psi = c_{HF}/c_{HF}^0$$

$$D_a^{(P,S)} = \frac{2\pi(1 - \phi_0)H^3 \alpha_S V_S^0 E_f^S}{i_p}$$

where the superscript P is intended to denote values specifically related to the geometry of a perforation. Substituting these dimensionless groups, Eq. (15.87) reduces to a form exactly like Eq. (15.72) except that the Damkohler number defined for a perforation applies. The solution is therefore the same.

$$\psi = \exp(-D_a^{(P,S)}\epsilon^{(P)}) \tag{15.89}$$

Using this equation, the acid concentration in a zone surrounding a perforation behind the reaction front can be calculated. Before demonstrating the application of these results, we need to determine the position of the reaction front.

Position of the Fast-Reaction Front in a Perforation

Equation (15.63) applies here

$$v_f^{II} = \frac{u_w}{\phi_0}\left[\frac{a_c^{(F)}(\epsilon_f^{(P)})}{1 + a_c^{(F)}(\epsilon_f^{(P)})}\right] \tag{15.63}$$

where u_w is given by Eq. (15.86) and

$$a_c^{(F)}(\epsilon_f^P) = A_c^{(F)} \exp(-D_a^{(P,S)}\epsilon_f^{(P)}) \tag{15.90}$$

The velocity of the front must be expressed in terms of the dimensionless quantity $\epsilon_f^{(P)}$. Imagine that a particle is displaced from one surface of constant w to another along a normal to both surfaces. The distance moved is $d\lambda$, say, and the velocity v_f^{II} is $d\lambda/dt$. However, λ depends on w; thus

$$\frac{d\lambda}{dt} = \frac{dw_f}{dt}\cdot\frac{d\lambda}{dw} = \frac{u_w}{\phi_0}\left[\frac{a_c^{(F)}(\epsilon_f^{(P)})}{1 + a_c^{(F)}(\epsilon_f^{(P)})}\right] \tag{15.91}$$

Noting that $d\lambda = \sqrt{(dx)^2 + (dy)^2 + (dz)^2}$ we can write

$$\frac{d\lambda}{dw} = \left[\left(\frac{\partial x}{\partial w}\right)^2 + \left(\frac{\partial y}{\partial w}\right)^2 + \left(\frac{\partial z}{\partial w}\right)^2\right]^{1/2} = H(\cosh^2 w - \cos^2 v)^{1/2} \tag{15.92}$$

Also, from the chain-rule of differentiation we have the following identity:

$$\frac{dw_f}{dt} = \frac{d\epsilon_f^{(P)}}{d\theta^{(P)}}\cdot\frac{dw_f}{d\epsilon_f^{(P)}}\cdot\frac{d\theta^{(P)}}{dt}$$

Using the definitions provided by Eq. (15.88), we find

$$\frac{dw_f}{dt} = \frac{i_p}{2\pi H^3\phi_0 \sinh w}\left\{\frac{1}{(\cosh^2 w - \cos^2 v}\right\}\frac{d\epsilon_f^{(P)}}{d\theta^{(P)}} \tag{15.93}$$

Substituting Eqs. (15.86), (15.92), and (15.93) into Eq. (15.91) we find the identical result as given by Eq. (15.63) for a linear flow system and Eq. (15.76) for a radial flow system, namely that

$$\frac{d\epsilon_f^{(P)}}{d\theta^{(P)}} = a_C^{(F)}(\epsilon_f^{(P)}) \tag{15.94}$$

This equation can be integrated as before to give

$$\theta^{(P)} = \frac{\exp(D_a^{(P,S)}\epsilon_f^{(P)}) - 1}{A_C^{(F)}D_a^{(P,S)}} + \epsilon_f^{(P)} \tag{15.95}$$

Example 15.12 Acidizing through Perforations

A small acid treatment, sometimes called an *acid wash*, may apply a small quantity of HF acid mixed with HCl to remove damage from the perforations. Suppose that acid is injected into a perforation at a rate of 10^{-4} m^3/sec. If the perforation has a radius of 0.5 cm at the wellbore and a depth of 20 cm, calculate the position of the fast reaction front after acid has been injected for 5 min. Take the formation composition to be the same as that considered in Examples 15.6 and 15.7.

Solution The perforation Damkohler number can be obtained from the linear value found in the laboratory as follows:

$$D_a^{(P)} = D_a\left(\frac{u}{L}\right)\left(\frac{2\pi H^3}{i_p}\right)$$

From Eq. (15.83), we see that

$$(0.5)^2 = H^2 \sinh^2 w_p$$

$$(20)^2 = H^2 \cosh^2 w_p$$

Solving for H and w_p it is found that

$$H = 20 \text{ cm} = 0.2 \text{ m} \qquad \text{and} \qquad w_p = 2.5 \times 10^{-2}$$

Thus, using the results obtained in Example 15.6

$$D_a^{(P,S)} = 0.62\left(\frac{1.24 \times 10^{-4}}{0.1}\right)\frac{2\pi(0.2)^3}{10^{-4}} = 0.386$$

After 5 min of acid injection

$$\theta^{(P)} = \frac{5(60 \text{ sec})(10^{-4} \text{ m}^3/\text{sec})}{2\pi(0.2 \text{ m})^3(0.23)} = 2.6$$

The acid capacity number is the same as that given in Example 15.7 and is equal to

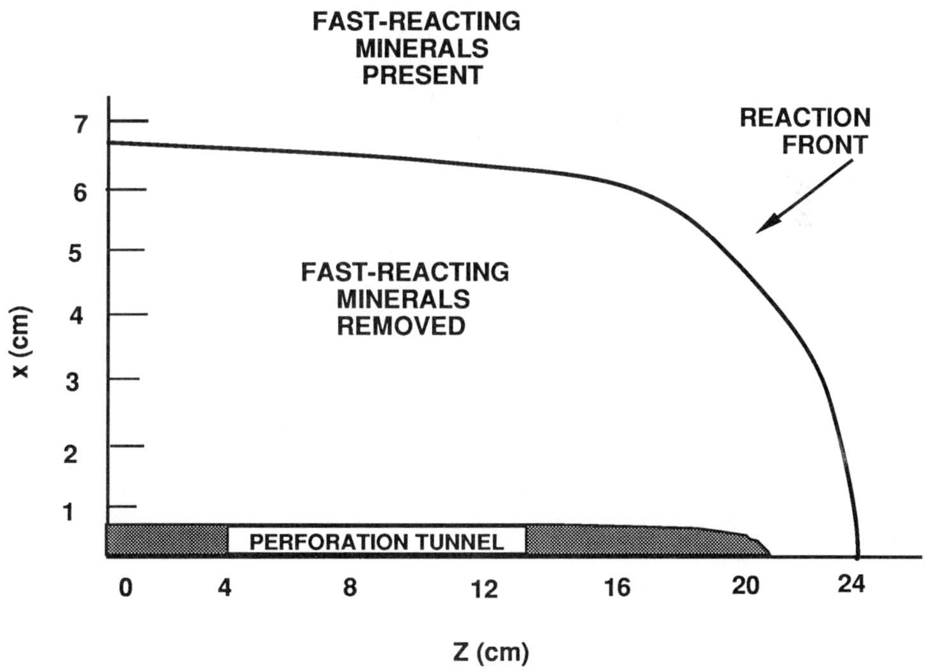

Figure 15.13 Graph showing position of the fast reaction front in the (x, z) plane after an acid wash (see Example 15.10 for details).

2.43×10^{-2}. Thus, from Eq. (15.90)

$$2.6 = \frac{\exp(0.386\epsilon_f^{(P)}) - 1}{(2.43 \times 10^{-2})(0.386)} + \epsilon_f^{(P)}$$

Solving yields $\epsilon_f^{(P)} = 6.09 \times 10^{-2}$. This dimensionless quantity is related to a distance of penetration through Eq. (15.88). However, $\epsilon_f^{(P)}$ depends on the angle v as well as w. The relationship between (w, v, γ) and (x, y, z) is given by Eq. (15.82). It is observed that $v = 0$ corresponds to $x = y = 0$. Substituting $v = 0$, $w_p = 2.5 \times 10^{-2}$, and $\epsilon_f^P = 6.09 \times 10^{-2}$ into Eq. (15.88) gives

$$0.0609 = \tfrac{1}{3}\cosh^3 w_f - \cosh w_f - \tfrac{1}{3} + 1 \qquad \text{or} \qquad w_f = 0.6765$$

For $v = 0$ and $w_f = 0.6765$, $x = y = 0$ and $z = 24.2$ cm is one point at the reaction front. This point is shown in Fig. 15.13. Another point on the reaction front can be found by setting $v = \pi/2$, $w_p = 2.5 \times 10^{-2}$ and $\epsilon_f^{(P)} = 0.0609$ into Eq. (15.88) gives $w_f = 0.3376$. This corresponds to another point $x = 6.88$ cm and $z = 0$, at the reaction front. This point is also shown in Fig. 15.13.

By selecting values of v intermediate between 0 and $\pi/2$, the entire reaction front can be calculated. The final shape is shown in Fig. 15.13. It can be observed that the front has penetrated into a radial region about 7 cm in radius around the perforation tunnel.

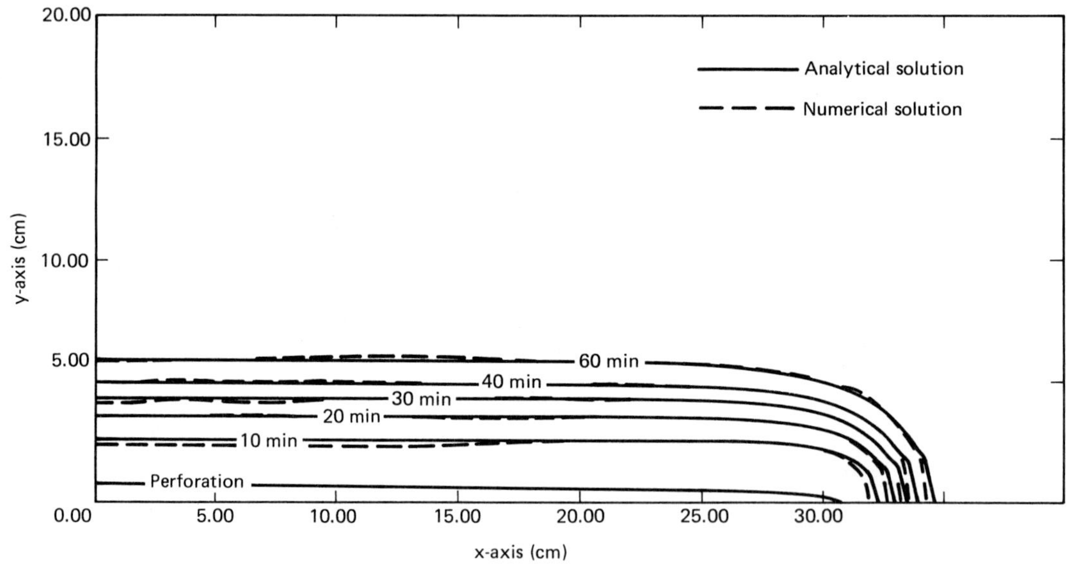

Conditions Simulated

Acid concentration: $c_{HF} = 1.619$ kg moles/m^3

Injection rate: $i_p = 1 \times 10^{-5}$ m^3/sec

Fast-reacting mineral: $c_F^0 = 0.884$ kg moles/m^3

Forward reaction rate: $E_f^{(F)} = 6.232 \times 10^{-6}$

Forward reaction rate: $E_f^{(S)} = 3.522 \times 10^{-9}$

Figure 15.14 Position of the reaction front in the zone surrounding the perforation shown as a function of time. The dashed line represents the approximate calculation [22].

Comparison with Precise Numerical Calculations

Several approximations have been applied in the development of a model of acidizing through perforations, and the extent to which these apply is not immediately obvious. Lea et al. [22] has reported numerical calculations showing the position of the reaction front under various conditions. The calculations consider a single perforation, which is ellipsoidal in shape, and start with the same fluxes used in the calculations presented here. Figure 15.14 shows the reaction front as a function of time, calculated by the more precise numerical methods. The numerical calculations are in good agreement with the method recommended here, thus providing justification for the approach.

Despite the fact that the approach recommended here is apparently quite accurate, there are a number of problems still to be resolved. For example, it is not known how the distribution of damage around the periphery of the perforation will alter the progress of the reaction front. Nor is it known to what extent this damage must be removed to restore much of the undamaged productivity of a perforation. These are matters to be resolved by more precise calculations.

REFERENCES

15.1. Schechter, R. S., and Gidley, J. L., *Amer. Inst. Chem. Eng. J., 15* (1969) 339.

15.2. Scheidegger, A. E., *The Physics of Flow through Porous Media*, New York: The Macmillan Co., 1960.

15.3. Purcell, W. R., *Trans. AIMME, 186* (1949) 39.

15.4. Rose, W. D., and Bruce, W. A., *Trans. AIMME, 186* (194) 127.

15.5. Burdine, N. T., Gournay, L. S., and Reichertz, P. P., *Trans. AIMME, 189* (1950) 195.

15.6. Hill, A. D., "Flow with Simultaneous Heterogeneous Reactions in Porous Media," PhD dissertation, The University of Texas at Austin, 1978.

15.7. Guin, J. A., Silberberg, I. H., and Schechter, R. S., *Indus. Eng. Chem. Fund., 10* (1971) 50.

15.8. Glover, M. C., and Guin, J. A., *Amer. Inst. Chem. Eng. J. 19* (1973) 1190.

15.9. Swift, S. T., and Fogler, H. S., *Chem. Eng. Sci., 32* (1977) 339.

15.10. Guin, J. A., and Schechter, R. S., *Soc. Pet. Eng. J., 11* (1971) 306; *Trans. AIMME, 251* (1971) 306.

15.11. Labrid, J. C., *Soc. Pet. Eng. J., 15* (1975) 117.

15.12. Lund, K., and Fogler, H. S., *Chem. Eng. Sci., 31* (1976) 381.

15.13. McCune, C. C., Fogler, H. S., and Kline, W. E., *Indust. Eng. Chem. Fund., 18* (1979) 188.

15.14. Walsh, M. P., Lake, L. W., and Schechter, R. S., *J. Pet. Tech., 34* (1982) 2097.

15.15. Lambert, M. E., "A Statistical Study of Reservoir Heterogeneity," MS thesis, The University of Texas at Austin, 1981.

15.16. Hill, A. D., Lindsay, D. M., Silberberg, I. H., and Schechter, R. S., *Soc. Pet. Eng. J., 21* (1981) 30.

15.17. Fogler, H. S., and Rege, S. D., *Chem. Eng. Comm., 42* (1986) 291.

15.18. Amundson, N. R., *J. Physical and Colloid Chem., 52* (1948) 1153.

15.19. Lindsay, D. M., "An Experimental Study of Sandstone Acidization," MS thesis, The University of Texas at Austin, 1976; Report No. U.T. 76-1, Texas Petroleum Research Committee, Austin, Texas, July 1976.

15.20. Hekim, Y., and Fogler, H. S., *Amer. Inst. Chem. Eng. J.*, *26* (1980) 403.
15.21. Sinex, W. E., Schechter, R. S., and Silberberg, I. H., *Indust. Eng. Chem. Fund.*, *11* (1972) 205.
15.22. Lea, C. M., Hill, A. D., and Sephnoori, K., "Simulation of Sandstone Acidizing of a Damaged Perforation," SPE 19419. Presented at the Formation Damage Control Symposium, Lafayette, Louisiana, Feb. 1990.

PROBLEMS

***15.1.** Given that the fraction of pores having an area between A and $A + dA$ is

$$\eta(A)\,dA = \beta^2 A \exp(-\beta A)\,dA$$

calculate the increase in the surface area, the porosity, and the permeability if β is decreased from $4 \times 10^{12}\ \text{m}^{-2}$ to $1 \times 10^{12}\ \text{m}^{-2}$.

****15.2.** Given that the fraction of pores having an area between A and $A + dA$ is

$$\eta(A)\,dA = \beta^2 A \exp(-\beta A)\,dA$$

and the porosity of the sample is ϕ, plot the capillary pressure as a function of saturation that would be expected for a porous medium represented by the following data:

$\beta = 1 \times 10^{12}\ \text{m}^{-2}$
$\phi = 17\%$
$\gamma_{\text{interfacial tension}} = 32\ \text{mN/m}$
The wetting is complete.

****15.3.** The pore density function, $\eta(A)$, of a set of parallel capillary tubes (see Fig. 15.1) can be determined from measurements in which the volume of mercury forced into a small core sample at a given pressure (p) is measured. If the porosity (ϕ) of the sample is known from other measurements and if the mercury is nonwetting so that the Laplace equation can be written as (assuming circular capillary pores)

$$p = \frac{2\sqrt{\pi}\,\gamma}{A^{1/2}}$$

where γ is the surface tension of mercury, then show that

$$l\eta(A) = \frac{\phi p^5}{32\pi^2\gamma^4}\frac{dS}{dp}$$

where $S(p)$ is the mercury saturation as a function of the applied pressure. Thus given $S(p)$, $l\eta(A)$ can be calculated (see Fig. 15.2).
[*Hint:* Start by showing that

$$\phi S(p) = \int_{A(p)}^{\infty} lA\eta(A)\,dA$$

and then differentiate both sides of this expression with respect to p].

****15.4.** A porous medium is composed of inert particles, each coated with a layer of a reactive mineral of thickness δ_0 so thin that the surface area per unit volume of porous medium (S) is not changed perceptibly as the mineral dissolves (see Fig.

15.3). The reaction rate between the flowing acid and the reactive mineral is first order, so that

$$r_A = -E_f^{(S)} c_A$$

where c_A is the acid concentration, which depends on position and time.

Assuming that the reaction rate constant $E_f^{(S)}$ is so small that the accumulation of acid ($\partial c_A/\partial t \cong 0$) can be neglected, show that the concentration of acid is given by

$$c_A = c_A^0 \exp\left[-\frac{E_f^{(S)} S}{u}(x - x_f(t))\right] \qquad (\text{for } x \geqslant, x_f)$$

$$c_A = c_A^0 \qquad (\text{for } x < x_f)$$

where c_A^0 is the inlet concentration and $x_f(t)$ is the position of the front separating the region where all of the reactive mineral has been dissolved from that where the reactive mineral coating still exists.

Prove that $x_f(t)$ must satisfy the equation

$$\delta_0 = E_f^{(S)} X \exp\left(-\frac{E_f^{(S)} S}{u} x_f\right) \int_0^t \exp\left[\frac{E_f^{(S)} S}{u} x_f(\tau)\right] d\tau$$

where X is the dissolving power of the injected acid. Show by differentiating this equation with respect to t, that

$$\frac{dx_f}{dt} = \frac{uX}{\delta_0 S}$$

and thus $$x_f = \frac{uX}{\delta_0 S}(t - t^*) \qquad (\text{for } t > t^*)$$

and $$x_f = 0 \qquad (\text{for } t < t^*)$$

where $$t^* = \frac{\delta_0}{E_f^{(S)} X}$$

***15.5.** A sandstone is composed of 88 wt% α-quartz, 5 wt% kaolinite rock fragments, and 7 wt% orthoclase. All of these minerals are detrital and they can be approximated as cubes. The α-quartz grains are 10 μm on a side, the feldspar (orthoclase) particles are 3 μm, and the kaolinite fragments are 0.8 μm on a side. Estimate the value of f_F for this sandstone. (*Note*: Compare this result with that found in Example 15.2).

***15.6.** Uranium dioxide (uraninite) is found deposited in certain sandstones as fine crystalline precipitates apparently formed when hexavalent uranium, which is water soluble, was reduced to U^{+4} by reaction with naturally occurring reducing compounds such as pyrite or organic materials. This uranium can be leached by in-situ processes. To be again made water soluble it must be oxidized to the hexavalent form. This step can be represented by the simple reaction

$$UO_2 + \tfrac{1}{2} O_2 \longrightarrow UO_3$$

If a core sample of sandstone containing 0.1 vol% uranium is flooded with an aqueous solution containing 15 ppm O_2, how many PV of the solution will have to be injected before the effluent oxygen concentration rises to 7.5 ppm O_2? Assume that uraninite is the only oxidizable mineral in the sandstone and that the sandstone porosity is 0.23. The density of uranium in the form of uraninite is 9780 kg of U/m^3 of rock.

***15.7.** To test the feasibility of recovering uranium from a sandstone deposit by an in-situ leaching process, a core 2 m in length is flooded with an aqueous solution containing 20 ppm O_2 and the effluent concentration is measured. Based on the data given in Table P15.7, estimate the parameter αE_f which defines the reaction rate of the uraninite (see Problem 15.6) with the oxygenated water where the rate of reaction with O_2 per unit of surface area obeys first-order kinetics as follows:

$$r_{O_2} = -E_f c_{O_2}$$

The importance of the parameter αE_f is that it will apply even if the length of the test core is increased or if the flow rate is changed.

TABLE P15.7 Coreflood Results

Property	Symbol	Value
Flux	u	6×10^{-3} cm/sec
Porosity	ϕ	0.23
Initial volume	V_0	5×10^{-4} m^3 of U/m^3 of rock
Uranium density	ρ_U	9780 kg of U/m^3 of rock

Time (hr)	Oxygen Effluent Concentration (ppm)
21.3	5.88
106.3	8.0
213.1	10.9
319.6	13.7
426.0	15.9
533.0	17.5

***15.8.** A core test for an acid treatment reveals that the Damkohler number characteristic of the fast reaction is 3.50 and the acid capacity number is 1.8×10^{-2} for a certain reservoir rock. Neglecting the effect of the slow reaction, plot the effluent acid concentration from the core as a function of PV acid injected. Express your answer in kg moles of HF/m^3. The fresh acid has a concentration of 3 wt% HF and 1 wt% HCl (density at 25°C 1070 kg/m^3).

Plot the effluent curve for a case in which the same experiment is repeated, except the acid flow is doubled.

***15.9.** A formation core sample (3 cm in diameter and 10 cm in length) is acidized using 3 wt% HF blended with 12 wt% HCl. The porosity of the sample (after an HCl preflush to remove carbonates) is 15%. The acid flux is 5×10^{-2} cm/sec. The effluent acid concentration is measured as a function of PV of acid injected with the results listed in Table P15.9.

Estimate the Damkohler numbers for both the fast and slow reactions and the acid capacity number for the fast-reacting minerals.

TABLE P15.9 Effluent Acid Concentration Measured as a Function of PV

Pore Volumes (of acid injected)	Effluent Acid Concentration (wt% HF)
5	0.19
10	0.30
15	0.46
20	0.68
25	0.96
30	1.30
35	1.67
40	1.68
45	1.69
50	1.70
55	1.70
60	1.71
65	1.71
70	1.72

***15.10.** The damaged zone surrounding a well having a 0.1 m in radius wellbore extends to a radius of 0.25 m. What volume of acid (expressed in m^3/m) will be required to remove the fast-reacting minerals from the damaged zone, given that for the particular injection rate the radial Damkohler number characteristic of the slow reactions is 0.5 and the acid capacity number for the fast-reacting minerals is 2×10^{-3}? Assume the reaction front to be sharp ($D_a^{(R,F)}$ is large).

Plot the acid concentration as a function of radius at a time when ψ is 0.5.

***15.11.** There is an optimum volume of acid and corresponding acid injection rate for each matrix acid treatment. Finding this optimum depends on knowledge of the depth to which the formation is damaged and a laboratory test acidizing a typical core sample which yields the three parameters $D_a^{(S)}$, $D_a^{(F)}$, and $A_C^{(F)}$. Consider a sandstone formation that is uncased and damaged. The wellbore radius is 0.1 m and the radius of the damaged zone is 0.3 m, how much acid (expressed in m^3/m of formation thickness) and at what rate should be injected to remove this damage? The thickness of the reaction front (position at which $\psi = 0.75$ to position at which $\psi = 0.25$) should be maintained at a value less than 20% of the depth of the damaged zone; i.e., $\delta < 0.2$.

The results of a linear core test made under conditions closely simulating the reservoir are as follows:

$u = 2 \times 10^{-2}$ cm/sec

$L = 20$ cm

$\phi = 0.15$

$D_a^{(F)} = 3.0$

$A_C^{(F)} = 1.6 \times 10^{-2}$

$D_a^{(S)} = 0.43$

**15.12. The acidization of a sandstone core 7 cm in length ($\phi = 0.15$) at an injected flux of 1.5×10^{-2} cm/sec reveals that when the ratio of the effluent concentration to the fresh acid concentration (ψ_e) is plotted as $\ln[(1/\psi_e) - 1]$ versus the PV injected, two straight line segments are observed. The one at small PV has a negative slope of 7.5 $\times 10^{-2}$ and an intercept of 3.0, whereas the line segment at large PV is practically horizontal so that $\ln(1/\psi_e - 1)$ has a value of approximately 0.2 independent of the number of PV injected.

These observations are important in the design of matrix acid treatments. Calculate the maximum damaged zone radius which can be treated if the maximum permitted injection rate (i/h) is 5.3×10^{-4} m^3/m of formation thickness, assuming the core sample tested is representative of the formation and the test temperature was the same as the formation temperature. The wellbore radius is 0.122 m.

**15.13. The acid balance equation representing the flow and simultaneous reaction of acid in the region where the fast-reacting minerals have been removed can be written as

$$\phi_0 \frac{\partial c_{HF}}{\partial t} + \mathbf{u}\cdot\nabla c_{HF} = -(1 - \phi_0)\alpha_S V_S^0 E_f^S c_{HF}$$

Starting with this expression, prove that Eq. (15.87), which applies to slow reactions in perforations, is correct. Start with the hypothesis that the pressure distribution about a perforation depends on w but not on v or γ.

16
Design Considerations in Matrix Acidizing of Sandstones

16.1 DESCRIPTION OF A SANDSTONE ACIDIZING TREATMENT

An acidizing treatment for a sandstone formation normally will consist of sequentially injecting three fluids: a preflush, the hydrofluoric acid–hydrochloric acid mixture (HF–HCl mixture), and an afterflush. Each of these fluids serves a definite purpose.

The preflush is usually HCl ranging in concentration from 5–15 wt% and containing corrosion inhibitors and other additives, as required. The preflush displaces connate water from the near-wellbore region, thereby minimizing direct contact between sodium and potassium ions in the formation brine and the HF or fluosilicate reaction products. Normally, this will reduce the possible redamaging of the formation by precipitation of insoluble sodium or potassium fluorides or fluosilicates. The HCl preflush also reacts with calcite or other calcareous materials, thereby reducing (or eliminating) reaction between HF and $CaCO_3$. This avoids waste of the more expensive HF and prevents precipitation of calcium fluoride (CaF_2). Finally, the preflush may serve to precool the formation.

Following the preflush, an HF–HCl mixture (usually 3 wt% HF and 12 wt% HCl) is then injected. The HF reacts with clays, sand, drilling mud, or cement to improve permeability near the wellbore. The HCl will not react with these materials but is needed to keep the pH low, thereby reducing the precipitation of HF reaction products.

An afterflush is generally applied to isolate the reacted HF from brine that may be used to flush the tubing, to displace spent acid into the formation, and to restore water wettability of the formation. Normally, one of three types of afterflush is used: (1) for oil wells, either a hydrocarbon afterflush, such as diesel oil or 15 wt% HCl is used; (2) for water injection wells, HCl is used; and (3) for gas wells, either acid or a gas (such as nitrogen or natural gas) is used. With a liquid afterflush chemicals are usually added to aid in removing treating fluids from the formation, restoring water wettability to formation solids and removing precipitated acid reaction products, and prevention of emulsion formation. When a gas is used as an afterflush, cleanup additives such as surfactants to reduce the surface tension are sometimes added to the HF–HCl stage of the treatment.

16.2 DAMAGE INDUCED BY ACID

Studies of the mechanism of acid attack in sandstone core samples were first published by Smith and Hendrickson [1]. Core plugs were acidized under carefully controlled conditions. The permeability was observed during injection of HF–HCl mixtures to monitor changes that occurred for various acid concentrations and applied pressure gradients (flow rates) in different sandstones. A plot of permeability change as a function of the amount of acid injected was subsequently called an *Acid Response Curve* [2] (Fig. 16.1).

These studies showed that upon continued injection of the HF–HCl mixture, the permeability initially decreases, reaches a minimum, and then increases as shown in Fig. 16.1. Smith et al. [2] reasoned that the initial permeability reduction

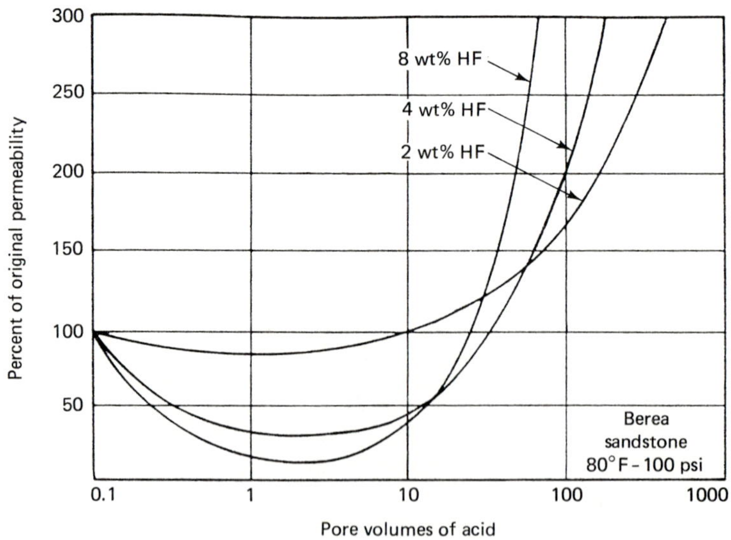

Figure 16.1 Graph showing effect of HF concentration on core response to HF–HCl mixtures [1]. (% of original permeability vs. pore volumes of acid.) (With permission of the Society of Petroleum Engineers.)

was caused by the partial disintegration of the sandstone matrix and the downstream migration of fines that plug flow channels. Continued exposure of the fines to unspent HF was thought eventually to result in their dissolution. Therefore, the subsequent permeability increase was thought to come from clearing the pore channels plugged by fines and the enlargement of other pore channels by the acid. Labrid [3, 4] has taken extensive equilibrium data for the sandstone–HF reaction and has proposed that the decrease in permeability observed during acidization is caused at least in part by the precipitation of orthosilicic acid or other reaction products (see Chap. 3).

Shaughnessy and Kunze [5] have observed that permeability damage to Berea cores will also result if HCl is used without HF. Their conclusion is that the observed reduction in permeability is not due to fines migration but is a result of the CO_2 produced by the reaction of acid with the carbonates present in Berea appearing as a separate CO_2-rich phase within the pore spaces. If this is the case, the rock is not at all damaged, but the relative permeability to acid is simply decreased by the presence of the additional fluid phase. This phase will, however, be readily removed when the well is put on production since CO_2 has a substantial solubility in both oil and water.

Because the permeability damage shown in Fig. 16.1 is severe, it is important to investigate those mechanisms responsible for it and to devise ways, if possible, to minimize its impact. Each of the possible mechanisms will be explored in detail.

Damage by Fines Migration

Figure 16.2 shows that as the acid flow rate through a Berea core is increased (the pressure gradient is correspondingly increased as shown on the curves), the

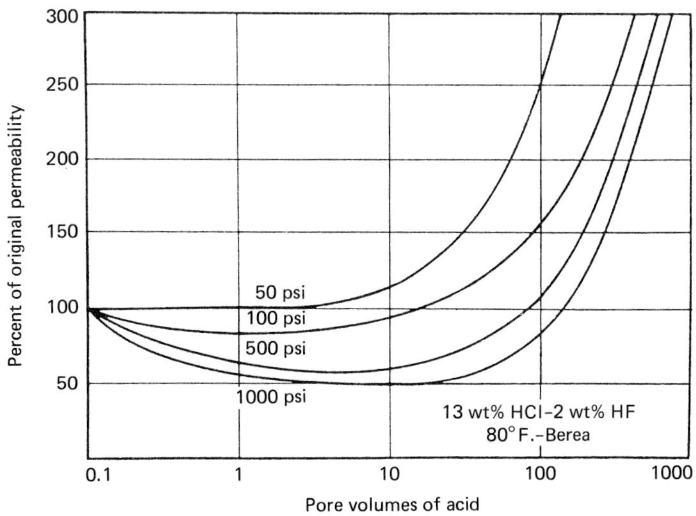

Figure 16.2 Graph showing effect of acid flow rate on Berea core response to 2 wt% HF–13 wt% HCl [1]. (% of original permeability vs. pore volumes of acid.) (With permission of the Society of Petroleum Engineers.)

permeability loss becomes more pronounced. Interestingly, larger volumes of acid are required to restore the core to its original permeability.

The increased permeability decline may be caused by an increase in the quantity of fines released because of the increased drag forces at high flow rates (see Chap. 4 on formation damage). Larger volumes of acid are probably required to achieve a given permeability increase because all the HF was not reacted when injected at higher flow rates. Since the effluent acid concentrations were not reported, this conjecture cannot be confirmed.

Carbon Dioxide Damage

Figure 16.3 shows that the solubility of CO_2 in spent acid solutions is exceeded when 15 wt% HCl is used to treat the formation. This assumes that all of the HCl reacts with carbonates and that the reaction products are calcium chloride and carbon dioxide ($CaCl_2$ and CO_2). If 15wt% HCl reacts to completion, the spent acid solution will contain about 20 wt% $CaCl_2$; thus, the lower two curves must be used to obtain the CO_2 solubility.

The dashed curves apply when 5 wt% HCl is used. In this case, the solubility exceeds the amount produced and a separate CO_2-rich phase is not expected to form.

2.2 kg moles/m^3 of CO_2 are formed when 15 wt% HCl is reacted, and if the formation pressure is 14 MPa, then about 0.4 kg mole/m^3 of CO_2 will reside in the acid solution and 1.8 kg moles/m^3 in a separate phase. At 76°C and 14 MPa the CO_2 saturation in the pore spaces will be about 22%. This large gas phase volume will certainly reduce the permeability to acid, especially since CO_2 would be expected to be nonwetting.

If gas evolution is responsible for part of the initial permeability damage

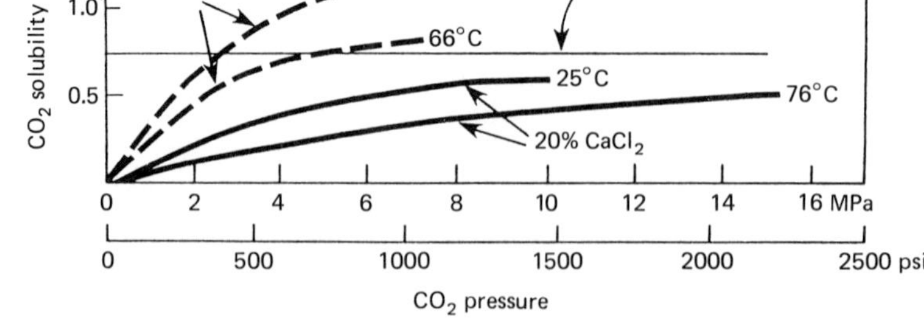

Figure 16.3 Graph showing solubility of CO_2 in spent acidizing solutions [5]. (CO_2 solubility vs. CO_2 pressure.) (With permission of the Society of Petroleum Engineers.)

observed when acidizing cores, then the damage should be apparent upon injection of plain HCl and should exhibit a strong pressure dependence. Figure 16.4 shows the percent permeability retained as a function of the pore volumes (PV) of HCl injected. (No HF was added.) The rate was held constant at 5 ml/min and all of the cores were 3.8 cm in diameter by 5.08 cm in length. Clearly, for this case, the permeability decrease is less pronounced at higher pressures than at lower ones. This is presumably due to the higher solubility of CO_2 at increased pressures and to the increased CO_2 phase density at higher pressures.

If the generation of a CO_2 gas saturation is responsible for the initial permeability loss, then removal of the gas will restore the initial permeability. Figure 16.4 clearly shows that continued injection of HCl, which would not be expected to remove clay fines responsible for pore clogging, does restore the permeability to a value equal to or greater than that of the untreated rock. Because, as shown in Fig. 16.3, CO_2 is soluble in HCl, continued acid injection dissolves and removes the CO_2 saturation, thereby increasing the relative permeability to the aqueous phase. In field treatments, CO_2 damage will similarly be removed and its presence is not of practical concern. It is not, however, certain that the formation of a separate CO_2-rich phase is the only damage mechanism involved. Indeed, at large back pressures the percentage of permeability reduction when HCl is applied alone is quite small, whereas the data of Smith and Hendrickson [1] shown in Fig. 16.1 indicate distinctly larger levels of permeability reduction when both HF and HCl are applied as a mixture. Thus it seems certain that although the formation of a separate CO_2 phase may contribute to the permeability reduction, it is not the only, nor perhaps even the primary, mechanism.

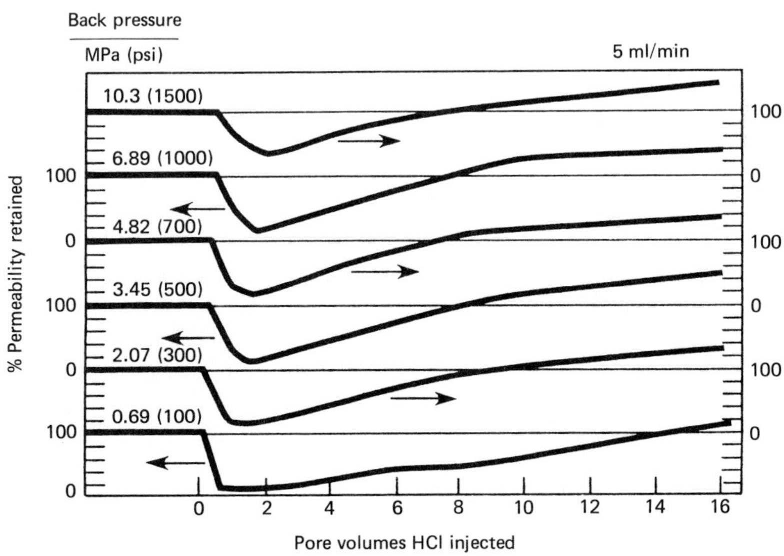

Figure 16.4 Graph showing Berea cores acidized with 15 wt% HCl at 22°C (72°F) [5]. (% permeability retained vs. pore volumes HCl injected.) (With permission of the Society of Petroleum Engineers.)

Precipitation

Aluminosilicates, such as clays and feldspars, all react with HF to form silicon and aluminum fluoride species in solution. Table 16.1 shows the possible aqueous complexes that can form when fluorine, silicon, aluminum, carbon, calcium, and chlorine are present. Under certain conditions, such as when fresh acid initially contacts the formation, the aqueous species formed will be those rich in fluorine and thus compounds such as AlF_3, AlF_2^{+1}, and SiF_6^{-2} will tend to dominate. As more and more mineral is dissolved, the proportions of fluorine in relation to aluminum and silicon are decreased. In this case AlF_2^{+1}, AlF^{+2}, and SiF_4 may be preferred, as discussed in Chapter 3.

When the concentrations of silicon and aluminum are quite high relative to that of HF, then there is a competition for the available fluorine. Aluminum has the greater affinity, and colloidal silica will tend to form and precipitate.

The competition between aluminum and silicon for fluorine can be easily observed by reacting clay, kaolinite [$Si_2Al_2O_5(OH)_4$] for example, with HF–HCl mixtures in a stirred reactor. If the aluminum and the silicon solution concentrations are equal, this means that equal amounts are dissolved and the reactions are congruent. Initially, it is expected that the reaction would be congruent, as noted in Chapter 14. This reaction is shown in Fig. 16.5. At early times the aluminum and silicon concentrations are equal. These data were obtained by mixing 40 g of kaolinite with 1000 cm^3 of solution containing 1 wt% HF and 4 wt% HCl.

After the reaction has progressed, the silicon concentration is observed in Fig. 16.5 to decrease indicating that it is being removed from solution by precip-

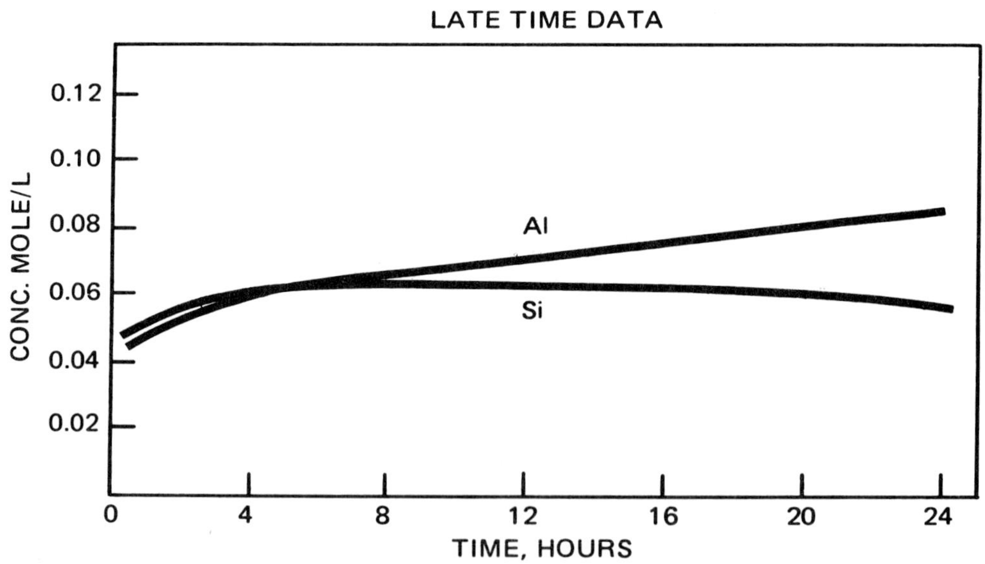

Figure 16.5 Graph showing reaction of kaolinite with 0.4 *M* HF/1.0 *M* HCl, 22°C (72°F) [5]. (Concentration vs. time.) (With permission of the Society of Petroleum Engineers.)

TABLE 16.1 Elements, Aqueous Species, and Minerals Considered in Model Calculations [6]

i	Elements	*i*	Elements
1	Hydrogen	5	Carbon
2	Fluorine	6	Calcium
3	Silicon	7	Chlorine
4	Aluminum	8	Oxygen

j	Aqueous Species	*j*	Aqueous Species
1	H^+	16	$Al(OH)_4^-$
2	F^-	17	HF
3	H_4SiO_4	18	HF_2^-
4	$H_3SiO_4^-$	19	OH^-
5	$H_2SiO_4^{-2}$	20	Cl^-
6	SiF_4	21	HCl
7	SiF_6^{-2}	22	CO_3^{-2}
8	AlF^{+2}	23	HCO_3^-
9	AlF_2^+	24	$CaCO_3$
10	AlF_3	25	$CaHCO_3^+$
11	AlF_4^-	26	$Ca(OH)^+$
12	AlF_5^{-2}	27	H_2CO_3
13	AlF_6^{-3}	28	Ca^{+2}
14	$AlOH^{+2}$	29	Al^+
15	$Al(OH)_2^+$	30	CaF_2

k	Mineral	Molecular Formula
1	Andalusite	Al_2SiO_5
2	Boehmite	$AlOOH$
3	Gibbsite	$Al_2(OH)_6$
4	Dickite	$Al_2Si_2O_5(OH)_4$
5	Halloysite	$Al_2Si_2O_5(OH)_4$
6	Kyanite	Al_2SiO_5
7	Quartz	SiO_2
8	Colloidal silica	$Si(OH)_4$
9	Calcite	$CaCO_3$
10	Fluorite	CaF_2
11	Fluellite	AlF_3
12	Kaolinite	$Al_2Si_2O_5(OH)_4$
13	Calcium hydroxide	$Ca(OH)_2$
14	Sillimanite	Al_2SiO_5

itation. Thus, it seems clear that colloidal silica will precipitate at some point during an acid treatment and tend to clog the pore spaces.

These data can be made quantitative if local equilibrium is assumed to apply during the flow process. Walsh et al. [6] have modeled the acidization process by considering the possibility that all of the aqueous compounds shown in Table 16.1 will exist while allowing any one or all of the solid species to exist. A typical result is shown in Fig. 16.6 in the form of a distance–time diagram. There are five different regions shown. In Region V all of the reactive solids have been dissolved. Part of the quartz was assumed to be nonreactive as a way of modeling the movement of the fast reaction front (see Chap. 15). Kaolinite is the reactive species that is initially present. In Region IV the solid phase is AlF_3. This solid, which does not exist prior to acidization, is formed by precipitation from the partially spent acid.

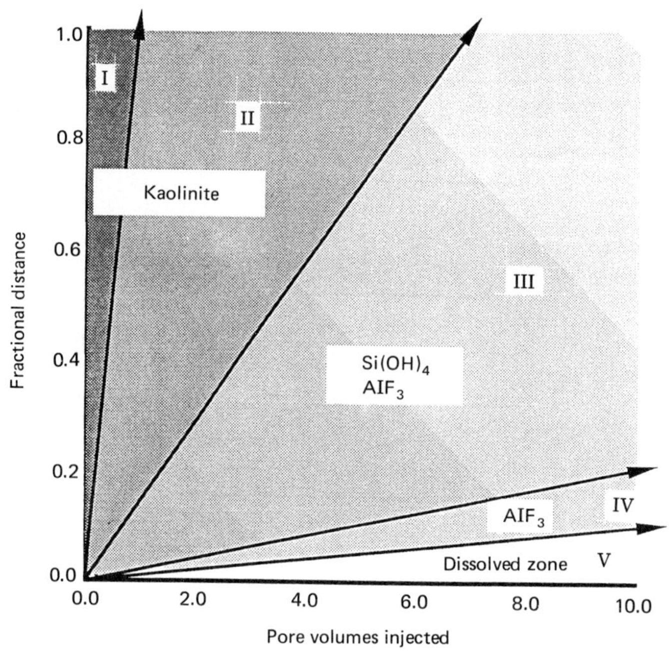

Region	Available Quartz (kmol/m^3)	Kaolinite (kmol/m^3)	Calcite (kmol/m^3)	Precipitated $Si(OH)_4$ (kmol/m^3)	Precipitated AlF_3 (kmol/m^3)	Precipitated $Al(OH)_3$ (kmol/m^3)	pH	pHF
I	0	2.5	0	0	0	0	6.2	–
II	0	2.5	0	0	0	0	5.2	5.4
III	0	0	0	8.4	4.9	0	1.0	−0.3
IV	0	0	0	0	4.9	0	1.0	−0.3
V	0	0	0	0	0	0	1.0	−0.3

Figure 16.6 Distance–time graph of a model-predicted acidization using 4 wt% HF for a sandstone formation containing 9 wt% kaolinite and 91 wt% inert quartz at 25°C. (Adapted from Reference [6].)

Both colloidal silica and AlF_3 have precipitated in Region III. This is a large region (it covers a large area in the distance–time diagram) and because the total amount of precipitate present in this zone is particularly large, the permeability in this region might be expected to be greatly reduced.

Region II contains spent acid solution in contact with a formation having the same solid composition as the initial formation. The regions shown in Fig. 16.6 are those which are predicted to exist when the aqueous phase is in local equilibrium with the solids that are contacted. In general, aqueous solutions tend to become supersaturated and so the precipitates are not likely to form as shown. Figure 16.6 represents a worst case. The question to be asked is: What can be done to minimize the damaging precipitation shown in Region III? Figure 16.7 provides a partial answer.

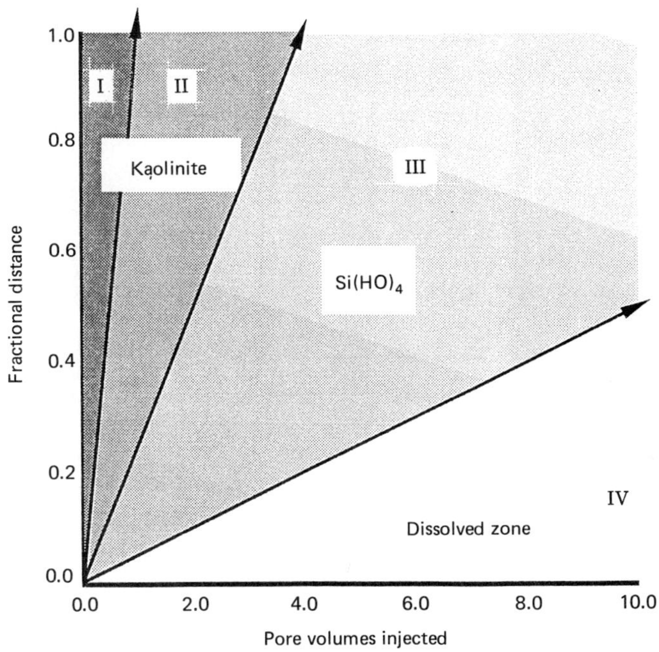

Region	Available Quartz (kmol/m^3)	Kaolinite (kmol/m^3)	Calcite (kmol/m^3)	Precipitated $Si(OH)_4$ (kmol/m^3)	Precipitated AlF_3 (kmol/m^3)	Precipitated $Al(OH)_3$ (kmol/m^3)	pH	pHF
I	0	2.5	0	0	0	0	6.2	–
II	0	2.5	0	0	0	0	2.7	5.8
III	0	0	0	6.2	0	0	−0.6	1.3
IV	0	0	0	0	0	0	−0.6	−0.3

Figure 16.7 Distance–time graph of a model-predicted acidization using 4 wt% HF/11 wt% HCl for a sandstone formation containing 9 wt% kaolinite and 91 wt% inert quartz at 25°C. (Adapted from Reference [6].)

Figure 16.7 shows a distance–time diagram for the same case depicted in Fig. 16.6, except that 11 wt% HCl has been injected together with the HF. In this case, AlF_3 does not precipitate at all during the process and furthermore the concentration of the colloidal silica precipitate is reduced. Thus, the presence of HCl mixed with HF is beneficial.

Figure 16.8 makes another important point. The process is repeated using a blend of 4 wt% HF and 11 wt% HCl. The formation now contains 12 wt% $CaCO_3$ in addition to the kaolinite. Note that $CaCO_3$ reacts rapidly with HCl and neutralizes it. Again AlF_3 appears as a precipitate and the solid concentration of $Si(OH)_4$ has increased. Thus failure to remove carbonates before contacting with the HF–HCl mixture can lead to serious problems. Not only will the compounds

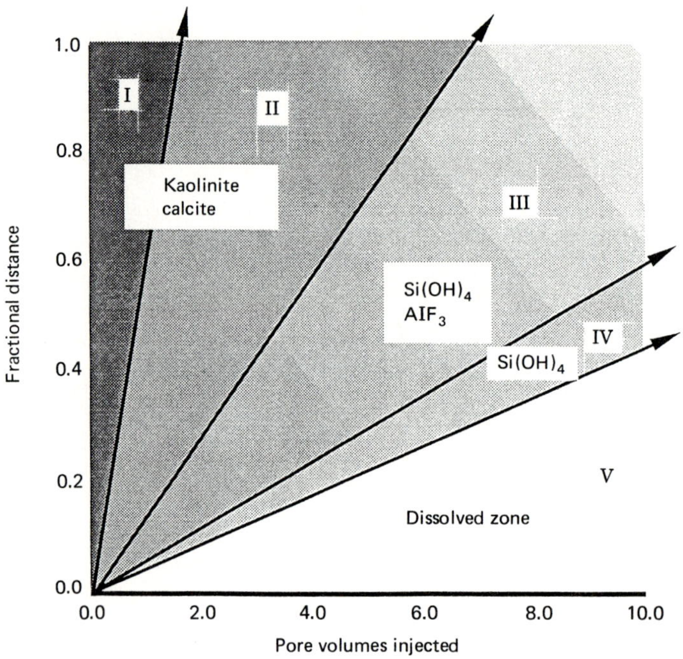

Region	Available Quartz (kmol/m³)	Kaolinite (kmol/m³)	Calcite (kmol/m³)	Precipitated $Si(OH)_4$ (kmol/m³)	Precipitated AlF_3 (kmol/m³)	Precipitated $Al(OH)_3$ (kmol/m³)	pH	pHF
I	0	2.5	3.0	0	0	0	6.2	–
II	0	2.5	3.0	0	0	0	3.7	5.4
III	0	0	0	8.2	4.9	0	−0.6	−0.3
IV	0	0	0	8.2	0	0	−0.6	−0.3
V	0	0	0	0	0	0	−0.6	−0.3

Figure 16.8 Distance–time graph of a model-predicted acidization using 4 wt%/11 wt% HCl for a sandstone formation containing 9 wt% kaolinite, 12 wt% calcite, and 79 wt% inert quartz at 25°C. (Adapted from Reference [6].)

shown here precipitate, but so will others, such as CaF_2. It is therefore imperative to employ an ample HCl preflush to ensure that all of the carbonates are removed from the wellbore region.

A preflush of 7.5 wt% HCl is generally preferred as compared to 15 wt% HCl if twice as much of the more dilute acid is used. The use of larger volumes will tend to ensure better contact with all of the near-wellbore minerals and may help to cool the formation.

Figure 16.9 shows another difficulty that may arise if the HF concentration is increased to levels much above those found in mud acids (3 wt% HF). Shown is the distance–time diagram representing the case in which the HF concentration

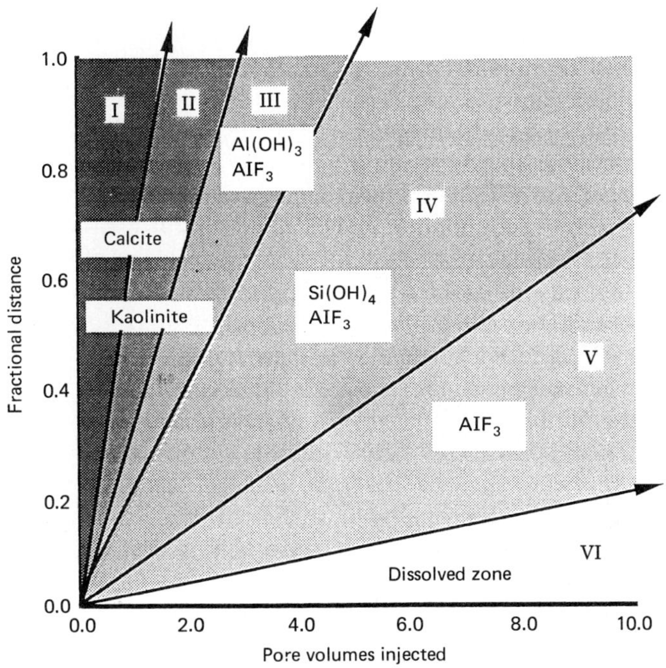

Region	Available Quartz (kmol/m³)	Kaolinite (kmol/m³)	Calcite (kmol/m³)	Precipitated $Si(OH)_4$ (kmol/m³)	Precipitated AlF_3 (kmol/m³)	Precipitated $Al(OH)_3$ (kmol/m³)	pH	pHF
I	0	2.5	3.0	0	0	0	6.1	–
II	0	2.5	3.0	0	0	0	3.6	4.8
III	0	2.5	0	0	2.3	1.9	2.3	4.8
IV	0	0	0	7.7	2.2	0	−0.6	−0.6
V	0	0	0	0	1.1	0	−0.6	−0.6
VI	0	0	0	0	0	0	−0.6	−0.6

Figure 16.9 Distance–time graph of a model-predicted acidization using 8 wt% HF/11 wt% HCl for a sandstone formation containing 9 wt% kaolinite, 4 wt% calcite, and 87 wt% inert quartz at 60°C. (Adapted from Reference [6].)

is increased to 8 wt%. In this case a new solid, Gibbsite [$Al(OH)_3$], appears as does AlF_3 even though the HCl concentration is 11 wt%. Thus the maximum HF concentrations that should be used in sandstone acidizing are those normally applied in mud acids—3–4 wt%. Even lower concentrations may prove beneficial.

The discussion relating to the formation of precipitates has been largely theoretical, but experimental evidence for precipitation damage in cores also exists. Shaughnessy and Kunze [5] have flooded a 3.8 cm × 76.2 cm Berea core first with HCl followed by a 3 wt% HF–10 wt% HCl solution. The test temperature was 66°C and the flow rate was 13.3 ml/min. A total of 2500 ml (13.5 PV) of HF–HCl was injected; however, midway into the treatment, injection was stopped for a period of 2 hr.

The spent acid effluent from the core test was analyzed for aluminum, silicon, and total fluorine. These results are plotted in Fig. 16.10. During the first 1000 ml of injection, the Si:Al ratio ranged from 0.2:1 to 0.5:1. This ratio is much lower than the 1.2:1 Si:Al ratio measured for the natural clays and indicates that the effluent acid has already reached the silica precipitation stage. Precipitation waves such as those shown in Figs. 16.6–16.8 have denuded the spent acid of a portion of its silica, which is precipitated as $Si(OH)_4$.

Following the 2 hr shut-in period, the effluent profile in Fig. 16.10 is characterized by a sharply rising aluminum concentration along with a nearly zero silicon content. The silicon in solution just prior to shut-in has totally precipitated. These results illustrate the hazard of allowing spent HF to remain in the formation rock, even for a few hours.

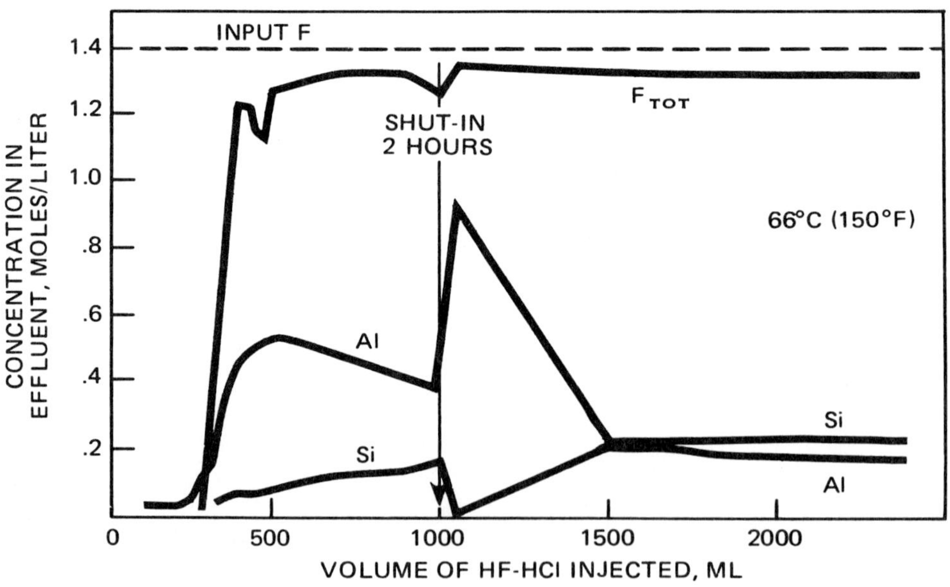

Figure 16.10 Graph showing effluent profiles from 3 wt% HF–10 wt% HCl stimulation [5]. (With permission of the Society of Petroleum Engineers.)

The precipitates tend to reduce the formation permeability. Walsh et al. [6] have calculated this permeability loss and shown that the results reported by Smith and Hendrickson [1] can be predicted by accounting for the change in the porosity resulting from the presence of the precipitate. Shaughnessy and Kunze have measured this permeability loss. After acidizing a Berea core with HCl, they partially treated the core with a 3 wt% HF–10 wt% HCl mixture. The HF–HCl spent acid front was allowed to advance only halfway into the core. Following a 2 hr shut-in period, flow was reversed and the spent acid was flushed out with 3 wt% NH_4Cl brine. The core was cut into sections and the permeability of each was measured. Three distinct permeability regions were found, as shown in Fig. 16.11.

Immediately adjacent to the injection face, the core showed a substantial gain in permeability due to the removal of both the carbonates (by HCl) and the silicates (by HF). Second, a low-permeability region lies between 5 and 45 cm from the injection face. This is the region penetrated by the spent HF–HCl mixture. Precipitation of silica caused by contact of the partially spent acid with aluminum-bearing minerals reduced the permeability to less than the HCl acidized value. The amount of permeability loss is indicated by the cross-hatched region in Fig. 16.11.

Finally, the region not invaded by the HF–HCl mixture has experienced a modest increase in permeability as a result of the HCl acidization.

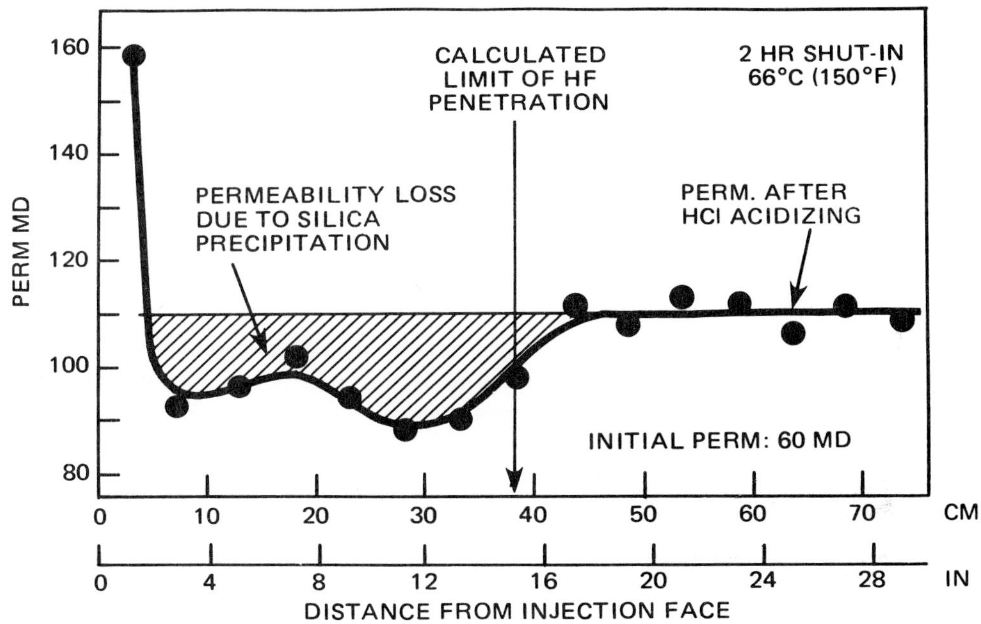

Figure 16.11 Graph showing permeability profile of a core shut-in after injecting 0.5 PV of 3 wt% HF–10 wt% HCl mixture [5]. (With permission of the Society of Petroleum Engineers.)

16.3 CORE MECHANICAL PROPERTIES

The volumes of acid used in a treatment are most often selected on the basis of economic considerations, but there is also a physical limitation on the quantity of acid that the formation can tolerate before becoming unconsolidated. The acid tends to dissolve the cementing material; therefore as acid is injected the formation becomes progressively weaker and finally disintegrates.

The effect of elevated temperatures, pressures, and confining load on the response to HF–HCl mixtures has been studied by Farley et al [7]. They found that the competence of core samples (measured by rock compressive strength) is drastically reduced after acid treatment. This can lead to a decrease in core permeability if excessive quantities of acid are applied. Although Farley et al. [7] have identified a potential problem, its reality has not been evaluated and is not, therefore, considered in the design of acid treatments.

16.4 PREFLUSH

The preflush is an important feature of a sandstone acid treatment. The main purposes of the preflush are to remove those formation materials which will react to a significant extent with HCl, to displace those formation waters containing ions (Na^+, Ca^{++}, etc.) which will tend to precipitate with HF, and to cool the formation thereby permitting deeper acid penetration.

To accomplish these purposes, a large volume of preflush seems to be desirable. Larger volumes should ensure better contact of the preflush with the formation materials, better removal of undesirable reaction products from the near-wellbore region, and more efficient cooling of the formation near the wellbore. By diluting 15 wt% HCl acids to obtain 7.5 wt% solutions, one can increase the volumes of fluid used without substantially increasing the cost and perhaps ease the task of inhibiting the corrosion.

If corrosion is a substantial problem, preflushing with formic acid is acceptable. Because of incomplete reactions, higher molecular weight organic acids are not recommended (see Chap. 3).

To ascertain the volume of preflush that should be used in a given situation, all HCl-soluble components must be removed from the near-wellbore region that is to be treated by the HF–HCl mixture. Thus, the volume of preflush is a function of the penetration depth of the main HF–HCl acid treatment and the volume fraction of acid-soluble minerals residing in that zone.

Example 16.1 The Preflush Volume

If a sandstone contains about 6 wt% HCl-soluble components, estimate the volume of preflush composed of 7.5 wt% HCl which is required if it is desired to remove the carbonates from a region 0.5 m in radius about a wellbore having a radius of 0.12 m. The formation has a porosity of 15%.

Solution The volume of carbonate per meter of formation is given by $\pi(r^2 - r_w^2)(1 - \phi)f$, where f is the volume fraction of acid-soluble material. For the present case, $f = 0.06$ since volume fractions and weight fractions do not differ markedly.

Thus, the volume of carbonates per meter of formation is

$$\pi[0.5^2 - (0.12)^2](0.85)(0.06) = 0.038 \text{ m}^3 \text{ of acid-soluble minerals/m of formation}$$

From Table 3.9, the dissolving power of 15 wt% HCl is 0.082. Thus, 7.5 wt% HCl will have a dissolving power of about 0.041.

$$\text{Volume of preflush} = \left(\frac{0.038 \text{ m}^3 \text{ of soluble minerals}}{\text{m of formation}}\right)\left(\frac{\text{m}^3 \text{ of preflush}}{0.041 \text{ m}^3 \text{ of soluble minerals}}\right)$$

$$\frac{\text{Volume of preflush}}{\text{m}} = \frac{0.92 \text{ m}^3 \text{ of acid}}{\text{m}}$$

The use of a larger HCl preflush also has the advantage of displacing the iron present as the minerals siderite and ankerite. Iron can precipitate from spent acid solutions to yield a damaging amorphous gel.

The precooling of the formation is one other benefit of a preflush.

Example 16.2 Precooling of the Near-Wellbore Region

Suppose the preflush enters the perforations at 50°C and the formation temperature is 95°C. To what distance is the formation cooled to 50°C, if the volume preflush used is 1 m^3/m of formation. The wellbore radius is 0.1 m and the formation porosity is 0.1.

Solution Equation (4.33) gives the distance to which any wave retarded by solid/liquid interaction will penetrate. For the case of thermal interaction, one has, according to Table 4.3,

$$\frac{\Delta A}{\Delta c} = \left(\frac{\rho_{\text{rock}} C_{\text{p rock}}}{\rho_{\text{fluid}} C_{\text{p fluid}}}\right)\left(\frac{1-\phi}{\phi}\right)$$

Since $\rho_{\text{rock}} C_{\text{p rock}}/\rho_{\text{fluid}} C_{\text{p fluid}} \cong 2/3$ for many sandstones (see Chap. 2), then

$$\frac{\Delta A}{\Delta c} = \frac{2}{3}\frac{(0.9)}{(0.1)} = 6$$

Substituting into Eq. (4.33) yields

$$(0.1)\pi(r_f^2 - 0.1^2) = \frac{1}{1+6} \qquad \text{or} \qquad r_f = 0.658 \text{ m}$$

In this case the reaction temperature in the near-wellbore region will be 50°C, rather than the original formation temperature of 95°C. Because the Damkohler number for the slow-reacting mineral will thereby be decreased, then the acid penetration distance will be increased.

16.5 ACID TREATMENT

Strategy

There is little advantage to be gained by acidizing an undamaged formation (see Example 13.2). Thus it seems self-evident that given a volume of acid, the optimal application will be to dissolve rock in the damaged zone. Acid reacted in the regions beyond the damaged zone will be of little utility. This general concept can be made mathematically precise using the calculus of variations [8], but the conclusion is essentially as stated. Thus, the injection rate should be adjusted to ensure that the overwhelming bulk of the acid is spent within the damaged region.

In addition to restricting the reaction of acid to the damaged zone, the design strategy imposed here also requires that the amount of acid expended in removing the slow-reacting minerals be minimized because if the damage is caused by debris dislodged from the rock surfaces it is almost certainly composed of fast-reacting minerals. Furthermore, fines from drilling, completion, or cementing fluids that have entered the pore spaces and clogged the pore throats are also likely to be highly reactive with HF if they react at all. Thus, the strategy of most efficiently removing the fast-reacting minerals from the damaged zone appears to be a logical one even though experimental evidence proving this to be the case does not now exist.

The fast-reacting minerals are removed from a Berea core upon injection of about 60 pore volumes of 3 wt% HF. This is observed as the intersection of the two line segments plotted in Fig. 15.8. Figure 16.1 shows for Berea sandstone that application of 60 PV of 4 wt% HF yields roughly 150% of the original permeability. Now it must be noted that the results displayed in Fig. 15.8 are not really correlatable to those given in Fig. 16.1 since the conditions of the experiments differed; however, a strategy requiring 60 PV to stimulate a damage Berea core would appear to be adequate.

There is an optimum acid injection rate that will be different for each acid treatment. An optimum value of a dependent variable exists as a compromise between competing forces. In this case a slow injection rate will yield a sharp reaction front, as shown by Fig. 16.12(a) and little of the fast-reacting minerals will be dissolved beyond the damaged zone. Because the injection rate is slow, however, the acid experiences a long contact time with the slow-reacting minerals behind the front and much of it is spent in unfruitful reactions with these minerals.

If the injection rate is increased, then the amount of acid reacting with the slow-reacting minerals is correspondingly decreased as shown by Fig. 16.12(b), but now the reaction front may become diffuse and unwanted reaction with the fast-reacting minerals beyond the damage zone may take place.

The amount of acid required to remove most of the minerals from the damaged zone will be a minimum when the injection rate is fixed at the optimum value. It should be noted that in some cases the optimum injection rate will exceed the maximum one permitted and therefore these treatments are simply designed based on the maximum allowable rate (see Example 13.3). In this case it would be helpful to measure the bottomhole wellbore pressure during the treatment so that as damage is removed and the wellbore pressure declines, the rate can be adjusted upward to more nearly approach the optimum value.

The depth of the damaged zone is rarely known, and most acid treatments are designed by first selecting a given volume of acid to be used. Associated with this volume is an optimum injection rate. According to the strategy defined here, it can be argued that small acid volumes correspond to small injection rates. This argument is similar to the one advanced by Hill et al. [9].

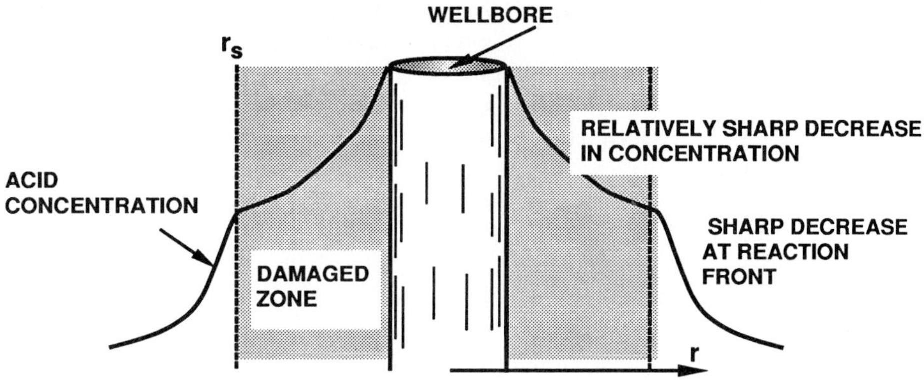

(a) CASE A SLOW INJECTION RATE

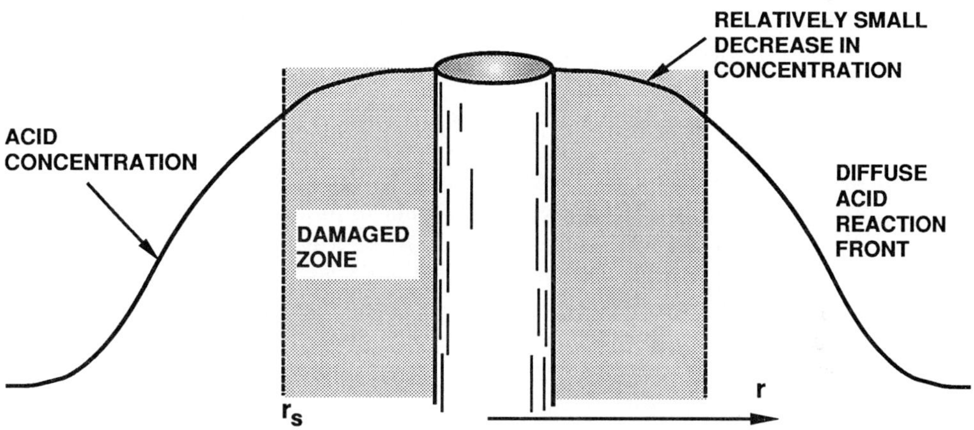

(b) CASE B FAST INJECTION RATE

Figure 16.12 Schematic diagram showing effect acid injection rate has upon acid concentration and reaction front. The profile of acid composition will depend on the injection rate. **(a)** Slow Injection Rate: Large decrease in concentration in the zone behind the front and sharp front. **(b)** Fast Injection Rate: Small decrease in concentration in the zone behind the front and diffuse front.

Acid Composition

Sandstone treatments should generally be carried out by using a mixture of HF and HCl acids, unless there is a substantial HCl-soluble fraction. The HCl alone may be effective (see Table 13.1) if the soluble fraction exceeds 25%. To minimize the degree of mineral precipitation from the spent or partially spent acid, it is necessary to maintain a large ratio of HCl to HF. Generally, the 3 wt% HF–12 wt% HCl ratio is adequate unless large amounts of carbonate have been by-passed during the preflush. Increasing the HF concentration should be applied together with an increased HCl concentration, rather than a decreased one as is often practiced.

In some treatments, such as the SIGMA process, the HCl concentration is limited and although greater depths of penetration may be achieved with this process, care should be taken to compensate, if possible, for the added risk of formation damage by precipitation. This process is discussed further in Section 16.7.

In some high-temperature wells (>100°C) precipitation of the reaction products may be quite rapid, making it difficult to push the spent acid deep into the formation using an afterflush before the supersaturated spent acid solutions precipitate. In high-temperature wells large preflush volumes are beneficial for cooling the formation, but it might also be advantageous to reduce the concentration of HF so as to reduce the volume of precipitate formed [5, 6]. This concept has been applied and the results appear encouraging [10].

Thus, for most cases the mud acid composition is adequate (3 wt% HF and 12 wt% HCl); however, reduced HF concentrations may be justified in formations that are hot or that contain large clay or feldspar concentrations.

Penetration Distance for Radial Flow

If at all possible, it is desirable to have the results of laboratory tests in which representative core samples of the formation to be treated are acidized preferably at low injection rates so that the acid effluent can be sharply divided into distinct curves, such as the one shown in Fig. 15.8. In many cases such information is, unfortunately, not available. Core samples are expensive to obtain and are rarely taken. Thus it will often be necessary to estimate the slow-reaction Damkohler number $D_a^{(R,S)}$, the acid capacity number $A_C^{(F)}$, and the fast-reaction Damkohler number, $D_a^{(R,F)}$ (see Chapter 15).

One approach that can be used in the absence of laboratory tests, is to assume that the Damkohler number representing the slow reaction in a Berea core will be representative of all systems. The dangers inherent in this assumption are evident. Grain sizes different from those found in Berea will cause the surface area available for reaction to differ from Berea. Furthermore, the slow-reacting minerals in other systems may differ from those in Berea, which include both quartz and kaolinite fragments. Thus, $E_f^{(S)}$ will differ from those found in Berea.

Despite these difficulties, as a first approximation consider

$$D_a^{(R,S)} = D_a^{(S)} \left(\frac{u}{L}\right) \frac{E_f^{(S)}(T)}{E_f^{(S)}(298)} \left(\frac{\pi r_w^2}{i/h}\right) \tag{16.1}$$

where $D_a^{(S)}$ is the Damkohler number measured in a linear core of length L at flux u with a reaction temperature of 298°K. The other variables in Eq. (16.1) refer to treatment variables. In the absence of acidization experiments, it is recommended to take (see Example 15.10):

$$D_a^{(S)} = 0.62 \qquad \text{and} \qquad \frac{u}{L} = 1.24 \times 10^{-3}\ \text{sec}^{-1}$$

and from Eq. (14.18)

$$\frac{E_f^{(S)}(T)}{E_f^{(S)}(298)} = \exp\left[-1150\left(\frac{1}{T} - \frac{1}{298}\right)\right] \tag{16.2}$$

Example 16.3 Estimation of the Radial Damkohler Number of the Slow-Reacting Minerals

Given a formation that is precooled to 60°C in the near-wellbore region, estimate the radial Damkohler number characteristic of the slow reaction. The wellbore radius is 0.12 m and the injection rate is to be 10^{-4} m^3/sec-m. Find the depth to which live acid will penetrate. Consider the acid to be spent when the acid concentration is reduced to 5% of its injected value.

Solution Equation (16.1) yields

$$D_a^{(R,S)} = \left[\frac{(0.62)(1.24 \times 10^{-3})\pi(0.12)^2}{10^{-4}}\right] \exp\left(\frac{1150}{298} - \frac{1150}{333}\right)$$

For this case, $D_a^{(R,S)} = 0.522$

The dimensionless penetration distance is given by Eq. (15.73) as

$$\psi = \exp(-D_a^{R,S)}\epsilon_f^{(R)})$$

which for $\psi = 0.05$ gives

$$\epsilon_f^{(R)} = 5.74 \qquad \text{or} \qquad \left(\frac{r_f}{r_w}\right)^2 - 1 = 5.74$$

Thus $r_f = 0.31$ m

Damage at depths greater than 0.3 m cannot be removed unless it is possible to either reduce the temperature or to increase i without fracturing the formation.

Thus, Example 16.3 shows that injection rates and reaction temperatures are critical variables in determining the maximum possible acid penetration depth. The penetration depth is not particularly sensitive to the formation mineralogy. This remarkable conclusion stems from the fact that slow-reacting materials are not substantially dissolved during treatment.

Volume of Acid Treatment in Radial Flow

The volume of acid to be used is that amount required to cause the acid front to move beyond the damaged zone [see Eq. (15.77)]. This volume of acid depends on the acid capacity number, which in turn, is a strong function of the quantity of fast-reacting mineral, V_F^0. This number cannot be estimated without a knowledge of the minerals present in the formation and their origin. To repeat briefly the conclusions reached in Chapter 15, detrital quartz, chert, and detrital clays are all included as slow-reacting minerals. Authogenic clays, detrital feldspars, and authogenic amorphous silica are all considered to be fast-reacting minerals. This assumes that carbonates have been essentially removed by the acid preflush.

Example 16.4 Acid Treatment Design

Design an appropriate acid treatment for an openhole well with radial damage extending to a radius of 0.2 m about a wellbore radius of 0.1 m. Estimate the increase in productivity that will result from a successful treatment. The formation temperature following the preflush is estimated to be 50°C. The following data are given:

Formation permeability:	50 mD	(from well test)
Formation skin factor:	5	(from well test)
Formation porosity:	0.13	(from well logs)
Formation thickness:	5 m	(from well logs)
Completion:	Openhole	
Well spacing:	200 m	

A mineral analysis of the formation to be acidized is presented in Table 16.2.

Solution The acid preflush must be designed to remove the carbonates, here the calcite and siderite, from the zone to be treated (see Example 16.1). It is observed that the volume of orthoclase and authogenic montmorillonite is substantial, and therefore it may be wise to decrease the HF concentration to reduce the density of $Si(OH)_4$ precipitate that will inevitably form. A concentration of 1.5 wt% HF will therefore be applied, rather the 3 wt% HF normally used; however, the HCl concentration will not be correspondingly reduced. There is field evidence to support this procedure [10]. At the treatment temperature of 50°C it may be possible to rapidly overflush the spent acid into the formation before $Si(OH_4)$ precipitates. Furthermore, there is some debate as to the damage that will result even if precipitation does occur [11]. The damage that can be directly attributable to $Si(OH)_4$ precipitation is not known and until this is better defined, it seems prudent to take precautions to minimize its impact.

The radial Damkohler number of the slow-reacting minerals can be estimated by applying Eq. (16.1):

TABLE 16.2 Mineral Analysis of the Formation to be Acidized (Example 16.3)

Mineral Name	Geological Origin	Vol% or Wt%
Calcite	Authogenic	4
Siderite	Authogenic	4
Quartz	Detrital	80
Orthoclase	Detrital	5
Kaolinite	Detrital	4
Montmorillonite	Authogenic	3
	Total	100

There are no further data available.

$$D_a^{(R,S)} = (0.62)(1.24 \times 10^{-3})\left(\frac{(0.1)^2\pi}{i/h}\right)\exp\left(\frac{1150}{298} - \frac{1150}{323}\right)$$

or

$$D_a^{(R,S)} = \frac{h}{i}(3.26 \times 10^{-5})$$

The optimal injection rate also depends on the radial Damkohler number characteristic of the fast-reacting minerals. An accurate estimate of this quantity is quite difficult to obtain without experimental data. Unfortunately, these data are rarely available. Damkohler numbers for Berea (see Example 15.8) and for a Phacoides sandstone [12] are available. As a crude estimate (see Fogler and McCune [13]) a scaling procedure is recommended as follows:

$$D_a^{(R,F)} = \frac{(D_a^{(F)})_B V_F^0(1 - \phi_0)E_f(T)u_B \pi r_w^2 h}{(V_F^0)_B(1 - \phi_0)_B E_f(T_B)L_B i}$$

The subscript B denotes values obtained using Berea or any other standard linear core test. Petrographic and X-ray analyses of representative samples of the Berea cores that were acidized to obtain the results shown in Figs. 15.7 and 15.8 and discussed in Example 15.8 showed that the only fast-reacting mineral is feldspar [14]. Clay (illite and chlorite) is present as rock fragments and thus is classified together with quartz as a slow-reacting mineral [14]. Once the carbonates are removed, then petrographic analysis shows $(V_F^0)_B = 0.052$ and $(1 - \phi_0)_B = 0.79$ [9].* Furthermore, from Example 15.8 $(D_a^{(F)})_B = 2.76$.

The rate expression $E_f(T)$ should be selected from those given in Chapter 14 for the most prevalent fast-reacting mineral. Table 16.2 shows that for the example treated here the mineral is orthoclase, so that Eq. (14.16) applies. Thus

$$\frac{E_f(T)}{E_f(298)} = \exp\left[-4680\left(\frac{1}{323} - \frac{1}{298}\right)\right] = 3.37$$

Noting that once the calcite and siderite are removed by the preflush, then $\phi_0 = 0.2$ and $V_F^0 = 0.08/0.92 = 0.087$.

An estimate for the fast-reacting radial Damkohler number is

$$D_a^{(R,F)} = \frac{(2.76)(0.087)(0.80)(3.37)(1.24 \times 10^{-3})\pi(0.1)^2 h}{(0.052)(0.79)i}$$

or

$$D_a^{(R,F)} = \frac{h}{i}(6.14 \times 10^{-4})$$

Since the radius of the damaged zone is given as 0.2 m, then the treatment strategy requires that the reaction front move to a position

$$\epsilon_f^{(R)} = \left(\frac{r_s}{r_w}\right)^2 - 1 = \left(\frac{0.2}{0.1}\right)^2 - 1 = 3$$

To maintain a sharp front, Eq. (15.81) shows that for $\delta = 0.3$ and $\psi_1 = 0.75$, then the required Damkohler number is

$$D_a^{(R,F)} = \frac{-\ln(1/\psi_1 - 1)}{\delta(\epsilon_f^{(R)} + 1)} = \frac{-\ln(1/0.75 - 1)}{0.3(3 + 1)} = 0.916$$

* A value for $V_F^0 = 0.098$ was found in Example 15.9, but the mineral analysis is used here.

which defines the injection rate since

$$\frac{i}{h} = \frac{6.14 \times 10^{-4}}{D_a^{(R,F)}} = 6.7 \times 10^{-4} \text{ m}^3/\text{m-sec}$$

This is the desired injection rate. Any faster injection of acid will result in a diffuse reaction front, as shown by Fig. 16.12(b). Let us suppose that this rate does not result in bottomhole pressures exceeding the fracture pressure (see Example 13.3) and is acceptable with regard to this constraint. However, the rate may not be acceptable in terms of the amount of acid that reacts with the slow-reacting mineral.

The slow-reacting Damkohler number is

$$D_a^{(R,S)} = \frac{h}{i}(3.26 \times 10^{-5}) = 4.86 \times 10^{-2}$$

and from Eq. (15.73) the acid concentration as it arrives at the reaction front is

$$\psi = \exp(-D_a^{(R)}\epsilon_f^{(R)}) = \exp[(-4.86 \times 10^{-2})(3)] = 0.86$$

Thus, only 14% of the acid will be consumed by the slow-reacting minerals at the time the reaction front arrives at the damaged zone radius. This is certainly an acceptable value. In fact, it might be wise to accept a greater value and further reduce the injection rate to sharpen the reaction front. This is an optimization problem best treated by numerically solving Eqs. (15.21), (15.22), and (15.23) in radial coordinates to minimize the amount of acid required to remove all of the fast-reacting minerals from within the damaged zone [15]. The approach shown here is an approximation to the more precise calculation.

Taking the acceptable injection rate to be $i/h = 6.7 \times 10^{-4}$ m^3/m-sec, to complete the design, it is now necessary to estimate the acid volume. An estimate of the acid capacity number is needed. This quantity is defined in Chapter 15 as

$$A_C^{(F)} = \frac{\phi_0 \beta_F c_{HF}^0 M_F}{(1 - \phi_0) V_0^F \rho_F}$$

Based on the available data, an estimate for the value of this dimensionless group can be obtained. The appropriate values for β_F are listed in Table 15.1. The formation mineralogy is complex, and an average value of the terms is dictated. Here we take a volume average so that

$$\frac{V_0^F \rho_F}{\beta_F M_F} = \sum_i \frac{V_{0,i}^F \rho_{F,i}}{\beta_{F,i} M_{F,i}}$$

where i represents a sum over the fast-reacting minerals. The original mineral analysis must be modified to account for the calcite and siderite that are removed by the acid preflush.

Thus the required number of moles of acid per unit volume of formation is found as a sum of the values in the last column of Table 16.3. These values are obtained using $\phi_0 = 0.2$. Table 16.3 reveals that 12 kg moles of HF/m^3 of formation are required to remove the fast-reacting minerals. Since the HF concentration is 1.5 wt%, then $c_{HF}^0 = 0.81$ kg mole/m^3. Thus

$$A_C^{(F)} = \frac{(0.2)(0.81)}{(1 - 0.2)12} = 1.69 \times 10^{-2}$$

TABLE 16.3 Calculation of the Acid Capacity Number for Example 16.3

Mineral (Name and Structure)	Density (kg/m^3)	Volume (m^3/m^3 of formation)	Mass (kg/m^3 of formation)	Moles (kg/m^3 of formation)	β_F† Moles Mineral / Mole HF	Moles of HF / m^3 of formation
α-Quartz (SiO_2)	2650	0.70	1860	30.9	0	0
* Montmorillonite ($Al_4Si_8O_{20}(OH)_4 \cdot 10H_2O$)	2400	0.026	62.4	0.06	0.018	3.83
Kaolinite ($Al_4Si_4O_{10}(OH)_8$)	2650	0.035	92.8	0.18	0	0
Ortholase ($KAlSi_3O_8$)	2600	0.044	114	0.414	0.050	8.28

* Estimated formula. Montmorillonite is generally quite complex with a high degree of isomorphic substitution.
† From Table 15.1

Equation (15.77) relates the pore volumes for the radial system $\theta^{(R)}$ to the penetration depth. Thus

$$\theta^{(R)} = \frac{\exp(D_a^{(R,S)}\epsilon_f^{(R)}) - 1}{A_C^F D_a^{(R,S)}} + \epsilon_f^{(R)} = \frac{\exp[(4.86 \times 10^{-2})(3)] - 1}{(1.69 \times 10^{-2})(4.86 \times 10^{-2})} + 3$$

or

$$\theta^{(R)} = 194$$

It follows from the definition of $\theta^{(R)}$ given by Eq. (15.71) that

$$\frac{it}{h} = \pi(0.1)^2(0.2)(194) = 1.22 \text{ m}^3\text{/m of acid}$$

This is a reasonable treatment volume.

Thus, the damaged well is to be treated using a total of 1.22 m^3/m of acid having a concentration of 1.5 wt% HF and 13.5 wt% HCl injected at a rate of 6.7×10^{-4} m^3/m-sec. The total treatment time will be about 30 min plus the time required to inject both the preflush and the afterflush. These latter two stages are to be injected as rapidly as possible.

The new well productivity that may be expected as a result of this acid treatment can now be estimated. The permeability in the damaged zone can be calculated using Eq. (6.4):

$$S = \left(\frac{k}{k_s} - 1\right) \ln \frac{r_s}{r_w}$$

From the data given

$$5 = \left(\frac{k}{k_s} - 1\right) \ln 2$$

or

$$\frac{k}{k_s} = 8.2$$

The stimulation results in a final porosity, given by

$$\phi_{\text{final}} = 0.13 + 0.87(0.08 + 0.08) = 0.27$$

Using Eq. (15.14)

$$\frac{k_A}{k_s} = \exp\left[7.5\left(\frac{0.27 - 0.20}{0.08}\right)\right] = 708$$

Thus, the new skin factor following the treatment is

$$S = \left(\frac{8.2}{708} - 1\right) \ln 2 = -0.685$$

To determine the stimulation ratio, first, calculate the effective wellbore radius following acidization using Eq. (13.3) as follows:

$$r_w'^{(A)} = 0.1 \exp(+0.685) = 0.198$$

Thus, the effect of the treatment is to apparently enlarge the wellbore. The effective wellbore radius prior to acidization (the damaged value) is

$$r_w'^{(D)} = 0.1 \exp(-5) = 6.74 \times 10^{-4}$$

Finally, using Eq. (11.4) for semi-steady-state production, the stimulation ratio is found to be

$$\frac{J_s}{J_d} = \frac{\ln\left(\frac{r_e}{r_w'^{(D)}}\right) - \frac{3}{4}}{\ln\left(\frac{r_e}{r_w'^{(A)}}\right) - \frac{3}{4}} = \frac{\ln\left(\frac{200}{6.73 \times 10^{-4}}\right) - \frac{3}{4}}{\ln\left(\frac{200}{0.198}\right) - \frac{3}{4}} = 1.92$$

This treatment may be expected to double current production.

Most often, the depth of the damage is unknown. In this case, an acid volume is selected from experience or perhaps because of economic limitations. Given an acid volume, then a corresponding injection rate has to be determined. This rate must ensure that the advancing acid front remains sharp but that the reaction with the slow-reacting minerals is not excessive. This is an optimization problem requiring a trial and error solution.

Example 16.5 Selection of the Injection Rate Given the Acid Volume

The well described in Example 16.3 is to be acidized with 1.0 m^3/m of acid having a concentration of 1.5 wt% HF and 13.5 wt% HCl. The depth of the damaged zone is unknown. What is the best injection rate?

Solution It is known from previous calculations (Example 16.4) that the Damkohler number for the fast reaction is related to the injection rate by

$$D_a^{(R,F)} = \frac{h}{i}(6.14 \times 10^{-4})$$

Let the value of the damaged zone dimensionless radius be

$$\epsilon_f^{(R)} = 2.5$$

The validity of this assumption can be tested as follows. The injection rate must not be too high, lest the reaction front be diffuse. Thus, applying the same criteria used in Example 16.4

$$D_a^{(R,F)} = \frac{-\ln(1/\psi_1 - 1)}{\delta(\epsilon_f^{(R)} + 1)} = \frac{-\ln(1/0.75 - 1)}{0.3(2.5 + 1)} = 1.05$$

The injection rate is (see Example 16.4)

$$\frac{i}{h} = \frac{6.14 \times 10^{-4}}{D_a^{(R,F)}} = 5.85 \times 10^{-4} \text{ m}^3\text{/m-sec}$$

The radial Damkohler number for the slow reaction is

$$D_a^{(R,S)} = \frac{h}{i}(3.26 \times 10^{-5}) = 5.57 \times 10^{-2}$$

The dimensionless injected total volume is

$$\theta^{(R)} = \frac{it}{\pi r_w^2 h \phi_0} = \frac{1.0}{\pi(0.1)^2(0.2)} = 1.59 \times 10^2$$

The assumed penetration distance of the reaction front can now be checked using Eq. (15.77) and $A_C^{(F)} = 1.69 \times 10^{-2}$. Thus

$$159 = \frac{\exp(5.57 \times 10^{-2}\, \epsilon_f^{(R)}) - 1}{(1.69 \times 10^{-2})(5.57 \times 10^{-2})} + \epsilon_f^{(R)} \qquad \text{or} \qquad \epsilon_f^{(R)} = 2.48$$

This value closely corresponds to the assumed one. The optimum injection rate corresponding to the given acid volume is 5.85×10^{-4} m^3/m-sec.

The results obtained in Example 16.5 show that there exists an optimum injection rate associated with each acid volume so that even if the depth of the damaged zone is unknown, then an optimum rate exists based on the strategy proposed here. Other strategies have been proposed. For example, Williams et al. [16] utilize a strategy that requires injection at the maximum possible rate, irrespective of the total acid volume to be utilized in the treatment. The concept underlying this strategy is simply to obtain the deepest acid penetration depth.

16.6 AFTERFLUSH

The precipitation of $Si(OH)_4$ may limit the success of sandstone acidizing treatments. To minimize the impact of precipitation, it is recommended that an afterflush be employed and that the well be returned to production immediately. The afterflush can be diesel oil, nitrogen, HCl, or ammonium chloride brine, and its use will tend to ensure that precipitation from spent acid solutions will occur away from the critical region very near the wellbore. For deep, hot formations the precipitation is rapid and the use of an afterflush may not be helpful. To have the most beneficial effect, the afterflushing process should occur immediately after acid injection so that there is no delay in displacing the spent acid [5, 6].

Aqueous afterflushes are the most effective in displacing spent acid from the near-wellbore region. They also serve to dilute the acid, thereby reducing the precipitation tendency. When practical, dilute HCl solutions are the afterflush of choice.

In some cases, however, the use of an aqueous afterflush is not desirable, thus necessitating that diesel oil or nitrogen be used. These cases occur when low reservoir permeability requires reestablishment of an oil or gas saturation around the well to allow rapid cleanup, or when low reservoir pressure requires that swabbing be employed to lift large aqueous volumes to the surface.

Shaughnessy and Kunze [5] report that at higher temperatures supersaturated solutions may precipitate in less than 30 min. Thus to some extent, there is a tradeoff between afterflush volume and time to initiate production. Generally, afterflush volumes applied immediately after the HF–HCl treatment would appear to be the safest procedure.

Additives

In some cases, it may be advantageous to add a mutual solvent to the afterflush for the purpose of removing unwanted corrosion inhibitors, which may have entered the formation and adsorbed on the minerals or blocked pore spaces [17]. These inhibitors may tend to reduce water wettability and therefore reduce the relative permeability to oil or to gas.

Ethylene glycol monobutyl ether (EGMBE) is often preferred as an additive in the afterflush [18, 19, 20]. A common concentration in 15 wt% HCl is 10 vol% EGMBE in the afterflush. Low molecular weight alcohols can also be added to aqueous phase afterflushes to help remove adsorbed corrosion inhibitors.

EGMBE can also be added to diesel oils when oil is used as an afterflush. It should not be added to the acid treatment nor to the preflush, since its presence may reduce effectiveness of the corrosion inhibitor.

16.7 ALTERNATIVE ACID FORMULATIONS

SIGMA

Because the depth of permeability alteration is severely limited in HF–HCl treatments much research has been devoted to finding ways to extend the action of acid deeper into the formation. One technique generates the HF acid in situ by injecting methyl formate and ammonium fluoride [21]. Methyl formate ($HCOOCH_3$) hydrolyzes in the presence of formation water to produce formic acid ($HCOOH$), which then reacts with ammonium fluoride (NH_4F) to yield HF. The reactions operable are as follows:

$$\begin{gathered}\underset{\substack{\text{methyl}\\\text{formate}}}{HCOOCH_3} + \underset{\text{water}}{H_2O} \rightleftharpoons \underset{\substack{\text{formic}\\\text{acid}}}{HCOOH} + \underset{\substack{\text{methyl}\\\text{alcohol}}}{CH_3OH}\\ \underset{\substack{\text{formic}\\\text{acid}}}{HCOOH} + \underset{\substack{\text{ammonium}\\\text{fluoride}}}{NH_4F} \rightleftharpoons \underset{\substack{\text{ammonium}\\\text{ion}}}{NH_4^+} + \underset{\substack{\text{formate}\\\text{ion}}}{HCOO^-} + \underset{\substack{\text{hydrofluoric}\\\text{acid}}}{HF}\end{gathered} \tag{16.3}$$

This acid system is retarded because of the slow rate of hydrolysis of methyl formate to produce formic acid. Typical hydrolysis rate determinations reported by Templeton et al. [21] show that 90% hydrolysis occurs in the presence of Wyoming bentonite after about 6 hr at 158°F, 18 hr at 140°F, and 30 hr at 122°F.

Extensive field results using this system have been reported and compared with conventional mud acid treatments. Self-generating mud acid (SIGMA) treatments in the field are reported to involve injection of the following chemicals: (1) a xylene preflush; (2) a 10 wt% HCl preflush; (3) a 7.5 wt% HCl, 1.5 wt% HF preflush; (4) a 3 wt% ammonium chloride spacer; and (5) the self-generating mud acid. Fluid volumes pumped and the composition of the self-generating mud acid were not specified.

After chemical injection, the wells are shut in long enough to allow complete acid reaction. They are then brought on production slowly over a period of several weeks to provide minimal disturbance to formation fines and gravel packs.

Higher pH Acids

Abrams et al. [22] report that an organic acid buffered with its ammonium salt and containing added NH_4F will dissolve clay at a relatively slow rate. These acids provide deep penetration. These authors report that increasing the reaction pH from the low values (less than zero) associated with the HCl–HF mixture to a buffer-regulated hydrofluoric acid–acetic acid mixture (pH $\cong$ 5.2) will reduce the reaction rate with clays by almost two orders of magnitude.

The buffered HF system was used to stimulate a deep sour-gas well (formation temperature 185°C) without significant corrosion problems and the field tests indicated that an effective stimulation was achieved.

Fluoroboric Acid

Recently, it has been proposed the fluoroboric acid (HBF_4) may provide deeper penetration into the formation than is possible using HF because HBF_4 must first hydrolyze before reacting with the silicates [23]. The reactions are thought to be as follows:

$$\begin{aligned} HBF_4 + H_2O &\rightleftharpoons HBF_3OH + HF && \text{(slow)} \\ HBF_3OH + H_2O &\rightleftharpoons HBF_2(OH)_2 + HF && \text{(rapid)} \\ HBF_2(OH)_2 + H_2O &\rightleftharpoons HBF(OH)_3 + HF && \text{(rapid)} \\ HBF(OH)_3 &\rightleftharpoons H_3BO_3 + HF && \text{(rapid)} \end{aligned} \tag{16.4}$$

This stepwise hydrolysis reaction is observed to produce 4 moles of HF per each mole of HBF_4 initially present. This HF is available to react with the silicates. Since the first reaction step is slow, it is thought that HBF_4 will provide deeper penetration because the HF can only react as it becomes available.

The rate of the first hydrolysis reaction has been shown to be given by

$$\frac{d}{dt}[HBF_4] = k_H[H^+][BF_4^-] \tag{16.5}$$

where k_H is the hydrolysis constant. It increases rapidly with temperature as follows [22]:

$$k_H = 1.44 \times 10^{17} e^{-26,183/RT}$$

where T is the temperature in °K, R is the gas constant (1.987 kcal/kg mole-°K), and k_H is expressed in m^3/mole-min. Based on this equation it appears that k_H will increase from 9×10^{-3} m^3/mole-min to 1.7 when the temperature is increased from 25°C to 65°C, an increase of 200-fold.

Thus, one would expect the penetration distance to be sensitive to the temperature. A second issue, and one of some importance, is that if HCl is added to the HBF_4 solution to accomplish the same purpose as that served by adding HCl to HF, then the hydrolysis reaction is greatly accelerated since this would effectively increase H^+ in the rate expression.

Thus, there are two aspects of HBF_4 to be considered. One is the temperature and the other is the hydrogen ion concentration. Kunze and Shaughnessy [23] have shown that at elevated temperatures (65°C) or in the presence of HCl, the reactions of HBF_4 with bentonite proceed at essentially the same rate as HF. Thus, there appeared to be no advantage to be gained using HBF_4 under these conditions. Core floods confirmed the conclusions.

It would appear that the use of HBF_4 would be most advantageous in shallow low-temperature formations or in formations precooled by the use of a substantial HCl preflush and perhaps even an HF–HCl flush following the HCl preflush. An advantage of HBF_4 is that it may reduce water sensitivity of sandstones [24, 25].

Sequential Hydrofluoric Acid System

Holden et al. [26] have reported successful stimulation using a sequential injection of HCl followed by NH_4F. The concept underlying this method, which has been reported by Hall [27], relies on cation exchange (see Chaps. 1 and 4). Clays generally have a fixed negative charge balanced by cations from the solution that adhere to the clay. These cations associated with the clay are exchangeable (see Chap. 1). This process can be represented as a chemical reaction in the following manner:

$$\text{clay}^- \cdot Na^+ + H^+\text{(solution)} \rightleftharpoons \text{clay}^- \cdot H^+ + Na^+\text{(solution)} \qquad (16.6)$$

Thus if a clay associated with sodium cations is brought into contact with an acid solution, say HCl, then the hydrogen ions will adsorb onto the clays thereby displacing the sodium cation. This process was discussed in considerable detail in Chapter 1. The sodium displaced from the surface will then form sodium chloride in solution.

The sequential process uses the cation-exchange properties of clays to generate HF in situ. The process consists of pumping alternate stages of HCl and NH_4F solutions. During the first stage, hydrogen ions are adsorbed by the ion-exchange process. When the NH_4F solution contacts the hydrogen-rich clays, the reverse exchange reaction occurs. Thus

$$\text{clay}^{-}\cdot\text{H}^{+} + \text{NH}_4^{+}(\text{solution}) \rightleftharpoons \text{clay}^{-}\cdot\text{NH}_4^{+} + \text{H}^{+}(\text{solution}) \qquad (16.7)$$

The H^+ then associates with F^- to form active HF, which can then dissolve clay.

In laboratory tests Hall [27] showed that the sequential acid process gave deeper penetration than did a 3 wt% HF–15 wt% HCl mixture. It seems reasonable that this would be the case.

It was also claimed that the process is very selective in that reaction with clay is much greater than with sand. This claim is true for the experiments reported by Hall [27], since the surface areas of the clays were generally large. However, once HF is formed in situ, its reactivity would be expected to conform to the rates described in Chapter 14 and only authogenic clays would be expected to react rapidly.

This procedure suffers from the disadvantage of not being able to blend HCl with HF. Precipitation problems might arise. This, however, is mitigated by the fact that the HF concentrations generated are usually quite small.

REFERENCES

16.1. Smith, C. F., and Hendrickson, A. R., *J. Pet. Tech., 17* (1965) 215.

16.2. Smith, C. F., Ross, W. M., and Hendrickson, A. R., "Hydrofluoric Acid Stimulation—Developments for Field Application," SPE 1284, presented at the 1965 Society of Petroleum Engineers Annual Meeting, Denver, Colorado, 1965.

16.3. Labrid, J. C., *Revue L. Institut Francais du Petrole, 26* (1971) 855.

16.4. Labrid, J. C., *Soc. Pet. Eng. J., 15* (1975) 117.

16.5. Shaughnessy, C. M., and Kunze, K. R., "Understanding Sandstone Acidizing Leads to Improved Field Practices," SPE 9388, presented at the 55th Fall Technical Conference and Exhibition of the Society of Petroleum Engineers, Dallas, Texas, 1980.

16.6. Walsh, M. P., Lake, L. M., and Schechter, R. S., *J. Pet. Tech., 34* (1982) 2097.

16.7. Farley, J. T., Miller, B. M., and Schoettle, V., *J. Pet. Tech., 22* (1970) 433.

16.8. Champion, L. S., "The Ultimate Yield from Oil Well Stimulation with Acids," MS thesis, The University of Texas at Austin, 1970.

16.9. Hill, A. D., Lindsay, D. M., Silberberg, I. H., and Schechter, R. S., *Soc. Pet. Eng. J., 21* (1981) 30.

16.10. Brannon, Dan H., Netters, C. K., Grimmer, P. J., *J. Pet. Tech., 39* (1987) 931.

16.11. Crowe, C. W., *J. Pet. Tech., 38* (1986) 1234.

16.12. Hekim, Y., and Fogler, H. S., *Amer. Inst. Chem. Eng. J., 26* (1980) 403.

16.13. Fogler, H. F., and McCune, C. C., *Amer. Inst. Chem. Eng. J., 22* (1978) 799.

16.14. Lindsay, D. M., "An Experimental Study of Sandstone Acidization" MS Thesis, The University of Texas at Austin, 1976.

16.15. Schechter, R. S., daMotta, E. Ponce, and Plavnik, B. "Optimizing Sandstone Acidization," SPE 19426, presented at the SPE Formation Damage Control Symposium, Lafayette, Louisiana, 1990.

16.16. Williams, B B., Gidley, J. L., and Schechter, R. S., *Acidizing Fundamentals*, Mono. Ser., *6*, Society of Petroleum Engineers, Richardson, Texas, 1979.

16.17. Crowe, C. W., and Minor, S. S., J. Petr. Tech., 37, (1985) 1857.

16.18. Gidley, J. L., *J. Pet. Tech., 23* (1971) 551.
16.19. Gidley, J. L., Ryan, T. P., and Mayhill, T. H., *J. Pet. Tech., 28* (1976) 1289.
16.20. Hall, B. E., *J. Pet. Tech., 27* (1975) 1439.
16.21. Templeton, C. C., Richardson, E. A., Karnes, G. T., and Lybarger, J. H., *J. Pet. Tech., 27* (1975) 1199.
16.22. Abrams, A., Scheuerman, R. F., Templeton, C. C., and Richardson, F. A., *J. Pet. Tech., 35* (1983) 2175.
16.23. Kunze, K. R., and Shaughnessy, C. M., "Acidizing Sandstone Formations with Fluoroboric Acid," SPE 9387, 55th Fall Technical Conference and Exhibition of the Society of Petroleum Engineers, Dallas, Texas, 1980.
16.24. Thomas, R. L., and Crowe, C. W., *J. Pet. Tech., 33* (1981) 1491.
16.25. Veley, C. D., *J. Pet. Tech., 21* (1969) 1111.
16.26. Holden, W. W., Prihoda, C. H., and Hall, B. E., *J. Pet. Tech., 33* (1981) 1485.
16.27. Hall, B. E., *J. Pet. Tech., 30* (1978) 1220.

PROBLEMS

****16.1.** When clays are dissolved by HF–HCl mixtures, invariably colloidal silica [$Si(OH)_4$] will precipitate. This precipitation following the dissolution of clays (or other aluminosilicates) can result in formation damage. The primary chemical reactions that occur (there are many others) when kaolinite [$Al_2Si_2O_5(OH)_4$] is dissolved are listed in Table P16.1. Suppose that 1 kg mole of kaolinite is added to a 1 m^3 of H_2O containing 0.5 kg mole of HF and the concentration of hydrogen ions is adjusted by adding HCl to maintain the final equilibrium pH of the solution as -0.5. How many kg moles of kaolinite will dissolve? How many kg moles of colloidal silica will precipitate? What will be the fraction of unreacted HF?

TABLE P16.1

Intraqueous Reactions	Logarithm of Equilibrium Constant
$SiF_4 \rightleftharpoons H_4SiO_4 + 4F^- + 4H^+ - 4H_2O$	−21.75
$AlF^{++} \rightleftharpoons Al^{+++} + F^-$	−6.13
$HF \rightleftharpoons H^+ + F^-$	−3.18
Mineral Solubility Reactions	**Logarithm of Equilibrium Constant**
(Solid) $Al_2Si_2O_5(OH)_4 \rightleftharpoons 2H_4SiO_4 + 2Al^{+++} + H_2O - 6H^+$	11.30
(Solid) $Si(OH)_4 \rightleftharpoons H_4SiO_4$	−2.36

In making the calculations, assume that it is permissible to replace activities by the concentrations expressed as kg moles/m^3. Thus, from the first reaction one can write

$$\frac{c_{H_4SiO_4} c_F^4 c_H^4}{c_{SiF_4}} = 10^{-21.75}$$

(*Note:* The activity of water can be taken as unity. So can the activities of the minerals.)

***16.2.** If a sandstone contains about 4 wt% HCl-soluble components, estimate the volume of preflush composed of 15 wt% HCl required to remove these acid-soluble minerals from a region 0.5 m in radius about a wellbore having a radius of 0.1 m. The formation has a porosity of 11%.

****16.3.** If a sandstone contains about 5 wt% HCl-soluble minerals, mostly carbonates, estimate the volume of 7.5 wt% HCl preflush required to remove these carbonates from a region around each perforation in a perforated interval that is to be acidized with an HF–HCl mixture. Each perforation is an oblate spheroid having a radius of 0.3 cm at the wellbore and penetrating a distance of 30 cm into the formation. There are 18 perforations/m, and the region surrounding each perforation from which the carbonates are to be removed is also an oblate spheroid concentric to the perforations and having a radius of 10 cm at the wellbore.

****16.4.** The acid preflush is intended to remove carbonates from the region near the wellbore, which is to be contacted with the HF–HCl mixture; however, a second positive effect is the cooling of the formation around the perforation. If a perforation is an oblate spheroid having a radius at the wellbore of 0.25 cm and a penetration depth of 32 cm, map the position of the temperature front surrounding the perforation after a total of 3×10^{-2} m^3 of preflush fluid has been injected. The temperature wave can be taken to be sharp so that in the region surrounding the perforation the temperature is that of the injected fluid, whereas across the wave it is the initial formation temperature. Also show the position of a tracer front. Note that a substantial region about the perforation is cooled to the injected fluid temperature by the preflush. The data are presented in Table P16.4.

TABLE P16.4 Treatment Data

Property	Symbol	Value
Formation porosity	ϕ	0.18
Formation rock density	ρ_{rock}	2650 kg/m^3
Formation heat capacity	$C_{P_{rock}}$	0.087 kcal/kg-°C
Preflush density	ρ	1030 kg/m^3
Preflush heat capacity	C_P	1 kcal/kg-°C

***16.5.** A sandstone formation is damaged to a depth of 0.3 m about a 0.1 m uncased wellbore. The formation is 3000 m deep and its temperature is 84°C. If the fracture gradient and the formation permeability restrict the acid injection rate to a maximum value of 5×10^{-4} m^3/m-sec, determine whether or not the damage can be fully removed by an HF–HCl treatment. Consider the treatment to be feasible if the HF concentration at the reaction front can be maintained above 20% of its injected concentration.

No other information regarding the composition or nature of the formation is available. State clearly all approximations used in your analysis.

***16.6.** Design a matrix acid treatment for a sandstone containing about 5 wt% authogenic clays and feldspars and about 9 wt% carbonates. The formation permeability is 75 mD, but the well is producing at a rate that would indicate an effective or average permeability of 10 mD. Thus, substantial damage is suspected and considerable benefit may be derived by a modest treatment. Because the depth of the damaged zone is unknown, consider a treatment of 1.8 m^3/m of 3 wt% HF and 12 wt% HCl.

The following data apply (no other data are available):

Net formation thickness:	10 m
Formation depth:	1500 m
Formation temperature:	66°C
Reservoir pressure:	15,500 kPa
Fracture gradient:	20 kPa/m
Porosity:	0.15
Wellbore radius:	0.1 m (uncased)
Drainage radius:	200 m

Your design should define the total volume of preflush to be used, the rate at which the preflush is to be injected, the rate at which the HF–HCl mixture is to be injected, and the volume of the afterflush. The afterflush is to consist of 5 wt% HCl. Predict the stimulation ratio that will result from your treatment.

17 Fundamentals and Design of Matrix Acidizing of Carbonates

The earliest acid treatments consisted of dumping hydrochloric acid into the wellbore and then applying pressure to force the acid into the formation. Because of equipment limitations, early treatments were conducted at pressures below the formation fracture pressure. As high-pressure pumping equipment became available, treatment pressure was increased until most carbonate acid treatments today are conducted above the formation fracture pressure. Nonetheless, some damaged wells remain where acid treatments must be performed below the fracture pressure to be effective. This situation is especially true when the treatment is limited to small acid volumes per unit of formation thickness. In this chapter the characteristics of this treatment process are investigated and design procedures recommended. It becomes evident that the mechanism of acid attack in carbonates is quite different from that found in sandstones. The fundamental difference is that the rate of acid reaction with carbonates is much faster than it is with most of the minerals found in sandstones (see Chap. 14). This difference in reaction rate produces a profound change in the mechanism of acid attack.

17.1 ACID CHANNELING IN CARBONATES: WORMHOLING

Flow and reaction in porous media can, at the extreme limits of reaction rate, take place by one of two mechanisms. In one limit, reaction between the fluid and the solid is such that a relatively sharp reaction front moves through the porous medium. This case is characteristic of matrix acidizing of sandstones using mixtures of hydrofluoric and hydrochloric acids (see Chap. 15). The second dissolution mechanism involves severe channeling. This condition results because sedimentary rocks are often composed of interconnected pores having a wide range of different sizes. More acid tends to flow through the larger pores than through the smaller ones. If the acid reaction rate is very fast, then a significant proportion of the acid entering both the larger pores and the smaller pores will react. However, since the larger pores receive more acid, they tend to enlarge at a faster rate causing even more acid to enter the pore. Thus, larger pores enlarge more rapidly than smaller ones. This unstable process tends to produce a single larger pore extending from the inlet face of a core of reservoir rocks to the outlet face. This large channel, which accepts virtually all of the acid, is known as a *wormhole* [1]. Figure 17.1 shows a cast of a wormhole [2]. This cast was prepared

Figure 17.1 Illustration showing Wood's metal casting of wormhole in Indiana limestone core (diameter = 1.5 in., length = 5 in.) acidized with HCl. (Photograph courtesy of H.S. Fogler)

by first acidizing an Indiana limestone core until acid breakthrough and then injecting molten Wood's alloy at 100°C. The Wood's alloy is allowed to solidify within the wormhole and then the core is totally dissolved in acid, leaving a reproduction of the etched channels.

The central channel is observed to be connected to somewhat smaller branches, which are in turn connected to even smaller branches, thus giving the entire geometric arrangement the appearance of being a fractal [3] or a structure very similar to patterns associated with diffusion-limited aggregation produced by a variety of other processes [4, 5].

When the diameter of the central pore becomes sufficiently large, the overall rate of reaction becomes diffusion limited, as described in Chapter 14, thus accounting for the somewhat uniform appearance of the central channel. The effective radius of the channel is sufficiently large to slow the reaction because acid must diffuse from the central region of the channel to the walls.

Calculations show that once the effective wormhole radius reaches a few mm, then long penetration distances are possible. These distances can range to tens of meters, depending on the acid injection rate. In the laboratory such long wormholes are not observed [6]. This limitation on wormhole length is attributed to fluid loss and is also suggested by the appearance of the cast shown in Fig. 17.1. The branches extending outward from the main channel are due to fluid loss. Thus, two of the main issues relating to the mechanism of matrix acidization in carbonate formations are severe channeling or wormholes and the limitation imposed on the length of the channels by fluid loss. The former issue is addressed first.

Wormhole formation occurs when larger pores tend to grow in cross-sectional area at a rate that greatly exceeds the growth rate of the smaller pores. The condition for wormhole growth can be made definite if one notes that the rate of change of pore area (A) with time (t) can, in certain limits to be discussed, be written in the form

$$\frac{dA}{dt} = \Psi A^{1-n} \tag{17.1}$$

where Ψ depends on parameters such as the reaction rate constant and the acid diffusion coefficient but not on pore area or time. Thus, Eq. (17.1) can be integrated as follows:

$$A = (A_0^n + n\Psi t)^{1/n} \tag{17.2}$$

where A_0 is the cross-sectional area when $t = 0$. Two extreme cases will arise. For $n > 0$ smaller pores tend to approach the same ultimate size as the larger pores. In a sense, the pores created by the acidization process do not remember which pores were the largest at the start of the process. This condition is characteristic of uniform acid attack. Acid channeling does not occur.

If $n < 0$ not only do the larger pores grow much more rapidly than the smaller ones, but according to Eq. (17.2) the pores grow to infinite size in a finite time. The larger the pore area is initially the more rapidly this limit is reached. Pores cannot, of course, become infinite and n cannot, therefore, be negative for all A.

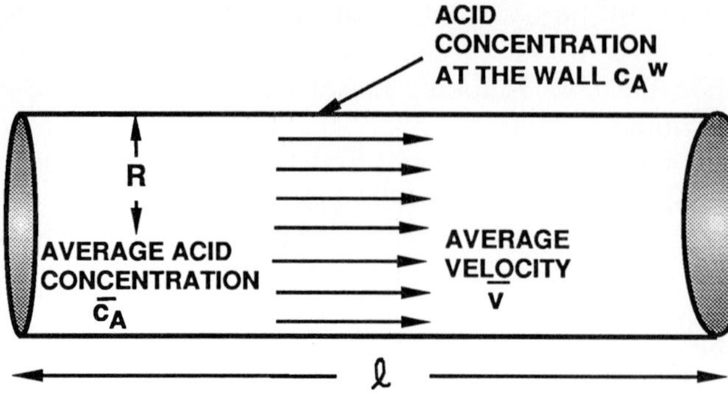

Figure 17.2 Sketch depicting acid flowing through cylindrical pore with reaction at the wall.

At some point when A is large, Eq. (17.1) will not apply. Nevertheless, the characteristic parameter for wormhole formation is $n < 0$.

To find the pore area growth law, the details of the interaction between the reactive fluid and solid pore wall must be investigated. The particular system of interest, a circular pore, is depicted by Fig. 17.2. The acid concentration throughout the pore is governed by Eq. (14.25). This equation is complex and since our goal is to obtain qualitative information about the tendency to wormhole, a simplified, but instructive equation will be considered. Thus, the average acid concentration at any axial position is defined by

$$\pi R^2 \, \bar{v} \, \frac{\partial \bar{c}_A}{\partial z} = -2\pi R K_M(\bar{c}_A - c_A^w) \tag{17.3}$$

The volumetric flow rate is $\pi R^2 \bar{v}$ and $2\pi R$ is the pore perimeter. The rate of acid transfer to the wall is approximated by $K_M(\bar{c}_A - c_A^w)$ where K_M* is for laminar flow approximately $(3D_A^2\bar{v}/2Rl)^{1/3}$ [7] and c_A^w is the acid wall concentration. Equation (17.3), although approximate, does contain all of the essential features of the process. To eliminate the acid concentration at the wall, an additional equation is needed. Since the rate of acid transfer to the pore surface must equal the rate of acid reaction, then for first-order kinetics (see Prob. 17.6 for more complex reaction mechanisms)

$$K_M(\bar{c}_A - c_A^w) = E_f c_A^w \tag{17.4}$$

Substituting for c_A^w in Eq. (17.3) and integrating

$$\bar{c}_A(l) = c_0 \exp\left(-\frac{2K_M E_f l}{R\bar{v}(K_M + E_f)}\right) \tag{17.5}$$

where l is the pore length and c_0 is the inlet acid concentration.

The fraction of acid entering a pore which reacts is $(c_0 - \bar{c}_A(l))/c_0$. Furthermore, if X is the acid dissolving power (see Chap. 3), then the volume of rock dissolved per unit of time is $AX\bar{v}[c_0 - \bar{c}_A(l)]/c_0$. The volume dissolved effectively

* This expression for K_M assumes that the diffusion boundary layer does not fully develop over the length of the tube. If it does another expression for K_M is required (see Prob. 17.4).

increases the pore area so that

$$l \frac{dA}{dt} = \frac{AX\bar{v}\,[c_0 - \bar{c}_A(l)]}{c_0} \tag{17.6}$$

Substituting Eq. (17.5), the pore growth rate is found to be

$$\frac{dA}{dt} = \frac{AX\bar{v}}{l}\left[1 - \exp\left(-\frac{2K_M E_f l}{R\bar{v}\,(K_M + E_f)}\right)\right] \tag{17.7}$$

There are two limits of interest. For slow reactions, $E_f \longrightarrow 0$ and Eq. (17.7) reduces to

$$\frac{dA}{dt} = 2\sqrt{\pi A}\,E_f X = \Psi A^{1/2} \tag{17.8}$$

Comparing this rate law with that given by Eq. (17.1), it is easy to see that $n = 1/2 > 0$. This result then gives uniform invasion and no channels will form.

For rapid reactions, $E_f \longrightarrow \infty$ and $K_M = (3D_A^2\bar{v}/2Rl)^{1/3}$

$$\frac{dA}{dt} = \frac{A\bar{v}X}{l}\left\{1 - \exp\left[-\left(\frac{12D_A^2 l^2}{R^4\bar{v}^2}\right)^{1/3}\right]\right\} \tag{17.9}$$

To further expand this equation recall that the average velocity for laminar flow in a capillary is given by

$$\bar{v} = \frac{A}{8\pi\mu}\left(-\frac{dp}{dz}\right)$$

so that

$$\bar{v} \cong \epsilon A \tag{17.10}$$

Equation (17.9) becomes

$$\frac{dA}{dt} = \frac{\epsilon A^2 X}{l}\left\{1 - \exp\left[-2.29\left(\frac{\pi D_A l}{\epsilon A^2}\right)^{2/3}\right]\right\} \tag{17.11}$$

For most reasonable size pores $\pi D_A l/\epsilon A^2 \gg 1$ and therefore

$$\frac{dA}{dt} = \frac{\epsilon X A^2}{l} \tag{17.12}$$

For these pores, $n = -1 < 0$ and severe channeling is to be expected. Thus between the limits of slow reaction rates and fast ones, the tendency for severe channeling appears gradually. Experiments conducted by W. Jazauri appear to confirm this prediction [8]. Acid was injected into a small hole in the center of a cylindrical core and allowed to flow radially outward. X-ray photographs were then taken to determine changes caused by the acid. Multiple small wormholes appeared when the reaction rate was retarded compared to the single large wormhole observed when HCl was used.

The tendency to form wormholes depends on the flow rate as well as the reaction rate. Thus by substituting K_M proportional to $\bar{v}^{1/3}$ into Eq. (17.7) and taking the limit as $\bar{v} \longrightarrow \infty$, one finds that the pore growth model defined by

Eq. (17.8) again applies. At high flow rates the system will behave as though the reaction rate is small. Thus, as the flow rate is increased the tendency for wormholes to form may also decrease. The influence of flow rate on wormhole formation has been studied by Daccord [9]. The results of his radial experiments in which plaster (calcium sulfate) is dissolved by water are shown in Fig. 17.3. In each of the three cases the experiment was stopped after 4 hr so that more water was injected at the higher flow rates. Note that when the injection rate is 48 cm³/hr

(a)

(b)

(c)

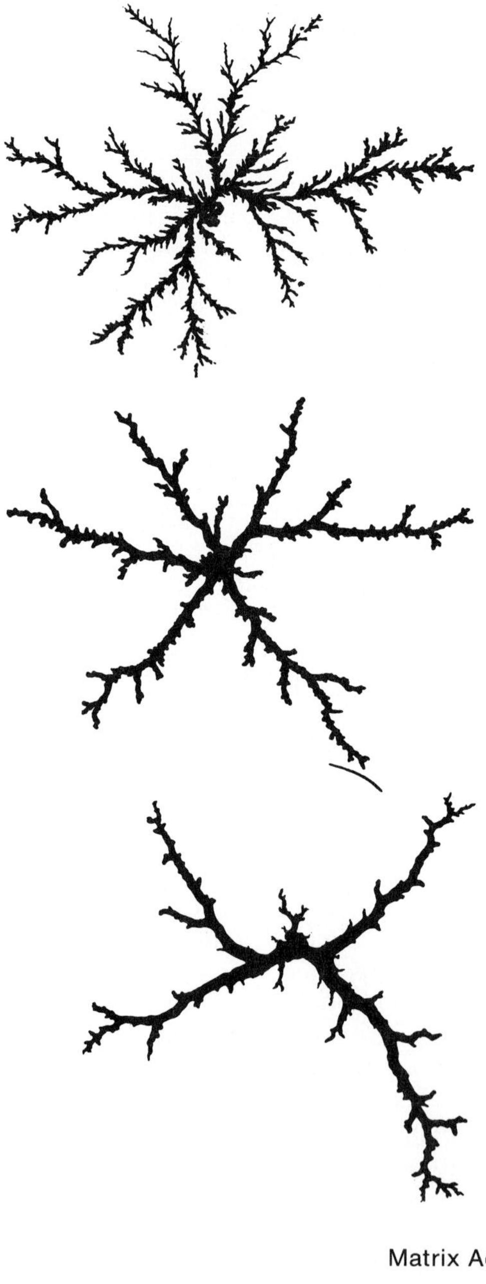

Figure 17.3 Typical wormhole patterns created in a radial flow geometry. The flow is stopped after 4 hr using three different injection rates. (a) 48 cm³/hr [6]. 4 cm³/hr [9]. (c) 2 cm³/hr. [From G. Daccord, *Physical Review Letters*, *58* (1987), 479. Used with permission.]

the wormhole structure is finer and more highly ramified, as compared to the patterns produced using slower injection rates. These experiments demonstrate the decreased tendency for wormholes to form as the fluid velocity is increased.

Example 17.1 Diffusion-Controlled Pore Growth

When the reaction rate is large, Eq. (17.11) applies and pore growth is diffusion controlled. Consider four different pore areas: 25×10^{-12}, 10×10^{-12}, 2×10^{-12}, and 0.4×10^{-12} m^2. Given that $\epsilon = 10^8$ m^{-1}-sec^{-1}, $D_A = 4 \times 10^{-9}$ m^2/sec, and $l = 3 \times 10^{-6}$ m, how long will it take the pore which is initially 25×10^{-12} m^2 to double in area if the acid is 2.5 wt% HCl and the rock is limestone? What will be the size of the other three pores at this time?

Solution For 2.5 wt% HCl, $X = 0.013$ (see Chap. 3). Equation (17.11) is then

$$\frac{dA}{dt} = (4.33 \times 10^{11})A^2 \left[1 - \exp - 2.29\left(\frac{3.77 \times 10^{-22}}{A^2}\right)^{2/3}\right]$$

This equation cannot be integrated, but $A(t)$ can be calculated numerically by considering small increments of time. A finite difference calculation will show that after $t = 0.085$ sec, the results shown in Table 17.1 apply. The largest pore has doubled in size, whereas the smaller ones have grown at a substantially slower pace. This is typical of reaction systems that exhibit wormhole formation [10].

TABLE 17.1 Pore Growth

Area at $t = 0$ (m^2)	Area at $t = 0.085$ sec (m^2)
25×10^{-12}	50×10^{-12}
10×10^{-12}	14.4×10^{-12}
2×10^{-12}	2.16×10^{-12}
0.4×10^{-12}	0.41×10^{-12}

17.2 FLUID LOSS FROM CIRCULAR PORES

The fact that the fluid lost from a growing wormhole may still be reactive (contain unreacted acid) explains the "tree-like" appearance shown in Fig. 17.1 and is certainly responsible for the discrepancy between the theoretical length and that observed in the laboratory. To model this fluid loss, consider the wormhole to have a cylindrical shape of radius R and length l as shown in Fig. 17.2. Although wormholes are certainly not cylindrical, predictions based on this model can be helpful in understanding some of the observations.

The approach taken here is identical to that previously applied to characterize the fluid loss from hydraulic fractures (see Chap. 8). An overall fluid-loss coefficient (C_R) having the form (the subscript R denotes a radial system)

$$u_r = \frac{C_R}{\sqrt{t}} \tag{17.13}$$

will be defined for the geometry shown by Fig. 17.4. The u_r represents the radial flux or fluid loss.

Start by considering the pressure in the formation at a point beyond the invaded zone. For a slightly compressible fluid this pressure is described by the differential equation in cylindrical coordinates.

$$\frac{1}{r}\frac{\partial}{\partial r}\left(r\frac{\partial p}{\partial r}\right) = \frac{\mu_f \phi \kappa_f}{k}\frac{\partial p}{\partial t} \tag{17.14}$$

where the subscript f is appended to denote properties of the reservoir fluid. This equation is to be solved subject to the following boundary and initial conditions:

$$\begin{aligned} p &= p_R \qquad (\text{at } t = 0) \\ p &= p_I \qquad (\text{at } r = r_I) \\ p &= p_R \qquad (\text{as } r \longrightarrow \infty) \end{aligned} \tag{17.15}$$

The solution to this problem is given by Crank [11]. For small contact times, the fluid loss per unit of wormhole length (q_l) is given by

$$q_l = u_r(2\pi r_I) = \left(\frac{2\pi^2 k}{\mu_f}\right)\left(\frac{p_I - p_D}{\sqrt{t_D}}\right) \tag{17.16}$$

where t_D is a dimensionless time expressed as

$$t_D = \frac{kt}{\mu_f \phi \kappa_f r_I^2} \cong \frac{kt}{\mu_f \phi \kappa_f R^2} \tag{17.17}$$

Here for simplicity the invaded zone radius r_I is taken to be approximately the wormhole radius. This is a good approximation for small contact times.

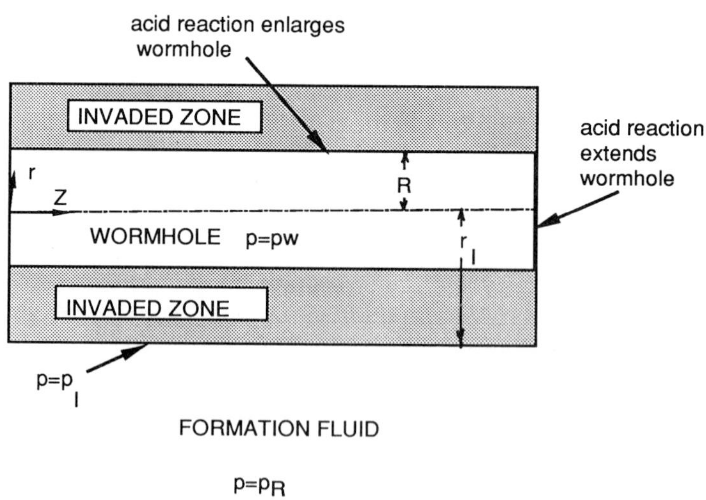

Figure 17.4 Sketch depicting the geometry of a wormhole and surrounding invaded zone.

As in Chapter 8 where the fluid loss from a fracture is considered, the flow in the invaded zone surrounding the hole is essentially an incompressible flow. Since the acid lost tends to increase the permeability in the surrounding region and since the acid has a small viscosity, the pressure drop across this zone is small and can be neglected. Thus $p_I \approx p_W$. Then the fluid loss per unit length (q_l) becomes

$$q_l = 2\pi R C_R/\sqrt{t} \tag{17.18}$$

where
$$C_R = \sqrt{\frac{\pi^2 \phi k \kappa_f}{\mu_f}}\,(p_w - p_R)$$

This equation can be compared with C_c, the compressibility coefficient, developed for fractures in Chapter 8.

Fluid-Loss Limit

To couple all of the mass transfer and hydrodynamic effects together to develop an expression for the wormhole geometry as a function of time is complex and requires a numerical solution of the equations. This sort of calculation is described in the next section. Here our goal is to describe the wormhole geometry when fluid loss is the critical feature of the process. For high fluid-loss rates, it is established for acid fracturing (see Chap. 10) that diffusion of the acid can be neglected. Acid is transported to the rock surface primarily by convection, not by diffusion. This same condition also holds for flow and reaction in wormholes. Thus the model to be considered is the circular wormhole of uniform radius R and length l depicted by Fig. 17.2. Both R and l depend on time. It is these two quantities that define the geometry of a wormhole. If q^* is the volume of acid per unit of time arriving at the end of the wormhole and X is the dissolving power of the acid, then the following volume balance must hold:

$$(1 - \phi)\pi(R^2 - R_p^2)\frac{dl}{dt} = q^* X \tag{17.19}$$

where R_p is the initial pore radius.

This balance is written assuming that the acid which arives at the end of the wormhole does so at full strength (this is true if diffusion of acid is negligible) and that the reaction takes place immediately to increase the wormhole length (acid does not penetrate into the rock matrix).

The volume of acid entering the wormhole per unit of time is q_0, but this amount does not reach the end of the wormhole because of fluid loss. Thus

$$q^* = q_0 - 2\pi R \int_0^l \frac{C_R dz}{\sqrt{t - \tau(z)}} \tag{17.20}$$

In writing this equation R is assumed to be a slowly varying function of time. Finally, the total rock volume dissolved at time t must equal the dissolving power of all of the acid used. Thus

$$q_0 t X = \pi(1 - \phi)(R^2 - R_p^2)l \tag{17.21}$$

Equations (17.19) and (17.21) can be substituted into Eq. (17.20) to yield the differential equation

$$\frac{q_0 t}{l}\frac{dl}{dt} = q_0 - 2\pi R C_R \int_0^l \frac{dz}{\sqrt{t - \tau(z)}} \tag{17.22}$$

If R is treated as practically independent of time, this equation can be solved to give

$$l = \frac{q_0 t^{1/2}}{2\pi^2 R C_R} \tag{17.23}$$

Substituting for l in Eq. (17.21) yields the following expression for R:

$$\frac{R^2 - R_p^2}{R} = \frac{2\pi X C_R t^{1/2}}{(1 - \phi)} \tag{17.24}$$

These equations permit estimation of the wormhole radius and length at a given time in the limit of high fluid-loss rates. If the diffusion of acid to the pore wall provides a significant contribution to the total acid flux to the wall, then these two equations will provide an inaccurate picture of the process.

Example 17.2 Approximate Wormhole Dimensions

What will be the dimensions of a pore initially 25 μm in radius after 15 min if the volume of 15 wt% HCl entering the pore per unit of time is 5×10^{-3} cm^3/sec? The fluid-loss coefficient is estimated to be 10^{-2} cm/sec$^{1/2}$. The limestone formation has a porosity of 10%.

Solution The dissolving power of 15 wt% HCl, according to Table 3.9, is 0.082.

Thus
$$\frac{R^2 - (25 \times 10^{-4})^2}{R} = \frac{2\pi(900)^{1/2}(0.082)(1 \times 10^{-2})}{(1 - 0.1)}$$

Solving yields

$$R = 0.17 \text{ cm} \quad \text{and} \quad l = \frac{5 \times 10^{-3}(900)^{1/2}}{2\pi^2(0.17)(1 \times 10^{-2})} = 4.5 \text{ cm}$$

For this example a rather short wormhole is produced. An examination of Eq. (17.23) reveals that for a given volume of acid entering a wormhole, the length can be increased by increasing the flow rate, q_0. Figure 17.5 shows the permeability increase of an Indiana limestone as a function of the pore volumes of acid injected. The abrupt increase in the permeability corresponds to the acid volume required to create a wormhole through the entire length of the core. As predicted by Eq. (17.23) the wormhole length increases more rapidly as the acid injection rate increases. (Ignore the experiment in which the acid was present as a microemulsion). This is in agreement with predictions based on Eqs. (17.23) and (17.24).

It is important to recognize the severe limitations imposed in the derivation of Eqs. (17.23) and (17.24). The acid lost from the wormhole is assumed to increase the wormhole radius without producing the tree-like branches shown in Fig.

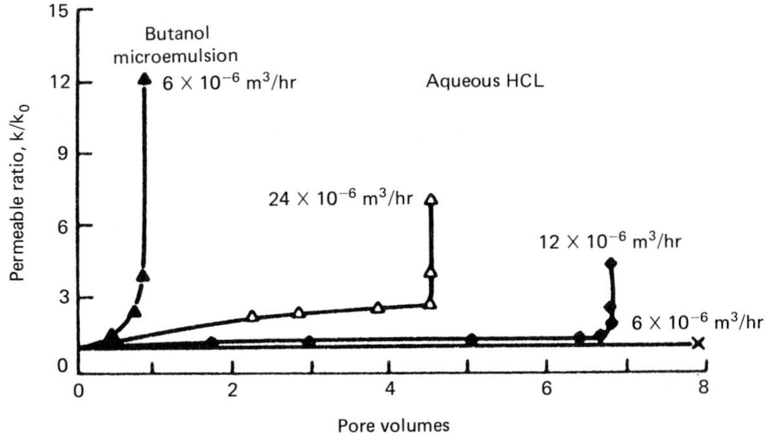

Figure 17.5 Graph of coreflood results for Indiana limestone showing permeability response as a function of fluid injection rate (core diameter = 0.025 m, core length = 0.054 m, 0.13 *N* aqueous HCl) [12]. (With permission of American Institute of Chemical Engineers.)

17.3(a). This restriction applies when the reaction rate is large. The analysis also neglects diffusion, which is a valid approximation for large fluid-loss rates. Thus, there is a limited range of conditions for which wormholes resemble "pipes" [Fig. 17.3(c)] rather than highly ramified structures. It is not possible at the present time to quantitatively define these limits, and thus great care should be exercised in formulating conclusions based on Eqs. (17.23) and (17.24).

As an example of the difficulty that can arise, consider the results shown by Fig. 17.6. The acidization of dolomite is observed to be more efficient at low injection rates. A smaller total volume of acid is required to achieve a sharp increase in the permeability when the injection rate is 6×10^{-6} m^3/hr than is needed at higher rates. This behavior is not predicted by Eqs. (17.23) and (17.24) nor is it in agreement with the results for the acid treatment of Indiana limestone.

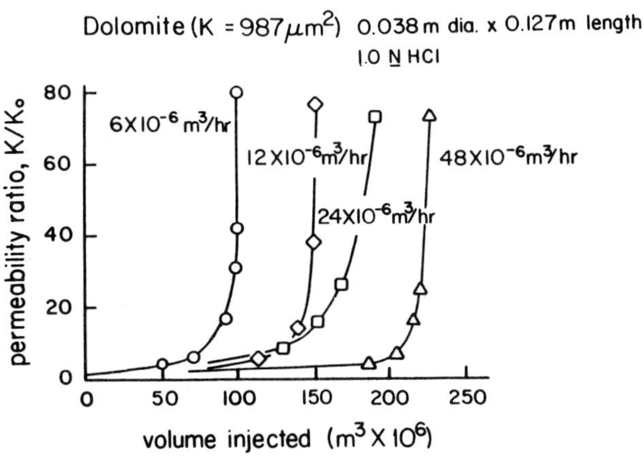

Figure 17.6 Graph of coreflood data for dolomite showing permeability response as a function of fluid injection rate (dolomite core k = 1 mD, diameter = 0.038 m, length = 0.127 m, 1.0 *N* HCl) [12]. (With permission of American Institute of Chemical Engineers.)

At room temperature the rate of HCl reaction with dolomite is several orders of magnitude smaller than it is with limestone (see Chap. 14). The tendency for wormhole formation is thus reduced in dolomites, and one will expect a number of highly branched structures in dolomite rather than a single channel. Furthermore, as the rate is increased the acid invasion is expected to become more uniform with increasing flow rate. Wood's alloy casts of the channels produced in dolomite confirm this conclusion [12].

Although it may seem counterintuitive, for a fixed acid volume deeper penetration is likely to be achieved with acids that react rapidly as compared to ones that react more slowly as long as both acids react rapidly enough to create wormholes. Thus using retarded acids for matrix acid treatments as has been often thought desirable, may in fact, be counterproductive. The most favorable approach for increasing acid penetration appears to be either reducing the acid diffusion rate [12] (which is effectively increasing the reaction rate) or reducing the fluid loss by using additives [6].

It should be noted that the reaction of HCl with dolomite increases markedly with temperature (see Chap. 14). Thus the behavior under reservoir conditions may more closely approximate the trends shown by Fig. 17.5 than by Fig. 17.6. This is, however, conjecture. Experiments delineating the effect of temperature on dolomite acidization have not yet been reported.

17.3 MODELING OF MATRIX ACIDIZING PROCESSES

In the previous sections the mechanisms occurring in isolated pores or wormholes have been considered. It is, of course, necessary to consider a large collection of such pores all interacting with flowing acid and as well as with other pores. Schechter and Gidley [1] and Guin et al. [13, 14] have considered a collection of essentially parallel pores and calculated the evolution of the pore sizes based on a stochastic equation, which includes a mechanism of pore "collisions." The concept being that as a pore expands in area, it incorporates other pores thus depleting their total number. They determined that this is an important mechanism contributing to change in permeability as rock is dissolved. This approach has not, however, produced models which can be used in the design of matrix treatments.

Hoefner and Fogler [2, 12] have used network models in which connected pores are allowed to increase in area according to certain rules designed to simulate the acid attack on pores. This approach indicates some of the qualitative features of the acidizing process but does not give guidance for the design of matrix treatments. The network approach taken alone cannot model the permeability changes that occur in laboratory core experiments, because pore collisions are ignored [1] and fluid loss is not really properly accounted for. However, network models are apparently useful for evaluating certain characteristics of acidized carbonates. These models are considered in the next section [15]. Important questions, such as the number of wormholes that form at each perforation, still remain unresolved. Both pore and network modeling are, however, quite instructive because the results add to our fundamental understanding of the process.

Model Results for a Collection of Parallel Pores

As noted in earlier sections, the large pores are the ones that dominate the acidizing process when the reaction is rapid. Since the largest pores are the most difficult to characterize by mercury injection, this method normally used to characterize the pore size distribution is not very informative as a starting point for modeling matrix acidizing. Limestones having very similar pore size distributions can, unfortunately, exhibit quite different responses to acid treatments. This fact leads us to conclude that detailed modeling based on specific pore size distributions will not likely be productive. Hung [16] has suggested that microscopic examinations of thin sections may be of greater utility. Optical imaging techniques may be helpful in automating this analysis [17]. Pore sizes are measured in terms of Ferret's diameter, which is the maximum spacing between tangents drawn parallel to a particular feature. The Ferret diameter has been arbitrarily considered to be a pore diameter by Hung [16] who has studied the acidizing of a collection of pores. Table 17.2 shows an example. The pore count refers to the number of pores in that diameter group found per unit area of thin section as revealed by a microscopic analysis. This particular thin section had pores ranging in size from 2 to 2000 μm in diameter.

Hung has modeled the simultaneous acidization of the collection of pores listed in Table 17.2. His calculation allowed acid to flow preferentially into the

TABLE 17.2 Data of a Carbonate Pore Size Distribution [16]

Group (no.)	Pore Radius (μm)	Pore Number (counts)
1	1	54,000
2	2	32,500
3	3	25,000
4	4	16,800
5	6	7,100
6	8	4,800
7	10	4,200
8	15	1,600
9	20	700
10	28	540
11	38	700
12	50	500
13	72	180
14	104	91
15	149	67
16	186	58
17	215	43
18	267	17
19	332	9
20	450	5
21	578	2
22	1029	1

larger pores and he took into account both fluid-loss and diffusion-limited reaction rates. Depending on the amount of acid entering into a given pore, both its area and its length were increased.

Using this computer simulation, Hung showed that the growth of pores with initial radii less than 30 μm was slight as compared to the growth of the larger ones. For example, in one case studied, at a time when the length of the largest pore had increased to 5 cm, pores with radii less than 40 μm had not yet attained a length of 1 cm. Decreasing the fluid loss or increasing the injection rate tended to increase the disparity in growth rate and the larger pores played an even more dominant role.

These results emphasize the difficulty that one encounters when attempting to develop a deterministic model of those processes which take place during matrix acidization with fast reacting acids. It does not appear that any method which attempts to model all of the processes occurring will lead to useful design procedures. Their primary value will continue to be the fundamental understanding that they provide. This is, of course, significant because the processes involved are extremely complicated and highly interrelated. The effect of changing one of the parameters cannot be understood without having a model.

It appears that some sort of stochastic model, one which incorporates characteristics of a large group of acid treatments without delving into the specific details of the wormhole geometry, will be of the greatest utility in so far as designing an acid treatment is concerned. Such an approach is described in the next section.

17.4 FRACTAL DESCRIPTION OF WORMHOLES

Fractals

Figures 17.1 and 17.3 show wormholes and their stochastic geometry. Daccord and Lenormand [18] have demonstrated that these structures are fractals and based on this characterization, design procedures for acid treatments in carbonates can be developed [15]. To appreciate this approach, let us make a few comments about fractals [3].

We all know that a smooth curve in space, such as the one shown by Fig. 17.7(a), is unidimensional. It can be represented parametrically by a single parameter. A surface, on the other hand, is described by two parameters and is, therefore, bidimensional. There are some curves which describe natural phenomena that have an extremely ragged appearance, such as the one shown in Fig. 17.7(b). This curve might be thought of as being the trajectory of a particle undergoing Brownian motion. Note that as the curve becomes increasingly irregular it takes on the appearance of a surface, even though we know that the trajectory of the center of mass of a particle is in fact still a curve and not a surface at all. One is, however, tempted to suggest that because of the very ragged appearance, this curve has a dimensionality greater than unity. It is not quite a surface, but it is difficult to characterize it as a curve.

The question then arises as to how to characterize the motion of a Brownian

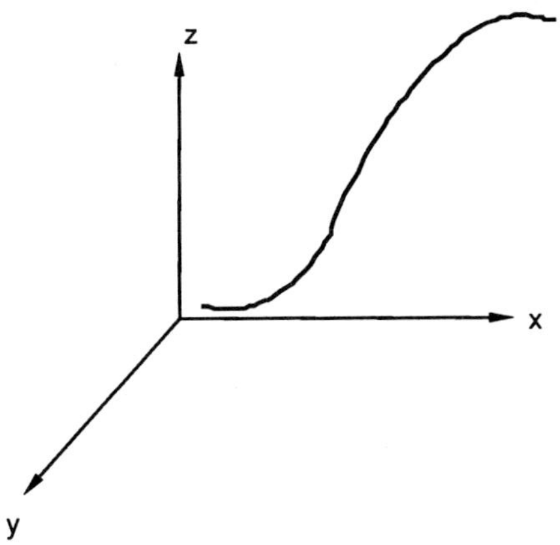

(a) CURVE

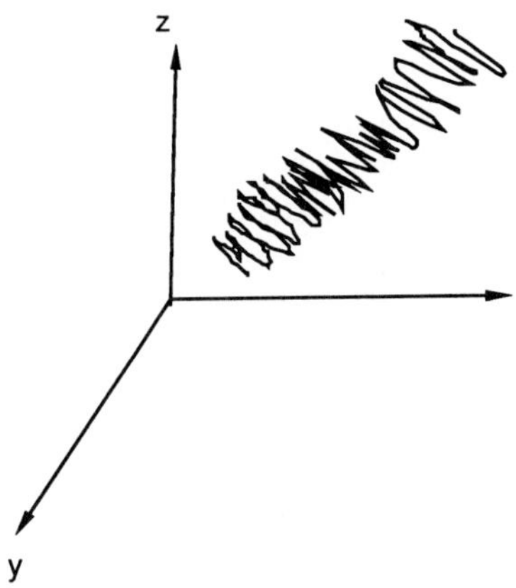

(b) IRREGULAR CURVE

Figure 17.7 Two unidimensional curves in space. **(a)** Smooth curve. **(b)** Irregular curve. [*Note:* **(b)** has the appearance of a surface because of its highly irregular and discontinous trajectory.]

particle, or in our case, the geometry of wormholes. One property that some of these irregular geometries exhibit is selfsimilarity. Some irregular curves can be usefully described based on this property. Selfsimilarity can be thought of as follows. Suppose that you take a ruler that is 1 m in length and measure the length of a segment of the curve shown in Fig. 17.7(b). Since the meter stick is rather long, many of the small (compared to the length of the meter stick) undulations of the curve are not sampled. To remedy this, let us take a smaller ruler—one having a length of 1 cm. Remeasurement of the same segment of curve will reveal a curve length different from that found with the meter stick, because now our measurement scale is smaller. If we continue to reduce the size ruler that is used to measure the length of the line, then over some range of ruler lengths (d) one might find the curve length L_{curve} to be given by

$$L_{curve} \approx d^{n_f}$$

provided we restrict d to a certain range of lengths. Over the range of ruler lengths for which this equation applies, the system is selfsimilar and is called a *fractal*. The parameter n_f defines the fractal "dimensionality" of the system.

Some wormholes are fractals. This behavior has been studied extensively by Daccord and his colleagues [9, 18]. Several techniques have been used to establish the fractal dimension. Many of the techniques used are restricted to analyzing two-dimensional wormholes, such as those depicted in Fig. 17.3. One interesting approach is to machine digitize a photograph of the wormholes using very fine spacing (Daccord et al. [19] used a lattice having 2^{16} points!). About every lattice, one forms an imaginary $L \times L$ square box. The number, $N(L)$, of wormhole points in this box is then counted and averaged over all possible lattice points as center. A log–log plot of $N(L)$ versus L has a slope n_f.

Using this technique and others, it has been determined that for wormholes $n_f = 1.6 \pm 0.1$. This is an important result for it establishes a general (so far as can be ascertained from both physical and computer experiments) characteristic of wormholes which applies within a certain range of acid injection rates, chemical reaction rates, and initial pore size distributions. Daccord et al. [15] believe that the fractal behavior is representative of the conditions normally encountered in matrix acid treatments using HCl in carbonate formations. There are, however, limitations and not all dissolution processes produce fractal wormholes.

The limitations on flow rates that bound fractal behavior are readily understood. If the flow is very slow, then the acid will simply dissolve rock at the face of the wellbore and only spent acid will penetrate into the rock. This very slow flow rate can be thought to create a definite front, the fluid/rock interface, which slowly recedes as acid dissolves rock. If X is the dissolving power of the acid, then iX must be the volume of rock dissolved per unit of time. This dissolution results in an increase in wellbore radius given by the simple volume balance equation

$$(1 - \phi)2\pi rh \frac{dr}{dt} = iX$$

or if v_c is the velocity of the front, then

$$v_c = \frac{iX}{2(1 - \phi)\pi rh} \tag{17.26}$$

Very slow injection rates result in the dissolution of rock at the wellbore surface. In this case wormholes do not form and hence cannot be fractals.

If the flow rate is extremely high, then the acid has little time to react and the invasion will be more or less homogeneous. Wormholes will not form. This can be made mathematically precise by examining Eq. (17.7) for the case of large velocities. As $\bar{v} \longrightarrow \infty$, then since $K_M \longrightarrow \infty$, one finds the following result:

$$\frac{dA}{dt} = 2\sqrt{\pi}\, XE_f A^{1/2} \tag{17.27}$$

Comparing this result with Eq. (17.1), it is observed that $n = \frac{1}{2}$ and wormholes do not form.

Thus at high velocities the dissolution becomes reaction controlled, the acid front moves more or less uniformly through porous media, and wormholes do not form. For field operations, the injection rate is controlled by the fracture gradient and the upper limit restricting wormhole formation is probably never reached. For matrix acidizing in carbonates, the range of flow rates is such that wormholes will generally form. Let us see if this fact can help us to design acid treatments.

Radial Flow Model

The concept that is applied to help design carbonate acid treatments is that there exists an effective radius (r_A), around the wellbore to which wormholes have penetrated after injection of a certain volume of acid. Within this effective radius, the pressure drop can be neglected because the wormholes are large conductive tubes. Thus, for incompressible flow into the formation from the wellbore

$$p_w - p_R = \frac{\mu i}{2\pi kh} \ln \frac{r_A(t)}{r_e} \tag{17.28}$$

where $r_A(t)$ is the depth of wormhole penetration into the formation. Define $v_{WH} = dr_A/dt$ and consider the dimensionless velocity $v_D = v_{WH}/v_c$. Clearly it is desirable to maximize v_{WH} since this effectively creates the largest r_A for a given volume of acid thereby maximizing the stimulation ratio.

Daccord et al. [15] claim that based on the fractal dimension and experiments at various injection rates

$$v_D = b r_A^{(2-n_f)} N_{Pe}^{-1/3} \tag{17.29}$$

with

$$N_{Pe} = \frac{i}{D_A h}$$

and $b = 1.7 \times 10^4 \text{ m}^{n_f-2}$. The penetration distance achieved by a given volume of acid is found by integrating Eq. (17.29).

$$r_A^{n_f} - r_w^{n_f} = n_f b \frac{V_A X}{2(1 - \phi)\pi h} N_{Pe}^{-1/3} \tag{17.30}$$

This is our final design equation. It shows that the fractal dimension n_f is an important quantity in determining the wormhole penetration distance and that high injection rates are undesirable because increasing the Peclet number (N_{Pe}) reduces the penetration distance; however, injection rates that are too slow ($N_{Pe} < 1$) will not yield wormholes. Thus Eq. (17.30) only applies whenever $N_{Pe} > 1$. Equation (17.30) will also fail at very high injection rates (large Peclet numbers) but according to Daccord et al. [15], this limit is of no practical concern.

Equation (17.30) has been derived in a slightly different form by Daccord et al. [15] in which r_w was assumed to be small in comparison to r_A. This is generally a good approximation.

Since r_A is the radius over which there is essentially no pressure drop, then a stimulation ratio can be defined as

$$\frac{J_s}{J_d} = \frac{\ln\left(\frac{r'_w}{r_e}\right) - \frac{3}{4}}{\ln\left(\frac{r_A}{r_e}\right) - \frac{3}{4}} \tag{17.31}$$

where r'_w is the effective wellbore radius including the skin factor prior to stimulation.

The optimum acid treatment found using Eq. (17.30) presents an interesting contrast to that obtained based on Eq. (17.23), which shows that wormhole lengths grow in proportion to the injection rate, and thus a given volume of acid is best applied at the highest possible injection rates. One the other hand, Eq. (17.30) suggests that moderate or even small injection rates are preferable. The difference between the two is fundamental.

Laboratory core flooding experiments in dolomite (Fig. 17.6) support the trends predicted by Eq. (17.30), whereas those in Indiana limestones (Fig. 17.5) appear to support Eq. (17.23). If the difference between these two trends is indeed a function of the reaction rate as is suspected [12], then Eq. (17.23) is likely to be the most useful at reservoir conditions since the reaction rates are increased as the temperature increases.

Clearly, further experimentation is required to ascertain which of the equations best represents the conditions within a reservoir. At the present time it is not possible to reject one approach in favor of the other. Therefore, both will be used in the examples to follow.

17.5 DESIGN OF MATRIX ACIDIZATION OF CARBONATES

The design of a matrix acid treatment for a carbonate formation consists of specifying acid type and volume, the injection rate and pressure that can be used without fracturing the formation. Two different design approaches will be used. These two different methods yield entirely different designs. It is not known which is the best one.

Example 17.3 Design of a Matrix Treatment

Design a matrix treatment for the formation whose relevant properties are given in Table 17.3. It is estimated that the high skin factor found by well tests results from formation damage that extends no more than a few meters about the wellbore. The treatment should extend well beyond this damaged zone. Specify the total acid volume to be used and the rate at which it should be applied.

Solution The maximum injection rate is determined by the fracture pressure and Eq. (13.2)

$$p_{wf} - p_R = \frac{i\mu_f}{2\pi kh}\left[\ln\frac{r_e}{r'_w} - \frac{3}{4}\right]$$

where $\quad r'_w = r_w \exp(-S) = 7.6 \times 10^{-2} \exp(-10) = 3.45 \times 10^{-6}$ m

TABLE 17.3 Well and Treatment Data Supporting Example 17.3

Well Property	Symbol	Value
Estimated skin factor	S	10
Formation depth	D	2300 m
Fracture gradient	FG	17 kPa/m
Formation permeability	k	8.93×10^{-15} m^2
Formation porosity	ϕ	0.12
Perforated interval	h	7 m
Perforation density	—	6 shots/m
Reservoir pressure	p_R	2.12×10^7 Pa
Wellbore radius	r_w	7.6×10^{-2} m
Drainage radius	r_e	200 m
Oil viscosity	μ_f	2×10^{-3} kg/m-sec
Oil compressibility	κ_f	1.45×10^{-6} kPa^{-1}
Formation temperature	T	175°F
Treatment Property	**Symbol**	**Value**
Acid type	—	15 wt% HCl
Dissolving power	X	0.082
Acid diffusivity	D_A	7×10^{-9} m^2/sec

The maximum allowable bottomhole well pressure is

$$p_{wf} = (FG)(D) = (17{,}000)(2300) = 3.91 \times 10^7 \text{ Pa}$$

In this case to ensure that the formation breakdown pressure is not exceeded, the bottomhole well-flowing pressure will be limited to 3.6×10^7 Pa. Thus

$$i_{max} = \frac{2\pi(3.6 \times 10^7 - 2.12 \times 10^7)(7)(8.93 \times 10^{-15})}{2 \times 10^{-3}} \left[\ln\left(\frac{200}{3.45 \times 10^{-6}}\right) - \frac{3}{4} \right]^{-1}$$

$$= 1.7 \times 10^{-4} \text{ m}^3/\text{sec}$$

This value is the maximum rate permitted. Further calculations must be made to ensure that the surface pressure corresponding to this rate can be tolerated. If not, the maximum rate must be further reduced. As the treatment progresses, it may be possible to increase the rate without exceeding the fracture pressure.

In this approach the fracture gradient will not represent a limitation on the process since slow rather than rapid injection rates are preferred. In fact, a design method based on Eq. (17.30) would yield optimum injection rates so slow that the total treatment time would be months. We therefore begin our design by specifying a reasonable treatment time. In this case, it is thought that injecting acid over a period of 4 hr is not unreasonable since the well is severely damaged and the injection rate, at least the initial one, is greatly restricted.

Consider an injection rate of 1×10^{-4} m^3/sec

Then
$$N_{Pe} = \frac{i}{hD_A} = \frac{1 \times 10^{-4}}{7(7 \times 10^{-9})} = 2.04 \times 10^3$$

During the 4 hr an acid volume of 1.44 m^3 would be injected. Based on this number, Eq. (17.30) yields

$$r_A^{1.67} - (7.6 \times 10^{-2})^{1.67} = (1.7 \times 10^4)(1.67)\frac{1.44(0.082)}{2\pi(1 - 0.12)}\left(\frac{1}{2.04 \times 10^5}\right)^{1/3}$$

or the acid penetration distance is

$$r_A = 3.1 \text{ m}$$

The stimulation ratio is

$$\frac{J_s}{J_d} = \frac{\ln\left(\dfrac{200}{3.45 \times 10^{-6}}\right) - \dfrac{3}{4}}{\ln\left(\dfrac{200}{3.1}\right) - \dfrac{3}{4}} = 5.0$$

This calculation can be repeated for several different injection rates and the results shown in Table 17.4 obtained.

Inspecting this table we see that the highest injection rate produces the greatest stimulation ratio. As the injection rate is decreased, according to this approach a more efficient use of the acid is realized. Smaller acid volumes produce acceptable stimulation ratios. It is clear that the optimum design of an acid treatment will require an economic analysis, but an acceptable treatment would appear to be a total acid volume of 1.44 m^3 injected at a rate of 10^{-4} m^3/sec. This volume would be expected to remove most of the damage since wormholes penetrate a radius of 3.1 m about the wellbore.

The final job specification would require corrosion inhibitors, diverting agents, and other additives (e.g., surfactants) that may be thought needed.

The acid in the tubing should be flushed into the formation with water containing potassium chloride, and the well can be immediately returned to production. Unless an organic acid is used together with hydrochloric or instead of hydrochloric, complete acid reaction is expected.

TABLE 17.4 Acidization Response Based on a 2-hr Treatment Time

i (m^3/sec)	V_A (m^3)	N_{Pe}	r_A (m)	$\frac{J_s}{J_d}$
1.5×10^{-4}	2.2	3×10^3	3.7	5.3
1.0×10^{-4}	1.4	2×10^3	3.1	5.0
5.0×10^{-5}	0.7	1×10^3	2.4	4.7
2.5×10^{-5}	0.4	5×10^2	1.8	4.3

Example 17.4 Design of a Matrix Treatment (Second Approach)

Determine the wormhole geometry that would be expected if 2.4 m^3 of 15 wt% HCl is applied at a rate near the maximum rate. Assume that only one wormhole forms at each perforation.

Solution Assuming that each perforation receives an equal proportion of the acid and that there is only one wormhole created at each perforation, then the dimensions of the wormhole can be estimated by considering that they are limited by fluid loss. This calculation is intended to be instructive, but is in fact unreliable for several reasons. There is no assurance that all perforations will receive equal volumes of acid. The use of diverting agents, discussed in Chapter 18, will help to distribute the acid evenly. Furthermore, although laboratory experiments tend to indicate the formation of a single wormhole, there is no evidence that this is true when acidizing an essentially infinite mass of rock through a perforation. Finally, the role of diffusion, neglected in the fluid-loss calculation, may well be important. Hung [13] has included this in his computer simulation, but because of the uncertainties in the process such a detailed calculation will not be helpful at this time.

If the acid enters all perforations then

$$q_0 = \frac{1.7 \times 10^{-4}}{7(6)} = 4.1 \times 10^{-6} \text{ m}^3/\text{sec}$$

and

$$C_R = \sqrt{\frac{\pi^2(0.12)(4.93 \times 10^{-15})(1.45 \times 10^{-9})}{2 \times 10^{-3}}(3.6 \times 10^7 - 2.12 \times 10^7)}$$

$$= 1.29 \times 10^{-3} \text{ m/sec}^{1/2}$$

Since in most practical cases $R(t) \gg R_p$, Eq. (17.24) yields

$$R = \frac{2\pi(0.082)(1.29 \times 10^{-3})[4(3600)]}{0.88} = 9.1 \times 10^{-2} \text{ m}$$

and

$$l = \frac{(4.05 \times 10^{-6})[4(3600)]^{1/2}}{2\pi^2(9.1 \times 10^{-2})(1.29 \times 10^{-3})} = 0.21 \text{ m}$$

Thus, the wormholes, one at each perforation, resemble a pipe about 9 cm in radius and 21 cm in length. This large radius, as compared to the wormhole length, is a result of the low acid injection rate. To reduce the radius and extend the length, the injection rate must be increased. Four hours to inject a rather small acid volume appears excessive. However, to increase the rate it will be necessary to fracture initially or to increase it gradually as damage is removed. According to this design approach, in contrast to Example 17.3, only near-wellbore damage will be removed by the treatment.

It may be thought that fluid-loss control additives in the acid solution can help extend the wormhole length, but because of the extremely low injection rate, this approach is unlikely to be unproductive. Furthermore, if multiple wormholes form at each perforation rather than the single one as assumed in the calculation, the penetration depth of each wormhole will be further restricted.

The final job specification requires the injection of 2.4 m^3 of 15 wt% HCl at a rate of 1.7×10^{-4} m^3/sec. Corrosion inhibitors, diverting agents, and other additives specific to the formation of interest (see Chap. 18) should be considered.

The reader can compare the results of the two different design approaches by examining Examples 17.3 and 17.4. There are no data available which can be used to ascertain the best design. Based on the results shown by Fig. 17.5, reasonable assumptions are imposed in the development of Eqs. (17.23) and (17.24). High rates are best. Moreover, the benefits that accrue from rapidly injecting acid into a naturally fractured reservoir and at formation temperatures, are significant. Thus the design presented in Example 17.3 would not appear to be the best choice. This assertion may change when additional data become available.

If the carbonate reservoir is vugular or highly fractured, then neither of the designs will represent the observed results. Generally when treating a number of wells in the same field, one should experiment to optimize the treatment.

17.6 NOVEL MATRIX ACID TREATMENTS FOR CARBONATES

Because hydrochloric acid reacts rapidly when contacted with either calcite or dolomite, a number of patents have been issued disclosing methods for reducing the reaction rate so as to reduce the tendency for wormhole formation. The proposed techniques can be divided into two general classifications:

1. One classification includes techniques that involve the injection of chemicals (not acids) that react slowly, but at a controlled rate to produce an acid. The acid product will then react rapidly with carbonates. These processes limit the reaction rate by producing in-situ acid at a controlled rate. They may have the further advantage of reducing corrosion problems whenever the chemicals from which the acid is derived are less corrosive than acid.

2. A second classification is represented by methods for retarding the effective acid reaction rate. This can be accomplished by reducing the rate of acid diffusion to the pore walls or by coating the rock surface with a chemical that slows the reaction of the acid with the rock.

In-situ Acid Formation

Hydrochloric acid. Halogen derivatives of hydrocarbons may in some circumstances slowly hydrolyze in the presence of water to yield as products an alcohol and an inorganic acid. Many of these compounds are liquid at room temperature and, therefore, one might imagine a process whereby HCl is produced at a controlled rate by injecting a mixture of water and the halogen. There is, however, one difficulty; namely, the solubility of halogen compounds in water is generally quite small—too small to produce an in-situ acid amount capable of dissolving enough rock to increase productivity.

Dilgren [20, 21] has proposed the use of a mutual solvent to enhance the miscibility of the halogen compound and water. One example is a solution of allyl chloride and mixed with a sufficient amount of isopropyl alcohol to form a single phase. The reaction is

$$\underset{\text{allyl chloride}}{C_3H_5Cl} + H_2O \rightleftharpoons \underset{\text{allyl alcohol}}{C_3H_5OH} + HCl \tag{17.32}$$

The reaction time can be varied widely by changing the organic portion of the halogen (normal propylchloride and tertiary butylchloride are also possible halogens according to Dilgren [20, 21], the water-to-alcohol ratio, the type of alcohol used as the mutual solvent, and the temperature.

Dilgren and Newman [22] modified this process to eliminate the expensive mutual solvent. They proposed that the water and the halogen be blended with a suitable surfactant to form a stable water external emulsion. This water external emulsion then is injected into the formation. The water and the organic phases are intimately mixed within the pore structure and react at a controlled rate to produce HCl.

Organic acids. Perrine [23] has patented a concept in which a selected organic compound capable of producing an organic acid upon contacting water is dissolved in a hydrocarbon phase. The hydrocarbon solution is injected into the carbonate formation, whereupon the organic compound diffuses from the hydrocarbon phase into the immobile water phase that remains in intimate contact with the hydrocarbon phase. One system proposed is a liquid hydrocarbon obtained by the extraction of a petroleum fraction with a polar solvent such as sulfur dioxide and mixed with acetic anhydride. Upon contacting water, acetic anhydride reacts to produce acetic acid. The reaction of acetic acid with carbonates was discussed in Chapter 3.

Weak acids. Weak organic acids which react more slowly than inorganic acids have been proposed for use in matrix acidizing of carbonate formations (see Chapter 14 for a discussion of this point). Fatt and Chittum [24] proposed using single phase solutions of organic acids with hydrocarbons for acidizing, Bombardieri and Martin [25] suggested using high concentrations of acetic acid, and Gidley [26] has claimed the use of a concentrated propionic acid which reacts with calcium carbonate to form an insoluble calcium propionate that subsequently dissolves in a salt-water afterflush.

All of these organic acid formulation have a relatively small dissolving power, especially because of the thermodynamic limits imposed on their reaction (see Chap. 3), and are not frequently used.

Reduced acid diffusivity. Hoefner et al. [27] have proposed an acid-in-oil microemulsion [28] for matrix acidizing. Since the acid is present in the form of suspended submicroscopic drops, acid diffusion occurs by Brownian motion of the drops rather than by diffusion of the molecules. Diffusivities of drops can range to more than one order of magnitude smaller than molecular diffusion coefficients [see Eq. (4.30)]. The influence of the diffusion coefficient on the creation of wormholes is observed by examining Eq. (17.11). Decreasing D_A substantially will reduce the value of A that satisfies the inequality

$$\frac{\pi D_A l}{\epsilon A^2} \gg 1$$

thus effectively limiting the growth of larger wormholes. Here one would expect wormholes to form, but to have a smaller characteristic size. For this reason, acid breakthrough in core flood tests may occur with smaller acid volumes as shown by Fig. 17-5 (see curve labelled butanol microemulsion).

Because the acid is present in small drops, microemulsions will necessarily have small dissolving power and a treatment will generally require greater fluid volumes than hydrochloric acid. This disadvantage may, however, be offset by the effect of greater acid penetration distances.

REFERENCES

17.1. Schechter, R. S., and Gidley, J. L., *Amer. Inst. Chem. Eng. J., 15* (1969) 339.

17.2. Hoefner, M. L., and Fogler, H. S., "Reaction Rate vs. Transport Limited Dissolution During Carbonate Acidizing: Application of Network Model," SPE 15573, presented at the 61st Fall Technical Conference and Exhibition of the Society of Petroleum Engineers, New Orleans, Louisiana, 1986.

17.3. Mandelbrot, B. B., *Fractal Geometry of Nature*, New York: W. H. Freeman and Co., 1977.

17.4. Witten, T. A., Sander, L. M., *Phys. Rev. Lett., 47* (1981) 1400.

17.5. Daccord, G., "Dissolution, Evaporations and Etchings," in *The Fractual Approach to Heterogeneous Chemistry*, D. Avnir (ed.), New York: John Wiley, 1989.

17.6. Nierode, D. E., and Williams, B. B., *Soc. Pet. Eng. J., 11* (1971) 406.

17.7. Levich, V. G., *Physicochemical Hydrodynamics*, Englewood Cliffs: Prentice Hall, 1962.

17.8. Williams, B. B., Gidley, J. L., Schechter, R. A., *Acidizing Fundamentals*, Richardson: Society of Petroleum Engineers (1979).

17.9. Daccord, G., *Phy. Rev. Lett., 58* (1987), 479.

17.10. Guin, J. A., and Schechter, R. S., *Soc. Pet. Eng. J., 11* (1971) 390.

17.11. Crank, J., *The Mathematics of Diffusion*, 2nd Ed., London: Clarendon Press, 1975.

17.12. Hoefner, M. L., and Fogler, H. S., *Amer. Inst. Chem. Eng. J., 34* (1988) 45.

17.13. Guin, J. A., Schechter, R. S., and Silberberg, I. H., *Indus. Eng. Chem. Fund., 10* (1971) 50.

17.14. Glover, M. C., and Guin, J. A., *Amer. Inst. Chem. Eng. J., 19* (1973) 1190.

17.15. Daccord, G., Touboulo, E., and Lenormand, R., *Soc. Pet. Eng. Prod. Eng.* (1989), 623.

17.16. Hung, K. C. M., "Modeling of Wormhole Behavior in Carbonate Acidizing," PhD dissertation, The University of Texas at Austin, 1987.

17.17. Ruzyla, K., *Soc. Pet. Eng. Form. Eval., 1* (1986) 389.

17.18. Daccord, G. and Lenormand, R., *Nature, 325* (1987) 41.

17.19. Daccord, G., Nittman, J., and Stanley, H. E., *Phys. Rev. Lett., 56* (1986) 336.

17.20. Dilgren, R. E., "Acidizing Oil Formation," U.S. Patent No. 3,215,199, November 2, 1965.

17.21. Dilgren, R.E., "Acidizing Oil Formations," U.S. Patent No. 3,297,090, January 10, 1967.

17.22. Dilgren, R. E., and Newman, F. M., "Acidizing Oil Formations," U.S. Patent No. 3,307,630, March 7, 1967.

17.23. Perrine, R. L., "Method of Acidizing Petroliferous Formations," U.S. Patent No. 2,863,832, December 9, 1958.

17.24. Fatt, I., and Chittum, J. F., "Method of Treating Wells with Acid," U.S. Patent No. 2,910,436, October 27, 1959.

17.25. Bombardieri, C. C., and Martin, T. H., "Acid Treating Process," U.S. Patent No. 3,251,415, May 17, 1966.

17.26. Gidley, J. L., "Method for Treating Carbonate Formations," U.S. Patent No. 3,441,085, April 19, 1969.
17.27. Hoefner, M. L., Fogler, H. S., Stenius, P., and Sjöblom, J., *J. Pet. Tech., 39* (1987) 203.
17.28. Bourrel, M., and Schechter, R. S., *Microemulsions and Related Systems*, New York: Marcel Dekker, 1988.

PROBLEMS

****17.1.** Suppose that a pore can be modeled as a narrow slit between parallel walls. The pore is depicted by Fig. P17.1. Furthermore, consider the fluid velocity in the slit to be constant and equal to $\bar{v}_x$. If the rate of erosion of the walls of the slit by the acid is a slow process because the acid inlet concentration (c_0) is small, then to a good approximation the acid concentration in the slit is described by the equation (see also Eq. (14.25)):

$$\bar{v}_x \frac{\partial c}{\partial x} = D_A \frac{\partial^2 c}{\partial y^2}$$

If the acid reaction rate is fast then

$$c = 0 \qquad \left(\text{at } y = \pm \frac{w}{2}\right)$$

(a) Prove that the acid concentration in the slit can be expressed as follows:

$$\frac{c_0 - c}{c_0} = 1 - \frac{4}{\pi} \sum_{n=0}^{\infty} \frac{(-1)^n}{2n + 1} \exp\left[-\frac{D_A(2n + 1)^2\pi^2 x}{w^2 \bar{v}_x}\right] \cos\left(\frac{2n + 1}{w} \pi y\right)$$

(Show that the differential equation and the boundary conditions are satisfied.)

(b) Find the *average* rate of increase of the pore area ($A = wl$) and show that it can be written as

$$\frac{dA}{dt} = \frac{8\bar{v}_x XA}{\pi l^2} \sum_{n=0}^{\infty} \frac{1}{(2n + 1)^2} \left\{1 - \exp\left[-\frac{D_A(2n + 1)^2\pi^2 l}{w^2 \bar{v}_x}\right]\right\}$$

Compare this result with the approximation defined by Eq. (17.9). Comment on the differences.

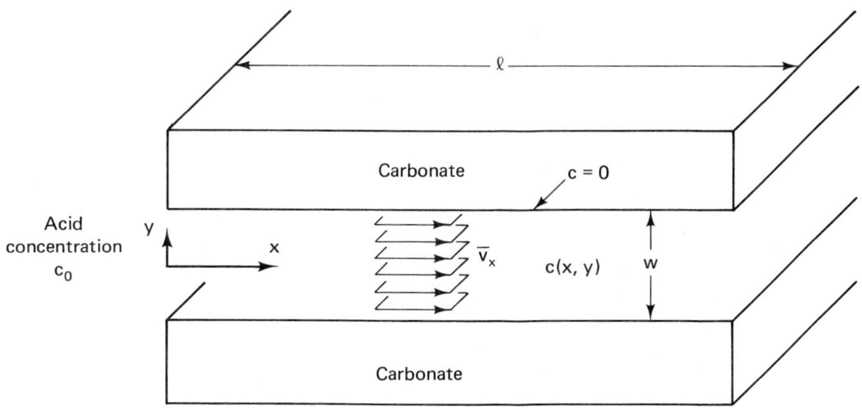

Figure P17.1 Sketch depicting a long, narrow pore.

***17.2.** Consider two pores, one having an area of 10 $(\mu m)^2$ and the other an area of 20 $(\mu m)^2$. Map the growth of these two pores as a function of t^* for two cases, where $t\,\Psi = t^*$. First use $n = \frac{1}{2}$ [see Eq. (17.2)] and then $n = -\frac{1}{2}$. Comment on the difference between the trajectories. In particular, indicate which pore growth function is most likely to lead to wormhole formation.

****17.3.** Equation (17.1) shows the pore growth function to be given as a function of two parameters: Ψ and n. For example, pores having an initial area A_0 will at time t have an area A. Thus for $n = \frac{1}{2}$ and $t^* = \Psi t$

$$A = \left(A_0^{1/2} + \frac{t^*}{2}\right)^2$$

relates A to A_0. Using the concept that the number of pores having an area A_0 at $t^* = 0$ will have an area A at t^*, plot the pore size distribution at $t^* = 0.1\ \mu m$ if the initial distribution is given by

$$\eta(A,0) = \beta^2 A \exp(-\beta A)$$

To make a comparison be certain that $\eta(A, t^*)$ satisfies

$$\int_0^\infty \eta(A,t^*)\, dA = 1 \text{ at } t^* = 0.1\ \mu m$$

****17.4.** The rate of mass transfer from a fluid flowing through a circular tube to the surface of the tube depends on a mass transfer coefficient K_M, which for laminar flow and a fully developed concentration boundary layer (long tube) is approximated by

$$\frac{RK_M}{D_A} = 1.62\left(\frac{R\bar{v}_z}{D_A}\right)^{0.8}$$

[E. L. Cussler, *Diffusion*, p. 230, Cambridge University Press, Cambridge, England (1984)]. For an acid flowing through a circular tube of radius 5×10^{-2} cm at an average velocity of 0.1 cm/sec, plot both the average acid concentration and the acid concentration at the wall divided by the inlet concentration as a function of position along the wormhole. The total length of the wormhole is 2 cm. The forward reaction rate constant is 2.2×10^{-3} cm/sec (assume first-order reaction) and the acid diffusion coefficient is 1.4×10^{-5} cm^2/sec.

Decide whether the overall reaction is diffusion or reaction rate controlled.

***17.5.** Acid flows into a wormhole at a rate of 1.5×10^{-3} cm^3/sec. If $E_f = 2.5 \times 10^{-3}$ cm/sec for a first-order reaction and $D_A = 4 \times 10^{-5}$ cm^2/sec, find the wormhole radius for which the reaction with flow is mass transfer controlled. If the acid concentration at the wall is less than 10% of the bulk concentration, then the overall reaction rate is mass transfer controlled. The flow is laminar and the mass transfer coefficient is given by the expression used in Problem 17.4.

****17.6.** The reaction of hydrochloric acid with dolomite is not first order but is, in fact, represented by Eq. (14.15). For this reaction rate expression, show for acid flowing through a circular pore that the acid concentration at the wall (c_A^W) can be represented by the equation

$$\frac{(c_A^W)^{1-\alpha} - (c_{A,\text{inlet}}^W)^{1-\alpha}}{1-\alpha} + \frac{mE_f}{K_M}\ln\frac{c_A^W}{c_{A,\text{inlet}}^W} = -\frac{E_f}{R\bar{v}_z}\,z$$

neglecting both fluid loss and the enlargement of the pore radius. In this equation $c_{A,\text{inlet}}^W$ must be selected so that it satisfies the equation

$$c_A^0 = c_{A,\text{inlet}}^W + \frac{E_f}{K_M}(c_{A,\text{inlet}}^W)^m$$

For the conditions listed in Table P17.6 show the acid concentration at the wall and the average acid concentration as a function of axial position along the pore. (*Note:* The flow is laminar so that the correlation given for the mass transfer coefficient in Prob. 17.4 applies.) The diffusion coefficient for HCl in water may be approximated based on the results plotted in Fig. 14.2

TABLE P17.6

Property	Symbol	Value
Fluid velocity in the wormhole	$\bar{v}_z$	0.05 cm/sec
Wormhole length	l	3 cm
Wormhole radius	R	0.15 cm
Inlet acid concentration	c_A^0	4 kg moles/m^3
Temperature	T	50°C

***17.7.** It is desired to create a wormhole that is at least 20 cm in length in a limestone formation having a 12% porosity. If the volume per unit of time of 15 wt% HCl entering the pore is 5×10^{-3} cm^3/sec and if this rate is maintained for 20 min, what is the maximum fluid-loss coefficient allowed that will provide the desired wormhole penetration?

***17.8.** Design a matrix treatment for the formation whose relevant properties are listed in Tables P17.8A and P17.8B. Specify the total acid volume and the rate at which it should be applied. The total pumping time should be limited to 90 min.

TABLE P17.8A Well Data

Property	Symbol	Value
Estimated skin factor	S	10
Formation depth	D	1500 m
Fraction gradient	FG	17 kPa/m
Formation permeability	k	2.7×10^{-15} m^2
Formation porosity	ϕ	0.15
Perforated interval	h	8 m
Perforation density	—	14 shots/m
Reservoir pressure	p_R	2.6×10^7 Pa
Wellbore radius	r_w	0.08 m
Drainage radius	r_e	200 m
Oil viscosity	μ_f	3×10^{-3} kg/m-sec
Oil compressibility	κ_f	1.45×10^{-6} kPa^{-1}
Formation temperature	T	180°F

TABLE P17.8B Treatment Data

Property	Symbol	Value
Acid type	—	15 wt% HCl
Dissolving power	X	0.086
Acid diffusivity	D_A	6.5×10^{-9} m^2/sec

18
Acid Additives

Acid fracturing or matrix acidizing are both complex chemical processes. Invariably, it is necessary to include additional chemicals in the acid solution to help achieve the desired goals without deleterious side effects. These additives are intended (1) to divert acid flow from more permeable, undamaged zones to less permeable ones thereby allowing a more uniform treatment; (2) to prevent corrosion of wellbore tubular goods; (3) to form complexes with iron, to prevent its precipitation; (4) to reduce friction drop through tubular goods; and (5) to eliminate emulsions and sludging in certain highly asphaltic oils. This chapter describes the function of some of the additives and describes the materials used.

18.1 DIVERTING AGENTS FOR MATRIX ACIDIZING

To obtain maximum benefit from a matrix or fracture acidizing treatment, it is necessary to treat the total productive interval. When several sands are open to the wellbore or the section to be treated is massive, it may be necessary to divide the treatment into stages. For example, flow can often be diverted effectively using downhole equipment such as packers [1]. However, since using a workover rig to move packers can significantly increase job costs, techniques for separating the treatment into stages without using packers are often preferred. One method of placing the acid, which applies to wells completed in zones of limited thickness with relative few perforations, is the use of ball sealers.

In this method small balls designed to lodge in a perforation and seal it shut as long as the wellbore pressure exceeds the reservoir pressure are dropped into the well. These balls travel into those perforations which are accepting fluid, thereby shutting off the flow. The fluid is then diverted to other less active perforations. In recent years, the development of neutrally buoyant sealers has apparently greatly improved their effectiveness [2].

The use of diverting agents provides another alternative to mechanically isolating and treating specific zones one at a time. Diverting agents are small particles that can either be applied batch-wise between fluid stages to temporarily plug the zone just treated or added continuously to the injected fluids. In either case the same materials are used, the difference being in their mode of application [3].

Advantages of Diverting Agents

This important issue has been discussed by Pye et al. [4], Cooper and Bolland [3], and Taha et al. [5]. When there are a number of zones to be treated, the injected acid will preferentially enter the most permeable zones rather than the less permeable more highly damaged zones where it is most needed. If the flow into each zone of a multizone system is described by Eq. (13.2), then the ratio of flow into any two zones is

$$\frac{i_j}{i_m} = \left(\frac{k_j h_j}{k_m h_m}\right) \frac{\ln\left(\dfrac{r_e}{r'_{wm}}\right) - \dfrac{3}{4}}{\ln\left(\dfrac{r_e}{r'_{wj}}\right) - \dfrac{3}{4}} \tag{18.1}$$

where $$r'_{wj} = r_w \exp(-S_j)$$

This equation assumes that the layers are noncommunicating; however, since acid treatments generally involve the injection of relatively small volumes of liquid, the opportunity for crossflow is restricted and Eq. (18.1) is approximately valid even if the layers are in communication.

Example 18.1 Preferential Flow into Thief Zones

Consider the acidization of a formation composed of two zones each having the same thickness. One zone has a permeability of 4.93×10^{-15} m^2 (5 mD) and is highly damaged, $S = 20$. The second zone has the same permeability as the first but is undamaged. Estimate the total volume of acid that must be injected so that each zone receives at least a volume of acid, V_A^*. The wellbore radius is 0.1 m and the drainage radius is 200 m.

Solution Neglecting the damage removed by the acid as it enters Zone 1, Eq. (18.1) gives

$$\frac{V_1}{V_2} = \frac{i_1 t}{i_2 t} = \frac{\ln\left(\dfrac{r_e}{r_w}\right) - \dfrac{3}{4}}{\ln\left(\dfrac{r_e}{r'_{w1}}\right) - \dfrac{3}{4}}$$

Since $r'_{w1} = 0.1 \exp(-20) = 2.06 \times 10^{-10}$ m,

then the volume ratio is found as follows:

$$\frac{V_1}{V_2} = \frac{\ln\left(\dfrac{200}{0.1}\right) - \dfrac{3}{4}}{\ln\left(\dfrac{200}{2.06 \times 10^{-10}}\right) - \dfrac{3}{4}} = \frac{1}{3.9}$$

Thus when $V_1 = V_A^*$, $V_2 = 3.9\ V_A^*$ and a total acid volume of $4.9\ V_A^*$ is required to treat both zones.

Example 18.1 illustrates the difficulties inherent in treating multiple producing zones or a single thick producing zone. It is crucial to spend acid in the places where production will be most influenced. This is the purpose of diverting agents.

Materials for Acid Diversion

Diverting agents are designed to bridge across the formation pores and form a filter cake on the sand surface. Chemicals used for diversion during matrix acidizing include finely ground inert organic resins, solid organic acids, deformable solids, mixtures of waxes, oil-soluble polymers, acid-swellable polymers, and mixtures of inert solids such as silica flour and rock salt with a water-soluble polymer. Perhaps other solids have been used but not reported.

The primary properties a good diverting material should exhibit are as follows:

1. The material should be essentially insoluble in concentrated acids;
2. It should form a dense filter cake of low permeability on the sand face without penetrating into the formation;
3. It should not have a high density so that settling in the wellbore is a problem; and finally
4. It should be easily removed from the sand face during production.

The materials most often used with acid treatments of oil wells are finely ground organic resins. These materials are sometimes a high molecular weight hydrocarbon resin that is primarily aliphatic in nature. Such resins exhibit considerable oil solubility and should be removed during production.

The organic resins should consist of a mixture of large particles mixed together with smaller ones. This mixture will permit the formation of a low-permeability filter cake. Using small particles alone may permit substantial penetration into the formation, which is, of course, undesirable since the particles may then be difficult to remove following the treatment. On the other hand, particles large enough to bridge at the sand face may not provide a low-permeability filter cake. The compromise appears to be the use of a mixture of sizes. One reported material contained larger particles ranging in size from 37 to 44 μm with a mean of 40 μm mixed with small particles ranging in size from 5 to 20 μm with a mean average of 15 μm [6]. Others have also reported oil-soluble resins to be effective acid diverting agents [7, 8, 9].

The importance of the small particles is shown by the Blake–Kozeny equation

$$k_{\text{cake}} = B \frac{\phi_{\text{cake}}^3}{(1 - \phi_{\text{cake}})^2} \left(\frac{V_{\text{p}}}{S_{\text{p}}}\right)^2 \tag{18.2}$$

where values of B vary as a function of the shape and size of the particles. For uniform spheres $B = 1/150$ [10]. In this equation S_{p} and V_{p} can be considered to be the specific areas and the specific volumes of the particles, respectively. This ratio is clearly enhanced by using smaller rather than larger particles. Furthermore, the porosity of the cake (ϕ_{cake}) is another important factor that is reduced by using a binodal rather than a monodisperse distribution of diverting material particle sizes. Thus, effective diverting systems are most often bimodal in size.

A Simple Theoretical Model of Fluid Diversion

In this section a simplified model in which diverting agents are added to the acid continuously is considered. Since the equations are solvable, we can, based on these results, develop a better understanding of the behavior of diverting agents in more complex systems and can appreciate some of the difficulties involved in modeling them (e.g., see [11, 12]). The system considered is represented by Fig. 18.1. Two zones are shown, one having a permeability much larger than the other. For example, $k_1/k_2 \gg 1$. The volumetric injection rate is i, which is assumed constant. In more realistic calculations, it may not be possible to sustain a particular injection rate without exceeding the fracture pressure. The rate will have to be reduced as the diverting agent builds up on the sand surface, but this difficulty is ignored here. Denote the total volume of fluid that has entered Zone 1 at time t as $V_1(t)$. Similarly, $V_2(t)$ stands for that volume entering Zone 2. Since i is a constant injection rate, then

$$V_1(t) + V_2(t) = it \tag{18.3}$$

As the fluid containing diverting agent enters the formation, a filter cake of increasing thickness builds up on the surface of the sand. If the filter cake is assumed incompressible (not always a good assumption [8]), then the permeability of the filter cake can be considered to be a constant irrespective of the pressure difference $p_{\text{w}}(t) - p_{\text{s}}(t)$ where p_{w} is the pressure in the wellbore and p_{s} is the pressure at the interface between the filter cake and the original formation. Both pressures depend on time. Let us call k_{cake} the *filter-cake permeability*. It will be observed that for a diverting agent to be effective, k_{cake} must be much smaller than either k_1 or k_2.

The injection into Zone 1 then depends on the pressure drop across the filter cake as follows:

$$\frac{i_j}{h_j} = \frac{1}{h_j}\frac{dV_j}{dt} = \frac{2\pi r_{\text{w}} k_{\text{cake}}}{\mu}\left(\frac{p_{\text{w}} - p_{\text{s}}^{(j)}}{l_j}\right) \tag{18.4}$$

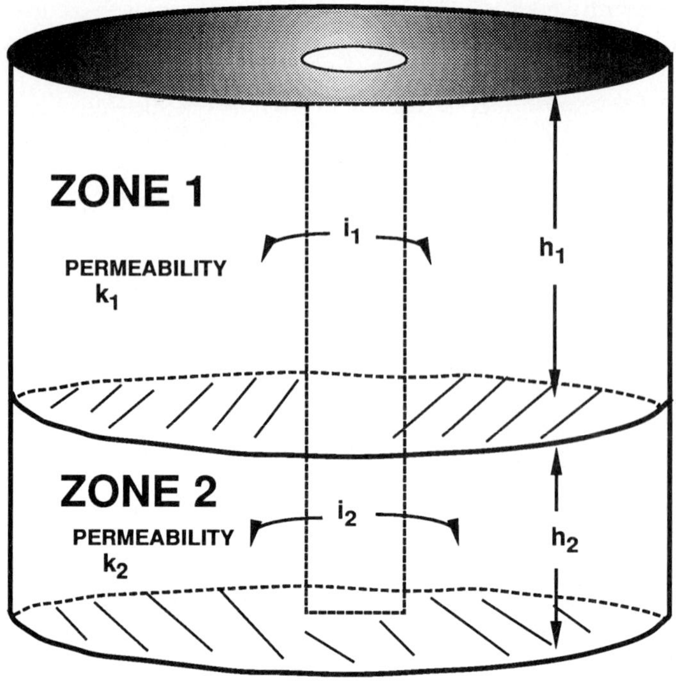

Figure 18.1 Sketch depicting acidization of a formation having two open zones.

where r_w is the wellbore radius and l_j is the thickness of the filter cake. The l_j must be proportional to the total volume of fluid that has been filtered. Thus

$$l_j = \frac{c_0 V_j}{(1 - \phi_{\text{cake}})(2\pi r_w h_j)} \qquad \text{for } j = 1, 2 \tag{18.5}$$

In this equation ϕ_{cake} is the porosity of the cake and c_0 is the volume of particulates per unit volume of solution excluding the diverting agent. We have now introduced the two most important design parameters: the concentration of particulates and the cake permeability. If the concentration c_0 is too large, then the flow in low- and high-permeability zones will be rapidly equalized but the pressure in the wellbore will also rapidly rise because the cake resistance in all zones will be large. If, on the other hand, c_0 is too small, then the diversion of fluid from the more permeable zones will not be effective.

Substituting Eq. (18.5) into Eq. (18.4) to eliminate l_j yields

$$\overline{V}_j \frac{d\overline{V}_j}{dt} = \frac{1}{2}\frac{d\overline{V}_j^2}{dt} = \beta(p_w - p_s^{(j)}) \tag{18.6}$$

where $$\beta = \frac{k_{\text{cake}}}{\mu}\frac{(2\pi r_w)^2}{c_0}(1 - \phi_{\text{cake}})$$

and $$\overline{V}_j = V_j/h_j$$

$\overline{V}_j$ is the treatment volume per unit of formation thickness. It is this value that we desire to equalize; that is, the goal of adding diverting agents is to cause $\overline{V}_1$ to equal $\overline{V}_2$ by the end of the treatment. Thus, it is convenient to consider $\overline{V}_j$ rather than V_j.

Equation (18.6) contains too many unknowns. There are two equations, but there are five unknowns. Equation (18.3) applies. Thus, two more equations are required. To establish the values of $p_s^{(i)}$, we use here the simple semisteady-state radial flow Eq. (13.2). Thus

$$\frac{d\overline{V}_j}{dt} = \frac{2\pi(p_s^{(j)} - p_R)\, k_j}{\mu}\left[\frac{1}{\ln\left(\dfrac{r_e}{r'_{wj}}\right) - \dfrac{3}{4}}\right] \tag{18.7}$$

where p_R is the reservoir pressure. On the right-hand side of Eq. (18.7), $p_s^{(j)}$ and r'_{wj} both depend on time; however, the dependence of r'_{wj} on time will be neglected in the analysis presented here. To include the time-dependent results of the skin factor requires a coupling of the movement of the acid front and the consequent removal of formation damage with the buildup of diverting agents on the formation face. Such comprehensive studies have been reported, but the methods used have involved numerical simulation [5, 8, 11, 12]. For simplicity, the influence of acid on the skin factor as the treatment progresses will be neglected here. This concept was discussed in Chapter 16.

Equation (18.7) can be solved for $p_s^{(j)}$ to yield

$$p_s^{(j)} = \frac{1}{\alpha_j}\frac{d\overline{V}_j}{dt} + p_R \tag{18.8}$$

where

$$\alpha_j = \frac{2\pi k_j}{\mu}\left[\frac{1}{\ln\left(\dfrac{r_e}{r'_{wj}}\right) - \dfrac{3}{4}}\right]$$

Substituting Eq. (18.8) into Eq. (18.6), the pressures $p_s^{(i)}$ are eliminated to give:

$$\frac{1}{2}\frac{d\overline{V}_j^2}{dt} = \beta(p_w - p_R) - \frac{\beta}{\alpha_j}\frac{d\overline{V}_j}{dt} \tag{18.9}$$

Subtract the equation for $j = 2$ from that for $j = 1$ and integrate. The pressures are eliminated to give

$$\overline{V}_1^2 - \overline{V}_2^2 = -2\beta\left(\frac{\overline{V}_1}{\alpha_1} - \frac{\overline{V}_2}{\alpha_2}\right) \tag{18.10}$$

From Eq. (18.3) we note

$$\overline{V}_2 = \frac{it}{h_2} - \frac{\overline{V}_1 h_1}{h_2} \tag{18.11}$$

which when substituted into Eq. (18.10) gives as the final result a quadratic equation for $\overline{V}_1$:

$$\left[1 - \left(\frac{h_1}{h_2}\right)^2\right]\overline{V}_1^2 + \left(\frac{2ith_1}{h_2^2} + \frac{2\beta}{\alpha_1} + \frac{2\beta h_1}{\alpha_2 h_2}\right)\overline{V}_1 - \left(\frac{i^2t^2}{h_2^2} + \frac{2\beta it}{\alpha_2 h_2}\right) = 0 \qquad (18.12)$$

$\overline{V}_2$ is determined from Eq. (18.10) once $\overline{V}_1$ is known.

Example 18.2 An Inadequate Diversion

To treat two zones of equal thickness, 0.25 m^3/m of fluid is used. It is desired that equal volumes of treating fluid enter each zone, even though the permeability of Zone 1 is 100 times that of Zone 2. A concentration of diverting agent 5×10^{-3} m^3/m^3 of solution is included as a part of the well treatment. Is this concentration sufficient? Take $(1 - \phi_{\text{cake}})k_{\text{cake}}/k_1 = 10^{-4}$, $r_e = 200$ m, and $r_w = 0.15$ m.

Solution From the information given

$$\frac{it}{2h} = 0.25$$

where h is the thickness of each zone.

Furthermore, from the definitions of β and α_j

$$\frac{2\beta}{\alpha_1} = \frac{4\pi r_w^2\,(1 - \phi_{\text{cake}})k_{\text{cake}}}{k_1 c_0}\left[\ln\left(\frac{r_e}{r_w}\right) - \frac{3}{4}\right]$$

or

$$\frac{2\beta}{\alpha_1} = \frac{4\pi(0.15)^2\,(10^{-4})}{5 \times 10^{-3}}\left[\ln\left(\frac{200}{0.15}\right) - \frac{3}{4}\right] = 3.64 \times 10^{-2}\ \text{m}^2$$

Similarly

$$\frac{2\beta}{\alpha_2} = 3.64\ \text{m}^2$$

Substituting into Eq. (18.11)

$$\{1.0 + 3.64 \times 10^{-2} + 3.64\}\overline{V}_1 - [0.25 + 3.64(0.5)] = 0$$

or

$$\overline{V}_1 = 0.443 \qquad \text{and} \qquad \overline{V}_2 = 0.057$$

Thus, about 85% of the injected fluid enters the more permeable zone. This result is a distinct improvement as compared to the distribution that would be obtained without using a diverting agent, but it is still not satisfactory. To further equalize the fluid distribution, one must either decrease k_{cake} or increase c_0.

Simultaneous Fluid Diversion and Acid Reactions

The simple model considered here neglects change in skin factor as a function of time. It is clear that as acid enters a zone, it creates permeability which tends to increase flow while the diverting agent tends to suppress flow. It is not, therefore, obvious that a design found to be effective when the influence of the acid in removing the damage is neglected will actually be effective.

Taha et al. [5] have simulated the acidization of a formation composed of two separate zones both damaged. For comparison, they first studied a vertically homogeneous reservoir; that is, both layers have precisely the same thickness and permeability of a damaged zone as well as undamaged permeability. There is, therefore, no need for diverting agents. Their simulation is complex. It takes into account the effect of acid reactions by simultaneously solving the acid balance equations defined in Chapter 15, allowing for the permeability of the damaged zone to be progressively increased as acid invades that zone and accounting for the growing filter-cake resistance.

For the vertically homogeneous case, a total acid volume of 1.86 m^3/m of fluid was required when injected at a rate (i/h) of 8.7×10^{-3} m^3/m-sec to remove a 15 cm damaged zone. The resulting stimulation ratio is 4.8, as shown in Table 18.1.

If the same formation is now considered to be heterogeneous with Zone 2, having a permeability ten times that of Zone 1 and if this formation is now treated with the same 1.86 m^3/m of acid volume, then as shown in Table 18.1, the stimulation ratio of the lower permeability zone has not been greatly improved and 91% of the production will be derived from the more permeable zone. However, even if the less permeable zone is highly stimulated, the contribution of the more permeable zone to the total production will not change substantially and neither will the total production. This example points to an important principle. Given a fixed amount of acid, it is of great importance to remove damage from the *most* permeable zones. These are after all, potentially the most productive zones. As shown in Table 18.1, even though 98% of the acid entered Zone 2, the stimulation ratio exceeds that of the base case where equal volumes of acid entered the two zones.

Taha [12] has considered even more complex problems and has shown that the benefits which accrue when diverting agents are applied can be substantial. This is not always the case, however. For example, the bottomhole treatment pressure is limited by the formation breakdown pressure. The presence of the filter cake adds an additional pressure drop, which will require restricting the acid injection rate so as not to exceed the breakdown pressure. A slower injection rate reduces the depth to which live acid penetrates (see Chap. 15). Thus damaged zones, which might be removed in the absence of diverting agents, may not be penetrated when diverting agents are present. Hence, great care should be exercised in the use of diverting agents. Too much can be detrimental to the treatment and they should, therefore, be used sparingly.

TABLE 18.1* Effect of Vertical Heterogeneity on Stimulation Results

	Zone 1 Permeability (mD)	Zone 2 Permeability (mD)	Zone 1 Stimulation Ratio	Zone 2 Stimulation Ratio	Overall Stimulation Ratio	** $\frac{V_1}{V_T}$
Base Case	10	10	4.89	4.89	4.89	0.50
Heterogeneous Case	10	100	1.25	5.56	5.18	0.02

* Each zone is damaged into a depth of 15 cm and the permeability in the damaged zone is $\frac{1}{10}$ of the permeability of the zones.

** Fraction of the total injected acid entering zone.

There are many variables that can be adjusted, such as the schedule of diverting agent injection, the concentration of diverting agent, the size distribution, the type of agent, and the injection rate. There is unfortunately as yet no proven method of optimization. Each particular case must be considered separately using an approach similar to that given by Taha [12] or by Doerler and Prouvost [8].

Cake Compressibility

In a study of organic resins as diverting agents, Hill and Galloway [11] reported the pressure drop across the face of the rock core as a function of time during a constant rate injection period. Equation (18.7) predicts a linear relationship. Figure 18.2 shows the observed response. Initially, $\Delta p_{\text{face}} = p_w - p_s$ increases slowly as the diverting agent distributes across the sand face. When the entire surface is covered, Δp_{face} increases rapidly and then rises linearly with volume injected. Thus once the cake has formed it is, under the particular test conditions, essentially imcompressible. This, however, is not always the case [8]. If a simulation of an acid treatment complete with diverting agents is contemplated, then tests must be performed under pressure differences which nearly approach those anticipated in the field.

Another issue to be addressed relates to the need for diverting agents to assure proper distribution of the preflush which, as described in Chapter 16, is crucial for satisfactory performance. Houchin et al. [6] have considered this question and concluded that a diverting agent should be applied in a nonreactive (3

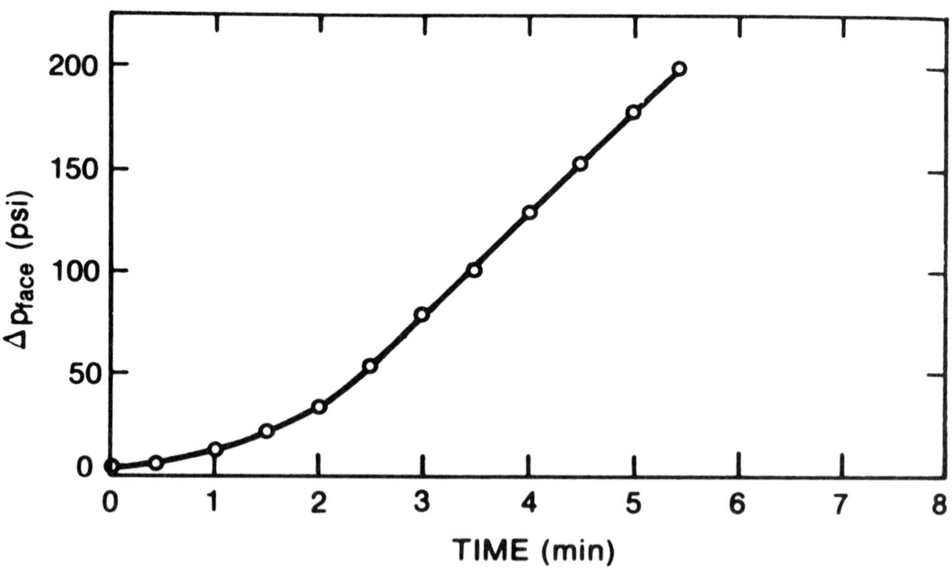

Figure 18.2 Graph showing pressure buildup due to organic resin diverting agent attending flow into a core (Δp_{face} versus time). Rate is 3 cc/min and the core is Berea (0.0254 m in diameter × 0.0762 m in length) [11]. (With permission of the Society of Petroleum Engineers.)

wt% NH_4Cl solution) injected ahead of the preflush to help assure uniform distribution of the preflush. Diverting agents can again be applied, if desired, in the same nonreactive solution after the preflush has been injected but before the HF/HCl treatment.

Are Core Tests Misleading?

It is perhaps important to note that the efficiency of a given diverting agent depends strongly on a large proportion of the pressure drop occurring across the filter cake. In the laboratory, short cores are normally used for testing and the pressure drop across the filter cake is the dominate one. In the field the pressure drops in the formation may be much larger for a given injection rate, thereby reducing the effectiveness of a diverting material applied at a certain concentration effectively in the laboratory. Care should therefore be used in translating from laboratory to field data. This is best done by modeling procedures, which extract the essential filter-cake properties from laboratory results and then use these in simulating the field process.

18.2 CORROSION INHIBITORS

The most important acid additives are corrosion inhibitors. Not only is the cost of the corrosion inhibitor often a significant proportion of the total treatment cost, especially if high bottomhole temperatures are encountered or long acid–pipe contact times are anticipated, but in some cases, the choice of the acid itself is governed by the selection of the most economical means for controlling corrosion and still accomplishing the goals established for the treatment. An explanation of the mechanism of acid corrosion inhibition and techniques for evaluating inhibitor performance are briefly described next. Factors important in inhibitor evaluation and practical suggestions for their selection are also described.

Corrosion of Metals

Any metal surface is a composite of electrodes electrically short circuited through the body of the metal itself (Fig. 18.3). So long as the metal remains dry, local action currents and corrosion are not observed. But on exposure of the metal to

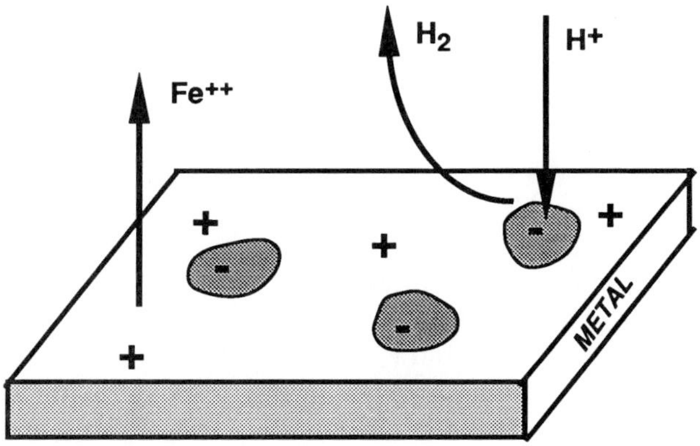

Figure 18.3 Sketch depicting a metal surface, which is composed of both anodic and cathodic sites.

aqueous salt, alkali, or acid solutions, local action cells are able to function and are accompanied by chemical conversion of the metal to corrosion products.

Many years ago it was thought that the two types of domains shown in Fig. 18.3 were entirely a result of impurities in the metal, thus accounting for the corrosion-resistant properties of pure aluminum. However, this theory is not correct. With iron and steel, the most important components of wellbore tubular goods, the negative electrodes are commonly portions of the iron surface itself covered perhaps by iron oxides (rust) and the positive electrodes are exposed areas, the positive and negative sites shifting from place to place as the corrosion reaction proceeds. Accordingly, high-purity iron in air-saturated water corrodes at essentially the same rate as impure or commercial iron. A difference in rates is observed in acids because impurities now enter predominately as electrodes of local action cells.

The reaction which takes place at the cathode (electrode where positive ions enter from the aqueous solution) is

$$2H^+ + 2e^- \rightleftharpoons H_2 \text{ (gas)} \tag{18.13}$$

Hydrogen ions that adsorb on the metal surface in the cathodic area are reduced to hydrogen atoms, which combine to form H_2 and then desorb as a gas. This reaction appears to become vigorous in hydrochloric acid solutions below about pH 4. This is approximately the pH at which $Fe(OH)_2$, a reaction product, becomes relatively soluble. In acetic or formic acids the reaction becomes pronounced at even higher pH values [13].

The oxidation reaction (or anode reaction) is

$$Fe^0 \rightleftharpoons 2\,e^- + Fe^{++} \tag{18.14}$$

that is, iron is oxidized to form ferrous ions which are soluble in acidic solutions. This reaction takes place at the anode (electrode where positive ions leave and enter aqueous solution). The electrons flow from the anodic regions of the metal to the cathodic regions through the metal. The acid solution completes the circuit.

To be effective, a corrosion inhibitor may reduce the reaction rate at the anode, at the cathode, or at both sites. Basically, two classes of inhibitors (distinguished by the manner in which they inhibit corrosion) are found to be useful. One class (the anodic type) functions by sharing electrons from the inhibitor molecule with anodic sites on the metal surfaces. The bond thus established terminates the reaction at that site. The second class (the cathodic type) forms a protective film (often a submonolayer) by attachment to the cathodic area of the metal surface through electrostatic attraction. In botn mechanisms of inhibition the effectiveness is determined by the ability of the inhibitor to adsorb on the steel surface. The typical effect of concentration of an inhibitor on the reaction between steel and 5 wt% H_2SO_4 is shown in Fig. 18.4 [14] showing that above a relatively low concentration, presumably the amount needed to form an adsorbed monolayer, an additional amount of inhibitor has little effect on further reducing the rate. Thus, factors that reduce the number of inhibitor molecules adsorbed will reduce inhibitor effectiveness. Perhaps the most important limiting factor is temperature. At a high temperature, the acid corrosion reaction rate increases and the ability

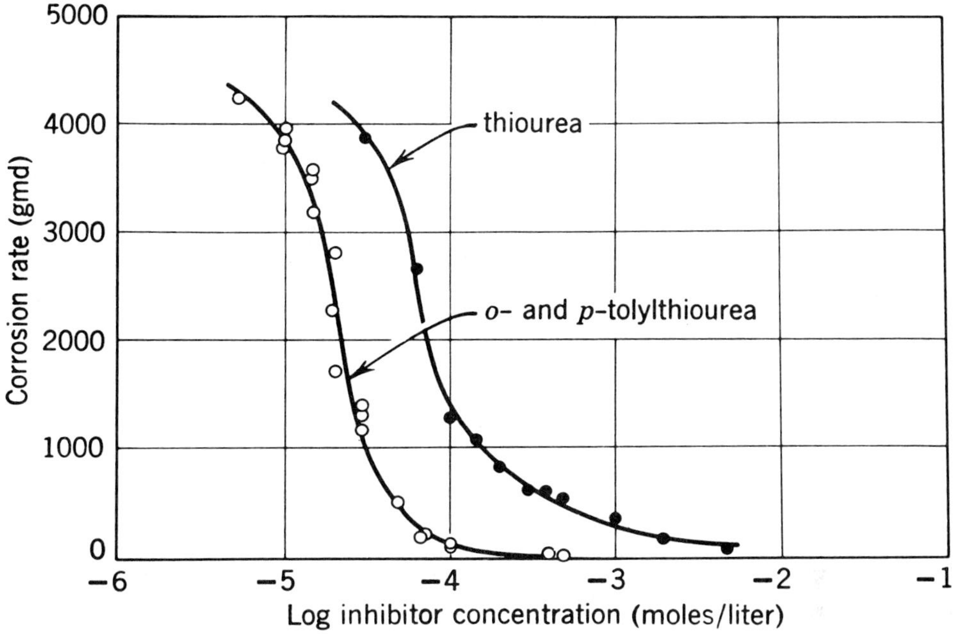

Figure 18.4 Graph showing effect of inhibitor concentration on corrosion rate of 0.1 wt% C steel in 5 wt% H_2SO_4 at 70°C [14]. [From T. Hoar and R. Holiday, *Journal of Applied Chemistry*, *3* (1953) 502. Used with permission of SCI.]

of the inhibitor to adsorb on the steel surface decreases. For these reasons, inhibitors for high-strength acids at temperatures above 50°C are expensive and difficult to formulate.

Compounds serving as inhibitors require, by and large, a favorable polar group or groups by which the molecule can attach itself to the metal surface. These include the organic N, S, OH, and amine groups. The size, orientation, shape, and the electric charge of the molecule all play a part in the effectiveness of inhibition. For example, the corrosion of iron in 1.6 wt% HCl is found to be inhibited by derivatives of thioglycolic acid and 3-mercaptopropionic acid to an extent that varies systematically with chain length [15]. Commercially available inhibitors are usually complex mixtures of organic compounds. Those inhibitors used by service companies are generally proprietary. For the purpose of the discussion here, it is not necessary to know the precise structure of these compounds. It is important, however, to be aware of the methods of testing corrosion inhibitors and to examine the relevance of these procedures to actual field results.

The laboratory evaluation of a corrosion inhibitor generally involves subjecting a coupon of the metal to be protected to the acid to be used [16]. The coupon is normally inserted into a heated pressure vessel containing the acid and the corrosion inhibitor to be evaluated. The amount of corrosion is determined by weighing the coupon before and after the test. Inhibitor effectiveness is expressed in terms of metal loss per unit area of metal exposed per unit of time. Typical units are grams per square meter per day (g/m^2-d), pounds per square foot per day (lb/ft^2-d), or millimeters of penetration per year (mm/y). Also to be noted is pitting, which is a localized type of attack. If appreciable attack is confined to a relatively small fixed area of the metal acting as anode, the resultant pits may be deep. A pitting factor defined as the ratio of the deepest metal penetration to the average penetration as determined by the weight loss of the specimen is sometimes reported. A pitting factor of unity represents uniform attack.

Many factors influence the corrosion rate measured in such tests [16]. Of major importance are (1) the degree of agitation, (2) metal type, (3) exposure time, (4) temperature, (5) acid type and concentration, (6) inhibitor type and concentration, (7) metal-to-acid volume ratio, (8) gas composition if an interface is present, (9) pressure, and (10) the presence of other additives such as surfactants.

Agitation. Since the hydrogen ions must diffuse to the surface and the reaction products diffuse away from the surface or small H_2 bubbles must be swept from the surface, agitation may be expected to increase the corrosion rate. If mass transfer is a limiting factor, then the effect of agitation may be pronounced. If oxygen is carefully excluded from the system and at high levels of inhibition, the effect of velocity should not be important. If on the other hand oxygen is present dissolved in the acid solution, then a remarkable effect of agitation may be observed. At low levels of agitation, the corrosion rate may decrease whereas at high levels it may increase [17]. It appears that oxygen in small concentrations at the metal surface increases the cathode polarization, thereby decreasing the corrosion rate; in higher concentrations it acts mainly as a depolarizer, increasing the corrosion rate. McDougall has reported similar effects of agitation [18] (see Fig. 18.5).

It is recommended that oxygen be excluded from the test by initially bubbling nitrogen through the acid. This may eliminate the effect of agitation [17] and provide assurance that the corrosion rate will not depend on the degree of turbulence in the borehole tubular goods.

Metal type. The effectiveness of a given corrosion inhibitor will depend on the metal. McDougall reported tests of five different commercial corrosion inhibitors (not identified) on N-80, J-55, and P-105 tubular goods [18]. These results clearly show the necessity of matching the inhibitor to the steel because a wide range of responses were obtained for a given inhibitor. Smith has pointed out that the API designations for steels are inadequate to distinguish corrosion rates because they identify yield strength, whereas corrosion characteristics are determined primarily by chemical composition [19]. Because yield strength may be achieved either by physical or chemical modifications in manufacturing, it is essential that the coupons tested be from a representative sample of the tubular goods protected.

Acid type and concentration. Acetic or formic acids are often selected for acidizing deep hot wells (>50°C) because it is difficult to inhibit the corrosive action of hydrochloric acid. Similarly, it is more difficult to protect against 28

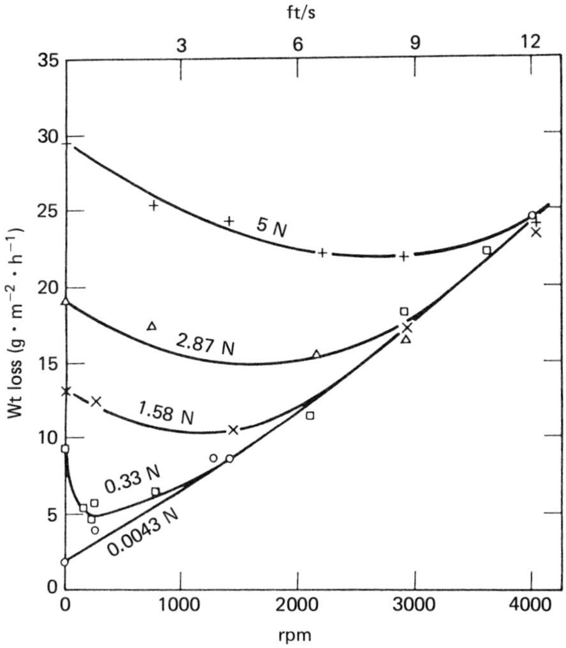

Figure 18.5 Graph showing effect of agitation on acid corrosion inhibitor effectiveness [17]. [Reprinted from W. Whitman et al., *Industrial and Engineering Chemistry*, *15* (1923) 672.]

wt% HCl than 15 wt%. In the presence of some inhibitors a 100-fold increase in corrosion rate at 35°C is observed when the acid concentration is increased from 15 to 28 wt% [18]. For other inhibitors a smaller increase was found.

Surface-to-volume ratio. Clearly, the concentration of inhibitor is reduced as it adsorbs onto the metal surface. Since the inhibitor is present in large excess in a batch test, then the decrease in inhibitor concentration during this test will not be perceptible. If the surface-to-volume ratio is greatly increased in field application, then the batch test may not be representative of downhole conditions. Smith et al. have examined this matter critically and indicate that volumes of acid together with inhibitor compared to coupon surface areas in excess of 12 ml/cm^2 are excessive compared with the common range of conditions found downhole [19].

Acid composition. Any additive that alters the tendency of the corrosion inhibitor to adsorb will also change its effectiveness. For example, surfactants added to the acid for various purposes may form micelles that solubilize the inhibitor, thereby decreasing the tendency for the inhibitor to adsorb on the metal surface (see Chap. 3 for further discussion of solubilization). Inorganic salts and mutual solvents can also alter the tendency for inhibitor adsorption. Batch tests to determine the extent of corrosion should use the acid solution including all of the additives to be used in the actual treatment. If possible, those additives which reduce the effectiveness of inhibitors should be included in a preflush or in the afterflush rather than in the acid solution.

Suggestions for Inhibitor Selection

Evidently, from the preceding discussion some corrosive loss from wellbore tubular goods must be expected when acidizing and the primary issue is what level can be tolerated. Most service company information is based on the assumption that a metal loss of 0.10 kg/m^2 can be tolerated during the treatment if no pitting occurs [20]. Sometimes an amount as high as 0.25 kg/m^2 is assumed allowable. If these metal losses cannot be sustained without adverse effects, a more effective inhibitor must be found.

If the inhibitor costs become prohibitive, it might be possible to reduce them by careful treatment design. Precooling the tubing by injecting a large very dilute hydrochloric acid preflush will, for example, be helpful. Using formic rather than hydrochloric acid will help to reduce the corrosion problem. Also, reducing the contact time to the smallest possible value will reduce the inhibitor requirements.

The most important aspect is the proper testing of representative metal samples using the precise acid formulation to be applied at the most adverse conditions of temperature and pressure. This test should be carried out in an oxygen-free environment noting that oxygen can both enhance and retard the corrosion rate, depending on the degree of agitation. Only under these conditions can a decision about the adequacy of the corrosion inhibitor be made.

18.3 COMPLEXING AGENTS

When appreciable quantities of iron in the form Fe^{+++} (ferric ions) rather than the usual Fe^{++} (ferrous ions) are present as mineral constituents in the near wellbore region of the formation, then ferric ions released by mineral dissolution during the acid treatment will reprecipitate as a gelatinous, damaging, ferric hydroxide whenever the pH of the spent acid solution rises above about 2.2. Since ferrous ions precipitate at a much higher pH, their precipitation is not considered to be a problem. In fact, properly conducted oil and gas well sandstone acidizing treatments will minimize the possibility of iron reprecipitation. Most of the minerals that contain ferric ions will be reacted during the acidizing stage when hydrofluoric acid is present, but a proper preflush (see Chap. 16) will ensure that the hydrochloric acid injected together with the hydrofluoric acid remains essentially unreacted, thereby maintaining a low solution pH and precluding the reprecipitation of iron.

Thus, the precipitation of ferric ions during acidizing is by no means a common problem. Ferric ions, in the presence of iron and steel-pipe and hydrochloric acid, are rapidly reduced to ferrous ions; therefore, the normal dissolution of pipe rust and scales does not necessarily lead to iron precipitation in the formation. From an operational standpoint, the precipitation of iron is a problem primarily in water injection wells where oxygenated water will have converted a significant fraction of the iron in the near wellbore region to the ferric form.

The precipitation of gelatinous ferric hydroxide can be prevented by adding certain complexing or sequestering agents to the acid. Several organic acids (citric, lactic, acetic, and gluconic) and their derivatives ethylene-diaminetetraacetic acid

(EDTA) and nitrilotriacetic acid (NTA) are considered useful sequestering agents. These acids have been evaluated by Smith et al. [21]. They found that each material has certain advantages and limitations. Both cost and performance vary widely. Performance is affected by both temperature and the presence of other metal ions. Only citric acid, EDTA, and NTA can hold as much as 3000 ppm ferric iron in spent acid solutions for more than four hours at temperatures above 50°C. Lactic, acetic, and gluconic acids may prevent ferric hydroxide precipitation from spent acids at lower temperatures and iron concentrations or for shorter periods of time. Lactic acid, in particular, is ineffective at temperatures above 80°C. Studies have indicated that several chemicals can be effective complexing agents when used in the correct application [21]. There is no one best agent for all applications.

EDTA is one of the most effective agents commonly available because large quantities may be used without precipitating calcium salts and it can be used to temperatures of 38°C. Since EDTA is rather expensive, citric acid may be the sequestering agent of choice; however, it must be applied in small quantities (1.7 kg/m^3) otherwise the calcium salt of citric acid will precipitate in the absence of ferric ions. NTA appears to be a useful compromise between citric acid and EDTA. It is less expensive than EDTA and can be used in substantial quantities without calcium salts precipitating and is effective up to 95°C.

The use of a sequestering agent is seldom recommended. If required, the agent should be added to the preflush in sandstone acidizing treatments or to the hydrochloric acid when treating carbonate formations. For further details in selecting the appropriate agent, the reader should consult the paper by Smith et al. [21].

More recently, Hall and Dill [22] have evaluated iron-control additives and have noted special problems that may arise in sour wells; that is, wells that contain hydrogen sulfides. They have proposed a technique for treating such wells to minimize the dangers of iron precipitation.

18.4 SURFACTANTS

Surface active agents are used in acid treating to demulsify spent acid and oil, to reduce interfacial tension, to speed cleanup, and to prevent sludge formation. As noted in Chapter 3, surfactants should never be added to treatment fluids without a full understanding of their effect evaluated at reservoir conditions and using live crude. Unless an extensive evaluation program has been conducted, one is best served by omitting them altogether from the formulation. This is especially true when adding surfactants to break or prevent emulsions.

Surfactants sometimes are added to acid to develop so-called low tension acids although this application has greatly diminished. Generally, an anionic or nonionic surfactant is used. Since the acid solutions are aι a pH well below the point of zero charge of mineral oxides (see Chap. 1), the mineral surfaces will be highly charged positively. It would be anticipated that certainly anionic surfactants and perhaps nonionic surfactants will also be strongly adsorbed on reservoir minerals under the low pH conditions and not, therefore, be available to lower the interfacial tension. Hall [23] has confirmed that adsorption will limit the effectiveness of surfactants for lowering interfacial tension during acid treatment.

Another issue that must be addressed is whether or not the surfactants added to the acid solution reduce the effectiveness of the corrosion inhibitor. In general, one would anticipate that both anionic and nonionic surfactants will interact strongly with corrosion inhibitors and reduce their tendency to adsorb. The best choice for surfactants would appear to be cationic. These surfactants will not under acidic conditions interact strongly with corrosion inhibitors (until micelles form) nor would they be strongly adsorbed onto reservoir rock so long as low pH conditions are maintained. To minimize the interaction with corrosion inhibitors, if this becomes a problem, cationic fluorocarbon surfactants may be the best choice provided that the desired result can be achieved by using them.

18.5 MUTUAL SOLVENTS

Acid corrosion inhibitors provide protection by adsorbing on metal surfaces. Upon entering the formation, the inhibitor is often strongly adsorbed on clays and other minerals especially as the acid pH increases. Crowe and Minor [24, 25] have studied the interaction of corrosion inhibitors with sandstone formations and found that in some cases corrosion inhibitors contain acid-insoluble residues that can cause plugging on the face of the formation and also that inhibitor adsorption can change the formation wettability to oil-wet conditions. They found that the damage resulting from corrosion inhibitors was more pronounced at lower temperatures and that mutual solvents tend to reduce the magnitude of the problem. A mutual solvent being a compound that exhibits significant solubility in both oil and water helps to dissolve both the adsorbed inhibitor and the acid-insoluble residue.

Many chemicals, including alcohols, ketones, and ethers, can be classified as mutual solvents. The specific compound normally referred to is a glycol ether. For sandstone acidizing a compound often applied is ethylene glycol monobutyl ether (EGMBE). In addition to its mutual solubility EGMBE also somewhat reduces the interfacial tension between oil and water and helps to remove adsorbed films from reservoir rock.

Gidley [26] reported that the productivity of oil wells in sandstone formations treated with HF–HCl increased five- to sixfold over that found with regular HF–HCl treatments if as much as 10 wt% EGMBE is used in the diesel oil afterflush of the treatment. A mutual solvent should not be added to either the preflush nor to the HF–HCl mixture since it will reduce the effectiveness of the corrosion inhibitor.

Taylor and Plummer [27] have also demonstrated improved results for gas wells when a mutual solvent was used. Sutton and Lasater [28] and Hall [23] both found that a mutual solvent reduces surfactant adsorption, again supporting the findings reported by Crowe and Minor [24, 25].

Thus in some cases, apparently lower reservoir temperatures, a reasonable afterflush might consist of diesel oil containing EGMBE to help speed the removal of adsorbed corrosion inhibitor and surfactant from the minerals. This mixture will help to maintain water wettability and to break Pickering emulsions (see Chap. 3).

REFERENCES

18.1. Harrison, N. W., *J. Pet. Tech., 24* (1972) 593.

18.2. Erbstoesser, S. R., *J. Pet. Tech., 32* (1980) 1903.

18.3. Cooper, R. E., and Bolland, J. A., "Effective Diversion During Matrix Acidization of Water Injection Wells," OTC 4795, presented at the Offshore Technology Conference, Houston, Texas, 1984.

18.4. Pye, D. S., Gallus, J. P., and Kemp, J. D., *Oil and Gas J.* (November 9, 1970) 76.

18.5. Taha, R. M., Hill, A. D., and Sepehrnoori, K., *J. Pet. Tech., 38* (1986) 753.

18.6. Houchin, L. R., Dunlap, D. D., Hudson, L. M., and Begnaud, P. C., "Evaluation of Oil Soluble Resin as an Acid-diverting Agent," SPE 15574, presented at the 61st Fall Technical Conference and Exhibition of the Society of Petroleum Engineers, New Orleans, Louisiana, 1986.

18.7. Crowe, C. W., "Evaluation of Oil Soluble Resin Mixture as Diverting Agents for Matrix Acidizing," SPE 3505, presented at the 1971 Society of Petroleum Engineers Annual Meeting, New Orleans, Louisiana, 1971.

18.8. Doerler, N., and Prouvost, L. P., "Diverting Agents: Laboratory Study and Meeting of Resultant Zone Injectivities, SPE 16250, presented at the International Symposium on Oilfield Chemistry of the Society of Petroleum Engineers, San Antonio, Texas, 1987.

18.9. King, G. E., and Hollingsworth, F. H., "Evaluation of Diverting Agent Effectiveness and Cleanup Characteristics Using a Dynamic Laboratory Model—High Permeability Case," SPE 8400, presented at the 54th Fall Technical Conference and Exhibition of the Society of Petroleum Engineers, Las Vegas, Nevada, 1979.

18.10. Bird, R. B., Stewart, W. E., and Lightfoot, E. N., *Transport Phenomena*, New York: John Wiley, 1960.

18.11. Hill, A. D., and Galloway, P. J., *J. Pet. Tech., 36* (1984) 1157.

18.12. Taha, R. M., "Single Well Model for Field Studies of Sandstone Matrix Acidizing," PhD dissertation, The University of Texas at Austin, 1987.

18.13. Uhlig, H. H., and Revie, R. H., *Corrosion and Corrosion Control*, 3rd ed., New York: John Wiley, 1985.

18.14. Hoar, T., and Holliday, R., *J. App. Chem., 3* (1953) 502.

18.15. Carroll, M., Travis, M., and Noggle, J., *Corrosion, 31* (1975) 123.

18.16. *Corrosion Inhibitors*, C. C. Nathan (ed.), National Association of Corrosion Engineers, Houston, Texas (1973) 156.

18.17. Whitman, W., Russell, R., Welling, C., and Cochrane, J., *Indust. and Eng. Chem., 15* (1923) 672.

18.18. McDougall, L. A., *Materials Protection, 8* (1969) 31.

18.19. Smith, C. F., Dollarhide, F. E., and Byth, N. J., *J. Pet. Tech., 30* (1978) 737.

18.20. Williams, B. B., Gidley, J. L., and Schechter, R. S., *Acidizing Fundamentals*, Mono. Ser. *6*, Society of Petroleum Engineers, Richardson, Texas, 1979.

18.21. Smith, C. E., Crowe, C. E., and Nocan, J. T., "Secondary Deposition of Iron Compounds Following Acidizing Treatments," SPE 2358, presented at the Society of Petroleum Engineers of the American Institute of Mining and Metallurgical Engineers Regional Meeting, Charleston, West Virginia, 1968.

18.22. Hall, B. E., and Dill, W. R., "Iron Control Additives for Limestone and Sandstone Acidizing of Sweet and Sour Wells," SPE 17157, presented at the Society of Petroleum Engineers Formation Damage Control Symposium, Bakersfield, California, 1988.
18.23. Hall, B. E., *J. Pet. Tech., 27* (1975) 1439.
18.24. Crowe, C. W., and Minor, S. S., "Acid Corrosion Inhibitor Adsorption and Its Effect on Matrix Stimulation," SPE 10650, presented at the Society of Petroleum Engineers Formation Damage Control Symposium, Lafayette, Louisiana, 1982.
18.25. Crowe, C. W., and Minor, S. S., "Effect of Acid Corrosion Inhibitors Upon Matrix Stimulation Results," SPE 11119, presented at the 57th Fall Technical Conference and Exhibition of the Society of Petroleum Engineers, New Orleans, Louisiana, 1982.
18.26. Gidley, J. L., *J. Pet. Tech., 23* (1971) 551.
18.27. Taylor, D. B., and Plummer, R. A., "Gas Well Stimulation Using Coiled Tubing and Acid with a Mutual Solvent," SPE 4115, presented at the 1972 Society of Petroleum Engineers Annual Meeting, San Antonio, Texas, 1972.
18.28. Sutton, G. D., and Lasater, R. M., "Aspect of Acid Additive Selection in Sandstone Acidizing," SPE 4114, presented at the 1972 Society of Petroleum Engineers Annual Meeting, San Antonio, Texas, 1972.

PROBLEMS

***18.1.** Three zones are to be acidized. Neglecting the effect of the acid on the formation permeability and the skin factor, determine the fraction of the total injected fluid that will enter each zone. The wellbore radius is 0.1 m and the drainage radius is 200 m. Additional data are listed in Table P18.1.

TABLE P18.1

Location (Zone No.)	Permeability (mD)	Thickness (m)	Skin Factor (S)
1	5	2	15
2	12	10	20
3	17	13	10

***18.2.** Two zones are to be simultaneously acidized. Diverting agents are to be added to equalize the acid volumes entering the two zones. It is decided that adequate diversion will require $V_1 = (1/4)V_2$ at the point in the treatment when 0.7 m^3/m of acid has been injected. Given the data listed in Table P18.2, state the concentration (c_0) of diverting agent recommended for use in the acid treatment.

TABLE P18.2

Location (Zone No.)	Permeability (mD)	Thickness (m)	Skin Factor (S)
1	15	10	10
2	200	6	10

Property	Symbol	Value
Welbore radius	r_w	0.07 m
Drainage radius	r_c	200 m
Diverting cake permeability	k_{cake}	0.1 mD
Diverting cake porosity	ϕ_{cake}	0.38

****18.3.** Two zones are to be acidized at an injection rate of 5×10^{-4} m^3/sec-m, and diverting agents are to be used to help equalize the acid flow into each zone. The total treatment time is to be 40 min. The concentration of diverting is to be selected to be as large as possible with, however, the restriction that the wellbore pressure not exceed 3.75×10^4 kPa because pressures in excess of this pressure may create fractures. If the maximum diverting agent concentration is applied, what fraction of the total acid volume will have entered Zone 1? Is this an improvement compared to acidization without diverting agent? If the improvement is not significant, what can be done? The data are listed in Table P18.3.

TABLE P18.3

Location (Zone No.)	Permeability (mD)	Thickness (m)	Skin Factor (S)
1	7	7	12
2	220	7	12

Property	Symbol	Value
Wellbore radius	r_w	0.1 m
Drainage radius	r_e	200 m
Diverting cake permeability	k_{cake}	0.3 mD
Diverting cake porosity	ϕ_{cake}	0.39
Reservoir pressure	p_R	1.1×10^4 kPa
Acid viscosity	μ	1.8 cp

****18.4.** The protonation reactions of many ligands, which chelate with metal ions as anions, determine the stability of the chelates that form with the metal ion. For example, ethylenediaminetetraacetic acid (EDTA) is most effective in the partially dissociated

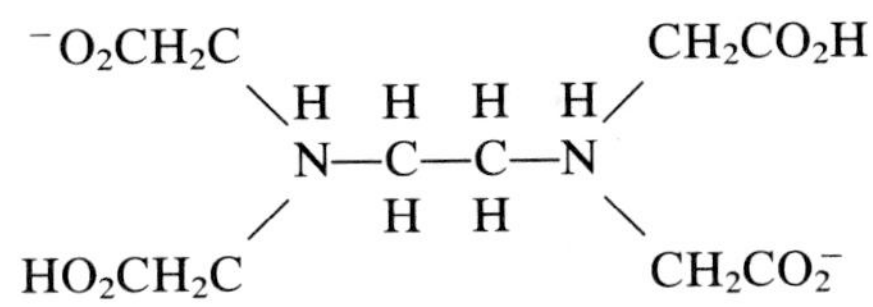

form which can be abbreviated as YH_2^{2-}. Clearly other possible forms are YH^{3-}, Y^{4-}, YH_3^{1-}, and YH_4. The dissociation constants of 25°C are as follows:

$$YH_4 \longrightarrow H^+ + YH_3^- \qquad K_1 = 1.175 \times 10^2$$

$$YH_3^- \longrightarrow H^+ + YH^{2-} \qquad K_2 = 5.632 \times 10^2$$

$$YH_2^{2-} \longrightarrow H^+ + YH^{3-} \qquad K_3 = 1.738 \times 10^6$$

$$YH^{3-} \longrightarrow H^+ + Y^{4-} \qquad K_4 = 2.188 \times 10^{10}$$

[The equilibrium constants are given by C. F. Bell, *Metal Chelation*, Clarendon Press, Oxford (1977).]

Determine the pH range over which ETDA is primarily in the YH_2^{2-} form for an EDTA concentration of 1 kg mole/m^3 in water.

PART 6
Sand Control Methods: Gravel Packing and Consolidation Techniques

19
Sand Control

19.1 PROBLEMS WITH PRODUCING UNCONSOLIDATED FORMATIONS

Control of formation sand is the principal production problem of oil and gas fields producing from recent clastic sediments. Sediments of the Pliocene and younger Tertiary ages are particularly troublesome, and sand production problems may be expected whenever wells are completed in unconsolidated reservoirs. Examples of areas where chronic sand problems occur in these geologically young sediments include U.S. Gulf Coast, California, Indonesia, Nigeria, Trinidad, and Venezuela. Although most of the formation control problems occur in these young sediments, failure may also occur in older formations when in-situ rock strength is reduced by poor completion and production practices.

Sand influx into producing wells causes a multitude of operating problems, many of which cannot be tolerated for safe and economical field operations. These problems include erosion of downhole tubulars and casing that can lead to the need for premature and costly workovers. In some cases, severe buckling can occur when lateral restraint provided by the reservoir sands is lost as the sands are produced [1].

Productivity loss caused by sand in the wellbore is the most common reason for downhole sand control. In some cases, this problem can be somewhat alleviated by increasing produced fluid velocities through the use of small tubing strings or by prematurely applying gas lift. Continuous sand production, however, runs the risk of erosion damage which can be hazardous [2].

The handling and disposing of produced sand is expensive and troublesome, especially offshore, where solids must be removed from surface facilities on platforms and transported to approved disposal sites.

It is, therefore, clearly necessary to assure maximum sand-free production rates. One way of accomplishing this goal is to limit production rates to levels that avoid sand production. In some cases this is the most cost-effective method of sand control, but in many cases low production rates are uneconomical. Offshore platforms, for example, have a limited lifetime. It is generally necessary to produce at high rates to justify platform construction. Thus the decision is often not whether or not to take sand control measures, but rather to select the procedure that is best in an economic sense. There are two approaches currently used for sand control. Gravel packing, as a method of sand control, is described in detail in this chapter. This approach has achieved a remarkable level of reliability in recent years, except perhaps in highly deviated wells, and is the most widely applied technique.

A second approach is to chemically consolidate the sand grains in a zone surrounding the wellbore by using a plastic material. The method is best applied to those formation intervals which are relatively free of clays and fines, have a uniform permeability, are thin (<5 m usually), and have no prior history of sand production. If these criteria are met, consolidation systems are attractive. High temperatures may, however, rule out use of chemical consolidation procedures. This technique is described in subsequent sections.

Gravel packs have a slight initial cost advantage over consolidation and a greater statistical chance for success in most types of completion [3]. These benefits of favorable reliability, cost, and versatility make gravel packing the preferred method of sand control in most, but certainly not all, cases.

19.2 SAND INFLUX AND WELL PRODUCTIVITY

Sand Failure Mechanisms

Formation stability under producing conditions is related to the stresses imposed on the formation matrix and the complex manner in which the matrix accommodates these stresses. Stresses imposed are a function of overburden pressure, pore pressure, flowing fluid pressure gradient near the wellbore, interfacial tension effects, and the drag forces on matrix grains caused by fluid flow. The manner by which these stresses are accommodated is determined by two interrelated factors, intrinsic formation strength and the formation's capability under a specific set of conditions to form stable arches around the wellbore face (see Fig. 19.1).

The strength of a formation is principally derived from the cementation between grains. In the case of unconsolidated sands, the cementation is primarily

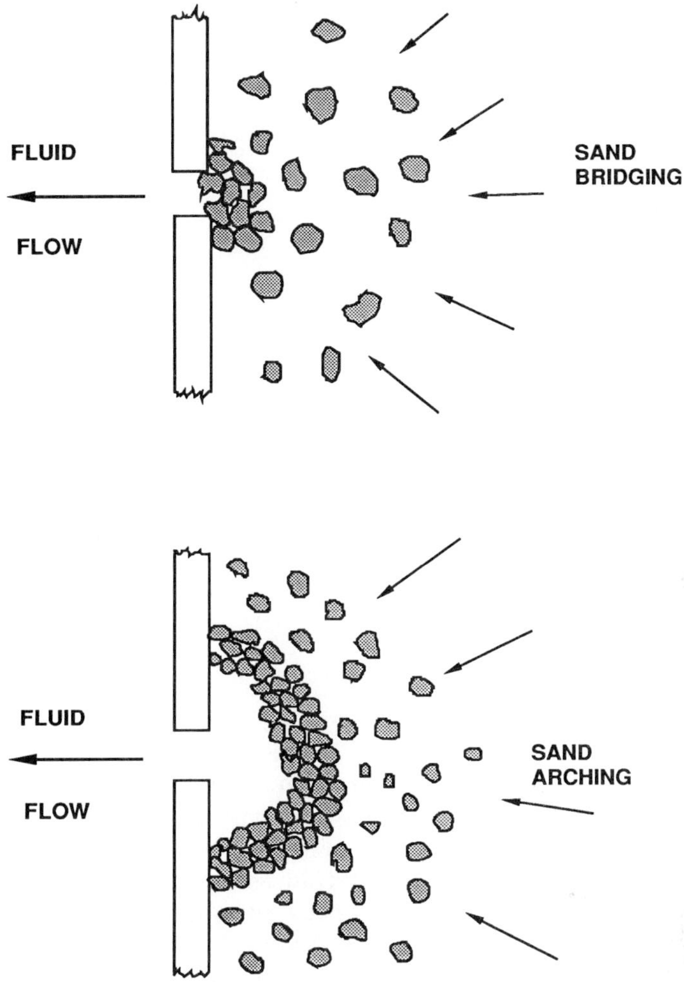

Figure 19.1 Sketch depicting the difference between sand bridging and sand arching at a perforation.

detrital clay and mud which have little or no strength. Sand formations cemented with precipitated minerals are generally much stronger and may resist destruction from formation stresses, or fail at a later date when the overburden stresses increase as reservoir pressure declines. In addition to cementation and the state of the confining stress (as determined by the overburden stress and pore pressure), intrinsic strength is also governed by grain shape and sorting.

For producing wells, the most important strength property of a formation is its resistance to shear. A widely used criterion for the shear failure of reservoir sands is the Mohr–Coulomb theory [4]. The shear strength of a sand may therefore be described in terms of its Mohr failure envelope (shear stress vs. normal stress). For unconsolidated sands, the failure envelope of this plot is not, as shown in Figure 19.2, a straight line. Curvature is caused by changes in the way the sand

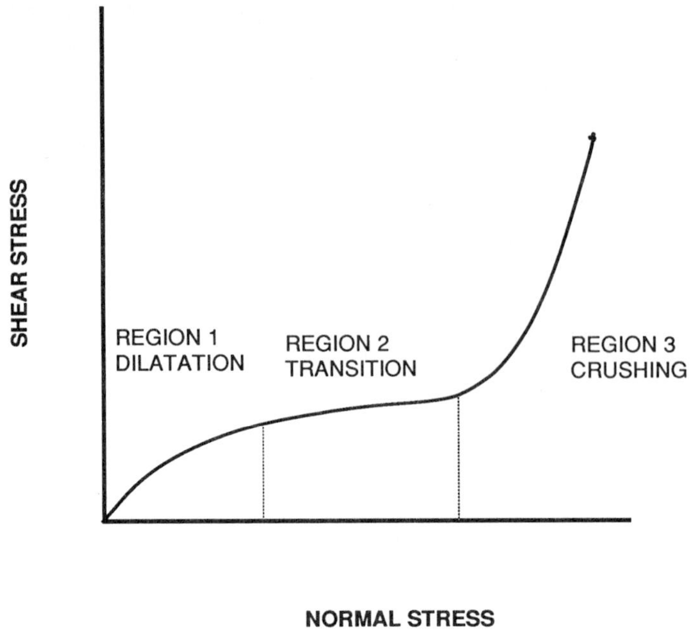

Figure 19.2 Mohr failure envelope for unconsolidated sands.

fails at different stress levels. The failure envelope is divided into three regions based on the dominating failure characteristic. At the low stress levels of Region 1, sand shears by the rolling and sliding of grains up over their neighbors. This is accompanied by an increase in bulk volume (dilatant behavior). At higher confining stresses (Region 2), some grains breakage begins to occur; and at sufficiently high stresses (Region 3), shear occurs almost entirely by grains breakage, accompanied by a decrease in bulk volume. Hall and Harrisberger demonstrated that higher pore pressure inhibits grains breakage [3]. This may be explained by the fact that pore pressure adds an equal increment to each of the principal stresses at every point within the pore spaces, thereby decreasing the stress ratio (see Chap. 2).

The reduced stress state at the wellbore face in the horizontal (radial) direction causes formations with low intrinsic strengths to form arches. For weakly cemented sands, resistance to sanding is dependent on the formation of stable, load-carrying arches spanning the producing cavities. An arch may be defined as a curved structure spanning an opening, serving to support a load by resolving the vertical stress (overburden load) into horizontal stresses. This resolution of forces sets up a pattern of shearing stresses within the arch structure, and thus arch behavior is related to the sand's shear strength. Through this arching effect, a part or most of the vertical loading due to the formation overburden at the wellbore face can be transferred to the formation adjacent to the arch structure.

It should be pointed out that arch formation is different from what is generally termed *sand bridging*. Sand bridging refers to blockage of an opening to sand movement through the opening by an interlocking of grains and transference of stress between grains within or at the mouth of the opening. As previously discussed, an arch is a curved structure entirely outside the opening. Loose, unstressed sand inside the arch is easily removed by fluid flow provided bridging across the opening does not occur. A cavity is then formed. Any loose sand remaining in this cavity does not contribute to the strength of the arch (Fig. 19.1).

Experimental work has demonstrated the effect of shear strength on the formation of arches and their stability [4, 5]. Hall and Harrisberger [3] conclude that two conditions are necessary for arch stability. These are dilatancy and cohesiveness or some other restraint on the surface grains of the arch. They reasoned that if shearing causes crushing, and thus a decrease in bulk volume at any location, the load must be shifted to an adjacent location, bringing its stress up to the level for crushing. Thus, a crushing failure is propagated. Conversely, shearing cannot propagate in a confined arch structure loaded within the dilatant range.

On the basis that radial effective stress (arch load) increases with depth, Suman makes the following assumptions regarding arch stability of unconsolidated sands as a function of depth [6]:

1. Shallow unconsolidated sands will either not arch over perforations or will form tenuous arches. Sand can readily flow (roll and slide) through the perforations.

2. At intermediate depths, rate-sensitive arches will form. Sand production then depends upon the rate of flow and the restraint provided by interfacial tension and sand angularity. Arch failure results in the rolling and sliding of sand through the perforations.

3. For somewhat greater depths, the arch load may be sufficient to provide for high rates of flow without failure.

4. Finally, for a given sand quality, the arch load may reach the level where arch failure through grain crushing will occur.

Obviously, precise evaluation of the multitude of variables affecting formation stability under dynamic well conditions is not possible. Many of the parameters, such as pore pressure, flowing fluid pressure gradient, etc., that determine formation stress at the wellbore face change over the life of the well. Moreover, the tendency for sand production increases when the flow is two phase. However, intrinsic strength is a direct indication of the formation's sanding tendency. Sand influx prediction techniques have been developed that evaluate in-situ formation strength.

Sand Influx Prediction Techniques

There is considerable evidence from laboratory measurements that a good correlation exists between formation intrinsic strength and the dynamic elastic constants determined from acoustic velocity and density measurements [7]. The most important of these dynamic elastic constants for predicting sanding problems is probably the shear modulus [8]. Other mechanical properties of importance include the bulk modulus and Poisson's ratio.

The most effective means for determining these in-situ mechanical properties of friable sands is through the use of well log data [8]. Since log measurements are made in-situ, they are fairly representative of the state of confining stress the formation experiences at completion. Therefore, the accuracy of these properties from log data should be better than the properties determined by direct strength tests with core samples. Weakly cemented materials would probably disaggregate during the core recovery operation, and no strength of cementation would be detected in direct strength measurements. Relief of overburden load results in core expansion and subsequent breaking of weak bonds.

In principle, shear and bulk moduli may be directly calculated from the sonic compressional and shear velocities and the bulk density (see Chap. 2). However, in many friable sands grain slippage readily occurs; and as a result, the shear wave is highly attenuated [9]. The velocity of the shear wave in these situations approaches the fluid velocity, and shear arrivals are often indistinct, buried among other wave arrivals. Techniques have been developed that make it possible to substitute conventional sonic log data in calculating the mechanical properties of friable sands [8, 9]. These mechanical properties, the dynamic moduli and Poisson's ratios, are then used as an indication of formation strength. There has been little published information on the reliability of these techniques.

The technique most frequently employed for prediction of sand influx is by analogy with other wells in the same horizon, field, or area. This approach implies a similarity of fluid types, rock characteristics, flow rates, and pressure drawdown for oil wells. A further refinement of this technique is the individual well test, which consists of producing a given well at gradually increasing rates until sand is produced, or the minimum acceptable production rate occurs. This may be a drillstem or a conventional production test. If sand is produced during the test, the well becomes a candidate for installation of sand control equipment. The results of this test can be projected to subsequent completions in the same formation.

Control by Perforating

Since the fluid velocity is a critical factor determining whether or not sand will be produced, it seems evident that the perforations should be designed to minimize this velocity. Sand problems are most often associated with highly permeable producing zones, and therefore, the perforation penetration depth is not an important parameter. The obvious design strategy to minimize the fluid velocity is to maximize the shot density of holes having large diameters. The selection of an appropriate jet gun should take these factors into consideration (see Chap. 7). Most likely tubing-conveyed or casing guns would be used and shot densities ranging from 25–30 shots/m would seem appropriate to help prevent the production of sand.

In some cases, even high shot densities with large perforation diameters will not prevent sand production at reasonable production rates. Then active sand control measures will be required. Gravel packing and sand consolidation are two methods now used. If it is known in advance that one of these methods is to be applied, high shot densities may not be desirable. Sand consolidation procedures, for example, are best applied when it can be assured that each and every perforation is treated. Thus as a prelude to using this technique, a limited number of perforations would be optimal.

19.3 GRAVEL PACKING

Principles

It is generally possible to prevent sand flux into the wellbore by reducing the production rates; however, in many cases a more economical approach is to take measures that positively restrain the movement of formation sand into the wellbore and at the same time permit the well to be produced at relatively high rates. One procedure that has achieved some success is gravel packing. The concept is simple. A screen or slotted pipe smaller in diameter than the casing or openhole diameter is positioned centrally within the wellbore. The annular space between the wellbore and the screen or pipe is then filled with gravel. The configuration of a completed gravel pack is shown in Fig. 19.3. The method of screen placement and subsequent filling of the annular space with gravel will be described in this chapter. Also discussed are the criteria for selection of gravel and the central screen or pipe designed to restrain the gravel. These design techniques have proven to be reliable in vertical wells even if the interval to be gravel packed is thick (200–300 m). Deviated wells such as those which predominate offshore are

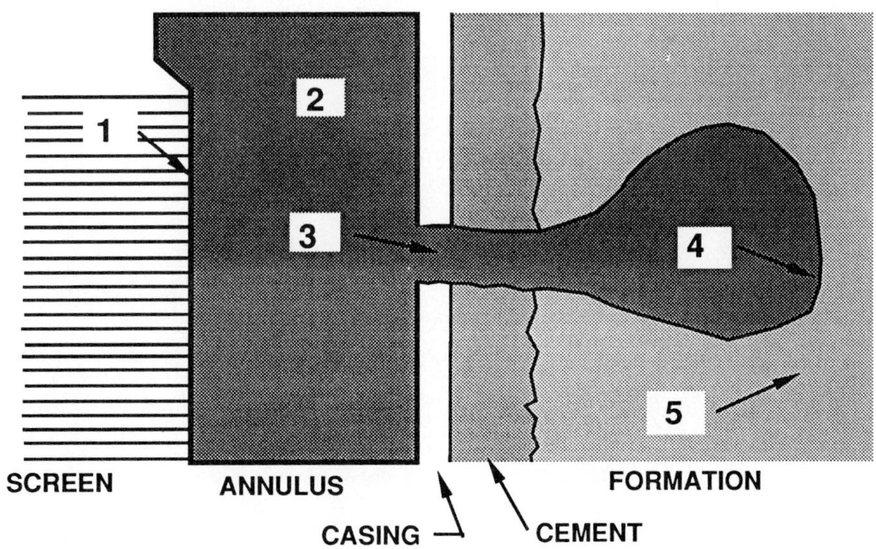

Figure 19.3 Sketch depicting the arrangement of a gravel pack. The screen holds against the casing wall.

more difficult to effectively gravel pack, and special consideration for deviated wells is required. This topic, which is still a subject of considerable research, is discussed in Section 19.4.

Basically, the gravel pack is intended to present a physical barrier to sand movement. The flow passages through the gravel must be small enough to prevent sand production but large enough to allow adequate well productivity. The passages should also be large enough to minimize plugging by clays, alsphaltenes, wax, or scale. The design of a gravel pack system involves three primary stages, all of which are considered in this chapter.

1. *Correctly sizing the gravel.* There is a relationship between the size and distribution of formation sand and the gravel size. As will be evident, the gravel is selected to prevent sand intrusion into the gravel pack. To size the gravel, an analysis of representative samples of formation sand is required. Once the gravel has been selected, the size of the openings in the screen or liner can then be determined.

2. *Correctly sizing the gravel pack thickness.* The liner or the screen must be large enough to accommodate all of the production equipment and to ensure that productivity is not restricted. On the other hand, the annular space filled with the gravel must be of sufficient thickness to ensure that the formation sand will be prevented from entering the wellbore.

3. *Correctly positioning the liner and placing the gravel.* The liner (or screen) must be centered and the gravel must completely fill the annular space between the liner and the formation wall or casing. For cased holes this includes the complete packing within the perforation tunnel to form a sharp interface between the gravel and the formation sand. The gravel pack arrangement must be mechanically stable, and movement due to the production of reservoir fluids should be restricted as much as possible.

Gravel Size Selection

Formation sampling and analysis. The selection of a gravel size is based primarily on the finest sand contained in the producing formation. Thus, obtaining a representative sample is critical. Types of samples available are

1. sands contained within the produced fluids;
2. samples obtained by bailing;
3. sidewall samples; or
4. samples obtained using rubber sleeve cores.

Generally, sand obtained from the production lines contains an excess of fines since the large grains will tend to settle back into the well. For this reason, bailed samples will generally be biased toward the larger sand grain sizes. When bailed samples are the only ones available, then a very conservative criterion for gravel size selection should be applied. This approach recognizes the biased nature of a bailed sample.

Rubber sleeve cores are generally considered to be more representative than are sidewall cores, which tend to contain some sand that has been crushed by the sampling process. Maly and Krueger [10] recommend at least one well in each field or area that produces sand from a given geological formation be sampled foot by foot through the producing intervals.

The formation samples should then be disaggregated, being careful not to crush the sand grains, and then passed through a series of sieves to separate the particulates into size ranges. To assure that the clays and the silt are separated from the sand grains, the sample is sometimes blended in water containing a dispersant. The dispersed particles are then washed through a 325-U.S.-mesh screen to separate most of the clay and the silt from the sand grains. The remaining sand fraction should then be dried and sieved [11, 12]. After shaking, the material retained on a progression of sieve sizes should be determined. The weight of the material passing through the small, wet sieve should also be measured and recorded.

Himes and Ruiz [13] considered the problem of analyzing sidewall cores, which are usually contaminated with drilling mud materials. This artificially introduced material is fine in particle size and can skew the sieve analysis of the sample, indicating a smaller median diameter than that of the true formation. Himes and Ruiz have used a particle size analyzer that detects particle size by the amount of light obstructed by the particle and have obtained a more accurate representation of the sand size by excluding the effects of mud contaminants. It seems reasonable that the current method of sand size determination recommended by the API [14] will eventually be replaced by more easily automated and perhaps more accurate procedures.

There are two different types of sieves in common use. The U.S. Standard Mesh series and the Tyler series are both used, but the U.S. series is the one most commonly used by the petroleum industry. Table 19.1 provides a comparison of these two designations. The data from a sieve analysis is generally presented as the cumulative weight fraction retained on succeeding screens, starting with the screens having the largest openings. The plot is generally presented on semilogarithmic paper, as shown in Fig. 19.4.

There are several numerical measures that can be used to characterize the size of the sand grains and their degree of sorting. A measure of the sand grain size is d_{40}, defined as the diameter corresponding to the 40th percentile. The d_{40} is read directly from the cumulative weight fraction plot obtained from the sieve analysis. The median, d_{50} (shown in Fig. 19.4), is also sometimes used as a measure of sand grain size.

Sorting is a term referring to the uniformity of the sand. A sand that includes grain sizes contained in a narrow range (small standard deviation) is considered to be well sorted. A measure of sorting found in the sand control literature is the uniformity coefficient U_c, which is defined as

$$U_c = \frac{d_{40}}{d_{90}} \tag{19.1}$$

TABLE 19.1 Comparison of Sieve Designations [5]

Standard Sieve Openings				Standard Sieve Openings			
Mesh	Sieve Opening	Mesh	Sieve Opening	Mesh	Sieve Opening	Mesh	Sieve Opening
U.S. Series	(mm)	Tyler Series	(mm)	U.S. Series	(mm)	Tyler Series	(mm)
$2\frac{1}{2}$	8.00	$2\frac{1}{2}$	7.925	35	0.50	32	0.495
3	6.73	3	6.68	40	0.42	35	0.417
$3\frac{1}{2}$	5.66	$3\frac{1}{2}$	5.613	45	0.351	42	0.351
4	4.76	4	4.699	50	0.297	48	0.295
5	4.00	5	3.962	60	0.250	60	0.246
6	3.36	6	3.327	70	0.210	65	0.208
7	2.83	7	2.794	80	0.177	80	0.175
8	2.38	8	2.362	100	0.149	100	0.147
10	2.00	9	1.981	120	0.124	115	0.124
12	1.68	10	1.651	140	0.104	150	0.104
14	1.41	12	1.397	170	0.088	170	0.088
16	1.19	14	1.168	200	0.074	200	0.074
18	1.00	16	0.991	230	0.062	250	0.062
20	0.84	20	0.833	270	0.053	270	0.053
25	0.71	24	0.701	325	0.044	325	0.044
30	0.589	28	0.589	400	0.037	400	0.037

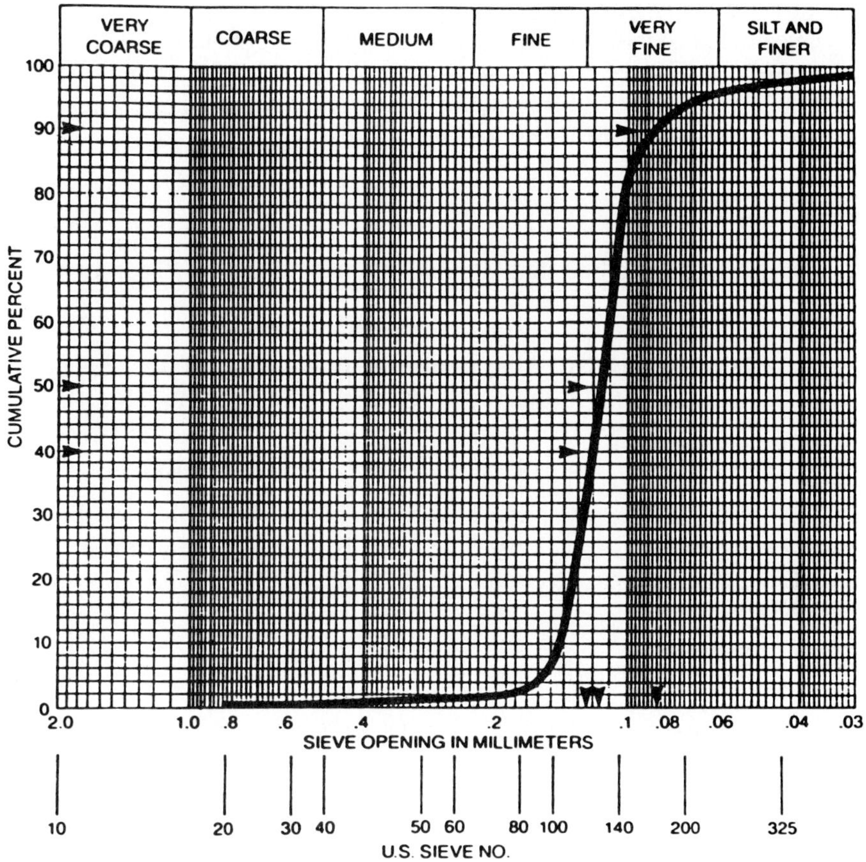

Figure 19.4 Sieve analysis plot of a disaggregated formation sand.

Example 19.1 Characteristics of Formation Sand

Based on the graph of the cumulative percent versus mesh size shown in Fig. 19.4, determine both the median and the uniformity coefficient of the sand.

Solution Reading from Fig. 19.4, we find

$$d_{40} = 0.122 \text{ mm}$$

$$d_{50} = 0.117 \text{ mm}$$

$$d_{90} = 0.086 \text{ mm}$$

The median is 0.117 mm and

$$U_c = \frac{0.122}{0.086} = 1.4$$

Schwartz [16] classifies sand as uniform (well sorted) if $U_c < 3$. For non-uniform or poorly sorted sand $U_c > 5$.

Porosity and permeability of gravel packs. The size of the gravel is to be selected so that the gravel pack is a physical barrier preventing sand movement. Shown in Fig. 19.5 are spheres packed in a cubic array—the diameter of the gravel is D and the diameter of the largest opening is d. These two dimensions are easily shown to be related by the following equation:

$$D = \frac{d}{\sqrt{2} - 1} = 2.41d \tag{19.2}$$

The porosity of a bed that is cubically packed is easily seen to be (see Prob 19.4)

$$\phi_{\text{cubic}} = 0.476 \tag{19.3}$$

Based on this porosity, the permeability of a bed of spheres packed in a cubic array is, according to the Blake–Kozeny equation [Eq. (10.2)],

$$k = \left(\frac{D^2}{150}\right) \frac{\phi^3}{(1 - \phi)^2} = (2.62 \times 10^{-3})D^2 \tag{19.4}$$

The cubic array is the loosest packing. The tightest packing that can occur is hexagonal close packing. This case is also depicted in Fig. 19.5. The centers of the spheres form an equilateral triangle, and the dashed lines bisect the angles. Thus

$$\frac{D}{2} = \left(\frac{D + d}{2}\right) \cos 30° \tag{19.5}$$

Solving for D we find

$$D = \frac{\sqrt{3}}{2 - \sqrt{3}} = 6.46d \tag{19.6}$$

The porosity of a bed hexagonally close packed is (see Prob 19.5)

$$\phi_{\text{close}} = 0.26$$

as compared with 0.476 for the loose cubic packing. The permeability as given by the Blake–Kozeny equation shown in Eq. (19.4) is

$$k = (2.14 \times 10^{-4})D^2 \tag{19.7}$$

where D is the diameter of the gravel.

Example 19.2 Packed Bed Permeability

Table 19.2 shows permeabilities for various sands. Compare the measured permeability with that calculated using both Eq. (19.5) and Eq. (19.7). Comment on whether the experimental packed bed was more nearly cubically or hexagonally packed.

Solution From Eq. (19.5) for the 6/10-mesh sand, a cubic packed permeability

$$k = (2.62 \times 10^{-3})D^2 = 0.0188 \text{ mm}^2 = 1.91 \times 10^4 \text{ D}$$

Similarly, for the 6/10-mesh sand using Eq. (19.7), we find the permeability of the hexagonal closed packed arrangement. The values for the other gravels are shown in Table 19.2. Clearly, the close packing is the better estimate of the actual experimental packing than is cubic packing.

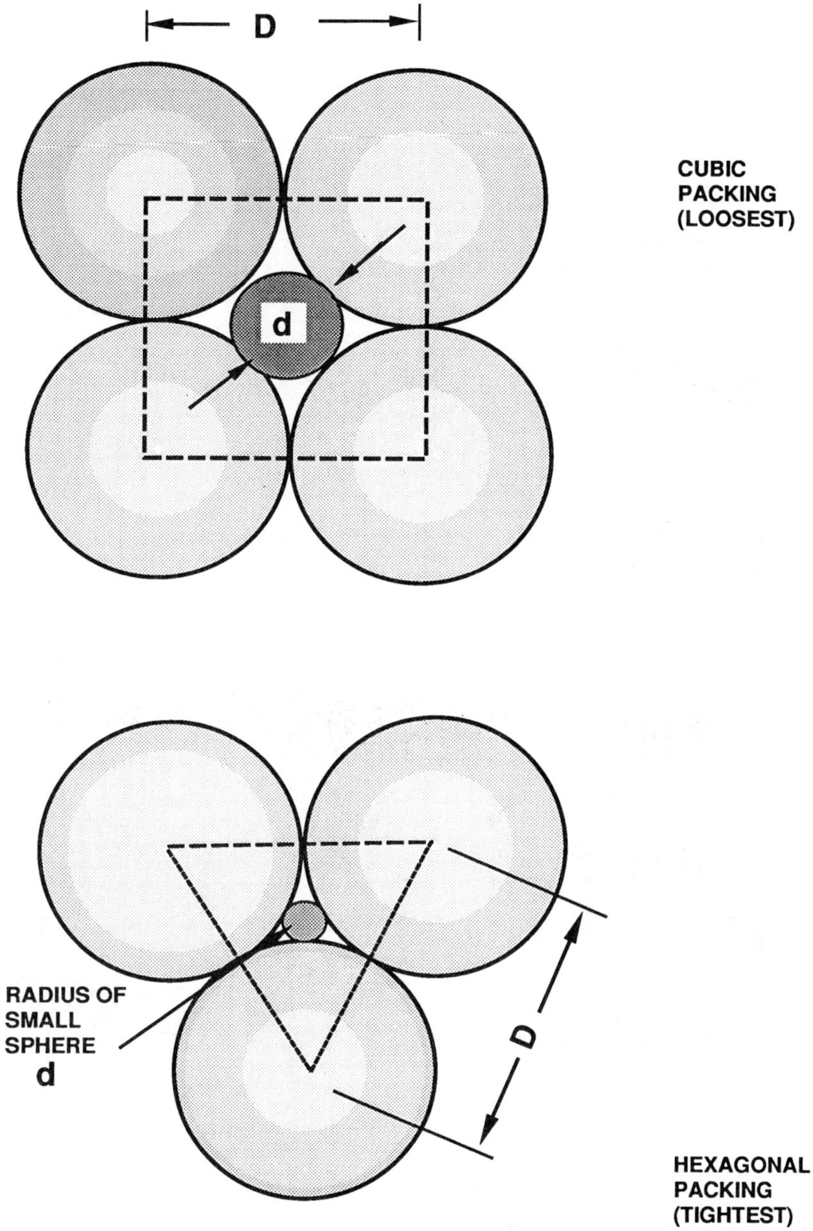

Figure 19.5 Two possible geometrical arrangements of a gravel pack.

The geometric arguments that have resulted in Eqs. (19.2) and (19.6) are central to the following design approach. Early approaches to gravel sizing were based on a desire to maintain large-size gravel diameters to provide a high permeability (see Table 19.2). Different investigators found that if large-diameter gravel is used, then formation sands can form stable bridges over openings within the

TABLE 19.2 Gravel Permeability

Sample Material	Median Diameter (mm)	Measured Permeability [17] (D)	Cubic Packing Permeability (D)	Hexagonal Packing Permeability (D)
6/10 Mesh	2.68	2703	19,100	1580
8/12 Mesh	2.03	1969	10,900	893
10/20 Mesh	1.42	652	5,350	437
20/40 Mesh	0.64	71	1,090	89
40/60 Mesh	0.36	69	344	28

gravel pack provided that the openings are no more than three times larger than the particle's size. This was the basis for the first relationships suggested for gravel sizing, and the choice allows the formation sand to migrate into the gravel pack until bridging occurs. At this point, formation sand no longer intrudes into the gravel pack.

Later it became evident that for maximum productivity, the formation sand must be stopped at the outer face of the gravel pack and not be allowed to migrate into the pack. This was necessary because it was proven that the mixing of the sand with the gravel could significantly reduce the permeability of the gravel pack [17], as shown in Fig. 19.6. The curves plotted represent the permeabilities of a

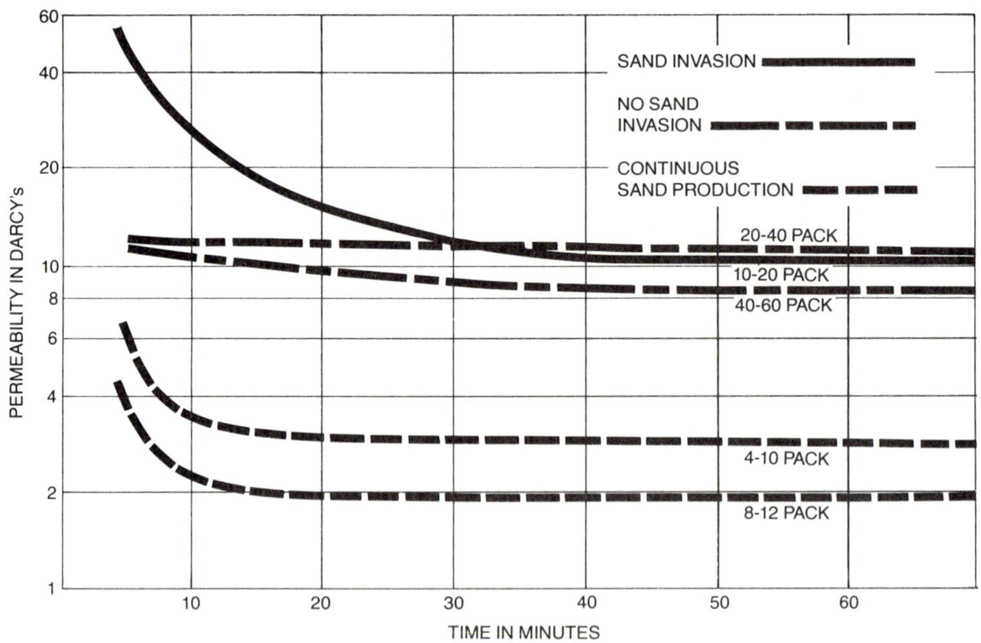

Figure 19.6 The permeability of a gravel pack shown as a function of time. The solid line denotes a case in which sand is continuously produced [18].

gravel pack consisting of the gravels tabulated in Table 19.2 placed in contact with the formation sand whose size distribution is that represented by the sieve analysis shown in Fig. 19.4. An oil is forced to flow first through the formation sand and then into the gravel pack. The permeability of the gravel pack will be a function of time as formation sand enters the gravel pack, reducing the permeability. Ultimately, however, the permeability will attain a steady-state plateau value.

The 4/10-mesh and 8/12-mesh gravels have a very high initial permeability as shown in Table 19.2; however, when these gravels were used in the gravel pack the permeabilities dropped to about 2 D (see Fig. 19.6) and furthermore, sand was continuously produced. The production of sand and the low final permeability make these two gravels unacceptable for gravel packing for the formation sand.

The 10/20-mesh gravel permeability declined to 9 D due to the invasion of formation sand into the gravel pack. In this case, the sand was ultimately restrained and not continuously produced.

Newer gravel sizing procedures as proposed by Schwartz [16], Saucier [19], and Stein [20] all call for a smaller-size gravel than the earlier approaches. Every effort is made to prevent formation sand from entering into the gravel pack consistent with the need to maintain high pack permeability. All of the methods for sizing gravel in current use yield about the same size requirement and generally these methods will, for example, select that gravel size shown in Fig. 19.6 to be the best. Since nearly all methods for selecting gravel yield approximately the same answer, we will restrict our attention to only one of the several methods now used.

Gravel selection. Schwartz [16] has suggested criteria for use in selecting gravel pack material based on formation sand analysis and the fluid velocity through the slots of the screen or the slotted pipe used to restrain the gravel. Schwartz defines uniform sands as those with uniformity coefficients <3, and nonuniform sands are defined to be those with coefficients >5. He notes that sand control using a relatively uniform gravel becomes more difficult as the uniformity coefficient of the formation sand increases. To allow for this difficulty, Schwartz proposed the following steps in selecting the gravel.

1. Estimate the gross production rate and the producing interval. Next estimate a slot size and pattern using manufacturers' guides. The rules governing the choice of an appropriate screen or liner are discussed in the next section. Gravel sizing is an iterative procedure since the choice of the slot size depends on the size of the gravel and vice versa. Calculate the entrance velocity into the screen or liner by dividing the production rate by 50% of the slot area.

2. Determine a critical sand grain size (d_c), which may also be called the *design point*, as follows:

- For sands with $U_c < 5$ and entrance velocities < 0.015 m/sec (0.05 ft/sec),

$$d_c(10) = d_{10}$$

- For $U_c > 5$ *or* entrance velocities > 0.015 m/sec,

$$d_c(40) = d_{40}$$

- For extremely nonuniform sands ($U_c > 10$) and velocities near or above 0.03 m/sec (0.1 ft/sec)

$$d_c(70) = d_{70}$$

3. Determine the critical gravel diameter (D_c), where

$$D_c(x) = 6d_c(x)$$

Note that this considers that the gravel pack will very nearly assume the hexagonal close packed configuration since the choice of 6 as a multiplying factor is dictated by Eq. (19.6).

4. Select the gravel size range by assuring that $D_c(x)$ is obtained and that $U_c < 1.5$. The sieve analysis of the gravel should satisfy these two conditions.

Example 19.3 Selection of Gravel Size

Given the formation sand sieve analysis as shown by Fig. 19.7, select the proper gravel size for a well that is expected to produce at a rate such that the fluid velocity through half of the open screen area is about 0.02 m/sec.

Solution Since the uniformity coefficient of the sand is <5 (see Example 19.1), but the fluid velocity exceeds 0.015 m/sec, the critical sand grain diameter is $d_c(40)$ and therefore

$$D_c(40) = 6d_c(40) = 6(0.122 \text{ mm}) = 0.73 \text{ mm}$$

This point is indicated on Fig. 19.7, which also shows the sand sieve analysis. To select the gravel size, we assume that the gravel size analysis is approximately a straight line and insist that $U_c = 1.5$ or less for the gravel. Let us try $U_c = 1.5$ or

$$\frac{D_{40}}{D_{90}} = 1.5$$

Solving we find

$$D_{90} = \frac{D_{40}}{1.5} = \frac{D_c(40)}{1.5} = \frac{(0.73)}{1.5} = 0.49 \text{ mm}$$

This point is shown in Fig. 19.7 and a straight line (dashed) is drawn between the two points $D_c(90)$ and $D_c(40)$. The intersection of this line with the 0 and 100 cumulative percents determines the range of gravel sizes. For the large gravel, we see that a size slightly greater than 20 mesh is permissible, but such a size is not available commercially. Select 20 mesh. At the smaller end, 40 mesh is found. Thus, the choice would be 20/40 mesh. This is a commercially available gravel. Note that this is the same gravel size range found to be best in the experimental study shown in Fig. 19.6.

Liner or Screen Selection

Slotted liners and screens are used to support the gravel of the pack (see Fig. 19.3). There are a great variety of them available. Their successful use in gravel packing depends mainly on the proper sizing of their openings.

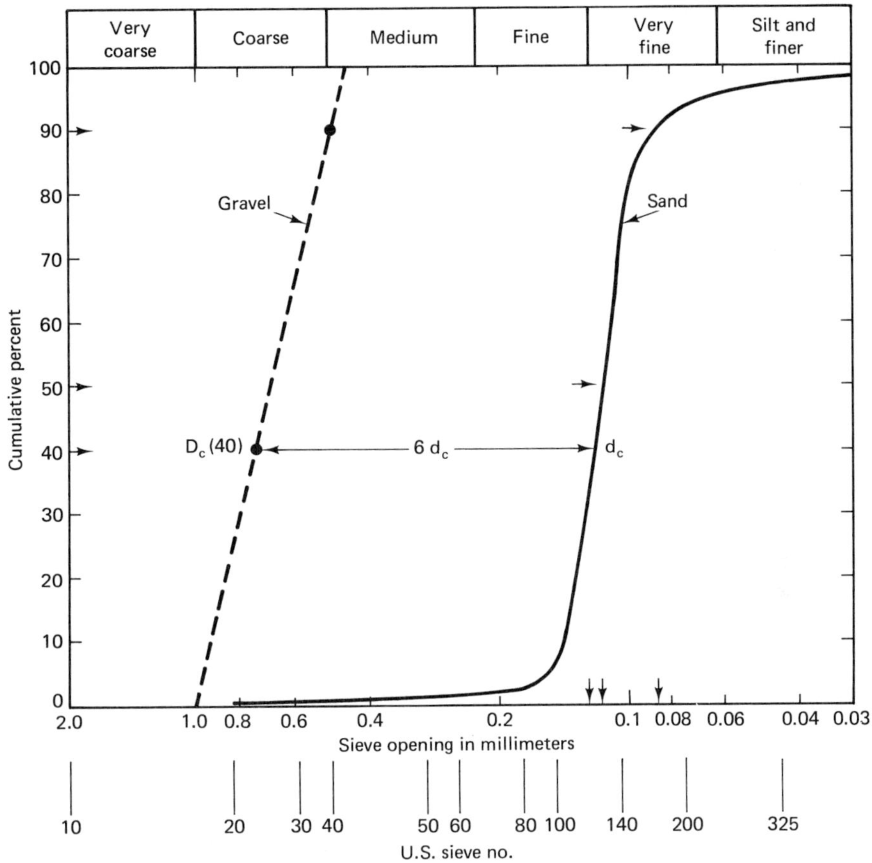

Figure 19.7 Sieve analysis plot and corresponding gravel sizes.

Slotted pipe for liners is milled in various patterns. The following patterns are available: staggered vertical slotted pipe, nonstaggered vertical slotted pipe, multiple vertical slotted pipe, and horizontal slotted pipe. The slots can be machined to have an undercut v-shape, narrow towards the formation and widening in the direction of flow toward the inside of the liner to minimize particle bridging.

Vertical slot widths are available from 0.03–1.25 cm straight cut or undercut. Horizontal slotted pipe slot widths can be 0.025–2.3 cm undercut.

Wire-wrapped screens are made by wrapping wire (which is normally stainless steel) on a pipe having previously machined holes or large slots. Wire-wrapped screen types available include: grooved type, wrapped-on pipe, ribbed type, and all welded.

Comparison of slotted liners to wire-wrapped screens. Although the slotted liners are less expensive than the wire-wrapped screens, they have some disadvantages as compared with the screens:

1. They cannot be constructed with slots of very small width, and the smallest available width is too large for certain applications.
2. They are more subject to corrosion and erosion than the screens.

Slot size. The choice of the slot opening is dictated by the need for absolute stoppage of the gravel. Thus, the slot size should be about the same size as the *smallest* gravel grain size [1]. This stringent requirement often dictates a slot opening that is smaller than the smallest sizes available in liners. Thus, screens are then the gravel restraint of choice. When 20/40-mesh gravels are used, the slot size required would be less than 0.45 mm. In this case, either a slotted pipe liner or a screen can be used. Screens often have the advantage of presenting a greater area of flow and are generally more stable to chemicals than are liners; however, they are more expensive.

Gravel pack thickness. An effective gravel pack should have a minimum thickness equal to four or five times the gravel grain diameter [21]; however, in practice much larger pack thicknesses are used. The minimum accepted thickness is about 0.7 cm, but 2–2.5 cm is recommended if at all possible.

Quality of Gravel

Gravel is commonly available in a number of sizes, such as those shown in Table 19.2. Gravel quality, angularity, size distribution, compaction, and permeability are all important to a successful pack.

Sparlin [17] recommends that no more than 10 wt% of the gravel should be either larger or smaller than the specified size range. Schwartz [16] recommends that no more than 12 wt% be retained on the coarsest screen and not more than 0.2 wt% pass the finest screen. Size should be controlled so that the proper formation-to-gravel-size ratio is maintained and absolute permeability is not adversely affected. Fines and dirt sometimes result from crushing in sacks during transport.

Acid solubility should be determined to prevent voids in the pack that may be created during acid stimulation. Gravel strength is important in some applications. In such cases, material containing high concentrations of weak mineral grains should not be used. Rodgers [23] recommends that the gravel consist of not less than 95 wt% quartz or siliceous minerals. Soft or ''earthy'' material like shale, gypsum, or anhydrite should not be included.

Gravel angularity and size distribution affect gravel permeability and compaction. For the same gravel-to-sand ratio, angular gravel is less likely to be invaded by sand because sand bridges more readily on the angular surfaces. More consistent packing is obtained with rounded gravel.

The Gravel Packing Process

The gravel packing tool and its operation. Gravel packs can be placed in both cased and open holes. If casing is required to permit production from selected zones or to exclude undesirable shale streaks, the gravel is then placed in the annular space between the liner or screen and the casing. In the open hole, casing is usually set above the producing interval, the hole below the casing is underreamed, and the gravel is packed in the annular space between the slotted liner or screen and the formation wall. While not recommended practice, multiple gravel packs positioned across different producing intervals are technically possible [24].

There are three main components of the screen assembly—the telltale screen, production screen, and a blank liner (see Fig. 19.8). The blank pipe provides a gravel reserve that will settle as gravel around the screen becomes compacted during production. The telltale screen is the part of the system that detects the arrival of gravel and filling of the lower portion of the annular space. A sudden pressure increase occurs when gravel completely covers the telltale screen. This indicates that the first stage of the operation is complete.

A very important component of the gravel packing tool is the crossover which permits fluid to flow from the drillpipe to the annular space and vice versa. A sketch showing the path of flow through the crossover is shown in Fig. 19.9. This flow arrangement is crucial because it permits the gravel packing operation to be carried out in a single trip without having to flow downward through the annular space between the casing and the drillstring, which is undesirable because larger volumes of fluid are required and it is difficult to prevent contamination of the fluid with dirt, scale pipe dope, or other extraneous materials commonly found in this space.

Another important component of the gravel packing tool is the washpipe, which is used to help direct the flow. It is shown explicitly in Fig. 19.9 and can be identified in Fig. 19.8(a) as the pipe that is seated within the O-ring sub. In this position circulated fluid is forced to flow through the lower telltale screen, which is called *lower circulation*. If the washpipe is raised so that it is no longer seated in the O-ring sub, then fluid can circulate through both the telltale screen and the production screen. This is called *upper circulation*.

The gravel packing tool is a mechanically complex, remarkably conceived device designed to permit the entire operation to be carried out in one trip. It can be set into various flow configurations based on commands transmitted by simply raising, lowering, or rotating the drillstring or by dropping ball sealers into the drillstring.

The entire assembly is run into the hole in the configuration shown by Fig. 19.8(a). When the production screen is positioned opposite the perforations, the sump packer is set by rotating the drillstring and the production packer is set by the application of fluid pressure once a ball sealer dropped into the drillstring is seated (see Fig. 19.8(b)). At this point a mixture of gravel and fluid is pumped down the drillstring, through the crossover, and into the annular space to be filled with gravel. When the washpipe is positioned as shown in Fig. 19.8(b), the fluid containing gravel circulates through the telltale screen where the gravel is separated from the fluid and begins to "dehydrate" forming a compacted gravel pack. The fluid stripped of the gravel returns through crossover and up the annulus (see Fig. 19.8(c) or 19.9). This phase of the operation is designed to enhance the placement of the gravel at the bottom of the production screen.

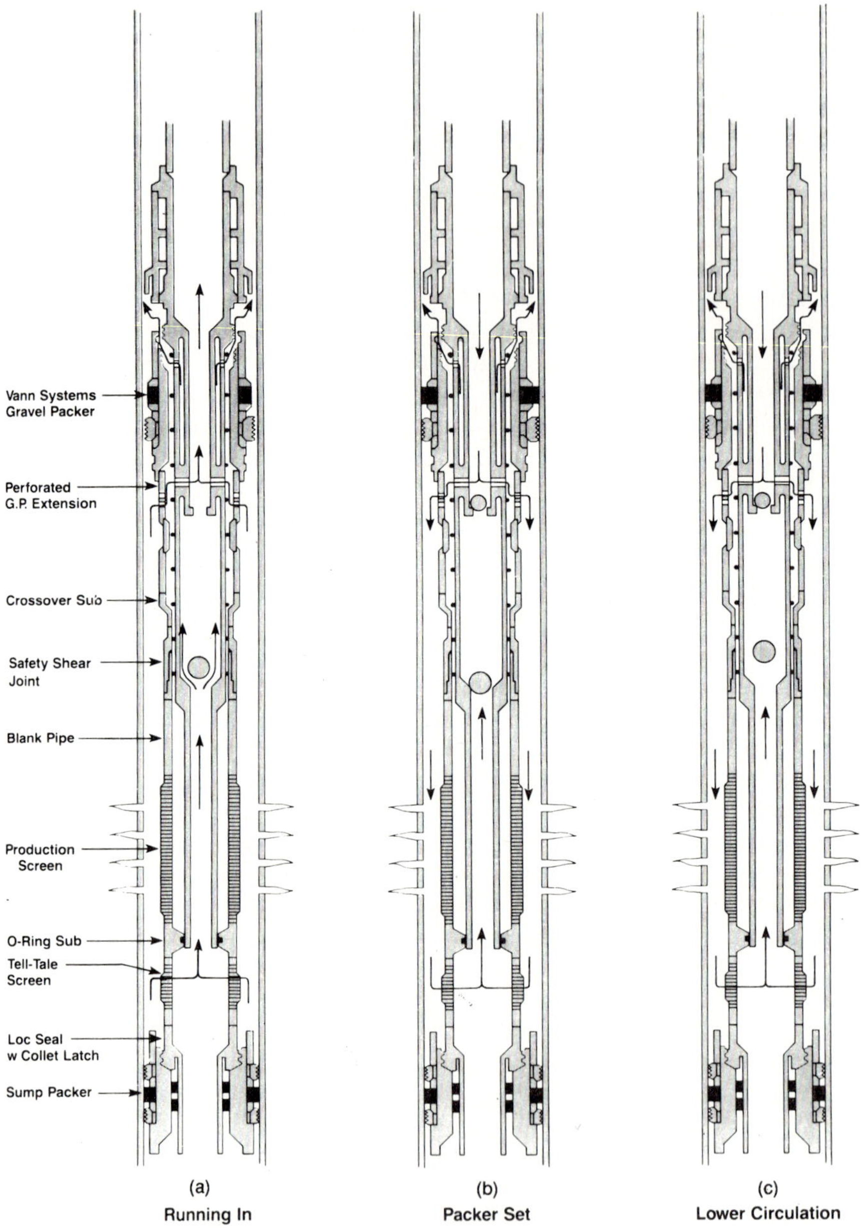

Figure 19.8 The various configurations of a gravel packing tool (a) in the running, (b) packer set and (c) lower circulation.

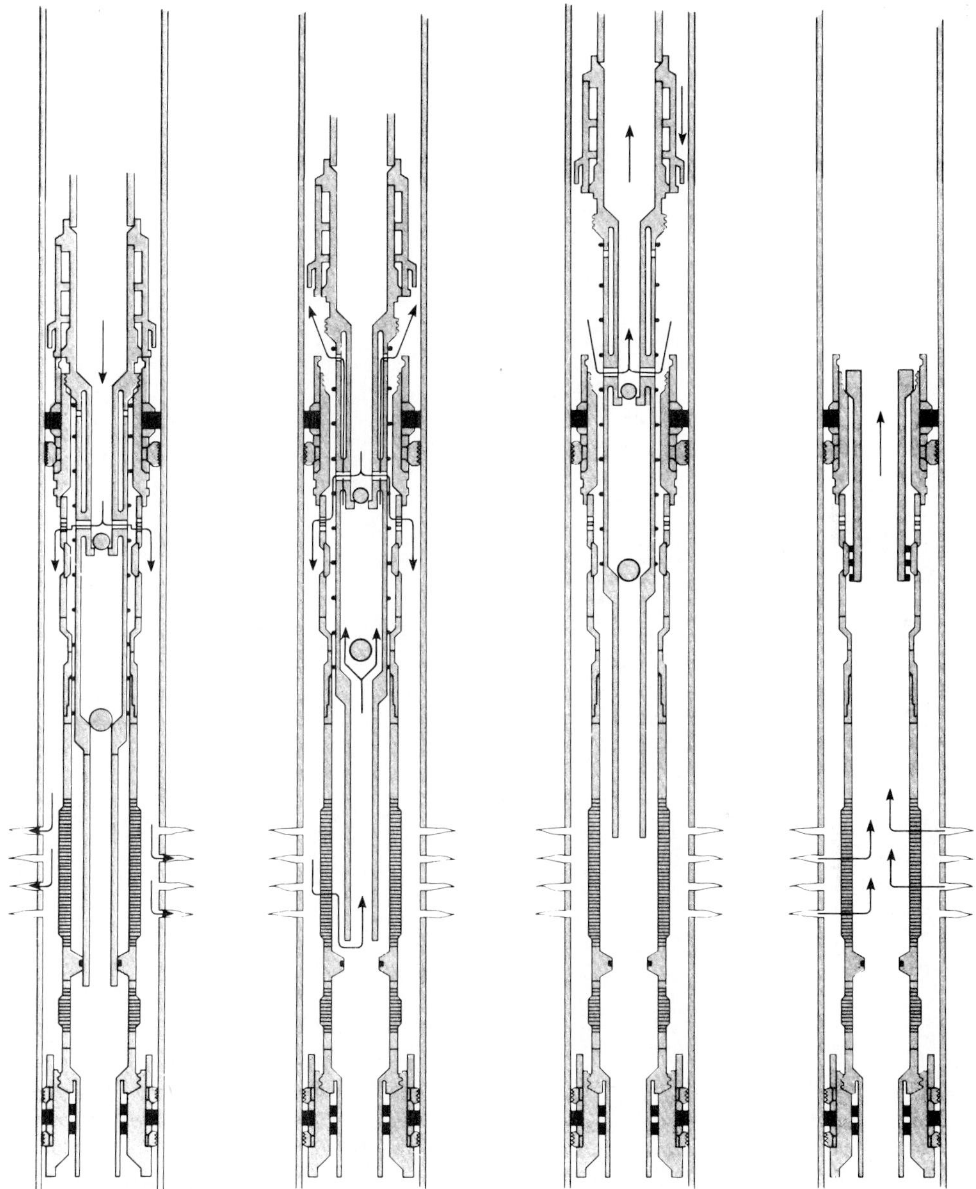

Figure 19.8 (continued) Various configurations of a gravel packing tool (d) squeeze position, (e) upper circulation, (f) reversing out, (g) production.

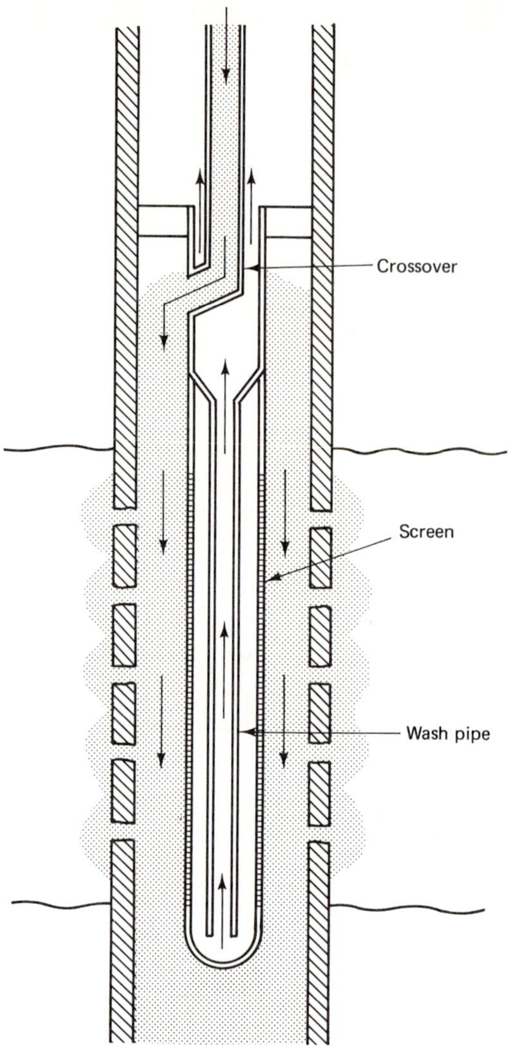

Figure 19.9 Simplified sketch of gravel packing tool showing the operation of the crossover.

Once the pressure drop across the telltale screen becomes significant, signalling the compaction of gravel around this screen, the tool is placed into the "squeeze" position. In this configuration [Fig. 19.8(d)] the fluid is not allowed to circulate back to the surface, but is, in fact, forced into the producing formation. This is an important phase of the operation because it is important to pack the perforations with gravel. Clearly, if the perforations fill with low-permeability sand instead of gravel the pressure loss attending production will be higher than if the perforation is filled with gravel [25]. This additional pressure drop can be significant.

It is also important to pack gravel into any void spaces that may exist outside of the casing. Thus, fluid injection with the tool in the squeeze configuration should be continued until the surface pressure reaches its maximum value. The flow should then be stopped until the pressure declines and then it should be restarted again. This surging procedure can be continued until the volume injected becomes small.

Once the perforations are fully packed with gravel, the washpipe is raised to permit fluid to be circulated through the production screen [Fig. 19.8(e)]. This upper circulation stage is intended to ensure that the production screen is completely filled with gravel.

At the final stage, the gravel packing tool is set into the reverse mode and gravel-laden fluid that remains in the drill string is forced upward by fluid injected into the annulus and flow through the crossover. The flow during the reverse stage does not disturb the gravel pack [see Fig. 19.8(f)].

That amount of gravel contained in the blank liner above the screen at the time the tool is fixed into the reverse configuration settles to form a gravel reservoir. The height of the settled gravel above the top of the production screen should amount to 10–15 m. Thus, the height of the blank pipe will depend on the gravel concentration in the injected fluid.

Finally, by rotating the drill string, the crossover and washpipe arrangement is detached from the screen and is removed. The gravel pack is then in the production configuration [Fig. 19.8(g)].

Fluid selection. Prior to the late 1960s the only methods used for gravel packing were the low-viscosity fluids such as acid, crude oil, or diesel with small gravel concentrations and using high pumping rates. This combination of low-viscosity fluids, small gravel concentrations, and high pumping rates was necessary to prevent gravel from bridging in tubes or in gravel packing tools. To improve success and the productivity of a gravel pack during this period emphasis was placed on improved completion techniques, i.e., carefully filtered completion fluids, good perforating techniques, and prudent gravel sizing. Every effort was made to avoid mixing of the gravel with any foreign solid particles (mud, perforating debris, etc.) by carefully cleaning the well and particularly with the formation sand by correctly sizing the gravel, because such mixing was known to damage the permeability of gravel [17].

In many cases, although clean fluids had been used, the perforation practice was excellent, and the gravel had been carefully sized, a surprisingly low productivity was often observed. To explain the difficulty, Sparlin and Copeland [22] suggested that the low permeability is due to mixing of gravel with the formation sand during the gravel placement process. As shown by Fig. 19.6 mixing of the formation sand with gravel can severely reduce the permeability of the gravel pack [17].

During the gravel packing operation, the main source of gravel and sand mixing was during the squeezing stage while the gravel is being pumped outside the perforations. When the gravel particles enter the perforations they can be traveling at high velocities and essentially sand blast the formation, thereby promoting gravel mixing with the formation sand. Also because gravel concentrations were low, large volumes of low-viscosity fluid were circulated thereby increasing the opportunity for mixing.

These difficulties have led to the widespread application of high-viscosity fluids for transporting gravel. These fluids are formulated to carry gravel in relatively high concentrations (1200 kg/m^3 of polymer solution is a typical value) without excessive settling. Thus, gravel can be transported through the perforations even at the modest fluid injection rates needed to control the injection pressures during the squeeze operation.

The fluids used for transporting gravel require a high viscosity to suspend sand but must readily flow into the formation to permit efficient transport of gravel into the perforations and cause the formation of a compact gravel/sand interface. Thus, we have two competing factors that come into play and a compromise is required. To perfectly suspend proppant highly cross-linked fluids are used in fracturing, but in general, these are unsuitable for gravel packing because high pressures would be required to inject them into the formation. A popular fluid used for gravel packing is hydroxyethyl cellulose (see Chap. 3) used at a concentration of 6–12 kg/m^3. A hydroxyethyl cellulose (HEC) solution containing 10 kg/m^3 has an apparent viscosity of 3000 cp at 0.03 sec^{-1} and 120°F when it contains 2 wt% NH_4Cl [26]. Scheuerman has shown that as the apparent viscosity at 0.03 sec^{-1} drops to 1000 cp, the fluid no longer has the capacity to carry sand [26]. Thus, the low shear rate apparent viscosity is important since it provides a measure of the minimum value that can be tolerated at any time during the gravel packing operation.

In addition to the need to transport gravel and be readily injected into the formation, the gel must also be easily produced when the well is put on production. To facilitate the recovery of the fluid, breakers are always included. For HEC, acids are often added to reduce the final viscosity [26]. The broken fluid should contain little residual to minimize formation damage.

By the 1980s a derivatized HEC was available, which possessed a nominal residue level of practically zero [27]. Despite these residual low levels, it is found that cores treated with this HEC gel would still leave a highly damaged permeability even after breaking and being displaced by brine [28]. Furthermore, difficulties were sometimes found in injecting the polymer solution into the cores without applying excess pressure. In addition, it appeared that field results from field jobs, which were apparently carried out without any difficulty, were not as good as expected. Thus, there appears to be an additional problem with HEC gel which is unrelated to the residual that remains upon breaking.

Chauveteau and Kohler [29] report that microgels, which are sometimes called *microfisheyes*, present in polysaccharide solutions can build up on the surface of the formation as well as invade the formation and clog the pores. Microgels are found in HEC solutions even when the polymer has been carefully purified and does not form a residue upon breaking [27]. Apparently, microgels are small crystals of the polymer ranging in size from 10–20 μm. This is in the size range to be damaging. Microgels are bits of the original polymer that have not dissolved in the water. It has been reasoned that these undissolved microcrystals are responsible for the high-injectivity pressures and the incomplete permeability recovery that cause the poorer-than-expected performance of gravel packs [30].

It has been reported that the microgels can be filtered from the polymer solution by passing it through a 10 μm filter. The difficulty with implementing this concept in the field is that like the formation rock, the filter will apparently become rapidly clogged. Cole et al. [27] have measured the concentration of microgels in HEC solutions. Although the number varies from sample to sample and in commercial products other nonhydratable impurities are found, they reported concentrations of 100–200 particles/(mm)^3.

It has been shown that the concentration of microgel particles can be reduced by subjecting the polymer solution to mechanical shear [31, 32]. Cole et al. [27] found that shearing the polymer solution by passing the solution through an orifice could reduce the concentration of microgels by a factor of about 3. This observation forms the basis of a two-stage filtration process to first dissolve and then to remove microgels—a prefilter [31] and a final filter. The gel prefilter is a high-pressure housing where hydraulic shearing of the polymer solution occurs. It contains a stack of several perforated modules with pressure-sensitive circular springs inserted between adjacent modules to provide adjustable gaps between the modules and the spring. The perforations are intended to disperse the microgels, and the variable-width gap causes the gel to be sheared to further reduce the microgel size. This system is intended to shear the solution in a controlled fashion to render the gel solution filterable but not to impart so much shear as to degrade the gel solution to the extent that it cannot carry gravel. There is an optimum degree of shear. Too much will reduce the gel's ability to carry gravel and too little will not produce a filterable solution.

The last stage of the process is a 10 μm filter which removes most of the microgels. The gels produced by this combination of shearing and then filtering seem to have acceptable properties. Figure 19.10 shows that the sheared and filtered polymer gel is easier to recover from the rock and leaves smaller residual damage than the same gel that has not been treated.

Figure 19.11 shows that the fluid-loss rate into a core is increased after the concentration of microgels has been decreased by shearing and filtration. Thus, the buildup of a compressible microgel cake on the face of the formation is retarded by the combination of shearing and filtration.

In addition to hydroxyethyl cellulose, hydroxypropyl cellulose, methyl cellulose, and xanthan have all been used for gravel packing. These gels are all improved by the shearing and filtration processes [31]. Xanthan was particularly interesting in that the shearing and filtration process did not appear to affect its ability to carry gravel. Thus, xanthan may become the polymer of choice in cases where the final gravel pack productivity is critical [31].

Gravel buildup patterns. When fluids having a low viscosity are used to transport gravel, the buildup of gravel is more or less from the bottom up. The gravel settles rapidly even when partially dehydrated against either the formation or the production screen. Polymer gels of even moderate apparent viscosities, such as those described in the previous section, are designed to limit the settling rate. A limit of 0.2 cm/min is sometimes imposed on the gravel settling rate [27]. When the gravel settling rate is smaller than this, the gravel does not settle per-

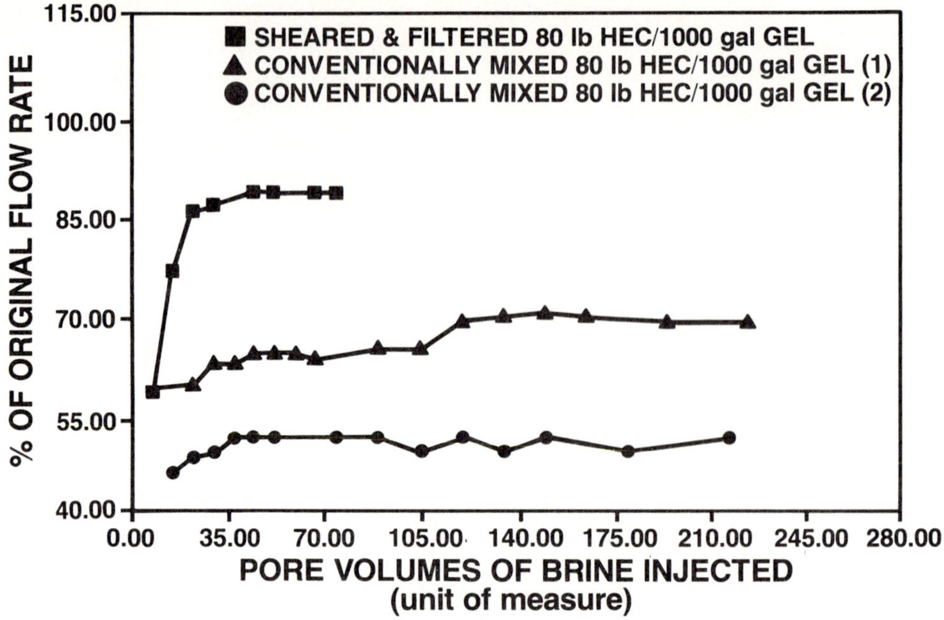

Figure 19.10 Flow rate recovery test. Hydroxyethyl cellulose injected into a Berea core followed by a brine. The vertical axis represents the flow rate after displacing the hydroxyethyl cellulose compared to the original rate [27]. [From C. Cole et al., SPE 17480, presented at the 58th Annual California Regional SPE Meeting, Long Beach (1988). With permission of Halliburton Services.]

ceptibly as it is separated from the gel. Figure 19.12 shows the gravel buildup during the squeeze phase of the gravel packing operation. Note the gravel separates from the gel at the formation interface, compacts within the perforation tunnel, and then builds up as nodes about perforations. These nodes can span the entire distance between the casing and the screen, preventing further gravel flow into the lower perforations. To ensure that all perforations are packed with gravel, it is necessary to repeat the squeeze several times. Each time the pressure reaches its limit, flow should be stopped to permit the pressure to relax and then the squeeze process should be repeated.

In visualizing the mechanical processes occurring during the gravel packing operation, note the buildup is from outside inward as shown by Fig. 19.12. This is in sharp contrast to the mechanics of sand transport using low-viscosity fluids.

19.4 GRAVEL PACKING DEVIATED WELLS

Low-Viscosity Fluids

Large offshore oil and gas fields are generally produced through deviated wells drilled from a central platform. Many of these wells penetrate the producing horizon at high angles ($> 60°$). Gravel packing these wells is not routinely successful.

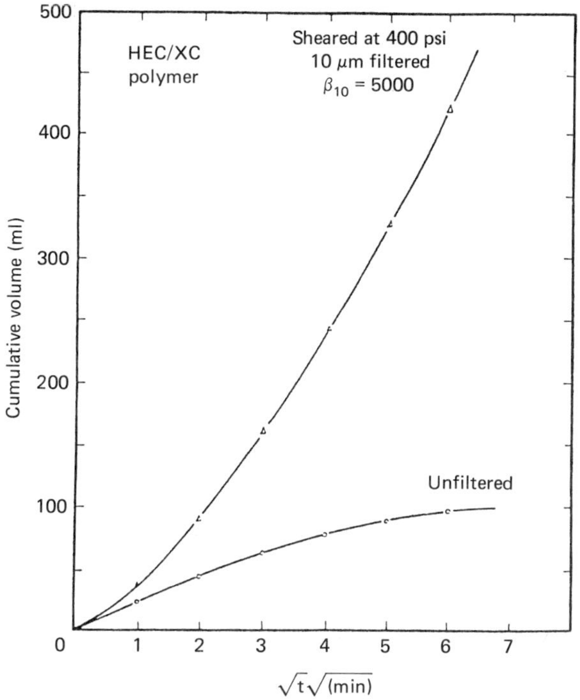

Figure 19.11 Graphical results of filtration test showing that sheared and filtered gel has a higher injectivity than untreated gel [31]. (With permission of the Society of Petroleum Engineers.)

The procedure becomes more difficult than packing near vertical wells, simply due to the influence of gravity. In high-angle holes, gravel conveyed in low-viscosity fluids tends to form dunes at the upper end of the screen and seal off the lower portion of the screen–wellbore annulus resulting in premature sandout. The lower portion of the interval to be gravel packed may be completely void of gravel. A sketch depicting premature sandout is shown in Fig. 19.13. Shryock [33] has demonstrated using a full-scale wellbore model that for deviations greater than 60° from the vertical (>60°, see Fig. 19.13), there are problems in obtaining complete packing of the system including difficulties in packing the perforations using either conventional low-viscosity fluids or viscous fluids to transport the gravel. Shryock observed that packing efficiency was significantly improved by increasing the resistance to flow in the annular space between the liner and the washpipe either by using larger diameter washpipes as proposed by Gruesbeck et al. [34] or by placing rubber baffles along the length of the washpipe as proposed by Maly et al. [35]. The reason for the improvement in gravel packing deviated wells as a result of increasing the flow resistance will be discussed in the next section; however, we should note that Shryock further reported that gravel packing efficiency can also be improved by using a gravel packing tool having a lower telltale screen combined with an intermediate viscosity carrier fluid (≈40 cp @ 100 sec^{-1}). This is similar to the fluids now used for packing vertical systems. His experiments included a study of highly deviated wells (≈80°).

When the telltale screen is located below the main production screen, the

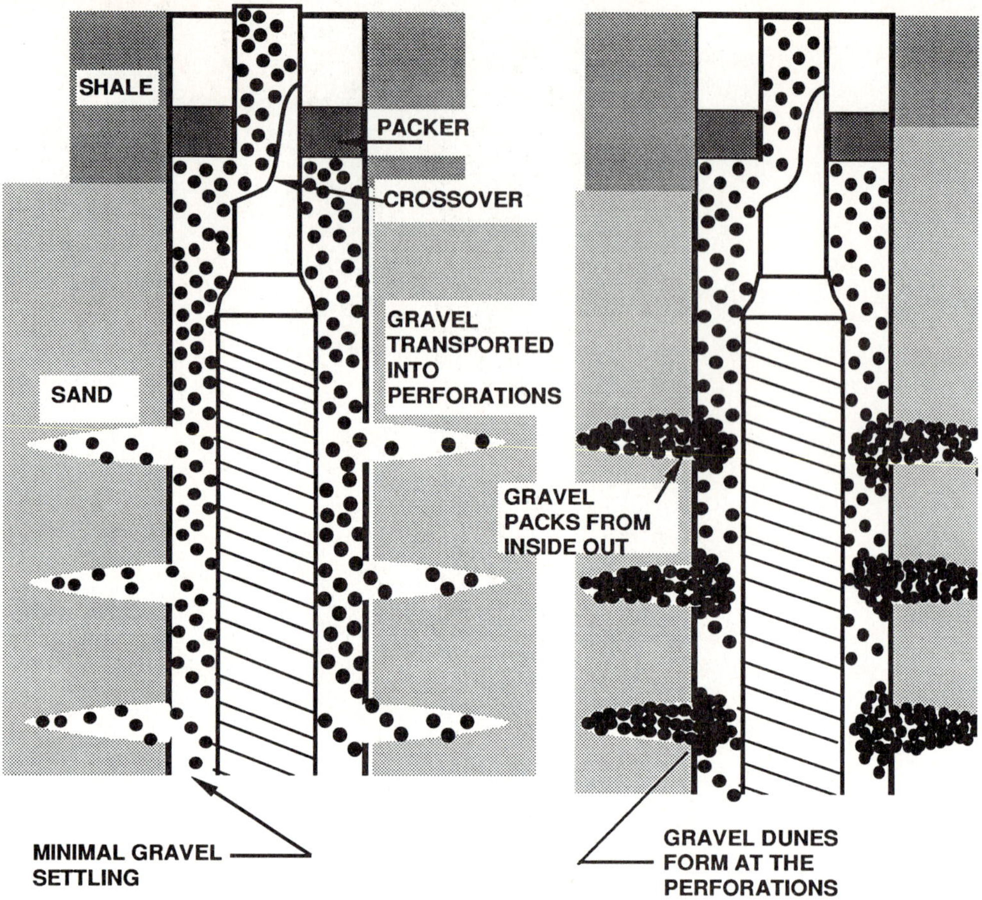

Figure 19.12 Sketch depicting gravel packing using viscous gels. Gravel is transported into the perforations and the gravel pack builds from outside inward.

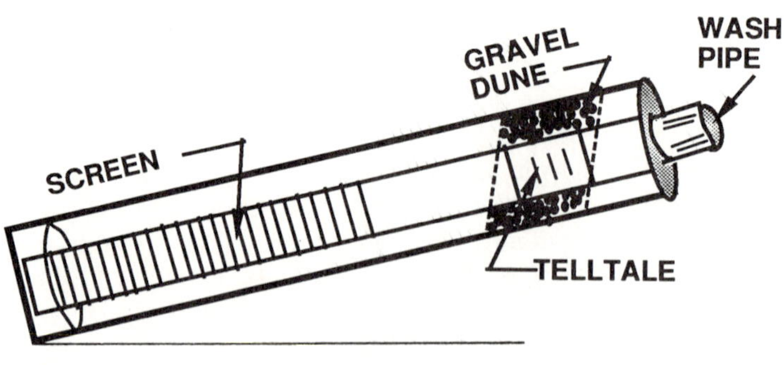

Figure 19.13 Sketch showing gravel dune bridging the annular space. No gravel can be transported to the lower part of the screen.

washpipe is inserted through an O-ring-sub between the telltale screen and the main screen as shown by Fig. 19.8(a). During the initial stages of gravel packing, all of the circulating fluids pass through the telltale screen (lower circulating configuration) initially filling the lower part of the pack. After this screen has been covered with gravel, thc washpipe can be raised mechanically out of the O-ring sub or a differential pressure valve can be opened allowing circulation through the main screen [Fig. 19.8(e)].

The use of a lower telltale screen permitted the full-scale model to be gravel packed, leaving only small voids in the channel over the dune and in the lower portion of the rathole. Thus, this procedure appears to be an alternative to restricting the flow in the washpipe–screen annulus. None of the proposed solutions appear to be totally satisfactory. Good gravel packing requires that both the perforation tunnels and void volume outside of the casing be well packed. This generally calls for squeezing carrier fluid into the formation during the initial stages of gravel placement, and this is best accomplished using a viscous carrier fluid. Once the perforations are packed, the tool can be placed in the circulating position. At this point, laboratory experiments indicate that a carrier fluid having an intermediate viscosity is advantageous. However, changing the carrier fluid viscosity during the course of the job is not an attractive procedure.

Hydrodynamic Considerations

Gruesbeck et al. [34] have shown in laboratory model experiments that increasing the diameter of the washpipe does reduce the tendency for premature sandout in deviated wells. In this section we will explore the reason for this seemingly tenuous relationship. The argument starts with the premise that the carrier fluid in its motion from the crossover port, where it enters the annular space between the screen and the casing, can effectively take only one of two flow paths. One flow path is the open space above the gravel dune, which will inevitably form in highly deviated wells when low viscosity fluids are used to transport the gravel, and a second path is within the annular space between the washpipe and the screen. Once a gravel dune forms, the flow of carrier fluid in that part of the annular region filled with gravel is quite small and can be neglected. The two possible flow paths are parallel and therefore

$$\left(\frac{\Delta p}{L}\right)_{\mathrm{o}} = \left(\frac{\Delta p}{L}\right)_{\mathrm{wp}} \tag{19.8}$$

where the subscript o denotes the pressure drop per unit length caused by the motion of the gravel-laden carrier fluid flowing over the top of the gravel dune and the subscript wp denotes the pressure drop per unit length resulting from the carrier flow in the washpipe–screen annulus. Equation (19.8) is instructive. It must always be satisfied for parallel paths. If the flow system is changed in such a way as to increase the flow resistance between the screen and the annulus, then fluid will be diverted and more will flow through the open area on top of the dune. The flows will adjust themselves so that Eq. (19.8) is satisfied.

Equation (19.8) shows in qualitative terms the advantage gained in increasing the washpipe diameter. Since the fluid velocity across the top of the dune is increased, then the tendency for gravel to be transported across the top of the dune rather than be deposited is correspondingly increased. Data presented by Gruesbeck et al. [34] verify this to be the case. The efficiency of gravel packing is increased dramatically by increasing the washpipe diameter. For the particular system studied, washpipe diameters in the range of 85–90% of the screen diameter provided good gravel packs even in highly deviated wells.

We can estimate the pressure drops given in Eq. (19.8) if the flows are in the fully developed turbulent regime by use of a friction factor (f) defined so that

$$\left(\frac{\Delta p}{L}\right)_o = \frac{2 f_o v_o^2 \rho}{R_o} + \psi \tag{19.9}$$

and

$$\left(\frac{\Delta p}{L}\right)_{wp} = \frac{2 f_{wp} v_{wp}^2 \rho}{R_{wp}} \tag{19.10}$$

Here v_o and v_{wp} are the velocities in the two parallel flow paths and R_o and R_{wp} are the respective hydraulic radii of the open area and the washpipe–screen annulus. The function ψ represents the additional contribution to the pressure drop due to the presence of the gravel particles contained within the fluid flowing through the open area above the gravel dune.

When the flow is turbulent, the friction factor can, with sufficient accuracy for our purposes, be represented as follows [35]:

$$f = \frac{0.0791}{\left(\dfrac{Rv\rho}{\mu}\right)^{0.25}} \tag{19.11}$$

If highly viscous fluids are used and the injection rates are small, the flow may be laminar. In such cases, the pressure drops would have to be determined by solving the hydrodynamic equations.

The hydraulic radius (R) used in Eq. (19.11) is defined [35] as

$$R = 4\,\frac{\text{area normal to flow}}{\text{wetted perimeter}} \tag{19.12}$$

For the washpipe–screen annulus the cross-sectional area is

$$A_{wp} = \pi(r_s^2 - r_{wp}^2) \tag{19.13}$$

where r_s is the inner radius of the screen and r_{wp} is the outer radius of the washpipe. The hydraulic radius is therefore

$$R_{wp} = 2(r_s - r_{wp}) \tag{19.14}$$

The calculation of A_o and R_o is slightly more complicated because the open area varies with the height of the gravel dune. Since most of the cases of interest to us will be for small open areas, the geometry of the screen will not matter. For simplicity, the geometry shown in Fig. 19.14 is assumed to represent the

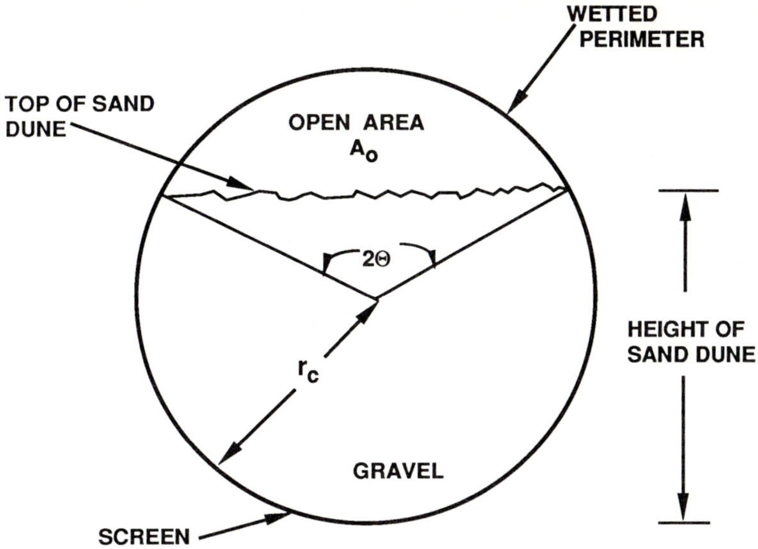

Figure 19.14 Sketch showing simplified gravel dune geometry.

gravel dune. It will be accurate when the screen is covered with gravel and the open area is small. In this case

$$A_o = r_c^2(\theta - \cos\theta \sin\theta) \tag{19.15}$$

and

$$R_o = 2r_c \left[\frac{\theta - \cos\theta \sin\theta}{\theta + \sin\theta} \right] \tag{19.16}$$

The additional pressure drop related to the presence of the gravel in the fluid flowing above the dune is obtained from the experimental results present by Gruesbeck et al. [34] and reproduced in Fig. 19.15. The concentration c_* appearing in Fig. 19.15 is a dimensionless quantity defined as the volume of gravel per unit volume of fluid and gravel. Furthermore, it is not necessarily the injected gravel concentration. Since that portion of the carrier fluid flowing through the washpipe–screen annulus does not contain gravel, the gravel concentration in the open area will be higher than the injected value. To determine this concentration, it is necessary to make material balances.

Overall Balance Equations

To determine whether or not a gravel dune that completely obstructs the transport of gravel will form, mass balances for both the gravel and the carrying fluid will be required. If i is the volumetric injection rate (gravel plus fluid), and if q_{wp} and q_* are the volumetric rates in the washpipe–screen annulus and in the open space, respectively, and assuming the fluid to be incompressible and neglecting the fluid loss to the formation, which should be relatively small when the gravel packing tool is in the circulating configuration, we find that

$$(1 - c_i)i = (1 - c_*)q_* + q_{wp} \tag{19.17}$$

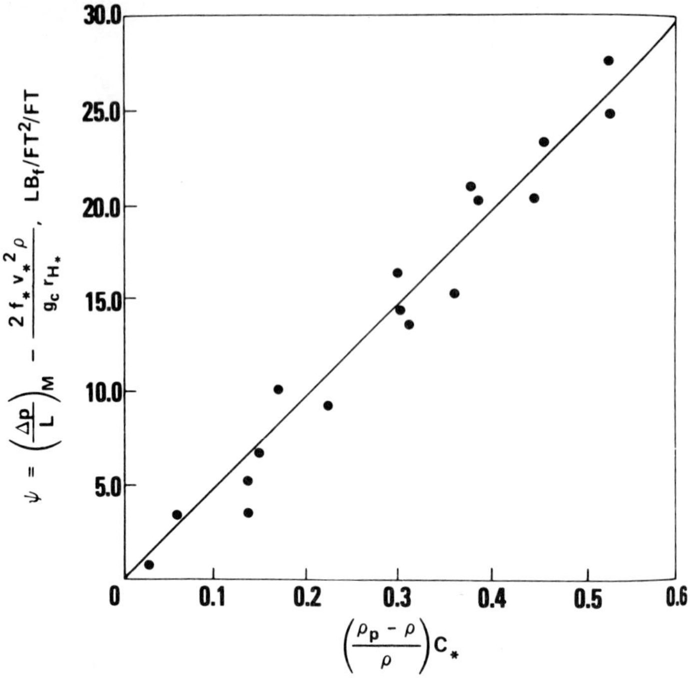

Figure 19.15 Plotted data for pressure gradient due to particles in a slurry over a horizontal equilibrium bank [34]. (With permission of the Society of Petroleum Engineers.)

Equation (19.17) is a volume balance on the carrying fluid. c_i is the injected gravel concentration expressed as volume of gravel per unit volume of gravel plus fluid. c_* is, as was previously noted, the concentration of gravel in the fluid flowing through the open area.

It is obvious that

$$q_* = v_o A_o \tag{19.18}$$

and

$$q_{wp} = v_{wp} A_{wp} \tag{19.19}$$

Similarly, a volume balance on the gravel takes the simple form

$$i c_i = q_* c_* \tag{19.20}$$

Equation (19.20) requires that all of the injected gravel be transported across the dune. This equation is, of course, only valid if the dune has attained an equilibrium height and in fact all of the gravel is transported across the dune. The condition for attaining an equilibrium height is an important factor in designing gravel packs. This will be discussed in the next section.

Example 19.4 Fluid Velocity in the Open Area

For $\Theta = 30°$ (see Fig. 19.14) calculate the fluid velocity (v_o) given the data presented in Table 19.3. Note that the overall injection rate is given and the velocities v_o and

TABLE 19.3 Parameters for Conventional or Low-Viscosity Gravel Pack

Quantity	Symbol	Value
Casing radius	r_c	0.075 m
Screen radius	r_s	0.024 m
Washpipe radius	r_{wp}	0.018 m
Injection rate	i	0.2 m^3/min
Gravel concentration	c_i	0.023
Gravel size	—	20/40 mesh
Gravel density	ρ_g	2.5×10^3 kg/m^3
Fluid viscosity	μ	10^{-3} kg/m-sec
Fluid density	ρ	10^3 kg/m^3

v_{wp} must be proportioned so that the pressure drops are the same as required by Eq. (19.8).

Solution For $\Theta = \pi/6$, Eq. (19.15) gives

$$A_o = (0.075)^2 \left[\frac{\pi}{6} - \cos\frac{\pi}{6}\sin\frac{\pi}{6}\right] = 5.1 \times 10^{-4} \text{ m}^2$$

and

$$R_o = 2(0.075)\left[\frac{\dfrac{\pi}{6} - \cos\dfrac{\pi}{6}\sin\dfrac{\pi}{6}}{\dfrac{\pi}{6} + \sin\dfrac{\pi}{6}}\right] = 1.33 \times 10^{-2} \text{ m}$$

Similarly, for the washpipe–screen annulus,

$$A_{wp} = \pi[(0.024)^2 - (0.018)^2] = 7.9 \times 10^{-4} \text{ m}^2$$

and

$$R_{wp} = 2(0.024 - 0.018) = 1.2 \times 10^{-2} \text{ m}$$

To initiate the calculation, let us assume that $q_{wp} = 0.12$ m^3/min. This leaves $q_* =$ 0.08 m^3/min. Given q_{wp}, the velocity v_{wp} is

$$v_{wp} = \frac{0.12}{60(7.9 \times 10^{-4})} = 2.53 \text{ m/sec}$$

The Reynolds number in the washpipe–screen annulus is

$$N_{Re}^{wp} = \frac{1.2 \times 10^{-2}(2.53)(1 \times 10^3)}{1 \times 10^{-3}} = 3.04 \times 10^4$$

From Eq. (19.11)

$$f_{wp} = \frac{0.0791}{(3.04 \times 10^4)^{0.25}} = 6 \times 10^{-3}$$

The pressure drop per unit length due to the flow in the washpipe–screen annulus is, therefore, from Eq. (19.10)

$$\left(\frac{\Delta p}{L}\right)_{wp} = \frac{2(6 \times 10^{-3})(2.53)^2\,(1 \times 10^3)}{1.2 \times 10^{-2}} = 6400 \text{ Pa/m}$$

The next step is to calculate the pressure drop attending the flow of gravel-laden fluid through the open area. The gravel concentration in the open area is given by Eq. (19.20) as follows:

$$c_* = \frac{(0.2)(0.023)}{0.08} = 0.0575$$

and therefore

$$\frac{\rho_g - \rho}{\rho} c_* = 1.5(0.0575) = 0.086$$

The quantity ψ can be read from Fig. 19.15. Thus

$$\psi = 3.5 \frac{\text{lb}_f}{\text{ft}^2\text{-ft}} \left(\frac{\text{ft}^2}{144 \text{ in}^2}\right)\left(\frac{6985 \text{ Pa}}{\text{lb}_f/\text{in}^2}\right)\left(\frac{\text{ft}}{0.3048 \text{ m}}\right) = 560 \text{ Pa/m}$$

This is the contribution to the pressure drop due to the presence of the gravel. The fluid velocity in the open area is

$$v_o = \frac{0.2 - 0.12}{60(5.1 \times 10^{-4})} = 2.61 \text{ m/sec}$$

and the Reynolds number in the space above the sand dune is

$$N_{Re} = \frac{(1.33 \times 10^{-2})(2.61)(1 \times 10^3)}{1 \times 10^{-3}} = 34{,}800$$

The friction factor is

$$f_o = \frac{0.0791}{(34{,}800)^{0.25}} = 5.79 \times 10^{-3}$$

Thus based on Eq. (19.9)

$$\left(\frac{\Delta p}{L}\right)_o = \frac{2(5.79 \times 10^{-3})(2.61)^2(1 \times 10^3)}{1.33 \times 10^{-2}} + 560 = 6496 \text{ Pa/m}$$

Since this pressure drop is about the same as that calculated for the washpipe–screen annulus, then the assumed q_{wp} is correct and the fluid velocity in the open area is 2.61 m/sec.

Equilibrium Dune Height

Experiments show that if the fluid velocity on top of the dune is high enough, then the dune attains an equilibrium height. Thus an equal number of gravel particles are lifted from the top of the dune as are deposited by settling. The fluid velocity for which this equilibrium is observed will be called a *critical fluid velocity* v^*. If the actual fluid velocity is greater than the critical value ($v_o > v^*$), then the height of the dune will decrease; that is, more gravel particles will be stripped from the top of the dune than are deposited. The height will decrease until $v_o = v^*$.

The critical velocity will certainly depend on the particle concentration as well as the viscous, inertial, gravitation, and buoyant forces that act on the gravel particles as they are transported over the equilibrium bank. Four dimensionless groups that include these forces are (1) the Reynolds number based on the particle settling velocity and the hydraulic radius of the open area; (2) the Reynolds number based on the particle settling velocity and the particle diameter; (3) the dimensionless density difference between the particles and the carrier fluid; and (4) the gravel concentration. Gruesbeck et al. [34] have reported the following empirical correlation verified by a series of experiments using a horizontal flow model:

$$\frac{v^*}{v_s} = 15 \left[\frac{R_o v_s \rho}{\mu}\right]^{0.39} \left[\frac{D v_s \rho}{\mu}\right]^{-0.73} \left[\frac{\rho_g - \rho}{\rho}\right]^{0.17} c_*^{0.14} \tag{19.21}$$

where v_s is the gravel particle settling velocity in a quiescent fluid at infinite dilution.

Equation (19.21) is an interesting and important result. It shows that the critical velocity increases approximately as $v_s^{2/3}$. This is a trend that makes considerable intuitive sense. One would expect that particles which settle faster require higher open area velocities to yield an equilibrium dune. An unexpected result is that the critical velocity increases as the open area hydraulic radius increases. This result may not be anticipated but is perhaps due to a greater drag force exerted on the particles resting on top of the dune when the area is smaller, thereby permitting equilibrium to be attained at smaller velocities.

The other trends, such as the variation of the critical velocity with gravel diameter or fluid viscosity, are not so obvious since changing these variables changes the settling velocity, as well as the terms appearing explicitly in Eq. (19.21).

Equation (19.21) applies to a horizontal system. The critical velocity decreases with decreasing inclination until an angle is reached at which particles will roll down an inclined gravel dune. At this angle the critical velocity vanishes. The angle at which the fluid velocity is zero depends upon the effective weight of the gravel particles in the fluid and the angularity of the particles. Gruesbeck et al. [34] found that for their system, the angle at which the equilibrium velocity vanished was between 45° and 55° from the vertical. No data were provided to indicate how the critical velocity varied at intermediate angles. The designer can take one of two approaches. One would be to interpolate between the two known endpoints and the other would be to use the critical value given for the horizontal case for all highly deviated wells. It is this latter approach which seems most reasonable. Designs based on the horizontal critical velocity will be conservative ones.

Example 19.5 The Critical Velocity

Determine if the dune specified in Example 19.4 is an equilibrium one. If not, will the dune height increase or decrease? Assume that the well is highly deviated.

Solution The median diameter of 20/40-mesh gravel (see Table 19.1) is (0.84 mm + 0.42 mm)/2 or 6.3×10^{-4} m. Since the fluid viscosity is nearly that of water, we can expect the settling velocity to exceed the Stokes's law limit. In this case, we can use a drag coefficient that applies to settling spheres for modest Reynolds numbers less than 200. Thus

$$C_D = \frac{18.5}{N_{Re}^{0.6}} \tag{19.22}$$

where the drag force is then $C_D(\pi D^2/8)\rho v_s^2$ [36]. Equating the drag force to the force exerted on the particle by gravity

$$g\left(\frac{\pi D^3}{6}\right)(\rho_g - \rho) = C_D\left(\frac{\pi D^2}{8}\right)\rho v_s^2 \tag{19.23}$$

where ρ_g is the density of the gravel which is about 2.5×10^3 kg/m³. Since $\rho \cong 1 \times 10^3$ kg/m³, then $\Delta\rho/\rho = 1.5$ and

$$C_D v_s^2 = \frac{8}{6}\frac{g\Delta\rho D}{\rho} = \frac{8}{6}(1.5)(9.8)(6.3 \times 10^{-4})$$

Thus

$$C_D v_s^2 = 0.0123 \text{ m}^2/\text{sec}^2$$

Substituting Eq. (19.22) for C_D

$$\frac{18.5 v_s^2}{\left[\dfrac{(6.3 \times 10^{-4})(1000)v_s}{1 \times 10^{-3}}\right]^{0.6}} = 0.0123$$

Solving we find

$$v_s = 0.088 \text{ m/sec}$$

The dimensionless groups appearing in Eq. (19.21) are

$$\frac{\rho D v_s}{\mu} = \frac{(1 \times 10^3)(6.3 \times 10^{-4})(0.088)}{1 \times 10^{-3}} = 55.9$$

$$\frac{R_o v_s \rho}{\mu} = \frac{(0.0133)(0.088)(1000)}{1 \times 10^{-3}} = 1170$$

$$\frac{\Delta\rho}{\rho} = 1.5$$

$$c_* = 0.0575 \quad \text{(from Example 19.4)}$$

Substituting into Eq. (19.21) yields

$$v^* = (15)(0.088)(1170)^{0.39}(55.9)^{-0.75}(1.5)^{0.17}(0.0575)^{0.14} \quad \text{or} \quad v^* = 0.79 \text{ m/sec}$$

Since from Example 19.4 the flow velocity in the annular space was found to be 2.61 m/sec, the height of the gravel dune will decrease. This also means that for the conditions selected, complete gravel packing of a highly deviated well will occur.

Decreasing the annular spacing between the washpipe and the screen is apparently an effective means for helping to prevent premature sandout when gravel packing in deviated wells using low-viscosity fluids. If, however, the annular spacing required to prevent dune formation becomes small, then difficulties in removing the washpipe after completing the job may be encountered. Thus, packing deviated wells using a low-viscosity fluid together with low sand concentrations leaves much to be desired. The emphasis placed here on the technique is not meant to imply an endorsement. The extensive calculations are presented to display clearly the difficulty involved. Because the approach presented by

Gruesbeck et al. [34] is the only technique having a mathematical, albeit empirical, basis, this makes it attractive for our consideration here even though the method suggested as a result of the modeling is not entirely satisfactory.

Intermediate-to-high-viscosity carrier fluids. Based on the discussion presented in the preceding section, it might be anticipated that high-viscosity fluids, which effectively prevent the formation of gravel dunes by reducing the settling velocity to small values, will effectively transport gravel into highly deviated wells and resolve the difficulties associated with low-fluid-viscosity gravel packing. Surprisingly, this turns out not to be the case.

In studies carried out in a full-scale model, Shyrock [37] observed that as the gravel slurry entered the annular space from the crossover port in a well inclined at an angle of 80° with respect to the vertical, the slurry being denser than the fluid initially present, settled and rolled down the low side of the wellbore in a layer roughly 5 cm deep. This layer gradually increased in thickness at the upper end of the wellbore and when about 50% of the slurry had been pumped, the gravel bridged at the top of the slotted liner causing a premature sandout! Only the upper 5 m of the annular space was completely filled.

Shyrock [37] observed other problems with viscous fluids, especially those that are cross-linked. He noted that the dehydration of the slurry is slow at points where either fluid is lost to the formation or at the screen. Thus whenever the cross-linked fluid breaks (additives ensure this, see Chap. 3), the pack will tend to settle leaving voids. Based on these experiments Shyrock recommends the use of an intermediate fluid viscosity such as the HEC gels described in the previous section and positioning a lower tell tale screen so as to initially force all of the sand-laden fluid to flow to the lower part of the wellbore.

Thus there is at the present time no totally satisfactory method for gravel packing offshore highly deviated wells, although new approaches have been described but not yet satisfactorily field tested [38].

19.5 SAND CONSOLIDATION

Sand problems occur in completing formations that are not naturally "consolidated" or cemented by mineral deposits precipitated at grain boundaries. Rather than attempting to restrain the movement of formation sand by gravel packs, an alternate procedure is to artificially consolidate the sand in the region near the wellbore. This will eliminate the need for screens or liners in the wellbore, which makes well operations in zones below sand-producing horizons more complex and adversely affects workovers. The consolidating medium is generally an organic resin that is polymerized at the contact point between sand grains. The strategy requires that the larger flow channels remain available for flow but also that a compressive strength be created that is sufficient to prevent sand movement when the well is produced. Values of the compressive strength in excess of 21 MPa have been deemed sufficient [1].

Resin Requirements

The important properties that a resin must possess are as follows:

1. The resin viscosity must not be excessive. Values less than 20 cp are desirable [41]. This will allow reasonable pumping times and permit good displacement of the excess resin by an overflush.

2. The resin must wet the formation solids. This is an essential requirement. It is necessary for the resin to be drawn by capillary forces into the tiny crevices between sand grains (see Fig. 19.16).

3. The final polymerized resin should be present in sufficient quantity and possess sufficient tensile and compressive strength to prevent sand movement during production.

4. The polymerization time must be controllable. Short times will make proper placement difficult. Excessive times are costly.

5. The final polymer must be chemically inert. It must tolerate long contact times with oil or brine, and in some cases, it must not be reactive with acids. Rensvold has shown that an epoxy resin was stable after exposure to more than 30 million pore volumes of hot flowing brine [39].

Consolidating Process

The resin systems now in use are epoxy and furan [1]. Inorganic cements have not yet found commercial application. The resins may be activated before or after they are in the formation. All consolidating processes begin with a series of preflushes. These preflushes are intended to create the desired wettability and to remove any compounds naturally present in the zone to be treated that may adversely affect the proper distribution of the resin or its ultimate strength.

Most often the absolute removal of water is desired, although a sand consolidation system that is placed with water is now in use [40]. Also, it is generally important to remove asphaltenes or waxes that may have precipitated and are adhering to the surface of the sand grains. To remove these compounds and to

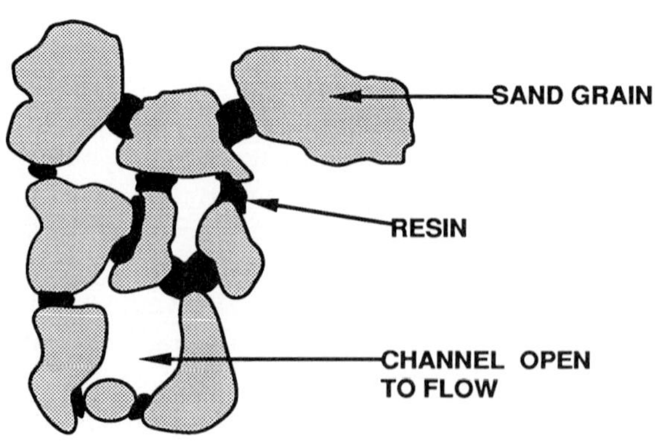

Figure 19.16 Sketch depicting consolidating process. The objective of formation sand consolidation is to cement sand grains together at the contact points.

displace the resident crude oil, several different preflush solutions may be injected sequentially. A diesel oil is commonly used as the initial preflush. This oil may include surfactants, which will help solubilize formation brine into the oil (see Chap. 3). In some cases this is followed by a mutual solvent, such as isopropyl alcohol, which is effective in dissolving both water and many oils [41]. A third oil flush may then be used. This preflush may include an adhesive enhancer that can react with the sand surface and increase the cohesion between the rock and the plastic. For epoxy resins, e.g., Shaughnessy et al. [42] report that an aminosilane coupling agent [3-(2-amino-ethylamino)propyltrimethoxysilane] to be effective.

Consolidating processes differ in use and composition of an overflush, temperature range, internal curing agent or catalyst, and resin solution.

Phase Separation Processes

These processes use dilute resin solutions in a hydrocarbon solvent combined with activator. After some time delay, polymerizing resin separates as a second liquid phase. Because of its wettability it is drawn into the grain–grain contact points by the capillary forces. The volume fraction of the resin phase must be substantial, in the range of 25–40% [1, 42], if the process is to be successful. Generally, an afterflush is not used with the phase separation processes since it is not desirable to displace resin from the region around the wellbore. Permeability is maintained by limiting the volume fraction of the separate resin phase.

Overflush Processes

These use a high-concentration resin solution. Permeability is established by injecting an overflush and displacing all but a residual resin saturation. The system should be designed to permit a stable displacement [43]. The overflush is, therefore, designed to control the thickness of the plastic film and thereby both the permeability and compressive strength.

Activators

These may initially be added to the resin solution or to the overflush. In some of the available processes, a two-step overflush is applied. The first is designed to restore permeability and the second to initiate polymerization. Generally, placing the activator in the overflush permits speeding of the curing time since it can be made to react rapidly. If the activator is mixed with the resin, then the reaction time must be carefully controlled.

Resin Placement

The interval to be treated must be isolated from the rest of the well to ensure injection into the perforations, prevent loss of process fluids, and prevent resin contamination. The treatment should be designed to treat a 1–1.5 m interval about the wellbore. Schroeder and Tucker [44] have made a study and show statistically that larger treatments last longer and are more reliable than smaller ones. Also, better results are obtained if the formation to be treated has not produced sand prior to treatment and if the treated intervals are not thick—generally less than 5–6 m.

If sand has been produced, then the voids in the formation outside of the casing must be refilled with gravel prior to initiating the consolidation treatment. In any event, the sand consolidation processes require extreme care to ensure success. Details such as mixing of fluids both in the tubing and in the formation must be carefully considered.

Consolidated Packs

In recent years, resin-coated gravel has been used in certain sand control applications [45, 46]. Once packed, the resin can be cured yielding a consolidated gravel pack having considerable chemical resistance which can prove useful, e.g., in steam injection wells where uncoated gravel will tend to dissolve [47]. Coated gravel also seems to have some advantages when repacking voids outside of the perforations after sand has been produced [1].

REFERENCES

19.1. Suman, G. O., *Sand Control Handbook*, Houston: World Oil, 1974.

19.2. Durrett, J. L., Golbin, W. T., Murray, J. W., and Tighe, R. E., *J. Pet. Tech., 29* (1977) 1664.

19.3. Hall, C. D., and Harrisberger, W. H., *J. Pet. Tech., 22* (1970) 821.

19.4. Tippie, D. B., and Kohlhaas, C. A., "Effect of Flow Rate on Stability of Unconsolidated Producing Sands," SPE 4533, presented at the 1973 Society of Petroleum Engineers Annual Meeting, Las Vegas, Nevada, 1973.

19.5. Bratli, R. K., and Risnes, R., "Stability and Failure of Sand Arches," SPE 8427, presented at the 54th Fall Technical Conference and Exhibition of the Society of Petroleum Engineers, Las Vegas, Nevada, 1979.

19.6. Suman, G. O., "Unconsolidated Sand Stabilization through Wellbore Stress State Control," SPE 5717, paper presented at the 50th Fall Technical Conference and Exhibition of the Society of Petroleum Engineers, Dallas, Texas, 1975.

19.7. Tixier, M. P., Loveless, G. W., and Anderson, R. A., *J. Pet. Tech., 27* (1975) 283.

19.8. Stein, N., and Hilchie, D. W., "Estimation of Maximum Production Rates Possible from Friable Sandstones without Using Sand Control Measures," SPE 3499, paper presented at the 1971 Society of Petroleum Engineers Annual Meeting, New Orleans, Louisiana, 1971.

19.9. Stein, N., *J. Pet. Tech., 28* (1976) 757.

19.10. Maly, G. P., and Krueger, R. F., *J. Pet. Tech., 23* (1971) 1403.

19.11. Krumbein, W. C., and Sloss, L. L., *Stratigraphy and Sedimentation*, San Francisco: W. H. Freeman and Company, 1963.

19.12. Griffiths, J. C., *Scientific Method in Analysis of Sediments*, New York: McGraw-Hill, 1967.

19.13. Himes, R. W., and Ruiz, S. J., "New Sidewall Core Analysis Techniques Improve Gravel Pack Design," SPE 14813, presented at the Seventh SPE Symposium on Formation Damage Control, Lafayette, Louisiana, 1986.

19.14. American Petroleum Institute, "Recommended Practices for Testing Sand Used in Gravel Packing Operations," RP 58, Dallas, Texas, 1986.
19.15. *Perry's Chemical Engineers' Handbook*, R. H. Perry and D. Green (eds.), New York: McGraw-Hill, 1984.
19.16. Schwartz, D. H., *J. Pet. Tech., 21* (1969) 1193.
19.17. Sparlin, D. D., "Sand and Gravel: A Study of Their Permeabilities," SPE 4772, presented at the Seventh SPE Symposium on Formation Damage Control, New Orleans, Louisiana, 1974.
19.18. B. J. Hughes Engineering Bulletin (BJ-476, 3/79), 1979.
19.19. Saucier, R. J., *J. Pet. Tech., 26* (1974) 205.
19.20. Stein, N., "Sand Control Method Using a Particulate Pack with External and Internal Particle Size Distribution Relationships," U.S. Patent No. 3,434,540, March 1969.
19.21. Sage, B. H., and Lacey, W. N., *Trans. AIME, 146* (1942) 89.
19.22. Sparlin, D. D., and Copeland, T., "Pressure Packing with Concentrated Gravel Slurry," SPE 4033, presented at the 1972 Society of Petroleum Engineers Annual Meeting, San Antonio, Texas, 1972.
19.23. Rodgers, E. B., Jr., *Oil and Gas J.* (November 1971) 54.
19.24. Shurtz, G. C., Breiner, W. G., and Comeaux, B. G., "New Thru-Tubing Gravel Pack Techniques," SPE 5660, presented at the 50th Fall Technical Conference and Exhibition of the Society of Petroleum Engineers, Dallas, Texas, 1975.
19.25. Saucier, R. J., *J. Pet. Tech., 26* (1974) 205.
19.26. Scheuerman, R. F., "New Look at Gravel Pack Carrier Fluid," SPE 12476, presented at the Seventh SPE Symposium on Formation Damage Control, Bakersfield, California, 1984.
19.27. Cole, C., Shah, S., Cavery, B., and Bellenger, B., "Monitoring HEC Gel Shearing to Optimize Improvements," SPE 17480, presented at the 58th Annual California Regional SPE Meeting, Long Beach, California, 1988.
19.28. Almond, S. W., "Factors Affecting Gelling Agent Residue Under Low Temperature Conditions," SPE 10658, presented at the Seventh SPE Symposium on Formation Damage Control, Lafayette, Louisiana, 1982.
19.29. Chauveteau, G., and Kohler, N., *Soc. Pet. Eng. J., 24* (1984) 361.
19.30. Torrest, R. S., "Deposit Buildup During Gravel Packing with Viscous Polymer Solutions and Water," *J. Pet. Tech., 35* (1983) 325.
19.31. Houchin, L. R., Hudson, L. W., Caothien, S., Daddazio, G., and Hashemi, R., "Reducing Formation Damage through Two-Stage Polymer Filtration," SPE 15408, presented at the 61st Annual Fall Technical Conference and Exhibition, New Orleans, Louisiana, 1986.
19.32. Ashton, J. P., and Nix, C. A., "Polymer Shear Mixer: A Device for Improving the Quality of Polymer Viscosified Brines," SPE 14829, presented at the Seventh SPE Symposium on Formation Damage Control, Lafayette, Louisiana, 1986.
19.33. Shryock, S. G., "Gravel Packing Studies in a Full-Scale, Deviated Wellbore," SPE 9421, presented at the 55th Fall Technical Conference and Exhibition of the Society of Petroleum Engineers, Dallas, Texas, 1980.
19.34. Gruesbeck, C., Salathiel, W. M., and Echols, E. E., *J. Pet. Tech., 31* (1979) 109.
19.35. Maly, G. P., Robinson, J. P., and Laurie, A. M., *J. Pet. Tech., 26* (1974) 19.
19.36. Bird, R. B., Stewart, W. E., and Lightfoot, E. N., *Transport Phenomena*, New York: John Wiley, 1960.
19.37. Shryock, S. G., *J. Pet. Tech., 35* (1983) 603.

19.38. Dickinson, W., and Anderson, R. R., "Gravel Packing of Horizontal Wells," SPE 16931, presented at the 62nd Fall Technical Conference and Exhibition of the Society of Petroleum Engineers, Dallas, 1987.
19.39. Rensvold, R. F., *Soc. Pet. Eng. J., 35* (1983) 238.
19.40. Murphey, J. R., Bila, V. J., and Totty, K., "Sand Consolidation Systems Placed with Water," SPE 5031, presented at the 1974 Society of Petroleum Engineers Annual Meeting, Houston, Texas, 1974.
19.41. Penberthy, W. L., Shaughnessy, C. M., Gruesbeck, C., Salathiel, W. M., *J. Pet. Tech., 30* (1978) 845.
19.42. Shaughnessy, C. M., Salathiel, W. M., and Penberthy, W. L., *J. Pet. Tech., 30* (1978) 1805.
19.43. Davies, D. R., Hagelaars, A. M. P. M., and Roberts, D. L., *J. Pet. Tech., 36* (1984) 1905.
19.44. Schroeder, R. H., and Tucker, M. J., "Evaluation of Completion Practices for Improved Sand Control Success and Longevity," SPE 5027, presented at the 1974 Society of Petroleum Engineers Annual Meeting, Houston, Texas, 1974.
19.45. Saunders, L. W., and McKinzie, H. L., *J. Pet. Tech., 33* (1981) 221.
19.46. Sinclair, A. R., and Graham, J. W., "An Effective Method of Sand Control," SPE 7004, presented at the Seventh SPE Symposium on Formation Damage Control, Lafayette, Louisiana, 1978.
19.47. Reed, M. G., "Gravel Pack and Formation Sandstone Dissolution During Steam Injection," SPE 8424, presented at the 54th Annual Fall Technical Conference and Exhibition, Las Vegas, Nevada, 1979.
19.48. Krumbein, W. C., *J. Sedimentary Petrology, 8* (1938) 84.

PROBLEMS

***19.1.** The weight fraction of a formation sand having a diameter greater than d is given by

$$w(d) = 1 - \exp\left(\frac{\beta}{d}\right)$$

where $w(d)$ is the cumulative weight fraction and β is a constant. Determine the uniformity coefficient.

****19.2.** Krumbein [48] has found that the grain-size distribution of formation sands can often be well represented by the phi-distribution

$$P(\Phi) = \frac{1}{\sqrt{2\pi}\,\sigma_\Phi} \exp - \frac{(\Phi - M_\Phi)^2}{2\sigma_\Phi^2}$$

where $P(\Phi)\,d\Phi$ represents the fraction of the sample having a phi-value between Φ and $\Phi + d\Phi$ and σ_Φ and M_Φ are characteristics of the particular reservoir sand. Here $\Phi = -\log_2 d_p$, where the particle diameter is expressed in mm.

(a) Prove the following identities:

$$\Phi_{50} = M_\Phi \qquad \text{and} \qquad \sigma_\Phi = \frac{\Phi_{84} - \Phi_{16}}{2}$$

(b) Consider the formation sand analysis presented in Table P19.2 and compare the predictions of the phi-distribution with the parameters M_Φ and σ_Φ determined by the equations derived in Part (a) with the actual experimental values.

TABLE P19.2 Formation Sand Analysis

Cumulative wt%	Grain Diameter (mm)	Cumulative wt%	Grain Diameter (mm)
5	0.74	55	0.14
10	0.56	60	0.13
15	0.46	65	0.12
20	0.37	70	0.11
25	0.30	75	0.097
30	0.24	80	0.089
35	0.22	85	0.076
40	0.19	90	0.069
45	0.17	95	0.061
50	0.15	—	—

***19.3.** A sieve analysis of a formation sand is given in Problem 19.2. Based on this analysis determine the median sand grain size, the geometric mean sand grain size, and the uniformity coefficient.

***19.4.** Show that a gravel pack composed of uniform spherical particles that are cubically packed will have a porosity of 47.6%.

****19.5.** Show that a gravel pack composed of uniform spherical particles that are hexagonally close packed will have a porosity of 25.4%.

***19.6.** Reduced production can be expected due to the additional pressure drop associated with the presence of gravel. Develop an equation for the skin factor resulting from gravel packing. Consider an uncased wellbore of radius r_w and a gravel pack of radius r_p. Let k be the formation permeability and k_p be that of the gravel pack. Based on this equation for the skin factor, what percentage reduction in production would you expect in an uncased well having a radius of 7.6×10^{-2} m when gravel packing with 40/60-mesh sand if the gravel pack screen has a radius of 4×10^{-2} m? The formation permeability is 0.45 D.

****19.7.** Sparlin [17] has studied the permeability of mixtures of sand and gravel to demonstrate the deleterious effect that sand intrusion into the gravel pack has on its permeability. In his studies Sparlin used the following approximation for the permeability of sand–gravel mixtures:

$$k = 5.1 \times 10^{-6}\, \phi^{5.1}\, d_{50}^2 \exp(-1.385\, \Delta\Phi)$$

where k is the permeability in darcies (D), ϕ is the percent porosity (porosity multiplied by 100), and

$$\Delta\Phi = \Phi_{84.1} - \Phi_{50}$$

The phi-function is defined in Problem 19.2. In all cases, the porosity of the gravel pack is 40%.

(a) Use Sparlin's equation for k to estimate the permeability of a 20/40-mesh gravel pack. Assume that the cumulative wt% for commercial gravel satisfies the equation

$$W_{gravel} = \alpha \ln(\beta D)$$

where α and β are constants and D is the diameter of a gravel grain. For this example, all of the gravel passes through the 20-mesh screen and none of it through the 40-mesh screen.

(b) Determine the permeability of a gravel pack composed of 75% gravel and 25% sand, where the sand is distributed as listed in Table P19.7.

TABLE P19.7

Cumulative wt%	Grain Diameter (mm)
10	0.216
20	0.193
30	0.180
40	0.168
50	0.152
60	0.142
70	0.135
80	0.117
90	0.089

***19.8.** The distribution of sand grain sizes of a given reservoir is known to satisfy the equation.

$$d = A \exp(-BW)$$

where W is the weight fraction of sand having a diameter larger than d and A and B are constants. Given that $U_c(=d_{40}/d_{90})$ is 4 and the median diameter ($= d_{50}$) is 0.2 mm

(a) Determine the constants A and B.

(b) Using the Schwartz criteria, specify the gravel size you would use to gravel pack to prevent sand production. Assume that the entrance velocities through one-half the screen area will be less than 0.015 m/sec.

Appendix: Units and Conversions

TABLE A.1 SI Units Used in This Book

Quantity	Name	Symbol
Length	Meter	m
Time	Second	s
Mass	Kilogram	kg
Temperature	Kelvin	°K
Amount of substance	Mole or Kilogram mole	mol kg mol
Amount of electrical charge	Coulomb or equivalent or kilogram equivalent	C equ kg equ

TABLE A.2 Special Names of SI Derived Units Used in This Book

Quantity	Name	Symbol	SI units
Force	Newton	N	$m\text{-}kg/s^2$
Pressure	Pascal	Pa	$kg/m\text{-}s^2$
Energy	Joule or calorie	J or cal	$m^2\text{-}kg/s^2$
Electrical potential	Volt	V	$m^2\text{-}kg/s^2\text{-}C$

TABLE A.3 SI Prefixes Used in This Book

Factor	Prefix	Symbol
10^6	mega	M
10^3	kilo	k
10^{-1}	deci	d
10^{-2}	centi	c
10^{-3}	milli	m
10^{-6}	micro	μ
10^{-9}	nano	n

TABLE A.4 Conversion from Field Units to SI Units

To convert	from	Field units	to	SI units	Multiply by
Length		Foot		m	0.304800
Mass		Pound		kg	0.453592
Time		Hour		s	3.60000×10^3
Rate		Barrels/day		m^3/s	1.840131×10^{-6}
Pressure		psia		Pa	6.894757×10^3
Viscosity		cp		kg/m-s	1.000000×10^{-3}
Permeability		darcy		m^2	9.86923×10^{-13}

Index

A

B

C

D

G

H

I

J

K

L

M

N

P

Q

R

S

T

V

W

X

Y

Z